Perwez Kalim PhD (KU)
Emeritus Professor of Mechanical Engineering
Wilkes University, Wilkes Barre, PA 18766
December 2023

Donated by:

A First Course
in the Finite
Element Method

THE PWS SERIES IN ENGINEERING

SECOND EDITION

A First Course

in the Finite

Element Method

DARYL L. LOGAN
Rose-Hulman Institute of Technology

PWS Publishing Company
BOSTON

PWS PUBLISHING COMPANY
20 Park Plaza, Boston, MA 02116-4324

To my parents, Shirley and John Logan

I(T)P™
International Thomson Publishing
The trademark ITP is used under license

Copyright © 1993 by PWS Publishing Company.
Copyright © 1992 by PWS-KENT Publishing Company. Copyright © 1986 by PWS Publishers.

PWS Publishing Company is a division of Wadsworth, Inc.

Library of Congress Cataloging-in-Publication Data

Logan, Daryl L.
    A first course in the finite element method / Daryl L. Logan. —
2nd ed.
        p.   cm.
    Includes bibliographical references and index.
    ISBN 0-534-92964-8
    1. Finite element method.   I. Title.
TA347.F5L64   1992
620'.001'515353—dc20   92-2857
                                                            CIP

Sponsoring Editor: Jonathan Plant
Assistant Editor: Mary Thomas
Production Editor: Helen Walden
Cover Design: Monique Calello
Text Design: Julia Gecha
Manufacturing Coordinator: Marcia Locke
Typesetter: Asco Trade Typesetting Ltd.
Printer/Binder: Maple-Vail Book Mfr. Group

Printed in the United States of America.
    93 94 95 96—10 9 8 7 6 5 4 3

Contents

3 Development of Truss Equations   60

4 Symmetry, Bandwidth, and a Computer Program for Truss Analysis   128

11 Isoparametric Formulation   **388**

12 Three-Dimensional Stress Analysis   **418**

13 Heat Transfer and Mass Transport   **434**

14 **Fluid Flow**   **485**

15 **Thermal Stress**   **512**

16 **Structural Dynamics and Time-Dependent Heat Transfer**   **538**

Appendix A   Matrix Algebra   588

Appendix B   Methods for Solution of Simultaneous Linear Equations   601

Appendix C   Equations from Elasticity Theory   616

Preface

The purpose of this second edition of the book is again to provide a simple, basic approach to the finite element method that can be understood by both undergraduate and graduate students without the usual prerequisites (such as structural analysis) required by most available texts in this area. The book is written primarily as a basic learning tool for the undergraduate student in civil and mechanical engineering whose main interest is in stress analysis and heat transfer. However, the concepts are presented in sufficiently simple form so that the book serves as a valuable learning aid for students of other backgrounds, as well as for practicing engineers. The text is geared toward those who want to apply the finite element method to solve practical physical problems.

General principles are presented for each topic, followed by traditional applications of these principles, which are in turn followed by computer applications where possible. This approach is taken to illustrate concepts used for computer analysis of large-scale problems.

The book proceeds from basic to advanced topics and can be suitably used in a two-course sequence. Topics include basic treatments of (1) simple springs and bars, leading to two- and three-dimensional truss analysis; (2) beam bending, leading to plane frame and grid analysis, and space frame analysis; (3) elementary plane stress/strain elements, leading to more advanced plane stress/strain elements; (4) axisymmetric stress; (5) isoparametric formulation of the finite element method; (6) three-dimensional stress; (7) heat transfer and fluid mass transport; (8) basic fluid mechanics; (9) thermal stress; and (10) time-dependent stress and heat transfer.

New topics/features include: how to handle inclined or skewed supports, beam element with nodal hinge, beam element arbitrarily located in space, the concept of substructure analysis, a completely new chapter on fluid mechanics, and a diskette including the source codes of six basic programs used in the text.

The direct approach, the principle of minimum potential energy, and Galerkin's residual method are introduced at various stages, as required, to develop the equations needed for analysis.

Appendices include: (1) basic matrix algebra used throughout the text, (2) solution methods for simultaneous equations, (3) basic theory of elasticity, and (4) the principle of virtual work.

More than 60 solved problems appear throughout the text. Most of the examples are solved "longhand" to illustrate the concepts; many of them are solved by digital computer to illustrate the use of the computer programs provided on the diskette enclosed in the back of the book. More than 300 end-of-chapter problems, including a number of new ones, are provided to reinforce concepts. Answers to many problems are included in the back of the book. Those end-of-chapter problems to be solved using a computer program are marked with a computer symbol.

Computer programs are incorporated directly into relevant places in the text to create a natural extension from basic principles to longhand examples and then to computer program examples. The programs are written specifically for instructional purposes and source codes written in FORTRAN language are included on the accompanying diskette. Each program solves a specific class or type of problem. These programs are easy for students to use. A single lecture is sufficient to explain how to use most of them.

To run any of the six special-purpose programs, you must create a ".EXE" file of the program. This is done using a FORTRAN compiler program such as the MS-DOS FORTRAN compiler.

Following is an outline of suggested topics for a first course (approximately 40 lectures; 50 minutes each) using this textbook.

Topic	*Number of Lectures*
Appendix A	1
Appendix B	1
Chapter 1	2
Chapter 2	3
Chapter 3, Sections 3.1–3.10	5
Exam 1	1
Chapter 4	4
Chapter 5, Sections 5.1–5.6	4
Chapter 6, Sections 6.1–6.3, 6.7	4
Exam 2	1
Chapter 7	4
Chapter 8	4
Chapter 13, Sections 13.1–13.8, 13.10–13.11	5
Exam 3	1

I use this outline in a one-quarter course for undergraduate and graduate students in civil and mechanical engineering. (If a total stress analysis emphasis is desired, Chapter 13 can be replaced, for instance, with material from

Chapters 9 and 10, Chapter 11, or some of Chapter 16.) I then finish the rest of the text in a second-quarter course.

I express my deepest appreciation to the staff at PWS-KENT Publishing Company, especially Jonathan Plant, Mary Thomas, and Helen Walden for their assistance in producing this second edition. I am also sincerely grateful to Linda Thompson for her editing of the manuscript. I also appreciate the initial reviews of the first manuscript by John O. Dow, University of Colorado at Boulder; Fernando Fagundo, University of Florida; Ronald Sack, University of Oklahoma; and Edward Ting, Purdue University, and the reviews of the second edition manuscript by Bruce A. DeVantier, Southern Illinois University; John O. Dow, University of Colorado; Joseph J. Rencis, Worcester Polytechnic Institute; and Stephen R. Swanson, University of Utah.

I am grateful to Ted Belytschko for his excellent teaching of the finite element method, which aided me in writing this text.

I thank the many students who used the notes that developed into this text. Special thanks to Ron Cenfetelli, Barry Davignon, Konstantinos Kariotos, Howard Koswara, Hidajat Harintho, Hari Salemganesan, Joe Kesari, Yanping Lu, and Khailan Zhang for checking and solving problems in the text. Thanks also to Peter Soller, David Danner, Ed Knoy, and Yanping Lu for "cleaning up" computer programs, and a special thank you to Professor Joseph Rencis and Chili Bao for getting computer program DFRAME running properly.

Finally, a very special thank you to my family, Diane, Kathy, Daryl Jr., and Paul, for their many sacrifices during the development of this second edition.

Daryl L. Logan

Notation

ENGLISH SYMBOLS

a_i	generalized coordinates (coefficients used to express displacement in general form)
A	cross-sectional area
$\underline{B}$	matrix relating strains to nodal displacements or relating temperature gradient to nodal temperatures
c	specific heat of a material
$\underline{C'}$	matrix relating stresses to nodal displacements
C	direction cosine in two dimensions
$C_x,\ C_y,\ C_z$	direction cosines in three dimensions
$\underline{d}$	element and structure nodal displacement matrix, both in global coordinates
$\underline{\hat{d}}$	local-coordinate element nodal displacement matrix
$\underline{D}$	matrix relating stresses to strains
e	exponential function
E	modulus of elasticity
$\underline{f}$	global-coordinate nodal force matrix
$\underline{\hat{f}}$	local-coordinate element nodal force matrix
$\underline{f_b}$	body force matrix
$\underline{f_h}$	heat transfer force matrix
$\underline{f_q}$	heat flux force matrix
$\underline{f_Q}$	heat source force matrix
$\underline{f_s}$	surface force matrix
$\underline{F}$	global-coordinate structure force matrix
$\underline{F_c}$	condensed force matrix
$\underline{F_i}$	global nodal forces

$\underline{F}_0$	equivalent force matrix
g	temperature gradient matrix or hydraulic gradient matrix
h	heat-transfer (or convection) coefficient
i, j, m	nodes of a triangular element
I	principal moment of inertia
$\underline{J}$	Jacobian matrix
k	spring stiffness
$\underline{k}$	global-coordinate element stiffness or conduction matrix
$\underline{k}_c$	condensed stiffness matrix, and conduction part of the stiffness matrix in heat-transfer problems
$\underline{\hat{k}}$	local-coordinate element stiffness matrix
$\underline{k}_h$	convective part of the stiffness matrix in heat-transfer problems
$\underline{K}$	global-coordinate structure stiffness matrix
K_{xx}, K_{yy}	thermal conductivities (or permeabilities, for fluid mechanics) in the x and y directions, respectively
L	length of a bar or beam element
m	maximum difference in node numbers in an element
$m(x)$	general moment expression
$\underline{\hat{m}}$	local mass matrix
$\hat{m}_i$	local nodal moments
$\underline{M}$	global mass matrix
$\underline{M}^*$	matrix used to relate displacements to generalized coordinates for a linear-strain triangle formulation
$\underline{M}'$	matrix used to relate strains to generalized coordinates for a linear-strain triangle formulation
n_b	bandwidth of a structure
n_d	number of degrees of freedom per node
$\underline{N}$	shape (interpolation or basis) function matrix
N_i	shape functions
p	surface pressure (or nodal heads in fluid mechanics)
p_r, p_z	radial and axial (longitudinal) pressures, respectively
P	concentrated load
$\underline{\hat{P}}$	concentrated local force matrix
q	heat flow (flux) per unit area
$\bar{q}$	rate of heat flow
q^*	heat flow per unit area on a boundary surface
Q	heat source generated per unit volume or internal fluid source
Q^*	line or point heat source
r, θ, z	radial, circumferential and axial coordinates, respectively
R	residual in Galerkin's integral

R_b	body force in the radial direction
R_{ix}, R_{iy}	nodal reactions in x and y directions, respectively
s, t, z'	natural coordinates attached to isoparametric element
s	surface area
t	thickness of a plane element
t_i, t_j, t_m	nodal temperatures of a triangular element
T	temperature function
T_∞	free-stream temperature
$\underline{T}$	displacement, force, and stiffness transformation matrix
$\underline{T_i}$	surface traction matrix in the i direction
u, v, w	displacement functions in the x, y, and z directions, respectively
U	strain energy
ΔU	change in stored energy
v	velocity of fluid flow
$\hat{V}$	shear force in a beam
w	distributed loading on a beam or along an edge of a plane element
W	work
x_i, y_i, z_i	nodal coordinates in the x, y, and z directions, respectively
$\hat{x}, \hat{y}, \hat{z}$	local element coordinate axes
x, y, z	structure global or reference coordinate axes
$\underline{X}$	body force matrix
X_b, Y_b	body forces in the x and y directions, respectively
Z_b	body force in longitudinal direction (axisymmetric case) or in the z direction (three-dimensional case)

GREEK SYMBOLS

α	coefficient of thermal expansion
$\alpha_i, \beta_i, \gamma_i, \delta_i$	used to express the shape functions defined by Eq. (7.2.10) and Eqs.(12.2.5)–(12.2.8)
δ	spring or bar deformation
ε	normal strain
$\underline{\varepsilon_T}$	thermal strain matrix
v	Poisson's ratio
ϕ_i	nodal angle of rotation or slope in a beam element
π_p	total potential energy
π_h	functional for heat-transfer problem
ρ	mass density of a material
ρ_w	weight density of a material

ω	angular velocity
Ω	potential energy of forces
ϕ	fluid head or potential
σ	normal stress
σ_T	thermal stress matrix
τ	shear stress
θ	angle between the x-axis and the local $\hat{x}$ axis for two-dimensional problems
θ_p	principal angle
$\theta_x, \theta_y, \theta_z$	angles between the global x, y, and z axes and the local $\hat{x}$ axis, respectively
$\underline{\Psi}$	general displacement function matrix

OTHER SYMBOLS

$\dfrac{d(\ )}{dx}$	derivative of a variable with respect to x
dt	time differential
$(\dot{\ })$	the dot over a variable denotes that the variable is being differential with respect to time
$[\ ]$	denotes a rectangular or a square matrix
$\{\ \}$	denotes a column matrix
$(\underline{\ })$	the underline of a variable denotes a matrix
$(\hat{\ })$	the hat over a variable denotes that the variable is being described in a local coordinate system
$[\ ]^{-1}$	denotes the inverse of a matrix
$[\ ]^{T}$	denotes the transpose of a matrix
$\dfrac{\partial(\ )}{\partial x}$	partial derivative with respect to x
$\dfrac{\partial(\ )}{\partial\{d\}}$	partial derivative with respect to each variable in $\{d\}$
$\blacksquare$	denotes the end of the solution of an example problem

Introduction

PROLOGUE

The finite element method is a numerical method for solving problems of engineering and mathematical physics. Typical problem areas of interest in engineering and mathematical physics that are solvable by use of the finite element method include structural analysis, heat transfer, fluid flow, mass transport, and electromagnetic potential.

For problems involving complicated geometries, loadings, and material properties, it is generally not possible to obtain analytical mathematical solutions. Analytical solutions are those given by a mathematical expression that yields the values of the desired unknown quantities at any location in a body (here total structure or physical system of interest) and are thus valid for an infinite number of locations in the body. These analytical solutions generally require the solution of ordinary or partial differential equations, which, because of the complicated geometries, loadings, and material properties, are not usually obtainable. Hence, we need to rely on numerical methods, such as the finite element method, for acceptable solutions. The finite element formulation of the problem results in a system of simultaneous algebraic equations for solution, rather than requiring the solution of differential equations. These numerical methods yield approximate values of the unknowns at discrete numbers of points in the continuum. Hence, this process of modeling a body by dividing it into an equivalent system of smaller bodies or units (finite elements) interconnected at points common to two or more elements (nodal points or nodes) and/or boundary lines and/or surfaces is called *discretization*. In the finite element method, instead of solving the problem for the entire body in one operation, one formulates the equations for each finite element and combines them to obtain the solution of the whole body.

Briefly, the solution for structural problems typically refers to determining the displacements at each node and the stresses within each element making up the structure that is subjected to applied loads. In nonstructural problems, the nodal unknowns may, for instance, be temperatures or fluid pressures due to thermal or fluid fluxes.

This chapter first presents a brief history of the development of the finite element method. You will see from this historical account that the method has only become a practical one for solving engineering problems in the past 35 years (paralleling the developments associated with the modern high-speed electronic digital computer). This historical account is followed by an introduction to matrix notation; then the need for matrix methods (as made practical by the development of the modern digital computer) in formulating the equations for solution is described. This section discusses both the role of the digital computer in solving the large systems of simultaneous algebraic equations associated with complex problems and the development of numerous computer programs based on the finite element method. Next, a general description of the steps involved to obtain a solution to a problem is provided. This description includes discussion of the types of elements available for a finite element method solution. Various representative applications are presented to illustrate the capacity of the method to solve problems, such as those involving complicated geometries, several different materials, and irregular loadings. Finally, Chapter 1 lists some of the advantages of the finite element method in solving problems of engineering and mathematical physics.

1.1   Brief History

This section presents a brief history of the finite element method as applied to both structural and nonstructural areas of engineering and to mathematical physics. References cited here are intended to augment this short introduction to the historical background.

The modern development of the finite element method began in the 1940s in the field of structural engineering with the work by Hrennikoff [1] in 1941 and McHenry [2] in 1943, who used a lattice of line (one-dimensional) elements (bars and beams) for the solution of stresses in continuous solids. In a paper published in 1943, but not widely recognized for many years, Courant [3] proposed setting up the solution of stresses in a variational form. Then he introduced piecewise interpolation (or shape) functions over triangular subregions making up the whole region as a method to obtain approximate numerical solutions. In 1947, Levy [4] developed the flexibility or force method, and, in 1953, his work [5] suggested that another method (the stiffness or displacement method) could be a promising alternative for use in analyzing statically redundant aircraft structures. However, his equations were cumbersome to solve by hand, and thus the method became popular only with the advent of the high-speed digital computer.

In 1954 Argyris and Kelsey [6, 7] developed matrix structural analysis methods using energy principles. This development illustrated the important role that energy principles would play in the finite element method.

The first treatment of two-dimensional elements was by Turner, Clough, Martin, and Topp [8] in 1956. They derived stiffness matrices for truss elements, beam elements, and two-dimensional triangular and rectangular elements in plane stress and outlined the procedure commonly known as the *direct stiffness method* for obtaining the total structure stiffness matrix. Along with the development of the high-speed digital computer in the early 1950s, the work of Turner, Clough, Martin, and Topp [8] prompted increased developments of finite element stiffness equations expressed in matrix notation. The first use of the phrase *finite element* was introduced by Clough [9] in 1960 when both triangular and rectangular elements were used for plane stress analysis.

A flat, rectangular-plate bending-element stiffness matrix was developed by Melosh [10] in 1961. This was followed by development of the curved-shell bending-element stiffness matrix for axisymmetric shells and pressure vessels by Grafton and Strome [11] in 1963.

The extension of the finite element method to three-dimensional problems with the development of a tetrahedral stiffness matrix was given by Martin [12] in 1961, by Gallagher, Padlog, and Bijlaard [13] in 1962, and by Melosh [14] in 1963. Additional three-dimensional elements were studied by Argyris [15] in 1964. The special case of axisymmetric solids was considered by Clough and Rashid [16] and Wilson [17] in 1965.

Most of the finite element work up to the early 1960s dealt with small strains and small displacements, elastic material behavior, and static loadings. However, large deflection and thermal analysis were considered by Turner, Dill, Martin, and Melosh [18] in 1960 and material nonlinearities by Gallagher, Padlog, and Bijlaard [13] in 1962, whereas buckling problems were initially treated by Gallagher and Padlog [19] in 1963. Extension of the method to visco-elasticity problems was done by Zienkiewicz, Watson, and King [20] in 1968.

In 1965, Archer [21] considered dynamic analysis in the development of the consistent-mass matrix, which is applicable to analysis of distributed-mass systems such as bars and beams in structural analysis.

With Melosh's [14] realization in 1963 that the finite element method could be set up in terms of a variational formulation, it began to be used to solve nonstructural applications. Field problems, such as the determination of the torsion of a shaft, fluid flow, and heat conduction, were solved by Zienkiewicz and Cheung [22] in 1965, Martin [23] in 1968, and Wilson and Nickel [24] in 1966.

Further extension of the method was made possible by the adaptation of weighted residual methods, first to derive the previously known elasticity equations used in structural analysis by Szabo and Lee [25] in 1969 and then for transient field problems by Zienkiewicz and Parekh [26] in 1970. It was then recognized that when direct formulations and variational formulations are difficult or not possible to use, the method of weighted residuals may at times be appropriate. For example, in 1977 Lyness, Owen, and Zienkiewicz [27] applied the method of weighted residuals to the determination of magnetic field.

Recently, problems associated with large-displacement nonlinear dynamic behavior, and improved numerical techniques for the solution of the resulting systems of equations have been considered by Belytschko [28, 29] in 1976.

A relatively new field of application of the finite element method is that of bioengineering [30, 31]. This field is still troubled by all the difficulties such as nonlinear materials, geometric nonlinearities, and other complexities still being discovered.

From the early 1950s to the present, enormous advances have been made in the application of the finite element method to solve complicated engineering problems. Engineers, applied mathematicians, and other scientists will undoubtedly continue to develop new applications. For an extensive bibliography on the finite element method, consult the work of Whiteman [32] or Norrie and de Vries [33].

1.2   Introduction to Matrix Notation

Matrix methods are a necessary tool used in the finite element method for purposes of simplifying the formulation of the element stiffness equations, for purposes of longhand solutions of various problems, and, most important, for use in programming the methods for high-speed electronic digital computers. Hence, matrix notation represents a simple and easy-to-use notation for writing and solving sets of simultaneous algebraic equations.

Appendix A presents a discussion of the significant matrix concepts used throughout the text. We will present here only a brief summary of the notation used in this text.

A **matrix** *is a rectangular array of quantities arranged in rows and columns that is often used to aid in expressing and solving a system of algebraic equations.* As examples of matrices that will be described in subsequent chapters, the force components $(F_{1x}, F_{1y}, F_{1z}, F_{2x}, F_{2y}, F_{2z}, \ldots, F_{nx}, F_{ny}, F_{nz})$ acting at the various nodes or points $(1, 2, \ldots, n)$ on a structure and the corresponding set of nodal displacements $(d_{1x}, d_{1y}, d_{1z}, d_{2x}, d_{2y}, d_{2z}, \ldots, d_{nx}, d_{ny}, d_{nz})$ can both be expressed as

$$\{F\} = \underline{F} = \begin{Bmatrix} F_{1x} \\ F_{1y} \\ F_{1z} \\ F_{2x} \\ F_{2y} \\ F_{2z} \\ \vdots \\ F_{nx} \\ F_{ny} \\ F_{nz} \end{Bmatrix} \qquad \{d\} = \underline{d} = \begin{Bmatrix} d_{1x} \\ d_{1y} \\ d_{1z} \\ d_{2x} \\ d_{2y} \\ d_{2z} \\ \vdots \\ d_{nx} \\ d_{ny} \\ d_{nz} \end{Bmatrix} \qquad (1.2.1)$$

The subscripts to the right of F and d identify the node and the direction of force or displacement, respectively. For instance, F_{1x} denotes the force at node 1 applied in the x direction. The matrices in Eqs. (1.2.1) are called *column matrices.* The brace notation { } will be used throughout the text to denote a column matrix. The whole set of force or displacement values in the column array is simply represented by $\{F\}$ or $\{d\}$. A more compact notation used throughout this text to represent any rectangular array is the underlining of the variable; that is, $\underline{F}$ or $\underline{d}$ underlined denotes these as general matrices (possibly column matrices or rectangular matrices—the type will become clear in the context of the discussion associated with the variable).

The more general case of a known rectangular matrix will be indicated by use of the bracket notation [ ]. For instance, the element and global structure stiffness matrices $[k]$ and $[K]$, respectively, developed throughout the text for various element types (such as those in Figure 1-1 on page 9), are represented by square matrices given as

$$[k] = \underline{k} = \begin{bmatrix} k_{11} & k_{12} & \cdots & k_{1n} \\ k_{21} & k_{22} & \cdots & k_{2n} \\ \vdots & \vdots & & \vdots \\ k_{n1} & k_{n2} & \cdots & k_{nn} \end{bmatrix} \qquad (1.2.2)$$

and

$$[K] = \underline{K} = \begin{bmatrix} K_{11} & K_{12} & \cdots & K_{1n} \\ K_{21} & K_{22} & \cdots & K_{2n} \\ \vdots & \vdots & & \vdots \\ K_{n1} & K_{n2} & \cdots & K_{nn} \end{bmatrix} \qquad (1.2.3)$$

where, in structural theory, the elements k_{ij} and K_{ij} are often referred to as *stiffness influence coefficients.*

You will learn that the global nodal forces $\underline{F}$ and the global nodal displacements $\underline{d}$ are related through use of the global stiffness matrix $\underline{K}$ by

$$\underline{F} = \underline{K}\underline{d} \qquad (1.2.4)$$

Equation (1.2.4) is called the *global stiffness equation.* It is the basic equation formulated in the stiffness or displacement method of analysis. Using the compact notation of underlining the variables, as in Eq. (1.2.4), should not cause you any difficulties in determining which matrices are column or rectangular arrays.

Subsequent chapters will discuss the element stiffness matrices $\underline{k}$ for various element types, such as bars, beams, and plane stress. They will also cover the procedure for obtaining the global stiffness matrices $\underline{K}$ for various structures and for solving Eq. (1.2.4) for the unknown displacements in matrix $\underline{d}$.

Using matrix concepts and operations will become routine with practice; they will be valuable tools for solving small problems longhand. However, matrix methods are crucial to the use of digital computers—necessary for the solution of complicated problems with their associated large number of simultaneous equations.

1.3   Role of the Computer

As mentioned, until the early 1950s, matrix methods and the associated finite element method were not readily adaptable for solving complicated problems because of the large number of algebraic equations that resulted. Hence, even though the finite element method was being used to describe complicated structures, the resulting large number of equations associated with the finite element method of structural analysis made the method extremely difficult and impractical to use. However, with the advent of the computer, the solution of thousands of equations in a matter of minutes became possible.

The development of the computer resulted in computational program development. Numerous special-purpose and general-purpose programs have been written to handle various complicated structural (and nonstructural) problems. The accompanying disk provides some special-purpose programs; they are simple to use and solidify the understanding of and elegance of the finite element method.

To use the computer, the analyst, having defined the finite element model, inputs the information into the computer. This formation may include the position of the element nodal coordinates, the manner in which elements are connected together, the material properties of the elements, the applied loads, boundary conditions, or constraints, and the kind of analysis to be performed. The computer then uses this information to generate and solve the equations necessary to carry out the analysis.

1.4   General Steps of the Finite Element Method

This section presents the general steps included in a finite element method formulation and solution to an engineering problem. We will use these steps as our guide in developing solutions for structural and nonstructural problems in subsequent chapters.

For simplicity's sake, for the presentation of the steps to follow, we will consider only the structural problem. The nonstructural heat-transfer and fluid mechanics problems and their analogies to the structural problem are considered in Chapters 13 and 14.

Typically, for the structural stress-analysis problem, the engineer seeks to determine displacements and stresses throughout the structure, which is in equilibrium and is subjected to applied loads. For many structures, it is difficult to determine the distribution of deformation using conventional methods, and thus the finite element method is necessarily used.

There are two general approaches associated with the finite element method. One approach, called the *force*, or *flexibility*, *method*, uses internal forces as the unknowns of the problem. To obtain the governing equations, first the equilibrium equations are used. Then necessary additional equations are found by introducing compatibility equations. The result is a set of algebraic equations for determining the redundant or unknown forces.

The second approach, called the *displacement,* or *stiffness, method,* assumes the displacements of the nodes as the unknowns of the problem. For instance, compatibility conditions requiring that elements connected at a common node, along a common edge, or on a common surface before loading remain connected at that node, edge, or surface after deformation takes place are initially satisfied. Then the governing equations are expressed in terms of nodal displacements using the equations of equilibrium and an applicable law relating forces to displacements.

These two approaches result in different unknowns (forces or displacements) in the analysis and different matrices associated with their formulations (flexibilities or stiffnesses). It has been shown [34] that, for computational purposes, the displacement (or stiffness) method is more desirable because its formulation is simpler for most structural analysis problems. Furthermore, a vast majority of general-purpose finite element programs have incorporated the displacement formulation for solving structural problems. Consequently, only the displacement method will be used throughout this text.

The finite element method involves modeling the structure using small interconnected elements called *finite elements.* A displacement function is associated with each finite element. Every interconnected element is linked, directly or indirectly, to every other element through common (or shared) interfaces, including nodes and/or boundary lines and/or surfaces. On making use of known stress/strain properties for the material making up the structure, one can determine the behavior of a given node in terms of the properties of every other element in the structure. The total set of equations describing the behavior of each node results in a series of algebraic equations best expressed in matrix notation.

We now present the steps, along with explanations necessary at this time, used in finite element method formulation and solution of a structural problem. The purpose of setting forth these general steps at this time is to expose you to the procedure generally followed in a finite element formulation of a problem. You will easily understand these steps as we illustrate them specifically for springs, bars, trusses, beams, plane frames, plane stress, axisymmetric stress, three-dimensional stress, heat transfer, and fluid flow in subsequent chapters. We suggest that you review this section periodically as we develop the specific element equations.

Keep in mind that the analyst must make decisions regarding dividing the structure or continuum into finite elements and selecting the element type or types to be used in the analysis (step 1) and the kinds of loads to be applied and the type of boundary conditions or supports to be applied. The other steps, 2, 3, 4, 5, 6, and 7, are carried out automatically by a computer program.

Step 1 Discretize and Select Element Types

Step 1 involves dividing the body into an equivalent system of finite elements with associated nodes and choosing the most appropriate element type. The total number of elements used and their variation in size and type within a given body are primarily matters of engineering judgment. The elements must be made small enough to give usable results and yet large enough to reduce computational effort. Small elements (and possibly higher-order elements) are generally desirable where the results are changing rapidly, such as where changes in geometry occur, whereas large elements can be used where results are relatively constant. More will be said regarding discretization guidelines in later chapters, particularly in Chapter 8, where the concept becomes quite significant. The discretized body or mesh is often created with mesh-generation programs or preprocessor programs available to the user.

The choice of elements used in a finite element analysis depends on the physical makeup of the body under actual loading conditions and how close to the actual behavior the analyst wants the results to be. Judgment concerning the appropriateness of one-, two-, or three-dimensional idealizations is necessary. Moreover, the choice of the most appropriate element for a particular problem is one of the major tasks that must be carried out by the designer/analyst. Elements that are commonly employed in practice—most of which are considered in this text—are shown in Figure 1-1.

The primary line elements, Figure 1-1(a), consist of bar (or truss) and beam elements. They have a cross-sectional area but are usually represented by line segments. In general, the cross-sectional area within the element can vary, but it will be considered to be constant throughout this text. These elements are often used to model trusses and frame structures (see Figure 1-2 on page 15, for instance). The simplest line element (called a *linear element*) has two nodes, one at each end, although higher-order elements having three nodes or more (called *quadratic, cubic,* etc. *elements*) also exist. The line elements are the simplest of elements to consider and will be discussed in Chapters 2, 3, 5, and 6 to illustrate many of the basic concepts of the finite element method.

The basic two-dimensional (or plane) elements, Figure 1-1(b), are loaded by forces in their own plane (plane stress or plane strain conditions). They are triangular or quadrilateral elements. The simplest two-dimensional elements have corner nodes only (linear elements) with straight sides or boundaries (see Chapter 7), although there are also higher-order elements, typically with midside nodes (called *quadratic elements*) and curved sides (see Chapters 9 and 11). The elements can have variable thicknesses throughout or be constant. They are often used to model a wide range of engineering problems (see Figures 1-3 and 1-4 on page 16).

The most common three-dimensional elements, Figure 1-1(c), are tetrahedral and hexahedral (or brick) elements; they are used when it becomes necessary to perform a three-dimensional stress analysis. The basic three-dimensional elements (see Chapter 12) have corner nodes only and straight sides, whereas higher-order elements with midedge nodes (and possible mid-face nodes) have curved surfaces for their sides.

The axisymmetric element, Figure 1-1(d), is developed by rotating a triangle or quadrilateral about a fixed axis located in the plane of the element through 360°. This element (described in Chapter 10) can be used when the geometry and loading of the problem are axisymmetric.

(a) Simple line element typically used to represent a bar or beam element

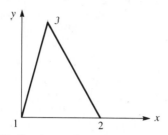
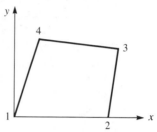

(b) Simple two-dimensional elements typically used to represent plane stress/strain

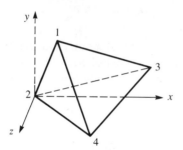
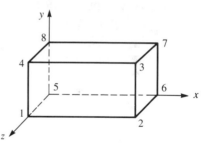

(c) Simple three-dimensional elements typically used to represent three-dimensional stress

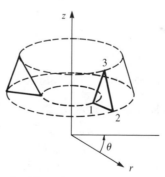

(d) Simple axisymmetric element used for axisymmetric problems

F I G U R E   1-1   Various types of finite elements

Step 2   **Select a Displacement Function**

Step 2 involves choosing a displacement function within each element. The function is defined within the element using the nodal values of the element. Linear, quadratic, and cubic polynomials are frequently used functions because they are simple to work with in finite element formulation. However, trigonometric series could also be used. For a two-dimensional element, the displacement function is a function of the coordinates in its plane (say, *x-y* plane). The functions are expressed in terms of the nodal unknowns (in the two-dimensional problem, in terms of an *x* and a *y* component). The same general displacement function can be used repeatedly for each element. Hence, the finite element method is one in which a continuous quantity, such as the displacement throughout the body, is approximated by a discrete model composed of a set of piecewise-continuous functions defined within each finite domain or finite element.

Step 3   **Define the Strain/Displacement and Stress/Strain Relationships**

Strain/displacement and stress/strain relationships are necessary for deriving the equations for each finite element. In the case of one-dimensional deformation, say, in the *x* direction, we have strain ε_x related to displacement *u* by

$$\varepsilon_x = \frac{du}{dx} \qquad (1.4.1)$$

for small strains. In addition, the stresses must be related to the strains through the stress/strain law—generally called the *constitutive law*. The ability to define the material behavior accurately is most important in obtaining acceptable results. The simplest of stress/strain laws, Hooke's law, often used in stress analysis, is given by

$$\sigma_x = E\varepsilon_x \qquad (1.4.2)$$

where σ_x = stress in the *x* direction and *E* = modulus of elasticity.

Step 4   **Derive the Element Stiffness Matrix and Equations**

Initially, the development of element stiffness matrices and element equations was based on the concept of stiffness influence coefficients, which presupposes a background in structural analysis. We now present alternative methods used in this text that do not require this special background.

Direct Equilibrium Method

According to this method, the stiffness matrix and element equations relating nodal forces to nodal displacements are obtained using force equilibrium conditions for a basic element, along with force/deformation relationships. Because this method is most easily adaptable to line or one-dimensional elements, Chapters 2, 3, and 5 illustrate this method for spring, bar, and beam elements, respectively.

Work or Energy Methods

To develop the stiffness matrix and equations for two- and three-dimensional elements, it is much easier to apply a work or energy method [35]. The principle of virtual work (using virtual displacements), the principle of minimum potential energy, and Castigliano's theorem are methods frequently used for the purpose of derivation of element equations.

The principle of virtual work outlined in Appendix E is applicable for any material behavior, whereas the principle of minimum potential energy and Castigliano's theorem are applicable only to elastic materials. Furthermore, the principle of virtual work can be used even when a potential function does not exist. However, all three principles yield identical element equations for linear-elastic materials; thus the method to be used for this kind of material in structural analysis is largely a matter of convenience and personal preference. We will present the principle of minimum potential energy—probably the most well known of the three energy methods mentioned here—in detail in Chapters 2 and 3 where it will be used to derive the spring and bar element equations. We will further generalize the principle and apply it to the beam element in Chapter 5 and to the plane stress/strain element in Chapter 7. Thereafter, the principle is routinely referred to as the basis for deriving all other stress-analysis stiffness matrices and element equations given in Chapters 9, 10, and 12.

For the purpose of extending the finite element method outside of the structural stress analysis field, a **functional** (defined to be a function of another function) analogous to the one to be used with the principle of minimum potential energy is quite useful in deriving the element stiffness matrix and equations (see Chapters 13 and 14 on heat transfer and fluid flow, respectively). For instance, letting π denote the functional and $f(x, y)$ denote a function f of two variables x and y, we then have $\pi = \pi(f(x, y))$, where π is a function of the function f.

Methods of Weighted Residuals

The methods of weighted residuals are useful for developing the element equations—particularly popular is Galerkin's method. These methods yield the same results as the energy methods, wherever the energy methods are applicable. They are particularly useful when a functional such as potential energy is not readily available. The weighted residual methods allow the finite element method to be applied directly to any differential equation.

Galerkin's method is introduced in Chapter 3, and then it is used to derive the bar element equations in Chapter 3, the beam element equations in Chapter 5, and then the combined heat-conduction/convection/mass transport problem in Chapter 13. For more information on the use of the methods of weighted residuals, see Reference [36]; for additional applications to the finite element method, consult References [37] and [38].

Using any of the methods just outlined will produce the equations to describe the behavior of an element. These equations are written conveniently in matrix form as

$$\begin{Bmatrix} f_1 \\ f_2 \\ f_3 \\ \vdots \\ f_n \end{Bmatrix} = \begin{bmatrix} k_{11} & k_{12} & k_{13} & \cdots & k_{1n} \\ k_{21} & k_{22} & k_{23} & \cdots & k_{2n} \\ k_{31} & k_{32} & k_{33} & \cdots & k_{3n} \\ \vdots & & & & \vdots \\ k_{n1} & & & \cdots & k_{nn} \end{bmatrix} \begin{Bmatrix} d_1 \\ d_2 \\ d_3 \\ \vdots \\ d_n \end{Bmatrix} \qquad (1.4.3)$$

or in compact matrix form as

$$\{f\} = [k]\{d\} \qquad (1.4.4)$$

where $\{f\}$ is the vector of element nodal forces, $[k]$ is the element stiffness matrix, and $\{d\}$ is the vector of unknown element nodal degrees of freedom or generalized displacements, n. Here generalized displacements may include such quantities as actual displacements, slopes, or even curvatures. The matrices in Eq. (1.4.4) will be developed and described in detail in subsequent chapters for specific element types, such as those in Figure 1-1.

Step 5 Assemble the Element Equations to Obtain the Global or Total Equations and Introduce Boundary Conditions

The individual element equations generated in Step 4 can now be added together using a method of superposition (called the *direct stiffness method*)— whose basis is nodal force equilibrium—to obtain the global equations for the whole structure. Implicit in the direct stiffness method is the concept of continuity, or compatibility, which requires that the structure remain together and that no tears occur anywhere in the structure.

The final assembled or global equation written in matrix form is

$$\{F\} = [K]\{d\} \qquad (1.4.5)$$

where $\{F\}$ is the vector of global nodal forces, $[K]$ is the structure global or total stiffness matrix, and $\{d\}$ is now the vector of known and unknown structure nodal degrees of freedom or generalized displacements. It can be shown that, at this stage, the global stiffness matrix $[K]$ is a singular matrix because its determinant is equal to zero. To remove this singularity problem we must invoke certain boundary conditions (or constraints or supports) so that the structure remains in place instead of moving as a rigid body. Further details and methods of invoking boundary conditions are given in subsequent chapters. At this time it is sufficient to note that invoking boundary or support conditions results in a modification of the global Eq. (1.4.5). We also emphasize that the applied known loads have been accounted for in the global force matrix $\{F\}$.

Step 6 Solve for the Unknown Degrees of Freedom (or Generalized Displacements)

Equation (1.4.5), modified to account for the boundary conditions, is a set of simultaneous algebraic equations that can be written in expanded matrix form as

$$
\begin{Bmatrix} F_1 \\ F_2 \\ \vdots \\ F_n \end{Bmatrix} = \begin{bmatrix} K_{11} & K_{12} & \cdots & K_{1n} \\ K_{21} & K_{22} & \cdots & K_{2n} \\ \vdots & & & \vdots \\ K_{n1} & K_{n2} & \cdots & K_{nn} \end{bmatrix} \begin{Bmatrix} d_1 \\ d_2 \\ \vdots \\ d_n \end{Bmatrix} \qquad (1.4.6)
$$

where now n is the structure total number of unknown nodal degrees of freedom. These equations can be solved for the d's by using an elimination method (such as Gauss's method) or an iterative method (such as Gauss–Seidel's method). These two methods are discussed in Appendix B. The d's are called the *primary unknowns* because they are the first quantities determined using the stiffness (or displacement) finite element method.

Step 7 **Solve for the Element Strains and Stresses**

For the structural stress-analysis problem, important secondary quantities of strain and stress (or moment and shear force) can be obtained because they can be directly expressed in terms of the displacements determined in Step 6. Typical relationships between strain and displacement and between stress and strain—such as Eqs. (1.4.1) and (1.4.2) for one-dimensional stress given in Step 3—can be used.

Step 8 **Interpret the Results**

The final goal is to interpret and analyze the results for use in the design/analysis process. Determination of locations in the structure where large deformations and large stresses occur is generally important in making design/analysis decisions. Postprocessor computer programs help the user to interpret the results by displaying them in graphical form.

1.5 Applications of the Finite Element Method

The finite element method can be used to analyze both structural and non-structural problems. Typical structural areas include

1. Stress analysis, including truss and frame analysis, and stress concentration problems typically associated with holes, fillets, or other changes in geometry in a body
2. Buckling
3. Vibration analysis

Nonstructural problems include

1. Heat transfer
2. Fluid flow, including seepage through porous media
3. Distribution of electric or magnetic potential

Finally, some biomechanical engineering problems (which may include stress analysis) typically include analyses of human

1. Spine
2. Skull
3. Hip joints
4. Jaw/gum tooth implants
5. Heart
6. Eye

We now present some typical applications of the finite element method. These applications will illustrate both the variety, size, and complexity of problems that can be solved using the method and the typical discretization process and kinds of elements used.

The first illustration, shown in Figure 1-2, is of a control tower for a railroad. The tower is a three-dimensional frame comprising a series of beam-type elements. The 48 elements are labeled by the circled numbers, whereas the 28 nodes are indicated by the uncircled numbers. Each node has three rotation and three displacement components associated with it. The rotations and displacements are called the *degrees of freedom*. Because of the loading conditions to which the tower structure is subjected, we have used a three-dimensional model.

The finite element method used for this frame enables the designer/analyst to quickly obtain displacements and stresses in the tower for typical load cases, as required by design codes. Before the development of the finite element method and the computer, even this relatively simple problem took many hours to solve.

The next illustration of the application of the finite element method to problem solving is the determination of displacements and stresses in an underground box culvert subjected to ground shock loading from a bomb explosion. Figure 1-3 shows the discretized model that included a total of 369 nodes, 40 one-dimensional bar or truss elements used to model the steel reinforcement in the box culvert, and 333 plane strain two-dimensional triangular and rectangular elements used to model the surrounding soil and concrete box culvert. With an assumption of symmetry, only half of the box culvert must be analyzed. This problem requires the solution of nearly 700 unknown nodal displacements. It illustrates that different kinds of elements (here bar and plane strain) can often be used in one finite element model.

Another problem, that of a hydraulic cylinder rod end shown in Figure 1-4, was modeled by 120 nodes and 297 plane strain triangular elements. Symmetry was also applied to the whole rod end so that only half of the rod end had to be analyzed, as shown. The purpose of this analysis was to locate areas of high stress concentration in the rod end.

Figure 1-5 shows a chimney stack section that is four form heights high (or a total of 32 ft high). In this illustration, 584 beam elements were used to model the vertical and horizontal stiffeners making up the formwork, whereas 252 flat-plate elements were used to model the inner wooden form and the

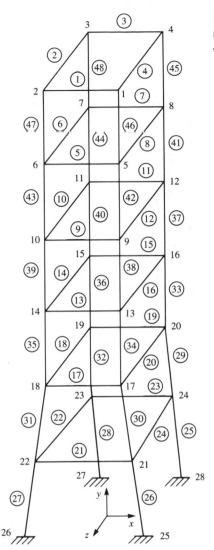

FIGURE 1-2 Discretized railroad control tower (28 nodes, 48 beam elements)

concrete shell. Owing to the irregular loading pattern on the structure, a three-dimensional model was necessary. Displacements and stresses in the concrete were of prime concern in this problem.

Figure 1-6 shows the finite element discretized model of a proposed steel die used in a plastic film-making process. The irregular geometry and associated potential stress concentrations necessitated use of the finite element method to obtain a reasonable solution. Two hundred forty axisymmetric elements were used to model the three-dimensional die.

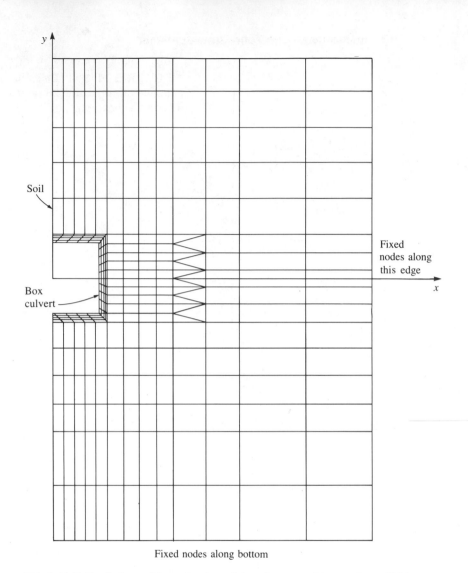

FIGURE 1-3   Discretized model underground box culvert (369 nodes, 40 bar elements, and 333 plane strain elements) [39]

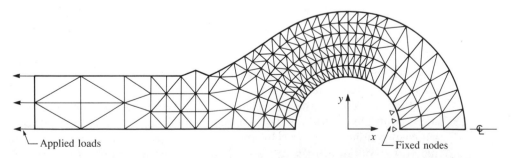

FIGURE 1-4   Two-dimensional analysis of hydraulic cylinder rod end (120 nodes, 297 plane strain triangular elements)

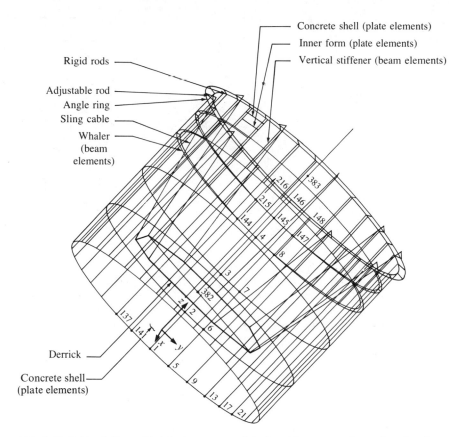

Concrete shell (plate elements)

Inner form (plate elements)

Vertical stiffener (beam elements)

Rigid rods

Adjustable rod

Angle ring

Sling cable

Whaler
(beam
elements)

Derrick

Concrete shell
(plate elements)

F I G U R E   1-5   Finite element model of a chimney stack section (end
view rotated 45°) (584 beam and 252 flat-plate
elements)

Figure 1-7 illustrates the use of a three-dimensional solid element to
model a swing casting for a backhoe frame. The three-dimensional hexahedral elements are necessary to model the irregularly shaped three-
dimensional casting. Two-dimensional models certainly would not yield accurate engineering solutions to this problem.

Figure 1-8 illustrates a two-dimensional heat-transfer model used to
determine the temperature distribution in earth subjected to a heat source—a
buried pipeline transporting a hot gas.

Finally, Figure 1-9 shows a three-dimensional finite element model of a
femur bone with an implant, used to study stresses in the bone and the cement
layer between bone and implant.

The preceding illustrations introduced you to the kinds of problems that
can be solved by the finite element method. Additional guidelines concerning
modeling techniques will be provided in Chapter 8.

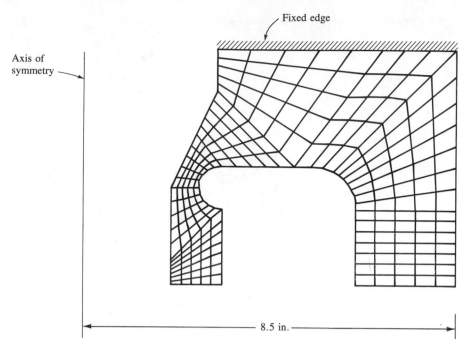

Axis of
symmetry

Fixed edge

8.5 in.

F I G U R E 1-6 Model of a high-strength steel die (240 axisymmetric
elements) used in plastic film industry [40]

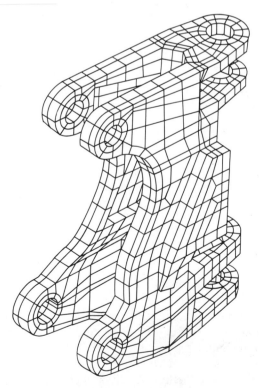

F I G U R E 1-7 Three-dimensional solid element model of swing casting
for backhoe frame

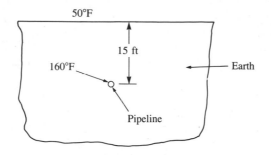

50°F

15 ft

160°F

Earth

Pipeline

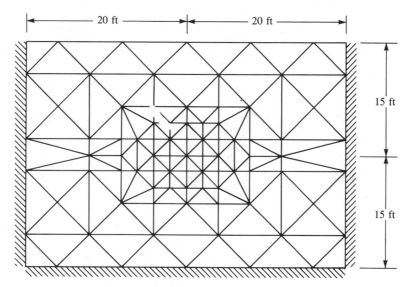

20 ft

20 ft

15 ft

15 ft

15 ft

FIGURE 1-8 Finite element model for two-dimensional temperature distribution in earth

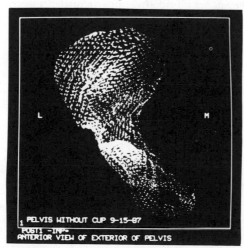

PELVIS WITHOUT CUP 9-15-87

POST1 -INP-
ANTERIOR VIEW OF EXTERIOR OF PELVIS

FIGURE 1-9 Finite element model of a femur bone with implant (nearly 5000 solid elements used in the model) (Courtesy of Harrington Arthritis Research Center, Phoenix, Arizona) [41]

1.6   Advantages of the Finite Element Method

As previously indicated, the finite element method has been applied to numerous problems, both structural and nonstructural. This method has a number of advantages that have made it very popular. They include the ability to

1. Model irregularly shaped bodies quite easily
2. Handle general load conditions without difficulty
3. Model bodies composed of several different materials because the element equations are evaluated individually
4. Handle unlimited numbers and kinds of boundary conditions
5. Vary the size of the elements to make it possible to use small elements where necessary
6. Alter the finite element model relatively easily and cheaply
7. Include dynamic effects
8. Handle nonlinear behavior existing with large deformations and nonlinear materials

The finite element method of structural analysis enables the designer to detect stress, vibration, and thermal problems during the design process and to evaluate design changes *before* the construction of a possible prototype. Thus, confidence in the acceptability of the prototype is enhanced. Moreover, the method, if used properly, can reduce the number of prototypes that need to be built.

Even though the finite element method was initially used for structural analysis, it has since been adapted to many other disciplines in engineering and mathematical physics, such as fluid flow, heat transfer, electromagnetic potentials, soil mechanics, and acoustics (see References [22, 23, 24, 27, 42, 43, 44]).

1.7   Computer Programs for the Finite Element Method

There are two general computer methods of approach to the solution of problems by the finite element method. One is to use large commercial programs, many of which have been scaled down to run on personal computers (PCs); these are called general-purpose programs and are designed to solve many types of problems. The other is to develop many small, special-purpose programs, such as those listed on the accompanying disk, to solve special problems. In this section, we will discuss the advantages and disadvantages of both methods. We will then list some of the available general-purpose programs and discuss some of their standard capabilities.

Some advantages of general-purpose programs are

1. The input is well organized and is developed with user ease in mind. Users do not need special knowledge of computer software or hardware. Preprocessors are readily available to help create the finite element model.

2. The programs are large systems that often can solve many types of problems of large or small size using the same input format.

3. Many of the programs can be expanded by adding new modules for new kinds of problems or new technology. Thus, they may be kept current with a minimum of effort.

4. Most large computer centers have one or more of the general-purpose programs available to users.

5. Many of the scaled-down programs have become very attractive in price and can solve a large range of problems [45].

Some disadvantages of general-purpose programs are

1. The initial cost of developing general-purpose programs is high.

2. General-purpose programs are less efficient than special-purpose programs because the computer must make many checks for each problem, some of which would not be necessary if a special-purpose program were used.

3. Many of the programs are proprietary. Hence, the user has little access to the logic of the program. If a revision must be made, it often has to be done by the developers.

4. Large computers (both core and off-line storage) are needed to handle general-purpose programs. (Scaled-down programs can be run on PCs.)

Some advantages of special-purpose programs are

1. The programs are usually relatively short, with low development costs.

2. Small computers are able to run the programs.

3. Additions can be made to the program quickly and at a low cost.

4. The programs are efficient in solving the problems they were designed to solve.

The major disadvantage of special-purpose programs is their inability to solve different classes of problems. Thus, one must have as many programs as there are different classes of problems to be solved.

There are numerous vendors supporting finite element programs, and the interested user should carefully consult the vendor before purchasing any software. However, to give you an idea about the various commercial personal computer programs now available for solving problems by the finite element method, we present a partial list of existing programs.

1. ALGOR [46]
2. ANSYS [47]
3. GIFTS [48]
4. IMAGES-3D [49]
5. MSC/PAL [50]
6. TAB/SAP86 [51]

Standard capabilities of many of the listed programs are provided in the preceding references and in Reference [45]. These capabilities include information on

1. Element types available, such as beam, plane stress, and three-dimensional solid
2. Type of analysis available, such as static and dynamic
3. Material behavior, such as linear-elastic and nonlinear
4. Load types, such as concentrated, distributed, thermal, and displacement (settlement)
5. Data generation, such as automatic generation of nodes, elements, and restraints (most programs have preprocessors to generate the mesh for the model)
6. Plotting, such as original and deformed geometry and stress and temperature contours (most programs have postprocessors to aid in interpreting results in graphical form)
7. Displacement behavior, such as small and large displacement and buckling
8. Selective output, such as at selected nodes, elements, and maximum or minimum values

All programs include at least the bar, beam, plane stress, plate-bending, and three-dimensional solid elements, and most now include heat-transfer analysis capabilities.

Complete capabilities of these programs are best obtained through program reference manuals, such as References [46–51].

REFERENCES

[1] Hrennikoff, A., "Solution of Problems in Elasticity by the Frame Work Method," *Journal of Applied Mechanics*, Vol. 8, No. 4, pp. 169–175, Dec. 1941.

[2] McHenry, D., "A Lattice Analogy for the Solution of Plane Stress Problems," *Journal of Institution of Civil Engineers*, Vol. 21, pp. 59–82, Dec. 1943.

[3] Courant, R., "Variational Methods for the Solution of Problems of Equilibrium and Vibrations," *Bulletin of the American Mathematical Society*, Vol. 49, pp. 1–23, 1943.

[4] Levy, S., "Computation of Influence Coefficients for Aircraft Structures with Discontinuities and Sweepback," *Journal of Aeronautical Sciences*, Vol. 14, No. 10, pp. 547–560, Oct. 1947.

[5] Levy, S., "Structural Analysis and Influence Coefficients for Delta Wings," *Journal of Aeronautical Sciences*, Vol. 20, No. 7, pp. 449–454, July 1953.

[6] Argyris, J. H., "Energy Theorems and Structural Analysis, *Aircraft Engineering*, Oct., Nov., Dec. 1954 and Feb., Mar., Apr., May 1955.

[7] J. H. Argyris and S. Kelsey, *Energy Theorems and Structural Analysis*, Butterworths, London, 1960 (collection of papers published in *Aircraft Engineering* in 1954 and 1955).

[8] Turner, M. J., Clough, R. W., Martin, H. C., and Topp, L. J., "Stiffness and Deflection Analysis of Complex Structures," *Journal of Aeronautical Sciences*, Vol. 23, No. 9, pp. 805–824, Sept. 1956.

[9] Clough, R. W., "The Finite Element Method in Plane Stress Analysis," *Proceedings*, American Society of Civil Engineers, 2nd Conference on Electronic Computation, Pittsburgh, Pa., pp. 345–378, Sept. 1960.

[10] Melosh, R. J., "A Stiffness Matrix for the Analysis of Thin Plates in Bending," *Journal of the Aerospace Sciences*, Vol. 28, No. 1, pp. 34–42, Jan. 1961.

[11] Grafton, P. E., and Strome, D. R., "Analysis of Axisymmetric Shells by the Direct Stiffness Method," *Journal of the American Institute of Aeronautics and Astronautics*, Vol. 1, No. 10, pp. 2342–2347, 1963.

[12] Martin, H. C., "Plane Elasticity Problems and the Direct Stiffness Method," *The Trend in Engineering*, Vol. 13, pp. 5–19, Jan. 1961.

[13] Gallagher, R. H., Padlog, J., and Bijlaard, P. P., "Stress Analysis of Heated Complex Shapes," *Journal of the American Rocket Society*, Vol. 32, pp. 700–707, May 1962.

[14] Melosh, R. J., "Structural Analysis of Solids," *Journal of the Structural Division*, Proceedings of the American Society of Civil Engineers, pp. 205–223, Aug. 1963.

[15] Argyris, J. H., "Recent Advances in Matrix Methods of Structural Analysis," *Progress in Aeronautical Science*, Vol. 4, Pergamon Press, New York, 1964.

[16] Clough, R. W., and Rashid, Y., "Finite Element Analysis of Axisymmetric Solids," *Journal of the Engineering Mechanics Division*, Proceedings of the American Society of Civil Engineers, Vol. 91, pp. 71–85, Feb. 1965.

[17] Wilson, E. L., "Structural Analysis of Axisymmetric Solids," *Journal of the American Institute of Aeronautics and Astronautics*, Vol. 3, No. 12, pp. 2269–2274, Dec. 1965.

[18] Turner, M. J., Dill, E. H., Martin, H. C., and Melosh, R. J., "Large Deflections of Structures Subjected to Heating and External Loads," *Journal of Aeronautical Sciences*, Vol. 27, No. 2, pp. 97–107, Feb. 1960.

[19] Gallagher, R. H., and Padlog, J., "Discrete Element Approach to Structural Stability Analysis," *Journal of the American Institute of Aeronautics and Astronautics*, Vol. 1, No. 6, pp. 1437–1439, 1963.

[20] Zienkiewicz, O. C., Watson, M., and King, I. P., "A Numerical Method of Visco-Elastic Stress Analysis," *International Journal of Mechanical Sciences,* Vol. 10, pp. 807–827, 1968.

[21] Archer, J. S., "Consistent Matrix Formulations for Structural Analysis Using Finite-Element Techniques," *Journal of the American Institute of Aeronautics and Astronautics,* Vol. 3, No. 10, pp. 1910–1918, 1965.

[22] Zienkiewicz, O. C., and Cheung, Y. K., "Finite Elements in the Solution of Field Problems," *The Engineer,* pp. 507–510, Sept. 24, 1965.

[23] Martin, H. C., "Finite Element Analysis of Fluid Flows," *Proceedings of the Second Conference on Matrix Methods in Structural Mechanics,* Wright-Patterson Air Force Base, Ohio, pp. 517–535, Oct. 1968. (AFFDL-TR-68-150, Dec. 1969; AD-703-685, N.T.I.S.)

[24] Wilson, E. L., and Nickel, R. E., "Application of the Finite Element Method to Heat Conduction Analysis," *Nuclear Engineering and Design,* Vol. 4, pp. 276–286, 1966.

[25] Szabo, B. A., and Lee, G. C., "Derivation of Stiffness Matrices for Problems in Plane Elasticity by Galerkin's Method," *International Journal of Numerical Methods in Engineering,* Vol. 1, pp. 301–310, 1969.

[26] Zienkiewicz, O. C., and Parekh, C. J., "Transient Field Problems: Two-Dimensional and Three-Dimensional Analysis by Isoparametric Finite Elements," *International Journal of Numerical Methods in Engineering,* Vol. 2, No. 1, pp. 61–71, 1970.

[27] Lyness, J. F., Owen, D. R. J., and Zienkiewicz, O. C., "Three-Dimensional Magnetic Field Determination Using a Scalar Potential. A Finite Element Solution," *Transactions on Magnetics,* Institute of Electrical and Electronics Engineers, pp. 1649–1656, 1977.

[28] Belytschko, T., "A Survey of Numerical Methods and Computer Programs for Dynamic Structural Analysis," *Nuclear Engineering and Design,* Vol. 37, No. 1, pp. 23–34, 1976.

[29] Belytschko, T., "Efficient Large-Scale Nonlinear Transient Analysis by Finite Elements," *International Journal of Numerical Methods in Engineering,* Vol. 10, No. 3, pp. 579–596, 1976.

[30] Huiskies, R., and Chao, E. Y. S., "A Survey of Finite Element Analysis in Orthopedic Biomechanics: The First Decade," *Journal of Biomechanics,* Vol. 16, No. 6, pp. 385–409, 1983.

[31] *Journal of Biomechanical Engineering,* Transactions of the American Society of Mechanical Engineers, (published quarterly) (1st issue published 1977).

[32] Whiteman, J. R., *A Bibliography for Finite Elements,* Academic Press, London, 1975.

[33] Norrie, D., and deVries, G., *Finite Element Bibliography,* IFI/Plenum, New York, 1976.

[34] Kardestuncer, H., *Elementary Matrix Analysis of Structures,* McGraw-Hill, New York, 1974.

[35] Oden, J. T., and Ripperger, E. A., *Mechanics of Elastic Structures*, 2nd ed., McGraw-Hill, New York, 1981.

[36] Finlayson, B. A., *The Method of Weighted Residuals and Variational Principles*, Academic Press, New York, 1972.

[37] Zienkiewicz, O. C., *The Finite Element Method*, 3rd ed., McGraw-Hill, London, 1977.

[38] Cook, R. D., Malkus, D. S. and Plesha, M. E., *Concepts and Applications of Finite Element Analysis*, 3rd ed., Wiley, New York, 1989.

[39] Koswara, H., *A Finite Element Analysis of Underground Shelter Subjected to Ground Shock Load*, M.S. Thesis, Rose-Hulman Institute of Technology, 1983.

[40] Greer, R. D., "The Analysis of a Film Tower Die Utilizing the ANSYS Finite Element Package," M.S. Thesis, Rose-Hulman Institute of Technology, Terre Haute, Indiana, May 1989.

[41] Koeneman, J. B., Hansen, T. M., and Beres, K., "The Effect of Hip Stem Elastic Modulus and Cement/Stem Bond on Cement Stresses," 36th Annual Meeting, Orthopaedic Research Society, Feb. 5–8, 1990, New Orleans, Louisiana.

[42] Girijavallabham, C. V., and Reese, L. C., "Finite-Element Method for Problems in Soil Mechanics," *Journal of the Structural Division*, American Society of Civil Engineers, No. Sm2, pp. 473–497, Mar. 1968.

[43] Young, C., and Crocker, M., "Transmission Loss by Finite-Element Method," *Journal of the Acoustical Society of America*, Vol. 57, No. 1, pp. 144–148, Jan. 1975.

[44] Silvester, P. P. and Ferrari, R. L., *Finite Elements for Electrical Engineers*, Cambridge University Press, Cambridge, 1983.

[45] Falk, H. and Beardsley, C. W., "Finite Element Analysis Packages for Personal Computers," *Mechanical Engineering*, pp. 54–71, Jan. 1985.

[46] Algor Interactive Systems, 260 Alpha Drive, Pittsburgh, PA 15238.

[47] Swanson, J. A., ANSYS-Engineering Analysis Systems User's Manual, Swanson Analysis Systems, Inc., Johnson Rd., P.O. Box 65, Houston, PA 15342.

[48] CASA/GIFTS, Inc., 7474 Greenway Center Dr., Suite 510, Greenbelt, MD 20770.

[49] Celestial Software, 125 University Ave., Berkeley, CA 94710.

[50] MacNeal-Schwendler Corporation, 815 Colorado Blvd., Los Angeles, CA 90041.

[51] Number Cruncher Microsystems, 1455 Hayes St., San Francisco, CA 94117.

PROBLEMS

1.1   Define the term *finite element*.

1.2   What does *discretization* mean in the finite element method?

1.3   In what year did the modern development of the finite element method begin?

1.4   In what year was the direct stiffness method introduced?

1.5   Define the term *matrix*.

1.6   What role did the computer play in the use of the finite element method?

1.7   List and briefly describe the general steps of the finite element method.

1.8   What is *the displacement method*?

1.9   List four common types of finite elements.

1.10   Name three commonly used methods for deriving the element stiffness matrix and element equations. Briefly describe each method.

1.11   To what does the term *degrees of freedom* refer?

1.12   List five typical areas of engineering where the finite element method is applied.

1.13   List five advantages of the finite element method.

Introduction to

the Stiffness

(Displacement) Method

INTRODUCTION

This chapter introduces some of the basic concepts on which the direct stiffness method is founded. The linear spring is introduced first because it provides a simple yet generally instructive tool to illustrate the basic concepts. We begin with a general definition of the stiffness matrix and then consider the derivation of the stiffness matrix for a linear-elastic spring element. We next illustrate how to assemble the total stiffness matrix for a structure comprising an assemblage of spring elements by using elementary concepts of equilibrium and compatibility. We then show how the total stiffness matrix for an assemblage can be obtained by superimposing the stiffness matrices of the individual elements in a direct manner. The term *direct stiffness method* evolved in reference to this method.

After establishing the total structure stiffness matrix, we illustrate how to impose boundary conditions—both homogeneous and nonhomogeneous. A complete solution including the nodal displacements and reactions is thus obtained. (The determination of internal forces is discussed in Chapter 3 in association with the bar element.)

We then introduce the principle of minimum potential energy, apply it to derive the spring element equations, and use it to solve a spring assemblage problem. We will illustrate this principle for the simplest of elements (those with small numbers of degrees of freedom) so that it will become a more easily understood concept when necessarily applied to elements with large numbers of degrees of freedom in subsequent chapters.

2.1   Definition of the Stiffness Matrix

The definition of the stiffness matrix is essential to understanding the stiffness method. We define the stiffness matrix as follows: *For an element, a **stiffness matrix** $\underline{\hat{k}}$ is a matrix such that $\underline{\hat{f}} = \underline{\hat{k}}\underline{\hat{d}}$, where $\underline{\hat{k}}$ relates local-coordinate $(\hat{x}, \hat{y}, \hat{z})$ nodal displacements $\underline{\hat{d}}$ to local forces $\underline{\hat{f}}$ of a single element.* (Throughout this text, the underline notation denotes a matrix, and the ^ symbol denotes quantities referred to a local-coordinate system set up to be convenient for the element.)

For a continuous medium or structure comprising a series of elements, a stiffness matrix $\underline{K}$ relates global-coordinate (x, y, z) nodal displacements $\underline{d}$ to global forces $\underline{F}$ of the whole medium or structure. (Lowercase letters such as x, y, and z without the ^ symbol denote global-coordinate variables.)

2.2   Derivation of the Stiffness Matrix for a Spring Element

Using the direct equilibrium approach, we will now derive the stiffness matrix for a one-dimensional linear spring—that is, a spring that obeys Hooke's law and resists forces only in the direction of the spring. Consider the linear spring element shown in Figure 2-1. Reference points 1 and 2 are located at the ends of the element. These reference points are called the *nodes* of the spring element. The local nodal forces are $\hat{f}_{1x}$ and $\hat{f}_{2x}$ for the spring element associated with the local axis $\hat{x}$. The local axis acts in the direction of the spring so as to be able to directly measure displacements and forces along the spring. The local nodal displacements are $\hat{d}_{1x}$ and $\hat{d}_{2x}$ for the spring element. These nodal displacements are called the *degrees of freedom* at each node. Positive directions for the forces and displacements at each node are taken in the positive $\hat{x}$ direction as shown in the figure. The symbol k is called the *spring constant* or *stiffness* of the spring.

Analogies to actual spring constants arise in numerous engineering problems. In Chapter 3, we see that a prismatic uniaxial bar has a spring constant $k = AE/L$, where A represents the cross-sectional area of the bar, E is the modulus of elasticity, and L is the bar length. Similarly, in Chapter 6 we show that a prismatic circular-cross-section bar in torsion has a spring constant $k = JG/L$, where J is the polar moment of inertia and G is the shear modulus of the material. For one-dimensional heat conduction (Chapter 13), $k =$

F I G U R E   2-1   Linear spring element with positive nodal displacement and force conventions

AK_{xx}/L, where K_{xx} is the thermal conductivity of the material and for one-dimensional fluid flow through a porous media (Chapter 14), $k = AK_{xx}/L$, where K_{xx} is the permeability coefficient of the material.

We now want to develop a relationship between nodal forces and nodal displacements for a spring element. This relationship will be the stiffness matrix. Therefore, we want to relate the nodal force matrix to the nodal displacement matrix as follows:

$$\begin{Bmatrix} \hat{f}_{1x} \\ \hat{f}_{2x} \end{Bmatrix} = \begin{bmatrix} k_{11} & k_{12} \\ k_{21} & k_{22} \end{bmatrix} \begin{Bmatrix} \hat{d}_{1x} \\ \hat{d}_{2x} \end{Bmatrix} \tag{2.2.1}$$

where the elements k_{ij} of the $\underline{\hat{k}}$ matrix in Eq. (2.2.1) are to be determined.

We now use the general steps outlined in Section 1.4 to derive the stiffness matrix for the spring element in this section (while keeping in mind that these same steps will be applicable later in the derivation of stiffness matrices of more general elements) and then to illustrate a complete solution of a spring assemblage in Section 2.3. Because our approach throughout this text is to derive various element stiffness matrices and then to illustrate how to solve engineering problems with the elements, step one now involves only selecting the element type.

Step 1   **Select Element Type**

Consider the linear spring (which can be an element in a system of springs) subjected to resulting nodal tensile forces T (which may result owing to the action of adjacent springs) directed along the spring axial direction $\hat{x}$ as shown in Figure 2-2, so as to be in equilibrium. The spring is represented by labeling nodes at each end and by labeling the element number. The original distance between nodes before deformation is denoted by L.

Step 2   **Select a Displacement Function**

A displacement function $\hat{u}$ is assumed. Here a linear displacement variation along the $\hat{x}$ axis of the spring is assumed because a linear function with specified endpoints has a unique path. Therefore,

$$\hat{u} = a_1 + a_2\hat{x} \tag{2.2.2}$$

In general, the total number of coefficients a is equal to the total number of degrees of freedom associated with the element. Here the total number of

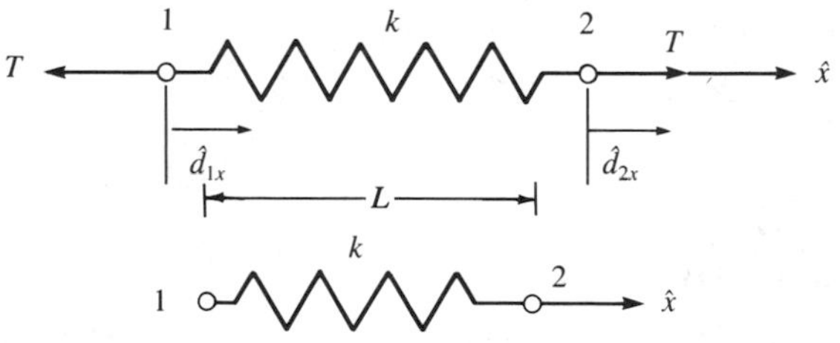

FIGURE   2-2   Linear spring subjected to tensile forces

degrees of freedom is two—an axial displacement at each of the two nodes of the element (we present further discussion regarding the choice of displacement functions in Section 3.2). In matrix form, Eq. (2.2.2) becomes

$$\hat{u} = [1 \quad \hat{x}]\begin{Bmatrix} a_1 \\ a_2 \end{Bmatrix} \tag{2.2.3}$$

We now want to express $\hat{u}$ as a function of the nodal displacements $\hat{d}_{1x}$ and $\hat{d}_{2x}$. We achieve this by evaluating $\hat{u}$ at each node and solving for a_1 and a_2 from Eq. (2.2.2) as follows:

$$\hat{u}(0) = \hat{d}_{1x} = a_1 \tag{2.2.4}$$

$$\hat{u}(L) = \hat{d}_{2x} = a_2 L + \hat{d}_{1x} \tag{2.2.5}$$

or, solving Eq. (2.2.5) for a_2,

$$a_2 = \frac{\hat{d}_{2x} - \hat{d}_{1x}}{L} \tag{2.2.6}$$

On substituting Eqs. (2.2.4) and (2.2.6) into Eq. (2.2.2), we have

$$\hat{u} = \left(\frac{\hat{d}_{2x} - \hat{d}_{1x}}{L}\right)\hat{x} + \hat{d}_{1x} \tag{2.2.7}$$

In matrix form, we express Eq. (2.2.7) as

$$\hat{u} = \left[1 - \frac{\hat{x}}{L} \quad \frac{\hat{x}}{L}\right]\begin{Bmatrix} \hat{d}_{1x} \\ \hat{d}_{2x} \end{Bmatrix} \tag{2.2.8}$$

or

$$\hat{u} = [N_1 \quad N_2]\begin{Bmatrix} \hat{d}_{1x} \\ \hat{d}_{2x} \end{Bmatrix} \tag{2.2.9}$$

Here

$$N_1 = 1 - \frac{\hat{x}}{L} \quad \text{and} \quad N_2 = \frac{\hat{x}}{L} \tag{2.2.10}$$

are called the *shape functions* because the N_i's express the shape of the assumed displacement function over the domain of the element when the ith element degree of freedom has unit value and all other degrees of freedom are zero. In this case, N_1 and N_2 are linear functions that have the properties that $N_1 = 1$ at node 1 and $N_1 = 0$ at node 2, whereas $N_2 = 1$ at node 2 and $N_2 = 0$ at node 1. Also, $N_1 + N_2 = 1$ for any axial coordinate along the bar. (Section 3.2 further explores this important relationship.) In addition, the N_i's are often called *interpolation functions* because we are interpolating to find the value of a function between given nodal values. The interpolation function may be different from the actual function except at the endpoints or nodes where the interpolation function and actual function must be equal to specified nodal values.

Step 3   **Define the Strain/Displacement and Stress/Strain Relationships**

The tensile forces T produce a total elongation (deformation) δ of the spring. For the linear spring, T and δ are related through Hooke's law by

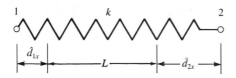

FIGURE 2-3   Deformed spring

$$T = k\delta \qquad (2.2.11)$$

where, because δ is the deformation of the spring, we have

$$\delta = \hat{u}(L) - \hat{u}(0) \qquad (2.2.12)$$

Making use of Eq. (2.2.7), Eq. (2.2.12) becomes

$$\delta = \hat{d}_{2x} - \hat{d}_{1x} \qquad (2.2.13)$$

The typical total elongation of the spring is shown in Figure 2-3. Here $\hat{d}_{1x}$ is a negative value because the direction of displacement is opposite the positive $\hat{x}$ direction, whereas $\hat{d}_{2x}$ is a positive value. From Eq. (2.2.13), we observe that the net displacement is the difference of the nodal displacements in the $\hat{x}$ direction.

Step 4 Derive the Element Stiffness Matrix and Equations

We now derive the spring element stiffness matrix as follows. By the sign convention for nodal forces, we have

$$\hat{f}_{1x} = -T \qquad \hat{f}_{2x} = T \qquad (2.2.14)$$

Using Eqs. (2.2.11), (2.2.13), and (2.2.14), we have

$$T = -\hat{f}_{1x} = k(\hat{d}_{2x} - \hat{d}_{1x})$$
$$T = \hat{f}_{2x} = k(\hat{d}_{2x} - \hat{d}_{1x}) \qquad (2.2.15)$$

or, rewriting Eqs. (2.2.15), we obtain

$$\hat{f}_{1x} = k(\hat{d}_{1x} - \hat{d}_{2x})$$
$$\hat{f}_{2x} = k(\hat{d}_{2x} - \hat{d}_{1x}) \qquad (2.2.16)$$

Now expressing Eqs. (2.2.16) in a single matrix equation yields

$$\begin{Bmatrix} \hat{f}_{1x} \\ \hat{f}_{2x} \end{Bmatrix} = \begin{bmatrix} k & -k \\ -k & k \end{bmatrix} \begin{Bmatrix} \hat{d}_{1x} \\ \hat{d}_{2x} \end{Bmatrix} \qquad (2.2.17)$$

This relationship holds for the spring along the $\hat{x}$ axis. From our basic definition of a stiffness matrix and use of Eq. (2.2.1) applied to Eq. (2.2.17), we obtain

$$\underline{\hat{k}} = \begin{bmatrix} k & -k \\ -k & k \end{bmatrix} \qquad (2.2.18)$$

as the stiffness matrix for a linear spring element. Here $\underline{\hat{k}}$ is called the *local stiffness matrix* for the element. We observe from Eq. (2.2.18) that $\underline{\hat{k}}$ is a symmetric (that is, $k_{ij} = k_{ji}$) square matrix (the number of rows equals the number of columns in $\hat{k}$). Appendix A gives more description and numerical examples of symmetric and square matrices.

Step 5 **Assemble the Element Equations to Obtain the Global Equations and Introduce Boundary Conditions**

The global stiffness matrix and global force matrix are assembled using nodal force equilibrium equations, force/deformation and compatibility equations from Section 2.3, and the direct stiffness method described in Section 2.4. This step applies for structures composed of more than one element such that

$$\underline{K} = [K] = \sum_{e=1}^{N} \underline{k}^{(e)} \quad \text{and} \quad \underline{F} = \{F\} = \sum_{e=1}^{N} \underline{f}^{(e)} \quad (2.2.19)$$

where $\underline{k}$ and $\underline{f}$ are now element stiffness and force matrices expressed in a global reference frame. (Throughout this text, the $\sum$ sign used in this context does not imply a simple summation of element matrices but denotes that these element matrices must be assembled properly according to the direct stiffness method described in Section 2.4.)

Step 6 **Solve for the Nodal Displacements**

The displacements are then determined by imposing boundary conditions and solving a system of equations, $\underline{F} = \underline{K}\underline{d}$, simultaneously.

Step 7 **Solve for the Element Forces**

Finally, the element forces are determined by back-substitution, applied to each element, into equations similar to Eqs. (2.2.16).

2.3 **Example of a Spring Assemblage**

Structures such as trusses, building frames, and bridges comprise basic structural components connected together to form the overall structures. To analyze these structures, we must determine the total structure stiffness matrix for an interconnected system of elements. Before considering the truss and frame, we will determine the total structure stiffness matrix for a spring assemblage by using the force/displacement matrix relationships derived in Section 2.2 for the spring element, along with considerations of fundamental concepts of nodal equilibrium and compatibility.

We will consider the specific example of the two-spring assemblage shown in Figure 2-4*. This example is general enough to illustrate the direct equilibrium approach for obtaining the total stiffness matrix of the spring assemblage. Here we fix node 1, and apply axial forces for F_{3x} at node 3 and F_{2x} at node 2. The stiffnesses of spring elements 1 and 2 are k_1 and k_2,

---

*Throughout this text, element numbers in figures are indicated by circles around them.

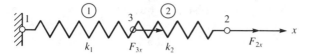

F I G U R E 2-4   Two-spring assemblage

respectively The nodes of the assemblage have been numbered 1, 3, and 2 for
further generalization because sequential numbering between elements generally does not occur in large problems. The x axis is the global axis of the
assemblage. For this example, the local $\hat{x}$ axis of each element coincides with
the global axis of the assemblage.
  For element 1, using Eq. (2.2.17), we have

$$\begin{Bmatrix} f_{1x} \\ f_{3x} \end{Bmatrix} = \begin{bmatrix} k_1 & -k_1 \\ -k_1 & k_1 \end{bmatrix} \begin{Bmatrix} d_{1x} \\ d_{3x} \end{Bmatrix} \qquad (2.3.1)$$

and for element 2, we have

$$\begin{Bmatrix} f_{3x} \\ f_{2x} \end{Bmatrix} = \begin{bmatrix} k_2 & -k_2 \\ -k_2 & k_2 \end{bmatrix} \begin{Bmatrix} d_{3x} \\ d_{2x} \end{Bmatrix} \qquad (2.3.2)$$

Furthermore, elements 1 and 2 must remain connected at common node 3
throughout the displacement. This is called the *continuity* or *compatibility*
requirement. The compatibility requirement yields

$$d_{3x}^{(1)} = d_{3x}^{(2)} = d_{3x} \qquad (2.3.3)$$

where, throughout this text, the superscript in parentheses above d refers to
the element number. Recall that the subscripts to the right identify the node
and the direction of displacement, respectively, and d_{3x} is the node 3 displacement of the total or global spring assemblage.
  Based on the sign conventions for element nodal forces given in Figure
2-1, we can write nodal equilibrium equations at nodes 3, 2, and 1 as

$$F_{3x} = f_{3x}^{(1)} + f_{3x}^{(2)} \qquad (2.3.4)$$

$$F_{2x} = f_{2x}^{(2)} \qquad (2.3.5)$$

$$F_{1x} = f_{1x}^{(1)} \qquad (2.3.6)$$

where F_{1x} results from the reaction at the fixed support. To further clarify the
resulting Eqs. (2.3.4)–(2.3.6), free-body diagrams of each element and node
(using the established sign conventions for element nodal forces) are shown
in Figure 2-5.

F I G U R E 2-5   Nodal forces consistent with element force sign convention

Here Newton's third law, of equal but opposite forces, is applied in moving from a node to an element associated with the node. Using Eqs. (2.3.1), (2.3.2), and (2.3.3) in Eqs. (2.3.4), (2.3.5), and (2.3.6), we obtain

$$F_{3x} = -k_1 d_{1x} + k_1 d_{3x} + k_2 d_{3x} - k_2 d_{2x}$$

$$F_{2x} = -k_2 d_{3x} + k_2 d_{2x} \qquad (2.3.7)$$

$$F_{1x} = k_1 d_{1x} - k_1 d_{3x}$$

In matrix form, Eqs. (2.3.7) are expressed by

$$\begin{Bmatrix} F_{3x} \\ F_{2x} \\ F_{1x} \end{Bmatrix} = \begin{bmatrix} k_1 + k_2 & -k_2 & -k_1 \\ -k_2 & k_2 & 0 \\ -k_1 & 0 & k_1 \end{bmatrix} \begin{Bmatrix} d_{3x} \\ d_{2x} \\ d_{1x} \end{Bmatrix} \qquad (2.3.8)$$

or, rearranging Eq. (2.3.8) in numerically increasing order of the nodal degrees of freedom, we have

$$\begin{Bmatrix} F_{1x} \\ F_{2x} \\ F_{3x} \end{Bmatrix} = \begin{bmatrix} k_1 & 0 & -k_1 \\ 0 & k_2 & -k_2 \\ -k_1 & -k_2 & k_1 + k_2 \end{bmatrix} \begin{Bmatrix} d_{1x} \\ d_{2x} \\ d_{3x} \end{Bmatrix} \qquad (2.3.9)$$

Equation (2.3.9) is now written as the single matrix equation

$$\underline{F} = \underline{K}\underline{d} \qquad (2.3.10)$$

where $\underline{F} = \begin{Bmatrix} F_{1x} \\ F_{2x} \\ F_{3x} \end{Bmatrix}$ is called the *global nodal force matrix*, $\underline{d} = \begin{Bmatrix} d_{1x} \\ d_{2x} \\ d_{3x} \end{Bmatrix}$ is

called the *global nodal displacement matrix*, and

$$\underline{K} = \begin{bmatrix} k_1 & 0 & -k_1 \\ 0 & k_2 & -k_2 \\ -k_1 & -k_2 & k_1 + k_2 \end{bmatrix} \qquad (2.3.11)$$

is called the *total* or *global stiffness matrix*.

In summary, to establish the stiffness equations and stiffness matrix, Eqs. (2.3.9) and (2.3.11), for a spring assemblage, we have used force/deformation relationships, Eqs. (2.3.1) and (2.3.2), compatibility relationship Eq. (2.3.3), and nodal force equilibrium Eqs. (2.3.4)–(2.3.6). We will consider the complete solution to this example problem after considering a more practical method of assembling the total stiffness matrix in Section 2.4 and discussing the support boundary conditions in Section 2.5.

2.4 Assembling the Total Stiffness Matrix by Superposition (Direct Stiffness Method)

We will now consider a more convenient method for constructing the total stiffness matrix. This method is based on proper superposition of the individual element stiffness matrices making up a structure (also see References [1] and [2]).

Referring to the two-spring assemblage of Section 2.3, the element stiffness matrices are given in Eqs. (2.3.1) and (2.3.2) as

$$
\underline{k}^{(1)} = \begin{array}{c} \quad d_{1x} \quad\;\; d_{3x} \\ \left[\begin{array}{cc} k_1 & -k_1 \\ -k_1 & k_1 \end{array} \right] \end{array} \qquad \underline{k}^{(2)} = \begin{array}{c} \quad d_{3x} \quad\;\; d_{2x} \\ \left[\begin{array}{cc} k_2 & -k_2 \\ -k_2 & k_2 \end{array} \right] \end{array} \qquad (2.4.1)
$$

Here the d_{ix}'s written above the columns in the $\underline{k}$'s indicate the degrees of freedom associated with each element.

The two element stiffness matrices, Eqs. (2.4.1), are not associated with the same degrees of freedom; that is, element 1 is associated with axial displacements at nodes 1 and 3, whereas element 2 is associated with axial displacements at nodes 2 and 3. Therefore, the element stiffness matrices cannot be directly added together (superimposed). To superimpose the element matrices, they must be expanded to the order of the total structure (spring assemblage) stiffness matrix so that each element matrix is associated with all the degrees of freedom of the structure. To expand each element stiffness matrix to the order of the total stiffness matrix, we simply add rows and columns of zeros for those displacements not associated with that particular element.

For element 1, we rewrite the stiffness matrix in expanded form so that Eq. (2.3.1) becomes

$$
k_1 \begin{bmatrix} 1 & 0 & -1 \\ 0 & 0 & 0 \\ -1 & 0 & 1 \end{bmatrix} \begin{Bmatrix} d_{1x}^{(1)} \\ d_{2x}^{(1)} \\ d_{3x}^{(1)} \end{Bmatrix} = \begin{Bmatrix} f_{1x}^{(1)} \\ f_{2x}^{(1)} \\ f_{3x}^{(1)} \end{Bmatrix} \qquad (2.4.2)
$$

where, from Eq. (2.4.2), we see that $d_{2x}^{(1)}$ and $f_{2x}^{(1)}$ are not associated with $\underline{k}^{(1)}$. Similarly, for element 2, we have

$$
k_2 \begin{bmatrix} 0 & 0 & 0 \\ 0 & 1 & -1 \\ 0 & -1 & 1 \end{bmatrix} \begin{Bmatrix} d_{1x}^{(2)} \\ d_{2x}^{(2)} \\ d_{3x}^{(2)} \end{Bmatrix} = \begin{Bmatrix} f_{1x}^{(2)} \\ f_{2x}^{(2)} \\ f_{3x}^{(2)} \end{Bmatrix} \qquad (2.4.3)
$$

Now considering force equilibrium at each node results in

$$
\begin{Bmatrix} f_{1x}^{(1)} \\ 0 \\ f_{3x}^{(1)} \end{Bmatrix} + \begin{Bmatrix} 0 \\ f_{2x}^{(2)} \\ f_{3x}^{(2)} \end{Bmatrix} = \begin{Bmatrix} F_{1x} \\ F_{2x} \\ F_{3x} \end{Bmatrix} \qquad (2.4.4)
$$

where Eq. (2.4.4) is really Eqs. (2.3.4)–(2.3.6) expressed in matrix form. Using Eqs. (2.4.2) and (2.4.3) in Eq. (2.4.4), we obtain

$$
k_1 \begin{bmatrix} 1 & 0 & -1 \\ 0 & 0 & 0 \\ -1 & 0 & 1 \end{bmatrix} \begin{Bmatrix} d_{1x}^{(1)} \\ d_{2x}^{(1)} \\ d_{3x}^{(1)} \end{Bmatrix} + k_2 \begin{bmatrix} 0 & 0 & 0 \\ 0 & 1 & -1 \\ 0 & -1 & 1 \end{bmatrix} \begin{Bmatrix} d_{1x}^{(2)} \\ d_{2x}^{(2)} \\ d_{3x}^{(2)} \end{Bmatrix} = \begin{Bmatrix} F_{1x} \\ F_{2x} \\ F_{3x} \end{Bmatrix}
$$

$$(2.4.5)$$

where, again, the superscripts on the d's indicate the element numbers. Simplifying Eq. (2.4.5) results in

$$
\begin{bmatrix}
k_1 & 0 & -k_1 \\
0 & k_2 & -k_2 \\
-k_1 & -k_2 & k_1 + k_2
\end{bmatrix}
\begin{Bmatrix}
d_{1x} \\
d_{2x} \\
d_{3x}
\end{Bmatrix}
=
\begin{Bmatrix}
F_{1x} \\
F_{2x} \\
F_{3x}
\end{Bmatrix}
\qquad (2.4.6)
$$

Here the superscripts indicating the element numbers associated with the nodal displacements have been dropped because $d_{1x}^{(1)}$ is really d_{1x}, $d_{2x}^{(2)}$ is really d_{2x}, and, by Eq. (2.3.3), $d_{3x}^{(1)} = d_{3x}^{(2)} = d_{3x}$, the node 3 displacement of the total assemblage. Equation (2.4.6), obtained through superposition, is identical to Eq. (2.3.9).

The expanded element stiffness matrices in Eqs. (2.4.2) and (2.4.3) could have been added directly to obtain the total stiffness matrix of the structure, given in Eq. (2.4.6). This reliable method of directly assembling individual element stiffness matrices to form the total structure stiffness matrix and the total set of stiffness equations is called the *direct stiffness method*. It is the most important step in the finite element method.

For this simple example, it is easy to expand the element stiffness matrices and then superimpose them to arrive at the total stiffness matrix. However, for problems involving a large number of degrees of freedom, it will become tedious to expand each element stiffness matrix to the order of the total stiffness matrix. To avoid this expansion of each element stiffness matrix, we suggest a direct, or shortcut, form of the direct stiffness method to obtain the total stiffness matrix. For the spring assemblage example, the columns of each element stiffness matrix are labeled according to the degrees of freedom associated with them as follows:

$$
\underline{k}^{(1)} =
\begin{array}{cc}
\begin{matrix} d_{1x} & d_{3x} \end{matrix} \\
\begin{bmatrix}
k_1 & -k_1 \\
-k_1 & k_1
\end{bmatrix}
\end{array}
\qquad
\underline{k}^{(2)} =
\begin{array}{cc}
\begin{matrix} d_{3x} & d_{2x} \end{matrix} \\
\begin{bmatrix}
k_2 & -k_2 \\
-k_2 & k_2
\end{bmatrix}
\end{array}
\qquad (2.4.7)
$$

$\underline{K}$ is then constructed simply by directly adding terms associated with degrees of freedom in $\underline{k}^{(1)}$ and $\underline{k}^{(2)}$ into their corresponding identical degree-of-freedom locations in $\underline{K}$ as follows:

$$
\underline{K} =
\begin{array}{ccc}
\begin{matrix} d_{1x} & d_{2x} & d_{3x} \end{matrix} \\
\begin{bmatrix}
k_1 & 0 & -k_1 \\
0 & k_2 & -k_2 \\
-k_1 & -k_2 & k_1 + k_2
\end{bmatrix}
\end{array}
\qquad (2.4.8)
$$

Here elements in $\underline{K}$ are located on the basis that degrees of freedom are ordered in increasing node numerical order for the total structure. Section 2.5 discusses the complete solution to the two-spring assemblage in conjunction with discussion of the support boundary conditions.

2.5 Boundary Conditions

We must specify boundary (or support) conditions for structure models such as the spring assemblage of Figure 2-4, or $\underline{K}$ will be singular; that is, the determinant of $\underline{K}$ will be zero and, therefore, its inverse will not exist. Without specifying adequate kinematic constraints or support conditions, the structure will be free to move as a rigid body.

Boundary conditions are of two general types: homogeneous boundary conditions—the most common—occur at locations that are completely prevented from movement; nonhomogeneous boundary conditions occur where finite nonzero values of displacement are specified, such as the settlement of a support.

In general, specified support conditions are treated mathematically by partitioning the global equilibrium equations as follows:

$$\left[\begin{array}{c|c} \underline{K}_{11} & \underline{K}_{12} \\ \hline \underline{K}_{21} & \underline{K}_{22} \end{array}\right] \left\{\begin{array}{c} \underline{d}_1 \\ \underline{d}_2 \end{array}\right\} = \left\{\begin{array}{c} \underline{F}_1 \\ \underline{F}_2 \end{array}\right\} \tag{2.5.1}$$

where we let $\underline{d}_1$ be the unconstrained or free displacements and $\underline{d}_2$ be the specified displacements. From Eq. (2.5.1), we have

$$\underline{K}_{11}\underline{d}_1 = \underline{F}_1 - \underline{K}_{12}\underline{d}_2 \tag{2.5.2}$$

and $$\underline{F}_2 = \underline{K}_{21}\underline{d}_1 + \underline{K}_{22}\underline{d}_2 \tag{2.5.3}$$

where $\underline{F}_1$ are the known nodal forces and $\underline{F}_2$ are the unknown nodal forces at the specified displacement nodes. $\underline{F}_2$ is found from Eq. (2.5.3) after determining $\underline{d}_1$ from Eq. (2.5.2). In Eq. (2.5.2), we assume that $\underline{K}_{11}$ is no longer singular, thus allowing for the determination of $\underline{d}_1$.

To illustrate the two general types of boundary conditions, let us consider Eq. (2.4.6), derived for the spring assemblage of Figure 2-4. We will first consider the case of homogeneous boundary conditions. Hence, all boundary conditions are such that the displacements are zero at certain nodes. Here we have $d_{1x} = 0$ because node 1 is fixed. Therefore, Eq. (2.4.6) can be written as

$$\begin{bmatrix} k_1 & 0 & -k_1 \\ 0 & k_2 & -k_2 \\ -k_1 & -k_2 & k_1 + k_2 \end{bmatrix} \left\{\begin{array}{c} 0 \\ d_{2x} \\ d_{3x} \end{array}\right\} = \left\{\begin{array}{c} F_{1x} \\ F_{2x} \\ F_{3x} \end{array}\right\} \tag{2.5.4}$$

Equation (2.5.4), written in expanded form, becomes

$$k_1(0) + (0)d_{2x} - k_1 d_{3x} = F_{1x}$$
$$0(0) + k_2 d_{2x} - k_2 d_{3x} = F_{2x} \tag{2.5.5}$$
$$-k_1(0) - k_2 d_{2x} + (k_1 + k_2)d_{3x} = F_{3x}$$

Consider the second and third of Eqs. (2.5.5) written in matrix form, we have

$$\begin{bmatrix} k_2 & -k_2 \\ -k_2 & k_1 + k_2 \end{bmatrix} \left\{\begin{array}{c} d_{2x} \\ d_{3x} \end{array}\right\} = \left\{\begin{array}{c} F_{2x} \\ F_{3x} \end{array}\right\} \tag{2.5.6}$$

We have now effectively partitioned off the first column and row of $\underline{K}$ and the first row of $\underline{d}$ and $\underline{F}$ to arrive at Eq. (2.5.6).

For homogeneous boundary conditions, Eq. (2.5.6) could have been obtained directly by deleting the row and column of Eq. (2.5.4) corresponding to the zero-displacement degrees of freedom. Here row 1 and column 1 are deleted because $d_{1x} = 0$. However, F_{1x} is not necessarily zero and must be determined as follows.

After solving Eq. (2.5.6) for d_{2x} and d_{3x}, we have

$$\begin{Bmatrix} d_{2x} \\ d_{3x} \end{Bmatrix} = \begin{bmatrix} k_2 & -k_2 \\ -k_2 & k_1 + k_2 \end{bmatrix}^{-1} \begin{Bmatrix} F_{2x} \\ F_{3x} \end{Bmatrix} = \begin{bmatrix} \dfrac{1}{k_2} + \dfrac{1}{k_1} & \dfrac{1}{k_1} \\ \dfrac{1}{k_1} & \dfrac{1}{k_1} \end{bmatrix} \begin{Bmatrix} F_{2x} \\ F_{3x} \end{Bmatrix} \qquad (2.5.7)$$

Using Eq. (2.5.7) in the first of Eqs. (2.5.5), we obtain the reaction F_{1x} as

$$F_{1x} = -k_1 d_{3x} \qquad (2.5.8)$$

We can express the unknown nodal force at node 1 (also called the *reaction*) in terms of the applied nodal forces F_{2x} and F_{3x} by using Eq. (2.5.7) for d_{3x} substituted into Eq. (2.5.8). The result is

$$F_{1x} = -F_{2x} - F_{3x} \qquad (2.5.9)$$

Therefore, for all homogeneous boundary conditions, we can delete the rows and columns corresponding to the zero-displacement degrees of freedom from the original set of equations and then solve for the unknown displacements. This procedure is useful for hand calculations. (However, Chapter 4 presents a more practical, computer-assisted scheme for solving the system of simultaneous equations.)

We now consider the case of nonhomogeneous boundary conditions. Hence, some of the specified displacements are nonzero. For simplicity's sake, let $d_{1x} = \delta$, where δ is a known displacement, in Eq. (2.4.6). We now have

$$\begin{bmatrix} k_1 & 0 & -k_1 \\ 0 & k_2 & -k_2 \\ -k_1 & -k_2 & k_1 + k_2 \end{bmatrix} \begin{Bmatrix} \delta \\ d_{2x} \\ d_{3x} \end{Bmatrix} = \begin{Bmatrix} F_{1x} \\ F_{2x} \\ F_{3x} \end{Bmatrix} \qquad (2.5.10)$$

Equation (2.5.10) written in expanded form becomes

$$k_1 \delta + 0 d_{2x} - k_1 d_{3x} = F_{1x}$$
$$0 \delta + k_2 d_{2x} - k_2 d_{3x} = F_{2x} \qquad (2.5.11)$$
$$-k_1 \delta - k_2 d_{2x} + (k_1 + k_2) d_{3x} = F_{3x}$$

Considering the second and third of Eqs. (2.5.11) because they have known right-side nodal forces F_{2x} and F_{3x}, we obtain

$$0 \delta + k_2 d_{2x} - k_2 d_{3x} = F_{2x}$$
$$-k_1 \delta - k_2 d_{2x} + (k_1 + k_2) d_{3x} = F_{3x} \qquad (2.5.12)$$

Transforming the known δ terms to the right side of Eqs. (2.5.12) results in

$$k_2 d_{2x} - k_2 d_{3x} = F_{2x}$$
$$-k_2 d_{2x} + (k_1 + k_2)d_{3x} = +k_1 \delta + F_{3x}$$

$(2.5.13)$

Rewriting Eqs. (2.5.13) in matrix form, we have

$$\begin{bmatrix} k_2 & -k_2 \\ -k_2 & k_1 + k_2 \end{bmatrix} \begin{Bmatrix} d_{2x} \\ d_{3x} \end{Bmatrix} = \begin{Bmatrix} F_{2x} \\ k_1 \delta + F_{3x} \end{Bmatrix}$$

$(2.5.14)$

Therefore, when dealing with nonhomogeneous boundary conditions, one cannot initially delete row 1 and column 1 of Eq. (2.5.10), corresponding to the nonhomogeneous boundary condition, as indicated by the resulting Eq. (2.5.14). Had we done so, the $k_1 \delta$ term in Eq. (2.5.14) would have been neglected, resulting in an error in the solution for the displacements. For nonhomogeneous boundary conditions, we must, in general, transform the terms associated with the known displacements to the right-side force matrix before solving for the unknown nodal displacements. This was illustrated by transforming the $k_1 \delta$ term of the second of Eqs. (2.5.12) to the right side of the second of Eqs. (2.5.13).

We could now solve for the displacements in Eq. (2.5.14) in a manner similar to that used to solve Eq. (2.5.6). However, we will not further pursue the solution of Eq. (2.5.14) because no new information is to be gained.

At this point, we summarize some properties of the stiffness matrix in Eq. (2.5.10) that are also applicable to the generalization of the finite element method.

1. $\underline{K}$ is symmetric, as is each of the element stiffness matrices. If you are familiar with structural mechanics, this symmetry property is not surprising and can be proved using the reciprocal laws described in such References as [3] and [4].

2. $\underline{K}$ is singular and thus no inverse exists until sufficient boundary conditions are imposed to remove the singularity and prevent rigid body motion.

3. The main diagonal terms of $\underline{K}$ are always positive. Otherwise, a positive nodal force F_i could produce a negative displacement d_i— a behavior contrary to the physical behavior of any actual structure.

To illustrate the stiffness method for the solution of spring assemblages we now present the following examples.

EXAMPLE 2.1

For the spring assemblage with arbitrarily numbered nodes shown in Figure 2-6, obtain (a) the global stiffness matrix, (b) the displacements of nodes 3 and 4, (c) the reaction forces at nodes 1 and 2, and (d) the forces in each spring. A

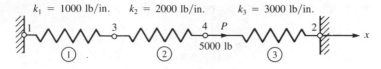

FIGURE 2-6   Spring assemblage for solution

force of 5000 lb is applied at node 4 in the x direction. The spring constants are given in the figure. Nodes 1 and 2 are fixed.

(a) We begin by making use of Eq. (2.2.18) to express each element stiffness matrix as follows:

$$\underline{k}^{(1)} = \begin{array}{cc} 1 & 3 \\ \begin{bmatrix} 1000 & -1000 \\ -1000 & 1000 \end{bmatrix} \end{array} \qquad \underline{k}^{(2)} = \begin{array}{cc} 3 & 4 \\ \begin{bmatrix} 2000 & -2000 \\ -2000 & 2000 \end{bmatrix} \end{array} \qquad (2.5.15)$$

$$\underline{k}^{(3)} = \begin{array}{cc} 4 & 2 \\ \begin{bmatrix} 3000 & -3000 \\ -3000 & 3000 \end{bmatrix} \end{array}$$

where the numbers above the columns indicate the nodal degrees of freedom associated with each element. For instance, element 1 is associated with degrees of freedom d_{1x} and d_{3x}.

Using the concept of superposition (the direct stiffness method), we obtain the global stiffness matrix as

$$\underline{K} = \underline{k}^{(1)} + \underline{k}^{(2)} + \underline{k}^{(3)}$$

or

$$\underline{K} = \begin{array}{cccc} 1 & 2 & 3 & 4 \\ \begin{bmatrix} 1000 & 0 & -1000 & 0 \\ 0 & 3000 & 0 & -3000 \\ -1000 & 0 & 1000 + 2000 & -2000 \\ 0 & -3000 & -2000 & 2000 + 3000 \end{bmatrix} \end{array} \qquad (2.5.16)$$

(b) The global stiffness matrix, Eq. (2.5.16), relates global forces to global displacements as follows:

$$\begin{Bmatrix} F_{1x} \\ F_{2x} \\ F_{3x} \\ F_{4x} \end{Bmatrix} = \begin{bmatrix} 1000 & 0 & -1000 & 0 \\ 0 & 3000 & 0 & -3000 \\ -1000 & 0 & 3000 & -2000 \\ 0 & -3000 & -2000 & 5000 \end{bmatrix} \begin{Bmatrix} d_{1x} \\ d_{2x} \\ d_{3x} \\ d_{4x} \end{Bmatrix} \qquad (2.5.17)$$

Applying the homogeneous boundary conditions $d_{1x} = 0$ and $d_{2x} = 0$ to Eq. (2.5.17), substituting applied nodal forces, and partitioning the first two equations of Eq. (2.5.17) (or deleting the first two rows of $\{F\}$ and $\{d\}$ and the first two rows and columns of $\underline{K}$ corresponding to the zero-displacement boundary conditions), we obtain

$$\left\{\begin{array}{c} 0 \\ 5000 \end{array}\right\} = \left[\begin{array}{cc} 3000 & -2000 \\ -2000 & 5000 \end{array}\right] \left\{\begin{array}{c} d_{3x} \\ d_{4x} \end{array}\right\} \tag{2.5.18}$$

Solving Eq. (2.5.18), we obtain

$$d_{3x} = \frac{10}{11}\text{in.} \qquad d_{4x} = \frac{15}{11}\text{in.} \tag{2.5.19}$$

(c) To obtain the global nodal forces (which include the reactions at nodes 1 and 2) we back-substitute Eqs. (2.5.19) and the boundary conditions $d_{1x} = 0$ and $d_{2x} = 0$ into Eq. (2.5.17). This substitution yields

$$\left\{\begin{array}{c} F_{1x} \\ F_{2x} \\ F_{3x} \\ F_{4x} \end{array}\right\} = \left[\begin{array}{cccc} 1000 & 0 & -1000 & 0 \\ 0 & 3000 & 0 & -3000 \\ -1000 & 0 & 3000 & -2000 \\ 0 & -3000 & -2000 & 5000 \end{array}\right] \left\{\begin{array}{c} 0 \\ 0 \\ \frac{10}{11} \\ \frac{15}{11} \end{array}\right\} \tag{2.5.20}$$

Multiplying matrices in Eq. (2.5.20) and simplifying, we obtain

$$F_{1x} = \frac{-10,000}{11}\text{lb} \qquad F_{2x} = \frac{-45,000}{11}\text{lb} \qquad F_{3x} = 0$$

$$F_{4x} = \frac{55,000}{11}\text{lb} \tag{2.5.21}$$

From these results, we observe that the sum of the reactions F_{1x} and F_{2x} is equal in magnitude but opposite in direction to the applied force F_{4x}. This result verifies equilibrium of the whole spring assemblage.

(d) Next we use local element Eq. (2.2.17) to obtain the forces in each element.

Element 1

$$\left\{\begin{array}{c} \hat{f}_{1x} \\ \hat{f}_{3x} \end{array}\right\} = \left[\begin{array}{cc} 1000 & -1000 \\ -1000 & 1000 \end{array}\right] \left\{\begin{array}{c} 0 \\ \frac{10}{11} \end{array}\right\} \tag{2.5.22}$$

Simplifying Eq. (2.5.22), we obtain

$$\hat{f}_{1x} = \frac{-10,000}{11}\text{lb} \qquad \hat{f}_{3x} = \frac{10,000}{11}\text{lb} \tag{2.5.23}$$

A free-body diagram of spring element 1 is shown in Figure 2-7. The spring is subjected to tensile forces given by Eqs. (2.5.23).

F I G U R E 2-7   Free-body diagram of element 1

Element 2

$$\begin{Bmatrix} \hat{f}_{3x} \\ \hat{f}_{4x} \end{Bmatrix} = \begin{bmatrix} 2000 & -2000 \\ -2000 & 2000 \end{bmatrix} \begin{Bmatrix} \frac{10}{11} \\ \frac{15}{11} \end{Bmatrix} \tag{2.5.24}$$

Simplifying Eq. (2.5.24), we obtain

$$\hat{f}_{3x} = \frac{-10,000}{11} \text{ lb} \qquad \hat{f}_{4x} = \frac{10,000}{11} \text{ lb} \tag{2.5.25}$$

A free-body diagram of spring element 2 is shown in Figure 2-8. The spring is subjected to tensile forces given by Eqs. (2.5.25).

F I G U R E   2-8   Free-body diagram of element 2

Element 3

$$\begin{Bmatrix} \hat{f}_{4x} \\ \hat{f}_{2x} \end{Bmatrix} = \begin{bmatrix} 3000 & -3000 \\ -3000 & 3000 \end{bmatrix} \begin{Bmatrix} \frac{15}{11} \\ 0 \end{Bmatrix} \tag{2.5.26}$$

Simplifying Eq. (2.5.26) yields

$$\hat{f}_{4x} = \frac{45,000}{11} \text{ lb} \qquad \hat{f}_{2x} = \frac{-45,000}{11} \text{ lb} \tag{2.5.27}$$

A free-body diagram of spring element 3 is shown in Figure 2-9. The spring is subjected to compressive forces given by Eqs. (2.5.27).

F I G U R E   2-9   Free-body diagram of element 3   ■

E X A M P L E   2.2

For the spring assemblage shown in Figure 2-10, obtain (a) the global stiffness matrix, (b) the displacements of nodes 2, 3, and 4, (c) the global nodal forces, and (d) the local element forces. Node 1 is fixed while node 5 is given a fixed, known displacement $\delta = 20.0$ mm. The spring constants are all equal to $k = 200$ kN/m.

  (a) We use Eq. (2.2.18) to express each element stiffness matrix as

$$\underline{k}^{(1)} = \underline{k}^{(2)} = \underline{k}^{(3)} = \underline{k}^{(4)} = \begin{bmatrix} 200 & -200 \\ -200 & 200 \end{bmatrix} \tag{2.5.28}$$

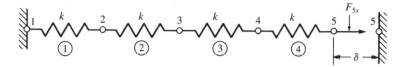

FIGURE 2-10 Spring assemblage for solution

Again using superposition, we obtain the global stiffness matrix as

$$\underline{K} = \begin{bmatrix} 200 & -200 & 0 & 0 & 0 \\ -200 & 400 & -200 & 0 & 0 \\ 0 & -200 & 400 & -200 & 0 \\ 0 & 0 & -200 & 400 & -200 \\ 0 & 0 & 0 & -200 & 200 \end{bmatrix} \qquad (2.5.29)$$

(b) The global stiffness matrix, Eq. (2.5.29), relates the global forces to the global displacements as follows:

$$\begin{Bmatrix} F_{1x} \\ F_{2x} \\ F_{3x} \\ F_{4x} \\ F_{5x} \end{Bmatrix} = \begin{bmatrix} 200 & -200 & 0 & 0 & 0 \\ -200 & 400 & -200 & 0 & 0 \\ 0 & -200 & 400 & -200 & 0 \\ 0 & 0 & -200 & 400 & -200 \\ 0 & 0 & 0 & -200 & 200 \end{bmatrix} \begin{Bmatrix} d_{1x} \\ d_{2x} \\ d_{3x} \\ d_{4x} \\ d_{5x} \end{Bmatrix} \qquad (2.5.30)$$

Applying the boundary conditions $d_{1x} = 0$ and $d_{5x} = 20$ mm $(=0.02$ m$)$, substituting known global forces $F_{2x} = 0$, $F_{3x} = 0$, and $F_{4x} = 0$, and partitioning the first and fifth equations of Eq. (2.5.30) corresponding to these boundary conditions, we obtain

$$\begin{Bmatrix} 0 \\ 0 \\ 0 \end{Bmatrix} = \begin{bmatrix} -200 & 400 & -200 & 0 & 0 \\ 0 & -200 & 400 & -200 & 0 \\ 0 & 0 & -200 & 400 & -200 \end{bmatrix} \begin{Bmatrix} 0 \\ d_{2x} \\ d_{3x} \\ d_{4x} \\ 0.02 \end{Bmatrix} \qquad (2.5.31)$$

Equation (2.5.31) is now rewritten, transposing the product of the appropriate stiffness coefficient multiplied by the known displacement to the left side.

$$\begin{Bmatrix} 0 \\ 0 \\ 4 \end{Bmatrix} = \begin{bmatrix} 400 & -200 & 0 \\ -200 & 400 & -200 \\ 0 & -200 & 400 \end{bmatrix} \begin{Bmatrix} d_{2x} \\ d_{3x} \\ d_{4x} \end{Bmatrix} \qquad (2.5.32)$$

Solving Eq. (2.5.32), we obtain

$$d_{2x} = 0.005 \text{ m} \qquad d_{3x} = 0.01 \text{ m} \qquad d_{4x} = 0.015 \text{ m} \qquad (2.5.33)$$

(c) The global nodal forces are obtained by back-substituting the boundary condition displacements and Eqs. (2.5.33) into Eq. (2.5.30). This substitution yields

$$F_{1x} = (-200)(0.005) = -1.0 \text{ kN}$$

$$F_{2x} = (400)(0.005) - (200)(0.01) = 0$$

$$F_{3x} = (-200)(0.005) + (400)(0.01) - (200)(0.015) = 0 \qquad (2.5.34)$$

$$F_{4x} = (-200)(0.01) + (400)(0.015) - (200)(0.02) = 0$$

$$F_{5x} = (-200)(0.015) + (200)(0.02) = 1.0 \text{ kN}$$

The results of Eqs. (2.5.34) yield the reaction F_{1x} opposite that of the nodal force F_{5x} required to displace node 5 by $\delta = 20.0$ mm. This result verifies equilibrium of the whole spring assemblage.

(d) Next, we make use of local element Eq. (2.2.17) to obtain the forces in each element.

Element 1

$$\left\{ \begin{matrix} \hat{f}_{1x} \\ \hat{f}_{2x} \end{matrix} \right\} = \left[\begin{matrix} 200 & -200 \\ -200 & 200 \end{matrix} \right] \left\{ \begin{matrix} 0 \\ 0.005 \end{matrix} \right\} \qquad (2.5.35)$$

Simplifying Eq. (2.5.35) yields

$$\hat{f}_{1x} = -1.0 \text{ kN} \qquad \hat{f}_{2x} = 1.0 \text{ kN} \qquad (2.5.36)$$

Element 2

$$\left\{ \begin{matrix} \hat{f}_{2x} \\ \hat{f}_{3x} \end{matrix} \right\} = \left[\begin{matrix} 200 & -200 \\ -200 & 200 \end{matrix} \right] \left\{ \begin{matrix} 0.005 \\ 0.01 \end{matrix} \right\} \qquad (2.5.37)$$

Simplifying Eq. (2.5.37) yields

$$\hat{f}_{2x} = -1 \text{ kN} \qquad \hat{f}_{3x} = 1 \text{ kN} \qquad (2.5.38)$$

Element 3

$$\left\{ \begin{matrix} \hat{f}_{3x} \\ \hat{f}_{4x} \end{matrix} \right\} = \left[\begin{matrix} 200 & -200 \\ -200 & 200 \end{matrix} \right] \left\{ \begin{matrix} 0.01 \\ 0.015 \end{matrix} \right\} \qquad (2.5.39)$$

Simplifying Eq. (2.5.39), we have

$$\hat{f}_{3x} = -1 \text{ kN} \qquad \hat{f}_{4x} = 1 \text{ kN} \qquad (2.5.40)$$

Element 4

$$\left\{ \begin{matrix} \hat{f}_{4x} \\ \hat{f}_{5x} \end{matrix} \right\} = \left[\begin{matrix} 200 & -200 \\ -200 & 200 \end{matrix} \right] \left\{ \begin{matrix} 0.015 \\ 0.02 \end{matrix} \right\} \qquad (2.5.41)$$

Simplifying Eq. (2.5.41), we obtain

$$\hat{f}_{4x} = -1 \text{ kN} \qquad \hat{f}_{5x} = 1 \text{ kN} \qquad (2.5.42)$$

∎

Finally, to review the major concepts presented in this chapter, we solve the following example problem.

E X A M P L E   2.3

(a) Using the ideas presented in Section 2.3 for the system of linear elastic springs shown in Figure 2-11, express the boundary conditions, the compatibility or continuity condition similar to Eq. (2.3.3), and the nodal equilibrium conditions similar to Eqs. (2.3.4)–(2.3.6). Then formulate the global stiffness matrix and equations for solution of the unknown global displacement and forces. The spring constants for the elements are k_1, k_2, and k_3; P is an applied force at node 2.

(b) Using the direct stiffness method, formulate the same global stiffness matrix and equation as in Part (a).

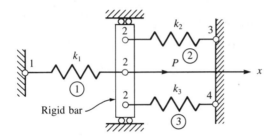

F I G U R E   2-11   Spring assemblage for solution

(a) The boundary conditions are

$$d_{1x} = 0 \qquad d_{3x} = 0 \qquad d_{4x} = 0 \tag{2.5.43}$$

The compatibility condition at node 2 is

$$d_{2x}^{(1)} = d_{2x}^{(2)} = d_{2x}^{(3)} = d_{2x} \tag{2.5.44}$$

The nodal equilibrium conditions are

$$F_{1x} = f_{1x}^{(1)}$$
$$P = f_{2x}^{(1)} + f_{2x}^{(2)} + f_{2x}^{(3)}$$
$$F_{3x} = f_{3x}^{(2)} \tag{2.5.45}$$
$$F_{4x} = f_{4x}^{(3)}$$

where the sign convention for positive element nodal forces given by Figure 2-1 was used in writing Eqs. (2.5.45). Figure 2-12 shows the element and nodal force free-body diagrams.

Using the local stiffness matrix Eq. (2.2.17) applied to each element, and compatibility condition Eq. (2.5.44), we obtain the total or global equilibrium

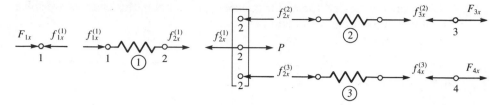

FIGURE 2-12   Free-body diagrams of elements and nodes of spring assemblage of Figure 2-11

equations as

$$F_{1x} = k_1 d_{1x} - k_1 d_{2x}$$

$$P = -k_1 d_{1x} + k_1 d_{2x} + k_2 d_{2x} - k_2 d_{3x} + k_3 d_{2x} - k_3 d_{4x}$$

$$F_{3x} = -k_2 d_{2x} + k_2 d_{3x}$$

$$F_{4x} = -k_3 d_{2x} + k_3 d_{4x}$$

$$(2.5.46)$$

In matrix form, we express Eqs. (2.5.46) as

$$\begin{Bmatrix} F_{1x} \\ P \\ F_{3x} \\ F_{4x} \end{Bmatrix} = \begin{bmatrix} k_1 & -k_1 & 0 & 0 \\ -k_1 & k_1 + k_2 + k_3 & -k_2 & -k_3 \\ 0 & -k_2 & k_2 & 0 \\ 0 & -k_3 & 0 & k_3 \end{bmatrix} \begin{Bmatrix} d_{1x} \\ d_{2x} \\ d_{3x} \\ d_{4x} \end{Bmatrix} \qquad (2.5.47)$$

Therefore, the global stiffness matrix is the square, symmetric matrix on the right side of Eq. (2.5.47). Making use of the boundary conditions, Eqs. (2.5.43), and then considering the second equation of Eqs. (2.5.46) or (2.5.47), we solve for d_{2x} as

$$d_{2x} = \frac{P}{k_1 + k_2 + k_3} \qquad (2.5.48)$$

We could have obtained this same result by deleting rows 1, 3, and 4 in the $\underline{F}$ and $\underline{d}$ matrices and rows and columns 1, 3, and 4 in $\underline{K}$, corresponding to zero displacement, as previously described in Section 2.4, and then solving for d_{2x}.

Using Eqs. (2.5.46), we now solve for the global forces as

$$F_{1x} = -k_1 d_{2x} \qquad F_{3x} = -k_2 d_{2x} \qquad F_{4x} = -k_3 d_{2x} \qquad (2.5.49)$$

The forces given by Eqs. (2.5.49) can be interpreted as the global reactions in this example. The negative signs in front of these forces indicate they are directed to the left (opposite the x axis).

(b) Using the direct stiffness method, we formulate the global stiffness matrix. First, using Eq. (2.2.18), we express each element stiffness matrix as

$$\underline{k}^{(1)} = \begin{matrix} d_{1x} & d_{2x} \\ \begin{bmatrix} k_1 & -k_1 \\ -k_1 & k_1 \end{bmatrix} \end{matrix} \qquad \underline{k}^{(2)} = \begin{matrix} d_{2x} & d_{3x} \\ \begin{bmatrix} k_2 & -k_2 \\ -k_2 & k_2 \end{bmatrix} \end{matrix} \qquad \underline{k}^{(3)} = \begin{matrix} d_{2x} & d_{4x} \\ \begin{bmatrix} k_3 & -k_3 \\ -k_3 & k_3 \end{bmatrix} \end{matrix}$$

$$(2.5.50)$$

where the degrees of freedom associated with each element are listed in the columns above each matrix. Using the direct stiffness method as outlined in Section 2.4, we add terms from each element stiffness matrix into the appropriate corresponding row and column in the global stiffness matrix to obtain

$$\underline{K} = \begin{matrix} d_{1x} & d_{2x} & d_{3x} & d_{4x} \\ \begin{bmatrix} k_1 & -k_1 & 0 & 0 \\ -k_1 & k_1 + k_2 + k_3 & -k_2 & -k_3 \\ 0 & -k_2 & k_2 & 0 \\ 0 & -k_3 & 0 & k_3 \end{bmatrix} \end{matrix} \qquad (2.5.51)$$

We observe that each element stiffness matrix $\underline{k}$ has been added into the location in the global $\underline{K}$ corresponding to the identical degree of freedom associated with the element $\underline{k}$. For instance, element 3 is associated with degrees of freedom d_{2x} and d_{4x}; hence, its contributions to $\underline{K}$ are in the 2–2, 2–4, 4–2, and 4–4 locations of $\underline{K}$ as indicated in Eq. (2.5.51) by the k_3 terms.

Having assembled the global $\underline{K}$ by the direct stiffness method, the global equations are then formulated in the usual manner by making use of the general Eq. (2.3.10), $\underline{F} = \underline{K}\underline{d}$. These equations have been previously obtained by Eq. (2.5.47) and therefore are not repeated.   ∎

2.6 Potential Energy Approach to Derive Spring Element Equations

One of the alternative methods often used to derive the element equations and the stiffness matrix for an element is based on the principle of *minimum potential energy*. (The use of this principle in structural mechanics is fully described in Reference [4].) This method has the advantage of being more general than the method given in Section 2.2 involving nodal and element equilibrium equations, along with the stress/strain law for the element. Thus, the principle of minimum potential energy is more adaptable for the determination of element equations for complicated elements (those with large numbers of degrees of freedom) such as the plane stress/strain element, the axisymmetric stress element, the plate bending element, and the three-dimensional solid stress element.

Again, we state that the principle of virtual work (see Appendix E) is applicable for any material behavior, whereas the principle of minimum potential energy is applicable only for elastic materials. However, both principles yield the same element equations for linear-elastic materials, which are the only kind considered in this text. Moreover, the principle of minimum potential energy, being included in the general category of *variational methods* (as is the principle of virtual work), leads to other variational functions (or functionals) similar to potential energy that can be formulated for other classes of problems, primarily of the nonstructural type. These other problems are generally classified as *field problems* and include, among others, torsion of a bar, heat transfer (see Chapter 13), fluid flow (Chapter 14), and electric potential.

Still other classes of problems, for which a variational formulation is not clearly definable, can be formulated by *weighted residual methods*, of which Galerkin's method is the most often used. We will describe Galerkin's method and its application to a bar element in Section 3.11. (For more information on weighted residual methods, also consult References [5], [6], and [7].)

Here we present the principle of minimum potential energy as used to derive the bar element equations. We will illustrate this seemingly complicated concept (complicated possibly because of the lack of physical insight associated with this approach) applied to the simplest of elements in hope that it will become a more comfortable concept when applied out of necessity to more complicated element types in subsequent chapters.

The total potential energy π_p of a structure is expressed in terms of displacements. In the finite element formulation, these will generally be nodal displacements such that $\pi_p = \pi_p(d_1, d_2, \ldots, d_n)$. When π_p is minimized with respect to these displacements, equilibrium equations result. For the spring element, we will show that the same nodal equilibrium equations $\underline{k}\underline{d} = \hat{f}$ result as previously derived in Section 2.2.

We first state the principle of minimum potential energy as follows:

> Of all the displacements that satisfy the given boundary conditions of a structure, those that satisfy the equations of equilibrium are distinguishable by a stationary value of the potential energy. If the stationary value is a minimum, the equilibrium state is stable.

To explain this principle, we must first explain the concepts of potential energy and of a stationary value of a function. We will now discuss these two concepts.

Total potential energy *is defined as the sum of the internal strain energy U and the potential energy of the external forces Ω; that is,*

$$\pi_p = U + \Omega \qquad (2.6.1)$$

Strain energy is the capacity of internal forces (or stresses) to do work through deformations (strains) in the structure; Ω is the capacity of forces such as body forces, surface traction forces, and applied nodal forces to do work through deformation of the structure.

Recall that a linear spring has force related to deformation by $F = kx$, where k is the spring constant and x is the deformation of the spring (Figure 2-13).

The differential internal work (or strain energy) dU in the spring is the internal force multiplied by the change in displacement through which the force moves, given by

$$dU = F\,dx \qquad (2.6.2)$$

Now we express F as $\qquad\qquad F = kx \qquad\qquad (2.6.3)$

Using Eq. (2.6.3) in Eq. (2.6.2), the differential strain energy becomes

$$dU = kx\,dx \qquad (2.6.4)$$

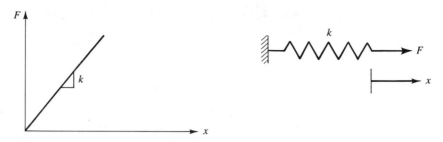

FIGURE 2-13 Force-deformation curve for linear spring

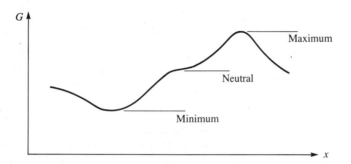

FIGURE 2-14 Stationary values of a function

The total strain energy is then given by

$$U = \int_0^x kx \, dx \tag{2.6.5}$$

Upon explicit integration of Eq. (2.6.5), we obtain

$$U = \tfrac{1}{2}kx^2 \tag{2.6.6}$$

Using Eq. (2.6.3) in Eq. (2.6.6), we have

$$U = \tfrac{1}{2}(kx)x = \tfrac{1}{2}Fx \tag{2.6.7}$$

Equation (2.6.7) indicates that the strain energy is the area under the force-deformation curve.

The potential energy of the external force, being opposite in sign from the external work expression because the potential energy of the external force is lost when the work is done by the external force, is given by

$$\Omega = -Fx \tag{2.6.8}$$

Therefore substituting Eqs. (2.6.6) and (2.6.8) into (2.6.1), the total potential energy becomes

$$\pi_p = \tfrac{1}{2}kx^2 - Fx \tag{2.6.9}$$

The concept of a *stationary value* of a function G (used in the definition of the principle of minimum potential energy) is shown in Figure 2-14. Here G is expressed as a function of the variable x. The stationary value can be a maximum, a minimum, or a neutral point of $G(x)$. To find a value of x yielding a stationary value of $G(x)$, we use differential calculus to differentiate G with respect to x and set the expression equal to zero, as follows:

$$\frac{dG}{dx} = 0 \qquad (2.6.10)$$

An analogous process will subsequently be used to replace G with π_p and x with discrete values (nodal displacements) d_i. With an understanding of variational calculus (see Reference [8]), the first variation of π_p (denoted by $\delta\pi_p$) could be used to minimize π_p. However, we will avoid the details of variational calculus and show that we can really use the familiar differential calculus to perform the minimization of π_p. To apply the principle of minimum potential energy—that is, to minimize π_p—we take the *variation* of π_p, defined in general as

$$\delta\pi_p = \frac{\partial\pi_p}{\partial d_1}\delta d_1 + \frac{\partial\pi_p}{\partial d_2}\delta d_2 + \cdots + \frac{\partial\pi_p}{\partial d_n}\delta d_n \qquad (2.6.11)$$

The principle states that equilibrium exists when the d_i define a structure state such that $\delta\pi_p = 0$ for arbitrary admissible variations δd_i from the equilibrium state. An *admissible variation* is one in which the displacement field still satisfies the boundary conditions and interelement continuity. Figure 2-15 shows the hypothetical actual axial displacement and an admissible one for a bar with specified boundary displacements $\hat{u}_1$ and $\hat{u}_2$. Here $\delta\hat{u}$ represents the variation in $\hat{u}$. In the general finite element formulation, $\delta\hat{u}$ would be replaced by δd_i. This implies that any of the δd_i might be nonzero. Hence, to satisfy $\delta\pi_p = 0$, all coefficients associated with the δd_i must be zero independently. Thus,

$$\frac{\partial\pi_p}{\partial d_i} = 0 \quad (i = 1, 2, 3, \ldots, n) \quad \text{or} \quad \frac{\partial\pi_p}{\partial\{d\}} = 0 \qquad (2.6.12)$$

where n equations must be solved for the n values of d_i that define the static equilibrium state of the structure. Equation (2.6.12) shows that for our purposes throughout this text, we can interpret *the variation of π_p* as a compact

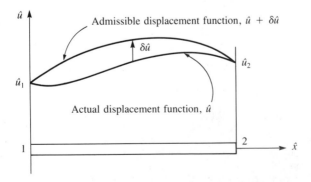

F I G U R E 2-15   Actual and admissible displacement functions

notation equivalent to differentiation of π_p with respect to the unknown nodal displacements for which π_p is expressed. For linear-elastic materials in equilibrium, the fact that π_p is a minimum is shown, for instance, in Reference [4].

Before discussing the formulation of the spring element equations, we now illustrate the concept of the principle of minimum potential energy by analyzing a single-degree-of-freedom spring subjected to an applied force, as given in Example 2.4. In this example, we will show that the equilibrium position of the spring corresponds to the minimum potential energy.

EXAMPLE 2.4

For the linear-elastic spring subjected to a force of 1000 lb shown in Figure 2-16, evaluate the potential energy for various displacement values and show that the minimum potential energy also corresponds to the equilibrium position of the spring.

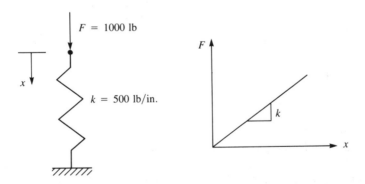

FIGURE 2-16 Spring subjected to force; load/displacement curve

We evaluate the total potential energy as

$$\pi_p = U + \Omega$$

where
$$U = \tfrac{1}{2}(kx)x \quad \text{and} \quad \Omega = -Fx$$

We now illustrate the minimization π_p through standard mathematics. Taking the variation of π_p with respect to x, or, equivalently, taking the derivative of π_p with respect to x as in Eqs. (2.6.11) and (2.6.12), we have

$$\delta\pi_p = \frac{\partial\pi_p}{\partial x}\,\delta x = 0$$

or, because δx is arbitrary and might not be zero,

$$\frac{\partial\pi_p}{\partial x} = 0$$

Using our previous expression for π_p, we obtain

T A B L E   2-1   Total potential energy
for various spring
deformations

Deformation x, in.	Total Potential Energy π_p, lb-in.
−4.00	8000
−3.00	5250
−2.00	3000
−1.00	1250
0.00	0
1.00	−750
2.00	−1000
3.00	−750
4.00	0
5.00	1250

$$\frac{\partial \pi_p}{\partial x} = 500x - 1000 = 0$$

or $x = 2.00$ in.

This value for x is then back-substituted into π_p to yield

$$\pi_p = 250(2)^2 - 1000(2) = -1000 \text{ lb-in.}$$

which corresponds to the minimum potential energy obtained in Table 2-1 by a searching technique. Here $U = (1/2)(kx)x$ is the strain energy or the area under the load/displacement curve shown in Figure 2-16, and $\Omega = -Fx$ is the potential energy of load F. For the given values of F and k, we then have

$$\pi_p = \tfrac{1}{2}(500)x^2 - 1000x = 250x^2 - 1000x$$

We now search for the minimum value of π_p for various values of spring deformation x. The results are shown in Table 2-1. A plot of π_p versus x is shown in Figure 2-17, where we observe that π_p has a minimum value at $x = 2.00$ in. This deformed position also corresponds to the equilibrium position because $(\partial \pi_p / \partial x) = 500(2) - 1000 = 0$. ∎

We now derive the spring element equations and stiffness matrix using the principal of minimum potential energy. Consider the linear spring subjected to nodal forces shown in Figure 2-18. Using Eq. (2.6.9), the total potential energy becomes

$$\pi_p = \tfrac{1}{2}k(\hat{d}_{2x} - \hat{d}_{1x})^2 - \hat{f}_{1x}\hat{d}_{1x} - \hat{f}_{2x}\hat{d}_{2x} \tag{2.6.13}$$

where $\hat{d}_{2x} - \hat{d}_{1x}$ is the deformation of the spring in Eq. (2.6.9). Simplifying Eq. (2.6.13), we obtain

$$\pi_p = \tfrac{1}{2}k(\hat{d}_{2x}^2 - 2\hat{d}_{2x}\hat{d}_{1x} + \hat{d}_{1x}^2) - \hat{f}_{1x}\hat{d}_{1x} - \hat{f}_{2x}\hat{d}_{2x} \tag{2.6.14}$$

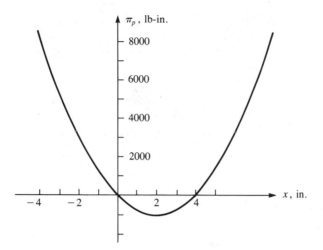

FIGURE 2-17   Variation of potential energy with spring deformation

FIGURE 2-18   Linear spring subjected to nodal forces

The minimization of π_p with respect to each nodal displacement requires that

$$\frac{\partial \pi_p}{\partial \hat{d}_{1x}} = \frac{1}{2}k(-2\hat{d}_{2x} + 2\hat{d}_{1x}) - \hat{f}_{1x} = 0$$

$$\frac{\partial \pi_p}{\partial \hat{d}_{2x}} = \frac{1}{2}k(2\hat{d}_{2x} - 2\hat{d}_{1x}) - \hat{f}_{2x} = 0$$

(2.6.15)

Simplifying Eqs. (2.6.15), we have

$$k(-\hat{d}_{2x} + \hat{d}_{1x}) = \hat{f}_{1x}$$

$$k(\hat{d}_{2x} - \hat{d}_{1x}) = \hat{f}_{2x}$$

(2.6.16)

In matrix form, we express Eq. (2.6.16) as

$$\begin{bmatrix} k & -k \\ -k & k \end{bmatrix} \begin{Bmatrix} \hat{d}_{1x} \\ \hat{d}_{2x} \end{Bmatrix} = \begin{Bmatrix} \hat{f}_{1x} \\ \hat{f}_{2x} \end{Bmatrix}$$

(2.6.17)

Since $\{\hat{f}\} = [\hat{k}]\{\hat{d}\}$, we have the stiffness matrix for the spring element obtained from Eq. (2.6.17):

$$[\hat{k}] = \begin{bmatrix} k & -k \\ -k & k \end{bmatrix}$$

(2.6.18)

As expected, Eq. (2.6.18) is identical to the stiffness matrix obtained in Section 2.2, Eq. (2.2.18).

We considered the equilibrium of a single spring element by minimizing the total potential energy with respect to the nodal displacements (See Example 2.4). We also developed the finite-element spring element equations by minimizing the total potential energy with respect to the nodal displacements. We now show that the total potential energy of an entire structure (here an assemblage of spring elements) can be minimized with respect to each nodal degree of freedom and this minimization results in the same finite element equations used for the solution as those obtained by the direct stiffness method.

E X A M P L E   2.5

Obtain the total potential energy of the spring assemblage (Figure 2-19) for Example 2.1 and find its minimum value. The procedure of assembling element equations can then be seen to be obtained from the minimization of the total potential energy.

Using Eq. (2.6.10) for each element of the spring assemblage, the total potential energy is given by

$$
\pi_p = \sum_{e=1}^{3} \pi_p^{(e)} = \tfrac{1}{2}k_1(d_{3x} - d_{1x})^2 + \tfrac{1}{2}k_2(d_{4x} - d_{3x})^2
$$
$$
+ \tfrac{1}{2}k_3(d_{2x} - d_{4x})^2 - f_{1x}^{(1)}d_{1x} - f_{3x}^{(1)}d_{3x} \qquad (2.6.19)
$$
$$
- f_{3x}^{(2)}d_{3x} - f_{4x}^{(2)}d_{4x} - f_{4x}^{(3)}d_{4x} - f_{2x}^{(3)}d_{2x}
$$

Upon minimizing π_p with respect to each nodal displacement, we obtain

$$
\frac{\partial \pi_p}{\partial d_{1x}} = -k_1 d_{3x} + k_1 d_{1x} - f_{1x}^{(1)} = 0
$$

$$
\frac{\partial \pi_p}{\partial d_{2x}} = k_3 d_{2x} - k_3 d_{4x} - f_{2x}^{(3)} = 0
$$

$$ \qquad (2.6.20) $$

$$
\frac{\partial \pi_p}{\partial d_{3x}} = k_1 d_{3x} - k_1 d_{1x} - k_2 d_{4x} + k_2 d_{3x} - f_{3x}^{(1)} - f_{3x}^{(2)} = 0
$$

$$
\frac{\partial \pi_p}{\partial d_{4x}} = k_2 d_{4x} - k_2 d_{3x} - k_3 d_{2x} + k_3 d_{4x} - f_{4x}^{(2)} - f_{4x}^{(3)} = 0
$$

In matrix form, Eqs. (2.6.20) become

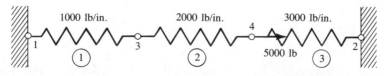

F I G U R E   2-19   Spring assemblage solved by energy principle

$$\begin{bmatrix} k_1 & 0 & -k_1 & 0 \\ 0 & k_3 & 0 & -k_3 \\ -k_1 & 0 & k_1+k_2 & -k_2 \\ 0 & -k_3 & -k_2 & k_2+k_3 \end{bmatrix} \begin{Bmatrix} d_{1x} \\ d_{2x} \\ d_{3x} \\ d_{4x} \end{Bmatrix} = \begin{Bmatrix} f_{1x}^{(1)} \\ f_{2x}^{(3)} \\ f_{3x}^{(1)}+f_{3x}^{(2)} \\ f_{4x}^{(2)}+f_{4x}^{(3)} \end{Bmatrix} \qquad (2.6.21)$$

Using nodal force equilibrium similar to Eqs. (2.3.4)–(2.3.6), we have the following:

$$f_{1x}^{(1)} = F_{1x}$$

$$f_{2x}^{(3)} = F_{2x}$$

$$f_{3x}^{(1)} + f_{3x}^{(2)} = F_{3x} \qquad (2.6.22)$$

$$f_{4x}^{(2)} + f_{4x}^{(3)} = F_{4x}$$

Using Eqs. (2.6.22) in (2.6.21) and substituting numerical values for k_1, k_2, and k_3, we obtain

$$\begin{bmatrix} 1000 & 0 & -1000 & 0 \\ 0 & 3000 & 0 & -3000 \\ -1000 & 0 & 3000 & -2000 \\ 0 & -3000 & -2000 & 5000 \end{bmatrix} \begin{Bmatrix} d_{1x} \\ d_{2x} \\ d_{3x} \\ d_{4x} \end{Bmatrix} = \begin{Bmatrix} F_{1x} \\ F_{2x} \\ F_{3x} \\ F_{4x} \end{Bmatrix} \qquad (2.6.23)$$

Equation (2.6.23) is identical to Eq. (2.5.17) obtained through the direct stiffness method. The assembled Eqs. (2.6.23) are then seen to be obtained from the minimization of the total potential energy. Upon applying the boundary conditions and substituting $F_{3x} = 0$ and $F_{4x} = 5000$ lb into Eq. (2.6.23), the solution is identical to that of Example 2.1. ∎

REFERENCES

[1] Turner, M. J., Clough, R. W., Martin, H. C., and Topp, L. J., "Stiffness and Deflection Analysis of Complex Structures," *Journal of the Aeronautical Sciences*, Vol. 23, No. 9, pp. 805–824, Sept. 1956.

[2] Martin, H. C., *Introduction to Matrix Methods of Structural Analysis*, McGraw-Hill, New York, 1966.

[3] Hsieh, Y. Y., *Elementary Theory of Structures*, 2nd ed., Prentice-Hall, Englewood Cliffs, N.J., 1982.

[4] Oden, J. T., and Ripperger, E. A., *Mechanics of Elastic Structures*, 2nd ed., McGraw-Hill, New York, 1981.

[5] Finlayson, B. A., *The Method of Weighted Residuals and Variational Principles*, Academic Press, New York, 1972.

[6] Zienkiewicz, O. C., *The Finite Element Method*, 3rd ed., McGraw-Hill, London, 1977.

[7] Cook, R. D., Malkus, D. S., and Plesha, M. E., *Concepts and Applications of Finite Element Analysis*, 3rd ed., Wiley, New York, 1989.

[8] Forray, M. J., *Variational Calculus in Science and Engineering*, McGraw-Hill, New York, 1968.

PROBLEMS

2.1   *a.* Obtain the global stiffness matrix $\underline{K}$ of the assemblage shown in Figure P2-1 by superimposing the stiffness matrices of the individual springs. Here k_1, k_2, and k_3 are the stiffnesses of the springs as shown.

   b. If nodes 1 and 2 are fixed and a force P acts on node 4 in the positive x direction, find an expression for the displacements of nodes 3 and 4.

   c. Determine the reaction forces at nodes 1 and 2.

   (*Hint:* Do this problem by writing the nodal equilibrium equations and then making use of the force/displacement relationships for each element as done in the first part of Section 2.4. Then solve the problem by the direct stiffness method.)

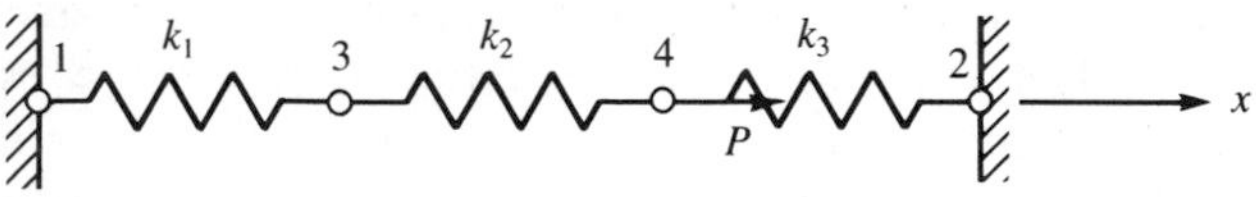

FIGURE P2-1

2.2   For the spring assemblage shown in Figure P2-2, determine the displacement at node 2 and the forces in each spring element. Also determine the force F_3. Given: node 3 displaces an amount $\delta = 1$ in. in the positive x direction due to the force F_3 and $k_1 = k_2 = 1000$ lb/in.

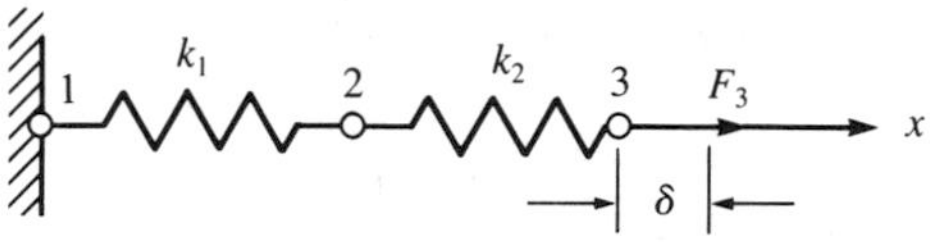

FIGURE P2-2

2.3   *a.* For the spring assemblage shown in Figure P2-3, obtain the global stiffness matrix by direct superposition.

   b. If nodes 1 and 5 are fixed and a force P is applied at node 3, determine the nodal displacements.

   c. Determine the reactions at the fixed nodes 1 and 5.

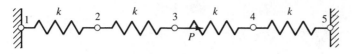

FIGURE P2-3

2.4 Resolve Problem 2.3 with $P = 0$ (no force applied at node 3) and with node 5 given a fixed, known displacement of δ as shown in Figure P2-4.

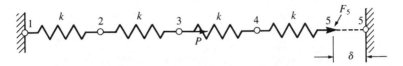

FIGURE P2-4

2.5–2.12 For the spring assemblages shown in Figures P2-5–P2-12, determine the nodal displacements, the forces in each element, and the reactions. Use the direct stiffness method for all problems.

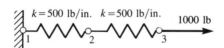

FIGURE P2-5

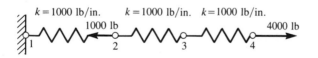

FIGURE P2-6

FIGURE P2-7

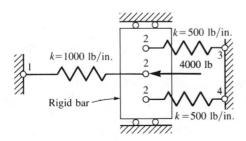

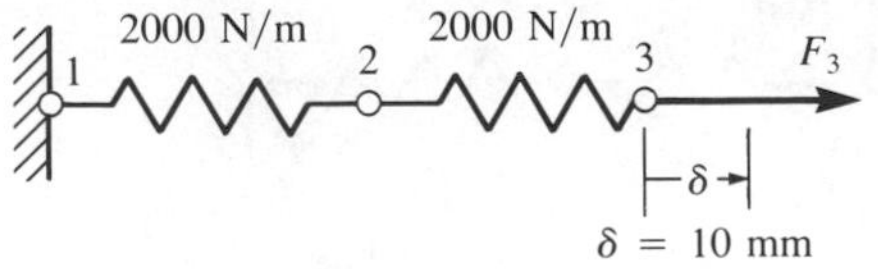

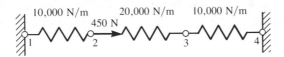

2000 N/m
2000 N/m
1
2
3
F_3
δ
δ = 10 mm

FIGURE P2-8

10,000 N/m
20,000 N/m
10,000 N/m
450 N
1
2
3
4

FIGURE P2-9

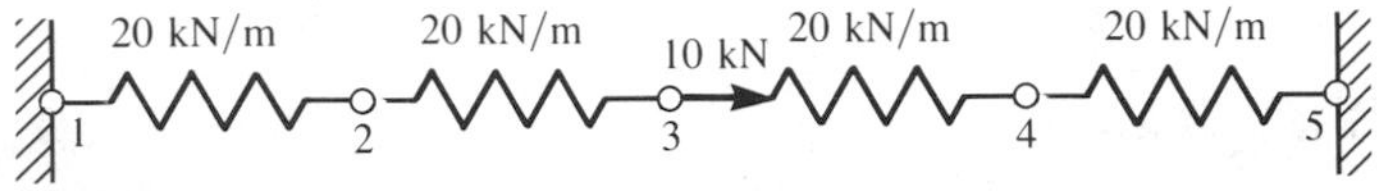

20 kN/m
20 kN/m
10 kN
20 kN/m
20 kN/m
1
2
3
4
5

FIGURE P2-10

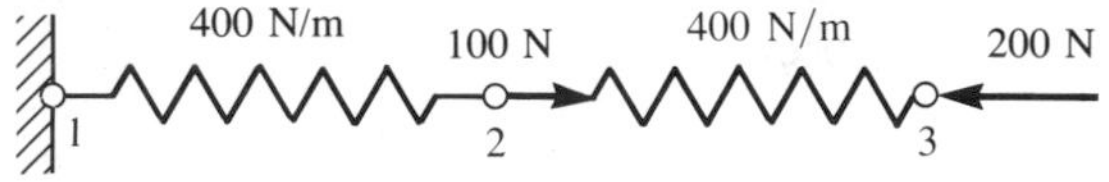

400 N/m
100 N
400 N/m
200 N
1
2
3

FIGURE P2-11

500 kN/m
1
3
1 kN
3
1000 kN/m
500 kN/m
2
3
4
1 kN

FIGURE P2-12

2.13 Use the principle of minimum potential energy developed in Section 2.6 to solve the spring problems shown in Figure P2-13; that is, plot the total potential energy for variations in the displacement of the free end of the spring to determine the minimum potential energy. Observe that the displacement that yields the minimum potential energy also yields the stable equilibrium position.

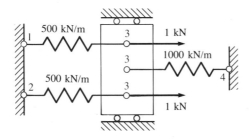

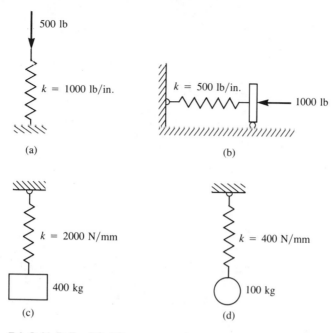

FIGURE P2-13

2.14 Reverse the direction of the load in Example 2.4 and recalculate the total potential energy and use this value to obtain the equilibrium value of displacement.

2.15 A nonlinear spring in Figure P2-15 has the force-deformation relationship $f = k\delta^2$. Express the total potential energy of the spring and use this potential energy to obtain the equilibrium value of displacement.

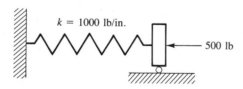

FIGURE P2-15

2.16–2.17 Solve problems 2.7 and 2.12 by the potential energy approach (see Example 2.5).

Development of Truss

Equations

INTRODUCTION

Having set forth the foundation on which the direct stiffness method is based, we will now derive the stiffness matrix for a linear-elastic bar (or truss) element using the general steps outlined in Chapter 1. We will include the introduction of both a local coordinate system, chosen with the element in mind, and a global or reference coordinate system, chosen to be convenient (for numerical purposes) with respect to the overall structure. We will also discuss the transformation of a vector from the local coordinate system to the global coordinate system, using the concept of transformation matrices to express the stiffness matrix of an arbitrarily oriented bar element in terms of the global system. We will solve two example plane truss problems to illustrate the procedure of establishing the total stiffness matrix and equations for solution of a structure.

Next we will describe how to handle inclined, or skewed, supports. We will then extend the stiffness method to include space trusses. We will develop the transformation matrix in three-dimensional space and analyze a space truss.

We will then use the principle of minimum potential energy and apply it to rederive the bar element equations. Finally, we will introduce Galerkin's residual method and then apply it to derive the bar element equations.

3.1 Derivation of the Stiffness Matrix for a Bar Element

We will now consider the derivation of the stiffness matrix for the linear-elastic, constant cross-sectional area (prismatic) bar element shown in Figure 3-1. The derivation here will be directly applicable to the solution of pin-connected trusses. The bar is subjected to tensile forces T directed along the

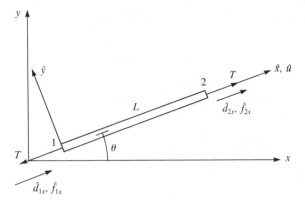

F I G U R E   3-1   Bar subjected to tensile forces T; positive nodal displacements and forces

axis of the bar and applied at nodes 1 and 2. Here we have introduced two coordinate systems, a local one $(\hat{x}, \hat{y})$ with $\hat{x}$ directed along the length of the bar, and a global one (x, y) assumed here to be best suited with respect to the total structure. Proper selection of global coordinate systems is best demonstrated through solution of two- and three-dimensional truss problems as illustrated in Sections 3.6 and 3.7. Both systems will be used extensively throughout this text.

The bar element is assumed to have constant cross-sectional area A, modulus of elasticity E, and initial length L. The nodal degrees of freedom are local axial displacements (longitudinal displacements directed along the length of the bar) represented by $\hat{d}_{1x}$ and $\hat{d}_{2x}$ at the ends of the element as shown in Figure 3-1.

From Hooke's law and the strain/displacement relationship, Eq. (1.4.1), we write

$$\sigma = E\varepsilon \tag{a}$$

$$\varepsilon = \frac{d\hat{u}}{d\hat{x}} \tag{b}$$

From force equilibrium, we have

$$A\sigma_x = T = \text{constant} \tag{c}$$

for no distributed load acting on the bar. (We will consider distributed loading in Section 3.9). Using Eq. (b) in (a) and then (a) in (c) and differentiating with respect to $\hat{x}$, we obtain the differential equation governing the linear-elastic bar behavior as

$$\frac{d}{d\hat{x}}\left(AE\frac{d\hat{u}}{d\hat{x}} \right) = 0 \tag{d}$$

where $\hat{u}$ is the axial displacement function in the $\hat{x}$ direction and A and E are written as though they were functions of $\hat{x}$ in the general form of the differential equation even though A and E will be assumed constant over the whole length of the bar in our derivations to follow.

The following assumptions are used in deriving the bar element stiffness matrix:

1. The bar cannot sustain shear force; that is, $\hat{f}_{1y} = 0$ and $\hat{f}_{2y} = 0$.
2. Any effect of transverse displacement is ignored.
3. Hooke's law applies; that is, axial stress σ_x is related to axial strain ε_x by $\sigma_x = E\varepsilon_x$.

The steps previously outlined in Chapter 1 are now used to derive the stiffness matrix for the bar element and then to illustrate a complete solution for a bar assemblage.

Step 1 Select Element Type

Represent the bar by labeling nodes at each end and in general by labeling the element number (see Figure 3-1).

Step 2 Select a Displacement Function

Assume a linear displacement variation along the $\hat{x}$ axis of the bar because a linear function with specified endpoints has a unique path. (Further discussion regarding the choice of displacement functions is provided in Section 3.2 and References [1]–[3].) Then

$$\hat{u} = a_1 + a_2\hat{x} \qquad (3.1.1)$$

with the total number of coefficients a_i always equal to the total number of degrees of freedom associated with the element. Here the total number of degrees of freedom is two—axial displacements at each of the two nodes of the element. Using the same procedure as in Section 2.2 for the spring element, we express Eq. (3.1.1) as

$$\hat{u} = \left(\frac{\hat{d}_{2x} - \hat{d}_{1x}}{L}\right)\hat{x} + \hat{d}_{1x} \qquad (3.1.2)$$

or, in matrix form, Eq. (3.1.2) becomes

$$\hat{u} = [N_1 \quad N_2]\begin{Bmatrix} \hat{d}_{1x} \\ \hat{d}_{2x} \end{Bmatrix} \qquad (3.1.3)$$

with shape functions given by

$$N_1 = 1 - \frac{\hat{x}}{L} \qquad N_2 = \frac{\hat{x}}{L} \qquad (3.1.4)$$

The linear displacement function $\hat{u}$ plotted over the length of the bar element is shown in Figure 3-2. The bar is shown with the same orientation as in Figure 3-1.

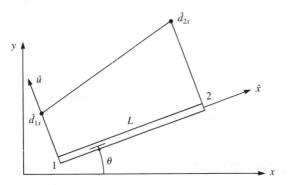

FIGURE 3-2   Displacement $\hat{u}$ plotted over the length of the element

Step 3   Define the Strain/Displacement and Stress/Strain Relationships

The strain/displacement relationship is

$$\varepsilon_x = \frac{d\hat{u}}{d\hat{x}} = \frac{\hat{d}_{2x} - \hat{d}_{1x}}{L} \qquad (3.1.5)$$

where Eqs. (3.1.3) and (3.1.4) have been used to obtain Eq. (3.1.5), and the stress/strain relationship is

$$\sigma_x = E\varepsilon_x \qquad (3.1.6)$$

Step 4   Derive the Element Stiffness Matrix and Equations

The element stiffness matrix is derived as follows. From elementary mechanics, we have

$$T = A\sigma_x \qquad (3.1.7)$$

Now using Eqs. (3.1.5) and (3.1.6) in Eq. (3.1.7), we obtain

$$T = AE\left(\frac{\hat{d}_{2x} - \hat{d}_{1x}}{L}\right) \qquad (3.1.8)$$

Also, by the nodal force sign convention of Figure 3-1,

$$\hat{f}_{1x} = -T \qquad (3.1.9)$$

or, by using Eq. (3.1.8), Eq. (3.1.9) becomes

$$\hat{f}_{1x} = \frac{AE}{L}(\hat{d}_{1x} - \hat{d}_{2x}) \qquad (3.1.10)$$

Similarly, $$\hat{f}_{2x} = T \qquad (3.1.11)$$

or, by Eq. (3.1.8), Eq. (3.1.11) becomes

$$\hat{f}_{2x} = \frac{AE}{L}(\hat{d}_{2x} - \hat{d}_{1x}) \qquad (3.1.12)$$

Expressing Eqs. (3.1.10) and (3.1.12) together in matrix form, we have

$$\begin{Bmatrix} \hat{f}_{1x} \\ \hat{f}_{2x} \end{Bmatrix} = \frac{AE}{L} \begin{bmatrix} 1 & -1 \\ -1 & 1 \end{bmatrix} \begin{Bmatrix} \hat{d}_{1x} \\ \hat{d}_{2x} \end{Bmatrix} \qquad (3.1.13)$$

Now, because $\hat{f} = \hat{k}\hat{d}$, we have, from Eq. (3.1.13),

$$\hat{k} = \frac{AE}{L} \begin{bmatrix} 1 & -1 \\ -1 & 1 \end{bmatrix} \qquad (3.1.14)$$

Equation (3.1.14) represents the stiffness matrix for a bar element. In Eq. (3.1.14), AE/L for a bar element is analogous to the spring constant k for a spring element.

Step 5   Assemble Element Equations to Obtain Global or Total Equations

Assemble the global stiffness and force matrices and global equations using the direct stiffness method described in Chapter 2 (see Section 3.6 for an example truss). This step applies for structures composed of more than one element such that (again)

$$\underline{K} = [K] = \sum_{e=1}^{N} \underline{k}^{(e)} \quad \text{and} \quad \underline{F} = \{F\} = \sum_{e=1}^{N} \underline{f}^{(e)} \qquad (3.1.15)$$

where now all local element stiffness matrices $\hat{k}$ must be transformed to global element stiffness matrices $\underline{k}$ before the direct stiffness method is applied as indicated by Eq. (3.1.15). (This concept of coordinate and stiffness matrix transformations is described in Sections 3.3 and 3.4.)

Step 6   Solve for the Nodal Displacements

Determine the displacements by imposing boundary conditions and simultaneously solving a system of equations, $\underline{F} = \underline{K}\underline{d}$.

Step 7   Solve for the Element Forces

Finally, determine the strains and stresses in each element by back-substitution of the displacements into equations similar to Eqs. (3.1.5) and (3.1.6).

We will now illustrate a solution for a one-dimensional bar problem.

E X A M P L E   3.1

For the three-bar assemblage shown in Figure 3-3, determine (a) the global stiffness matrix, (b) the displacements of nodes 2 and 3, and (c) the reactions at nodes 1 and 4. A force of 3000 lb is applied in the x direction at node 2. The length of each element is 30 in. Let $E = 30 \times 10^6$ psi and $A = 1$ in.^{2} for elements 1 and 2, and $E = 15 \times 10^6$ psi and $A = 2$ in.^{2} for element 3. Nodes 1 and 4 are fixed.

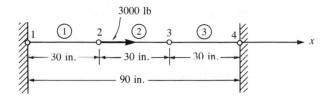

FIGURE 3-3   Three-bar assemblage

(a) Using Eq. (3.1.14), the element stiffness matrices are

$$
\underline{k}^{(1)} = \underline{k}^{(2)} = \frac{(1)(30 \times 10^6)}{30} \begin{matrix} & 1 & 2 \\ & 2 & 3 \end{matrix} \begin{bmatrix} 1 & -1 \\ -1 & 1 \end{bmatrix} = 10^6 \begin{bmatrix} 1 & -1 \\ -1 & 1 \end{bmatrix}
$$

$$(3.1.16)$$

$$
\underline{k}^{(3)} = \frac{(2)(15 \times 10^6)}{30} \begin{matrix} 3 & 4 \end{matrix} \begin{bmatrix} 1 & -1 \\ -1 & 1 \end{bmatrix} = 10^6 \begin{bmatrix} 1 & -1 \\ -1 & 1 \end{bmatrix}
$$

where, again, the numbers above the matrices in Eqs. (3.1.16) indicate the displacements associated with each matrix. Assembling the element stiffness matrices by the direct stiffness method, we obtain the global stiffness matrix as

$$
\underline{K} = 10^6 \begin{matrix} d_{1x} & d_{2x} & d_{3x} & d_{4x} \end{matrix} \begin{bmatrix} 1 & -1 & 0 & 0 \\ -1 & 1+1 & -1 & 0 \\ 0 & -1 & 1+1 & -1 \\ 0 & 0 & -1 & 1 \end{bmatrix}
$$

$$(3.1.17)$$

(b) Equation (3.1.17) relates global nodal forces to global nodal displacements as follows:

$$
\begin{Bmatrix} F_{1x} \\ F_{2x} \\ F_{3x} \\ F_{4x} \end{Bmatrix} = 10^6 \begin{bmatrix} 1 & -1 & 0 & 0 \\ -1 & 2 & -1 & 0 \\ 0 & -1 & 2 & -1 \\ 0 & 0 & -1 & 1 \end{bmatrix} \begin{Bmatrix} d_{1x} \\ d_{2x} \\ d_{3x} \\ d_{4x} \end{Bmatrix}
$$

$$(3.1.18)$$

Invoking the boundary conditions, we have

$$
d_{1x} = 0 \qquad d_{4x} = 0
$$

$$(3.1.19)$$

Using the boundary conditions, substituting known applied global forces into Eq. (3.1.18), and partitioning the first and fourth equations of Eq. (3.1.18), we solve the second and third equations of Eq. (3.1.18) to obtain

$$
\begin{Bmatrix} 3000 \\ 0 \end{Bmatrix} = 10^6 \begin{bmatrix} 2 & -1 \\ -1 & 2 \end{bmatrix} \begin{Bmatrix} d_{2x} \\ d_{3x} \end{Bmatrix}
$$

$$(3.1.20)$$

Solving Eq. (3.1.20) simultaneously for the displacements yields

$$d_{2x} = 0.002 \text{ in.} \qquad d_{3x} = 0.001 \text{ in.} \qquad (3.1.21)$$

(c) Back-substituting Eqs. (3.1.19) and (3.1.21) into Eq. (3.1.18), the global nodal forces, which include the reactions at nodes 1 and 4, are obtained as follows:

$$F_{1x} = 10^6(d_{1x} - d_{2x}) = 10^6(0 - 0.002) = -2000 \text{ lb}$$

$$F_{2x} = 10^6(-d_{1x} + 2d_{2x} - d_{3x}) = 10^6[0 + 2(0.002) - 0.001] = 3000 \text{ lb}$$

$$F_{3x} = 10^6(-d_{2x} + 2d_{3x} - d_{4x}) = 10^6[-0.002 + 2(0.001) - 0] = 0$$

$$F_{4x} = 10^6(-d_{3x} + d_{4x}) = 10^6(-0.001 + 0) = -1000 \text{ lb} \qquad (3.1.22)$$

The results of Eqs. (3.1.22) show that the sum of the reactions F_{1x} and F_{4x} is equal in magnitude but opposite in direction to the applied nodal force of 3000 lb at node 2. Equilibrium of the bar assemblage is thus verified. Furthermore, Eqs. (3.1.22) show that $F_{2x} = 3000$ lb and $F_{3x} = 0$ are merely the applied nodal forces at nodes 2 and 3, respectively, which further enhances the validity of our solution.   ■

3.2 Selecting Approximation Functions for Displacements

Consider the following guidelines, as they relate to the one-dimensional bar element, when selecting a displacement function. (Further discussion regarding selection of displacement functions and other kinds of approximation functions (such as temperature functions) will be provided in Chapter 5 for the beam element, in Chapter 7 for the constant-strain triangular element, in Chapter 9 for the linear-strain triangular element, and in Chapter 13 for the heat-transfer problem, and is also provided in References [1]–[3].)

1. Common approximation functions are usually polynomials such as that given by Eq. (3.1.1) or equivalently by Eq. (3.1.3), where the function is expressed in terms of the shape functions.

2. The approximation function should be continuous within the bar element. The simple linear function of Eq. (3.1.1) certainly is continuous within the element.

3. The approximating function should provide interelement continuity for all degrees of freedom at each node for discrete line elements, and along common boundary lines and surfaces for two- and three-dimensional elements. For the bar element, we must ensure that nodes common to two or more elements remain common to these elements upon deformation and thus prevent overlaps or voids between elements. For example, consider the two-bar structure

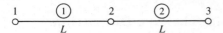

FIGURE 3-4   Interelement continuity of a two-bar structure

shown in Figure 3-4. For the two-bar structure, the linear function for $\hat{u}$ within each element will ensure that elements 1 and 2 remain connected; that is, the displacement at node 2 for element 1 will equal the displacement at the same node 2 for element 2. This rule was also illustrated by Eq. (2.3.3). The linear function is then called a *conforming* (or *compatible*) *function* for the bar element because it ensures both the satisfaction of continuity between adjacent elements and of continuity within the element.

4. The approximation function should allow for rigid-body displacement and for a state of constant strain within the element. The one-dimensional displacement function, Eq. (3.1.1), satisfies these criteria because the a_1 term allows for rigid-body motion (constant motion of the body without straining) and the $a_2\hat{x}$ term allows for constant strain since $\varepsilon_x = d\hat{u}/d\hat{x} = a_2$ is a constant. (This state of constant strain in the element can, in fact, occur if elements are chosen small enough.) The simple polynomial Eq. (3.1.1) satisfying this fourth guideline is then said to be *complete* for the bar element. Completeness of a function is a necessary condition for convergence to the exact answer, for instance, for displacements and stresses (see Reference [3]).

The idea that the interpolation (approximation) function must allow for a rigid-body displacement means that the function must be capable of yielding a constant value (say, a_1), because such a value can, in fact, occur. Therefore, we must consider the case

$$\hat{u} = a_1 \tag{3.2.1}$$

or
$$a_1 = \hat{d}_{1x} = \hat{d}_{2x} \tag{3.2.2}$$

Using Eq. (3.2.2) in Eq. (3.1.3), we have

$$\hat{u} = N_1\hat{d}_{1x} + N_2\hat{d}_{2x} = (N_1 + N_2)a_1 \tag{3.2.3}$$

From Eqs. (3.2.1) and (3.2.3), we then have

$$\hat{u} = a_1 = (N_1 + N_2)a_1 \tag{3.2.4}$$

Therefore, by Eq. (3.2.4), we obtain

$$N_1 + N_2 = 1 \tag{3.2.5}$$

Thus, Eq. (3.2.5) shows that the displacement interpolation functions must add to unity at every point within the element so that $\hat{u}$ will yield a constant value when a rigid-body displacement occurs.

3.3 Transformation of Vectors in Two Dimensions

In many problems it is convenient to introduce both local and global (or reference) coordinates. Local coordinates are always chosen to conveniently represent the individual element. Global coordinates are chosen to be convenient for the whole structure.

Given the nodal displacement of an element, represented by the vector **d** in Figure 3-5, we want to relate the components of this vector in one coordinate system to components in another. For general purposes, we will assume in this section that **d** is not coincident with either the local or global axes. In this case, we want to relate global displacement components to local ones. In so doing, we will develop a transformation matrix that will subsequently be used to develop the global stiffness matrix for a bar element. We define the angle θ to be positive when measured counterclockwise from x to $\hat{x}$. We can express vector displacement **d** in both global and local coordinates by

$$\mathbf{d} = d_x\mathbf{i} + d_y\mathbf{j} = \hat{d}_x\hat{\mathbf{i}} + \hat{d}_y\hat{\mathbf{j}} \tag{3.3.1}$$

where **i** and **j** are unit vectors in the x and y directions, and $\hat{\mathbf{i}}$ and $\hat{\mathbf{j}}$ are unit vectors in the $\hat{x}$ and $\hat{y}$ directions. We will now relate **i** and **j** to $\hat{\mathbf{i}}$ and $\hat{\mathbf{j}}$ through use of Figure 3-6.

Using Figure 3-6 and vector addition, we obtain

$$\mathbf{a} + \mathbf{b} = \mathbf{i} \tag{3.3.2}$$

Also, from the law of cosines,

$$|\mathbf{a}| = |\mathbf{i}| \cos \theta \tag{3.3.3}$$

and because **i** is, by definition, a unit vector, its magnitude is given by

$$|\mathbf{i}| = 1 \tag{3.3.4}$$

Therefore, we obtain $|\mathbf{a}| = 1 \cos \theta$ $\tag{3.3.5}$

Similarly, $|\mathbf{b}| = 1 \sin \theta$ $\tag{3.3.6}$

Now **a** is in the $\hat{\mathbf{i}}$ direction and **b** is in the $-\hat{\mathbf{j}}$ direction. Therefore,

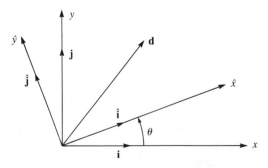

F I G U R E 3-5   General displacement vector **d**

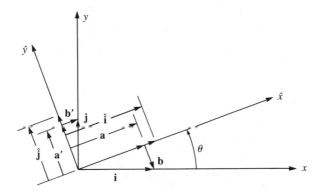

F I G U R E 3-6   Relationship between local and global unit vectors

$$\mathbf{a} = |\mathbf{a}|\hat{\mathbf{i}} = (\cos \theta)\hat{\mathbf{i}} \qquad (3.3.7)$$

and
$$\mathbf{b} = |\mathbf{b}|(-\hat{\mathbf{j}}) = (\sin \theta)(-\hat{\mathbf{j}}) \qquad (3.3.8)$$

Using Eqs. (3.3.7) and (3.3.8) in Eq. (3.3.2) yields

$$\mathbf{i} = \cos \theta \hat{\mathbf{i}} - \sin \theta \hat{\mathbf{j}} \qquad (3.3.9)$$

Similarly, from Figure 3-6, we obtain

$$\mathbf{a}' + \mathbf{b}' = \mathbf{j} \qquad (3.3.10)$$

$$\mathbf{a}' = \cos \theta \hat{\mathbf{j}} \qquad (3.3.11)$$

$$\mathbf{b}' = \sin \theta \hat{\mathbf{i}} \qquad (3.3.12)$$

Using Eqs. (3.3.11) and (3.3.12) in Eq. (3.3.10), we have

$$\mathbf{j} = \sin \theta \hat{\mathbf{i}} + \cos \theta \hat{\mathbf{j}} \qquad (3.3.13)$$

Now, using Eqs. (3.3.9) and (3.3.13) in Eq. (3.3.1), we have

$$d_x(\cos \theta \hat{\mathbf{i}} - \sin \theta \hat{\mathbf{j}}) + d_y(\sin \theta \hat{\mathbf{i}} + \cos \theta \hat{\mathbf{j}}) = \hat{d}_x \hat{\mathbf{i}} + \hat{d}_y \hat{\mathbf{j}} \qquad (3.3.14)$$

Combining like coefficients of $\hat{\mathbf{i}}$ and $\hat{\mathbf{j}}$ in Eq. (3.3.14), we obtain

$$d_x \cos \theta + d_y \sin \theta = \hat{d}_x$$

and
$$-d_x \sin \theta + d_y \cos \theta = \hat{d}_y \qquad (3.3.15)$$

In matrix form, Eqs. (3.3.15) are written as

$$\begin{Bmatrix} \hat{d}_x \\ \hat{d}_y \end{Bmatrix} = \begin{bmatrix} C & S \\ -S & C \end{bmatrix} \begin{Bmatrix} d_x \\ d_y \end{Bmatrix} \qquad (3.3.16)$$

where $C = \cos \theta$ and $S = \sin \theta$.

Equation (3.3.16) relates the global displacement $\underline{d}$ to the local displacement $\underline{\hat{d}}$. The matrix

$$\begin{bmatrix} C & S \\ -S & C \end{bmatrix} \qquad (3.3.17)$$

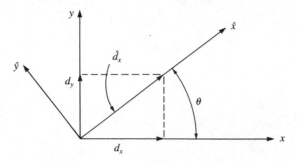

FIGURE 3-7   Relationship between local and global displacements

is called the *transformation matrix*. It will be used in Section 3.4 to develop the global stiffness matrix for an arbitrarily oriented bar element and to transform global nodal displacements and forces to local ones.

Now, for the case of $\hat{d}_y = 0$, we have, from Eq. (3.3.1),

$$d_x\mathbf{i} + d_y\mathbf{j} = \hat{d}_x\hat{\mathbf{i}} \qquad (3.3.18)$$

Figure 3-7 shows $\hat{d}_x$ expressed in terms of global x and y components. Using trigonometry and Figure 3-7, we then obtain the magnitude of $\hat{d}_x$ as

$$\hat{d}_x = Cd_x + Sd_y \qquad (3.3.19)$$

Equation (3.3.19) is equivalent to the first equation of Eq. (3.3.16).

EXAMPLE 3.2

The global nodal displacements at node 2 have been determined to be $d_{2x} = 0.1$ in. and $d_{2y} = 0.2$ in. for the bar element shown in Figure 3-8. Determine the local $\hat{x}$ displacement at node 2.

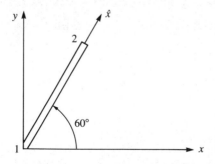

FIGURE 3-8   Bar element

Using Eq. (3.3.19), we obtain

$$\hat{d}_{2x} = (\cos 60°)(0.1) + (\sin 60°)(0.2) = 0.223 \text{ in.} \qquad \blacksquare$$

3.4   Global Stiffness Matrix

We will now use the transformation relationship Eq. (3.3.16) to obtain the global stiffness matrix for a bar element. We need the global stiffness matrix of each element to assemble the total global stiffness matrix of the structure. We have shown in Eq. (3.1.13) that for a bar element in the local coordinate system,

$$\begin{Bmatrix} \hat{f}_{1x} \\ \hat{f}_{2x} \end{Bmatrix} = \frac{AE}{L} \begin{bmatrix} 1 & -1 \\ -1 & 1 \end{bmatrix} \begin{Bmatrix} \hat{d}_{1x} \\ \hat{d}_{2x} \end{Bmatrix} \qquad (3.4.1)$$

or

$$\underline{\hat{f}} = \underline{\hat{k}}\underline{\hat{d}} \qquad (3.4.2)$$

We now want to relate the global element nodal forces f to the global nodal displacements d for a bar element arbitrarily oriented with respect to the global axes as was shown in Figure 3-1. This relationship will yield the global stiffness matrix $\underline{k}$ of the element. That is, we want to find a matrix $\underline{k}$ such that

$$\begin{Bmatrix} f_{1x} \\ f_{1y} \\ f_{2x} \\ f_{2y} \end{Bmatrix} = \underline{k} \begin{Bmatrix} d_{1x} \\ d_{1y} \\ d_{2x} \\ d_{2y} \end{Bmatrix} \qquad (3.4.3)$$

or, in simplified matrix form, Eq. (3.4.3) becomes

$$\underline{f} = \underline{k}\underline{d} \qquad (3.4.4)$$

We observe from Eq. (3.4.3) that a total of four components of force and four of displacement arise when global coordinates are used. However, a total of two components of force and two of displacement appear for the local-coordinate representation of a spring or a bar, as shown by Eq. (3.4.1). By using relationships between local and global force components and between local and global displacement components, we will be able to obtain the global stiffness matrix. We know from transformation relationship Eq. (3.3.15) that

$$\hat{d}_{1x} = d_{1x} \cos \theta + d_{1y} \sin \theta$$
$$\hat{d}_{2x} = d_{2x} \cos \theta + d_{2y} \sin \theta \qquad (3.4.5)$$

In matrix form, Eqs. (3.4.5) can be written as

$$\begin{Bmatrix} \hat{d}_{1x} \\ \hat{d}_{2x} \end{Bmatrix} = \begin{bmatrix} C & S & 0 & 0 \\ 0 & 0 & C & S \end{bmatrix} \begin{Bmatrix} d_{1x} \\ d_{1y} \\ d_{2x} \\ d_{2y} \end{Bmatrix} \qquad (3.4.6)$$

or as

$$\underline{\hat{d}} = \underline{T}^*\underline{d} \qquad (3.4.7)$$

where

$$T^* = \begin{bmatrix} C & S & 0 & 0 \\ 0 & 0 & C & S \end{bmatrix} \qquad (3.4.8)$$

Similarly, since forces transform in the same manner as displacements, we have

$$\left\{ \begin{matrix} \hat{f}_{1x} \\ \hat{f}_{2x} \end{matrix} \right\} = \begin{bmatrix} C & S & 0 & 0 \\ 0 & 0 & C & S \end{bmatrix} \left\{ \begin{matrix} f_{1x} \\ f_{1y} \\ f_{2x} \\ f_{2y} \end{matrix} \right\} \qquad (3.4.9)$$

or, using Eq. (3.4.8), we can write Eq. (3.4.9) as

$$\hat{f} = T^*f \qquad (3.4.10)$$

Now, substituting Eq. (3.4.7) into Eq. (3.4.2), we obtain

$$\hat{f} = \hat{k}T^*d \qquad (3.4.11)$$

and using Eq. (3.4.10) in (3.4.11) yields

$$T^*f = \hat{k}T^*d \qquad (3.4.12)$$

However, to write the final expression relating global nodal forces to global nodal displacements for an element, we must invert T^* in Eq. (3.4.12). This is not immediately possible because T^* is not a square matrix. Therefore, we must expand $\hat{d}$, $\hat{f}$, and $\hat{k}$ to the order that is consistent with the use of global coordinates even though $\hat{f}_{1y}$ and $\hat{f}_{2y}$ are zero. Using Eq. (3.3.16) for each nodal displacement, we thus obtain

$$\left\{ \begin{matrix} \hat{d}_{1x} \\ \hat{d}_{1y} \\ \hat{d}_{2x} \\ \hat{d}_{2y} \end{matrix} \right\} = \begin{bmatrix} C & S & 0 & 0 \\ -S & C & 0 & 0 \\ 0 & 0 & C & S \\ 0 & 0 & -S & C \end{bmatrix} \left\{ \begin{matrix} d_{1x} \\ d_{1y} \\ d_{2x} \\ d_{2y} \end{matrix} \right\} \qquad (3.4.13)$$

or

$$\hat{d} = Td \qquad (3.4.14)$$

where

$$T = \begin{bmatrix} C & S & 0 & 0 \\ -S & C & 0 & 0 \\ 0 & 0 & C & S \\ 0 & 0 & -S & C \end{bmatrix} \qquad (3.4.15)$$

Similarly, we can write

$$\hat{f} = Tf \qquad (3.4.16)$$

because forces are like displacements—both are vectors. Also, $\hat{k}$ must be expanded to a 4 × 4 matrix. Therefore, Eq. (3.4.1) in expanded form becomes

$$\begin{Bmatrix} \hat{f}_{1x} \\ \hat{f}_{1y} \\ \hat{f}_{2x} \\ \hat{f}_{2y} \end{Bmatrix} = \frac{AE}{L} \begin{bmatrix} 1 & 0 & -1 & 0 \\ 0 & 0 & 0 & 0 \\ -1 & 0 & 1 & 0 \\ 0 & 0 & 0 & 0 \end{bmatrix} \begin{Bmatrix} \hat{d}_{1x} \\ \hat{d}_{1y} \\ \hat{d}_{2x} \\ \hat{d}_{2y} \end{Bmatrix} \qquad (3.4.17)$$

In Eq. (3.4.17), since $\hat{f}_{1y}$ and $\hat{f}_{2y}$ are zero, rows of zeros corresponding to the row numbers $\hat{f}_{1y}$ and $\hat{f}_{2y}$ appear in $\hat{\underline{k}}$. Now using Eqs. (3.4.14) and (3.4.16) in Eq. (3.4.2), we obtain

$$T\underline{f} = \hat{\underline{k}}T\underline{d} \qquad (3.4.18)$$

Equation (3.4.18) is Eq. (3.4.12) expanded. Now, premultiplying both sides of Eq. (3.4.18) by $\underline{T}^{-1}$, we have

$$\underline{f} = \underline{T}^{-1}\hat{\underline{k}}T\underline{d} \qquad (3.4.19)$$

where $\underline{T}^{-1}$ is the *inverse* of $\underline{T}$. However, it can be shown (see Problem 3.26) that

$$\underline{T}^{-1} = \underline{T}^T \qquad (3.4.20)$$

where $\underline{T}^T$ is the *transpose* of $\underline{T}$. The property of square matrices such as $\underline{T}$ given by Eq. (3.4.20) defines $\underline{T}$ to be an orthogonal matrix. The transformation matrix $\underline{T}$ between rectangular coordinate frames is orthogonal. This property of $\underline{T}$ is used throughout this text. Substituting Eq. (3.4.20) into Eq. (3.4.19), we obtain

$$\underline{f} = \underline{T}^T\hat{\underline{k}}T\underline{d} \qquad (3.4.21)$$

Equating Eqs. (3.4.4) and (3.4.21), we obtain the global stiffness matrix for an element as

$$\underline{k} = \underline{T}^T\hat{\underline{k}}\underline{T} \qquad (3.4.22)$$

Substituting Eq. (3.4.15) for $\underline{T}$ and the expanded form of $\hat{\underline{k}}$ given in Eq. (3.4.17) into Eq. (3.4.22), we obtain $\underline{k}$ given in explicit form by

$$\underline{k} = \frac{AE}{L} \begin{bmatrix} C^2 & CS & -C^2 & -CS \\ & S^2 & -CS & -S^2 \\ & & C^2 & CS \\ \text{Symmetry} & & & S^2 \end{bmatrix} \qquad (3.4.23)$$

Now, since the trial displacement function Eq. (3.1.1) was assumed piecewise-continuous element by element, the stiffness matrix for each element can be summed using the direct stiffness method to obtain

$$\sum_{e=1}^{N} \underline{k}^{(e)} = \underline{K} \qquad (3.4.24)$$

where $\underline{K}$ is the total stiffness matrix and N is the total number of elements. Similarly, each element global nodal force matrix can be summed such that

$$\sum_{e=1}^{N} \underline{f}^{(e)} = \underline{F} \qquad (3.4.25)$$

$\underline{K}$ now relates the global nodal forces $\underline{F}$ to the global nodal displacements $\underline{d}$ for the whole structure by

$$\underline{F} = \underline{K}\underline{d} \qquad (3.4.26)$$

E X A M P L E   3.3

For the bar element shown in Figure 3-9, evaluate the global stiffness matrix with respect to the $x - y$ coordinate system. Let the bar's cross-sectional area equal 2 in.^{2}, length equal 60 in., and modulus of elasticity equal 30×10^6 psi. The angle the bar makes with the x axis is $30°$.

To evaluate the global stiffness matrix $\underline{k}$ for a bar, we use Eq. (3.4.23) with angle θ defined to be positive when measured counterclockwise from x to $\hat{x}$. Therefore,

$$\theta = 30° \qquad C = \cos 30° = \frac{\sqrt{3}}{2} \qquad S = \sin 30° = \frac{1}{2}$$

$$\underline{k} = \frac{(2)(30 \times 10^6)}{60}\begin{bmatrix} \dfrac{3}{4} & \dfrac{\sqrt{3}}{4} & \dfrac{-3}{4} & \dfrac{-\sqrt{3}}{4} \\[2mm] & \dfrac{1}{4} & \dfrac{-\sqrt{3}}{4} & \dfrac{-1}{4} \\[2mm] & & \dfrac{3}{4} & \dfrac{\sqrt{3}}{4} \\[2mm] \text{Symmetry} & & & \dfrac{1}{4} \end{bmatrix} \qquad (3.4.27)$$

Simplifying Eq. (3.4.27), we have

$$\underline{k} = 10^6 \begin{bmatrix} 0.75 & 0.433 & -0.75 & -0.433 \\ & 0.25 & -0.433 & -0.25 \\ & & 0.75 & 0.433 \\ \text{Symmetry} & & & 0.25 \end{bmatrix} \qquad (3.4.28) \quad \blacksquare$$

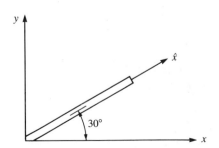

FIGURE 3-9   Bar element for stiffness matrix evaluation

3.5   Computation of Stress for a Bar in the *x-y* Plane

We will now consider the determination of the stress in a bar element. For a bar, the local forces are related to the local displacements by Eq. (3.1.13). This equation is repeated here for convenience.

$$\left\{\begin{matrix} \hat{f}_{1x} \\ \hat{f}_{2x} \end{matrix}\right\} = \frac{AE}{L}\begin{bmatrix} 1 & -1 \\ -1 & 1 \end{bmatrix}\left\{\begin{matrix} \hat{d}_{1x} \\ \hat{d}_{2x} \end{matrix}\right\} \tag{3.5.1}$$

The usual definition of axial tensile stress is

$$\sigma = \frac{\hat{f}_{2x}}{A} \tag{3.5.2}$$

where $\hat{f}_{2x}$ is used because it pulls on the bar as shown in Figure 3-10. By Eq. (3.5.1),

$$\hat{f}_{2x} = \frac{AE}{L}[-1 \quad 1]\left\{\begin{matrix} \hat{d}_{1x} \\ \hat{d}_{2x} \end{matrix}\right\} \tag{3.5.3}$$

Therefore, combining Eqs. (3.5.2) and (3.5.3) yields

$$\underline{\sigma} = \frac{E}{L}[-1 \quad 1]\hat{\underline{d}} \tag{3.5.4}$$

Now, using Eq. (3.4.7), we obtain

$$\underline{\sigma} = \frac{E}{L}[-1 \quad 1]\underline{T}^*\underline{d} \tag{3.5.5}$$

Equation (3.5.5) can be expressed in simpler form as

$$\underline{\sigma} = \underline{C}'\underline{d} \tag{3.5.6}$$

where, using Eq. (3.4.8),

$$\underline{C}' = \frac{E}{L}[-1 \quad 1]\begin{bmatrix} C & S & 0 & 0 \\ 0 & 0 & C & S \end{bmatrix} \tag{3.5.7}$$

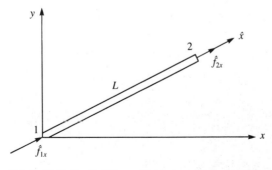

F I G U R E   3-10   Basic bar element with positive nodal forces

After multiplying the matrices in Eq. (3.5.7), we have

$$\underline{C'} = \frac{E}{L}[-C \quad -S \quad C \quad S]$$ (3.5.8)

E X A M P L E   3.4

For the bar shown in Figure 3-11, determine the axial stress. Let $A = 4 \times 10^{-4}$ m², $E = 210$ GPa, $L = 2$ m, and the angle between x and $\hat{x}$ equal 60°. Assume the global displacements have been previously determined to be $d_{1x} = 0.25$ mm, $d_{1y} = 0.0$, $d_{2x} = 0.50$ mm, and $d_{2y} = 0.75$ mm.

We can use Eq. (3.5.6) to evaluate the axial stress. Therefore, we first calculate $\underline{C'}$ from Eq. (3.5.8) as

$$\underline{C'} = \frac{210 \times 10^6 \text{ kN/m}^2}{2 \text{ m}} \left[-1 \quad -\frac{\sqrt{3}}{2} \quad \frac{1}{2} \quad \frac{\sqrt{3}}{2} \right]$$ (3.5.9)

where we have used $C = \cos 60° = \frac{1}{2}$ and $S = \sin 60° = \sqrt{3}/2$ in Eq. (3.5.9). Now $\underline{d}$ is given by

$$\underline{d} = \begin{Bmatrix} d_{1x} \\ d_{1y} \\ d_{2x} \\ d_{2y} \end{Bmatrix} = \begin{Bmatrix} 0.25 \times 10^{-3} \text{ m} \\ 0.0 \\ 0.50 \times 10^{-3} \text{ m} \\ 0.75 \times 10^{-3} \text{ m} \end{Bmatrix}$$ (3.5.10)

Using Eqs. (3.5.9) and (3.5.10) in Eq. (3.5.6), we obtain the bar axial stress as

$$\sigma = \frac{210 \times 10^6}{2} \left[-1 \quad -\frac{\sqrt{3}}{2} \quad \frac{1}{2} \quad \frac{\sqrt{3}}{2} \right] \begin{Bmatrix} 0.25 \\ 0.0 \\ 0.50 \\ 0.75 \end{Bmatrix} \times 10^{-3}$$

$$\sigma = 81.32 \times 10^3 \text{ kN/m}^2 = 81.32 \text{ MPa}$$ ■

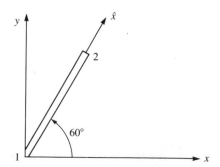

F I G U R E   3-11   Bar element for stress evaluation

3.6   Solution of a Plane Truss

We will now illustrate the use of equations developed in Sections 3.4 and 3.5, along with the direct stiffness method of assembling the total stiffness matrix and equations, to solve the following plane truss example problems. *A plane truss is a structure composed of bar elements all lying in a common plane that are connected together by frictionless pins.* The plane truss also must have loads acting only in the common plane.

EXAMPLE 3.5

For the plane truss composed of the three elements shown in Figure 3-12 subjected to a downward force of 10,000 lb applied at node 1, determine the x and y displacements at node 1 and the stresses in each element. Let $E = 30 \times 10^6$ psi and $A = 2$ in.^{2} for all elements. The lengths of the elements are shown in the figure.

First, we determine the global stiffness matrices for each element by using Eq. (3.4.23). This requires the determination of the angle θ between the global x axis and the local $\hat{x}$ axis for each element. In this example, the direction of the $\hat{x}$ axis for each element is taken in the direction *from* node 1 *to* the other node. The node numbering is arbitrary for each element. However, once the direction is chosen, the angle θ is then established as positive when measured counterclockwise from positive x to $\hat{x}$. For element 1, the local $\hat{x}$ axis is directed from node 1 to node 2; therefore, $\theta^{(1)} = 90°$. For element 2, the local $\hat{x}$ axis is directed from node 1 to node 3 and $\theta^{(2)} = 45°$. For element 3, the local $\hat{x}$ axis is directed from node 1 to node 4 and $\theta^{(3)} = 0°$. It is convenient to construct Table 3-1 to aid in determining each element stiffness matrix.

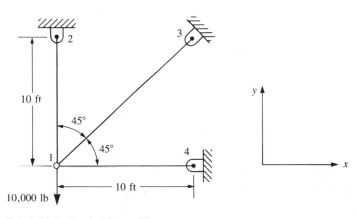

FIGURE 3-12   Plane truss

T A B L E 3-1   Data for the truss of Figure 3-12

Element	$\theta°$	C	S	C^2	S^2	CS
1	90°	0	1	0	1	0
2	45°	$\sqrt{2}/2$	$\sqrt{2}/2$	$\frac{1}{2}$	$\frac{1}{2}$	$\frac{1}{2}$
3	0°	1	0	1	0	0

There are a total of eight nodal components of displacement, or degrees of freedom, for the truss before boundary constraints are imposed. Thus, the order of the total stiffness matrix must be 8 × 8. We could then expand the $\underline{k}$ matrix for each element to the order 8 × 8 by adding rows and columns of zeros as explained in the first part of Section 2.4. Alternatively, we could label the rows and columns of each element stiffness matrix according to the displacement components associated with it as explained in the latter part of Section 2.4. Using this latter approach, the total stiffness matrix $\underline{K}$ is constructed simply by adding terms from the individual element stiffness matrices into their corresponding locations in $\underline{K}$. This approach will be used here and throughout this text.

For element 1, using Eq. (3.4.23), along with Table 3-1 for the direction cosines, we obtain

$$
\underline{k}^{(1)} = \frac{(30 \times 10^6)(2)}{120}
\begin{array}{c}
\begin{matrix} d_{1x} & d_{1y} & d_{2x} & d_{2y} \end{matrix} \\
\begin{bmatrix}
0 & 0 & 0 & 0 \\
0 & 1 & 0 & -1 \\
0 & 0 & 0 & 0 \\
0 & -1 & 0 & 1
\end{bmatrix}
\end{array}
\qquad (3.6.1)
$$

Similarly, for element 2, we have

$$
\underline{k}^{(2)} = \frac{(30 \times 10^6)(2)}{120 \times \sqrt{2}}
\begin{array}{c}
\begin{matrix} d_{1x} & d_{1y} & d_{3x} & d_{3y} \end{matrix} \\
\begin{bmatrix}
0.5 & 0.5 & -0.5 & -0.5 \\
0.5 & 0.5 & -0.5 & -0.5 \\
-0.5 & -0.5 & 0.5 & 0.5 \\
-0.5 & -0.5 & 0.5 & 0.5
\end{bmatrix}
\end{array}
\qquad (3.6.2)
$$

and for element 3, we have

$$
\underline{k}^{(3)} = \frac{(30 \times 10^6)(2)}{120}
\begin{array}{c}
\begin{matrix} d_{1x} & d_{1y} & d_{4x} & d_{4y} \end{matrix} \\
\begin{bmatrix}
1 & 0 & -1 & 0 \\
0 & 0 & 0 & 0 \\
-1 & 0 & 1 & 0 \\
0 & 0 & 0 & 0
\end{bmatrix}
\end{array}
\qquad (3.6.3)
$$

The common factor of $30 \times 10^6 \times 2/120 \, (= 500{,}000)$ can be taken from each of Eqs. (3.6.1)–(3.6.3). After adding terms from the individual element stiffness matrices into their corresponding locations in $\underline{K}$, we obtain the total stiffness matrix as

$$\underline{K} = (500{,}000)\begin{array}{c}\begin{array}{cccccccc} d_{1x} & \;\; d_{1y} & d_{2x} & d_{2y} & \;\; d_{3x} & \;\;\;\; d_{3y} & d_{4x} & d_{4y}\end{array}\\ \begin{bmatrix} 1.354 & 0.354 & 0 & 0 & -0.354 & -0.354 & -1 & 0 \\ 0.354 & 1.354 & 0 & -1 & -0.354 & -0.354 & 0 & 0 \\ 0 & 0 & 0 & 0 & 0 & 0 & 0 & 0 \\ 0 & -1 & 0 & 1 & 0 & 0 & 0 & 0 \\ -0.354 & -0.354 & 0 & 0 & 0.354 & 0.354 & 0 & 0 \\ -0.354 & -0.354 & 0 & 0 & 0.354 & 0.354 & 0 & 0 \\ -1 & 0 & 0 & 0 & 0 & 0 & 1 & 0 \\ 0 & 0 & 0 & 0 & 0 & 0 & 0 & 0 \end{bmatrix}\end{array}$$

$$(3.6.4)$$

The global $\underline{K}$ matrix, Eq. (3.6.4), relates the global forces to the global displacements. We thus write the total structure stiffness equations, accounting for the applied force at node 1 and the boundary constraints at nodes 2, 3, and 4 as follows:

$$\begin{Bmatrix} 0 \\ -10{,}000 \\ F_{2x} \\ F_{2y} \\ F_{3x} \\ F_{3y} \\ F_{4x} \\ F_{4y} \end{Bmatrix} = (500{,}000)\begin{bmatrix} 1.354 & 0.354 & 0 & 0 & -0.354 & -0.354 & -1 & 0 \\ 0.354 & 1.354 & 0 & -1 & -0.354 & -0.354 & 0 & 0 \\ 0 & 0 & 0 & 0 & 0 & 0 & 0 & 0 \\ 0 & -1 & 0 & 1 & 0 & 0 & 0 & 0 \\ -0.354 & -0.354 & 0 & 0 & 0.354 & 0.354 & 0 & 0 \\ -0.354 & -0.354 & 0 & 0 & 0.354 & 0.354 & 0 & 0 \\ -1 & 0 & 0 & 0 & 0 & 0 & 1 & 0 \\ 0 & 0 & 0 & 0 & 0 & 0 & 0 & 0 \end{bmatrix}$$
$$\times \begin{Bmatrix} d_{1x} \\ d_{1y} \\ d_{2x} = 0 \\ d_{2y} = 0 \\ d_{3x} = 0 \\ d_{3y} = 0 \\ d_{4x} = 0 \\ d_{4y} = 0 \end{Bmatrix} \qquad (3.6.5)$$

We could now use the partitioning scheme described in the first part of Section 2.5 to obtain the equations used to determine unknown displacements d_{1x} and d_{1y}—that is, partition the first two equations from the third through eighth in Eq. (3.6.5). Alternatively, we could eliminate rows and columns in the total stiffness matrix corresponding to zero displacements as previously described in the latter part of Section 2.5. Here we will use the latter approach; that is, we eliminate rows and columns 3 through 8 in Eq. (3.6.5) because those rows and columns correspond to zero displacements. (Remember, this direct approach must be modified for nonhomogeneous boundary conditions as was indicated in Section 2.5.) We then obtain

$$\begin{Bmatrix} 0 \\ -10,000 \end{Bmatrix} = (500,000) \begin{bmatrix} 1.354 & 0.354 \\ 0.354 & 1.354 \end{bmatrix} \begin{Bmatrix} d_{1x} \\ d_{1y} \end{Bmatrix} \qquad (3.6.6)$$

Equation (3.6.6) can now be solved for the displacements by multiplying both sides of the matrix equation by the inverse of the 2×2 stiffness matrix, or by solving the two equations simultaneously. Using either procedure for solution, the resulting displacements are

$$d_{1x} = 0.414 \times 10^{-2} \text{ in.} \qquad d_{1y} = -1.59 \times 10^{-2} \text{ in.}$$

The minus sign in the d_{1y} result indicates that the displacement component in the y direction at node 1 is in the direction opposite that of the positive y direction; that is, a downward displacement occurs at node 1.

Using Eq. (3.5.6) and Table 3-1, we determine the stresses in each element as follows:

$$\sigma^{(1)} = \frac{30 \times 10^6}{120} [0 \ -1 \ 0 \ 1] \begin{Bmatrix} d_{1x} = 0.414 \times 10^{-2} \\ d_{1y} = -1.59 \times 10^{-2} \\ d_{2x} = 0 \\ d_{2y} = 0 \end{Bmatrix} = 3965 \text{ psi}$$

$$\sigma^{(2)} = \frac{30 \times 10^6}{120\sqrt{2}} \left[-\frac{\sqrt{2}}{2} \ -\frac{\sqrt{2}}{2} \ \frac{\sqrt{2}}{2} \ \frac{\sqrt{2}}{2} \right] \begin{Bmatrix} d_{1x} = 0.414 \times 10^{-2} \\ d_{1y} = -1.59 \times 10^{-2} \\ d_{3x} = 0 \\ d_{3y} = 0 \end{Bmatrix}$$

$$= 1471 \text{ psi}$$

and

$$\sigma^{(3)} = \frac{30 \times 10^6}{120} [-1 \ 0 \ 1 \ 0] \begin{Bmatrix} d_{1x} = 0.414 \times 10^{-2} \\ d_{1y} = -1.59 \times 10^{-2} \\ d_{4x} = 0 \\ d_{4y} = 0 \end{Bmatrix} = -1035 \text{ psi}$$

We now verify our results by examining force equilibrium at node 1; that is, summing forces in the global x and y directions, we obtain

$$\sum F_x = 0 \qquad (1471 \text{ psi})(2 \text{ in.}^2)\frac{\sqrt{2}}{2} - (1035 \text{ psi})(2 \text{ in.}^2) = 0$$

$$\sum F_y = 0 \qquad (3965 \text{ psi})(2 \text{ in.}^2) + (1471 \text{ psi})(2 \text{ in.}^2)\frac{\sqrt{2}}{2} - 10,000 = 0 \qquad \blacksquare$$

E X A M P L E 3.6

For the two-bar truss shown in Figure 3-13, determine the displacement in the y direction of node 1, and the axial force in each element. A force of $P = 1000$ kN is applied at node 1 in the positive y direction while node 1

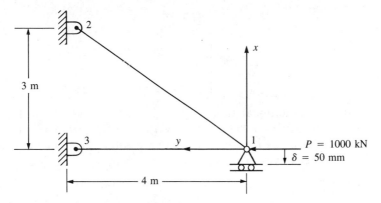

FIGURE 3-13   Two-bar truss

settles an amount $\delta = 50$ mm in the negative x direction. Let $E = 210$ GPa and $A = 6.00 \times 10^{-4}$ m^2 for each element. The lengths of the elements are shown in the figure.

We begin by using Eq. (3.4.23) to determine each element stiffness matrix.

Element 1

$$\cos \theta^{(1)} = \frac{3}{5} = 0.60 \qquad \sin \theta^{(1)} = \frac{4}{5} = 0.80$$

$$\underline{k}^{(1)} = \frac{(6.0 \times 10^{-4}\,\text{m}^2)(210 \times 10^6\,\text{kN/m}^2)}{5\,\text{m}} \begin{bmatrix} 0.36 & 0.48 & -0.36 & -0.48 \\ & 0.64 & -0.48 & -0.64 \\ & & 0.36 & 0.48 \\ \text{Symmetry} & & & 0.64 \end{bmatrix}$$

$$(3.6.7)$$

Simplifying Eq. (3.6.7), we obtain

$$\underline{k}^{(1)} = (25,200) \begin{matrix} d_{1x} \quad\; d_{1y} \quad\;\; d_{2x} \quad\;\; d_{2y} \\ \begin{bmatrix} 0.36 & 0.48 & -0.36 & -0.48 \\ & 0.64 & -0.48 & -0.64 \\ & & 0.36 & 0.48 \\ \text{Symmetry} & & & 0.64 \end{bmatrix} \end{matrix} \quad (3.6.8)$$

Element 2

$$\cos \theta^{(2)} = 0.0 \qquad \sin \theta^{(2)} = 1.0$$

$$\underline{k}^{(2)} = \frac{(6.0 \times 10^{-4})(210 \times 10^6)}{4} \begin{bmatrix} 0 & 0 & 0 & 0 \\ & 1 & 0 & -1 \\ & & 0 & 0 \\ \text{Symmetry} & & & 1 \end{bmatrix} \quad (3.6.9)$$

$$\underset{k^{(2)}}{=} (25{,}200) \begin{array}{c} \begin{array}{cccc} d_{1x} & d_{1y} & d_{3x} & d_{3y} \end{array} \\ \begin{bmatrix} 0 & 0 & 0 & 0 \\ & 1.25 & 0 & -1.25 \\ & & 0 & 0 \\ \text{Symmetry} & & & 1.25 \end{bmatrix} \end{array} \qquad (3.6.10)$$

where, for computational simplicity, Eq. (3.6.10) is written with the same factor (25,200) in front of the matrix as Eq. (3.6.8). Superimposing the element stiffness matrices, Eqs. (3.6.8) and (3.6.10), we obtain the global $\underline{K}$ matrix and relate the global forces to global displacements by

$$\begin{Bmatrix} F_{1x} \\ F_{1y} \\ F_{2x} \\ F_{2y} \\ F_{3x} \\ F_{3y} \end{Bmatrix} = (25{,}200) \begin{bmatrix} 0.36 & 0.48 & -0.36 & -0.48 & 0 & 0 \\ & 1.89 & -0.48 & -0.64 & 0 & -1.25 \\ & & 0.36 & 0.48 & 0 & 0 \\ & & & 0.64 & 0 & 0 \\ & & & & 0 & 0 \\ \text{Symmetry} & & & & & 1.25 \end{bmatrix} \begin{Bmatrix} d_{1x} \\ d_{1y} \\ d_{2x} \\ d_{2y} \\ d_{3x} \\ d_{3y} \end{Bmatrix}$$

$$(3.6.11)$$

We can again partition equations with known displacements and then simultaneously solve those associated with unknown displacements. To do this partitioning, we consider the boundary conditions given by

$$d_{1x} = \delta \qquad d_{2x} = 0 \qquad d_{2y} = 0 \qquad d_{3x} = 0 \qquad d_{3y} = 0 \qquad (3.6.12)$$

Therefore, using Eqs. (3.6.12), we partition equation 2 from equations 1, 3, 4, 5, and 6 of Eq. (3.6.11) and are left with

$$P = 25{,}200(0.48\delta + 1.89 d_{1y}) \qquad (3.6.13)$$

where $F_{1y} = P$ and $d_{1x} = \delta$ were substituted into Eq. (3.6.13). Expressing Eq. (3.6.13) in terms of P and δ allows these two influences on d_{1y} to be clearly separated. Solving Eq. (3.6.13) for d_{1y}, we have

$$d_{1y} = 0.000021 P - 0.254 \delta \qquad (3.6.14)$$

Now substituting the numerical values $P = 1000$ kN and $\delta = -0.05$ m into Eq. (3.6.14), we obtain

$$d_{1y} = 0.0337 \text{ m} \qquad (3.6.15)$$

where the positive value indicates horizontal displacement to the left.

The local element forces are obtained by using Eq. (3.4.11). We then have the following.

Element 1

$$\begin{Bmatrix} \hat{f}_{1x} \\ \hat{f}_{2x} \end{Bmatrix} = (25{,}200) \begin{bmatrix} 1 & -1 \\ -1 & 1 \end{bmatrix} \begin{bmatrix} 0.60 & 0.80 & 0 & 0 \\ 0 & 0 & 0.60 & 0.80 \end{bmatrix} \begin{Bmatrix} d_{1x} = -0.05 \\ d_{1y} = 0.0337 \\ d_{2x} = 0 \\ d_{2y} = 0 \end{Bmatrix}$$

$$(3.6.16)$$

Performing the matrix triple product in Eq. (3.6.16), we obtain

$$\hat{f}_{1x} = -76.6 \text{ kN} \qquad \hat{f}_{2x} = 76.6 \text{ kN} \qquad (3.6.17)$$

Element 2

$$\left\{ \begin{matrix} \hat{f}_{1x} \\ \hat{f}_{3x} \end{matrix} \right\} = (31,500) \begin{bmatrix} 1 & -1 \\ -1 & 1 \end{bmatrix} \begin{bmatrix} 0 & 1 & 0 & 0 \\ 0 & 0 & 0 & 1 \end{bmatrix} \left\{ \begin{matrix} d_{1x} = -0.05 \\ d_{1y} = 0.0337 \\ d_{3x} = 0 \\ d_{3y} = 0 \end{matrix} \right\} \qquad (3.6.18)$$

Performing the matrix triple product in Eq. (3.6.18), we obtain

$$\hat{f}_{1x} = 1061 \text{ kN} \qquad \hat{f}_{3x} = -1061 \text{ kN} \qquad (3.6.19)$$

Verification of the computations by checking that equilibrium is satisfied at node 1 is left to your discretion. ∎

3.7 Transformation Matrix and Stiffness Matrix for a Bar in Three-Dimensional Space

We will now derive the transformation matrix necessary to obtain the general stiffness matrix of a bar element arbitrarily oriented in three-dimensional space as shown in Figure 3-14. Let the coordinates of node 1 be taken as x_1, y_1, and z_1, and those of node 2 be taken as x_2, y_2, and z_2. Also, let θ_x, θ_y, and θ_z be the angles measured from the global x, y, and z axes, respectively, to the local $\hat{x}$ axis. Here $\hat{x}$ is directed along the element from node 1 to node 2. We must now determine $\underline{T}^*$ such that $\hat{\underline{d}} = \underline{T}^* \underline{d}$. We begin the derivation of $\underline{T}^*$ by considering the vector $\hat{\mathbf{d}} = \mathbf{d}$ expressed in three dimensions as

$$\hat{d}_x \hat{\mathbf{i}} + \hat{d}_y \hat{\mathbf{j}} + \hat{d}_z \hat{\mathbf{k}} = d_x \mathbf{i} + d_y \mathbf{j} + d_z \mathbf{k} \qquad (3.7.1)$$

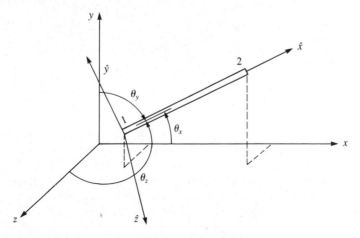

F I G U R E   3-14   Bar in three-dimensional space

where $\hat{\mathbf{i}}$, $\hat{\mathbf{j}}$, and $\hat{\mathbf{k}}$ are unit vectors associated with the local $\hat{x}$, $\hat{y}$, and $\hat{z}$ axes, respectively, and $\mathbf{i}$, $\mathbf{j}$, and $\mathbf{k}$ are unit vectors associated with the global x, y, and z axes. Taking the dot product of Eq. (3.7.1) with $\hat{\mathbf{i}}$, we have

$$\hat{d}_x + 0 + 0 = d_x(\hat{\mathbf{i}} \cdot \mathbf{i}) + d_y(\hat{\mathbf{i}} \cdot \mathbf{j}) + d_z(\hat{\mathbf{i}} \cdot \mathbf{k}) \qquad (3.7.2)$$

and, by definition of the dot product,

$$\hat{\mathbf{i}} \cdot \mathbf{i} = \frac{x_2 - x_1}{L} = C_x$$

$$\hat{\mathbf{i}} \cdot \mathbf{j} = \frac{y_2 - y_1}{L} = C_y \qquad (3.7.3)$$

$$\hat{\mathbf{i}} \cdot \mathbf{k} = \frac{z_2 - z_1}{L} = C_z$$

where $\qquad L = [(x_2 - x_1)^2 + (y_2 - y_1)^2 + (z_2 - z_1)^2]^{1/2}$

and $\qquad C_x = \cos \theta_x \qquad C_y = \cos \theta_y \qquad C_z = \cos \theta_z \qquad (3.7.4)$

Here C_x, C_y, and C_z are the projections of $\hat{\mathbf{i}}$ on $\mathbf{i}$, $\mathbf{j}$, and $\mathbf{k}$, respectively. Therefore, using Eqs. (3.7.3) in Eq. (3.7.2), we have

$$\hat{d}_x = C_x d_x + C_y d_y + C_z d_z \qquad (3.7.5)$$

For a vector in space directed along the $\hat{x}$ axis, Eq. (3.7.5) gives the components of that vector in the global x, y, and z directions. Now usng Eq. (3.7.5), $\underline{\hat{d}} = \underline{T}^* \underline{d}$ can be written in explicit form as

$$\left\{ \begin{matrix} \hat{d}_{1x} \\ \hat{d}_{2x} \end{matrix} \right\} = \begin{bmatrix} C_x & C_y & C_z & 0 & 0 & 0 \\ 0 & 0 & 0 & C_x & C_y & C_z \end{bmatrix} \left\{ \begin{matrix} d_{1x} \\ d_{1y} \\ d_{1x} \\ d_{2x} \\ d_{2y} \\ d_{2z} \end{matrix} \right\} \qquad (3.7.6)$$

where

$$\underline{T}^* = \begin{bmatrix} C_x & C_y & C_z & 0 & 0 & 0 \\ 0 & 0 & 0 & C_x & C_y & C_z \end{bmatrix} \qquad (3.7.7)$$

is the transformation matrix, which enables the local displacement matrix $\underline{\hat{d}}$ to be expressed in terms of displacement components in the global coordinate system.

We have shown in Section 3.4 that the global stiffness matrix (the stiffness matrix for a bar element referred to global axes) is given in general by $\underline{k} = \underline{T}^T \underline{\hat{k}} \underline{T}$. This equation will now be used to express the general form of the stiffness matrix of a bar arbitrarily oriented in space. In general, we must expand the transformation matrix in a manner analogous to that done in expanding $\underline{T}^*$ to $\underline{T}$ in Section 3.4. However, the same result will be obtained here by simply using $\underline{T}^*$, defined by Eq. (3.7.7), in place of $\underline{T}$. Then $\underline{k}$ is obtained using the equation $\underline{k} = (\underline{T}^*)^T \underline{\hat{k}} \underline{T}^*$ as follows:

$$k = \begin{bmatrix} C_x & 0 \\ C_y & 0 \\ C_z & 0 \\ 0 & C_x \\ 0 & C_y \\ 0 & C_z \end{bmatrix} \frac{AE}{L} \begin{bmatrix} 1 & -1 \\ -1 & 1 \end{bmatrix} \begin{bmatrix} C_x & C_y & C_z & 0 & 0 & 0 \\ 0 & 0 & 0 & C_x & C_y & C_z \end{bmatrix} \qquad (3.7.8)$$

Simplifying Eq. (3.7.8), we obtain the explicit form of k as

$$k = \frac{AE}{L} \begin{bmatrix} C_x^2 & C_x C_y & C_x C_z & -C_x^2 & -C_x C_y & -C_x C_z \\ & C_y^2 & C_y C_z & -C_x C_y & -C_y^2 & -C_y C_z \\ & & C_z^2 & -C_x C_z & -C_y C_z & -C_z^2 \\ & & & C_x^2 & C_x C_y & C_x C_z \\ & & & & C_y^2 & C_y C_z \\ \text{Symmetry} & & & & & C_z^2 \end{bmatrix} \qquad (3.7.9)$$

You should verify Eq. (3.7.9). First, expand T^* to a 6×6 square matrix in a manner similar to that done in Section 3.4 for the two-dimensional case. Then expand $\hat{k}$ to a 6×6 matrix by adding appropriate rows and columns of zeros (for the $\hat{d}_z$ terms) to Eq. (3.4.17), and finally, perform the matrix triple product $k = T^T \hat{k} T$ (see Problem 3.44).

Equation (3.7.9) is the basic form of the stiffness matrix for a bar element arbitrarily oriented in three-dimensional space. We will now analyze a simple space truss to illustrate the concepts developed in this section. We will show that the direct stiffness method provides a simple procedure for solving space truss problems.

EXAMPLE 3.7

Analyze the space truss shown in Figure 3-15. The truss is composed of four nodes, whose coordinates (in inches) are shown in the figure, and three elements with cross-sectional areas given in the figure. The modulus of elasticity $E = 1.2 \times 10^6$ psi for all elements. A load of 1000 lb is applied at node 1 in the negative z direction. Nodes 2, 3, and 4 are supported by ball-and-socket joints and thus constrained from movement in the x, y, and z directions. Node 1 is constrained from movement in the y direction by the roller shown in Figure 3-15.

Using Eq. (3.7.9), we will now determine the stiffness matrices of the three elements in Figure 3-15. To simplify the numerical calculations, we first express k for each element, given by Eq. (3.7.9), in the following form:

$$k = \frac{AE}{L} \begin{bmatrix} \lambda & -\lambda \\ -\lambda & \lambda \end{bmatrix} \qquad (3.7.10)$$

where λ is a 3×3 submatrix defined by

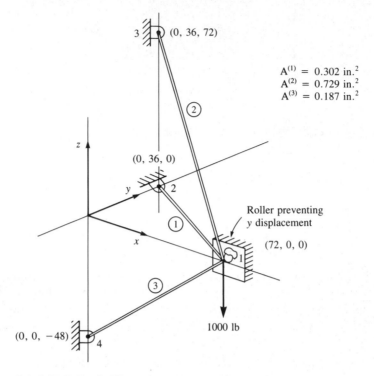

FIGURE 3-15   Space truss

$$
\underset{\sim}{\lambda} = \begin{bmatrix} C_x^2 & C_x C_y & C_x C_z \\ C_y C_x & C_y^2 & C_y C_z \\ C_z C_x & C_z C_y & C_z^2 \end{bmatrix}
\tag{3.7.11}
$$

Therefore, determining $\underset{\sim}{\lambda}$ will sufficiently describe $\underset{\sim}{k}$.

Element 3

The direction cosines of element 3 are given, in general, by

$$
C_x = \frac{x_4 - x_1}{L^{(3)}} \qquad C_y = \frac{y_4 - y_1}{L^{(3)}} \qquad C_z = \frac{z_4 - z_1}{L^{(3)}}
\tag{3.7.12}
$$

where the notation x_i, y_i, and z_i is used to denote the coordinates of each node, and $L^{(e)}$ denotes the element length. From the coordinate information given in Figure 3-15, we obtain the length and the direction cosines as

$$
L^{(3)} = [(-72.0)^2 + (-48.0)^2]^{1/2} = 86.5 \text{ in.}
$$

$$
C_x = \frac{-72.0}{86.5} = -0.833 \qquad C_y = 0 \qquad C_z = \frac{-48.0}{86.5} = -0.550
\tag{3.7.13}
$$

Using the results of Eqs. (3.7.13) in Eq. (3.7.11),

$$\underline{\lambda} = \begin{bmatrix} 0.69 & 0 & 0.46 \\ 0 & 0 & 0 \\ 0.46 & 0 & 0.30 \end{bmatrix} \tag{3.7.14}$$

and, from Eq. (3.7.10),

$$\underline{k}^{(3)} = \frac{(0.187)(1.2 \times 10^6)}{86.5} \begin{matrix} \begin{matrix} d_{1x}\,d_{1y}\,d_{1z} & d_{4x}\,d_{4y}\,d_{4z} \end{matrix} \\ \begin{bmatrix} \underline{\lambda} & -\underline{\lambda} \\ -\underline{\lambda} & \underline{\lambda} \end{bmatrix} \end{matrix} \tag{3.7.15}$$

Element 1

Similarly, for element 1, we obtain

$$L^{(1)} = 80.5 \text{ in.}$$
$$C_x = -0.89 \qquad C_y = 0.45 \qquad C_z = 0$$

$$\underline{\lambda} = \begin{bmatrix} 0.79 & -0.40 & 0 \\ -0.40 & 0.20 & 0 \\ 0 & 0 & 0 \end{bmatrix}$$

and $$\underline{k}^{(1)} = \frac{(0.302)(1.2 \times 10^6)}{80.5} \begin{matrix} \begin{matrix} d_{1x}\,d_{1y}\,d_{1z} & d_{2x}\,d_{2y}\,d_{2z} \end{matrix} \\ \begin{bmatrix} \underline{\lambda} & -\underline{\lambda} \\ -\underline{\lambda} & \underline{\lambda} \end{bmatrix} \end{matrix} \tag{3.7.16}$$

Element 2

Finally, for element 2, we obtain

$$L^{(2)} = 108 \text{ in.}$$
$$C_x = -0.667 \qquad C_y = 0.33 \qquad C_z = 0.667$$

$$\underline{\lambda} = \begin{bmatrix} 0.45 & -0.22 & -0.45 \\ -0.22 & 0.11 & 0.45 \\ -0.45 & 0.45 & 0.45 \end{bmatrix}$$

and $$\underline{k}^{(2)} = \frac{(0.729)(1.2 \times 10^6)}{108} \begin{matrix} \begin{matrix} d_{1x}\,d_{1y}\,d_{1z} & d_{3x}\,d_{3y}\,d_{3z} \end{matrix} \\ \begin{bmatrix} \underline{\lambda} & -\underline{\lambda} \\ -\underline{\lambda} & \underline{\lambda} \end{bmatrix} \end{matrix} \tag{3.7.17}$$

Using the zero-displacement boundary conditions $d_{1y} = 0$, $d_{2x} = d_{2y} = d_{2z} = 0$, $d_{3x} = d_{3y} = d_{3z} = 0$, and $d_{4x} = d_{4y} = d_{4z} = 0$, we can cancel the corresponding rows and columns of each element stiffness matrix. After canceling appropriate rows and columns in Eqs. (3.7.15), (3.7.16), and (3.7.17), and then superimposing the resulting element stiffness matrices, we have the total stiffness matrix for the truss as

$$\underline{K} = \begin{matrix} \begin{matrix} d_{1x} & \quad d_{1z} \end{matrix} \\ \begin{bmatrix} 9000 & -2450 \\ -2450 & 4450 \end{bmatrix} \end{matrix} \tag{3.7.18}$$

The global stiffness equations are then expressed by

$$\begin{Bmatrix} 0 \\ -1000 \end{Bmatrix} = \begin{bmatrix} 9000 & -2450 \\ -2450 & 4550 \end{bmatrix} \begin{Bmatrix} d_{1x} \\ d_{1z} \end{Bmatrix} \qquad (3.7.19)$$

Solving Eq. (3.7.19) for the displacements, we obtain

$$d_{1x} = -0.072 \text{ in.}$$
$$d_{1z} = -0.264 \text{ in.} \qquad (3.7.20)$$

where the minus signs in the displacements indicate these displacements to be in the negative x and z directions.

We will now determine the stress in each element. The stresses are determined by using Eq. (3.5.6) expanded to three dimensions. Thus, for an element with nodes i and j, Eq. (3.5.6) expanded to three dimensions becomes

$$\sigma = \frac{E}{L}[-C_x \quad -C_y \quad -C_z \quad C_x \quad C_y \quad C_z] \begin{Bmatrix} d_{ix} \\ d_{iy} \\ d_{iz} \\ d_{jx} \\ d_{jy} \\ d_{jz} \end{Bmatrix} \qquad (3.7.21)$$

Derive Eq. (3.7.21) in a manner similar to that used to derive Eq. (3.5.6) (see Problem 3.45, for instance). For element 3, using Eqs. (3.7.13) for the direction cosines, along with the proper length and modulus of elasticity, we obtain the stress as

$$\sigma^{(3)} = \frac{1.2 \times 10^6}{86.5}[0.83 \quad 0 \quad 0.55 \quad -0.83 \quad 0 \quad -0.55] \begin{Bmatrix} -0.072 \\ 0 \\ -0.264 \\ 0 \\ 0 \\ 0 \end{Bmatrix}$$

$$(3.7.22)$$

Simplifying Eq. (3.7.22), the result is

$$\sigma^{(3)} = -2850 \text{ psi}$$

where the negative sign in the answer indicates a compressive stress. The stresses in the other elements can be determined in a manner similar to that used for element 3. For brevity's sake, we will not show the calculations but merely list these stresses:

$$\sigma^{(1)} = -945 \text{ psi} \qquad \sigma^{(2)} = 1440 \text{ psi} \qquad \blacksquare$$

3.8   Inclined, or Skewed, Supports

In the preceding sections the supports were oriented such that the resulting boundary conditions on the displacements were in the global directions.

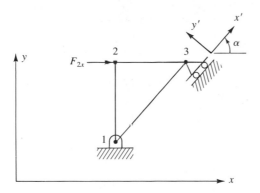

F I G U R E 3-16   Plane truss with inclined boundary conditions at node 3

However if a support is inclined, or skewed, at an angle α from the global x axis, as shown at node 3 in the plane truss of Figure 3-16, the resulting boundary conditions on the displacements are not in the global x-y directions but are in the local x'-y' directions.

To account for inclined boundary conditions, we must perform a transformation of the global displacements at node 3 only into the local nodal coordinate system x'-y', while keeping all other displacements in the x-y global system. We can then enforce the zero-displacement boundary condition d'_{3y} in the force-displacement equations and, finally, solve the equations in the usual manner.

The transformation used is analogous to that for transforming a vector from local to global coordinates. For the plane truss we use Eq. (3.3.16) applied to node 3 as follows:

$$\begin{Bmatrix} d'_{3x} \\ d'_{3y} \end{Bmatrix} = \begin{bmatrix} \cos \alpha & \sin \alpha \\ -\sin \alpha & \cos \alpha \end{bmatrix} \begin{Bmatrix} d_{3x} \\ d_{3y} \end{Bmatrix} \qquad (3.8.1)$$

or rewriting Eq. (3.8.1), we have

$$\{d'_3\} = [t_3]\{d_3\} \qquad (3.8.2)$$

where

$$[t_3] = \begin{bmatrix} \cos \alpha & \sin \alpha \\ -\sin \alpha & \cos \alpha \end{bmatrix} \qquad (3.8.3)$$

We now write the transformation for the entire nodal displacement vector as

$$\{d'\} = [T_1]\{d\} \qquad (3.8.4)$$

or

$$\{d\} = [T_1]^T\{d'\} \qquad (3.8.5)$$

where the transformation matrix for the entire truss is

$$[T_1] = \begin{bmatrix} [I] & & \\ & [I] & \\ & & [t_3] \end{bmatrix} \qquad (3.8.6)$$

The identity matrix $[I]$ and $[t_3]$ have the same 2×2 order, that order in general being equal to the number of degrees of freedom at each node.

On considering Eqs. (3.8.5) and (3.8.6), we observe that only node 3 components of $\{d\}$ are really transformed to local (skewed) axes components. This transformation is indeed necessary whenever the local axes x'-y' fixity directions are known.

Furthermore, the global force vector can also be transformed by using the same transformation as for $\{d'\}$:

$$\{f'\} = [T_1]\{f\} \qquad (3.8.7)$$

In global coordinates, we then have

$$\{f\} = [K]\{d\} \qquad (3.8.8)$$

On premultiplying Eq. (3.8.8) by $[T_1]$, we have

$$[T_1]\{f\} = [T_1][K]\{d\} \qquad (3.8.9)$$

For the truss in Figure 3-16, the left side of Eq. (3.8.9) is

$$\begin{bmatrix} [I] & [0] & [0] \\ [0] & [I] & [0] \\ [0] & [0] & [t_3] \end{bmatrix} \begin{Bmatrix} f_{1x} \\ f_{1y} \\ f_{2x} \\ f_{2y} \\ f_{3x} \\ f_{3y} \end{Bmatrix} = \begin{Bmatrix} f_{1x} \\ f_{1y} \\ f_{2x} \\ f_{2y} \\ f'_{3x} \\ f'_{3y} \end{Bmatrix} \qquad (3.8.10)$$

where the fact that local forces transform similarly to Eq. (3.8.2) as

$$\{f'_3\} = [t_3]\{f_3\} \qquad (3.8.11)$$

has been used in Eq. (3.8.10). Also, each submatrix in Eq. (3.8.10) is of order 2×2, with the null, or zero, matrix denoted by $[0]$. From Eq. (3.8.10), we see that only the node 3 components of $\{f\}$ have been transformed to the local axes components, as desired.

To obtain the desired displacement vector with global displacement components at nodes 1 and 2 and local displacement components at node 3, we use Eq. (3.8.5) to obtain

$$\begin{Bmatrix} d_{1x} \\ d_{1y} \\ d_{2x} \\ d_{2y} \\ d_{3x} \\ d_{3y} \end{Bmatrix} = \begin{bmatrix} [I] & [0] & [0] \\ [0] & [I] & [0] \\ [0] & [0] & [t_3]^T \end{bmatrix} \begin{Bmatrix} d'_{1x} \\ d'_{1y} \\ d'_{2x} \\ d'_{2y} \\ d'_{3x} \\ d'_{3y} \end{Bmatrix} \qquad (3.8.12)$$

In Eq. (3.8.12), we observe that only the node 3 global components are transformed, as indicated by the placement of the $[t_3]^T$ matrix. We denote the square matrix in Eq. (3.8.12) by $[T_1]^T$. In general, we place a 2×2 $[t]$ matrix in $[T_1]$ wherever the transformation from global to local displacements is needed (where skewed supports exist). Using Eq. (3.8.12) in Eq. (3.8.9), we have

$$[T_1]\{f\} = [T_1][K][T_1]^T\{d'\} \tag{3.8.13}$$

Using Eq. (3.8.10), the form of Eq. (3.8.13) becomes

$$\begin{Bmatrix} F_{1x} \\ F_{1y} \\ F_{2x} \\ F_{2y} \\ F'_{3x} \\ F'_{3y} \end{Bmatrix} = [T_1][K][T_1]^T \begin{Bmatrix} d_{1x} \\ d_{1y} \\ d_{2x} \\ d_{2y} \\ d'_{3x} \\ d'_{3y} \end{Bmatrix} \tag{3.8.14}$$

as $d_{1x} = d'_{1x}$, $d_{1y} = d'_{1y}$, $d_{2x} = d'_{2x}$, and $d_{2y} = d'_{2y}$ from Eq. (3.8.12). Equation (3.8.14) is the desired form that allows all known global and inclined boundary conditions to be enforced. The global forces now result in the left side of Eq. (3.8.14). To solve Eq. (3.8.14), first perform the matrix triple product $[T_1][K][T_1]^T$. Then invoke the following boundary conditions (for the truss in Figure 3-16):

$$d_{1x} = 0, \qquad d_{1y} = 0, \qquad d'_{3y} = 0 \tag{3.8.15}$$

Then substitute the known value of the applied force F_{2x} along with $F_{2y} = 0$ and $F'_{3x} = 0$ into Eq. (3.8.14). Finally, partition the equations with known displacements (here Eqs. 1, 2, and 6 of Eq. (3.8.14)) and then simultaneously solve those associated with the unknown displacements d_{2x}, d_{2y}, and d'_{3x}.

After solving for the displacements, return to Eq. (3.8.14) to obtain the global reactions F_{1x} and F_{1y} and the inclined roller reaction F'_{3y}.

EXAMPLE 3.8

For the plane truss shown in Figure 3-17, determine the displacements and reactions. Let $E = 210$ GPa, $A = 6.00 \times 10^{-4}$ m² for elements 1 and 2, and $A = 6\sqrt{2} \times 10^{-4}$ m² for element 3.

We begin by using Eq. (3.4.23) to determine each element stiffness matrix as follows:

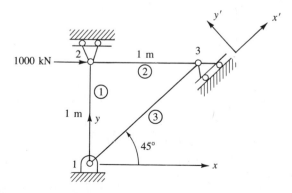

FIGURE 3-17   Plane truss with inclined support

Element 1

$$\cos \theta = 0 \qquad \sin \theta = 1$$

$$
k^{(1)} = \frac{(6.0 \times 10^{-4})(210 \times 10^9)}{1 \text{ m}}
\begin{array}{cccc}
d_{1x} & d_{1y} & d_{2x} & d_{2y}
\end{array}
\begin{bmatrix}
0 & 0 & 0 & 0 \\
 & 1 & 0 & -1 \\
 & & 0 & 0 \\
\text{Symmetry} & & & 1
\end{bmatrix}
\qquad (3.8.16)
$$

Element 2

$$\cos \theta = 1 \qquad \sin \theta = 0$$

$$
k^{(2)} = \frac{(6.0 \times 10^{-4})(210 \times 10^9)}{1 \text{ m}}
\begin{array}{cccc}
d_{2x} & d_{2y} & d_{3x} & d_{3y}
\end{array}
\begin{bmatrix}
1 & 0 & -1 & 0 \\
 & 0 & 0 & 0 \\
 & & 1 & 0 \\
\text{Symmetry} & & & 0
\end{bmatrix}
\qquad (3.8.17)
$$

Element 3

$$\cos \theta = \frac{\sqrt{2}}{2} \qquad \sin \theta = \frac{\sqrt{2}}{2}$$

$$
k^{(3)} = \frac{(6\sqrt{2} \times 10^{-4})(210 \times 10^9)}{\sqrt{2}}
\begin{array}{cccc}
d_{1x} & d_{1y} & d_{3x} & d_{3y}
\end{array}
\begin{bmatrix}
0.5 & 0.5 & -0.5 & -0.5 \\
 & 0.5 & -0.5 & -0.5 \\
 & & 0.5 & 0.5 \\
\text{Symmetry} & & & 0.5
\end{bmatrix}
\qquad (3.8.18)
$$

Using the direct stiffness method on Eqs. (3.8.16) through (3.8.18), we obtain the global K matrix as

$$
K = 1260 \times 10^5 \text{ N/m}
\begin{bmatrix}
0.5 & 0.5 & 0 & 0 & -0.5 & -0.5 \\
 & 1.5 & 0 & -1 & -0.5 & -0.5 \\
 & & 1 & 0 & -1 & 0 \\
 & & & 1 & 0 & 0 \\
 & & & & 1.5 & 0.5 \\
\text{Symmetry} & & & & & 0.5
\end{bmatrix}
\qquad (3.8.19)
$$

Next we obtain the transformation matrix T_1 using Eq. (3.8.6) to transform the global displacements at node 3 into local nodal coordinates x'-y'. In using Eq. (3.8.6) the angle α is 45°.

$$
[T_1] =
\begin{bmatrix}
1 & 0 & 0 & 0 & 0 & 0 \\
0 & 1 & 0 & 0 & 0 & 0 \\
0 & 0 & 1 & 0 & 0 & 0 \\
0 & 0 & 0 & 1 & 0 & 0 \\
0 & 0 & 0 & 0 & \sqrt{2}/2 & \sqrt{2}/2 \\
0 & 0 & 0 & 0 & -\sqrt{2}/2 & \sqrt{2}/2
\end{bmatrix}
\qquad (3.8.20)
$$

Next we use Eq. (3.8.14) (in general, we would use Eq. (3.8.13)) to express the assembled equations. First define $\underline{K}^* = \underline{T_1}\underline{K}\underline{T_1}^T$ and evaluate in steps as follows:

$$\underline{T_1}\underline{K} = 1260 \times 10^5 \begin{bmatrix} 0.5 & 0.5 & 0 & 0 & -0.5 & -0.5 \\ 0.5 & 1.5 & 0 & -1 & -0.5 & -0.5 \\ 0 & 0 & 1 & 0 & -1 & 0 \\ 0 & -1 & 0 & 1 & 0 & 0 \\ -0.707 & -0.707 & -0.707 & 0 & 1.414 & 0.707 \\ 0 & 0 & 0.707 & 0 & -0.707 & 0 \end{bmatrix}$$

$$(3.8.21)$$

and

$$\underline{T_1}\underline{K}\underline{T_1}^T = 1260 \times 10^5 \begin{array}{cccccc} d_{1x} & d_{1y} & d_{2x} & d_{2y} & d'_{3x} & d'_{3y} \\ \begin{bmatrix} 0.5 & 0.5 & 0 & 0 & -0.707 & 0 \\ 0.5 & 1.5 & 0 & -1 & -0.707 & 0 \\ 0 & 0 & 1 & 0 & -0.707 & 0.707 \\ 0 & -1 & 0 & 1 & 0 & 0 \\ -0.707 & -0.707 & -0.707 & 0 & 1.500 & -0.500 \\ 0 & 0 & 0.707 & 0 & -0.500 & 0.500 \end{bmatrix} \end{array}$$

$$(3.8.22)$$

Applying the boundary conditions, $d_{1x} = d_{1y} = d_{2y} = d'_{3y} = 0$, to Eq. (3.8.22), we obtain

$$\begin{Bmatrix} F_{2x} = 1000 \text{ kN} \\ F'_{3x} = 0 \end{Bmatrix} = (1260 \times 10^5) \begin{bmatrix} 1 & -0.707 \\ -0.707 & 1.50 \end{bmatrix} \begin{Bmatrix} d_{2x} \\ d'_{3x} \end{Bmatrix} \qquad (3.8.23)$$

Solving Eq. (3.8.23) for the displacements, we obtain

$$d_{2x} = 11.91 \text{ mm}$$
$$d'_{3x} = 5.613 \text{ mm} \qquad (3.8.24)$$

Postmultiplying the known displacement vector times Eq. (3.8.22) (see Eq. (3.8.14), we obtain the reactions as

$$F_{1x} = -500 \text{ kN}$$
$$F_{1y} = -500 \text{ kN}$$
$$F_{2y} = 0 \qquad (3.8.25)$$
$$F'_{3y} = 707 \text{ kN}$$

The free-body diagram of the truss with the reactions is shown in Figure 3-18. You can easily verify that the truss is in equilibrium. ∎

3.9 Potential Energy Approach to Derive Bar Element Equations

We now present the principle of minimum potential energy to derive the bar element equations. Recall from Section 2.6 that the total potential energy π_p

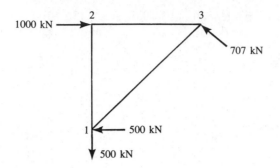

FIGURE 3-18   Free-body diagram of the truss of Figure 3-17

was defined as the sum of the internal strain energy U and the potential energy of the external forces Ω as

$$\pi_p = U + \Omega \qquad (3.9.1)$$

To evaluate the strain energy for a bar, we consider only the work done by the internal forces during deformation. Because we are dealing with a one-dimensional bar, the internal force doing work is given in Figure 3-19 as $\sigma_x(\Delta y)(\Delta z)$, due only to normal stress σ_x. The displacement of the x face of the element is $\Delta x(\varepsilon_x)$; the displacement of the $x + \Delta x$ face is $\Delta x(\varepsilon_x + d\varepsilon_x)$. The change in displacement is then $\Delta x\, d\varepsilon_x$, where $d\varepsilon_x$ is the differential change in strain occurring over length Δx. The differential internal work (or strain energy) dU is the internal force multiplied by the displacement through which the force moves, given by

$$dU = \sigma_x(\Delta y)(\Delta z)(\Delta x)\, d\varepsilon_x \qquad (3.9.2)$$

Rearranging and letting the volume of the element approach zero, we obtain, from Eq. (3.9.2),

$$dU = \sigma_x\, d\varepsilon_x\, dV \qquad (3.9.3)$$

For the whole bar, we then have

$$U = \int\int\int_V \left\{ \int_0^{\varepsilon_x} \sigma_x\, d\varepsilon_x \right\} dV \qquad (3.9.4)$$

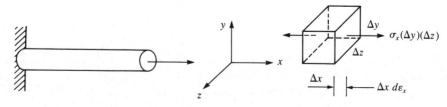

FIGURE 3-19   Internal force in a one-dimensional bar

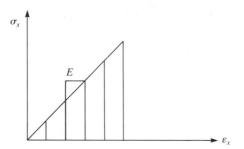

FIGURE 3-20   Linear-elastic (Hooke's law) material

Now, for a linear-elastic (Hooke's law) material as shown in Figure 3-20, we see that $\sigma_x = E\varepsilon_x$. Hence, substituting this relationship into Eq. (3.9.4), integrating with respect to ε_x, and then resubstituting σ_x for $E\varepsilon_x$, we have

$$U = \frac{1}{2} \iiint_V \sigma_x \varepsilon_x \, dV \qquad (3.9.5)$$

as the expression for the strain energy for one-dimensional stress.

The potential energy of the external forces, being opposite in sign from the external work expression because the potential energy of external forces is lost when the work is done by the external forces, is given by

$$\Omega = - \iiint_V \hat{X}_b \hat{u} \, dV - \iint_{S_1} \hat{T}_x \hat{u} \, dS - \sum_{i=1}^{M} \hat{f}_{ix} \hat{d}_{ix} \qquad (3.9.6)$$

where the first, second, and third terms on the right side of Eq. (3.9.6) represent the potential energy of (1) body forces $\hat{X}_b$ (in units of force per unit volume), (2) surface loading or traction $\hat{T}_x$ (in units of force per unit surface area), and (3) nodal concentrated forces $\hat{f}_{ix}$. The forces $\hat{X}_b$, $\hat{T}_x$, and $\hat{f}_{ix}$ are considered to act in the local $\hat{x}$ direction of the bar as shown in Figure 3-21. In Eqs. (3.9.5) and (3.9.6), V is the volume of the body and S_1 is the part of the surface S on which surface loading acts. For a bar element with two nodes and one degree of freedom per node, $M = 2$.

We are now ready to describe the finite element formulation of the bar element equations using the principle of minimum potential energy.

The finite element process seeks a minimum in the potential energy within the constraint of an assumed displacement pattern within each element. The greater the number of degrees of freedom associated with the element (usually meaning increasing the number of nodes), the more closely will the solution approximate the true one and ensure complete equilibrium (provided the true displacement can, in the limit, be approximated). An approximate finite element solution using the stiffness method will always provide an approximate value of potential energy greater than or equal to the correct one. This method also results in a structure behavior that is predicted

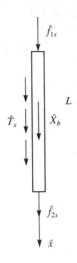

FIGURE 3-21   General forces acting on a one-dimensional bar

to be physically stiffer than, or at best to have the same stiffness as, the actual one. This is explained by the fact that the structure model is allowed to displace only into shapes defined by the terms of the assumed displacement field within each element of the structure. The correct shape is usually only approximated by the assumed field, although the correct shape can be the same as the assumed field. The assumed field effectively constrains the structure from deforming in its natural manner. This constraint effect stiffens the predicted behavior of the structure.

Apply the following steps when using the principle of minimum potential energy to derive the finite element equations.

1. Formulate an expression for the total potential energy.

2. Assume the displacement pattern to vary with a finite set of undetermined parameters (here these are the nodal displacements d_{ix}), which are substituted into the expression for total potential energy.

3. Obtain a set of simultaneous equations minimizing the total potential energy with respect to these nodal parameters. These resulting equations represent the element equations.

The resulting equations are the approximate (or possibly exact) equilibrium equations whose solution for the nodal parameters seeks to minimize the potential energy when back-substituted into the potential energy expression. The preceding three steps will now be followed to derive the bar element equations and stiffness matrix.

Consider the bar element of length L, with constant cross-sectional area A, shown in Figure 3-21. Using Eqs. (3.9.5) and (3.9.6), the total potential energy, Eq. (3.9.1), becomes

$$\pi_p = \frac{A}{2} \int_0^L \sigma_x \varepsilon_x \, d\hat{x} - \hat{f}_{1x}\hat{d}_{1x} - \hat{f}_{2x}\hat{d}_{2x} - \iint_S \hat{u}\hat{T}_x \, dS - \iiint_V \hat{u}\hat{X}_b \, dV$$

$$(3.9.7)$$

since A is a constant and variables σ_x and ε_x at most vary with $\hat{x}$.

From Eqs. (3.1.3) and (3.1.4), we have the axial displacement function expressed in terms of the shape functions and nodal displacements by

$$\hat{u} = [N]\{\hat{d}\} \qquad (3.9.8)$$

where

$$[N] = \left[1 - \frac{\hat{x}}{L} \quad \frac{\hat{x}}{L} \right] \qquad (3.9.9)$$

and

$$\{\hat{d}\} = \begin{Bmatrix} \hat{d}_{1x} \\ \hat{d}_{2x} \end{Bmatrix} \qquad (3.9.10)$$

Then, using the strain/displacement relationship $\varepsilon_x = d\hat{u}/d\hat{x}$, the axial strain can be written as

$$\{\varepsilon_x\} = \left[-\frac{1}{L} \quad \frac{1}{L} \right]\{\hat{d}\} \qquad (3.9.11)$$

or

$$\{\varepsilon_x\} = [B]\{\hat{d}\} \qquad (3.9.12)$$

where we define

$$[B] = \left[-\frac{1}{L} \quad \frac{1}{L} \right] \qquad (3.9.13)$$

The axial stress/strain relationship is given by

$$\{\sigma_x\} = [D]\{\varepsilon_x\} \qquad (3.9.14)$$

where

$$[D] = [E] \qquad (3.9.15)$$

for the one-dimensional stress/strain relationship and E is the modulus of elasticity. Now, by Eq. (3.9.12), we can express Eq. (3.9.14) as

$$\{\sigma_x\} = [D][B]\{\hat{d}\} \qquad (3.9.16)$$

Using Eq. (3.9.17) expressed in matrix notation form, we have the total potential energy given by

$$\pi_p = \frac{A}{2} \int_0^L \{\sigma_x\}^T\{\varepsilon_x\} \, d\hat{x} - \{\hat{d}\}^T\{P\} - \iint_S \{\hat{u}\}^T\{\hat{T}_x\} \, dS - \iiint_V \{\hat{u}\}^T\{\hat{X}_b\} \, dV$$

$$(3.9.17)$$

where $\{P\}$ now represents the concentrated nodal loads and where in general both $\underline{\sigma}_x$ and $\underline{\varepsilon}_x$ are column matrices. For proper matrix multiplication we must place the transpose on $\{\sigma_x\}$. Similarly, $\{\hat{u}\}$ and $\{\hat{T}_x\}$ in general are column matrices, so for proper matrix multiplication $\{\hat{u}\}$ is transposed in Eq. (3.9.17).

Using Eqs. (3.9.11), (3.9.12), and (3.9.16) in Eq. (3.9.17), we obtain

$$\pi_p = \frac{A}{2} \int_0^L \{\hat{d}\}^T [B]^T [D]^T [B] \{\hat{d}\} \, d\hat{x} - \{\hat{d}\}^T \{P\} - \int\int_S \{\hat{d}\}^T [N]^T \{\hat{T}_x\} \, dS$$

$$- \int\int\int_V \{d\}^T [N]^T \{\hat{X}_b\} \, dV \qquad (3.9.18)$$

In Eq. (3.9.18), π_p is seen to be a function of $\{\hat{d}\}$; that is, $\pi_p = \pi_p(\hat{d}_{1x}, \hat{d}_{2x})$. However, $[B]$ and $[D]$, Eqs. (3.9.13) and (3.9.15), and the nodal degrees of freedom $\hat{d}_{1x}$ and $\hat{d}_{2x}$ are not functions of $\hat{x}$. Therefore, integrating Eq. (3.9.18) with respect to $\hat{x}$ yields

$$\pi_p = \frac{AL}{2} \{\hat{d}\}^T [B]^T [D]^T [B] \{\hat{d}\} - \{\hat{d}\}^T \{\hat{f}\} \qquad (3.9.19)$$

where

$$\{\hat{f}\} = \{P\} + \int\int_S [N]^T \{\hat{T}_x\} \, dS + \int\int\int_V [N]^T \{\hat{X}_b\} \, dV \qquad (3.9.20)$$

From Eq. (3.9.20), we observe three separate types of load contributions from body forces, surface tractions, and concentrated nodal forces. We define these surface tractions and body-force matrices as

$$\{\hat{f}_s\} = \int\int_S [N]^T \{\hat{T}_x\} \, dS \qquad (3.9.20a)$$

$$\{\hat{f}_b\} = \int\int\int_V [N]^T \{\hat{X}_b\} \, dV \qquad (3.9.20b)$$

The minimization of π_p with respect to each nodal displacement requires that

$$\frac{\partial \pi_p}{\partial \hat{d}_{1x}} = 0 \quad \text{and} \quad \frac{\partial \pi_p}{\partial \hat{d}_{2x}} = 0 \qquad (3.9.21)$$

Now we explicitly evaluate π_p given by Eq. (3.9.19) to apply Eq. (3.9.21). We define the following for convenience:

$$\{U^*\} = \{\hat{d}\}^T [B]^T [D]^T [B] \{\hat{d}\} \qquad (3.9.22)$$

Using Eqs. (3.9.10), (3.9.13), and (3.9.15) in Eq. (3.9.22) yields

$$\{U^*\} = [\hat{d}_{1x} \quad \hat{d}_{2x}] \begin{Bmatrix} -\dfrac{1}{L} \\ \dfrac{1}{L} \end{Bmatrix} [E] \begin{bmatrix} -\dfrac{1}{L} & \dfrac{1}{L} \end{bmatrix} \begin{Bmatrix} \hat{d}_{1x} \\ \hat{d}_{2x} \end{Bmatrix} \qquad (3.9.23)$$

Simplifying Eq. (3.9.23), we obtain

$$U^* = \frac{E}{L^2}(\hat{d}_{1x}^2 - 2\hat{d}_{1x}\hat{d}_{2x} + \hat{d}_{2x}^2) \qquad (3.9.24)$$

Also, the explicit expression for $\{\hat{d}\}^T\{\hat{f}\}$ is

$$\{\hat{d}\}^T\{\hat{f}\} = \hat{d}_{1x}\hat{f}_{1x} + \hat{d}_{2x}\hat{f}_{2x} \qquad (3.9.25)$$

Therefore, using Eqs. (3.9.24) and (3.9.25) in Eq. (3.9.19) and then applying Eqs. (3.9.21), we obtain

$$\frac{\partial \pi_p}{\partial \hat{d}_{1x}} = \frac{AL}{2}\left[\frac{E}{L^2}(2\hat{d}_{1x} - 2\hat{d}_{2x}) \right] - \hat{f}_{1x} = 0$$

and $\hspace{8cm} (3.9.26)$

$$\frac{\partial \pi_p}{\partial \hat{d}_{2x}} = \frac{AL}{2}\left[\frac{E}{L^2}(-2\hat{d}_{1x} + 2\hat{d}_{2x}) \right] - \hat{f}_{2x} = 0$$

In matrix form, we express Eqs. (3.9.26) as

$$\frac{\partial \pi_p}{\partial \{\hat{d}\}} = \frac{AE}{L}\begin{bmatrix} 1 & -1 \\ -1 & 1 \end{bmatrix}\begin{Bmatrix} \hat{d}_{1x} \\ \hat{d}_{2x} \end{Bmatrix} - \begin{Bmatrix} \hat{f}_{1x} \\ \hat{f}_{2x} \end{Bmatrix} = \begin{Bmatrix} 0 \\ 0 \end{Bmatrix} \qquad (3.9.27)$$

or, since $\{\hat{f}\} = [\hat{k}]\{\hat{d}\}$, we have the stiffness matrix for the bar element obtained from Eq. (3.9.27) as

$$[\hat{k}] = \frac{AE}{L}\begin{bmatrix} 1 & -1 \\ -1 & 1 \end{bmatrix} \qquad (3.9.28)$$

As expected, Eq. (3.9.28) is identical to the stiffness matrix obtained in Section 3.1.

Finally, instead of the cumbersome process of explicitly evaluating π_p, we can use the matrix differentiation as given by Eq. (2.6.12) and apply it directly to Eq. (3.9.19) to obtain

$$\frac{\partial \pi_p}{\partial \{\hat{d}\}} = AL[B]^T[D][B]\{\hat{d}\} - \{\hat{f}\} = 0 \qquad (3.9.29)$$

where $[D]^T = [D]$ has been used in writing Eq. (3.9.29). The result of the evaluation of $AL[B]^T[D][B]$ is then equal to $[\hat{k}]$ given by Eq. (3.9.28). Throughout this text, we will use this matrix differentiation concept (also see Appendix A), which greatly simplifies the task of evaluating $[\hat{k}]$.

E X A M P L E   3.9

A bar of length L is subjected to a linearly distributed axial loading, which varies from zero at node 1 to a maximum at node 2 (Figure 3-22). Determine the equivalent nodal loads.

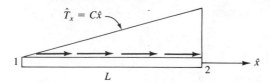

FIGURE   3-22   Element subjected to linearly varying axial load

Using Eq. (3.9.20a), we solve for the energy equivalent nodal forces of the distributed loading as follows:

$$\{\hat{f}_0\} = \begin{Bmatrix} \hat{f}_{1x} \\ \hat{f}_{2x} \end{Bmatrix} = \int_0^L \begin{Bmatrix} 1 - \dfrac{\hat{x}}{L} \\ \dfrac{\hat{x}}{L} \end{Bmatrix} \{C\hat{x}\} \, d\hat{x} \qquad (3.9.30)$$

$$= \begin{Bmatrix} \dfrac{C\hat{x}^2}{2} - \dfrac{C\hat{x}^3}{3L} \\ \dfrac{C\hat{x}^3}{3L} \end{Bmatrix}_0^L$$

$$= \begin{Bmatrix} \dfrac{CL^2}{6} \\ \dfrac{CL^2}{3} \end{Bmatrix} \qquad (3.9.31)$$

where the integration was carried out over the length of the bar, since $\hat{T}_x$ is in units of force/length.

Note that the total load is the area under the load distribution given by

$$F = \frac{1}{2}(L)(CL) = \frac{CL^2}{2} \qquad (3.9.32)$$

Therefore, comparing Eq. (3.9.31) with (3.9.32), the equivalent nodal loads for a linearly varying load are

$$\hat{f}_{1x} = \frac{1}{3}F = \frac{1}{3} \text{ of the total load}$$

$$\qquad (3.9.33)$$

$$\hat{f}_{2x} = \frac{2}{3}F = \frac{2}{3} \text{ of the total load}$$

In summary, for the simple two-noded bar element subjected to a linearly varying load (triangular loading), place $\frac{1}{3}$ of the total load at the node where the distributed loading begins (zero end of the load) and $\frac{2}{3}$ of the total load at the node where the peak value of the distributed load ends.   ∎

E X A M P L E   3.10

For the rod loaded axially as shown in Figure 3-23, determine the axial displacement and axial stress. Let $E = 30 \times 10^6$ psi, $A = 2$ in.^{2}, and $L = 60$ in. Use (a) one and (b) two elements in the finite element solutions. (In Section 3.10, one-, two-, four-, and eight-element solutions will be presented from the computer program TRUSS.) (TRUSS is described in Section 4.5, and the listing is found on the diskette in the back of the text.)

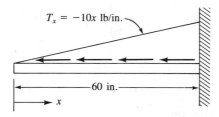

FIGURE 3-23   Rod subjected to triangular load distribution

(a) One-element solution (see Figure 3-24).

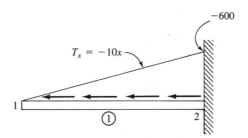

FIGURE 3-24   One-element model

From Eq. (3.9.20a), the distributed load matrix is evaluated as follows:

$$\{F_0\} = \int_0^L [N]^T \{T_x\} \, dx \qquad (3.9.34)$$

where T_x is a line load in units of pounds per inch and $\hat{f}_0 = F_0$ as $x = \hat{x}$. Therefore, using Eq. (3.1.4) for $[N]$ in Eq. (3.9.34), we obtain

$$\{F_0\} = \int_0^L \left\{ \begin{array}{c} 1 - \dfrac{x}{L} \\[2mm] \dfrac{x}{L} \end{array} \right\} \{-10x\} \, dx \qquad (3.9.35)$$

or

$$\begin{Bmatrix} F_{1x} \\ F_{2x} \end{Bmatrix} = \begin{Bmatrix} \dfrac{-10L^2}{2} + \dfrac{10L^2}{3} \\ \dfrac{-10L^2}{3} \end{Bmatrix} = \begin{Bmatrix} \dfrac{-10L^2}{6} \\ \dfrac{-10L^2}{3} \end{Bmatrix} = \begin{Bmatrix} \dfrac{-10(60)^2}{6} \\ \dfrac{-10(60)^2}{3} \end{Bmatrix}$$

or $\qquad\qquad F_{1x} = -6000 \text{ lb}, \qquad F_{2x} = -12,000 \text{ lb} \qquad\qquad (3.9.36)$

Using Eq. (3.9.33), we could have determined the same forces at nodes 1 and 2—that is, $\frac{1}{3}$ of the total load is at node 1 and $\frac{2}{3}$ of the total load is at node 2. Using Eq. (3.9.28), the stiffness matrix is given by

$$k^{(1)} = 10^6 \begin{bmatrix} 1 & -1 \\ -1 & 1 \end{bmatrix}$$

The element equations are then

$$10^6 \begin{bmatrix} 1 & -1 \\ -1 & 1 \end{bmatrix} \begin{Bmatrix} d_{1x} \\ 0 \end{Bmatrix} = \begin{Bmatrix} -6000 \\ R_{2x} - 12,000 \end{Bmatrix} \qquad\qquad (3.9.37)$$

Solving Eq. 1 of Eq. (3.9.37), we obtain

$$d_{1x} = -0.006 \text{ in.} \qquad\qquad (3.9.38)$$

The stress is obtained from Eq. (3.9.14) as

$$
\begin{aligned}
\{\sigma_x\} &= [D]\{\varepsilon_x\} \\
&= E[B]\{d\} \\
&= E\begin{bmatrix} -\dfrac{1}{L} & \dfrac{1}{L} \end{bmatrix} \begin{Bmatrix} d_{1x} \\ d_{2x} \end{Bmatrix} \\
&= E\left(\dfrac{d_{2x} - d_{1x}}{L}\right) \\
&= 30 \times 10^6 \left(\dfrac{0 + 0.006}{60}\right) \\
&= 3000 \text{ psi } (T) \qquad\qquad (3.9.39)
\end{aligned}
$$

(b) Two-element solution (see Figure 3-25).

We first obtain the element forces. For element 2, we divide the load into a uniform part and a triangular part. For the uniform part, one-half the total

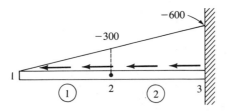

F I G U R E   3-25   Two-element model

uniform load is placed at each node associated with the element. Therefore, the total uniform part is

$$(30 \text{ in.})(300 \text{ lb/in.}) = -9000 \text{ lb}$$

and using Eq. (3.9.33) for the triangular part of the load, we have, for element 2,

$$\begin{Bmatrix} f_{2x}^{(2)} \\ f_{3x}^{(2)} \end{Bmatrix} = \begin{Bmatrix} -[\frac{1}{2}(9000) + \frac{1}{3}(4500)] \\ -[\frac{1}{2}(9000) + \frac{2}{3}(4500)] \end{Bmatrix} = \begin{Bmatrix} -6000 \text{ lb} \\ -7500 \text{ lb} \end{Bmatrix} \qquad (3.9.40)$$

For element 1, the total force is from the triangular-shaped distributed load only and is given by

$$\tfrac{1}{2}(30 \text{ in.})(300 \text{ lb/in.}) = -4500 \text{ lb}$$

Based on Eq. (3.9.33), this load is separated into nodal forces as shown:

$$\begin{Bmatrix} f_{1x}^{(1)} \\ f_{2x}^{(1)} \end{Bmatrix} = \begin{Bmatrix} \frac{1}{3}(-4500) \\ \frac{2}{3}(-4500) \end{Bmatrix} = \begin{Bmatrix} -1500 \text{ lb} \\ -3000 \text{ lb} \end{Bmatrix} \qquad (3.9.41)$$

The final nodal force matrix is then

$$\begin{Bmatrix} F_{1x} \\ F_{2x} \\ F_{3x} \end{Bmatrix} = \begin{Bmatrix} -1500 \\ -6000 - 3000 \\ R_{3x} - 7500 \end{Bmatrix} \qquad (3.9.42)$$

The element stiffness matrices are now

$$\underline{k}^{(1)} = \underline{k}^{(2)} = \frac{AE}{L/2} \begin{array}{cc} 1 & 2 \\ 2 & 3 \\ \begin{bmatrix} 1 & -1 \\ -1 & 1 \end{bmatrix} \end{array} = 2 \times 10^6 \begin{array}{cc} 1 & 2 \\ 2 & 3 \\ \begin{bmatrix} 1 & -1 \\ -1 & 1 \end{bmatrix} \end{array} \qquad (3.9.43)$$

The assembled global stiffness matrix is

$$\underline{K} = 2 \times 10^6 \begin{bmatrix} 1 & -1 & 0 \\ -1 & 2 & -1 \\ 0 & -1 & 1 \end{bmatrix} \qquad (3.9.44)$$

The assembled global equations are then

$$2 \times 10^6 \begin{bmatrix} 1 & -1 & 0 \\ -1 & 2 & -1 \\ 0 & -1 & 1 \end{bmatrix} \begin{Bmatrix} d_{1x} \\ d_{2x} \\ d_{3x} = 0 \end{Bmatrix} = \begin{Bmatrix} -1500 \\ -9000 \\ R_{3x} - 7500 \end{Bmatrix} \qquad (3.9.45)$$

where the boundary condition $d_{3x} = 0$ has been substituted into Eq. (3.9.45). Now solving Eqs. 1 and 2 of Eq. (3.9.45), we obtain

$$d_{1x} = -0.006 \text{ in.}$$
$$d_{2x} = -0.00525 \text{ in.} \qquad (3.9.46)$$

The element stresses are as follows:

Element 1

$$\sigma_x = E\left[-\frac{1}{30} \quad \frac{1}{30}\right]\begin{Bmatrix} d_{1x} = -0.006 \\ d_{2x} = -0.00525 \end{Bmatrix}$$

$$= 750 \text{ psi } (T) \tag{3.9.47}$$

Element 2

$$\sigma_x = E\left[-\frac{1}{30} \quad \frac{1}{30}\right]\begin{Bmatrix} d_{2x} = -0.00525 \\ d_{3x} = 0 \end{Bmatrix}$$

$$= 5250 \text{ psi } (T) \tag{3.9.48}$$

3.10   Comparison of Finite Element Solution to Exact Solution

We will now compare the finite element solutions for Example 3.10 using one, two, four, and eight elements to model the bar element and the exact solution. The exact solution for displacement is obtained by solving the following equation:

$$\delta = \frac{1}{AE}\int_0^x P(x)\, dx \tag{3.9.49a}$$

where, using the following free-body digram, we have

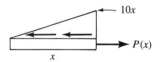

$$P(x) = \tfrac{1}{2}x(10x) = 5x^2 \tag{3.9.49b}$$

Therefore, substituting Eq. (3.9.49b) into (3.9.49a), we have

$$\delta = \frac{1}{AE}\int_0^x 5x^2\, dx$$

$$= \frac{5x^3}{3AE} + C_1 \tag{3.9.49c}$$

Now applying the boundary condition at $x = L$, we obtain

$$\delta(L) = 0 = \frac{5L^3}{3AE} + C_1$$

or

$$C_1 = -\frac{5L^3}{3AE} \tag{3.9.49d}$$

Substituting Eq. (3.9.49d) into (3.9.49c), the final expression for displacement is then

$$\delta = \frac{5}{3AE}(x^3 - L^3) \qquad\qquad (3.9.50)$$

The exact solution for axial stress is obtained by solving the following equation:

$$\sigma(x) = \frac{P(x)}{A} = \frac{5x^2}{2 \text{ in.}^2} = 2.5x^2 \qquad\qquad (3.9.51)$$

Figure 3-26 shows a plot of Eq. (3.9.50) along with the finite element solutions (part of which were obtained in Example 3.10). Some conclusions from these results are

1. The finite element solutions match the exact solution at the node points. The reason these nodal values are correct is due to the fact that the element nodal forces were calculated based on being energy-equivalent to the distributed load based on the assumed linear displacement field within each element.

2. Although the node values for displacement match the exact solution, the values at locations between the nodes is poor using few elements (see one- and two-element solutions) as we used a linear displacement function within each element, whereas the exact solution, Eq. (3.9.50), is a cubic function. However, as we use increasing numbers of elements, the finite element solution

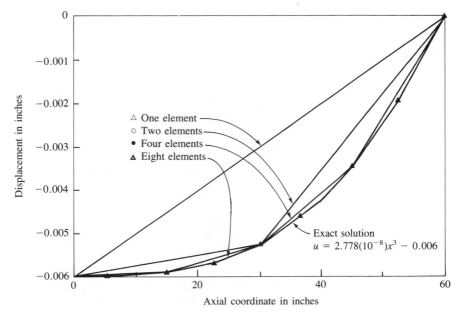

FIGURE   3-26   Comparison of exact and finite element solutions for axial displacement

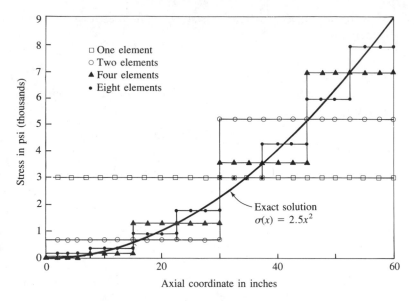

FIGURE 3-27 Comparison of exact and finite element solutions for axial stress

converges to the exact solution (see the four- and eight-element solutions in Figure 3-26).

3. The stress is derived from the slope of the displacement curve as $\sigma = E\varepsilon = E(du/dx)$. Therefore, by the finite element solution, since u is a linear function in each element, axial stress is constant in each element. It then takes even more elements to model the first derivative of the displacement function or, equivalently, the axial stress. This is shown in Figure 3-27, where the best results occur for the eight-element solution.

4. The best approximation of the stress occurs at the midpoint of the element, not at the nodes (Figure 3-27). This is because the derivative of displacement is better predicted between the nodes than at the nodes.

5. The stress is not continuous across element boundaries. Therefore, equilibrium is not satisfied across element boundaries. Also, equilibrium within each element is, in general, not satisfied. This is shown in Figure 3-28 for element 1 in the two-element solution and element 1 in the eight-element solution (in the eight-element solution the forces are obtained from the computer code TRUSS (see Chapter 4) solution). As the number of elements used increases, the discontinuity in the stress decreases across element boundaries and the approximation of equilibrium improves.

Finally, in Figure 3-29, we show the convergence of axial stress at the fixed end $(x = L)$ as the number of elements increases.

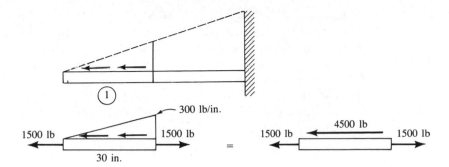

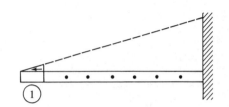

Two-element solution

Eight-element solution

FIGURE 3-28   Free-body diagram of element 1 in both two- and eight-element models showing that equilibrium is not satisfied

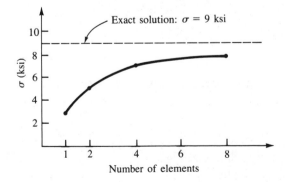

FIGURE 3-29   Axial stress at fixed end as number of elements increases

However, if we formulate the problem in a customary general way, as described in detail in Chapter 5 for beams subjected to distributed loading, we can obtain the exact stress distribution with any of the models used. That is, letting $\hat{f} = \underline{\hat{k}}\underline{\hat{d}} - \hat{f}_0$, where $\hat{f}_0$ is the initial nodal replacement force system of the distributed load on each element, we subtract the initial replacement force system from the $\underline{\hat{k}}\underline{\hat{d}}$ result. This yields the nodal forces in each element. For example, considering element 1 of the two-element model, we have (see also Eqs. (3.9.33) and (3.9.41))

$$\underline{\hat{f}}_0 = \left\{ \begin{matrix} -1500\,\text{lb} \\ -3000\,\text{lb} \end{matrix} \right\}$$

Using $\underline{\hat{f}} = \underline{\hat{k}}\underline{\hat{d}} - \hat{f}_0$, we obtain

$$\underline{\hat{f}} = \frac{2(30 \times 10^6)}{(30\,\text{in.})} \begin{bmatrix} 1 & -1 \\ -1 & 1 \end{bmatrix} \left\{ \begin{matrix} -0.006\,\text{in.} \\ -0.00525\,\text{in.} \end{matrix} \right\} - \left\{ \begin{matrix} -1500\,\text{lb} \\ -3000\,\text{lb} \end{matrix} \right\}$$

$$= \left\{ \begin{matrix} -1500 + 1500 \\ 1500 + 3000 \end{matrix} \right\} = \left\{ \begin{matrix} 0 \\ 4500 \end{matrix} \right\}$$

as the actual nodal forces. Drawing a free-body diagram of the element 1, we have

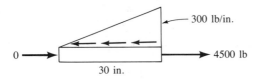

$$\Sigma F_x = 0: \quad -\frac{1}{2}(300\,\text{lb/in.})(30\,\text{in.}) + 4500\,\text{lb} = 0$$

For other kinds of elements (other than beams), this adjustment is ignored in practice. The adjustment is less important for plane and solid elements than for beams. Also, these adjustments are more difficult to formulate for an element of general shape.

3.11   Galerkin's Residual Method and Its Application to a One-Dimensional Bar

General Formulation

We have developed the bar finite element equations by the direct method in Section 3.1 and by the potential energy method (one of a number of variational methods) in Section 3.9. In fields other than structural/solid mechanics, it is quite probable that a variational principle, analogous to the principle of minimum potential energy, for instance, may not be known or even exist. In some flow problems in fluid mechanics and in mass transport problems (see Chapter 13), we often have only the differential equation and boundary conditions available. However, the finite element method can still be applied.

The methods of weighted residuals applied directly to the differential equation can be used to develop the finite element equations. In this section, we describe Galerkin's residual method in general and then apply it to the bar element. This development provides the basis for later applications of Galerkin's method to the beam element in Chapter 5 and to the nonstructural heat-transfer element (specifically, the one-dimensional combined conduction, convection, and mass transport element described in Chapter 13). Owing to the mass transport phenomena, the variational formulation is not known (or certainly is difficult to obtain) and, hence, Galerkin's method is necessarily applied to develop the finite element equations.

There are a number of other residual methods. Among these are collocation, least squares, and least squares collocation. (For more on these methods, see Reference [5].) However, since Galerkin's method is more well known than the other residual methods, it is the only one described in this text.

In weighted residual methods, a trial or approximate function is chosen to approximate the independent variable, such as a displacement or a temperature, in a problem defined by a differential equation. This trial function will not, in general, satisfy the governing differential equation. Thus, the substitution of the trial function into the differential equation results in a residual over the whole region of the problem as follows:

$$\iiint_V R \, dV = \text{minimum} \qquad (3.11.1)$$

In the residual method, we require that a weighted value of the residual be a minimum over the whole region. The weighting functions allow the weighted integral of residuals to go to zero. Denoting the weighting function by W, the general form of the weighted residual integral is

$$\iiint_V RW \, dV = 0 \qquad (3.11.2)$$

Using Galerkin's method, we choose the interpolation function, such as Eq. (3.1.3), in terms of N_i shape functions for the independent variable in the differential equation. In general, this substitution yields the residual $R \neq 0$. By the Galerkin criterion, the shape functions N_i are chosen to play the role of the weighting functions W. Thus, for each i we have

$$\iiint_V RN_i \, dV = 0 \qquad (i = 1, 2, \ldots, n) \qquad (3.11.3)$$

Equation (3.11.3) results in a total of n equations. Equation (3.11.3) applies to points within the region of a body without reference to boundary conditions such as specified applied loads or displacements. To obtain boundary conditions, we apply integration by parts to Eq. (3.11.3), which yields integrals applicable for the region and its boundary.

Bar Element Formulation

We now illustrate Galerkin's method to formulate the bar element stiffness equations. We begin with the basic differential equation, without distributed load, derived in Section 3.1 as

$$\frac{d}{d\hat{x}}\left(AE\frac{d\hat{u}}{d\hat{x}}\right) = 0 \qquad (3.11.4)$$

where constants A and E are now assumed. The residual R is now defined to be Eq. (3.11.4). Applying Galerkin's criterion, Eq. (3.11.3), to Eq. (3.11.4), we have

$$\int_0^L \frac{d}{d\hat{x}}\left(AE\frac{d\hat{u}}{d\hat{x}}\right)N_i \, d\hat{x} = 0 \qquad (i = 1, 2) \qquad (3.11.5)$$

We now apply integration by parts to Eq. (3.11.5). Integration by parts is given in general by

$$\int u \, dv = uv - \int v \, du \qquad (3.11.6)$$

where u and v are simply variables in the general equation. Letting

$$u = N_i \qquad du = \frac{dN_i}{d\hat{x}} d\hat{x}$$

$$dv = \frac{d}{d\hat{x}}\left(AE\frac{d\hat{u}}{d\hat{x}}\right)d\hat{x} \qquad v = AE\frac{d\hat{u}}{d\hat{x}} \qquad (3.11.7)$$

in Eq. (3.11.5) and integrating by parts according to Eq. (3.11.6), Eq. (3.11.5) becomes

$$\left(N_i AE\frac{d\hat{u}}{d\hat{x}}\right)\Bigg|_0^L - \int_0^L AE\frac{d\hat{u}}{d\hat{x}}\frac{dN_i}{d\hat{x}} d\hat{x} = 0 \qquad (3.11.8)$$

where the integration by parts introduces the boundary conditions.

Recall that, because $\hat{u} = [N]\{\hat{d}\}$, we have

$$\frac{d\hat{u}}{d\hat{x}} = \frac{dN_1}{d\hat{x}}\hat{d}_{1x} + \frac{dN_2}{d\hat{x}}\hat{d}_{2x} \qquad (3.11.9)$$

or, by using Eqs. (3.1.4) for N_1 and N_2,

$$\frac{d\hat{u}}{d\hat{x}} = \left[-\frac{1}{L} \quad \frac{1}{L}\right]\begin{Bmatrix}\hat{d}_{1x}\\ \hat{d}_{2x}\end{Bmatrix} \qquad (3.11.10)$$

Using Eq. (3.11.10) in Eq. (3.11.8), we then express Eq. (3.11.8) as

$$AE\int_0^L \frac{dN_i}{d\hat{x}}\left[-\frac{1}{L} \quad \frac{1}{L}\right]d\hat{x}\begin{Bmatrix}\hat{d}_{1x}\\ \hat{d}_{2x}\end{Bmatrix} = \left(N_i AE\frac{d\hat{u}}{d\hat{x}}\right)\Bigg|_0^L \qquad (i = 1, 2) \qquad (3.11.11)$$

Equation (3.11.11) is really two equations (one for $N_i = N_1$ and one for $N_i = N_2$). First, using the weighting function $N_i = N_1$, we have

$$AE \int_0^L \frac{dN_1}{d\hat{x}} \left[-\frac{1}{L} \quad \frac{1}{L} \right] d\hat{x} \begin{Bmatrix} \hat{d}_{1x} \\ \hat{d}_{2x} \end{Bmatrix} = \left(N_1 AE \frac{d\hat{u}}{d\hat{x}} \right)\Big|_0^L \qquad (3.11.12)$$

Substituting for $dN_1/d\hat{x}$, we obtain

$$AE \int_0^L \left[-\frac{1}{L} \right] \left[-\frac{1}{L} \quad \frac{1}{L} \right] d\hat{x} \begin{Bmatrix} \hat{d}_{1x} \\ \hat{d}_{2x} \end{Bmatrix} = \hat{f}_{1x} \qquad (3.11.13)$$

where $\hat{f}_{1x} = EA(d\hat{u}/d\hat{x})$ because $N_1 = 1$ at $x = 0$ and $N_1 = 0$ at $x = L$. Evaluating Eq. (3.11.13) yields

$$\frac{AE}{L}(\hat{d}_{1x} - \hat{d}_{2x}) = \hat{f}_{1x} \qquad (3.11.14)$$

Similarly, using $N_i = N_2$, we obtain

$$AE \int_0^L \left[\frac{1}{L} \right] \left[-\frac{1}{L} \quad \frac{1}{L} \right] d\hat{x} \begin{Bmatrix} \hat{d}_{1x} \\ \hat{d}_{2x} \end{Bmatrix} = \left(N_2 AE \frac{d\hat{u}}{d\hat{x}} \right)\Big|_0^L \qquad (3.11.15)$$

Simplifying Eq. (3.11.15) yields

$$\frac{AE}{L}(\hat{d}_{2x} - \hat{d}_{1x}) = \hat{f}_{2x} \qquad (3.11.16)$$

where $\hat{f}_{2x} = EA(d\hat{u}/d\hat{x})$ because $N_2 = 1$ at $x = L$ and $N_2 = 0$ at $x = 0$. Equations (3.11.14) and (3.11.16) are then seen to be the same as Eqs. (3.1.13) and (3.9.27) derived, respectively, by the direct and variational methods.

REFERENCES

[1] Turner, M. J., Clough, R. W., Martin, H. C., and Topp, L. J., "Stiffness and Deflection Analysis of Complex Structures," *Journal of the Aeronautical Sciences,* Vol. 23, No. 9, Sept. 1956, pp. 805–824.

[2] Martin, H. C., "Plane Elasticity Problems and the Direct Stiffness Method," *The Trend in Engineering,* Vol. 13, Jan. 1961, pp. 5–19.

[3] Melosh, R. J., "Basis for Derivation of Matrices for the Direct Stiffness Method," *Journal of the American Institute of Aeronautics and Astronautics,* Vol. 1, No. 7, July 1963, pp. 1631–1637.

[4] Oden, J. T., and Ripperger, E. A., *Mechanics of Elastic Structures,* 2nd ed., McGraw-Hill, New York, 1981.

[5] Finlayson, B. A., *The Method of Weighted Residuals and Variational Principles,* Academic Press, New York, 1972.

[6] Zienkiewicz, O. C., *The Finite Element Method,* 3rd ed., McGraw-Hill, London, 1977.

[7] Cook, R. D., Malkus, D. S. and Plesha, M. E., *Concepts and Applications of Finite Element Analysis,* 3rd ed., Wiley, New York, 1989.

[8] Forray, M. J., *Variational Calculus in Science and Engineering,* McGraw-Hill, New York, 1968.

PROBLEMS

3.1   *a.* Compute the total stiffness matrix $\underline{K}$ of the assemblage shown in Figure P3-1 by superimposing the stiffness matrices of the individual bars. Note $\underline{K}$ should be in terms of $A_1, A_2, A_3, E_1, E_2, E_3, L_1, L_2,$ and L_3. Here A, E, and L are generic symbols used for cross-sectional area, modulus of elasticity, and length, respectively.

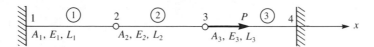

FIGURE   P3-1

b. Now let $A_1 = A_2 = A_3 = A$, $E_1 = E_2 = E_3 = E$, and $L_1 = L_2 = L_3 = L$. If nodes 1 and 4 are fixed and a force P acts at node 3 in the positive x direction, find expressions for the displacement of nodes 2 and 3 in terms of A, E, L, and P.

c. Now let $A = 1$ in.^{2}, $E = 10 \times 10^6$ psi, $L = 10$ in., and $P = 1000$ lb.

   i. Determine the numerical values of the displacements of nodes 2 and 3.

   ii. Determine the numerical values of the reactions at nodes 1 and 4.

   iii Determine the stresses in elements 1, 2, and 3.

3.2–3.11   For the bar assemblages shown in Figures P3-2–P3-11, determine the nodal displacements, the forces in each element, and the reactions. Use the direct stiffness method for these problems.

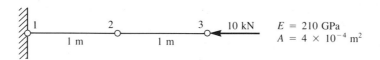

FIGURE   P3-2

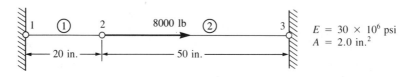

FIGURE   P3-3

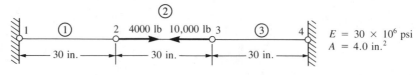

FIGURE   P3-4

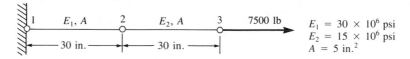

F I G U R E   P3-5

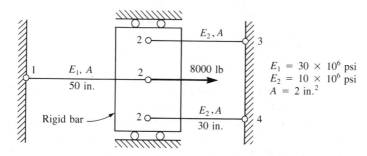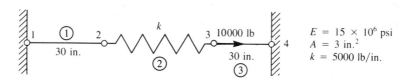

F I G U R E   P3-6

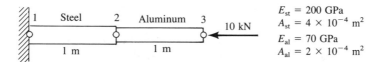

F I G U R E   P3-7

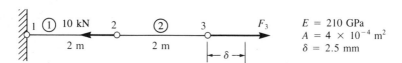

F I G U R E   P3-8

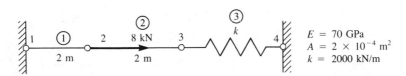

F I G U R E   P3-9

F I G U R E   P3-10

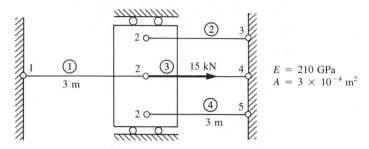

FIGURE   P3-11

3.12   Solve for the axial displacement and stress in the tapered bar shown in Figure P3-12 using one and then two constant-area elements. Evaluate the area at the center of each element length. Use that area for each element. Let $A_0 = 2$ in.^{2}, $L = 20$ in., $E = 10 \times 10^6$ psi, and $P = 1000$ lb. Compare your finite element solutions with the exact solution.

FIGURE   P3-12

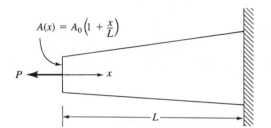

3.13   Determine the stiffness matrix for the bar element with end nodes and mid-length node shown in Figure P3-13. Let axial displacement $u = a_1 + a_2 x + a_3 x^2$. (This is a higher-order element in that strain now varies linearly through the element.)

FIGURE   P3-13

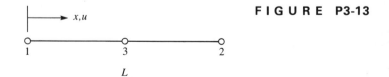

3.14   Given the following displacement function for the two-noded bar element:

$$u = a + bx^2$$

Is this a valid displacement function? Discuss why or why not.

3.15   For each of the bar elements shown in Figure P3-15, evaluate the global x-y stiffness matrix.

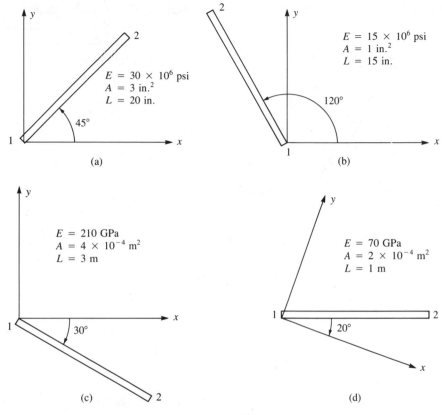

FIGURE P3-15

3.16   For the bar elements shown in Figure P3-16, the global displacements have been determined to be $d_{1x} = 0.5$ in., $d_{1y} = 0.0$, $d_{2x} = 0.25$ in., and $d_{2y} = 0.75$ in. Determine the local $\hat{x}$ displacements at each end of the bars. Let $E = 12 \times 10^6$ psi, $A = 0.5$ in.^{2}, and $L = 60$ in. for each element.

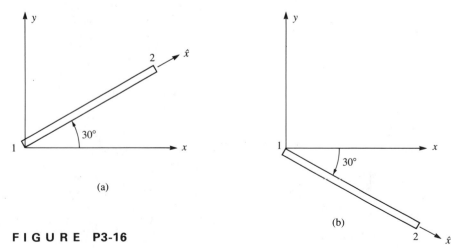

FIGURE P3-16

3.17 For the bar elements shown in Figure P3-17, the global displacements have been determined to be $d_{1x} = 0.0$, $d_{1y} = 2.5$ mm, $d_{2x} = 5.0$ mm, and $d_{2y} = 3.0$ mm. Determine the local $\hat{x}$ displacements at the ends of each bar. Let $E = 210$ GPa, $A = 10 \times 10^{-4}$ m^2, and $L = 3$ m for each element.

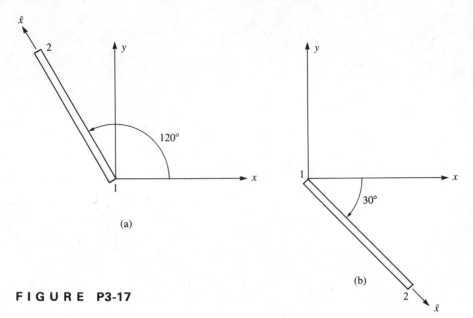

FIGURE P3-17

3.18 Using the method of Section 3.5, determine the axial stress in each of the bar elements shown in Figure P3-18.

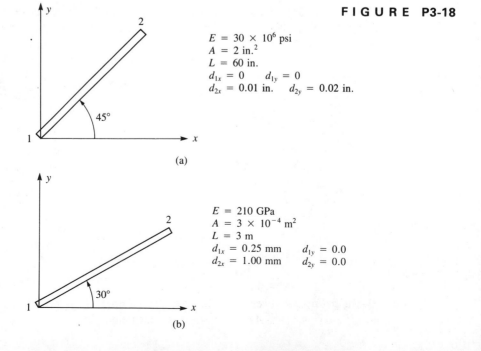

FIGURE P3-18

$E = 30 \times 10^6$ psi
$A = 2$ in.^{2}
$L = 60$ in.
$d_{1x} = 0$ $d_{1y} = 0$
$d_{2x} = 0.01$ in. $d_{2y} = 0.02$ in.

$E = 210$ GPa
$A = 3 \times 10^{-4}$ m^2
$L = 3$ m
$d_{1x} = 0.25$ mm $d_{1y} = 0.0$
$d_{2x} = 1.00$ mm $d_{2y} = 0.0$

3.19   **a.** Assemble the stiffness matrix for the assemblage shown in Figure P3-19 by superimposing the stiffness matrices of the springs. Here k is the stiffness of each spring.
         b. Find the x and y components of deflection of node 1.

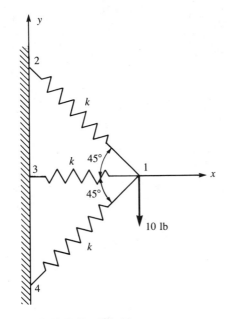

F I G U R E   P3-19

3.20   For the plane truss structure shown in Figure P3-20, determine the displacement of node 2 using the stiffness method. Also determine the stress in element 1. Let $A = 5$ in.^{2}, $E = 1 \times 10^6$ psi, and $L = 100$ in.

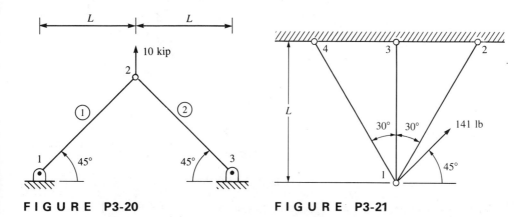

F I G U R E   P3-20   **F I G U R E   P3-21**

3.21   Find the horizontal and vertical displacements of node 1 for the truss shown in Figure P3-21. Assume AE is the same for each element.

3.22   For the truss shown in Figure P3-22 solve for the horizontal and vertical components of displacement at node 1 and determine the stress in each element. Also verify force equilibrium at node 1. All elements have $A_1 = 1$ in.^{2} and $E = 10 \times 10^6$ psi. Let $L = 100$ in.

FIGURE P3-22

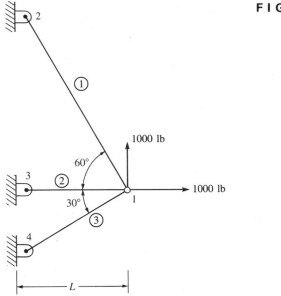

3.23   For the truss shown in Figure P3-23, solve for the horizontal and vertical components of displacement at node 1. Also determine the stress in element 1. Let $A = 1$ in.^{2}, $E = 10.0 \times 10^6$ psi, and $L = 100$ in.

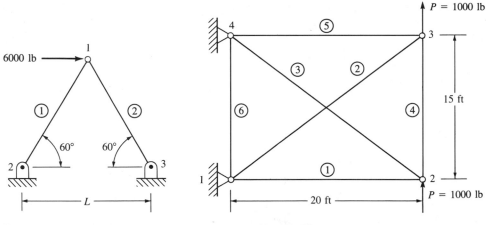

FIGURE P3-23   **FIGURE P3-24**

3.24   Determine the nodal displacements and the element forces for the truss shown in Figure P3-24. Assume all elements have the same AE.

3.25   Now remove the element connecting nodes 2 and 4 in Figure P3-24. Then determine the nodal displacements and element forces.

3.26 Now remove *both* cross elements in Figure P3-24. Can you determine the nodal displacements? If not, why?

3.27 Determine the displacement components at node 3 and the element forces for the plane truss shown in Figure P3-27. Let $A = 3$ in.^{2} and $E = 30 \times 10^6$ psi for all elements. Verify force equilibrium at node 3.

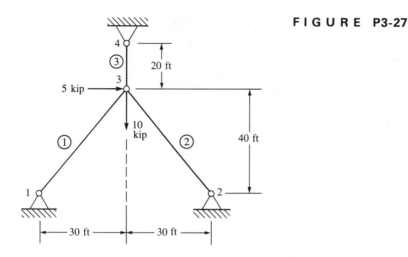

FIGURE P3-27

3.28 Show that for the transformation matrix T of Eq. (3.4.15), $T^T = T^{-1}$ and, hence, Eq. (3.4.21) is indeed correct, thus also illustrating that $k = T^T \hat{k} T$ is the expression for the global stiffness matrix for an element.

3.29–3.30 For the plane trusses shown in Figures P3-29 and P3-30, determine the horizontal and vertical displacements of node 1 and the stresses in each element. All elements have $E = 210$ GPa and $A = 4.0 \times 10^{-4}$ m^2.

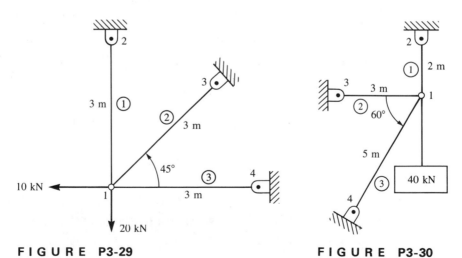

FIGURE P3-29

FIGURE P3-30

3.31 Remove element 1 from Figure P3-30 and resolve the problem. Compare the displacements and stresses to the results for Problem 3.30.

3.32   For the plane truss shown in Figure P3-32, determine the nodal displacements, the element forces and stresses, and the support reactions. All elements
have $E = 70$ GPa and $A = 3.0 \times 10^{-4}$ m². Verify force equilibrium at nodes
2 and 4. Use symmetry in your model.

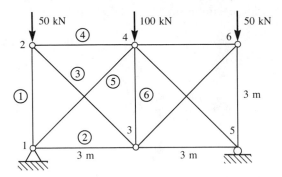

FIGURE P3-32

3.33   For the plane truss supported by the spring at node 1 in Figure P3-33,
determine the nodal displacements and the stresses in each element. Let
$E = 210$ GPa and $A = 5.0 \times 10^{-4}$ m² for both truss elements.

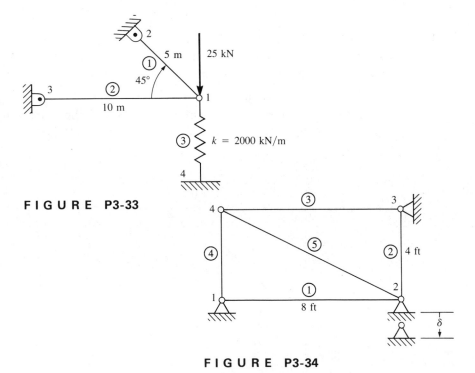

FIGURE P3-33

FIGURE P3-34

3.34   For the plane truss shown in Figure P3-34, node 2 settles an amount $\delta = 0.05$
in. Determine the forces and stresses in each element due to this settlement.
Let $E = 30 \times 10^6$ psi and $A = 2$ in.² for each element.

3.35 For the symmetric plane truss shown in Figure P3-35, determine (a) the deflection of node 1 and (b) the stress in element 1. The AE/L for element 3 is $2AE/L$ that of the other elements. Let $AE/L = 10^6$ lb/in. Then let $A = 1$ in.^{2}, $L = 10$ in., and $E = 10 \times 10^6$ psi to obtain numerical results.

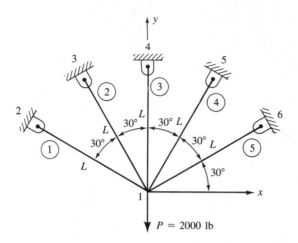

FIGURE P3-35

3.36–3.37 For the space truss elements shown in Figures P3-36 and P3-37, the global displacements at node 1 have been determined to be $d_{1x} = 0.1$ in., $d_{1y} = 0.2$ in., and $d_{1z} = 0.15$ in. Determine the displacement along the local $\hat{x}$ axis at node 1 of the elements. The coordinates, in inches, are shown in the figures.

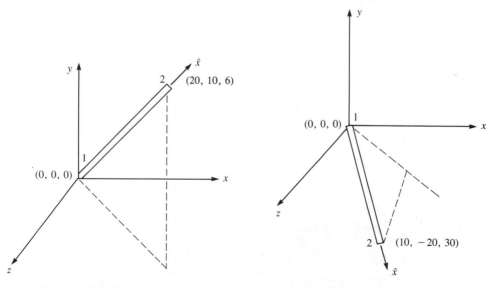

FIGURE P3-36   **FIGURE P3-37**

3.38–3.39   For the space truss elements shown in Figures P3-38 and P3-39, the global displacements at node 2 have been determined to be $d_{2x} = 5$ mm, $d_{2y} = 10$ mm, and $d_{2z} = 15$ mm. Determine the displacement along the local $\hat{x}$ axis at node 2 of the elements. The coordinates, in meters, are shown in the figures.

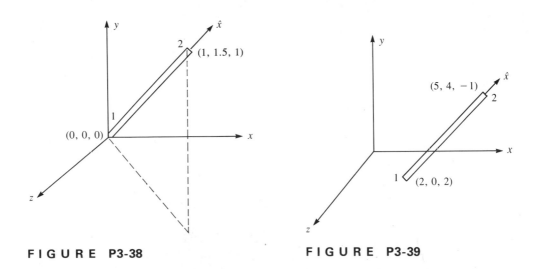

FIGURE P3-38   **FIGURE P3-39**

3.40–3.41   For the space trusses shown in Figures P3-40 and P3-41, determine the nodal displacements and the stresses in each element. Let $E = 210$ GPa and $A = 10 \times 10^{-4}$ m^2 for all elements. Verify force equilibrium at node 1. The coordinates of each node, in meters, are shown in the figure. All supports are ball-and-socket joints.

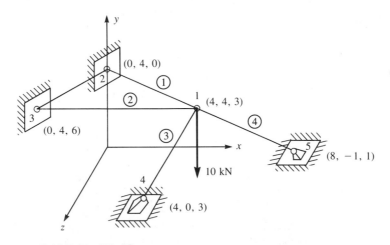

FIGURE P3-40

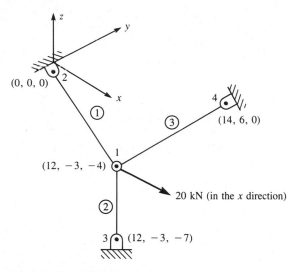

F I G U R E   P3-41

3.42   For the space truss subjected to a 1000-lb load in the x direction, as shown
in Figure P3-42, determine the displacement of node 5. Also determine the
stresses in each element. Let $A = 4$ in.^{2} and $E = 30 \times 10^6$ psi for all elements.
The coordinates of each node, in inches, are shown in the figure. Nodes 1, 2,
3, and 4 are supported by ball-and-socket joints (fixed supports).

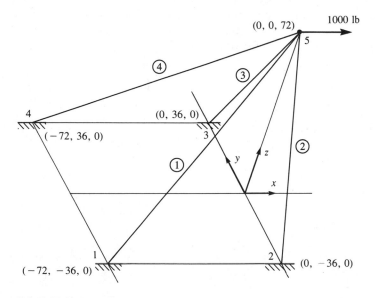

F I G U R E   P3-42

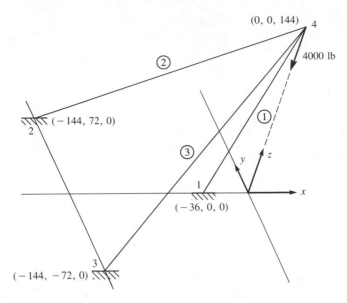

F I G U R E P3-43

3.43 For the space truss subjected to the 4000 lb load acting as shown in Figure P3-43, determine the displacement of node 4. Also determine the stresses in each element. Let $A = 6$ in.^{2} and $E = 30 \times 10^6$ psi for all elements. The coordinates of each node, in inches, are shown in the figure. Nodes 1, 2, and 3 are supported by ball-and-socket joints (fixed supports).

3.44 Verify Eq. (3.7.9) for $\underline{k}$ by first expanding $\underline{T}^*$, given by Eq. (3.7.7), to a 6×6 square matrix in a manner similar to that done in Section 3.4 for the two-dimensional case. Then expand $\hat{\underline{k}}$ to a 6×6 matrix by adding appropriate rows and columns of zeros (for the $\hat{d}_z$ terms) to Eq. (3.4.17), and finally, perform the matrix triple product $\underline{k} = \underline{T}^T \hat{\underline{k}} \underline{T}$.

3.45 Derive Eq. (3.7.21) for stress in space truss elements by a process similar to that used to derive Eq. (3.5.6) for stress in a plane truss element.

3.46–3.48 For the plane trusses with inclined supports shown in Figures P3-46–3-48, solve for the nodal displacements and element stresses in the bars. Let $A = 2$ in.^{2}, $E = 30 \times 10^6$ psi, and $L = 30$ in. for each truss.

F I G U R E P3-46

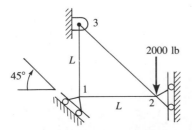

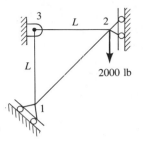

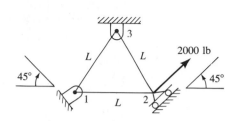

FIGURE P3-47   **FIGURE P3-48**

3.49   Use the principle of minimum potential energy developed in Section 3.8 to solve the bar problems shown in Figure P3-49, that is, plot the total potential energy for variations in the displacement of the free end of the bar to determine the minimum potential energy. Observe that the displacement that yields the minimum potential energy also yields the stable equilibrium position. Use displacement increments of 0.002 in., beginning with $x = -0.004$. Let $E = 30 \times 10^6$ psi and $A = 2$ in.^{2} for the bars.

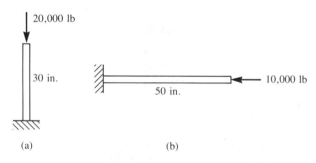

(a)   (b)

FIGURE P3-49

3.50   Derive the stiffness matrix for the nonprismatic bar shown in Figure P3-50 using the principle of minimum potential energy. Let E be constant.

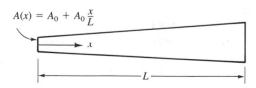

$A(x) = A_0 + A_0 \dfrac{x}{L}$

FIGURE P3-50

3.51   For the bar subjected to the linear varying axial load shown in Figure P3-51, determine the nodal displacements and axial stress distribution using (a) two equal-length elements and (b) four equal-length elements. Let $A = 2$ in.^{2} and $E = 30 \times 10^6$ psi. Compare the finite element solution with an exact solution.

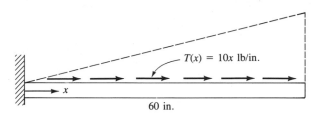

$T(x) = 10x$ lb/in.

x

60 in.

FIGURE   P3-51

3.52   For the bar subjected to the uniform line load in the axial direction shown in Figure P3-52, determine the nodal displacements and axial stress distribution using (a) two equal-length elements and (b) four equal-length elements. Compare the finite element results with an exact solution. Let $A = 2$ in.^{2} and $E = 30 \times 10^6$ psi.

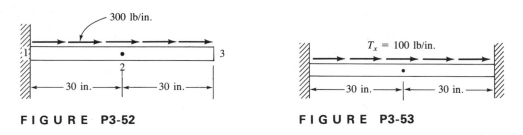

300 lb/in.

$T_x = 100$ lb/in.

— 30 in. — | — 30 in. —   — 30 in. — | — 30 in. —

FIGURE   P3-52   **FIGURE   P3-53**

3.53   For the bar fixed at both ends and subjected to the uniformly distributed loading shown in Figure P3-53, determine the displacement at the middle of the bar and the stress in the bar. Let $A = 2$ in.^{2} and $E = 30 \times 10^6$ psi.

3.54   For the bar hanging under its own weight shown in Figure P3-54, determine the nodal displacements using (a) two equal-length elements and (b) four equal-length elements. Let $A = 2$ in.^{2}, $E = 30 \times 10^6$ psi, and weight density $\rho_w = 0.283$ lb/in.^{3} (*Hint:* The internal force is a function of x. Use the potential energy approach.)

FIGURE   P3-54

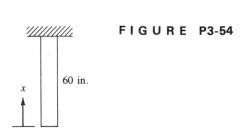

60 in.

x

3.55 Determine the energy equivalent nodal forces for the axial distributed loading shown acting on the bar elements in Figure P3-55.

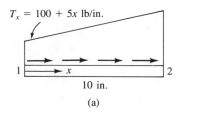

$T_x = 100 + 5x$ lb/in.

1 → x 2

10 in.

(a)

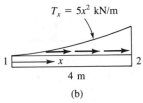

$T_x = 5x^2$ kN/m

1 → x 2

4 m

(b)

F I G U R E   P3-55

Symmetry, Bandwidth,

and a Computer Program

for Truss Analysis

INTRODUCTION

In this chapter, we will first introduce the use of symmetry to reduce the size of a problem, and thus facilitate its solution. We will use an example truss problem to illustrate the concept. We will solve the resulting set of stiffness equations by a process of partitioning the matrix equations.

Next, we will introduce the concepts of banded-symmetric matrices and bandwidth because we will implement these concepts in a computer program to analyze trusses.

Because the longhand solution of large problems is not feasible, we conclude this chapter with a description of a computer program implementation of the matrix stiffness method. We will give a general computer flowchart for solution of both two- and three-dimensional truss structures and discuss the manner in which data is input into the computer program listed on the accompanying disk.

4.1 Use of Symmetry in Structure and Partitioning for Solution of Equations

In many instances, we can use symmetry to facilitate the solution of a problem. **Symmetry** *means correspondence in size, shape, and position of loads; material properties; and boundary conditions that are on opposite sides of a dividing line or plane.* Use of symmetry allows us to consider a reduced problem instead of the actual problem. Thus, the order of the total stiffness matrix and total set of stiffness equations can be reduced. Longhand solution time is then reduced and computer solution time for large-scale problems is substantially decreased.

E X A M P L E   4.1

Solve the plane truss problem shown in Figure 4-1. The truss is composed of eight elements and five nodes as shown. A vertical load of $2P$ is applied at node 4. Nodes 1 and 5 are pin supports. Bar elements 1, 2, 7, and 8 have axial stiffnesses of $\sqrt{2}AE$ and bars 3, 4, 5, and 6 have axial stiffnesses of AE. Here, again, A and E represent the cross-sectional area and modulus of elasticity of a bar.

In this problem, we will use a plane of symmetry. The vertical plane perpendicular to the plane truss passing through nodes 2, 4, and 3 is the plane of symmetry because identical geometry, material, loading, and boundary conditions occur at the corresponding locations on opposite sides of this plane. For loads such as $2P$, occurring in the plane of symmetry, one-half of the total load must be applied to the reduced structure. For elements occurring in the plane of symmetry, one-half of the cross-sectional area must be used in the reduced structure. Furthermore, for nodes in the plane of symmetry, the displacement components normal to the plane of symmetry must be set to zero in the reduced structure; that is, we set $d_{2x} = 0$, $d_{3x} = 0$, and $d_{4x} = 0$. Figure 4-2 shows the reduced structure to be used to analyze the plane truss of Figure 4-1.

We begin the solution of the problem by determining the angles θ for each bar element. For instance, for element 1, assuming $\hat{x}$ to be directed from node 1 to node 2, we obtain $\theta^{(1)} = 45°$. Table 4-1 is used in determining each element stiffness matrix.

There are a total of eight nodal components of displacement for the truss before boundary constraints are imposed. Therefore, $\underline{K}$ must be of order 8×8. For element 1, using Eq. (3.4.23) along with Table 4-1 for the direction

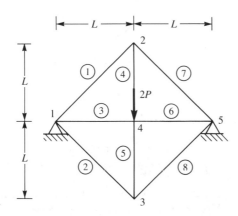

F I G U R E   4-1   Plane truss

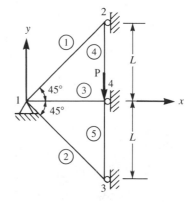

F I G U R E   4-2   Truss of Figure 4-1 reduced by symmetry

T A B L E 4-1   Data for the truss of Figure 4-2

Element	$\theta°$	C	S	C^2	S^2	CS
1	45°	$\sqrt{2}/2$	$\sqrt{2}/2$	1/2	1/2	1/2
2	315°	$\sqrt{2}/2$	$-\sqrt{2}/2$	1/2	1/2	$-1/2$
3	0°	1	0	1	0	0
4	90°	0	1	0	1	0
5	90°	0	1	0	1	0

cosines, we obtain

$$\underline{k}^{(1)} = \frac{\sqrt{2}AE}{\sqrt{2}L} \begin{array}{cccc} d_{1x} & d_{1y} & d_{2x} & d_{2y} \end{array} \begin{bmatrix} \frac{1}{2} & \frac{1}{2} & -\frac{1}{2} & -\frac{1}{2} \\ \frac{1}{2} & \frac{1}{2} & -\frac{1}{2} & -\frac{1}{2} \\ -\frac{1}{2} & -\frac{1}{2} & \frac{1}{2} & \frac{1}{2} \\ -\frac{1}{2} & -\frac{1}{2} & \frac{1}{2} & \frac{1}{2} \end{bmatrix} \qquad (4.1.1)$$

Similarly, for elements 2, 3, 4, and 5, we obtain

$$\underline{k}^{(2)} = \frac{\sqrt{2}AE}{\sqrt{2}L} \begin{array}{cccc} d_{1x} & d_{1y} & d_{3x} & d_{3y} \end{array} \begin{bmatrix} \frac{1}{2} & -\frac{1}{2} & -\frac{1}{2} & \frac{1}{2} \\ -\frac{1}{2} & \frac{1}{2} & \frac{1}{2} & -\frac{1}{2} \\ -\frac{1}{2} & \frac{1}{2} & \frac{1}{2} & -\frac{1}{2} \\ \frac{1}{2} & -\frac{1}{2} & -\frac{1}{2} & \frac{1}{2} \end{bmatrix} \qquad (4.1.2)$$

$$\underline{k}^{(3)} = \frac{AE}{L} \begin{array}{cccc} d_{1x} & d_{1y} & d_{4x} & d_{4y} \end{array} \begin{bmatrix} 1 & 0 & -1 & 0 \\ 0 & 0 & 0 & 0 \\ -1 & 0 & 1 & 0 \\ 0 & 0 & 0 & 0 \end{bmatrix} \qquad (4.1.3)$$

$$\underline{k}^{(4)} = \frac{AE}{L} \begin{array}{cccc} d_{4x} & d_{4y} & d_{2x} & d_{2y} \end{array} \begin{bmatrix} 0 & 0 & 0 & 0 \\ 0 & \frac{1}{2} & 0 & -\frac{1}{2} \\ 0 & 0 & 0 & 0 \\ 0 & -\frac{1}{2} & 0 & \frac{1}{2} \end{bmatrix} \qquad (4.1.4)$$

$$\underline{k}^{(5)} = \frac{AE}{L} \begin{array}{cccc} d_{3x} & d_{3y} & d_{4x} & d_{4y} \end{array} \begin{bmatrix} 0 & 0 & 0 & 0 \\ 0 & \frac{1}{2} & 0 & -\frac{1}{2} \\ 0 & 0 & 0 & 0 \\ 0 & -\frac{1}{2} & 0 & \frac{1}{2} \end{bmatrix} \qquad (4.1.5)$$

where, Eqs. (4.1.1)–(4.1.5), the column labels indicate the degrees of freedom associated with each element. Also, because elements 4 and 5 lie in the plane of symmetry, one half of their original areas have been used in Eqs. (4.1.4) and (4.1.5).

We will limit the solution to that of determining the displacement components. Therefore, considering the boundary constraints that result in zero displacement components, we can immediately obtain the reduced set of equations by eliminating rows and columns in each element stiffness matrix corresponding to a zero displacement component; that is, because $d_{1x} = 0$ and $d_{1y} = 0$ (owing to the pin support at node 1 in Figure 4-2) and $d_{2x} = 0$, $d_{3x} = 0$, and $d_{4x} = 0$ (owing to the symmetry condition), we can cancel rows and columns corresponding to these displacement components in each element stiffness matrix before assembling the total stiffness matrix. The resulting set of stiffness equations is

$$\frac{AE}{L} \begin{bmatrix} 1 & 0 & -\frac{1}{2} \\ 0 & 1 & -\frac{1}{2} \\ -\frac{1}{2} & -\frac{1}{2} & 1 \end{bmatrix} \begin{Bmatrix} d_{2y} \\ d_{3y} \\ d_{4y} \end{Bmatrix} = \begin{Bmatrix} 0 \\ 0 \\ -P \end{Bmatrix} \qquad (4.1.6)$$

Solution by Partitioning (Condensing) of Matrices

We will solve Eq. (4.1.6) by separating the matrices into submatrices by drawing dashed horizontal and vertical lines as shown in Eq. (4.1.6). The purpose of this partitioning is to expedite further calculations into which the matrix enters. We will use partitioning here to make one part of a matrix all zeros. (In Section 2.5, we used partitioning to separate unknown displacements from known ones in solving for $\underline{d}$.)

It is most appropriate to begin with the general set of stiffness equations partitioned as follows:

$$\begin{bmatrix} \underline{K}_{11} & \underline{K}_{12} \\ \underline{K}_{21} & \underline{K}_{22} \end{bmatrix} \begin{Bmatrix} \underline{d}_1 \\ \underline{d}_2 \end{Bmatrix} = \begin{Bmatrix} \underline{0} \\ \underline{F} \end{Bmatrix} \qquad (4.1.7)$$

These equations can then be applied to the specific example, Eq. (4.1.6). Equation (4.1.7) can be written as separate matrix equations as follows:

$$\underline{K}_{11}\underline{d}_1 + \underline{K}_{12}\underline{d}_2 = \underline{0} \qquad (4.1.8)$$

$$\underline{K}_{21}\underline{d}_1 + \underline{K}_{22}\underline{d}_2 = \underline{F} \qquad (4.1.9)$$

Solving Eq. (4.1.8) for $\underline{d}_1$, we obtain

$$\underline{d}_1 = -\underline{K}_{11}^{-1}\underline{K}_{12}\underline{d}_2 \qquad (4.1.10)$$

Using Eq. (4.1.10) in Eq. (4.1.9) yields

$$\underline{K}_{21}(-\underline{K}_{11}^{-1}\underline{K}_{12}\underline{d}_2) + \underline{K}_{22}\underline{d}_2 = \underline{F} \qquad (4.1.11)$$

Simplifying Eq. (4.1.11), we obtain

$$(\underline{K}_{22} - \underline{K}_{21}\underline{K}_{11}^{-1}\underline{K}_{12})\underline{d}_2 = \underline{F} \qquad (4.1.12)$$

Defining $\underline{k}_c$ as the condensed stiffness matrix

$$\underline{k}_c = \underline{K}_{22} - \underline{K}_{21}\underline{K}_{11}^{-1}\underline{K}_{12} \qquad (4.1.13)$$

we can express Eq. (4.1.12) as

$$\underline{k}_c\underline{d}_2 = \underline{F} \qquad (4.1.14)$$

Multiplying both sides of Eq. (4.1.14) by $(\underline{k}_c)^{-1}$, we obtain

$$\underline{d}_2 = (\underline{k}_c)^{-1}\underline{F} \qquad (4.1.15)$$

Using the definition of $\underline{k}_c$, Eq. (4.1.13), applied to partitioned Eq. (4.1.6), we obtain

$$\underline{k}_c = \frac{AE}{L}\left[[1] - \begin{bmatrix} -\frac{1}{2} & -\frac{1}{2} \end{bmatrix}\begin{bmatrix} 1 & 0 \\ 0 & 1 \end{bmatrix}^{-1}\begin{Bmatrix} -1/2 \\ -1/2 \end{Bmatrix}\right] \qquad (4.1.16)$$

Simplifying Eq. (4.1.16) yields

$$\underline{k}_c = \frac{AE}{L}\left\{[1] - \begin{bmatrix} 1 \\ 2 \end{bmatrix}\right\} = \frac{AE}{L}\begin{bmatrix} 1 \\ 2 \end{bmatrix} \qquad (4.1.17)$$

Then the inverse of $\underline{k}_c$ becomes

$$(\underline{k}_c)^{-1} = \begin{bmatrix} \dfrac{2L}{AE} \end{bmatrix} \qquad (4.1.18)$$

Using Eq. (4.1.18) in Eq. (4.1.15), we obtain

$$\underline{d}_2 = [d_{4y}] = \begin{bmatrix} \dfrac{-2PL}{AE} \end{bmatrix} \qquad (4.1.19)$$

where $\underline{F} = [-P]$ has also been used in Eq. (4.1.15). Now using partitioned Eq. (4.1.6) along with Eq. (4.1.19) in Eq. (4.1.10), we obtain the other displacements as

$$\begin{Bmatrix} d_{2y} \\ d_{3y} \end{Bmatrix} = -\begin{bmatrix} 1 & 0 \\ 0 & 1 \end{bmatrix}\begin{Bmatrix} -1/2 \\ -1/2 \end{Bmatrix}\begin{bmatrix} \dfrac{-2PL}{AE} \end{bmatrix} \qquad (4.1.20)$$

On simplifying Eq. (4.1.20), the final displacements are

$$\begin{Bmatrix} d_{2y} \\ d_{3y} \end{Bmatrix} = \begin{Bmatrix} \dfrac{-PL}{AE} \\ \dfrac{-PL}{AE} \end{Bmatrix} \qquad (4.1.21) \quad \blacksquare$$

4.2 Banded-Symmetric Matrices and Bandwidth

The coefficient matrix (stiffness matrix) for the linear equations that occur in structural analysis is always symmetric and banded. Because a meaningful analysis generally requires the use of a large number of variables, the implementation of compressed storage of the stiffness matrix is desirable both from the viewpoint of fitting into memory (immediate access portion of the computer) and computational efficiency. We will discuss the banded-symmetric format, which is not necessarily the most efficient format but is relatively simple to implement on the computer.

Another method, based on the concept of the skyline of the stiffness matrix, is often used to improve the efficiency in solving the equations. *The skyline is an envelope that begins with the first nonzero coefficient in each column of the stiffness matrix* (see Figure 4-4 on page 133). In skylining, only the coefficients between the main diagonal and the skyline are stored (normal-

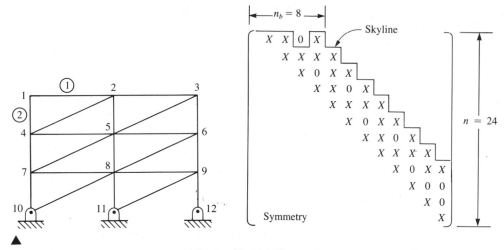

FIGURE 4-3   Plane truss for bandwidth illustration

FIGURE 4-4   Stiffness matrix for the plane truss of Figure 4-3, where X denotes, in general, blocks of 2 × 2 submatrices with nonzero coefficients

ly by successive columns) in a one-dimensional array. In general, this procedure takes even less storage space in the computer and is more efficient in terms of equation solving than the conventional banded format. However, the programs in this text are based on the banded-symmetric format, so we will discuss this concept in greater detail. (For more information on skylining, consult References [7], [8], and [9].)

A matrix is **banded** if the nonzero terms of the matrix are gathered about the main diagonal. To illustrate this concept, consider the plane truss of Figure 4-3.

From Figure 4-3, we see that element 2 connects nodes 1 and 4. Therefore, the 2 × 2 submatrices at positions 1-1, 1-4, 4-1, and 4-4 of Figure 4-4 will have nonzero coefficients. Figure 4-4 represents the total stiffness matrix of the plane truss. The X's denote nonzero coefficients. From Figure 4-4, we observe that the nonzero terms are within the band shown. Using a banded storage format, only the main diagonal and the nonzero upper codiagonals need be stored as shown in Figure 4-5. Note that any codiagonal with a nonzero term requires storage of the whole codiagonal *and* any codiagonals between it and the main diagonal. The use of banded storage is not only efficient, but permits direct use of the Scientific Subroutine Package (SSP) subroutine called MCHB. The description of this subroutine in the SSP manual gives a more detailed explanation of banded compressed storage (see Reference [1]).

We now define the semibandwidth n_b as $n_b = n_d(m + 1)$, where n_d is the number of degrees of freedom per node and m is the maximum difference in node numbers determined by calculating the difference in node numbers for each element of a finite element model. In the example for the plane truss of Figure 4-3, $m = 4 - 1 = 3$ and $n_d = 2$, so that $n_b = 2(3 + 1) = 8$.

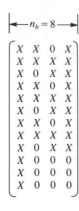

$\left|\!\!\longleftarrow n_b = 8 \longrightarrow\!\!\right|$

$$\begin{bmatrix} X & X & 0 & X \\ X & X & X & X \\ X & 0 & X & X \\ X & X & 0 & X \\ X & X & X & X \\ X & 0 & X & X \\ X & X & 0 & X \\ X & X & X & X \\ X & 0 & X & X \\ X & 0 & 0 & 0 \\ X & 0 & 0 & 0 \\ X & 0 & 0 & 0 \end{bmatrix}$$

FIGURE 4-5   Banded storage format of the stiffness matrix of Figure 4-4

Execution time (primarily, equation-solving time) is a function of the number of equations to be solved. Without using banded storage of global stiffness matrix $\underline{K}$, it has been shown (see Reference [2]) that execution time is proportional to $(1/3)n^3$, where n is the number of equations to be solved, or, equivalently, the size of $\underline{K}$. Using banded storage of $\underline{K}$, the execution time is proportional to $(n)n_b^2$. The ratio of time of execution without banded storage to that using banded storage is then $(1/3)(n/n_b)^2$. For the plane truss example, this ratio is $(1/3)(24/8)^2 = 3$. Therefore, it takes about three times as long to execute the solution of the example truss if banded storage is not used.

Hence, to reduce bandwidth we should number systematically and try to have a minimum difference between adjacent nodes. A small bandwidth is usually achieved by consecutive node numbering across the shorter dimension, as shown in Figure 4-3 on page 133. The computer program TRUSS uses the banded-symmetric format for storing the global stiffness matrix, $\underline{K}$. (We present a flowchart of TRUSS in Section 4.3 and discuss the use of the program in Section 4.4.)

Several automatic node renumbering schemes have been computerized (see Reference [3]). This option is available in most general-purpose computer programs. Alternatively, the wavefront or frontal method is becoming popular for optimizing equation solution time. In the **wavefront method**, elements, instead of nodes, are automatically renumbered.

In the wavefront method the assembly of the equations alternates with their solution by Gauss elimination. The sequence in which the equations are processed is determined by element numbering rather than by node numbering. The first equations eliminated are those associated with element 1 only. Next the contributions of stiffness coefficients of the adjacent element, element 2, are added to the system of equations. If any additional degrees of freedom are contributed by elements 1 and 2 only—that is, no other elements contribute stiffness coefficients to specific degrees of freedom—these equations are eliminated (condensed) from the system of equations. As one or more additional elements make their contributions to the system of equations and additional degrees of freedom are contributed only by these elements, those degrees of freedom are eliminated from the solution. This repetitive alternation between assembly and solution was initially seen as a wavefront that sweeps over the structure in a pattern determined by the element numbering. For greater efficiency of this method, consecutive element numbering should be done across the structure in a direction that spans the smallest number of nodes.

The wavefront method, although somewhat more difficult to understand and to program than the banded-symmetric method, is computationally more efficient. A banded solver stores and processes any blocks of zeros created in assembling the stiffness matrix. These blocks of zero coefficients are not stored or processed using the wavefront method. Many large-scale computer programs are now using the wavefront method to solve the system of equations. (For additional details of this method, see References [4]–[7].) Example 4.2 illustrates the wavefront method for solution of a truss problem.

E X A M P L E   4.2

For the plane truss shown in Figure 4-6, illustrate the wavefront solution procedure.

We will solve this problem in symbolic form. Merging k's for elements 1, 2, and 3 and enforcing boundary conditions at node 1, we have

$$
\begin{array}{cc}
d_{2x} & \qquad\qquad d_{2y} \\
\end{array}
$$

$$
\begin{bmatrix}
k_{33}^{(1)} + k_{11}^{(2)} + k_{11}^{(3)} & k_{34}^{(1)} + k_{12}^{(2)} + k_{12}^{(3)} & k_{13}^{(3)} & k_{14}^{(3)} & k_{13}^{(2)} & k_{14}^{(2)} \\
k_{43}^{(1)} + k_{21}^{(2)} + k_{21}^{(3)} & k_{44}^{(1)} + k_{22}^{(2)} + k_{22}^{(3)} & k_{23}^{(3)} & k_{24}^{(3)} & k_{23}^{(2)} & k_{24}^{(2)} \\
k_{31}^{(2)} & k_{32}^{(3)} & k_{33}^{(3)} & k_{34}^{(3)} & k_{33}^{(2)} & k_{34}^{(2)} \\
k_{41}^{(3)} & k_{42}^{(3)} & k_{43}^{(3)} & k_{44}^{(3)} & k_{43}^{(2)} & k_{44}^{(2)} \\
k_{31}^{(2)} & k_{32}^{(2)} & 0 & 0 & 0 & 0 \\
k_{41}^{(2)} & k_{42}^{(2)} & 0 & 0 & 0 & 0
\end{bmatrix}
\begin{Bmatrix}
d_{2x} \\
d_{2y} \\
d'_{3x} \\
d'_{3y} \\
d'_{4x} \\
d'_{4y}
\end{Bmatrix}
$$

$$
= \begin{Bmatrix}
0 \\
0 \\
0 \\
-P \\
0 \\
0
\end{Bmatrix}
\qquad (4.2.1)
$$

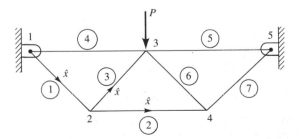

F I G U R E   4-6   Truss for wavefront solution

Eliminating d_{2x} and d_{2y} (as all stiffness contributions from node 2 degrees of freedom have been included from these elements; these contributions are from elements 1, 2, and 3) by static condensation or Gauss elimination yields

$$[k_c'] \begin{Bmatrix} d_{3x}' \\ d_{3y}' \\ d_{4x}' \\ d_{4y}' \end{Bmatrix} = \{F_c'\} \qquad (4.2.2)$$

where the condensed stiffness and force matrices are

$$[k_c'] = [K_{22}'] - [K_{21}'][K_{11}']^{-1}[K_{12}'] \qquad (4.2.3)$$

$$\{F_c'\} = \{F_2'\} - [K_{21}'][K_{11}']^{-1}\{F_1'\} \qquad (4.2.4)$$

where primes on the degrees of freedom, such as d_{3x}' in Eq. (4.2.1), indicate that all stiffness coefficients associated with that degree of freedom have not yet been included. Now include elements 4, 5, and 6 for degrees of freedom at node 3. The resulting equations are

$$\begin{array}{cccc} d_{3x} & d_{3y} & d_{4x} & d_{4y} \end{array}$$

$$\begin{bmatrix} k_{c11}' + k_{33}^{(4)} + k_{11}^{(5)} + k_{11}^{(6)} & k_{34}^{(4)} + k_{12}^{(5)} + k_{12}^{(6)} + k_{c12}' & k_{13}^{(6)} + k_{c13}' & k_{14}^{(6)} + k_{c14}' \\ k_{c21}' + k_{34}^{(4)} + k_{21}^{(5)} + k_{21}^{(6)} & k_{44}^{(4)} + k_{22}^{(5)} + k_{22}^{(6)} + k_{c22}' & k_{23}^{(6)} + k_{c23}' & k_{24}^{(6)} + k_{c24}' \\ \hline k_{c31}' + k_{31}^{(6)} & k_{c32}' + k_{32}^{(6)} & k_{c33}' + k_{33}^{(6)} & k_{c34}' + k_{34}^{(6)} \\ k_{c41}' + k_{41}^{(6)} & k_{c42}' + k_{42}^{(6)} & k_{c43}' + k_{43}^{(6)} & k_{c44}' + k_{44}^{(6)} \end{bmatrix}$$

$$\times \begin{Bmatrix} d_{3x} \\ d_{3y} \\ d_{4x} \\ d_{4y} \end{Bmatrix} = \begin{Bmatrix} 0 \\ -P \\ 0 \\ 0 \end{Bmatrix} \qquad (4.2.5)$$

Using static condensation, we eliminate d_{3x} and d_{3y} (as all contributions from node 3 degrees of freedom have been included from each element) to obtain

$$[k_c''] \begin{Bmatrix} d_{4x}' \\ d_{4y}' \end{Bmatrix} = \{F_c''\} \qquad (4.2.6)$$

where

$$[k_c''] = [K_{22}''] - [K_{21}''][K_{11}'']^{-1}[K_{12}''] \qquad (4.2.7)$$

$$\{F_c''\} = \{F_2''\} - [K_{21}''][K_{11}'']^{-1}\{F_1''\} \qquad (4.2.8)$$

Next we include element 7 contributions to the stiffness matrix. The condensed set of equations yield

$$[k_c'''] \begin{Bmatrix} d_{4x} \\ d_{4y} \end{Bmatrix} = \{F_c'''\} \qquad (4.2.9)$$

$$[k_c'''] = [K_{22}'''] - [K_{21}'''][K_{11}''']^{-1}[K_{12}'''] \qquad (4.2.10)$$

where

$$\{F_c'''\} = \{F_2'''\} - [K_{21}'''][K_{11}''']^{-1}\{F_1'''\} \qquad (4.2.11)$$

The elimination procedure is now complete, and we solve Eq. (4.2.9) for d_{4x} and d_{4y}. Then we back-substitute d_{4x} and d_{4y} into Eq. (4.2.5) to obtain d_{3x} and d_{3y}. Finally, we back-substitute d_{3x} through d_{4y} into Eq. (4.2.1) to obtain d_{2x}

and d_{2y}. Static condensation and Gauss elimination with back-substitution have been used to solve the set of equations for all the degrees of freedom. The solution procedure has then proceeded as if it were a wave sweeping over the structure, starting at node 2, engulfing node 2 and elements with degrees of freedom at node 2, and then sweeping through node 3 and finally node 4. ■

    We now describe a practical computer scheme that is often used for the solution of the resulting system of algebraic equations. The significance of this scheme is that it takes advantage of the fact that the stiffness method produces a banded $\underline{K}$ matrix in which the nonzero elements occur about the main diagonal in $\underline{K}$; while solving the equations this banded format is maintained.

EXAMPLE   4.3

We will now use a simple example to illustrate this computer scheme. Consider the three-spring assemblage shown in Figure 4-7. The assemblage is subjected to forces at node 2 of 100 lb in the x direction and 200 lb in the y direction. Node 1 is completely constrained from displacement in both the x and y directions, whereas node 3 is completely constrained in the y direction but is displaced a known amount δ in the x direction.

    Our purpose here is not to obtain the actual $\underline{K}$ for the assemblage, but to illustrate the scheme used for solution. The general solution can be shown to be given by

$$
\begin{bmatrix}
k_{11} & k_{12} & k_{13} & k_{14} & k_{15} & k_{16} \\
 & k_{22} & k_{23} & k_{24} & k_{25} & k_{26} \\
 & & k_{33} & k_{34} & k_{35} & k_{36} \\
 & & & k_{44} & k_{45} & k_{46} \\
 & & & & k_{55} & k_{56} \\
 \text{Symmetry} & & & & & k_{66}
\end{bmatrix}
\begin{Bmatrix}
d_{1x} \\ d_{1y} \\ d_{2x} \\ d_{2y} \\ d_{3x} \\ d_{3y}
\end{Bmatrix}
=
\begin{Bmatrix}
F_{1x} \\ F_{1y} \\ F_{2x} = 100 \\ F_{2y} = 200 \\ F_{3x} \\ F_{3y}
\end{Bmatrix}
\qquad (4.2.12)
$$

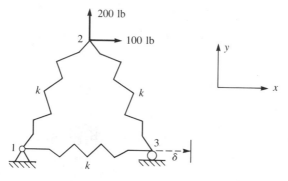

FIGURE 4-7   Three-spring assemblage

where $\underline{K}$ has been left in general form. On imposing the boundary conditions, the computer program transforms Eq. (4.2.12) to the following:

$$\begin{bmatrix} 1 & 0 & 0 & 0 & 0 & 0 \\ 0 & 1 & 0 & 0 & 0 & 0 \\ 0 & 0 & k_{33} & k_{34} & 0 & 0 \\ 0 & 0 & k_{43} & k_{44} & 0 & 0 \\ 0 & 0 & 0 & 0 & 1 & 0 \\ 0 & 0 & 0 & 0 & 0 & 1 \end{bmatrix} \begin{Bmatrix} d_{1x} \\ d_{1y} \\ d_{2x} \\ d_{2y} \\ d_{3x} \\ d_{3y} \end{Bmatrix} = \begin{Bmatrix} 0 \\ 0 \\ 100 - k_{35}\delta \\ 200 - k_{45}\delta \\ \delta \\ 0 \end{Bmatrix} \qquad (4.2.13)$$

From Eq. (4.2.13), we can see that $d_{1x} = 0$, $d_{1y} = 0$, $d_{3y} = 0$, and $d_{3x} = \delta$. These displacements are consistent with the imposed boundary conditions. The unknown displacements, d_{2x} and d_{2y}, can be determined routinely by solving Eq. (4.2.13).

We will now explain the computer scheme that is generally applicable to transform Eq. (4.2.12) to Eq. (4.2.13). First, the terms associated with the known displacement boundary condition(s) within each equation were transformed to the right side of those equations. In the third and fourth equations of Eq. (4.2.12), $k_{35}\delta$ and $k_{45}\delta$ were transformed to the right side, as shown in Eq. (4.2.13). Then the right-side force term corresponding to the known displacement row was equated to the known displacement. In the fifth equation of Eq. (4.2.12), where $d_{3x} = \delta$, the right-side, fifth-row force term F_{3x} was equated to the known displacement δ, as shown in Eq. (4.2.13). For the homogeneous boundary conditions, the affected rows of $\underline{F}$, corresponding to the zero-displacement rows, were replaced with zeros. Again, this is done in the computer scheme only to obtain the nodal displacements and does not imply that these nodal forces are zero. We obtain the unknown nodal forces by determining the nodal displacements and back-substituting these results into the original Eq. (4.2.12). Since $d_{1x} = 0$, $d_{1y} = 0$, and $d_{3y} = 0$ in Eq. (4.2.12), the first, second, and sixth rows of the force matrix of Eq. (4.2.13) were set to zero. Finally, for both nonhomogenous and homogeneous boundary conditions, the rows and columns of $\underline{K}$ corresponding to these prescribed boundary conditions were set to zero except the main diagonal, which was made unity; that is, the first, second, fifth, and sixth rows and columns of $\underline{K}$ in Eq. (4.2.12) were set to zero, except for the main diagonal terms, which were made unity. Although not necessary, setting the main diagonal terms equal to one facilitates the simultaneous solution of the six equations in Eq. (4.2.13) by an elimination method used in the computer program. This modification is shown in the $\underline{K}$ matrix of Eq. (4.2.13).   ■

4.3   Flowchart for the Solution of a Truss Problem by the Direct Stiffness Method

In Figure 4-8, we present a flowchart of a finite element program (called TRUSS) used for the analysis of two- and three-dimensional trusses.

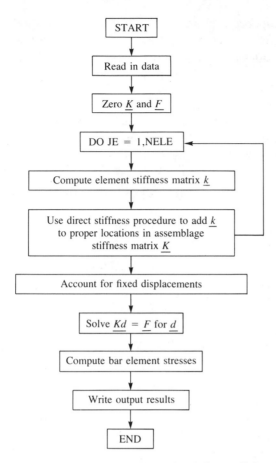

FIGURE 4-8 Flowchart of truss finite element program (NELE represents the number of elements)

4.4 Description of a Computer Program for Truss Analysis

We will now describe a computer program called TRUSS (the source code for TRUSS is provided on the enclosed disk) that can be used to solve two- and three-dimensional truss problems. The program is based on the flowchart of Section 4.3.

First, we list the general steps for modeling and for entering data into the program as follows:

1. Establish the global coordinate axes (x, y, and z). (At each node j, displacements will then be decomposed into x, y, and z components.)

2. Number the elements and the nodes. (Identify the total numbers of elements and nodes.)

3. Specify the nodal coordinates. (Then x_j, y_j, and z_j represent the coordinates of each node j.)

4. Specify the support or boundary conditions. (That is, specify the degrees of freedom that are to be zero at each support node.)

5. Specify the loads or forces in global-coordinate components at the nodes. (*Note:* In program TRUSS, F_{1j}, F_{2j}, and F_{3j} are the x, y, and z components of the possible forces acting at node j. This notation is different from that previously used in this text because the computer program was written with the first subscript on F denoting the global direction of the force and the second subscript denoting the node at which the force acts. The description of variables for program TRUSS further describes this notation.)

6. Specify the connectivity or topology. (Identify which nodes connect to which elements.)

7. Describe the element properties. (For each truss element, specify the modulus of elasticity and the cross-sectional area.)

To use the computer program TRUSS, we first describe the input data. Hence, the argument list in SUBROUTINE DATA, which reads data into the program, is described as follows:

SUBROUTINE DATA

(NELE,NNODE,MUD,IFIX,XC,YC,ZC,FORCE,NODE,E,A)

This subroutine reads in the data needed for the truss analysis program and transfers the values of the variables back to the calling program through the calling arguments. In addition, it computes the number of nonzero upper codiagonals needed in the stiffness matrix and stores the result in MUD. All data is free format; separate data by commas. Always include all data; that is, include zeros where they occur.

Description of Variables

I is a dummy variable representing the x, y, and z directions, $(I = 1, 2, 3)$.

J is a dummy variable that indicates the node number.

K is a dummy variable that indicates the element number.

NELE is the number of elements in the finite element model.

NNODE is the number of nodes in the finite element model.

IFIX(3,J) specifies whether a displacement component at node J is fixed with respect to global coordinates; for example, IFIX(1,J) specifies whether the x component of displacement is fixed or free. If IFIX(I,J) = 1, the Ith displacement component at node J is fixed.

If IFIX(I,J) = 0, the Ith displacement component is free.

XC(J), YC(J), and ZC(J) are the x, y, and z coordinates of node J.

FORCE(3,J) represents the three components of the external force applied at node J with respect to global coordinates.

NODE(2,K) represents the two nodes of element K.

E(K) and A(K) are the modulus of elasticity and the cross-sectional area of element K.

Input Data

The order of input data by sequence of parameters needed by TRUSS is as follows:

1. Title (columns 1–80).
   Title of problem; any alphanumeric data to identify the problem. This will be printed as a heading on the first page of output.

2. Basic parameters line

   NELE,NNODE

3. Node data (one line for each node)

   J,IFIX(1,J),IFIX(2,J),IFIX(3,J),XC(J),YC(J),ZC(J),

   FORCE(1,J),FORCE(2,J),FORCE(3,J)

4. Element data (one line for each element)

   K,NODE(1,K),NODE(2,K),E(K),A(K)

EXAMPLE 4.4

We now consider the space truss solved longhand in Section 3.7 to illustrate how to input data into TRUSS. For convenience' sake, the truss model shown in Figure 3-15 is repeated here as Figure 4-9.

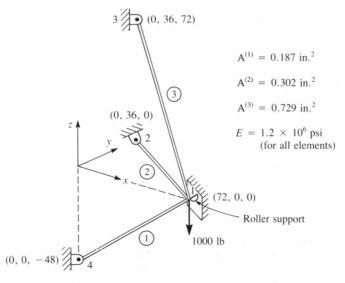

$A^{(1)} = 0.187$ in.^{2}

$A^{(2)} = 0.302$ in.^{2}

$A^{(3)} = 0.729$ in.^{2}

$E = 1.2 \times 10^6$ psi
(for all elements)

FIGURE 4-9   Space truss used to illustrate input data for computer program TRUSS

T A B L E 4-2   Input for the space truss of Figure 4-9

Line	Data on the Line (Beginning in Column 1)
1	SPACE TRUSS EXAMPLE OF SECTION 3.7
2	3,4
3	1,0,1,0,72.0,0.,0.,0.,0., − 1000.0
4	2,1,1,1,0.0,36.0,0.,0.,0.,0.
5	3,1,1,1,0.0,36.0,72.0,0.,0.,0.
6	4,1,1,1,0.0,0.0, − 48.0,0.,0.,0.
7	1,1,4,1.2E + 6,0.187
8	2,1,2,1.2E + 6,0.302
9	3,1,3,1.2E + 6,0.729

Table 4-2 shows a typical data file (using the space truss of Figure 4-8) based on the general description of the input data. In Table 4-2, line 1 is the identifying title of the problem being solved. On line 2 are the numbers of elements and nodes of the truss, lines 3–6 describe node information, and lines 7–9 describe element information. The solution to this problem given in Section 3.7 can be compared to the computer program results given in Table 4-3.

REFERENCES

[1] SYSTEM/360, Scientific Subroutine Package, IBM.

[2] Kardestuncer, H., *Elementary Matrix Analysis of Structures*, McGraw-Hill, New York, 1974.

[3] Collins, R. J., "Bandwidth Reduction by Automatic Renumbering," *International Journal For Numerical Methods in Engineering*, Vol. 6, pp. 345–356, 1973.

[4] Melosh, R. J., and Bamford, R. M., "Efficient Solution of Load-Deflection Equations," *Journal of the Structural Division*, American Society of Civil Engineers, No. ST4, pp. 661–676, April 1969.

[5] Irons, B. M., "A Frontal Solution Program for Finite Element Analysis," *International Journal for Numerical Methods in Engineering*, Vol. 2, No. 1, pp. 5–32, 1970.

[6] Meyer, C., "Solution of Linear Equations-State-of-the-Art," *Journal of the Structural Division*, American Society of Civil Engineers, Vol. 99, No. ST7, pp. 1507–1526, 1973.

[7] Jennings, A., *Matrix Computation for Engineers and Scientists*, Wiley, London, 1977.

[8] Cook, R. D., Malkus, D. S., and Plesha, M. E., *Concepts and Applications of Finite Element Analysis*, 3rd ed., Wiley, New York, 1989.

[9] Bathe, K. J., and Wilson, E. L., *Numerical Methods in Finite Element Analysis*, Prentice Hall, Englewood Cliffs, N.J., 1976.

T A B L E   4-3   Output for the space truss of Figure 4-9 from computer
program TRUSS

SPACE TRUSS EXAMPLE OF SECTION 3.7

NUMBER OF ELEMENTS(NELE) = 3
NUMBER OF NODES(NNODE)   = 4

NODE POINTS

K	IFIX			XC(K)	YC(K)	ZC(K)
1	0	1	0	7.200000E + 01	0.000000E + 00	0.000000E + 00
2	1	1	1	0.000000E + 00	3.600000E + 01	0.000000E + 00
3	1	1	1	0.000000E + 00	3.600000E + 01	7.200000E + 01
4	1	1	1	0.000000E + 00	0.000000E + 00	− 4.800000E + 01

FORCE(1, K)	FORCE(2, K)	FORCE(3, K)
0.000000E + 00	0.000000E + 00	− 1.000000E + 03
0.000000E + 00	0.000000E + 00	0.000000E + 00
0.000000E + 00	0.000000E + 00	0.000000E + 00
0.000000E + 00	0.000000E + 00	0.000000E + 00

ELEMENTS

K	NODE(1, K)		E(K)	A(K)
1	1	4	1.2000E + 06	1.8700E − 01
2	1	2	1.2000E + 06	3.0200E − 01
3	1	3	1.2000E + 06	7.2900E − 01

NUMBER OF NONZERO UPPER CODIAGONALS(MUD) = 11

DISPLACEMENTS	X	Y	Z
NODE NUMBER 1	−0.7111E − 01	0.0000E + 00	−0.2662E + 00
NODE NUMBER 2	0.0000E + 00	0.0000E + 00	0.0000E + 00
NODE NUMBER 3	0.0000E + 00	0.0000E + 00	0.0000E + 00
NODE NUMBER 4	0.0000E + 00	0.0000E + 00	0.0000E + 00

STRESSES IN ELEMENTS (IN CURRENT UNITS)

ELEMENT NUMBER	STRESS
1 =	− 0.28685E + 04
2 =	− 0.94819E + 03
3 =	0.14454E + 04

PROBLEMS

4.1   For the truss shown in Figure P4-1, use symmetry to determine the displacements of the nodes and the stresses in each element. All elements have $E = 30 \times 10^6$ psi. Elements 1, 2, 4, and 5 have $A = 10$ in.^{2} and element 3 has $A = 20$ in.^{2}

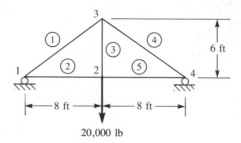

FIGURE   P4-1

4.2   All elements of the structure in Figure P4-2 have the same AE except element 1, which has an axial stiffness of $2AE$. Find the displacements of the nodes and the stresses in elements 2, 3, and 4 by using symmetry. Check equilibrium at node 4. You might want to use the results obtained from the stiffness matrix of Problem 3.24.

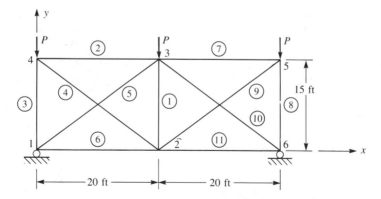

FIGURE   P4-2

4.3   For the roof truss shown in Figure P4-3, use symmetry to determine the displacements of the nodes and the stresses in each element. All elements have $E = 210$ GPa and $A = 10 \times 10^{-4}$ m^2.

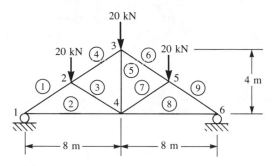

20 kN

20 kN ④ ③ 20 kN

⑤

⑥

4 m

② ③ ⑦ ⑧

④ ⑨

⑥

8 m   8 m

FIGURE P4-3

4.4 Determine the bandwidths of the plane trusses shown in Figure P4-4. What conclusions can you draw regarding labeling of nodes?

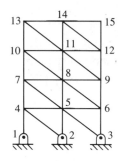
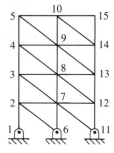

FIGURE P4-4

Use the computer code TRUSS from the disk or any other suitable computer program to solve Problems 4.5–4.13. (*Hint:* Don't forget to set IFIX = 1 in the z direction for all nodes of a plane truss.)

4.5 For the bridge truss shown in Figure P4-5, use symmetry to determine the displacements of the nodes and the stresses in each element. All elements have $E = 210$ GPa and $A = 30 \times 10^{-4}$ m² except element 7, which has $A = 60 \times 10^{-4}$ m².

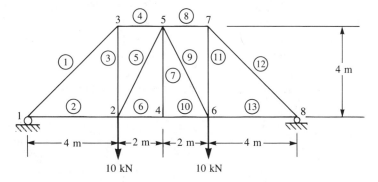

FIGURE P4-5

4.6 For the plane truss shown in Figure P4-6, determine the joint displacements and the element stresses. Let $E = 30 \times 10^6$ psi and $A = 4$ in.^{2} for all elements.

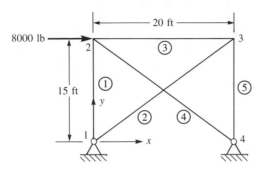

FIGURE P4-6

4.7 A typical truss made of structural steel for a mill building is shown in Figure P4-7. Design the truss members such that the maximum stress in any member is 15 ksi. Assume all elements have the same cross-sectional area.

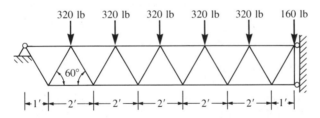

FIGURE P4-7

4.8 For the three-dimensional truss shown in Figure P4-8, determine the displacement components at node 1 and the stresses in each element. Let $A = 4$ in.^{2} and $E = 30 \times 10^6$ psi.

4.9 If the space truss shown in Figure P4-9 is to be constructed of 30,000 psi yield-strength steel and a safety factor of 2.0 is to be used, determine the cross-sectional area required for the elements. All elements must have the same final cross-sectional area. The loading on node 7 is a vertical force of 4500 lb and a horizontal force of 2800 lb. Nodes 1 and 2 are completely fixed, whereas node 3 is constrained from vertical displacement.

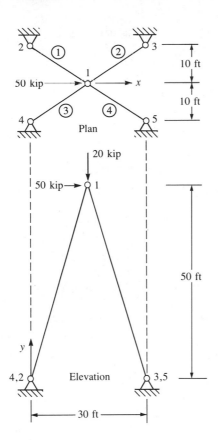

Plan

20 kip

50 kip → 1

50 ft

y

4,2   Elevation   3,5

← 30 ft →

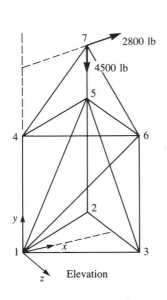

7   2800 lb

4500 lb

5

4   6

y

2

1   x   3

z   Elevation

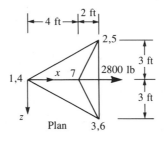

← 4 ft → 2 ft

2,5

3 ft

1,4   x 7   2800 lb

3 ft

z

Plan   3,6

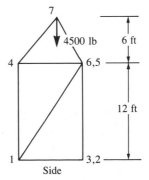

7

4500 lb   6 ft

4   6,5

12 ft

1   3,2

Side

FIGURE P4-9

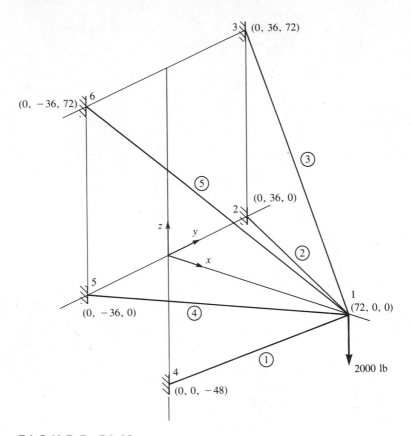

FIGURE   P4-10

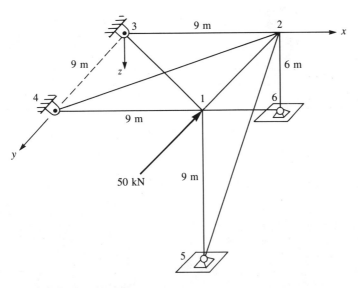

FIGURE   P4-11

4.10 For the space truss shown in Figure P4-10, all nodes other than node 1 are
 fixed (ball-and-socket joints). Determine the displacement components at
node 1 and the element stresses. Let the cross-sectional areas of elements 2
and 5 be 0.729 in.^{2}, that of elements 1 and 4 be 0.302 in.^{2}, and that of element
3 be 0.374 in.^{2} Let $E = 1.2 \times 10^6$ psi for all elements. A force of 2000 lb is
applied at node 1 in the negative z direction. The coordinates of each node
(in units of inches) are shown in the figure.

4.11 For the space truss shown in Figure P4-11, nodes 3, 4, 5, and 6 are ball-and-
socket joints. Determine the displacement components at nodes 1 and 2 and
the element stresses. Let $E = 70$ GPa and $A = 4 \times 10^{-4}$ m^2 for all elements.
A force of 50 kN is applied at node 1 in the negative y direction.

4.12 For the plane truss shown in Figure P4-12, determine the nodal displacements and the element stresses. Nodes 1 and 2 are pin joints. Let $E = 210$
GPa and $A = 3 \times 10^{-4}$ m^2 for all elements.

4.13 For the space truss shown in Figure P4-13, determine the nodal displacements and the element stresses. Let $E = 210$ GPa and $A = 1 \times 10^{-4}$ m^2 for
all elements. Nodes 1, 2, and 3 are ball-and-socket joints.

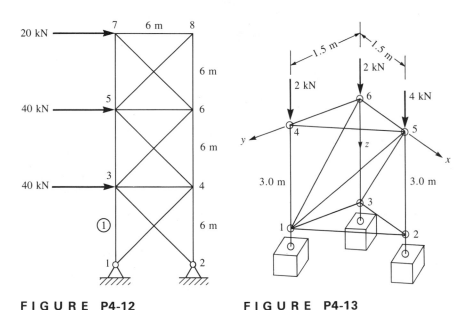

FIGURE P4-12   FIGURE P4-13

4.14   For the plane truss shown in Figure P4-14, determine the nodal displacements and element stresses. Let $A = 1 \times 10^{-4}$ m^2 and $E = 200$ GPa for each
element and $P = 10$ kN.

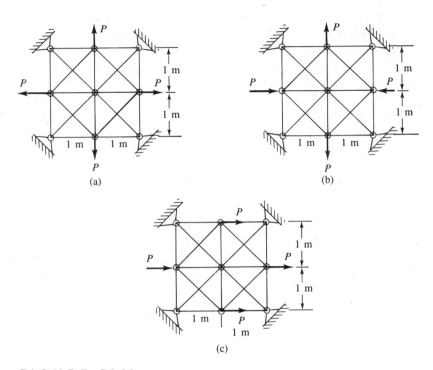

F I G U R E   P4-14

Development of

Beam Equations

INTRODUCTION

We begin this chapter by developing the stiffness matrix for the bending of a beam element, the most common of all structural elements as evidenced by its prominence in buildings, bridges, towers, and many other structures. The beam element is considered to be straight and to have constant cross-sectional area. We will first derive the beam element stiffness matrix by using the principles developed for simple beam theory.

We will then present simple examples to illustrate the assemblage of beam element stiffness matrices and the solution of beam problems by the direct stiffness method presented in Chapter 2. The solution of a beam problem illustrates that the degrees of freedom associated with a node are a transverse displacement and a rotation. We will include the nodal shear forces and bending moments and the resulting shear force and bending moment diagrams as part of the total solution.

Next, we will discuss procedures for handling distributed loading, since beams and frames are often subjected to distributed loading as well as concentrated nodal loading. We will follow the discussion with solutions of beams subjected to distributed loading.

We will then develop the beam element stiffness matrix for a beam element with a nodal hinge and illustrate the solution of a beam with an internal hinge.

To further acquaint you with the potential energy approach for developing stiffness matrices and equations, we will again develop the beam bending element equations using this approach. We hope to increase your confidence in this approach as it will be used throughout much of this text to develop stiffness matrices and equations for more complex elements, such as two-dimensional (plane) stress, axisymmetric, and three-dimensional stress.

Finally, the Galerkin residual method is applied to derive the beam element equations.

The concepts presented in this chapter are prerequisite to understanding the concepts for frame analysis presented in Chapter 6.

5.1 Beam Stiffness

In this section, we will derive the stiffness matrix for a simple beam element. A **beam** *is a long, slender structural member generally subjected to transverse loading that produces significant bending effects as opposed to twisting or axial effects.* This bending deformation is measured as a transverse displacement and a rotation. Hence, the degrees of freedom considered per node are a transverse displacement and a rotation (as opposed to only an axial displacement for the bar element of Chapter 3).

Consider the beam element shown in Figure 5-1. The beam is of length L with axial local coordinate $\hat{x}$ and transverse local coordinate $\hat{y}$. The local transverse nodal displacements are given by $\hat{d}_{iy}$'s and the rotations by $\hat{\phi}_i$'s. The local nodal forces are given by $\hat{f}_{iy}$'s and the bending moments by $\hat{m}_i$'s as shown. We initially neglect all axial effects.

At all nodes, the following sign conventions are used:

1. Moments are positive in the counterclockwise direction.
2. Rotations are positive in the counterclockwise direction.
3. Forces are positive in the positive $\hat{y}$ direction.
4. Displacements are positive in the positive $\hat{y}$ direction.

Figure 5-2 indicates the sign conventions used in simple beam theory for positive shear forces $\hat{V}$ and bending moments $\hat{m}$.

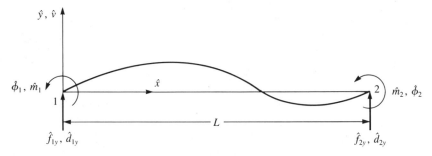

FIGURE 5-1   Beam element with positive nodal displacements, rotations, forces, and moments

FIGURE 5-2   Beam theory sign conventions for shear forces and bending moments

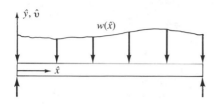
(a) Beam under load $w(\hat{x})$

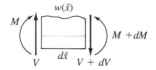
(b) Differential beam element

F I G U R E 5-3   Beam under distributed load

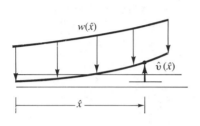
(a) Portion of deflected curve of beam

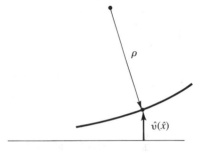
(b) Radius of deflected curve at $\hat{v}(\hat{x})$

F I G U R E 5-4   Deflected curve of beam

The differential equation governing elementary linear-elastic beam be-havior (see Reference [1]) is derived as follows. Consider the beam shown in Figure 5-3 subjected to a distributed loading $w(\hat{x})$ (force/length). From force and moment equilibrium of a differential element of the beam, we have

$$-w\, d\hat{x} + dV = 0 \quad \text{or} \quad w = -\frac{dV}{d\hat{x}} \qquad (5.1.1a)$$

$$V\, d\hat{x} + dM = 0 \quad \text{or} \quad V = \frac{dM}{d\hat{x}} \qquad (5.1.1b)$$

Also, the curvature κ of the beam is related to the moment by

$$\kappa = \frac{1}{\rho} = \frac{M}{EI} \qquad (5.1.1c)$$

where ρ is the radius of the deflected curve shown in Figure 5-4b, $\hat{v}$ is the transverse displacement function in the $\hat{y}$ direction (see Figure 5-4a), E is the modulus of elasticity, and I is the principal moment of inertia about the $\hat{z}$ axis (where the $\hat{z}$ axis is perpendicular to the $\hat{x}$ and $\hat{y}$ axes).

The curvature for small slopes $\theta = d\hat{v}/d\hat{x}$ is given by

$$\kappa = \frac{d^2\hat{v}}{d\hat{x}^2} \qquad (5.1.1d)$$

Using Eq. (5.1.1d) in (5.1.1c), we obtain

$$\frac{d^2\hat{v}}{d\hat{x}^2} = \frac{M}{EI} \qquad (5.1.1e)$$

Solving Eq. (5.1.1e) for M and substituting this result into (5.1.1b) and (5.1.1a), we obtain

$$\frac{d^2}{d\hat{x}^2}\left(EI\frac{d^2\hat{v}}{d\hat{x}^2}\right) = -w(\hat{x}) \qquad (5.1.1f)$$

For constant EI and only nodal forces and moments, Eq. (5.1.1f) becomes

$$EI\frac{d^4\hat{v}}{d\hat{x}^4} = 0 \qquad (5.1.1g)$$

We will now follow the steps outlined in Chapter 1 to develop the stiffness matrix and equations for a beam element and then to illustrate complete solutions for beams.

Step 1   Select Element Type

Represent the beam by labeling nodes at each end and in general by labeling the element number (see Figure 5-1).

Step 2   Select a Displacement Function

Assume the transverse displacement variation through the element length to be

$$\hat{v}(\hat{x}) = a_1\hat{x}^3 + a_2\hat{x}^2 + a_3\hat{x} + a_4 \qquad (5.1.2)$$

The complete cubic displacement function Eq. (5.1.2) is appropriate because there are four total degrees of freedom (a transverse displacement and a small rotation at each node). The cubic function also satisfies the basic beam differential equation—further justifying its selection. In addition, the cubic function also satisfies the conditions of displacement and slope continuity at nodes shared by two elements.

Using the same procedure as described in Section 2.2, we express $\hat{v}$ as a function of the nodal degrees of freedom $\hat{d}_{1y}$, $\hat{d}_{2y}$, $\hat{\phi}_1$, and $\hat{\phi}_2$ as follows:

$$\hat{v}(0) = \hat{d}_{1y} = a_4$$

$$\frac{d\hat{v}(0)}{d\hat{x}} = \hat{\phi}_1 = a_3$$

$$\hat{v}(L) = \hat{d}_{2y} = a_1 L^3 + a_2 L^2 + a_3 L + a_4 \qquad (5.1.3)$$

$$\frac{d\hat{v}(L)}{d\hat{x}} = \hat{\phi}_2 = 3a_1 L^2 + 2a_2 L + a_3$$

Solving Eqs. (5.1.3) for a_1 through a_4 in terms of the nodal degrees of freedom and substituting into Eq. (5.1.2), we have

$$\hat{v} = \left[\frac{2}{L^3}(\hat{d}_{1y} - \hat{d}_{2y}) + \frac{1}{L^2}(\hat{\phi}_1 + \hat{\phi}_2)\right]\hat{x}^3$$

$$+ \left[-\frac{3}{L^2}(\hat{d}_{1y} - \hat{d}_{2y}) - \frac{1}{L}(2\hat{\phi}_1 + \hat{\phi}_2)\right]\hat{x}^2 + \hat{\phi}_1\hat{x} + \hat{d}_{1y} \qquad (5.1.4)$$

In matrix form, we express Eq. (5.1.4) as

$$\hat{v} = [N]\{\hat{d}\} \qquad (5.1.5)$$

where

$$\{\hat{d}\} = \begin{Bmatrix} \hat{d}_{1y} \\ \hat{\phi}_1 \\ \hat{d}_{2y} \\ \hat{\phi}_2 \end{Bmatrix} \qquad (5.1.6a)$$

and where

$$[N] = [N_1 \quad N_2 \quad N_3 \quad N_4] \qquad (5.1.6b)$$

and

$$N_1 = \frac{1}{L^3}(2\hat{x}^3 - 3\hat{x}^2 L + L^3) \qquad N_2 = \frac{1}{L^3}(\hat{x}^3 L - 2\hat{x}^2 L^2 + \hat{x}L^3)$$

$$(5.1.7)$$

$$N_3 = \frac{1}{L^3}(-2\hat{x}^3 + 3\hat{x}^2 L) \qquad N_4 = \frac{1}{L^3}(\hat{x}^3 L - \hat{x}^2 L^2)$$

N_1, N_2, N_3, and N_4 are called the **shape functions** for a beam element. For the beam element, $N_1 = 1$ when evaluated at node 1 and $N_1 = 0$ when evaluated at node 2. Because N_2 is associated with $\hat{\phi}_1$, we have, from the second of Eqs. (5.1.7), $(dN_2/d\hat{x}) = 1$ when evaluated at node 1. Shape functions N_3 and N_4 have analogous results for node 2.

Step 3 **Define the Strain/Displacement and Stress/Strain Relationships**

Assume the following axial strain/displacement relationship to be valid:

$$\varepsilon_x(\hat{x}, \hat{y}) = \frac{d\hat{u}}{d\hat{x}} \qquad (5.1.8)$$

where $\hat{u}$ is the axial displacement function. From the deformed configuration of the beam shown in Figure 5-5, we relate the axial displacement to the transverse displacement by

$$\hat{u} = -\hat{y}\frac{d\hat{v}}{d\hat{x}} \qquad (5.1.9)$$

where we should recall from elementary beam theory (see Reference [2]) the basic assumption that cross sections of the beam (such as cross section $ABCD$) that are planar before bending deformation remain planar after deformation and, in general, rotate through an angle $(d\hat{v}/d\hat{x})$. Using Eq. (5.1.9) in Eq. (5.1.8), we obtain

$$\varepsilon_x(\hat{x}, \hat{y}) = -\hat{y}\frac{d^2\hat{v}}{d\hat{x}^2} \qquad (5.1.10)$$

From elementary beam theory, the bending moment and shear force are related to the transverse displacement function. Since we will use these relationships in the derivation of the beam element stiffness matrix, we now present them as

$$\hat{m}(\hat{x}) = EI\frac{d^2\hat{v}}{d\hat{x}^2} \qquad \hat{V} = EI\frac{d^3\hat{v}}{d\hat{x}^3} \qquad (5.1.11)$$

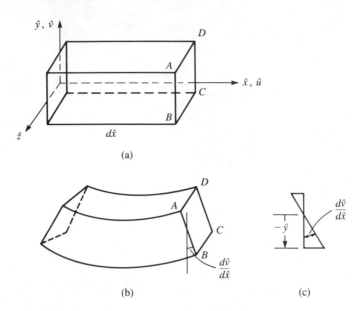

(a)

(b)   (c)

F I G U R E   5-5   Beam segment (a) before deformation and (b) after
deformation; (c) angle of rotation of cross-section *ABCD*

Step 4   Derive the Element Stiffness Matrix and Equations

First, derive the element stiffness matrix and equations using a direct equilibrium approach. We now use the nodal and beam theory sign conventions for
shear forces and bending moments, along with Eqs. (5.1.4) and (5.1.11), to
obtain

$$\hat{f}_{1y} = \hat{V} = EI\frac{d^3\hat{v}(0)}{d\hat{x}^3} = \frac{EI}{L^3}(12\hat{d}_{1y} + 6L\hat{\phi}_1 - 12\hat{d}_{2y} + 6L\hat{\phi}_2)$$

$$\hat{m}_1 = -\hat{m} = -EI\frac{d^2\hat{v}(0)}{d\hat{x}^2} = \frac{EI}{L^3}(6L\hat{d}_{1y} + 4L^2\hat{\phi}_1 - 6L\hat{d}_{2y} + 2L^2\hat{\phi}_2)$$

$$\hat{f}_{2y} = -\hat{V} = -EI\frac{d^3\hat{v}(L)}{d\hat{x}^3} = \frac{EI}{L^3}(-12\hat{d}_{1y} - 6L\hat{\phi}_1 + 12\hat{d}_{2y} - 6L\hat{\phi}_2)$$

$$\hat{m}_2 = \hat{m} = EI\frac{d^2\hat{v}(L)}{d\hat{x}^2} = \frac{EI}{L^3}(6L\hat{d}_{1y} + 2L^2\hat{\phi}_1 - 6L\hat{d}_{2y} + 4L^2\hat{\phi}_2)$$

(5.1.12)

where the minus signs in the second and third of Eqs. (5.1.12) are the result of
opposite nodal and beam theory positive bending moment conventions at
node 1 and opposite nodal and beam theory positive shear force conventions
at node 2 as seen by comparing Figures 5-1 and 5-2. Equations (5.1.12) relate
the nodal forces to the nodal displacements. In matrix form, Eqs. (5.1.12)
become

$$
\begin{Bmatrix} \hat{f}_{1y} \\ \hat{m}_1 \\ \hat{f}_{2y} \\ \hat{m}_2 \end{Bmatrix} = \frac{EI}{L^3} \begin{bmatrix} 12 & 6L & -12 & 6L \\ 6L & 4L^2 & -6L & 2L^2 \\ -12 & -6L & 12 & -6L \\ 6L & 2L^2 & -6L & 4L^2 \end{bmatrix} \begin{Bmatrix} \hat{d}_{1y} \\ \hat{\phi}_1 \\ \hat{d}_{2y} \\ \hat{\phi}_2 \end{Bmatrix} \tag{5.1.13}
$$

where the stiffness matrix is then

$$
\underline{\hat{k}} = \frac{EI}{L^3} \begin{bmatrix} 12 & 6L & -12 & 6L \\ 6L & 4L^2 & -6L & 2L^2 \\ -12 & -6L & 12 & -6L \\ 6L & 2L^2 & -6L & 4L^2 \end{bmatrix} \tag{5.1.14}
$$

Equation (5.1.13) indicates that $\underline{\hat{k}}$ relates transverse forces and bending moments to transverse displacements and rotations, whereas axial effects have been neglected.

5.2 Example of Assemblage of Beam Stiffness Matrices

Step 5 **Assemble the Element Equations to Obtain the Global Equations and Introduce Boundary Conditions**

Consider the beam in Figure 5-6 as an example to illustrate the procedure for assemblage of beam element stiffness matrices. Assume EI to be constant throughout the beam. A force of 1000 lb and a moment of 1000 lb-ft are applied to the beam at midlength. The left end is a fixed support and the right end is a pin support.

First, we discretize the beam into two elements with nodes 1, 2, and 3 as shown. We include a node at midlength because applied force and moment exist at midlength and, at this time, loads are assumed to be applied only at nodes. (Another procedure for handling loads applied on elements will be discussed in Section 5.4.)

Using Eq. (5.1.14), the global stiffness matrices for the two elements are now given by

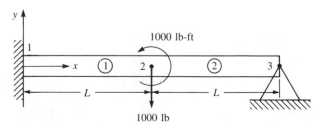

FIGURE 5-6   Fixed-hinged beam subjected to a force and a moment

$$
\underline{k}^{(1)} = \frac{EI}{L^3}
\begin{array}{cccc}
d_{1y} & \phi_1 & d_{2y} & \phi_2
\end{array}
\begin{bmatrix}
12 & 6L & -12 & 6L \\
6L & 4L^2 & -6L & 2L^2 \\
-12 & -6L & 12 & -6L \\
6L & 2L^2 & -6L & 4L^2
\end{bmatrix}
\qquad (5.2.1)
$$

and

$$
\underline{k}^{(2)} = \frac{EI}{L^3}
\begin{array}{cccc}
d_{2y} & \phi_2 & d_{3y} & \phi_3
\end{array}
\begin{bmatrix}
12 & 6L & -12 & 6L \\
6L & 4L^2 & -6L & 2L^2 \\
-12 & -6L & 12 & -6L \\
6L & 2L^2 & -6L & 4L^2
\end{bmatrix}
\qquad (5.2.2)
$$

where the degrees of freedom associated with each beam element are indicated by the usual labels above the columns in each element stiffness matrix. Here the local coordinate axes for each element coincide with the global x and y axes of the whole beam. Consequently, the local and global stiffness matrices are identical, so hats ($^\wedge$) are not needed in Eqs. (5.2.1) and (5.2.2).

The total stiffness matrix can now be assembled for the beam by using the direct stiffness method. When the total (global) stiffness matrix has been assembled, the external global nodal forces are related to the global nodal displacements. Using direct superposition and Eqs. (5.2.1) and (5.2.2), the governing equations for the beam are thus given by

$$
\begin{Bmatrix}
F_{1y} \\
M_1 \\
F_{2y} \\
M_2 \\
F_{3y} \\
M_3
\end{Bmatrix}
= \frac{EI}{L^3}
\begin{bmatrix}
12 & 6L & -12 & 6L & 0 & 0 \\
6L & 4L^2 & -6L & 2L^2 & 0 & 0 \\
-12 & -6L & 12+12 & -6L+6L & -12 & 6L \\
6L & 2L^2 & -6L+6L & 4L^2+4L^2 & -6L & 2L^2 \\
0 & 0 & -12 & -6L & 12 & -6L \\
0 & 0 & 6L & 2L^2 & -6L & 4L^2
\end{bmatrix}
$$

$$
\times
\begin{Bmatrix}
d_{1y} \\
\phi_1 \\
d_{2y} \\
\phi_2 \\
d_{3y} \\
\phi_3
\end{Bmatrix}
\qquad (5.2.3)
$$

Now considering the boundary conditions, or constraints, of the fixed support at node 1 and the hinge support at node 3, we have

$$
\phi_1 = 0 \qquad d_{1y} = 0 \qquad d_{3y} = 0 \qquad (5.2.4)
$$

On considering the third, fourth, and sixth equations of Eq. (5.2.3) corresponding to the rows with unknown degrees of freedom and using Eqs. (5.2.4), we obtain

$$\left\{\begin{array}{c} -1000 \\ 1000 \\ 0 \end{array}\right\} = \frac{EI}{L^3}\left[\begin{array}{ccc} 24 & 0 & 6L \\ 0 & 8L^2 & 2L^2 \\ 6L & 2L^2 & 4L^2 \end{array}\right]\left\{\begin{array}{c} d_{2y} \\ \phi_2 \\ \phi_3 \end{array}\right\} \qquad (5.2.5)$$

where $F_{2y} = -1000$ lb, $M_2 = 1000$ lb-ft, and $M_3 = 0$ have been substituted into the reduced set of equations. We could now solve Eq. (5.2.5) simultaneously for the unknown nodal displacement d_{2y} and the unknown nodal rotations ϕ_2 and ϕ_3. We leave the final solution for you to obtain. Section 5.3 provides complete solutions to beam problems.

5.3 Examples of Beam Analysis Using the Direct Stiffness Method

We will now perform complete solutions for beams with various boundary supports and loads to further illustrate the use of the equations developed in Section 5.1.

EXAMPLE 5.1

Using the direct stiffness method, solve the problem of the propped cantilever beam subjected to end load P in Figure 5-7. The beam is assumed to have constant EI and length $2L$. It is supported by a roller at midlength and is built in at the right end.

We have discretized the beam and established global coordinate axes as shown in Figure 5-7. We will determine the nodal displacements and rotations, the reactions, and the complete shear force and bending moment diagrams.

Using Eq. (5.1.14) for each element, along with superposition, we obtain the structure total stiffness matrix as

$$K = \frac{EI}{L^3}\begin{array}{c} \\ \\ \\ \\ \\ \\ \end{array}\overset{\begin{array}{cccccc} d_{1y} & \phi_1 & d_{2y} & \phi_2 & d_{3y} & \phi_3 \end{array}}{\left[\begin{array}{cccccc} 12 & 6L & -12 & 6L & 0 & 0 \\ & 4L^2 & -6L & 2L^2 & 0 & 0 \\ & & 12+12 & -6L+6L & -12 & 6L \\ & & & 4L^2+4L^2 & -6L & 2L^2 \\ & & & & 12 & -6L \\ \text{Symmetry} & & & & & 4L^2 \end{array}\right]} \qquad (5.3.1)$$

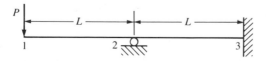

FIGURE 5-7   Propped cantilever beam

The governing equations for the beam are then given by

$$
\begin{Bmatrix} F_{1y} \\ M_1 \\ F_{2y} \\ M_2 \\ F_{3y} \\ M_3 \end{Bmatrix} = \frac{EI}{L^3} \begin{bmatrix} 12 & 6L & -12 & 6L & 0 & 0 \\ 6L & 4L^2 & -6L & 2L^2 & 0 & 0 \\ -12 & -6L & 24 & 0 & -12 & 6L \\ 6L & 2L^2 & 0 & 8L^2 & -6L & 2L^2 \\ 0 & 0 & -12 & -6L & 12 & -6L \\ 0 & 0 & 6L & 2L^2 & -6L & 4L^2 \end{bmatrix} \begin{Bmatrix} d_{1y} \\ \phi_1 \\ d_{2y} \\ \phi_2 \\ d_{3y} \\ \phi_3 \end{Bmatrix}
$$

$$(5.3.2)$$

On applying the boundary conditions

$$d_{2y} = 0 \qquad d_{3y} = 0 \qquad \phi_3 = 0 \tag{5.3.3}$$

and partitioning the equations associated with unknown displacements [the first, second, and fourth equations of Eq. (5.3.2)] from those equations associated with known displacements in the usual manner, we obtain the final set of equations for a longhand solution as

$$
\begin{Bmatrix} -P \\ 0 \\ 0 \end{Bmatrix} = \frac{EI}{L^3} \begin{bmatrix} 12 & 6L & 6L \\ 6L & 4L^2 & 2L^2 \\ 6L & 2L^2 & 4L^2 \end{bmatrix} \begin{Bmatrix} d_{1y} \\ \phi_1 \\ \phi_2 \end{Bmatrix} \tag{5.3.4}
$$

We will now solve Eq. (5.3.4) by the partitioning scheme described in Chapter 4. Rearranging Eq. (5.3.4) and partitioning (indicated by the dotted lines), we obtain

$$
\begin{Bmatrix} 0 \\ 0 \\ \hline -P \end{Bmatrix} = \frac{EI}{L^3} \begin{bmatrix} 4L^2 & 2L^2 & \vdots & 6L \\ 2L^2 & 8L^2 & \vdots & 6L \\ \hline 6L & 6L & \vdots & 12 \end{bmatrix} \begin{Bmatrix} \phi_1 \\ \phi_2 \\ \hline d_{1y} \end{Bmatrix} \tag{5.3.5}
$$

Using Eq. (4.1.13) for $\underline{k}_c$, we obtain

$$
\underline{k}_c = \frac{EI}{L^3} \left[[12] - [6L \quad 6L] \begin{bmatrix} 4L^2 & 2L^2 \\ 2L^2 & 8L^2 \end{bmatrix}^{-1} \begin{Bmatrix} 6L \\ 6L \end{Bmatrix} \right] \tag{5.3.6}
$$

Equation (5.3.6) can be simplified to

$$
\underline{k}_c = \frac{EI}{L^3} \left(\frac{12}{7} \right) \tag{5.3.7}
$$

Now using Eq. (4.1.15), we obtain the transverse displacement at node 1 as

$$
d_{1y} = -\frac{7PL^3}{12EI} \tag{5.3.8}
$$

where the minus sign indicates that the displacement of node 1 is downward. Substituting the appropriate partitioned parts of $\underline{K}$ from Eq. (5.3.5) and the result from Eq. (5.3.8) into Eq. (4.1.10), we have

$$
\underline{d}_1 = \begin{Bmatrix} \phi_1 \\ \phi_2 \end{Bmatrix} = \begin{bmatrix} \dfrac{2}{7L^2} & -\dfrac{1}{14L^2} \\[2ex] -\dfrac{1}{14L^2} & \dfrac{1}{7L^2} \end{bmatrix} \begin{Bmatrix} 6L \\ 6L \end{Bmatrix} \left[-\frac{7PL^3}{12EI} \right] \tag{5.3.9}
$$

Equation (5.3.9) can be simplified to

$$\phi_1 = \frac{3PL^2}{4EI} \qquad \phi_2 = \frac{PL^2}{4EI} \qquad (5.3.10)$$

where the positive signs indicate counterclockwise rotations at nodes 1 and 2.

We will now determine the global nodal forces. To do this, we substitute the known global nodal displacements and rotations, Eqs. (5.3.9) and (5.3.10), into Eq. (5.3.2). The resulting equations are

$$\begin{Bmatrix} F_{1y} \\ M_1 \\ F_{2y} \\ M_2 \\ F_{3y} \\ M_3 \end{Bmatrix} = \frac{EI}{L^3} \begin{bmatrix} 12 & 6L & -12 & 6L & 0 & 0 \\ 6L & 4L^2 & -6L & 2L^2 & 0 & 0 \\ -12 & -6L & 24 & 0 & -12 & 6L \\ 6L & 2L^2 & 0 & 8L^2 & -6L & 2L^2 \\ 0 & 0 & -12 & -6L & 12 & -6L \\ 0 & 0 & 6L & 2L^2 & -6L & 4L^2 \end{bmatrix} \begin{Bmatrix} -\dfrac{7PL^3}{12EI} \\[2mm] \dfrac{3PL^2}{4EI} \\[2mm] 0 \\[2mm] \dfrac{PL^2}{4EI} \\[2mm] 0 \\[2mm] 0 \end{Bmatrix}$$

$$(5.3.11)$$

Multiplying the matrices on the right-hand side of Eq. (5.3.11), we obtain the global nodal forces and moments as

$$F_{1y} = -P \qquad M_1 = 0 \qquad F_{2y} = \tfrac{5}{2}P$$

$$M_2 = 0 \qquad F_{3y} = -\tfrac{3}{2}P \qquad M_3 = \tfrac{1}{2}PL \qquad (5.3.12)$$

The results of Eqs. (5.3.12) can be interpreted as follows: The value of $F_{1y} = -P$ is the applied force at node 1, as it must be. The values of F_{2y}, F_{3y}, and M_3 are the reactions from the supports as felt by the beam. The moments M_1 and M_2 are zero because no applied or reactive moments are present on the beam at node 1 or node 2.

It is generally necessary to determine the local nodal forces associated with each element of a large structure to perform a stress analysis of the entire structure. We will thus consider the forces in element 1 of this example to illustrate this concept (element 2 can be treated similarly). Using Eqs. (5.3.8) and (5.3.10) in the $\hat{f} = \underline{k}\hat{d}$ equation for element 1 [also see Eq. (5.1.13)], we have

$$\begin{Bmatrix} \hat{f}_{1y} \\ \hat{m}_1 \\ \hat{f}_{2y} \\ \hat{m}_2 \end{Bmatrix} = \frac{EI}{L^3} \begin{bmatrix} 12 & 6L & -12 & 6L \\ 6L & 4L^2 & -6L & 2L^2 \\ -12 & -6L & 12 & -6L \\ 6L & 2L^2 & -6L & 4L^2 \end{bmatrix} \begin{Bmatrix} -\dfrac{7PL^3}{12EI} \\[2mm] \dfrac{3PL^2}{4EI} \\[2mm] 0 \\[2mm] \dfrac{PL^2}{4EI} \end{Bmatrix} \qquad (5.3.13)$$

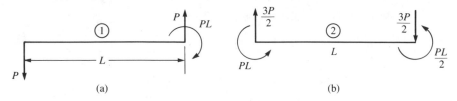

FIGURE 5-8   Free-body diagrams showing forces and moments on (a)
element 1 and (b) element 2

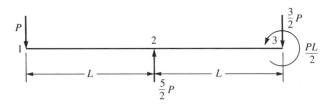

FIGURE 5-9   Nodal forces and moment on the beam

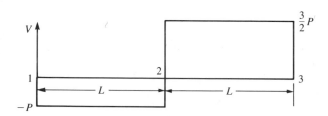

FIGURE 5-10   Shear force diagram for the beam of Figure 5-9

where, again, since the local coordinate axes of the element coincide with the
global axes of the whole beam, we have used the relationships $\underline{d} = \hat{\underline{d}}$ and $\underline{k} = \hat{\underline{k}}$
(that is, the local nodal displacements are also the global nodal displacements,
and so forth). Equation (5.3.13) yields

$$\hat{f}_{1y} = -P \qquad \hat{m}_1 = 0 \qquad \hat{f}_{2y} = P \qquad \hat{m}_2 = -PL \qquad (5.3.14)$$

A free-body diagram of element 1, shown in Figure 5-8(a), should help you to
understand the results of Eqs. (5.3.14). The figure shows a nodal transverse
force of negative P at node 1 and of positive P and negative moment PL at
node 2. These values are consistent with the results given by Eqs. (5.3.14). For
completeness, the free-body diagram of element 2 is shown in Figure 5-8(b).
We can easily verify the element nodal forces by writing an equation similar
to Eq. (5.3.13). From the results of Eqs. (5.3.12), the nodal forces and moments
for the whole beam are shown on the beam in Figure 5-9. Using the beam sign
conventions established in Section 5.1, we obtain the shear force V and
bending moment M diagrams as shown in Figures 5-10 and 5-11.   ∎

In general, for complex beam structures, we will use the element local
forces to determine the shear force and bending moment diagrams for each
element. We can then use these values for design purposes. Chapter 6 will
further discuss this concept as used in computer codes.

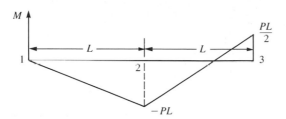

FIGURE 5-11   Bending moment diagram for the beam of Figure 5-9

EXAMPLE 5.2

Determine the nodal displacements and rotations, global nodal forces, and element forces for the beam shown in Figure 5-12. We have discretized the beam as indicated by the node numbering. The beam is fixed at nodes 1 and 5 and has a roller support at node 3. Vertical loads of 10,000 lb each are applied at nodes 2 and 4. Let $E = 30 \times 10^6$ psi and $I = 500$ in.[4] throughout the beam.

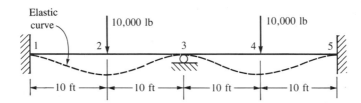

FIGURE 5-12   Beam example

We must have consistent units; therefore, the 10-ft lengths in Figure 5-12 will be converted to 120 in. during the solution. Using Eq. (5.1.14), along with superposition of the four beam element stiffness matrices, we obtain the global stiffness matrix and the global equations as given by Eq. (5.3.15) on p. 164. Here the lengths of each element are the same. Thus, we can factor an L out of the superimposed stiffness matrix.

For a longhand solution, we reduce Eq. (5.3.15) in the usual manner by application of the boundary conditions

$$d_{1y} = \phi_1 = d_{3y} = d_{5y} = \phi_5 = 0$$

The resulting equation is

$$\begin{Bmatrix} -10{,}000 \\ 0 \\ 0 \\ -10{,}000 \\ 0 \end{Bmatrix} = \frac{EI}{L^3} \begin{bmatrix} 24 & 0 & 6L & 0 & 0 \\ 0 & 8L^2 & 2L^2 & 0 & 0 \\ 6L & 2L^2 & 8L^2 & -6L & 2L^2 \\ 0 & 0 & -6L^2 & 24 & 0 \\ 0 & 0 & 2L^2 & 0 & 8L^2 \end{bmatrix} \begin{Bmatrix} d_{2y} \\ \phi_2 \\ \phi_3 \\ d_{4y} \\ \phi_4 \end{Bmatrix} \qquad (5.3.16)$$

The rotations (slopes) at nodes 2, 3, and 4 are equal to zero because of symmetry in loading, geometry, and material properties about a plane per-

$$
\begin{Bmatrix} F_{1y} \\ M_1 \\ F_{2y} \\ M_2 \\ F_{3y} \\ M_3 \\ F_{4y} \\ M_4 \\ F_{5y} \\ M_5 \end{Bmatrix}
= \frac{EI}{L^3}
\begin{bmatrix}
12 & 6L & -12 & 6L & 0 & 0 & 0 & 0 & 0 & 0 \\
6L & 4L^2 & -6L & 2L^2 & 0 & 0 & 0 & 0 & 0 & 0 \\
-12 & -6L & 12+12 & -6L+6L & -12 & 6L & 0 & 0 & 0 & 0 \\
6L & 2L^2 & -6L+6L & 4L^2+4L^2 & -6L & 2L^2 & 0 & 0 & 0 & 0 \\
0 & 0 & -12 & -6L & 12+12 & -6L+6L & -12 & 6L & 0 & 0 \\
0 & 0 & 6L & 2L^2 & -6L+6L & 4L^2+4L^2 & -6L & 2L^2 & 0 & 0 \\
0 & 0 & 0 & 0 & -12 & -6L & 12+12 & -6L+6L & -12 & 6L \\
0 & 0 & 0 & 0 & 6L & 2L^2 & -6L+6L & 4L^2+4L^2 & -6L & 2L^2 \\
0 & 0 & 0 & 0 & 0 & 0 & -12 & -6L & 12 & -6L \\
0 & 0 & 0 & 0 & 0 & 0 & 6L & 2L^2 & -6L & 4L^2
\end{bmatrix}
\begin{Bmatrix} d_{1y} \\ \phi_1 \\ d_{2y} \\ \phi_2 \\ d_{3y} \\ \phi_3 \\ d_{4y} \\ \phi_4 \\ d_{5y} \\ \phi_5 \end{Bmatrix}
$$

$$(5.3.15)$$

pendicular to the beam length and passing through node 3. Therefore, $\phi_2 = \phi_3 = \phi_4 = 0$ and we can further reduce Eq. (5.3.16) to

$$
\begin{Bmatrix} -10,000 \\ -10,000 \end{Bmatrix} = \frac{EI}{L^3} \begin{bmatrix} 24 & 0 \\ 0 & 24 \end{bmatrix} \begin{Bmatrix} d_{2y} \\ d_{4y} \end{Bmatrix} \tag{5.3.17}
$$

Solving for the displacements using $L = 120$ in., $E = 30 \times 10^6$ psi, and $I = 500$ in.^{4} in Eq. (5.3.17), we obtain

$$
d_{2y} = d_{4y} = -0.048 \text{ in.} \tag{5.3.18}
$$

as expected because of symmetry.

As observed from the solution of this problem, the greater the static redundancy (degrees of static indeterminacy or number of unknown forces and moments that cannot be determined by equations of statics), the smaller the kinematic redundancy (unknown nodal degrees of freedom, such as displacements or slopes)—hence, the fewer the number of unknown degrees of freedom to be solved for. Moreover, the use of symmetry, when applicable, reduces the number of unknown degrees of freedom even further. We can now back-substitute the results from Eq. (5.3.18), along with the numerical values for E, I, and L, into Eq. (5.3.15) to determine the global nodal forces as

$$F_{1y} = 5000 \text{ lb} \qquad M_1 = 25,000 \text{ lb-ft}$$

$$F_{2y} = 10,000 \text{ lb} \qquad M_2 = 0$$

$$F_{3y} = 10,000 \text{ lb} \qquad M_3 = 0 \tag{5.3.19}$$

$$F_{4y} = 10,000 \text{ lb} \qquad M_4 = 0$$

$$F_{5y} = 5000 \text{ lb} \qquad M_5 = -25,000 \text{ lb-ft}$$

Once again, the global nodal forces (and moments) at the support nodes (nodes 1, 3, and 5) can be interpreted as the reaction forces, and the global nodal forces at nodes 2 and 4 are the applied nodal forces.

However, for large structures we must obtain the local element shear force and bending moment at each node end of the element because these values are used in the design/analysis process. We will again illustrate this concept for the element connecting nodes 1 and 2 in Figure 5-12. Using the local equations for this element, for which all nodal displacements have now been determined, we obtain

$$
\begin{Bmatrix} \hat{f}_{1y} \\ \hat{m}_1 \\ \hat{f}_{2y} \\ \hat{m}_2 \end{Bmatrix} = \frac{EI}{L^3} \begin{bmatrix} 12 & 6L & -12 & 6L \\ 6L & 4L^2 & -6L & 2L^2 \\ -12 & -6L & 12 & -6L \\ 6L & 2L^2 & -6L & 4L^2 \end{bmatrix} \begin{Bmatrix} \hat{d}_{1y} = 0 \\ \hat{\phi}_1 = 0 \\ \hat{d}_{2y} = -0.048 \\ \hat{\phi}_2 = 0 \end{Bmatrix} \tag{5.3.20}
$$

Simplifying Eq. (5.3.20), we have

$$
\begin{Bmatrix} \hat{f}_{1y} \\ \hat{m}_1 \\ \hat{f}_{2y} \\ \hat{m}_2 \end{Bmatrix} = \begin{Bmatrix} 5000 \text{ lb} \\ 25,000 \text{ lb-ft} \\ -5000 \text{ lb} \\ 25,000 \text{ lb-ft} \end{Bmatrix} \tag{5.3.21}
$$

If you wish, you can draw a free-body diagram to confirm the equilibrium of the element. ∎

Finally, you should note that due to symmetry about a vertical plane passing through node 3, we could have initially considered one-half of this beam and used the following model.

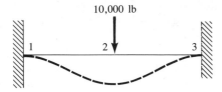

EXAMPLE 5.3

Determine the nodal displacements and rotations and the global and element forces for the beam shown in Figure 5-13. We have discretized the beam as shown by the node numbering. The beam is fixed at node 1, has a roller support at node 2, and has an elastic spring support at node 3. A downward vertical force of $P = 50$ kN is applied at node 3. Let $E = 210$ GPa and $I = 2 \times 10^{-4}$ m^4 throughout the beam, and let $k = 200$ kN/m.

Using Eq. (5.1.14) for each beam element and Eq. (2.2.18) for the spring element as well as the direct stiffness method, we obtain the structure stiffness matrix as

$$
\underline{K} = \frac{EI}{L^3}
\begin{array}{c}
\begin{array}{ccccccc}
d_{1y} & \phi_1 & d_{2y} & \phi_2 & d_{3y} & \phi_3 & d_{4y}
\end{array} \\
\begin{bmatrix}
12 & 6L & -12 & 6L & 0 & 0 & 0 \\
 & 4L^2 & -6L & 2L^2 & 0 & 0 & 0 \\
 & & 24 & 0 & -12 & 6L & 0 \\
 & & & 8L^2 & -6L & 2L^2 & 0 \\
 & & & & 12 + \dfrac{kL^3}{EI} & -6L & -\dfrac{kL^3}{EI} \\
 & & & & & 4L^2 & 0 \\
 & \text{Symmetry} & & & & & \dfrac{KL^3}{EI}
\end{bmatrix}
\end{array}
\qquad (5.3.22a)
$$

where the spring stiffness matrix $\underline{k}_s$ given below by Eq. (5.3.22(b)) has been directly added into the global stiffness matrix corresponding to its degrees of freedom at nodes 3 and 4.

$$
\underline{k}_s =
\begin{array}{c}
\begin{array}{cc} d_{3y} & d_{4y} \end{array} \\
\begin{bmatrix} k & -k \\ -k & k \end{bmatrix}
\end{array}
\qquad (5.3.22b)
$$

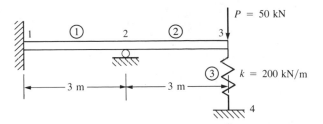

FIGURE 5-13   Beam example

It is easier to solve the problem using the general variables, later making numerical substitutions into the final displacement expressions. The governing equations for the beam are then given by

$$
\begin{Bmatrix} F_{1y} \\ M_1 \\ F_{2y} \\ M_2 \\ F_{3y} \\ M_3 \\ F_{4y} \end{Bmatrix} = \frac{EI}{L^3}
\begin{bmatrix}
12 & 6L & -12 & 6L & 0 & 0 & 0 \\
 & 4L^2 & -6L & 2L^2 & 0 & 0 & 0 \\
 & & 24 & 0 & -12 & 6L & 0 \\
 & & & 8L^2 & -6L & 2L^2 & 0 \\
 & & & & 12+k' & -6L & -k' \\
 & & & & & 4L^2 & 0 \\
 & \text{Symmetry} & & & & & k'
\end{bmatrix}
\begin{Bmatrix} d_{1y} \\ \phi_1 \\ d_{2y} \\ \phi_2 \\ d_{3y} \\ \phi_3 \\ d_{4y} \end{Bmatrix}
$$

$$(5.3.23)$$

where $k' = kL^3/(EI)$ is used to simplify the notation. We now apply the boundary conditions

$$d_{1y} = 0 \qquad \phi_1 = 0 \qquad d_{2y} = 0 \qquad d_{4y} = 0 \qquad (5.3.24)$$

We delete the first three equations and the seventh equation of Eq. (5.3.23), rearrange the remaining three equations, and then partition these equations to again reduce the matrix inversion associated with determining the unknown displacements. This partitioned set of equations is given by

$$
\begin{Bmatrix} 0 \\ 0 \\ -P \end{Bmatrix} = \frac{EI}{L^3}
\begin{bmatrix}
8L^2 & 2L^2 & -6L \\
2L^2 & 4L^2 & -6L \\
-6L & -6L & 12+k'
\end{bmatrix}
\begin{Bmatrix} \phi_2 \\ \phi_3 \\ d_{3y} \end{Bmatrix}
\qquad (5.3.25)
$$

Using Eq. (4.1.13) for $\underline{k}_c$ we obtain

$$
\underline{k}_c = \frac{EI}{L^3} \left[[12+k'] - [-6L \quad -6L]
\begin{bmatrix} 8L^2 & 2L^2 \\ 2L^2 & 4L^2 \end{bmatrix}^{-1}
\begin{Bmatrix} -6L \\ -6L \end{Bmatrix} \right]
\qquad (5.3.26)
$$

Performing the matrix inversion and multiplications in Eq. (5.3.26), we obtain

$$
\underline{k}_c = \frac{EI}{L^3} \left[\frac{12}{7} + k' \right]
\qquad (5.3.27)
$$

Using Eq. (4.1.15), we obtain the displacement at node 3 as

$$d_{3y} = \frac{-7PL^3}{EI}\left(\frac{1}{12 + 7k'}\right) \qquad (5.3.28)$$

Substituting appropriate partitioned parts of $\underline{K}$ from Eq. (5.3.25) and the result from Eq. (5.3.28) into Eq. (4.1.10), we have

$$\begin{Bmatrix} \phi_2 \\ \phi_3 \end{Bmatrix} = -\frac{\begin{bmatrix} 2 & -1 \\ -1 & 4 \end{bmatrix}}{14L^2}\begin{Bmatrix} -6L \\ -6L \end{Bmatrix}\left[\frac{-7PL^3}{EI}\left(\frac{1}{12 + 7k'}\right)\right] \qquad (5.3.29)$$

Equation (5.3.29) is simplified to

$$\phi_2 = -\frac{3PL^2}{EI}\left(\frac{1}{12 + 7k'}\right) \qquad \phi_3 = -\frac{9PL^2}{EI}\left(\frac{1}{12 + 7k'}\right) \qquad (5.3.30)$$

The influence of the spring stiffness on the displacements is easily seen in Eqs. (5.3.28) and (5.3.30). Solving for the numerical displacements using $P - 50$ kN, $L = 3$ m, $E = 210$ GPa $(= 210 \times 10^6$ kN/m$^2)$, $I = 2 \times 10^{-4}$ m^4, and $k' = 0.129$ in Eq. (5.3.28), we obtain

$$d_{3y} = \frac{-7(50 \text{ kN})(3 \text{ m})^3}{(210 \times 10^6 \text{ kN/m}^2)(2 \times 10^{-4} \text{ m}^4)}\left(\frac{1}{12 + 7(0.129)}\right) = -0.0174 \text{ m}$$

$$(5.3.31)$$

Similar substitutions into Eqs. (5.3.30) yield

$$\phi_2 = -0.00249 \text{ rad} \qquad \phi_3 = -0.00747 \text{ rad} \qquad (5.3.32)$$

We now back-substitute the results from Eqs. (5.3.31) and (5.3.32), along with numerical values for P, E, I, L, and k', into Eq. (5.3.23) to obtain the global nodal forces as

$$F_{1y} = -69.9 \text{ kN} \qquad M_1 = -69.7 \text{ kN·m}$$

$$F_{2y} = 116.4 \text{ kN} \qquad M_2 = 0.0 \text{ kN·m} \qquad (5.3.33a)$$

$$F_{3y} = -50.0 \text{ kN} \qquad M_3 = 0.0 \text{ kN·m}$$

For the beam-spring structure, an additional global force F_{4y} is determined at the base of the spring as follows:

$$F_{4y} = -d_{3y}k = (0.0174)200 = 3.5 \text{ kN} \qquad (5.3.33b)$$

This force provides the additional global y force for equilibrium of the structure.

A free-body diagram, including the forces and moments from Eqs. (5.3.33a) and (5.3.33b) acting on the beam, is shown here.

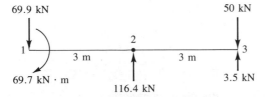

FIGURE 5-14   Free-body diagram of beam of Figure 5-13   ∎

5.4 Distributed Loading

Beam members can support distributed loading as well as concentrated nodal loading. Therefore, we must be able to account for distributed loading. Consider the fixed-fixed beam subjected to a uniformly distributed loading w shown in Figure 5-15. The reactions, determined from structural analysis theory (see Reference [2]), are shown in Figure 5-16. These reactions are called *fixed-end reactions*. In general, **fixed-end reactions** are those reactions at the ends of an element if the ends of the element are assumed to be fixed—that is, if displacements and rotations are prevented. (For those of you who are unfamiliar with the analysis of indeterminate structures, assume these reactions as given and proceed with the rest of the discussion, as we will develop these results in a subsequent presentation of the work equivalence method.) Therefore, guided by the results from structural analysis for the case of a uniformly distributed load, we replace the load by concentrated nodal forces and moments tending to have the same effect on the beam as the actual distributed load. Figure 5-17 illustrates this idea for a beam. We have replaced the uniformly distributed load by an equivalent force system consisting of a concentrated nodal force and moment at each end of the member carrying the distributed load. These equivalent forces are always of opposite

FIGURE 5-15   Fixed-fixed beam subjected to a uniformly distributed load

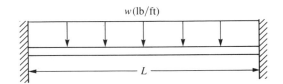

FIGURE 5-16   Fixed-end reactions for the beam of Figure 5-15

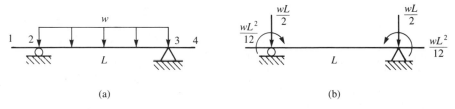

(a)   (b)

FIGURE 5-17   (a) Beam with a distributed load and (b) the equivalent nodal force system

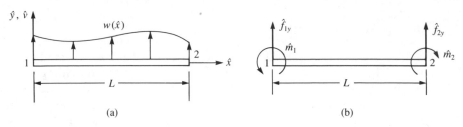

FIGURE 5-18   (a) Beam element subjected to a general load and (b) the equivalent nodal force system

sign from the fixed-end forces. If we want to analyze the behavior of loaded member 2–3 in better detail, we can place a node at midspan and use the same procedure just described for each of the two elements representing the horizontal member. That is, to determine the maximum deflection and maximum moment in the beam span, a node is needed at midspan of beam segment 2–3.

Work Equivalence Method

We can use the work equivalence method to replace a distributed load by a set of discrete loads. This method is based on the concept that the work of the distributed load is equal to that of the discrete load replacement for arbitrary nodal displacements. To illustrate the method, we consider the example shown in Figure 5-18. The work due to the distributed load is given by

$$W_{\text{distributed}} = \int_0^L w(\hat{x})\hat{v}(\hat{x})\, d\hat{x} \tag{5.4.1}$$

where $\hat{v}(\hat{x})$ is the transverse displacement given by Eq. (5.1.4). The work due to the discrete nodal forces is given by

$$W_{\text{discrete}} = \hat{m}_1\hat{\phi}_1 + \hat{m}_2\hat{\phi}_2 + \hat{f}_{1y}\hat{d}_{1y} + \hat{f}_{2y}\hat{d}_{2y} \tag{5.4.2}$$

We can then determine the nodal moments and forces $\hat{m}_1$, $\hat{m}_2$, $\hat{f}_{1y}$, and $\hat{f}_{2y}$ used to replace the distributed load by using the concept of work equivalence—that is, by setting $W_{\text{distributed}} = W_{\text{discrete}}$ for arbitrary displacements $\hat{\phi}_1$, $\hat{\phi}_2$, $\hat{d}_{1y}$, and $\hat{d}_{2y}$.

Example of Load Replacement

To illustrate more clearly the concept of work equivalence, we will now consider a beam subjected to a specified distributed load. Consider the uniformly loaded beam shown in Figure 5-19. The support conditions are not shown because they are not relevant to the replacement scheme. By letting $W_{\text{discrete}} = W_{\text{distributed}}$ and by assuming arbitrary $\hat{\phi}_1$, $\hat{\phi}_2$, $\hat{d}_{1y}$, and $\hat{d}_{2y}$, we will find equivalent nodal forces $\hat{m}_1$, $\hat{m}_2$, $\hat{f}_{1y}$, and $\hat{f}_{2y}$.

Using Eqs. (5.4.1) and (5.4.2) for $W_{\text{distributed}} = W_{\text{discrete}}$, we have

$$\int_0^L w(\hat{x})\hat{v}(\hat{x})\, d\hat{x} = \hat{m}_1\hat{\phi}_1 + \hat{m}_2\hat{\phi}_2 + \hat{f}_{1y}\hat{d}_{1y} + \hat{f}_{2y}\hat{d}_{2y} \tag{5.4.3}$$

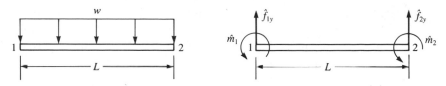

F I G U R E 5-19.   (a) Beam subjected to a uniformly distributed loading
and (b) the equivalent nodal forces to be determined

Evaluating the left-hand side of Eq. (5.4.3) by substituting $w(\hat{x}) = -w$ and $\hat{v}(\hat{x})$ from Eq. (5.1.4), we obtain the work due to the distributed load as

$$\int_0^L w(\hat{x})\hat{v}(\hat{x})\, d\hat{x} = -\frac{Lw}{2}(\hat{d}_{1y} - \hat{d}_{2y}) - \frac{L^2 w}{4}(\hat{\phi}_1 + \hat{\phi}_2) - Lw(\hat{d}_{2y} - \hat{d}_{1y})$$

$$+ \frac{L^2 w}{3}(2\hat{\phi}_1 + \hat{\phi}_2) - \hat{\phi}_1\left(\frac{L^2 w}{2}\right) - \hat{d}_{1y}(wL) \qquad (5.4.4)$$

Now using Eqs. (5.4.3) and (5.4.4) for arbitrary nodal displacements, we let $\hat{\phi}_1 = 1$, $\hat{\phi}_2 = 0$, $\hat{d}_{1y} = 0$, and $\hat{d}_{2y} = 0$ and then obtain

$$\hat{m}_1(1) = -\left(\frac{L^2 w}{4} - \frac{2}{3}L^2 w + \frac{L^2}{2}w\right) = -\frac{wL^2}{12} \qquad (5.4.5)$$

Similarly, letting $\hat{\phi}_1 = 0$, $\hat{\phi}_2 = 1$, $\hat{d}_{1y} = 0$, and $\hat{d}_{2y} = 0$ yields

$$\hat{m}_2(1) = -\left(\frac{L^2 w}{4} - \frac{L^2 w}{3}\right) = \frac{wL^2}{12} \qquad (5.4.6)$$

Finally, letting all nodal displacements equal zero except first $\hat{d}_{1y}$ and then $\hat{d}_{2y}$, we obtain

$$\hat{f}_{1y}(1) = -\frac{Lw}{2} + Lw - Lw = -\frac{Lw}{2}$$

$$\qquad (5.4.7)$$

$$\hat{f}_{2y}(1) = \frac{Lw}{2} - Lw = -\frac{Lw}{2}$$

We can conclude that, in general, for any given load function $w(\hat{x})$, we can multiply by $\hat{v}(\hat{x})$ and then integrate according to Eq. (5.4.3) to obtain the concentrated nodal forces (and/or moments) used to replace the distributed load. Moreover, we can obtain the load replacement by using the concept of fixed-end reactions from structural analysis theory. Tables of fixed-end reactions have been generated for numerous load cases and can be found in texts on structural analysis such as Reference [2]. A table of equivalent nodal forces has been generated in Appendix D of this text, guided by the fact that fixed-end reaction forces are of opposite sign from those obtained by the work equivalence method.

Hence, if a concentrated load is applied other than at the natural intersection of two elements, we can use the concept of equivalent nodal forces to replace the concentrated load by nodal concentrated values acting at the beam ends, instead of creating a node on the beam at the location where the load is applied. We provide examples of this procedure for handling concentrated loads on elements in beam Example 5.5 and in plane frame Example 6.3.

General Formulation

In general, we can account for distributed loads or concentrated loads acting on beam elements by starting with the following formulation application for a general structure

$$\underline{F} = \underline{K}\underline{d} - \underline{F}_o \qquad (5.4.8)$$

where $\underline{F}_o$ are called the *equivalent nodal forces*, now expressed in terms of global-coordinate components, which are of such magnitude that they yield the same displacements as would the distributed load. Using the table in Appendix D of equivalent nodal forces $\hat{f}_o$ expressed in terms of local-coordinate components, we can express $\underline{F}_o$ in terms of global-coordinate components.

Recall that $\underline{F}$ represents the global nodal concentrated forces, including the reactions. Since we now assume concentrated nodal forces are initially not present ($\underline{F} = 0$), we can rewrite Eq. (5.4.8) as

$$\underline{F}_o = \underline{K}\underline{d} \qquad (5.4.9)$$

On solving for $\underline{d}$ in Eq. (5.4.9) and then substituting the global displacements $\underline{d}$ and equivalent nodal forces $\underline{F}_o$ into Eq. (5.4.8), we obtain the actual global nodal forces $\underline{F}$. For example, using the definition of $\hat{f}_o$ and Eqs. (5.4.5)–(5.4.7) (or using load case 4 in Appendix D) for a uniformly distributed load w acting over a one-element beam, we have

$$\underline{F}_o = \left\{ \begin{array}{c} \dfrac{-wL}{2} \\[2mm] \dfrac{-wL^2}{12} \\[2mm] \dfrac{-wL}{2} \\[2mm] \dfrac{wL^2}{12} \end{array} \right\} \qquad (5.4.10)$$

This concept can be applied on a local basis to obtain the local nodal forces $\hat{f}$ in individual elements of structures by applying Eq. (5.4.8) locally as

$$\hat{f} = \hat{\underline{k}}\hat{\underline{d}} - \hat{f}_o \qquad (5.4.11)$$

where $\hat{f}_o$ are the equivalent local nodal forces.

Examples 5.4, 5.5, and 5.6 illustrate the method of equivalent nodal forces for solving beams subjected to distributed and concentrated loadings. We will use global-coordinate notation in Examples 5.4, 5.5, and 5.6—treating the beam as a general structure rather than as an element.

EXAMPLE 5.4

For the cantilever beam subjected to the uniform load w in Figure 5-20, solve for the right-end vertical displacement and rotation and then for the nodal forces. Assume the beam to have constant EI throughout its length.

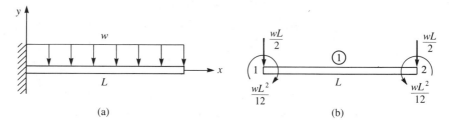

(a)   (b)

FIGURE 5-20   (a) Cantilever beam subjected to a uniformly distributed load and (b) the equivalent nodal force system

We begin by discretizing the beam. Here only one element will be used to represent the whole beam. Next, the distributed load is replaced by its work equivalent nodal forces as shown in Figure 5-20(b). These work equivalent nodal forces are those that result from the uniformly distributed load acting over the whole beam element. (Or see appropriate load case 4 in Appendix D). Using the beam element stiffness Eq. (5.1.13), we obtain

$$\frac{EI}{L^3}\begin{bmatrix} 12 & -6L \\ -6L & 4L^2 \end{bmatrix}\begin{Bmatrix} d_{2y} \\ \phi_2 \end{Bmatrix} = \begin{Bmatrix} \dfrac{-wL}{2} \\[2mm] \dfrac{wL^2}{12} \end{Bmatrix} \qquad (5.4.12)$$

where we have applied the nodal forces from Figure 5-20(b) and the boundary conditions $d_{1y} = 0$ and $\phi_1 = 0$ to reduce the number of matrix equations for the normal longhand solution. Solving Eq. (5.4.12) for the displacements, we obtain

$$\begin{Bmatrix} d_{2y} \\ \phi_2 \end{Bmatrix} = \frac{L}{6EI}\begin{bmatrix} 2L^2 & 3L \\ 3L & 6 \end{bmatrix}\begin{Bmatrix} \dfrac{-wL}{2} \\[2mm] \dfrac{wL^2}{12} \end{Bmatrix} \qquad (5.4.13)$$

Simplifying Eq. (5.4.13), we obtain the displacement and rotation as

$$\begin{Bmatrix} d_{2y} \\ \phi_2 \end{Bmatrix} = \begin{Bmatrix} \dfrac{-wL^4}{8EI} \\[2mm] \dfrac{-wL^3}{6EI} \end{Bmatrix} \qquad (5.4.14)$$

In this case, the method of replacing the distributed load by discrete concentrated loads gives exact solutions for the displacement and rotation.

We will now illustrate the procedure for obtaining the global nodal forces. For convenience, we first define the product $\underline{K}\underline{d}$ to be $\underline{F}^{(e)}$, where $\underline{F}^{(e)}$ are called the *effective global nodal forces*. On using Eq. (5.4.14) for $\underline{d}$, we then have

$$
\left\{
\begin{array}{c}
F_{1y}^{(e)} \\
M_1^{(e)} \\
F_{2y}^{(e)} \\
M_2^{(e)}
\end{array}
\right\}
=
\frac{EI}{L^3}
\left[
\begin{array}{cccc}
12 & 6L & -12 & 6L \\
6L & 4L^2 & -6L & 2L^2 \\
-12 & -6L & 12 & -6L \\
6L & 2L^2 & -6L & 4L^2
\end{array}
\right]
\left\{
\begin{array}{c}
0 \\
0 \\
\dfrac{-wL^4}{8EI} \\
\dfrac{-wL^3}{6EI}
\end{array}
\right\}
\qquad (5.4.15)
$$

Simplifying Eq. (5.4.15), we obtain

$$
\left\{
\begin{array}{c}
F_{1y}^{(e)} \\
M_1^{(e)} \\
F_{2y}^{(e)} \\
M_2^{(e)}
\end{array}
\right\}
=
\left\{
\begin{array}{c}
\dfrac{wL}{2} \\
\dfrac{5wL^2}{12} \\
\dfrac{-wL}{2} \\
\dfrac{wL^2}{12}
\end{array}
\right\}
\qquad (5.4.16)
$$

We then use Eqs. (5.4.10) and (5.4.16) in Eq. (5.4.8) to obtain the correct global nodal forces as

$$
\left\{
\begin{array}{c}
F_{1y} \\
M_1 \\
F_{2y} \\
M_2
\end{array}
\right\}
=
\left\{
\begin{array}{c}
\dfrac{wL}{2} \\
\dfrac{5wL^2}{12} \\
\dfrac{-wL}{2} \\
\dfrac{wL^2}{12}
\end{array}
\right\}
-
\left\{
\begin{array}{c}
\dfrac{-wL}{2} \\
\dfrac{-wL^2}{12} \\
\dfrac{-wL}{2} \\
\dfrac{wL^2}{12}
\end{array}
\right\}
=
\left\{
\begin{array}{c}
wL \\
\dfrac{wL^2}{2} \\
0 \\
0
\end{array}
\right\}
\qquad (5.4.17)
$$

In Eq. (5.4.17), F_{1y} is the vertical force reaction and M_1 is the moment reaction as applied by the clamped support at node 1. The results for displacement given by Eq. (5.4.14) and the global nodal forces given by Eq. (5.4.17) are sufficient to complete the solution of the cantilever beam problem. ■

We will solve the following example to illustrate the procedure for handling concentrated loads acting on beam elements at locations other than nodes.

EXAMPLE 5.5

For the cantilever beam subjected to the concentrated load P in Figure 5-21, solve for the right-end vertical displacement and rotation and the nodal forces, including reactions, by replacing the concentrated load with equivalent nodal forces acting at each end of the beam. Assume EI constant throughout the beam.

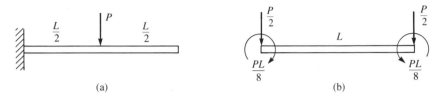

(a)   (b)

FIGURE 5-21   (a) Cantilever beam subjected to a concentrated load and (b) the equivalent nodal force replacement system

We begin by discretizing the beam. Here only one element is used with nodes at each end of the beam. We then replace the concentrated load as shown in Figure 5-21(b) by using appropriate loading case 1 in Appendix D. Using the beam element stiffness Eq. (5.1.14), we obtain

$$\frac{EI}{L^3}\begin{bmatrix} 12 & -6L \\ -6L & 4L^2 \end{bmatrix}\begin{Bmatrix} d_{2y} \\ \phi_2 \end{Bmatrix} = \begin{Bmatrix} \dfrac{-P}{2} \\ \dfrac{PL}{8} \end{Bmatrix} \qquad (5.4.18)$$

where we have applied the nodal forces from Figure 5-21(b) and the boundary conditions $d_{1y} = 0$ and $\phi_1 = 0$ to reduce the number of matrix equations for the usual longhand solution. Solving Eq. (5.4.18) for the displacements, we obtain

$$\begin{Bmatrix} d_{2y} \\ \phi_2 \end{Bmatrix} = \frac{L}{6EI}\begin{bmatrix} 2L^2 & 3L \\ 3L & 6 \end{bmatrix}\begin{Bmatrix} \dfrac{-P}{2} \\ \dfrac{PL}{8} \end{Bmatrix} \qquad (5.4.19)$$

Simplifying Eq. (5.4.19), we obtain the displacement and rotation as

$$\begin{Bmatrix} d_{2y} \\ \phi_2 \end{Bmatrix} = \begin{Bmatrix} \dfrac{-5PL^3}{48EI} \\ \dfrac{-PL^2}{8EI} \end{Bmatrix} \qquad (5.4.20)$$

To obtain the nodal forces, we begin by evaluating the effective nodal forces $\underline{F}^{(e)} = \underline{K}\underline{d}$ as

$$
\begin{Bmatrix} F_{1y}^{(e)} \\ M_1^{(e)} \\ F_{2y}^{(e)} \\ M_2^{(e)} \end{Bmatrix} = \frac{EI}{L^3} \begin{bmatrix} 12 & 6L & -12 & 6L \\ 6L & 4L^2 & -6L & 2L^2 \\ -12 & -6L & 12 & -6L \\ 6L & 2L^2 & -6L & 4L^2 \end{bmatrix} \begin{Bmatrix} 0 \\ 0 \\ \dfrac{-5PL^3}{48EI} \\ \dfrac{-PL^2}{8EI} \end{Bmatrix}
\qquad (5.4.21)
$$

Simplifying Eq. (5.4.21), we obtain

$$
\begin{Bmatrix} F_{1y}^{(e)} \\ M_1^{(e)} \\ F_{2y}^{(e)} \\ M_2^{(e)} \end{Bmatrix} = \begin{Bmatrix} \dfrac{P}{2} \\ \dfrac{3PL}{8} \\ -P \\ \dfrac{2}{PL} \\ \dfrac{PL}{8} \end{Bmatrix}
\qquad (5.4.22)
$$

Then using Eq. (5.4.22) and the equivalent nodal forces from Figure 5-21(b) in Eq. (5.4.8), we obtain the correct nodal forces as

$$
\begin{Bmatrix} F_{1y} \\ M_1 \\ F_{2y} \\ M_2 \end{Bmatrix} = \begin{Bmatrix} \dfrac{P}{2} \\ \dfrac{3PL}{8} \\ \dfrac{-P}{2} \\ \dfrac{PL}{8} \end{Bmatrix} - \begin{Bmatrix} \dfrac{-P}{2} \\ \dfrac{-PL}{8} \\ \dfrac{-P}{2} \\ \dfrac{PL}{8} \end{Bmatrix} = \begin{Bmatrix} P \\ \dfrac{PL}{2} \\ 0 \\ 0 \end{Bmatrix}
\qquad (5.4.23)
$$

We can see from Eq. (5.4.23) that F_{1y} is equivalent to the vertical reaction force and M_1 is the reaction moment as applied by the clamped support at node 1. ∎

Finally, to illustrate the procedure for handling concentrated nodal forces and distributed loads acting simultaneously on beam elements, we will solve the following example.

E X A M P L E   5.6

For the cantilever beam subjected to the concentrated free-end load P and the uniformly distributed load w acting over the whole beam as shown

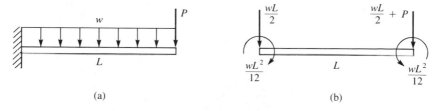

(a) (b)

FIGURE 5-22 (a) Cantilever beam subjected to a concentrated load and a distributed load and (b) the equivalent nodal force replacement system

in Figure 5-22, determine the free-end displacements and the nodal forces.

Once again, the beam is modeled using one element with nodes 1 and 2, and the distributed load is replaced as shown in Figure 5-22(b) using appropriate loading case 4 in Appendix D. Using the beam element stiffness Eq. (5.1.14), we obtain

$$
\frac{EI}{L^3}\begin{bmatrix} 12 & -6L \\ -6L & 4L^2 \end{bmatrix}\begin{Bmatrix} d_{2y} \\ \phi_2 \end{Bmatrix} = \begin{Bmatrix} \dfrac{-wL}{2} - P \\ \dfrac{wL^2}{12} \end{Bmatrix} \tag{5.4.24}
$$

where we have applied the nodal forces from Figure 5-22(b) and the boundary conditions $d_{1y} = 0$ and $\phi_1 = 0$ to reduce the number of matrix equations for the usual longhand solution. Solving Eq. (5.4.24) for the displacements, we obtain

$$
\begin{Bmatrix} d_{2y} \\ \phi_2 \end{Bmatrix} = \begin{Bmatrix} \dfrac{-wL^4}{8EI} - \dfrac{PL^3}{3EI} \\ \dfrac{-wL^3}{6EI} - \dfrac{PL^2}{2EI} \end{Bmatrix} \tag{5.4.25}
$$

Next, we obtain the effective nodal forces as

$$
\begin{Bmatrix} F_{1y}^{(e)} \\ M_1^{(e)} \\ F_{2y}^{(e)} \\ M_2^{(e)} \end{Bmatrix} = \frac{EI}{L^3}\begin{bmatrix} 12 & 6L & -12 & 6L \\ 6L & 4L^2 & -6L & 2L^2 \\ -12 & -6L & 12 & -6L \\ 6L & 2L^2 & -6L & 4L^2 \end{bmatrix}\begin{Bmatrix} 0 \\ 0 \\ \dfrac{-wL^4}{8EI} - \dfrac{PL^3}{3EI} \\ \dfrac{-wL^3}{6EI} - \dfrac{PL^2}{2EI} \end{Bmatrix} \tag{5.4.26}
$$

Simplifying Eq. (5.4.26), we obtain

$$
\begin{Bmatrix} F_{1y}^{(e)} \\ M_1^{(e)} \\ F_{2y}^{(e)} \\ M_2^{(e)} \end{Bmatrix} = \begin{Bmatrix} P + \dfrac{wL}{2} \\[2mm] PL + \dfrac{5wL^2}{12} \\[2mm] -P - \dfrac{wL}{2} \\[2mm] \dfrac{wL^2}{12} \end{Bmatrix} \qquad (5.4.27)
$$

Finally, subtracting the equivalent nodal force matrix [see Figure 5-22(b)] from the effective force matrix of Eq. (5.4.27), we obtain the correct nodal forces as

$$
\begin{Bmatrix} F_{1y} \\ M_1 \\ F_{2y} \\ M_2 \end{Bmatrix} = \begin{Bmatrix} P + \dfrac{wL}{2} \\[2mm] PL + \dfrac{5wL^2}{12} \\[2mm] -P - \dfrac{wL}{2} \\[2mm] \dfrac{wL^2}{12} \end{Bmatrix} - \begin{Bmatrix} \dfrac{-wL}{2} \\[2mm] \dfrac{-wL^2}{12} \\[2mm] \dfrac{-wL}{2} \\[2mm] \dfrac{wL^2}{12} \end{Bmatrix} = \begin{Bmatrix} P + wL \\[2mm] PL + \dfrac{wL^2}{2} \\[2mm] -P \\[2mm] 0 \end{Bmatrix} \qquad (5.4.28)
$$

From Eq. (5.4.28), we see that F_{1y} is equivalent to the vertical reaction force, M_1 is the reaction moment at node 1, and F_{2y} is equal to the applied downward force P at node 2. (Remember that only the equivalent nodal force matrix is subtracted, not the original concentrated load matrix. This is based on the general formulation, Eq. (5.4.8).]   ∎

In general, for any structure in which an equivalent nodal force replacement is made, the actual nodal forces acting on the structure are determined by first evaluating the effective nodal forces $\underline{F}^{(e)}$ for the structure and then subtracting off the equivalent nodal forces $\underline{F}_o$ for the structure, as indicated in Eq. (5.4.8). Similarly, for any element of a structure in which equivalent nodal force replacement is made, the actual local nodal forces acting on the element are determined by first evaluating the effective local nodal forces $\hat{f}^{(e)}$ for the element and then subtracting off the equivalent local nodal forces $\hat{f}_o$ associated only with the element, as indicated in Eq. (5.4.11). We provide other examples of this procedure in plane frame Examples 6.2 and 6.3.

5.5   Beam Element With Nodal Hinge

In some beams an internal hinge may be present. In general, this internal hinge causes a discontinuity in the slope of the deflection curve at the hinge.

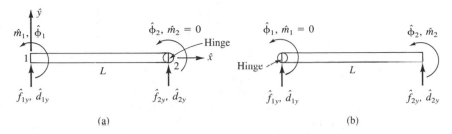

F I G U R E 5-23 Beam element with (a) hinge at right end and (b) hinge at left end

Also, the bending moment is zero at the hinge. We could construct other types of connections that release other generalized end forces; that is, connections can be designed to make the shear force or axial force zero at the connection. These special conditions can be treated by starting with the generalized unreleased beam stiffness matrix [Eq. (5.1.14)] and eliminating the known zero force or moment. This yields a modified stiffness matrix with the desired force or moment equal to zero and the corresponding displacement or slope eliminated.

We now consider the most common cases of a beam element with a nodal hinge at the right end or left end, as shown in Figure 5-23. For the beam element with a hinge at its right end, the moment $\hat{m}_2$ is zero and we partition the $\underline{\hat{k}}$ matrix [Eq. (5.1.14)] to eliminate the degree of freedom $\hat{\phi}_2$ associated with $\hat{m}_2 = 0$ as follows:

$$\underline{\hat{k}} = \frac{EI}{L^3} \left[\begin{array}{ccc:c} 12 & 6L & -12 & 6L \\ 6L & 4L^2 & -6L & 2L^2 \\ -12 & -6L & 12 & -6L \\ \hdashline 6L & 2L^2 & -6L & 4L^2 \end{array} \right] \qquad (5.5.1)$$

We condense out the degree of freedom $\hat{\phi}_2$ associated with $\hat{m}_2 = 0$. Using the concept of solution by partitioning in Section 4.1, the condensed stiffness matrix is now written in a form similar to Eq. (4.1.13) for $\underline{k}_c$ except the zero matrix is now the lower one in this case. The result is

$$\underline{k}_c = [K_{11}] - [K_{12}][K_{22}]^{-1}[K_{21}] \qquad (5.5.2)$$

$$= \frac{EI}{L^3} \begin{bmatrix} 12 & 6L & -12 \\ 6L & 4L^2 & -6L \\ -12 & -6L & 12 \end{bmatrix} - \frac{EI}{L^3} \left\{ \begin{array}{c} 6L \\ 2L^2 \\ -6L \end{array} \right\} \frac{1}{4L^2} [6L \quad 2L^2 \quad -6L]$$

$$= \frac{3EI}{L^3} \begin{bmatrix} 1 & L & -1 \\ L & L^2 & -L \\ -1 & -L & 1 \end{bmatrix} \qquad (5.5.3)$$

and the element equations (force-displacement equations) with the hinge at node 2 are

$$\begin{Bmatrix} \hat{f}_{1y} \\ \hat{m}_1 \\ \hat{f}_{2y} \end{Bmatrix} = \frac{3EI}{L^3} \begin{bmatrix} 1 & L & -1 \\ L & L^2 & -L \\ -1 & -L & 1 \end{bmatrix} \begin{Bmatrix} \hat{d}_{1y} \\ \hat{\phi}_1 \\ \hat{d}_{2y} \end{Bmatrix} \qquad (5.5.4)$$

The generalized rotation $\hat{\phi}_2$ has been eliminated from the equation and will not be calculated using this scheme. However, $\hat{\phi}_2$ is not zero in general. We can expand Eq. (5.5.4) to include $\hat{\phi}_2$ by adding zeros in the fourth row and column of the $\hat{k}$ matrix to maintain $\hat{m}_2 = 0$, as follows:

$$\begin{Bmatrix} \hat{f}_{1y} \\ \hat{m}_1 \\ \hat{f}_{2y} \\ \hat{m}_2 \end{Bmatrix} = \frac{3EI}{L^3} \begin{bmatrix} 1 & L & -1 & 0 \\ L & L^2 & -L & 0 \\ -1 & -L & 1 & 0 \\ 0 & 0 & 0 & 0 \end{bmatrix} \begin{Bmatrix} \hat{d}_{1y} \\ \hat{\phi}_1 \\ \hat{d}_{2y} \\ \hat{\phi}_2 \end{Bmatrix} \qquad (5.5.5)$$

For the beam element with a hinge at its left end, the moment $\hat{m}_1$ is zero, and we partition the $\hat{k}$ matrix [Eq. (5.1.14)] to eliminate the zero moment $\hat{m}_1$ and its corresponding rotation $\hat{\phi}_1$ to obtain

$$\begin{Bmatrix} \hat{f}_{2y} \\ \hat{f}_{2y} \\ \hat{m}_2 \end{Bmatrix} = \frac{3EI}{L^3} \begin{bmatrix} 1 & -1 & L \\ -1 & 1 & -L \\ L & -L & L^2 \end{bmatrix} \begin{Bmatrix} \hat{d}_{1y} \\ \hat{d}_{2y} \\ \hat{\phi}_2 \end{Bmatrix} \qquad (5.5.6)$$

The expanded form of Eq. (5.5.6) including $\hat{\phi}_1$ is

$$\begin{Bmatrix} \hat{f}_{1y} \\ \hat{m}_1 \\ \hat{f}_{2y} \\ \hat{m}_2 \end{Bmatrix} = \frac{3EI}{L^3} \begin{bmatrix} 1 & 0 & -1 & L \\ 0 & 0 & 0 & 0 \\ -1 & 0 & 1 & -L \\ L & 0 & -L & L^2 \end{bmatrix} \begin{Bmatrix} \hat{d}_{1y} \\ \hat{\phi}_1 \\ \hat{d}_{2y} \\ \hat{\phi}_2 \end{Bmatrix} \qquad (5.5.7)$$

EXAMPLE 5.7

Determine the displacement and rotation at node 2 and the element forces for the uniform beam with an internal hinge at node 2 shown in Figure 5-24. Let EI be a constant.

We can assume the hinge as part of element 1. Therefore, using Eq. (5.5.5), the stiffness matrix of element 1 is

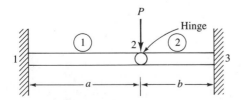

FIGURE 5-24   Beam with internal hinge

$$
\underline{k}^{(1)} = \frac{3EI}{a^3}
\begin{matrix}
\begin{matrix} d_{1y} & \phi_1 & d_{2y} & \phi_2 \end{matrix} \\
\begin{bmatrix}
1 & a & -1 & 0 \\
a & a^2 & -a & 0 \\
-1 & -a & 1 & 0 \\
0 & 0 & 0 & 0
\end{bmatrix}
\end{matrix}
\qquad (5.5.8)
$$

The stiffness matrix of element 2 is obtained from Eq. (5.1.14) as

$$
\underline{k}^{(2)} = \frac{EI}{b^3}
\begin{matrix}
\begin{matrix} d_{2y} & \phi_2 & d_{3y} & \phi_3 \end{matrix} \\
\begin{bmatrix}
12 & 6b & -12 & 6b \\
6b & 4b^2 & -6b & 2b^2 \\
-12 & -6b & 12 & -6b \\
6b & 2b^2 & -6b & 4b^2
\end{bmatrix}
\end{matrix}
\qquad (5.5.9)
$$

Superimposing Eqs. (5.5.8) and (5.5.9) and applying the boundary conditions

$$
d_{1y} = 0, \qquad \phi_1 = 0, \qquad d_{3y} = 0, \qquad \phi_3 = 0
$$

we obtain the total stiffness matrix and total set of equations as

$$
EI
\begin{bmatrix}
\dfrac{3}{a^3} + \dfrac{12}{b^3} & \dfrac{6}{b^2} \\[2ex]
\dfrac{6}{b^2} & \dfrac{4}{b}
\end{bmatrix}
\begin{Bmatrix} d_{2y} \\ \phi_2 \end{Bmatrix}
= \begin{Bmatrix} -P \\ 0 \end{Bmatrix}
\qquad (5.5.10)
$$

Solving Eq. (5.5.10), we obtain

$$
d_{2y} = \frac{-a^3 b^3 P}{3(b^3 + a^3)EI}
$$

$$
\phi_2 = \frac{a^3 b^2 P}{2(b^3 + a^3)EI}
$$

$$
\qquad (5.5.11)
$$

The value ϕ_2 is actually that associated with element 2—that is, ϕ_2 in Eq. (5.5.11) is actually $\phi_2^{(2)}$. The value of ϕ_2 at the right end of element 1 ($\phi_2^{(1)}$) is, in general, not equal to $\phi_2^{(2)}$. If we had chosen to assume the hinge to be part of element 2, then we would have used Eq. (5.1.14) for the stiffness matrix of element 1 and Eq. (5.5.7) for the stiffness matrix of element 2. This would allow us to obtain $\phi_2^{(1)}$, which is different than $\phi_2^{(2)}$.

Using Eq. (5.5.4) for element 1, we obtain the element forces as

$$
\begin{Bmatrix} \hat{f}_{1y} \\ \hat{m}_1 \\ \hat{f}_{2y} \end{Bmatrix}
= \frac{3EI}{a^3}
\begin{bmatrix}
1 & a & -1 \\
a & a^2 & -a \\
-1 & -a & 1
\end{bmatrix}
\begin{Bmatrix}
0 \\
0 \\
\dfrac{-a^3 b^3 P}{3(b^3 + a^3)EI}
\end{Bmatrix}
\qquad (5.5.12)
$$

Simplifying Eq. (5.5.12), we obtain the forces as

$$\hat{f}_{1y} = \frac{b^3 P}{b^3 + a^3}$$

$$\hat{m}_1 = \frac{ab^3 P}{b^3 + a^3} \qquad (5.5.13)$$

$$\hat{f}_{2y} = -\frac{b^3 P}{b^3 + a^3}$$

Using Eq. (5.5.9) and the results from Eq. (5.5.11), we obtain the element 2 forces as

$$
\begin{Bmatrix} \hat{f}_{2y} \\ \hat{m}_2 \\ \hat{f}_{3y} \\ \hat{m}_3 \end{Bmatrix} = \frac{EI}{b^3}
\begin{bmatrix}
12 & 6b & -12 & 6b \\
6b & 4b^2 & -6b & 2b^2 \\
-12 & -6b & 12 & -6b \\
6b & 2b^2 & -6b & 4b^2
\end{bmatrix}
\begin{Bmatrix}
-\dfrac{a^3 b^3 P}{3(b^3 + a^3)EI} \\[2mm]
\dfrac{a^3 b^2 P}{2(b^3 + a^3)EI} \\[2mm]
0 \\[2mm]
0
\end{Bmatrix}
$$

$$(5.5.14)$$

Simplifying Eq. (5.5.14), we obtain the element forces as

$$\hat{f}_{2y} = -\frac{a^3 P}{b^3 + a^3}$$

$$\hat{m}_2 = 0$$

$$\hat{f}_{3y} = \frac{a^3 P}{b^3 + a^3} \qquad (5.5.15)$$

$$\hat{m}_3 = -\frac{ba^3 P}{b^3 + a^3}$$

5.6   Potential Energy Approach to Derive Beam Element Equations

We will now derive the beam element equations using the principle of minimum potential energy. The procedure is similar to that used in Section 3.9 in deriving the bar element equations. Again, our primary purpose in applying the principle of minimum potential energy is to enhance your understanding of the principle, as it will be used routinely in subsequent chapters to develop element stiffness equations. We use the same notation here as in Section 3.9.

The total potential energy for a beam is

$$\pi_p = U + \Omega \qquad (5.6.1)$$

where the general one-dimensional expression for the strain energy U for a beam is given by

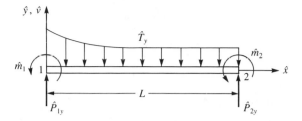

FIGURE 5-25 Beam element subjected to surface loading and concentrated nodal forces

$$U = \iiint_V \frac{1}{2} \sigma_x \varepsilon_x \, dV \qquad (5.6.2)$$

and for a single beam element subjected to both distributed and concentrated nodal loads, the potential energy of forces is given by

$$\Omega = -\iint_{S_1} \hat{T}_y \hat{v} \, dS - \sum_{i=1}^{2} \hat{P}_{iy} \hat{d}_{iy} - \sum_{i=1}^{2} \hat{m}_i \hat{\phi}_i \qquad (5.6.3)$$

where body forces are now neglected. The terms on the right-hand side of Eq. (5.6.3) represent the potential energy of (1) transverse surface loading $\hat{T}_y$ (in units of force per unit surface area, acting over surface S_1); (2) nodal concentrated forces $\hat{P}_{iy}$; and (3) moments $\hat{m}_i$. Again, $\hat{v}$ is the transverse displacement function for the beam element of length L shown in Figure 5-25.

Consider the beam element to have constant cross-sectional area A. The differential volume for the beam element can then be expressed as

$$dV = dA \, d\hat{x} \qquad (5.6.4)$$

and the differential area over which the surface loading acts is

$$dS = b \, d\hat{x} \qquad (5.6.5)$$

where b is the constant width. Using Eqs. (5.6.4) and (5.6.5) in Eqs. (5.6.1)–(5.6.3), the total potential energy becomes

$$\pi_p = \iint_{\hat{x}} \int_A \frac{1}{2} \sigma_x \varepsilon_x \, dA \, d\hat{x} - \int_0^L b\hat{T}_y \hat{v} \, d\hat{x} - \sum_{i=1}^{2} (\hat{P}_{iy} \hat{d}_{iy} + \hat{m}_i \hat{\phi}_i) \qquad (5.6.6)$$

Substituting Eq. (5.1.5) for $\hat{v}$ into the strain/displacement relationship Eq. (5.1.10), repeated here for convenience as

$$\varepsilon_x = -\hat{y} \frac{d^2 \hat{v}}{d\hat{x}^2} \qquad (5.6.7)$$

we express the strain in terms of nodal displacements and rotations as

$$\{\varepsilon_x\} = -\hat{y} \left[\frac{12\hat{x} - 6L}{L^3} \quad \frac{6\hat{x}L - 4L^2}{L^3} \quad \frac{-12\hat{x} + 6L}{L^3} \quad \frac{6\hat{x}L - 2L^2}{L^3} \right] \{\hat{d}\} \qquad (5.6.8)$$

or
$$\{\varepsilon_x\} = -\hat{y}[B]\{\hat{d}\} \qquad (5.6.9)$$

where we define

$$[B] = \left[\frac{12\hat{x} - 6L}{L^3} \quad \frac{6\hat{x}L - 4L^2}{L^3} \quad \frac{-12\hat{x} + 6L}{L^3} \quad \frac{6\hat{x}L - 2L^2}{L^3} \right] \qquad (5.6.10)$$

The stress/strain relationship is given by

$$\{\sigma_x\} = [D]\{\varepsilon_x\} \qquad (5.6.11)$$

where
$$[D] = [E] \qquad (5.6.12)$$

and E is the modulus of elasticity. Using Eq. (5.6.9) in Eq. (5.6.11), we obtain

$$\{\sigma_x\} = -\hat{y}[D][B]\{\hat{d}\} \qquad (5.6.13)$$

Next, the total potential energy Eq. (5.6.6) is expressed in matrix notation as

$$\pi_p = \int_{\hat{x}} \int_A \frac{1}{2}\{\sigma_x\}^T\{\varepsilon_x\}\, dA\, d\hat{x} - \int_0^L b\hat{T}_y[\hat{v}]^T\, d\hat{x} - \{\hat{d}\}^T\{\hat{P}\} \qquad (5.6.14)$$

Using Eqs. (5.6.9), (5.6.12), and (5.6.13), and defining $w = b\hat{T}_y$ as the line load (load per unit length) in the $\hat{y}$ direction, we express the total potential energy, Eq. (5.6.14), in matrix form as

$$\pi_p = \int_0^L \frac{EI}{2}\{\hat{d}\}^T[B]^T[B]\{\hat{d}\}\, d\hat{x} - \int_0^L w\{\hat{d}\}^T[N]^T\, d\hat{x} - \{\hat{d}\}^T\{\hat{P}\} \qquad (5.6.15)$$

where we have used the definition of the moment of inertia

$$I = \int_A \int y^2\, dA \qquad (5.6.16)$$

to obtain the first term on the right-hand side of Eq. (5.6.15). In Eq. (5.6.15), π_p is now expressed as a function of $\{\hat{d}\}$.

Differentiating π_p in Eq. (5.6.15) with respect to $\hat{d}_{1y}$, $\hat{\phi}_1$, $\hat{d}_{2y}$, and $\hat{\phi}_2$ and equating each term to zero to minimize π_p, we obtain four element equations, which are written in matrix form as

$$EI \int_0^L [B]^T[B]\, d\hat{x}\{\hat{d}\} - \int_0^L [N]^T w\, d\hat{x} - \{\hat{P}\} = 0 \qquad (5.6.17)$$

The derivation of the four element equations is left as an exercise (see Problem 5.32). Representing the nodal force matrix as the sum of those nodal forces resulting from distributed loading and concentrated loading, we have

$$\{\hat{f}\} = \int_0^L [N]^T w\, d\hat{x} + \{\hat{P}\} \qquad (5.6.18)$$

Using Eq. (5.6.18), the four element equations given by explicitly evaluating Eq. (5.6.17) are then identical to Eq. (5.1.13). Since $\{\hat{f}\} = [\hat{k}]\{\hat{d}\}$, we have from Eq. (5.6.17),

$$[\hat{k}] = EI \int_0^L [B]^T [B] \, d\hat{x} \qquad (5.6.19)$$

Using Eq. (5.6.10) in Eq. (5.6.19) and integrating, $[\hat{k}]$ is evaluated in explicit form as

$$[\hat{k}] = \frac{EI}{L^3} \begin{bmatrix} 12 & 6L & -12 & 6L \\ & 4L^2 & -6L & 2L^2 \\ & & 12 & -6L \\ \text{Symmetry} & & & 4L^2 \end{bmatrix} \qquad (5.6.20)$$

Equation (5.6.20) represents the local stiffness matrix for a beam element. As expected, Eq. (5.6.20) is identical to Eq. (5.1.14) developed previously.

5.7 Galerkin's Method to Derive Beam Element Equations

We will now illustrate Galerkin's method to formulate the beam element stiffness equations. We begin with the basic differential Eq. (5.1.1a) with transverse loading w now included; that is,

$$EI \frac{d^4 \hat{v}}{d\hat{x}^4} - w = 0 \qquad (5.7.1)$$

We now define the residual R to be Eq. (5.7.1). Applying Galerkin's criterion, Eq. (3.11.3), to Eq. (5.7.1), we have

$$\int_0^L \left(EI \frac{d^4 \hat{v}}{d\hat{x}^4} - w \right) N_i \, d\hat{x} = 0 \qquad (i = 1, 2, 3, 4) \qquad (5.7.2)$$

where the shape functions N_i are defined by Eqs. (5.1.7).

We now apply integration by parts twice to the first term in Eq. (5.7.2) to yield

$$\int_0^L EI(\hat{v}_{,\hat{x}\hat{x}\hat{x}\hat{x}}) N_i \, d\hat{x} = \int_0^L EI(\hat{v}_{,\hat{x}\hat{x}})(N_{i,\hat{x}\hat{x}}) \, d\hat{x} + EI[N_i(\hat{v}_{,\hat{x}\hat{x}\hat{x}}) - (N_{i,\hat{x}})(\hat{v}_{,\hat{x}\hat{x}})]_0^L$$

$$(5.7.3)$$

where the notation of the comma followed by the subscript $\hat{x}$ indicates differentiation with respect to $\hat{x}$. Again, integration by parts introduces the boundary conditions.

Since $\hat{v} = [N]\{\hat{d}\}$ as given by Eq. (5.1.5), we have

$$\hat{v}_{,\hat{x}\hat{x}} = \left[\frac{12\hat{x} - 6L}{L^3} \quad \frac{6\hat{x}L - 4L^2}{L^3} \quad \frac{-12\hat{x} + 6L}{L^3} \quad \frac{6\hat{x}L - 2L^2}{L^3} \right] \{\hat{d}\} \qquad (5.7.4)$$

or, using Eq. (5.6.10),

$$\hat{v}_{,\hat{x}\hat{x}} = [B]\{\hat{d}\} \qquad (5.7.5)$$

Substituting Eq. (5.7.5) into Eq. (5.7.3), and then Eq. (5.7.3) into Eq. (5.7.2), we obtain

$$\int_0^L (N_{i,\hat{x}\hat{x}})EI[B]\, d\hat{x}\{\hat{d}\} - \int_0^L N_i w\, d\hat{x} + [N_i\hat{V} - (N_{i,\hat{x}})\hat{m}]\{\hat{d}\}|_0^L = 0$$

$$(i = 1, 2, 3, 4) \quad (5.7.6)$$

where Eqs. (5.1.11) have been used in the boundary terms. Equation (5.7.6) is really four equations (one each for $N_i = N_1$, N_2, N_3, and N_4). Instead of directly evaluating Eq. (5.7.6) for each N_i, as was done in Section 3.11, we can express the four equations of Eq. (5.7.6) in matrix form as

$$\int_0^L [B]^T EI[B]\, d\hat{x}\{\hat{d}\} = \int_0^L [N]^T w\, d\hat{x} + ([N]^T_{,\hat{x}}\hat{m} - [N]^T\hat{V})|_0^L \quad (5.7.7)$$

where we have used the relationship $[N]_{,xx} = [B]$ in Eq. (5.7.7).

Observe that the integral term on the left side of Eq. (5.7.7) is identical to the stiffness matrix previously given by Eq. (5.6.19) and the first term on the right side of Eq. (5.7.7) represents the equivalent nodal forces due to distributed loading [also given in Eq. (5.6.18)]. The two terms in parentheses on the right side of Eq. (5.7.7) are the same as the concentrated force matrix $\{\hat{P}\}$ of Eq. (5.6.18). We explain this by evaluating $[N]_{,\hat{x}}$ and $[N]$, where $[N]$ is defined by Eq. (5.1.6), at the ends of the element as follows:

$$[N]_{,\hat{x}}|_0 = [0 \quad 1 \quad 0 \quad 0] \qquad [N]_{,\hat{x}}|_L = [0 \quad 0 \quad 0 \quad 1]$$
$$[N]|_0 = [1 \quad 0 \quad 0 \quad 0] \qquad [N]|_L = [0 \quad 0 \quad 1 \quad 0] \qquad (5.7.8)$$

Therefore, using Eqs. (5.7.8) in Eq. (5.7.7), the following terms result:

$$\begin{Bmatrix} 0 \\ 0 \\ 0 \\ 1 \end{Bmatrix} \hat{m}(L) - \begin{Bmatrix} 0 \\ 1 \\ 0 \\ 0 \end{Bmatrix} \hat{m}(0) - \begin{Bmatrix} 0 \\ 0 \\ 1 \\ 0 \end{Bmatrix} \hat{V}(L) + \begin{Bmatrix} 1 \\ 0 \\ 0 \\ 0 \end{Bmatrix} \hat{V}(0) \qquad (5.7.9)$$

These nodal shear forces and moments are illustrated in Figure 5-26.

Note that when element matrices are assembled, two shear forces and two moments from adjacent elements contribute to the concentrated force and concentrated moment at the node common to the adjacent elements as shown

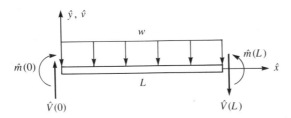

FIGURE 5-26   Beam element with shear forces, moments, and a distributed load

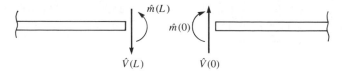

FIGURE 5-27 Shear forces and moments acting on adjacent elements meeting at a node

in Figure 5-27. These concentrated shear forces $\hat{V}(0) - \hat{V}(L)$ and moments $\hat{m}(L) - \hat{m}(0)$ are often zero; that is, $\hat{V}(0) = \hat{V}(L)$ and $\hat{m}(L) = \hat{m}(0)$ occur except when a concentrated nodal force or moment exists at the node. In the actual computations, we handle the expressions given by Eq. (5.7.9) by including them as concentrated nodal values making up the matrix $\{P\}$.

REFERENCES

[1] Logan, D. L., *Mechanics of Materials*, Harper-Collins, New York, 1991.

[2] Hsieh, Y. Y., *Elementary Theory of Structures*, 2nd ed., Prentice-Hall, Englewood Cliffs, N.J., 1982.

PROBLEMS

5.1 Use Eqs. (5.1.7) to plot the shape functions N_1 and N_3 and the derivatives $(dN_2/d\hat{x})$ and $(dN_4/d\hat{x})$, which represent the shapes (variations) of the slopes $\hat{\phi}_1$ and $\hat{\phi}_2$ over the length of the beam element.

5.2 Derive the element stiffness matrix for the beam element in Figure 5-1 if the rotational degrees of freedom are assumed positive clockwise instead of counterclockwise. Compare the two different nodal sign conventions and discuss.

Solve all problems using the finite element stiffness method.

5.3 For the beam shown in Figure P5-3, determine the rotation at pin support A, and the rotation and displacement under the load P. Determine the reactions. Draw the shear force and bending moment diagrams. Let EI be constant throughout the beam.

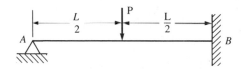

FIGURE P5-3

5.4   For the cantilever beam subjected to the free-end load *P* shown in Figure P5-4, determine the maximum deflection and the reactions. Let *EI* be constant throughout the beam.

F I G U R E   P5-4

5.5–5.11   For the beams shown in Figures P5-5–P5-11, determine the displacements and the slopes at the nodes, the forces in each element, and the reactions. Also, draw the shear force and bending moment diagrams.

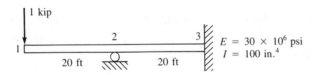

$E = 30 \times 10^6$ psi
$I = 100$ in.^{4}

F I G U R E   P5-5

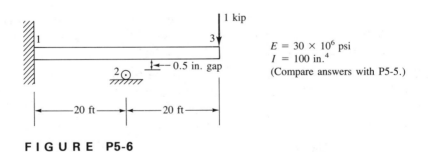

$E = 30 \times 10^6$ psi
$I = 100$ in.^{4}
(Compare answers with P5-5.)

F I G U R E   P5-6

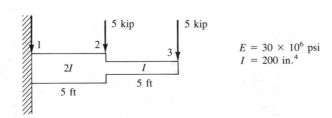

$E = 30 \times 10^6$ psi
$I = 200$ in.^{4}

F I G U R E   P5-7

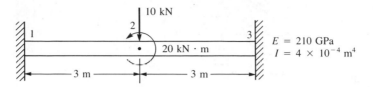

FIGURE P5-8

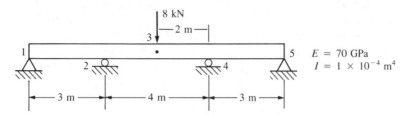

FIGURE P5-9

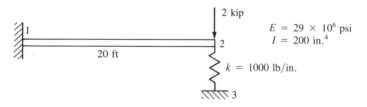

FIGURE P5-10

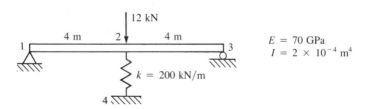

FIGURE P5-11

5.12 For the fixed-fixed beam subjected to the uniform load *w* shown in Figure P5-12, determine the midspan deflection and the reactions. Draw the shear force and bending moment diagrams. The middle section of the beam has a bending stiffness of 2*EI*; the other sections have bending stiffnesses of *EI*.

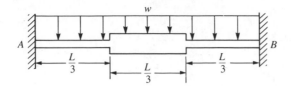

FIGURE   P5-12

5.13   Determine the midspan deflection and the reactions and draw the shear force and bending moment diagrams for the fixed-fixed beam subjected to uniformly distributed load *w* shown in Figure P5-13. Assume *EI* constant throughout the beam. Compare your answers with the classical solution (that is, with the appropriate equivalent joint forces given in Appendix D).

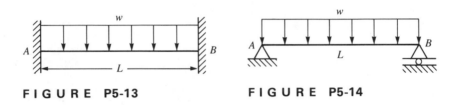

FIGURE   P5-13   **FIGURE   P5-14**

5.14   Determine the midspan deflection and the reactions and draw the shear force and bending moment diagrams for the simply supported beam subjected to the uniformly distributed load *w* shown in Figure P5-14. Assume *EI* constant throughout the beam.

5.15   For the beam loaded as shown in Figure P5-15, determine the free-end deflection and the reactions and draw the shear force and bending moment diagrams.

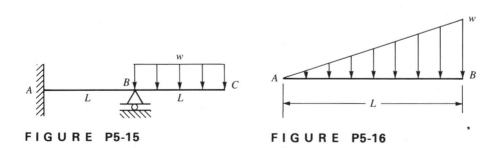

FIGURE   P5-15   **FIGURE   P5-16**

5.16   Using the concept of work equivalence, determine the nodal forces and moments (called *equivalent nodal forces*) used to replace the linearly varying distributed load shown in Figure P5-16.

5.17 Assume the beam of Figure P5-16 to be fixed at each end. Determine the reactions and draw the shear force and bending moment diagrams. Assume *EI* constant throughout the beam. Use two beam elements to discretize the beam. Compare your answers for the reactions with the appropriate equivalent joint forces given in Appendix D.

5.18 For the beam subjected to the linearly varying line load *w* shown in Figure P5-18, determine the right-end rotation and the reactions. Assume *EI* constant throughout the beam.

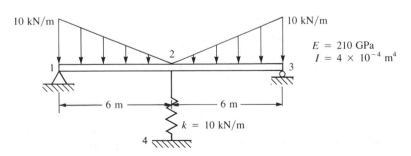

FIGURE   P5-18

5.19–5.24 For the beams shown in Figures P5-19–P5-24, determine the nodal displacements and slopes, the forces in each element, and the reactions.

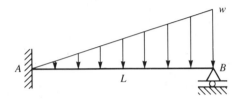

8 kN/m

$E = 70$ GPa
$I = 3 \times 10^{-4}$ m^4

4 m

4 m

FIGURE   P5-19

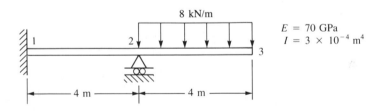

10 kN/m

10 kN/m

$E = 210$ GPa
$I = 4 \times 10^{-4}$ m^4

6 m

6 m

$k = 10$ kN/m

FIGURE   P5-20

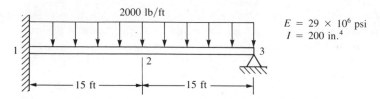

$E = 29 \times 10^6$ psi
$I = 200$ in.^{4}

F I G U R E   P5-21

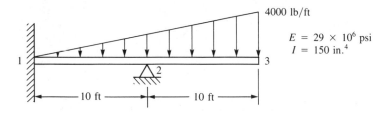

4000 lb/ft

$E = 29 \times 10^6$ psi
$I = 150$ in.^{4}

F I G U R E   P5-22

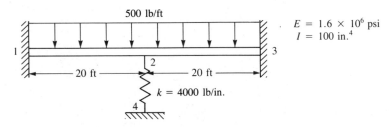

$E = 1.6 \times 10^6$ psi
$I = 100$ in.^{4}

F I G U R E   P5-23

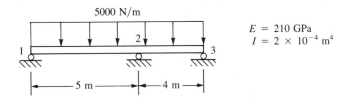

$E = 210$ GPa
$I = 2 \times 10^{-4}$ m^4

F I G U R E   P5-24

5.25   For the beam shown in Figure P5-25 subjected to the concentrated load P and distributed load w, determine the midspan displacement and the reactions. Let EI be constant throughout the beam.

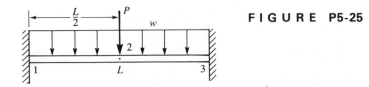

F I G U R E   P5-25

5.26   For the beam shown in Figure P5-26 subjected to the two concentrated loads *P*, determine the deflection at the midspan. Use the equivalent load replacement method. Let *EI* be constant throughout the beam.

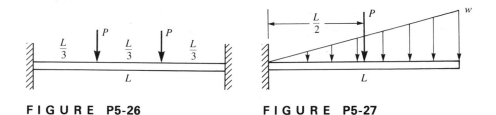

FIGURE P5-26   **FIGURE P5-27**

5.27   For the beam shown in Figure P5-27 subjected to the concentrated load *P* and the linearly varying line load *w*, determine the free-end deflection and rotation and the reactions. Use the equivalent load replacement method. Let *EI* be constant throughout the beam.

5.28–5.30   For the beams shown in Figures P5-28–P5-30, with internal hinge, determine the deflection at the hinge. Let $E = 210$ GPa and $I = 2 \times 10^{-4}$ m^4.

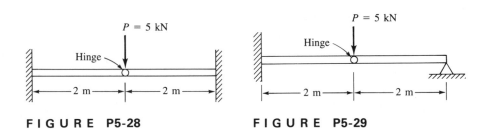

FIGURE P5-28   **FIGURE P5-29**

5.31   Derive the stiffness matrix for a beam element with a nodal linkage—that is, the shear is 0 at node *i*, but the usual shear and moment resistance are present at node *j* (see Figure P5-31).

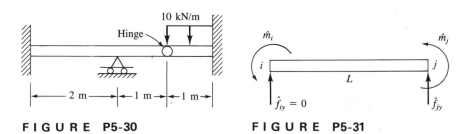

FIGURE P5-30   **FIGURE P5-31**

5.32   Explicitly evaluate π_p of Eq. (5.6.15); then differentiate π_p with respect to $\hat{d}_{1y}$, $\hat{\phi}_1$, $\hat{d}_{2y}$, and $\hat{\phi}_2$ and set each of these equations to zero (that is, minimize π_p) to obtain the four element equations for the beam element. Then express these equations in matrix form.

5.33   Determine the free-end deflection for the tapered beam shown in Figure P5-33. Here $I(x) = I_0(1 + nx/L)$ where I_0 is the moment of inertia at $x = 0$. Compare the exact beam theory solution with a 2-element finite element solution for $n = 2$.

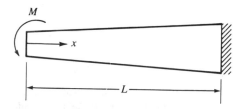
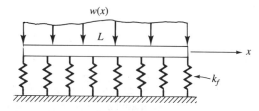

FIGURE P5-33   FIGURE P5-34

5.34   Derive the equations for the beam element on an elastic foundation (Figure P5-34) using the principle of minimum potential energy. Here k_f is the subgrade spring constant per unit length. The potential energy of the beam is

$$\pi_p = \int_0^L \frac{1}{2} EI(v'')^2 \, dx + \int_0^L \frac{k_f v^2}{2} \, dx - \int_0^L wv \, dx$$

See Figure P5-34.

5.35   Derive the equations for the beam element on an elastic foundation (see Figure P5-34) using Galerkin's method. The basic differential equation for the beam on an elastic foundation is

$$(EIv'')'' = w - k_f v$$

Plane Frame and

Grid Equations

INTRODUCTION

Many structures, such as buildings (Figure 6-1) and bridges, are composed of frames and/or grids. This chapter develops the equations and methods for solution of plane frames and grids.

First, we will develop the stiffness matrix for a beam element arbitrarily oriented in a plane. We will then include the axial nodal displacement degree of freedom in the local beam element stiffness matrix. Then we will combine these results to develop the stiffness matrix, including axial deformation effects, for an arbitrarily oriented beam element; thus making it possible to analyze plane frames. Specific examples of plane frame analysis follow. We will then consider frames with inclined or skewed supports.

Next, we will develop the grid element stiffness matrix. We will present the solution of a grid deck system to illustrate the application of the grid equations. We will then develop the stiffness matrix for a beam element arbitrarily oriented in space. We will also consider the concept of substructure analysis.

A discussion of the use of a computer program called PFRAME (listed on accompanying disk) follows. This program facilitates the solution of complex plane frame and grid problems with large numbers of degrees of freedom.

6.1 Two-Dimensional Arbitrarily Oriented Beam Element

We can derive the stiffness matrix for an arbitrarily oriented beam element, as shown in Figure 6-2, in a manner similar to that used for the bar element in Chapter 3. The local axes $\hat{x}$ and $\hat{y}$ are located along the beam element and transverse to the beam element, respectively, and the global axes x and y are located to be convenient for the total structure.

FIGURE 6-1   Rigid building frame

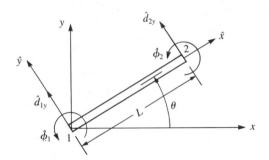

FIGURE 6-2   Arbitrarily oriented beam element

Recall that we can relate local displacements to global displacements by using Eq. (3.3.16), repeated here for convenience as

$$\begin{Bmatrix} \hat{d}_x \\ \hat{d}_y \end{Bmatrix} = \begin{bmatrix} C & S \\ -S & C \end{bmatrix} \begin{Bmatrix} d_x \\ d_y \end{Bmatrix} \qquad (6.1.1)$$

Using the second equation of Eq. (6.1.1) for the beam element, we relate local nodal degrees of freedom to global degrees of freedom by

$$\begin{Bmatrix} \hat{d}_{1y} \\ \hat{\phi}_1 \\ \hat{d}_{2y} \\ \hat{\phi}_2 \end{Bmatrix} = \begin{bmatrix} -S & C & 0 & 0 & 0 & 0 \\ 0 & 0 & 1 & 0 & 0 & 0 \\ 0 & 0 & 0 & -S & C & 0 \\ 0 & 0 & 0 & 0 & 0 & 1 \end{bmatrix} \begin{Bmatrix} d_{1x} \\ d_{1y} \\ \phi_1 \\ d_{2x} \\ d_{2y} \\ \phi_2 \end{Bmatrix} \qquad (6.1.2)$$

where, for a beam element, we define

$$
\underline{T} = \begin{bmatrix}
-S & C & 0 & 0 & 0 & 0 \\
0 & 0 & 1 & 0 & 0 & 0 \\
0 & 0 & 0 & -S & C & 0 \\
0 & 0 & 0 & 0 & 0 & 1
\end{bmatrix}
\tag{6.1.3}
$$

as the *transformation matrix*. The axial effects are not yet included. Equation (6.1.2) indicates that rotation is invariant with respect to either coordinate system. For example, $\hat{\phi}_1 = \phi_1$, and moment $\hat{m}_1 = m_1$ can be considered to be a vector pointing normal to the $\hat{x} - \hat{y}$ plane or to the $x - y$ plane by the usual right-hand rule. From either viewpoint, the moment is in the $\hat{z} = z$ direction. Therefore, moment is unaffected as the element changes orientation in the $x - y$ plane.

Substituting Eq. (6.1.3) for $\underline{T}$ and Eq. (5.1.14) for $\hat{\underline{k}}$ into Eq. (3.4.22), $\underline{k} = \underline{T}^T \hat{\underline{k}} \underline{T}$, we obtain

$$
\underline{k} = \frac{EI}{L^3}
\begin{bmatrix}
12S^2 & -12SC & -6LS & -12S^2 & 12SC & -6LS \\
 & 12C^2 & 6LC & 12SC & -12C^2 & 6LC \\
 & & 4L^2 & 6LS & -6LC & 2L^2 \\
 & & & 12S^2 & -12SC & 6LS \\
 & & & & 12C^2 & -6LC \\
\text{Symmetry} & & & & & 4L^2
\end{bmatrix}
\tag{6.1.4}
$$

where, again, $C = \cos\theta$ and $S = \sin\theta$. It is not necessary here to expand $\underline{T}$ given by Eq. (6.1.3) to make it a square matrix to be able to use Eq. (3.4.22). Since Eq. (3.4.22) is a generally applicable equation, the matrices used must merely be of the correct order for matrix multiplication (see Appendix A for more on matrix multiplication). The stiffness matrix Eq. (6.1.4) is the global stiffness matrix for a beam element that includes shear and bending resistance. Local axial effects are not yet included. The transformation from local to global stiffness by multiplying matrices $\underline{T}^T \hat{\underline{k}} \underline{T}$, as done in Eq. (6.1.4), is usually done on the computer by using such subroutines as GMPRD and GTPRD, which can be found in the Scientific Subroutine Package (see Reference [1]).

We will now include the axial effects in the element, as shown in Figure 6-3. For axial effects, we recall from Eq. (3.1.13),

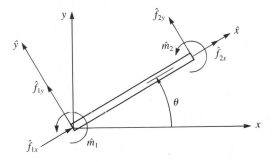

FIGURE   6-3   Local forces acting on a beam element

$$\begin{Bmatrix} \hat{f}_{1x} \\ \hat{f}_{2x} \end{Bmatrix} = \frac{AE}{L} \begin{bmatrix} 1 & -1 \\ -1 & 1 \end{bmatrix} \begin{Bmatrix} \hat{d}_{1x} \\ \hat{d}_{2x} \end{Bmatrix} \tag{6.1.5}$$

Combining the axial effects of Eq. (6.1.5) with the shear and principal bending moment effects of Eq. (5.1.13), we have, in local coordinates,

$$\begin{Bmatrix} \hat{f}_{1x} \\ \hat{f}_{1y} \\ \hat{m}_1 \\ \hat{f}_{2x} \\ \hat{f}_{2y} \\ \hat{m}_2 \end{Bmatrix} = \begin{bmatrix} C_1 & 0 & 0 & -C_1 & 0 & 0 \\ 0 & 12C_2 & 6C_2L & 0 & -12C_2 & 6C_2L \\ 0 & 6C_2L & 4C_2L^2 & 0 & -6C_2L & 2C_2L^2 \\ -C_1 & 0 & 0 & C_1 & 0 & 0 \\ 0 & -12C_2 & -6C_2L & 0 & 12C_2 & -6C_2L \\ 0 & 6C_2L & 2C_2L^2 & 0 & -6C_2L & 4C_2L^2 \end{bmatrix} \begin{Bmatrix} \hat{d}_{1x} \\ \hat{d}_{1y} \\ \hat{\phi}_1 \\ \hat{d}_{2x} \\ \hat{d}_{2y} \\ \hat{\phi}_2 \end{Bmatrix} \tag{6.1.6}$$

where

$$C_1 = \frac{AE}{L} \quad \text{and} \quad C_2 = \frac{EI}{L^3} \tag{6.1.7}$$

and, therefore,

$$\hat{k} = \begin{bmatrix} C_1 & 0 & 0 & -C_1 & 0 & 0 \\ 0 & 12C_2 & 6C_2L & 0 & -12C_2 & 6C_2L \\ 0 & 6C_2L & 4C_2L^2 & 0 & -6C_2L & 2C_2L^2 \\ -C_1 & 0 & 0 & C_1 & 0 & 0 \\ 0 & -12C_2 & -6C_2L & 0 & 12C_2 & -6C_2L \\ 0 & 6C_2L & 2C_2L^2 & 0 & -6C_2L & 4C_2L^2 \end{bmatrix} \tag{6.1.8}$$

The $\hat{k}$ matrix in Eq. (6.1.8) now includes axial effects (in the $\hat{x}$ direction), as well as shear force effects (in the $\hat{y}$ direction) and principal bending moment effects (about the $\hat{z} = z$ axis). Using Eqs. (6.1.1) and (6.1.2), we now relate the local to the global displacements by

$$\begin{Bmatrix} \hat{d}_{1x} \\ \hat{d}_{1y} \\ \hat{\phi}_1 \\ \hat{d}_{2x} \\ \hat{d}_{2y} \\ \hat{\phi}_2 \end{Bmatrix} = \begin{bmatrix} C & S & 0 & 0 & 0 & 0 \\ -S & C & 0 & 0 & 0 & 0 \\ 0 & 0 & 1 & 0 & 0 & 0 \\ 0 & 0 & 0 & C & S & 0 \\ 0 & 0 & 0 & -S & C & 0 \\ 0 & 0 & 0 & 0 & 0 & 1 \end{bmatrix} \begin{Bmatrix} d_{1x} \\ d_{1y} \\ \phi_1 \\ d_{2x} \\ d_{2y} \\ \phi_2 \end{Bmatrix} \tag{6.1.9}$$

where T has now been expanded to include local axial deformation effects as

$$T = \begin{bmatrix} C & S & 0 & 0 & 0 & 0 \\ -S & C & 0 & 0 & 0 & 0 \\ 0 & 0 & 1 & 0 & 0 & 0 \\ 0 & 0 & 0 & C & S & 0 \\ 0 & 0 & 0 & -S & C & 0 \\ 0 & 0 & 0 & 0 & 0 & 1 \end{bmatrix} \tag{6.1.10}$$

Substituting T from Eq. (6.1.10) and $\hat{k}$ from Eq. (6.1.8) into Eq. (3.4.22), we obtain the general transformed global stiffness matrix for a beam element that includes axial force, shear force, and bending moment effects as follows:

$$k = \frac{E}{L} \times$$

$$
\begin{bmatrix}
AC^2 + \dfrac{12I}{L^2}S^2 & \left(A - \dfrac{12I}{L^2}\right)CS & -\dfrac{6I}{L}S & -\left(AC^2 + \dfrac{12I}{L^2}S^2\right) & -\left(A - \dfrac{12I}{L^2}\right)CS & -\dfrac{6I}{L}S \\[2ex]
 & AS^2 + \dfrac{12I}{L^2}C^2 & \dfrac{6I}{L}C & -\left(A - \dfrac{12I}{L^2}\right)CS & -\left(AS^2 + \dfrac{12I}{L^2}C^2\right) & \dfrac{6I}{L}C \\[2ex]
 & & 4I & \dfrac{6I}{L}S & -\dfrac{6I}{L}C & 2I \\[2ex]
 & & & AC^2 + \dfrac{12I}{L^2}S^2 & \left(A - \dfrac{12I}{L^2}\right)CS & \dfrac{6I}{L}S \\[2ex]
 & & & & AS^2 + \dfrac{12I}{L^2}C^2 & -\dfrac{6I}{L}C \\[2ex]
\text{Symmetry} & & & & & 4I
\end{bmatrix}
$$

$$(6.1.11)$$

The analysis of a rigid plane frame can be undertaken by applying stiffness matrix Eq. (6.1.11). *A **rigid plane frame** is defined here as a series of beam elements rigidly connected to each other*; that is, the original angles made between elements at their joints remain unchanged after the deformation. Furthermore, moments are transmitted from one element to another at the joints. Hence, moment continuity exists at the rigid joints. In addition, the element centroids, as well as the applied loads, lie in a common plane. From Eq. (6.1.11), we observe that the element stiffnesses of a frame are functions of E, A, L, I, and the angle of orientation θ of the element with respect to the global-coordinate axes.

6.2   Rigid Plane Frame Examples

To illustrate the use of the equations developed in Section 6.1, we will now perform complete solutions for the following rigid plane frames.

E X A M P L E   6.1

As the first example of rigid plane frame analysis, solve the simple "bent" shown in Figure 6-4.

The frame is fixed at nodes 1 and 4 and subjected to a positive horizontal force of 10,000 lb applied at node 2 and to a positive moment of 5000 lb-in. applied at node 3. The global-coordinate axes and the element lengths are shown in Figure 6-4. Let $E = 30 \times 10^6$ psi and $A = 10$ in.^{2} for all elements, and let $I = 200$ in.^{4} for elements 1 and 3, and $I = 100$ in.^{4} for element 2.

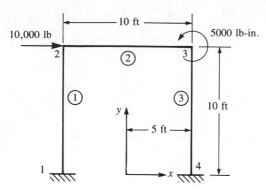

FIGURE 6-4   Plane frame for analysis

Using Eq. (6.1.11), we obtain the global stiffness matrices for each element.

Element 1

For element 1, the angle between the global x and the local $\hat{x}$ axes is $90°$ because $\hat{x}$ is assumed to be directed from node 1 to node 2. Therefore,

$$C = \cos 90° = \frac{x_2 - x_1}{L^{(1)}} = \frac{-5 - (-5)}{120} = 0$$

$$S = \sin 90° = \frac{y_2 - y_1}{L^{(1)}} = \frac{120 - 0}{120} = 1$$

Also,

$$\frac{12I}{L^2} = \frac{12(200)}{(10 \times 12)^2} = 0.167$$

$$\frac{6I}{L} = \frac{6(200)}{10 \times 12} = 10.0 \qquad\qquad (6.2.1)$$

$$\frac{E}{L} = \frac{30 \times 10^6}{10 \times 12} = 250,000$$

Then, using Eqs. (6.2.1) to help in evaluating Eq. (6.1.11) for element 1, we obtain

$$
\underline{k}^{(1)} = 250,000
\begin{array}{c}
\begin{array}{cccccc} d_{1x} & d_{1y} & \phi_1 & d_{2x} & d_{2y} & \phi_2 \end{array} \\
\begin{bmatrix}
0.167 & 0 & -10 & -0.167 & 0 & -10 \\
0 & 10 & 0 & 0 & -10 & 0 \\
-10 & 0 & 800 & 10 & 0 & 400 \\
-0.167 & 0 & 10 & 0.167 & 0 & 10 \\
0 & -10 & 0 & 0 & 10 & 0 \\
-10 & 0 & 400 & 10 & 0 & 800
\end{bmatrix}
\end{array}
$$

$$(6.2.2)$$

Element 2

For element 2, the angle between x and $\hat{x}$ is zero because $\hat{x}$ is directed from node 2 to node 3. Therefore,

$$C = 1 \qquad S = 0$$

Also,

$$\frac{12I}{L^2} = \frac{12(100)}{120^2} = 0.0835$$

$$\frac{6I}{L} = \frac{6(100)}{120} = 5.0 \qquad\qquad (6.2.3)$$

$$\frac{E}{L} = 250,000$$

Using the quantities obtained in Eqs. (6.2.3) in evaluating Eq. (6.1.11) for element 2, we obtain

$$
\underline{k}^{(2)} = 250,000
\begin{array}{c}
\begin{array}{cccccc} d_{2x} & d_{2y} & \phi_2 & d_{3x} & d_{3y} & \phi_3 \end{array} \\
\begin{bmatrix}
10 & 0 & 0 & -10 & 0 & 0 \\
0 & 0.0835 & 5 & 0 & 0.0835 & 5 \\
0 & 5 & 400 & 0 & -5 & 200 \\
-10 & 0 & 0 & 10 & 0 & 0 \\
0 & 0.0835 & -5 & 0 & 0.0835 & -5 \\
0 & 5 & 200 & 0 & -5 & 400
\end{bmatrix}
\end{array}
\qquad (6.2.4)
$$

Element 3

For element 3, the angle between x and $\hat{x}$ is 270° because $\hat{x}$ is directed from node 3 to node 4. Therefore,

$$C = 0 \qquad S = -1$$

Therefore, evaluating Eq. (6.1.11) for element 3, we obtain

$$
\underline{k}^{(2)} = 250,000
\begin{array}{c}
\begin{array}{cccccc} d_{3x} & d_{3y} & \phi_3 & d_{4x} & d_{4y} & \phi_4 \end{array} \\
\begin{bmatrix}
0.167 & 0 & 10 & -0.167 & 0 & 10 \\
0 & 10 & 0 & 0 & -10 & 0 \\
10 & 0 & 800 & -10 & 0 & 400 \\
-0.167 & 0 & -10 & 0.167 & 0 & -10 \\
0 & -10 & 0 & 0 & 10 & 0 \\
10 & 0 & 400 & -10 & 0 & 800
\end{bmatrix}
\end{array}
$$

$$(6.2.5)$$

Superposition of Eqs. (6.2.2), (6.2.4), and (6.2.5) and application of the boundary conditions $d_{1x} = d_{1y} = \phi_1 = 0$ and $d_{4x} = d_{4y} = \phi_4 = 0$ at nodes 1 and 4 yield the reduced set of equations for a longhand solution as

$$
\begin{Bmatrix} 10{,}000 \\ 0 \\ 0 \\ 0 \\ 0 \\ 5000 \end{Bmatrix} = 250{,}000
\begin{bmatrix}
10.167 & 0 & 10 & -10 & 0 & 0 \\
0 & 10.0835 & 5 & 0 & 0.0835 & 5 \\
10 & 5 & 1200 & 0 & -5 & 200 \\
-10 & 0 & 0 & 10.167 & 0 & 10 \\
0 & 0.0835 & -5 & 0 & 10.0835 & -5 \\
0 & 5 & 200 & 10 & -5 & 1200
\end{bmatrix}
$$

$$
\times
\begin{Bmatrix} d_{2x} \\ d_{2y} \\ \phi_2 \\ d_{3x} \\ d_{3y} \\ \phi_3 \end{Bmatrix}
\qquad (6.2.6)
$$

Solving Eq. (6.2.6) for the displacements and rotations, we have

$$
\begin{Bmatrix} d_{2x} \\ d_{2y} \\ \phi_2 \\ d_{3x} \\ d_{3y} \\ \phi_3 \end{Bmatrix}
=
\begin{Bmatrix}
0.211 \text{ in.} \\
0.00148 \text{ in.} \\
-0.00153 \text{ rad} \\
0.209 \text{ in.} \\
-0.00148 \text{ in.} \\
-0.00149 \text{ rad}
\end{Bmatrix}
\qquad (6.2.7)
$$

The element forces can now be obtained using $\hat{f} = \hat{k}\underline{T}\underline{d}$ for each element, as was previously done in solving truss and beam problems. We will illustrate this procedure only for element 1. For element 1, on using Eq. (6.1.10) for $\underline{T}$, and Eq. (6.2.7) for the displacements at node 2, we have

$$
\underline{T}\underline{d} =
\begin{bmatrix}
0 & 1 & 0 & 0 & 0 & 0 \\
-1 & 0 & 0 & 0 & 0 & 0 \\
0 & 0 & 1 & 0 & 0 & 0 \\
0 & 0 & 0 & 0 & 1 & 0 \\
0 & 0 & 0 & -1 & 0 & 0 \\
0 & 0 & 0 & 0 & 0 & 1
\end{bmatrix}
\begin{Bmatrix}
d_{1x} = 0 \\
d_{1y} = 0 \\
\phi_1 = 0 \\
d_{2x} = 0.211 \\
d_{2y} = 0.00148 \\
\phi_2 = -0.00153
\end{Bmatrix}
\qquad (6.2.8)
$$

On multiplying the matrices in Eq. (6.2.8), we obtain

$$
\underline{T}\underline{d} =
\begin{Bmatrix}
0 \\ 0 \\ 0 \\ 0.00148 \\ -0.211 \\ -0.00153
\end{Bmatrix}
\qquad (6.2.9)
$$

Then

$$\hat{f} = \hat{k}\underline{T}\underline{d} = 250{,}000 \begin{bmatrix} 10 & 0 & 0 & -10 & 0 & 0 \\ 0 & 0.167 & 10 & 0 & -0.167 & 10 \\ 0 & 10 & 800 & 0 & -10 & 400 \\ -10 & 0 & 0 & 10 & 0 & 0 \\ 0 & -0.167 & -10 & 0 & 0.167 & -10 \\ 0 & 10 & 400 & 0 & -10 & 800 \end{bmatrix}$$

$$\times \begin{Bmatrix} 0 \\ 0 \\ 0 \\ 0.00148 \\ -0.211 \\ -0.00153 \end{Bmatrix} \qquad (6.2.10)$$

Simplifying Eq. (6.2.10), we obtain the local forces acting on element 1 as

$$\begin{Bmatrix} \hat{f}_{1x} \\ \hat{f}_{1y} \\ \hat{m}_1 \\ \hat{f}_{2x} \\ \hat{f}_{2y} \\ \hat{m}_2 \end{Bmatrix} = \begin{Bmatrix} -3700 \text{ lb} \\ 4990 \text{ lb} \\ 376{,}000 \text{ lb-in.} \\ 3700 \text{ lb} \\ -4990 \text{ lb} \\ 223{,}000 \text{ lb-in.} \end{Bmatrix} \qquad (6.2.11)$$

A free-body diagram of element 1 is shown in Figure 6-5. In Figure 6-5, the $\hat{x}$ axis is directed from node 1 to node 2—consistent with the order of the nodal degrees of freedom used in developing the stiffness matrix for the element. Since the x-y plane was initially established as shown in Figure 6-4, the z axis is directed outward—consequently, so is the $\hat{z}$ axis (recall $\hat{z} = z$). The $\hat{y}$ axis is then established such that $\hat{x}$ cross $\hat{y}$ yields the direction of $\hat{z}$. The signs on the resulting element forces in Eq. (6.2.11) are thus consistently

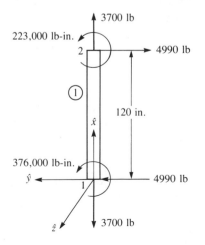

FIGURE 6-5 Free-body diagram of element 1

shown in Figure 6-5. The forces in elements 2 and 3 can be obtained in a manner similar to that used to obtain Eq. (6.2.11) for the nodal forces in element 1. Here we report only the final results for the forces in elements 2 and 3, and leave it to your discretion to perform the detailed calculations. The element forces are as follows:

Element 2

$$\hat{f}_{2x} = 5010\,\text{lb} \qquad \hat{f}_{2y} = -3700\,\text{lb} \qquad \hat{m}_2 = -223{,}000\,\text{lb-in.}$$
$$\hat{f}_{3x} = -5010\,\text{lb} \qquad \hat{f}_{3y} = 3700\,\text{lb} \qquad \hat{m}_3 = -221{,}000\,\text{lb-in.}$$

$$(6.2.12)$$

Element 3

$$\hat{f}_{3x} = 3700\,\text{lb} \qquad \hat{f}_{3y} = 5010\,\text{lb} \qquad \hat{m}_3 = 226{,}000\,\text{lb-in.}$$
$$\hat{f}_{4x} = -3700\,\text{lb} \qquad \hat{f}_{4y} = -5010\,\text{lb} \qquad \hat{m}_4 = 375{,}000\,\text{lb-in.}$$

Considering moment equilibrium at node 2, we see that on element 1, $\hat{m}_2 = 223{,}000$ lb-in., and the opposite value, $-223{,}000$ lb-in., occurs on element 2. Similarly, moment equilibrium is satisfied at node 3.   ■

E X A M P L E   6.2

To illustrate the procedure for solving frames subjected to distributed loads, solve the rigid plane frame shown in Figure 6-6. The frame is fixed at nodes 1 and 3 and subjected to a uniformly distributed load of 1000 lb/ft applied downward over element 2. The global-coordinate axes have been established at node 1. The element lengths are shown in the figure. Let $E = 30 \times 10^6$ psi, $A = 100$ in.^{2}, and $I = 1000$ in.^{4} for both elements of the frame.

We begin by replacing the distributed load acting on element 2 by nodal forces and moments acting at nodes 2 and 3. Using Eqs. (5.4.5)–(5.4.7) (or Appendix D), the equivalent nodal forces and moments are calculated as

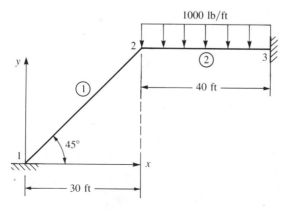

F I G U R E   6-6   Plane frame for analysis

$$f_{2y} = -\frac{wL}{2} = -\frac{(1000)40}{2} = -20{,}000 \text{ lb} = -20 \text{ k}$$

$$m_2 = -\frac{wL^2}{12} = -\frac{(1000)40^2}{12} = -133{,}333 \text{ lb-ft} = -1600 \text{ k-in.}$$

$$f_{3y} = -\frac{wL}{2} = -\frac{(1000)40}{2} = -20{,}000 \text{ lb} = -20 \text{ k}$$

$$(6.2.13)$$

$$m_3 = \frac{wL^2}{12} = \frac{(1000)40^2}{12} = 133{,}333 \text{ lb-ft} = 1600 \text{ k-in.}$$

Using Eq. (6.1.11), each element stiffness matrix is determined as follows.

Element 1

$$\theta^{(1)} = 45° \qquad C = 0.707 \qquad S = 0.707 \qquad L^{(1)} = 42.4 \text{ ft} = 509.0 \text{ in.}$$

$$\frac{E}{L} = \frac{30 \times 10^3}{509} = 58.93$$

$$\underline{k}^{(1)} = 58.93 \begin{bmatrix} 50.02 & 49.98 & 8.33 \\ 49.98 & 50.02 & -8.33 \\ 8.33 & -8.33 & 4000 \end{bmatrix} \qquad (6.2.14)$$

Simplifying Eq. (6.2.14), we obtain

$$\underline{k}^{(1)} = \begin{array}{ccc} d_{2x} & d_{2y} & \phi_2 \end{array} \\ \begin{bmatrix} 2948 & 2945 & 491 \\ 2945 & 2948 & -491 \\ 491 & -491 & 235{,}700 \end{bmatrix} \qquad (6.2.15)$$

where only the parts of the stiffness matrix associated with degrees of freedom at node 2 are included because node 1 is fixed.

Element 2

$$\theta^{(2)} = 0° \qquad C = 1 \qquad S = 0 \qquad L^{(2)} = 40 \text{ ft} = 480 \text{ in.}$$

$$\frac{E}{L} = \frac{30 \times 10^3}{480} = 62.50$$

$$\underline{k}^{(2)} = 62.50 \begin{bmatrix} 100 & 0 & 0 \\ 0 & 0.052 & 12.5 \\ 0 & 12.5 & 4000 \end{bmatrix} \qquad (6.2.16)$$

Simplifying Eq. (6.2.16), we obtain

$$\underline{k}^{(2)} = \begin{array}{ccc} d_{2x} & d_{2y} & \phi_2 \end{array} \\ \begin{bmatrix} 6250 & 0 & 0 \\ 0 & 3.25 & 781.25 \\ 0 & 781.25 & 250{,}000 \end{bmatrix} \qquad (6.2.17)$$

where, again, only the parts of the stiffness matrix associated with degrees of freedom at node 2 are included because node 3 is fixed. On superimposing the stiffness matrices of the elements, using Eqs. (6.2.15) and (6.2.17), and using Eq. (6.2.13) for the nodal forces and moments only at node 2 (because the structure is fixed at node 3), we have

$$
\left\{ \begin{array}{c} 0 \\ -20 \\ -1600 \end{array} \right\} = \left[\begin{array}{ccc} 9198 & 2945 & 491 \\ 2945 & 2951 & 290 \\ 491 & 290 & 485{,}700 \end{array} \right] \left\{ \begin{array}{c} d_{2x} \\ d_{2y} \\ \phi_2 \end{array} \right\} \qquad (6.2.18)
$$

Solving Eq. (6.2.18) for the displacements and the rotation at node 2, we obtain

$$
\left\{ \begin{array}{c} d_{2x} \\ d_{2y} \\ \phi_2 \end{array} \right\} = \left\{ \begin{array}{c} 0.0033 \text{ in.} \\ -0.0097 \text{ in.} \\ -0.0033 \text{ rad} \end{array} \right\} \qquad (6.2.19)
$$

The local forces in each element can now be determined. The procedure for elements that are subjected to a distributed load must be applied to element 2. Recall the local forces are given by $\hat{f} = \underline{\hat{k}} \underline{T} \underline{d}$. For element 1, we then have

$$
\underline{T}\underline{d} = \left[\begin{array}{cccccc} 0.707 & 0.707 & 0 & 0 & 0 & 0 \\ -0.707 & 0.707 & 0 & 0 & 0 & 0 \\ 0 & 0 & 1 & 0 & 0 & 0 \\ 0 & 0 & 0 & 0.707 & 0.707 & 0 \\ 0 & 0 & 0 & -0.707 & 0.707 & 0 \\ 0 & 0 & 0 & 0 & 0 & 1 \end{array} \right] \left\{ \begin{array}{c} 0 \\ 0 \\ 0 \\ 0.0033 \\ -0.0097 \\ -0.0033 \end{array} \right\} \qquad (6.2.20)
$$

Simplifying Eq. (6.2.20) yields

$$
\underline{T}\underline{d} = \left\{ \begin{array}{c} 0 \\ 0 \\ 0 \\ -0.00452 \\ -0.0092 \\ -0.0033 \end{array} \right\} \qquad (6.2.21)
$$

Using Eq. (6.2.21) and Eq. (6.1.8) for $\underline{\hat{k}}$, we obtain

$$
\left\{ \begin{array}{c} \hat{f}_{1x} \\ \hat{f}_{1y} \\ \hat{m}_1 \\ \hat{f}_{2x} \\ \hat{f}_{2y} \\ \hat{m}_2 \end{array} \right\} = \left[\begin{array}{cccccc} 5893 & 0 & 0 & -5893 & 0 & 0 \\ & 2.730 & 694.8 & 0 & -2.730 & 694.8 \\ & & 117{,}900 & 0 & -694.8 & 117{,}900 \\ & & & 5893 & 0 & 0 \\ & & & & 2.730 & -694.8 \\ \text{Symmetry} & & & & & 235{,}800 \end{array} \right] \left\{ \begin{array}{c} 0 \\ 0 \\ 0 \\ -0.00452 \\ -0.0092 \\ -0.0033 \end{array} \right\}
$$

$$
(6.2.22)
$$

Simplifying Eq. (6.2.22) yields the local forces in element 1 as

$$
\begin{array}{llll}
\hat{f}_{1x} = 26.64\,\text{k} & \hat{f}_{1y} = -2.268\,\text{k} & \hat{m}_{1x} = -389.1\,\text{k-in.} \\
\hat{f}_{2x} = -26.64\,\text{k} & \hat{f}_{2y} = 2.268\,\text{k} & \hat{m}_{2x} = -778.2\,\text{k-in.}
\end{array}
\qquad (6.2.23)
$$

For element 2, the local forces are given by Eq. (5.4.11) because a distributed load is acting on the element. From Eqs. (6.1.10) and (6.2.19), we then have

$$
\underline{T}\underline{d} =
\begin{bmatrix}
1 & 0 & 0 & 0 & 0 & 0 \\
0 & 1 & 0 & 0 & 0 & 0 \\
0 & 0 & 1 & 0 & 0 & 0 \\
0 & 0 & 0 & 1 & 0 & 0 \\
0 & 0 & 0 & 0 & 1 & 0 \\
0 & 0 & 0 & 0 & 0 & 1
\end{bmatrix}
\begin{Bmatrix}
0.0033 \\
-0.0097 \\
-0.0033 \\
0 \\
0 \\
0
\end{Bmatrix}
\qquad (6.2.24)
$$

Simplifying Eq. (6.2.24), we obtain

$$
\begin{Bmatrix}
0.0033 \\
-0.0097 \\
-0.0033 \\
0 \\
0 \\
0
\end{Bmatrix}
\qquad (6.2.25)
$$

Using Eq. (6.2.25) and Eq. (6.1.8) for $\underline{\hat{k}}$, we have

$$
\underline{\hat{k}}\underline{d} = \underline{\hat{k}}\underline{T}\underline{d} =
\begin{bmatrix}
6250 & 0 & 0 & -6250 & 0 & 0 \\
 & 3.25 & 781.1 & 0 & -3.25 & 781.1 \\
 & & 250{,}000 & 0 & -781.1 & 125{,}000 \\
 & & & 6250 & 0 & 0 \\
 & & & & 3.25 & -781.1 \\
\text{Symmetry} & & & & & 250{,}000
\end{bmatrix}
\times
\begin{Bmatrix}
0.0033 \\
-0.0097 \\
-0.0033 \\
0 \\
0 \\
0
\end{Bmatrix}
\qquad (6.2.26)
$$

Simplifying Eq. (6.2.26) yields

$$
\underline{\hat{k}}\underline{d} =
\begin{Bmatrix}
20.63 \\
-2.58 \\
-832.57 \\
-20.63 \\
2.58 \\
-412.50
\end{Bmatrix}
\qquad (6.2.27)
$$

To obtain the actual element local nodal forces, we apply Eq. (5.4.11); that is, we must subtract the equivalent nodal forces, Eqs. (6.2.13), from Eq. (6.2.27) to yield

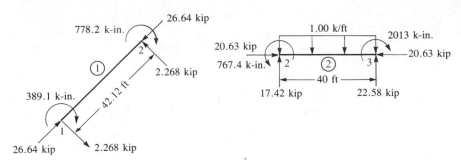

FIGURE 6-7   Free-body diagrams of elements 1 and 2

$$\begin{Bmatrix} \hat{f}_{2x} \\ \hat{f}_{2y} \\ \hat{m}_2 \\ \hat{f}_{3x} \\ \hat{f}_{3y} \\ \hat{m}_3 \end{Bmatrix} = \begin{Bmatrix} 20.63 \\ -2.58 \\ -832.57 \\ -20.63 \\ 2.58 \\ -412.50 \end{Bmatrix} - \begin{Bmatrix} 0 \\ -20 \\ -1600 \\ 0 \\ -20 \\ 1600 \end{Bmatrix} \qquad (6.2.28)$$

Simplifying Eq. (6.2.28), we obtain

$$\hat{f}_{2x} = 20.63\,\text{k} \qquad \hat{f}_{2y} = 17.42\,\text{k} \qquad \hat{m}_2 = 767.4\,\text{k-in.}$$
$$\hat{f}_{3x} = -20.63\,\text{k} \qquad \hat{f}_{3y} = 22.58\,\text{k} \qquad \hat{m}_3 = -2013\,\text{k-in.} \qquad (6.2.29)$$

Using Eqs. (6.2.23) and (6.2.29) for the local forces in each element, we can construct the free-body diagram for each element, as shown in Figure 6-7. From the free-body diagrams, one can confirm the equilibrium of each element, the total frame, and joint 2 as desired.   ∎

In Example 6.3, we will illustrate the equivalent joint force replacement method for a frame subjected to a load acting on an element instead of at one of the joints of the structure. Since no distributed loads are present, the point of application of the concentrated load could be treated as an extra joint in the analysis, and we could solve the problem in the same manner as Example 6.1.

This approach has the disadvantage of increasing the total number of joints, as well as the size of the total structure stiffness matrix $\underline{K}$. For small structures solved by computer, this does not pose a problem. However, for very large structures, this might reduce the maximum size of the structure that could be analyzed. Certainly, this additional node greatly increases the long-hand solution time for the structure. Hence, we will illustrate a standard procedure based on the concept of equivalent joint forces applied to the case of concentrated loads. We will again use Appendix D.

EXAMPLE 6.3

Solve the frame shown in Figure 6-8(a). The frame consists of the three elements shown and is subjected to a 15-kip horizontal load applied at midlength of element 1. Nodes 1, 2, and 3 are fixed, and the dimensions are shown in the figure. Let $E = 30 \times 10^6$ psi, $I = 800$ in.^{4}, and $A = 8$ in.^{2} for all elements.

    1. We first express the applied load in the element 1 local coordinate system (here $\hat{x}$ is directed from node 1 to node 4). This is shown in Figure 6-8(b).

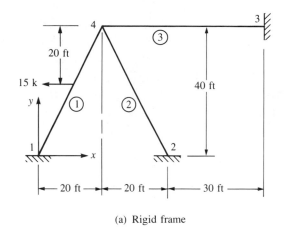

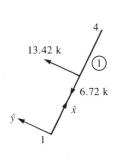

(a) Rigid frame

(b) Applied load expressed in element 1 local-coordinate system

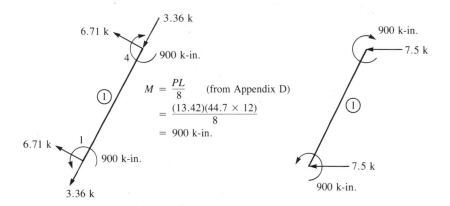

$$M = \frac{PL}{8} \quad \text{(from Appendix D)}$$
$$= \frac{(13.42)(44.7 \times 12)}{8}$$
$$= 900 \text{ k-in.}$$

(c) Equivalent joint forces expressed in local-coordinate system

(d) Final equivalent joint forces expressed in global-coordinate system

FIGURE 6-8   Rigid frame with a load applied on an element

2. Next, we determine the equivalent joint forces at each end of element 1, using the table in Appendix D. (These forces are of opposite sign from what are traditionally known as *fixed-end forces* in classical structural analysis theory.) These equivalent forces (and moments) are shown in Figure 6-8(c).

3. We then transform the equivalent joint forces from the present local-coordinate-system forces into the global-coordinate-system forces, using the equation $f = \underline{T}^T \hat{f}$, where $\underline{T}$ is defined by Eq. (6.1.10). These global joint forces are shown in Figure 6-8(d).

4. Then we analyze the structure in Figure 6-8(d), using the equivalent joint forces (plus actual joint forces, if any) in the usual manner.

5. We obtain the final internal forces developed at the ends of each element that has an applied load (here element 1 only) by subtracting Step 2 joint forces from Step 4 joint forces; that is, Eq. (5.4.11) is applied locally to all elements originally having loads acting on them.

The solution of the structure as shown in Figure 6-8(d) now follows. Using Eq. (6.1.11), we obtain the global stiffness matrix for each element.

Element 1

For element 1, the angle between the global x and the local $\hat{x}$ axes is $63.43°$ because $\hat{x}$ is assumed to be directed from node 1 to node 4. Therefore,

$$C = \cos 63.43° = \frac{x_4 - x_1}{L^{(1)}} = \frac{20 - 0}{44.7} = 0.447$$

$$S = \sin 63.43° = \frac{y_4 - y_1}{L^{(1)}} = \frac{40 - 0}{44.7} = 0.895$$

$$\frac{12I}{L^2} = \frac{12(800)}{(44.7 \times 12)^2} = 0.0334 \qquad \frac{6I}{L} = \frac{6(800)}{44.7 \times 12} = 8.95$$

$$\frac{E}{L} = \frac{30 \times 10^3}{44.7 \times 12} = 55.9$$

Using the preceding results in Eq. (6.1.11) for $\underline{k}$, we obtain

$$\underline{k}^{(1)} = \begin{array}{c} d_{4x} d_{4y} \phi_4 \\ \begin{bmatrix} 90.9 & 178 & 448 \\ 178 & 359 & -224 \\ 448 & -224 & 179,000 \end{bmatrix} \end{array} \qquad (6.2.30)$$

where only the parts of the stiffness matrix associated with degrees of freedom at node 4 are included because node 1 is fixed and, hence, not needed in the solution for the nodal displacements.

Element 3

For element 3, the angle between x and $\hat{x}$ is zero because $\hat{x}$ is directed from node 4 to node 3. Therefore,

$$C = 1 \qquad S = 0 \qquad \frac{12I}{L^2} = \frac{12(800)}{(50 \times 12)^2} = 0.0267$$

$$\frac{6I}{L} = \frac{6(800)}{50 \times 12} = 8.00 \qquad \frac{E}{L} = \frac{30 \times 10^3}{50 \times 12} = 50$$

Substituting these results into $\underline{k}$, we obtain

$$\underline{k}^{(3)} = \begin{array}{c} \begin{array}{ccc} d_{4x} & d_{4y} & \phi_4 \end{array} \\ \begin{bmatrix} 400 & 0 & 0 \\ 0 & 1.334 & 400 \\ 0 & 400 & 160{,}000 \end{bmatrix} \end{array} \qquad (6.2.31)$$

since node 3 is fixed.

Element 2

For element 2, the angle between x and $\hat{x}$ is 116.57° because $\hat{x}$ is directed from node 2 to node 4. Therefore,

$$C = \frac{20 - 40}{44.7} = -0.447 \qquad S = \frac{40 - 0}{44.7} = 0.895$$

$$\frac{12I}{L^2} = 0.0334 \qquad \frac{6I}{L} = 8.95 \qquad \frac{E}{L} = 55.9$$

since element 2 has the same properties as element 1. Substituting these results into $\underline{k}$, we obtain

$$\underline{k}^{(2)} = \begin{array}{c} \begin{array}{ccc} d_{4x} & d_{4y} & \phi_4 \end{array} \\ \begin{bmatrix} 90.9 & -178 & 448 \\ -178 & 359 & 224 \\ 448 & 224 & 179{,}000 \end{bmatrix} \end{array} \qquad (6.2.32)$$

since node 2 is fixed. On superimposing the stiffness matrices given by Eqs. (6.2.30), (6.2.31), and (6.2.32), and using the nodal forces given in Figure 6-8(d) at node 4 only, we have

$$\begin{Bmatrix} -7.50\,\mathrm{k} \\ 0 \\ -900\,\mathrm{k\text{-}in.} \end{Bmatrix} = \begin{bmatrix} 582 & 0 & 896 \\ 0 & 719 & 400 \\ 896 & 400 & 518{,}000 \end{bmatrix} \begin{Bmatrix} d_{4x} \\ d_{4y} \\ \phi_4 \end{Bmatrix} \qquad (6.2.33)$$

Simultaneously solving the three equations in Eq. (6.2.33), we obtain

$$d_{4x} = -0.0103 \text{ in.}$$

$$d_{4y} = 0.000956 \text{ in.} \qquad (6.2.34)$$

$$\phi_4 = -0.00172 \text{ rad}$$

Next, we determine the element forces by again using $\hat{\underline{f}} = \hat{\underline{k}}\underline{T}\underline{d}$. In general, we have

$$
\underline{T}\underline{d} = \begin{bmatrix} C & S & 0 & 0 & 0 & 0 \\ -S & C & 0 & 0 & 0 & 0 \\ 0 & 0 & 1 & 0 & 0 & 0 \\ 0 & 0 & 0 & C & S & 0 \\ 0 & 0 & 0 & -S & C & 0 \\ 0 & 0 & 0 & 0 & 0 & 1 \end{bmatrix} \begin{Bmatrix} d_{ix} \\ d_{iy} \\ \phi_i \\ d_{jx} \\ d_{jy} \\ \phi_j \end{Bmatrix}
$$

Thus, the preceding matrix multiplication yields

$$
\underline{T}\underline{d} = \begin{Bmatrix} Cd_{ix} + Sd_{iy} \\ -Sd_{ix} + Cd_{iy} \\ \phi_i \\ Cd_{jx} + Sd_{jy} \\ -Sd_{jx} + Cd_{jy} \\ \phi_j \end{Bmatrix} \tag{6.2.35}
$$

Element 1

$$
\underline{T}\underline{d} = \begin{Bmatrix} 0 \\ 0 \\ 0 \\ (0.447)(-0.0103) + (0.895)(0.000956) \\ (-0.895)(-0.0103) + (0.447)(0.000956) \\ -0.00172 \end{Bmatrix} = \begin{Bmatrix} 0 \\ 0 \\ 0 \\ -0.00374 \\ 0.00963 \\ -0.00172 \end{Bmatrix}
$$

$$\tag{6.2.36}$$

Using Eq. (6.1.8) for $\underline{\hat{k}}$ and Eq. (6.2.36), we obtain

$$
\underline{\hat{k}}\underline{T}\underline{d} = \begin{bmatrix} 447 & 0 & 0 & -447 & 0 & 0 \\ 0 & 1.868 & 500.5 & 0 & -1.868 & 500.5 \\ 0 & 500.5 & 179{,}000 & 0 & -500.5 & 89{,}490 \\ -447 & 0 & 0 & 447 & 0 & 0 \\ 0 & -1.868 & -500.5 & 0 & 1.868 & -500.5 \\ 0 & 500.5 & 89{,}490 & 0 & -500.5 & 179{,}000 \end{bmatrix} \times \begin{Bmatrix} 0 \\ 0 \\ 0 \\ -0.00374 \\ 0.00963 \\ -0.00172 \end{Bmatrix}
$$

$$\tag{6.2.37}$$

These values are now called *effective nodal forces*. Multiplying the matrices of Eq. (6.2.37) and using Eq. (5.4.11) to subtract the equivalent nodal forces in local coordinates for the element shown in Figure 6-8(c), we obtain the final nodal forces in element 1 as

$$
\underline{\hat{f}}^{(1)} = \begin{Bmatrix} 1.67 \\ -0.88 \\ -158 \\ -1.67 \\ 0.88 \\ -311 \end{Bmatrix} - \begin{Bmatrix} -3.36 \\ 6.71 \\ 900 \\ -3.36 \\ 6.71 \\ -900 \end{Bmatrix} = \begin{Bmatrix} 5.03 \text{ k} \\ -7.59 \text{ k} \\ -1058 \text{ k-in.} \\ 1.68 \text{ k} \\ -5.83 \text{ k} \\ 589 \text{ k-in.} \end{Bmatrix} \tag{6.2.38}
$$

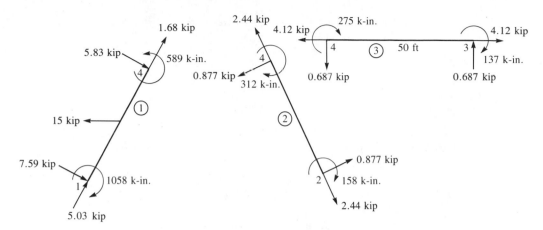

FIGURE 6-9 Free-body diagrams of each element of the frame in Figure 6-8(a)

Similarly, we can use Eqs. (6.2.35) and (6.1.8) for elements 3 and 2 to obtain the local nodal forces in these elements. Since these elements do not have any applied loads on them, the final nodal forces in local coordinates associated with each element are given by $\hat{\underline{f}} = \hat{\underline{k}} \underline{T} \underline{d}$. These forces have been determined as follows.

Element 3

$$\hat{f}_{4x} = -4.12 \text{ k} \qquad \hat{f}_{4y} = -0.687 \text{ k} \qquad \hat{m}_4 = -275 \text{ k-in.}$$
$$\hat{f}_{3x} = 4.12 \text{ k} \qquad \hat{f}_{3y} = 0.687 \text{ k} \qquad \hat{m}_3 = -137 \text{ k-in.} \qquad (6.2.39)$$

Element 2

$$\hat{f}_{2x} = -2.44 \text{ k} \qquad \hat{f}_{2y} = -0.877 \text{ k} \qquad \hat{m}_2 = -158 \text{ k-in.}$$
$$\hat{f}_{4x} = 2.44 \text{ k} \qquad \hat{f}_{4y} = 0.877 \text{ k} \qquad \hat{m}_4 = -312 \text{ k-in.} \qquad (6.2.40)$$

Free-body diagrams of each element are shown in Figure 6-9. Each element has been determined to be in equilibrium, as often occurs even if errors are made in the longhand calculations. However, equilibrium at node 4 and equilibrium of the whole frame are also satisfied. For instance, using the results of Eqs. (6.2.38), (6.2.39), and (6.2.40) to check equilibrium at node 4, which is implicit in the formulation of the global equations, we have

$$\sum M_4 = 589 - 275 - 312 = 2 \text{ k-in.} \qquad \text{(close to zero)}$$
$$\sum F_x = 1.68(0.447) + 5.83(0.895) - 2.44(0.447)$$
$$- 0.877(0.895) - 4.12 = -0.027 \text{ k} \qquad \text{(close to zero)}$$
$$\sum F_y = 1.68(0.895) - 5.83(0.447) + 2.44(0.895)$$
$$- 0.877(0.447) - 0.687 = 0.004 \text{ k} \qquad \text{(close to zero)}$$

Thus, the solution has been verified to be correct within the accuracy associated with a longhand solution.   ∎

To illustrate the solution of a problem involving both bar and frame elements, we will solve the following example.

E X A M P L E   6.4

The bar element 2 is used to stiffen the cantilever beam element 1, as shown in Figure 6-10. Determine the displacements at node 1 and the element forces. For the bar, let $A = 1.0 \times 10^{-3}$ m². For the beam, let $A = 2 \times 10^{-3}$ m², $I = 5 \times 10^{-5}$ m⁴, and $L = 3$ m. For both the bar and the beam elements, let $E = 210$ GPa. Let the angle between the beam and the bar be 45°. A downward force of 500 kN is applied at node 1.

For brevity's sake, since nodes 2 and 3 are fixed, we keep only the parts of $\underline{k}$ for each element that are needed to obtain the global $\underline{K}$ matrix necessary for solution of the nodal degrees of freedom. Using Eq. (3.4.23), we obtain $\underline{k}$ for the bar as

$$\underline{k}^{(2)} = \frac{(1 \times 10^{-3})(210 \times 10^{6})}{(3/\cos 45°)} \begin{bmatrix} 0.5 & 0.5 \\ 0.5 & 0.5 \end{bmatrix}$$

or, simplifying this equation, we obtain

$$\underline{k}^{(2)} = 70 \times 10^{3} \begin{array}{c} \\ \end{array} \overset{\displaystyle d_{1x} \quad d_{1y}}{\begin{bmatrix} 0.354 & 0.354 \\ 0.354 & 0.354 \end{bmatrix}} \qquad (6.2.41)$$

Using Eq. (6.1.11), we obtain $\underline{k}$ for the beam (including axial effects) as

$$\underline{k}^{(1)} = 70 \times 10^{3} \overset{\displaystyle d_{1x} \quad\; d_{1y} \quad\;\; \phi_{1}}{\begin{bmatrix} 2 & 0 & 0 \\ 0 & 0.067 & 0.10 \\ 0 & 0.10 & 0.20 \end{bmatrix}} \qquad (6.2.42)$$

where $(E/L) \times 10^{-3}$ has been factored out in evaluating Eq. (6.2.42).

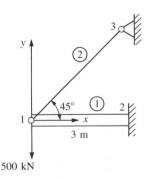

FIGURE 6-10   Cantilever beam with a bar element support

We assemble Eqs. (6.2.41) and (6.2.42) in the usual manner to obtain the global stiffness matrix as

$$K = 70 \times 10^3 \begin{bmatrix} 2.354 & 0.354 & 0 \\ 0.354 & 0.421 & 0.10 \\ 0 & 0.10 & 0.20 \end{bmatrix} \qquad (6.2.43)$$

The global equations are then written for node 1 as

$$\begin{Bmatrix} 0 \\ -500 \\ 0 \end{Bmatrix} = 70 \times 10^3 \begin{bmatrix} 2.354 & 0.354 & 0 \\ 0.354 & 0.421 & 0.10 \\ 0 & 0.10 & 0.20 \end{bmatrix} \begin{Bmatrix} d_{1x} \\ d_{1y} \\ \phi_1 \end{Bmatrix} \qquad (6.2.44)$$

Solving Eq. (6.2.44), we obtain

$$d_{1x} = 0.00338 \text{ m} \qquad d_{1y} = -0.0225 \text{ m} \qquad \phi_1 = 0.0113 \text{ rad} \qquad (6.2.45)$$

In general, the local element forces are obtained using $\hat{f} = \hat{k}T\underline{d}$. For the bar element, we then have

$$\begin{Bmatrix} \hat{f}_{1x} \\ \hat{f}_{3x} \end{Bmatrix} = \frac{AE}{L} \begin{bmatrix} 1 & -1 \\ -1 & 1 \end{bmatrix} \begin{bmatrix} C & S & 0 & 0 \\ 0 & 0 & C & S \end{bmatrix} \begin{Bmatrix} d_{1x} \\ d_{1y} \\ d_{3x} \\ d_{3y} \end{Bmatrix} \qquad (6.2.46)$$

The matrix triple product of Eq. (6.2.46) yields (as one equation)

$$\hat{f}_{1x} = \frac{AE}{L}(Cd_{1x} + Sd_{1y}) \qquad (6.2.47)$$

Substituting the numerical values into Eq. (6.2.47), we obtain

$$\hat{f}_{1x} = \frac{(1 \times 10^{-3} \text{ m}^2)(210 \times 10^6 \text{ kN/m}^2)}{4.24 \text{ m}} \left[\frac{\sqrt{2}}{2}(0.00338 - 0.0225) \right] \qquad (6.2.48)$$

Simplifying Eq. (6.2.48), we obtain

$$\hat{f}_{1x} = -670 \text{ kN} \qquad (6.2.49)$$

where the negative sign means $\hat{f}_{1x}$ is in the direction opposie $\hat{x}$ for element 1. Similarly, we obtain

$$\hat{f}_{3x} = 670 \text{ kN} \qquad (6.2.50)$$

Since the local and global axes are coincident for the beam element, we have $\hat{f} = f$ and $\hat{\underline{d}} = \underline{d}$. Therefore, from Eq. (6.1.6), we have at node 1,

$$\begin{Bmatrix} \hat{f}_{1x} \\ \hat{f}_{1y} \\ \hat{m}_1 \end{Bmatrix} = \begin{bmatrix} C_1 & 0 & 0 \\ 0 & 12C_2 & 6C_2L \\ 0 & 6C_2L & 4C_2L^2 \end{bmatrix} \begin{Bmatrix} d_{1x} \\ d_{1y} \\ \phi_1 \end{Bmatrix} \qquad (6.2.51)$$

where only the upper part of the stiffness matrix is needed because the displacements at node 2 are equal to zero. Substituting numerical values into

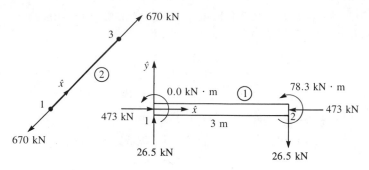

FIGURE 6-11   Free-body diagrams of the bar and beam elements of Figure 6-10

Eq. (6.2.51), we obtain

$$
\left\{ \begin{array}{c} \hat{f}_{1x} \\ \hat{f}_{1y} \\ \hat{m}_1 \end{array} \right\} = 70 \times 10^3 \begin{bmatrix} 2 & 0 & 0 \\ 0 & 0.067 & 0.10 \\ 0 & 0.10 & 0.20 \end{bmatrix} \left\{ \begin{array}{c} 0.00338 \\ -0.0225 \\ 0.0113 \end{array} \right\}
$$

The matrix product then yields

$$\hat{f}_{1x} = 473 \text{ kN} \qquad \hat{f}_{1y} = -26.5 \text{ kN} \qquad \hat{m}_1 = 0.0 \text{ kN·m} \qquad (6.2.52)$$

Similarly, using Eq. (6.1.6), we have at node 2,

$$
\left\{ \begin{array}{c} \hat{f}_{2x} \\ \hat{f}_{2y} \\ \hat{m}_2 \end{array} \right\} = 70 \times 10^3 \begin{bmatrix} -2 & 0 & 0 \\ 0 & -0.067 & -0.10 \\ 0 & 0.10 & 0.10 \end{bmatrix} \left\{ \begin{array}{c} 0.00338 \\ -0.0225 \\ 0.0113 \end{array} \right\}
$$

The matrix product then yields

$$\hat{f}_{2x} = -473 \text{ kN} \qquad \hat{f}_{2y} = 26.5 \text{ kN} \qquad \hat{m}_2 = -78.3 \text{ kN·m} \qquad (6.2.53)$$

To help interpret the results of Eqs. (6.2.49), (6.2.50), (6.2.52), and (6.2.53), free-body diagrams of each element are shown in Figure 6-11. To further verify the results, we can show a check on equilibrium of node 1 to be satisfied.   ∎

6.3   Inclined or Skewed Supports—Frame Element

For the frame element in Figure 6-12, the transformation matrix T used to transform global to local nodal displacements is given by Eq. (6.1.10).

In the example shown in Figure 6-12, we use T applied to node 3 as follows:

$$
\left\{ \begin{array}{c} d'_{3x} \\ d'_{3y} \\ \phi'_3 \end{array} \right\} = \begin{bmatrix} \cos \alpha & \sin \alpha & 0 \\ -\sin \alpha & \cos \alpha & 0 \\ 0 & 0 & 1 \end{bmatrix} \left\{ \begin{array}{c} d_{3x} \\ d_{3y} \\ \phi_3 \end{array} \right\}
$$

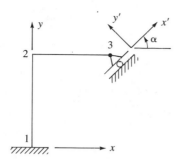

FIGURE 6-12   Frame with inclined support

The same steps as given in Section 3.8 then follow for the truss. The resulting equations for the plane frame in Figure 6-12 are

$$[T_i]\{f\} = [T_i][K][T_i]^T\{d\}$$

or

$$\begin{Bmatrix} F_{1x} \\ F_{1y} \\ M_1 \\ F_{2x} \\ F_{2y} \\ M_2 \\ F'_{3x} \\ F'_{3y} \\ M_3 \end{Bmatrix} = [T_i][K][T_i]^T \begin{Bmatrix} d_{1x} = 0 \\ d_{1y} = 0 \\ \phi_1 = 0 \\ d_{2x} \\ d_{2y} \\ \phi_2 \\ d'_{3x} \\ d'_{3y} = 0 \\ \phi'_3 = \phi_3 \end{Bmatrix}$$

where

$$[T_i] = \begin{bmatrix} [I] & [0] & [0] \\ [0] & [I] & [0] \\ [0] & [0] & [t_3] \end{bmatrix}$$

and

$$[t_3] = \begin{bmatrix} \cos\alpha & \sin\alpha & 0 \\ -\sin\alpha & \cos\alpha & 0 \\ 0 & 0 & 1 \end{bmatrix}$$

6.4 Grid Equations

A **grid** is a structure on which loads are applied perpendicular to the plane of the structure, as opposed to a plane frame where loads are applied in the plane of the structure. We will now develop the grid element stiffness matrix. The elements of a grid are assumed to be rigidly connected, so that the original angles between elements connected together at a node remain unchanged. Both torsional and bending moment continuity then exist at the node point

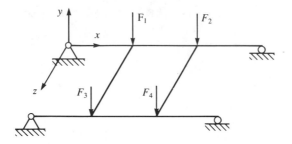

FIGURE 6-13   Typical grid structure

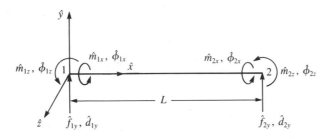

FIGURE 6-14   Grid element with nodal degrees of freedom and nodal forces

of a grid. Examples of grids include floor and bridge deck systems. A typical grid structure subjected to loads F_1, F_2, F_3, and F_4 is shown in Figure 6-13.

We will now consider the development of the grid element stiffness matrix and element equations. A representative grid element with the nodal degrees of freedom and nodal forces is shown in Figure 6-14. The degrees of freedom at each node for a grid are a vertical deflection $\hat{d}_{iy}$ (normal to the grid), a torsional rotation $\hat{\phi}_{ix}$ about the $\hat{x}$ axis, and a bending rotation $\hat{\phi}_{iz}$ about the $\hat{z}$ axis. The nodal forces consist of a transverse force $\hat{f}_{iy}$, a torsional moment $\hat{m}_{ix}$ about the $\hat{x}$ axis, and a bending moment $\hat{m}_{iz}$ about the $\hat{z}$ axis.

To develop the local stiffness matrix for a grid element, we need to include the torsional effects in the basic beam element stiffness matrix Eq. (5.1.14). Recall that Eq. (5.1.14) already accounts for the bending and shear effects.

We can derive the torsional bar element stiffness matrix in a manner analogous to that used for the axial bar element stiffness matrix in Chapter 3. In the derivation, we simply replace $\hat{f}_{ix}$ with $\hat{m}_{ix}$, $\hat{d}_{ix}$ with $\hat{\phi}_{ix}$, E with G (the shear modulus), A with J (the torsional constant, or stiffness factor), σ with τ (shear stress), and ε with γ (shear strain).

The actual derivation is briefly presented as follows. We assume a circular cross section with radius R for simplicity but without loss of generality.

Step 1

Figure 6-15 shows the sign conventions for nodal torque and angle of twist, and for element torque.

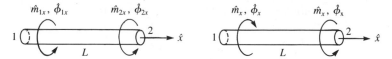

FIGURE 6-15 Nodal and element torque sign conventions

Step 2

We assume a linear angle-of-twist variation along the $\hat{x}$ axis of the bar such that

$$\hat{\phi} = a_1 + a_2\hat{x} \tag{6.4.1}$$

Using the usual procedure of expressing a_1 and a_2 in terms of unknown nodal angles of twist $\hat{\phi}_{1x}$ and $\hat{\phi}_{2x}$, we obtain

$$\hat{\phi} = \left(\frac{\hat{\phi}_{2x} - \hat{\phi}_{1x}}{L}\right)\hat{x} + \hat{\phi}_{1x} \tag{6.4.2}$$

or, in matrix form, Eq. (6.4.2) becomes

$$\hat{\phi} = [N_1 \quad N_2]\begin{Bmatrix}\hat{\phi}_{1x}\\\hat{\phi}_{2x}\end{Bmatrix} \tag{6.4.3}$$

with the shape functions given by

$$N_1 = 1 - \frac{\hat{x}}{L} \qquad N_2 = \frac{\hat{x}}{L} \tag{6.4.4}$$

Step 3

We obtain the shear strain γ/angle of twist $\hat{\phi}$ relationship by considering the torsional deformation of the bar segment shown in Figure 6-16. Assuming that all radial lines, such as OA, remain straight during twisting or torsional deformation, we observe that the arc length $\widehat{AB}$ is given by

$$\widehat{AB} = \gamma_{max}\, d\hat{x} = R\, d\hat{\phi}$$

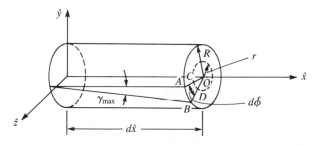

FIGURE 6-16 Torsional deformation of a bar segment

Solving for the maximum shear strain γ_{max}, we obtain

$$\gamma_{max} = \frac{R \, d\hat{\phi}}{d\hat{x}}$$

Similarly, at any radial position r, we then have, from similar triangles OAB and OCD,

$$\gamma = r\frac{d\hat{\phi}}{d\hat{x}} = \frac{r}{L}(\hat{\phi}_{2x} - \hat{\phi}_{1x}) \tag{6.4.5}$$

where we have used Eq. (6.4.2) to derive the final expression in Eq. (6.4.5).

The shear stress τ/shear strain γ relationship for linear-elastic isotropic materials is given by

$$\tau = G\gamma \tag{6.4.6}$$

where G is the shear modulus of the material.

Step 4

We derive the element stiffness matrix in the following manner. From elementary mechanics, we have the shear stress related to the applied torque by

$$\hat{m}_x = \frac{\tau J}{R} \tag{6.4.7}$$

where J is called the *polar moment of inertia* for the circular cross section or, generally, the *torsional constant* for noncircular cross sections. Using Eqs. (6.4.5) and (6.4.6) in Eq. (6.4.7), we obtain

$$\hat{m}_x = \frac{GJ}{L}(\hat{\phi}_{2x} - \hat{\phi}_{1x}) \tag{6.4.8}$$

By the nodal torque sign convention of Figure 6-15,

$$\hat{m}_{1x} = -\hat{m}_x \tag{6.4.9}$$

or, by using Eq. (6.4.8) in Eq. (6.4.9), we obtain

$$\hat{m}_{1x} = \frac{GJ}{L}(\hat{\phi}_{1x} - \hat{\phi}_{2x}) \tag{6.4.10}$$

Similarly,

$$\hat{m}_{2x} = \hat{m}_x \tag{6.4.11}$$

or

$$\hat{m}_{2x} = \frac{GJ}{L}(\hat{\phi}_{2x} - \hat{\phi}_{1x}) \tag{6.4.12}$$

Expressing Eqs. (6.4.10) and (6.4.12) together in matrix form, we have the resulting torsion bar stiffness matrix equation:

$$\begin{Bmatrix} \hat{m}_{1x} \\ \hat{m}_{2x} \end{Bmatrix} = \frac{GJ}{L} \begin{bmatrix} 1 & -1 \\ -1 & 1 \end{bmatrix} \begin{Bmatrix} \hat{\phi}_{1x} \\ \hat{\phi}_{2x} \end{Bmatrix}$$

(6.4.13)

Hence, the stiffness matrix for the torsion bar is

$$\underline{\hat{k}} = \frac{GJ}{L} \begin{bmatrix} 1 & -1 \\ -1 & 1 \end{bmatrix}$$

(6.4.14)

The cross sections of various structures, such as bridge decks, are often not circular. However, Eqs. (6.4.13) and (6.4.14) are still general; to apply them to other cross sections, we simply evaluate the torsional constant J for the particular cross section. For instance, for cross sections made up of thin rectangular shapes such as channels, angles, or I shapes, we approximate J by

$$J = \sum \tfrac{1}{3} b_i t_i^3$$

(6.4.15)

where b_i is the length of any element of the cross section and t_i is the thickness of any element of the cross section. In Table 6-1, we list values of J for various common cross sections. The first four cross sections are called *open sections*. Equation (6.4.15) applies only to these open cross sections. (For more information on the J concept, consult References [2] and [3], and for an extensive table of torsional constants for various cross sectional shapes, consult Reference [4].) We assume the loading to go through the shear center of these open cross sections in order to prevent twisting of the cross section. For more on the shear center consult [2] and [5].

On combining the torsional effects of Eq. (6.4.13) with the shear and bending effects of Eq. (5.1.13), we obtain the local stiffness matrix equation for a grid element as

$$\begin{Bmatrix} \hat{f}_{1y} \\ \hat{m}_{1x} \\ \hat{m}_{1z} \\ \hat{f}_{2y} \\ \hat{m}_{2x} \\ \hat{m}_{2z} \end{Bmatrix} = \begin{bmatrix} \dfrac{12EI}{L^3} & 0 & \dfrac{6EI}{L^2} & \dfrac{-12EI}{L^3} & 0 & \dfrac{6EI}{L^2} \\ & \dfrac{GJ}{L} & 0 & 0 & \dfrac{-GJ}{L} & 0 \\ & & \dfrac{4EI}{L} & \dfrac{-6EI}{L^2} & 0 & \dfrac{2EI}{L} \\ & & & \dfrac{12EI}{L^3} & 0 & \dfrac{-6EI}{L^2} \\ & & & & \dfrac{GJ}{L} & 0 \\ \text{Symmetry} & & & & & \dfrac{4EI}{L} \end{bmatrix} \begin{Bmatrix} \hat{d}_{1y} \\ \hat{\phi}_{1x} \\ \hat{\phi}_{1z} \\ \hat{d}_{2y} \\ \hat{\phi}_{2x} \\ \hat{\phi}_{2z} \end{Bmatrix}$$

(6.4.16)

where, from Eq. (6.4.16), the local stiffness matrix for a grid element is

TABLE 6-1 Torsional constants J and shear centers SC for various cross sections

Cross Section	*Torsional Constant*

1. Channel

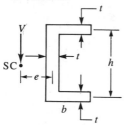

$$J = \frac{t^3}{3}(h + 2b)$$

$$e = \frac{h^2 b^2 t}{4I}$$

2. Angle

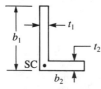

$$J = \tfrac{1}{3}(b_1 t_1^3 + b_2 t_2^3)$$

3. Z section

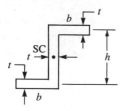

$$J = \frac{t^3}{3}(2b + h)$$

4. Wide-flanged beam with unequal flanges

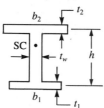

$$J = \tfrac{1}{3}(b_1 t_1^3 + b_2 t_2^3 + h t_w^3)$$

5. Solid circular

$$J = \frac{\pi}{2} r^4$$

6. Closed hollow rectangular

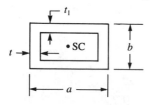

$$J = \frac{2 t t_1 (a - t)^2 (b - t_1)^2}{at + bt_1 - t^2 - t_1^2}$$

$$\hat{k}_G = \begin{bmatrix} \dfrac{12EI}{L^3} & 0 & \dfrac{6EI}{L^2} & -\dfrac{12EI}{L^3} & 0 & \dfrac{6EI}{L^2} \\[2mm] 0 & \dfrac{GJ}{L} & 0 & 0 & -\dfrac{GJ}{L} & 0 \\[2mm] \dfrac{6EI}{L^2} & 0 & \dfrac{4EI}{L} & -\dfrac{6EI}{L^2} & 0 & \dfrac{2EI}{L} \\[2mm] -\dfrac{12EI}{L^3} & 0 & -\dfrac{6EI}{L^2} & \dfrac{12EI}{L^3} & 0 & -\dfrac{6EI}{L^2} \\[2mm] 0 & -\dfrac{GJ}{L} & 0 & 0 & \dfrac{GJ}{L} & 0 \\[2mm] \dfrac{6EI}{L^2} & 0 & \dfrac{2EI}{L} & -\dfrac{6EI}{L^2} & 0 & \dfrac{4EI}{L} \end{bmatrix} \qquad (6.4.17)$$

(columns labeled $\hat{d}_{1y}$, $\hat{\phi}_{1x}$, $\hat{\phi}_{1z}$, $\hat{d}_{2y}$, $\hat{\phi}_{2x}$, $\hat{\phi}_{2z}$)

and the degrees of freedom are in the order (1) vertical deflection, (2) torsional rotation, and (3) bending rotation, as indicated by the notation used above the columns of Eq. (6.4.17).

The transformation matrix relating local to global degrees of freedom for a grid is given by

$$T_G = \begin{bmatrix} 1 & 0 & 0 & 0 & 0 & 0 \\ 0 & C & S & 0 & 0 & 0 \\ 0 & -S & C & 0 & 0 & 0 \\ 0 & 0 & 0 & 1 & 0 & 0 \\ 0 & 0 & 0 & 0 & C & S \\ 0 & 0 & 0 & 0 & -S & C \end{bmatrix} \qquad (6.4.18)$$

where θ is now positive, taken counterclockwise from x to $\hat{x}$ in the $x - z$ plane, and

$$C = \cos \theta = \frac{x_j - x_i}{L} \qquad S = \sin \theta = \frac{z_j - z_i}{L}$$

where L is the length of the element from node i to node j. As indicated by Eq. (6.4.18) for a grid, the vertical deflection $\hat{d}_y$ is invariant with respect to a coordinate transformation (that is, $y = \hat{y}$).

The global stiffness matrix for a grid element arbitrarily oriented in the $x = z$ plane is then given by using Eqs. (6.4.17) and (6.4.18) in

$$k_G = T_G^T \hat{k}_G T_G \qquad (6.4.19)$$

Having formulated the global stiffness matrix for the grid element, the procedure for solution then follows in the same manner as that for the plane frame.

To illustrate the use of the equations developed in Section 6.4, we will now solve the following grid structures.

EXAMPLE 6.5

Analyze the grid shown in Figure 6-17. The grid consists of three elements, is fixed at nodes 2, 3, and 4, and is subjected to a downward vertical force (perpendicular to the plane passing through the grid elements) of 100 kips. The global-coordinate axes have been established at node 3, and the element lengths are shown in the figure. Let $E = 30 \times 10^3$ ksi, $G = 12 \times 10^3$ ksi, $I = 400$ in.^{4}, and $J = 110$ in.^{4} for all elements of the grid.

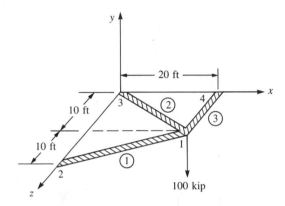

FIGURE 6-17 Grid for analysis

Substituting Eq. (6.4.17) for the local stiffness matrix and Eq. (6.4.18) for the transformation matrix into Eq. (6.4.19), we can obtain each element global stiffness matrix. To expedite the longhand solution, the boundary conditions at nodes 2, 3, and 4,

$$d_{2y} = \phi_{2x} = \phi_{2z} = 0 \qquad d_{3y} = \phi_{3x} = \phi_{3z} = 0 \qquad d_{4y} = \phi_{4x} = \phi_{4z} = 0$$
$$(6.4.20)$$

make it possible to use only the upper left-hand 3×3 partitioned part of the local stiffness and transformation matrices associated with the degrees of freedom at node 1. Therefore, the global stiffness matrices for each element are as follows.

Element 1

For element 1, we assume the local $\hat{x}$ axis to be directed from node 1 to node 2 for the formulation of the element stiffness matrix. We need the following expressions to evaluate the element stiffness matrix:

$$C = \cos\theta = \frac{x_2 - x_1}{L^{(1)}} = \frac{-20 - 0}{22.36} = -0.894$$

$$S = \sin\theta = \frac{z_2 - z_1}{L^{(1)}} = \frac{10 - 0}{22.36} = 0.447$$

$$\frac{12EI}{L^3} = \frac{12(30 \times 10^3)(400)}{(22.36 \times 12)^3} = 7.45$$

(6.4.21)

$$\frac{6EI}{L^2} = \frac{6(30 \times 10^3)(400)}{(22.36 \times 12)^2} = 1000$$

$$\frac{GJ}{L} = \frac{(12 \times 10^3)(110)}{(22.36 \times 12)} = 4920$$

$$\frac{4EI}{L} = \frac{4(30 \times 10^3)(400)}{(22.36 \times 12)} = 179,000$$

Considering the boundary condition Eqs. (6.4.20), using the results of Eqs. (6.4.21) in Eq. (6.4.17) for $\underline{k}_G$ and Eq. (6.4.18) for $\underline{T}_G$, and then applying Eq. (6.4.19), we obtain the upper left-hand 3×3 partitioned part of the global stiffness matrix for element 1 as

$$\underline{k}^{(1)} = \begin{bmatrix} 1 & 0 & 0 \\ 0 & -0.894 & -0.447 \\ 0 & 0.447 & -0.894 \end{bmatrix} \begin{bmatrix} 7.45 & 0 & 1000 \\ 0 & 4920 & 0 \\ 1000 & 0 & 179,000 \end{bmatrix}$$

$$\times \begin{bmatrix} 1 & 0 & 0 \\ 0 & -0.894 & 0.447 \\ 0 & -0.447 & -0.894 \end{bmatrix}$$

Performing the matrix multiplications, we obtain

$$\underline{k}^{(1)} = \begin{bmatrix} 7.45 & -447 & -894 \\ -447 & 39,700 & 69,600 \\ -894 & 69,600 & 144,000 \end{bmatrix}$$

(6.4.22)

Element 2

For element 2, we assume the local $\hat{x}$ axis to be directed from node 1 to node 3 for the formulation of the element stiffness matrix. We need the following expressions to evaluate the element stiffness matrix:

$$C = \frac{x_3 - x_1}{L^{(2)}} = \frac{-20 - 0}{22.36} = -0.894$$

(6.4.23)

$$S = \frac{z_3 - z_1}{L^{(2)}} = \frac{-10 - 0}{22.36} = -0.447$$

Other expressions used in Eq. (6.4.17) are identical to those in Eqs. (6.4.21) for element 1 because E, G, I, J, and L are identical. Evaluating Eq. (6.4.19) for the global stiffness matrix for element 2, we obtain

$$
\underline{k}^{(2)} = \begin{bmatrix} 1 & 0 & 0 \\ 0 & -0.894 & 0.447 \\ 0 & -0.447 & -0.894 \end{bmatrix} \begin{bmatrix} 7.45 & 0 & 1000 \\ 0 & 4920 & 0 \\ 1000 & 0 & 179,000 \end{bmatrix}
$$

$$
\times \begin{bmatrix} 1 & 0 & 0 \\ 0 & -0.894 & -0.447 \\ 0 & 0.447 & -0.894 \end{bmatrix}
$$

Simplifying, we obtain

$$
\underline{k}^{(2)} = \begin{bmatrix} 7.45 & 447 & -894 \\ 447 & 39,700 & -69,600 \\ -894 & -69,600 & 144,000 \end{bmatrix} \tag{6.4.24}
$$

Element 3

For element 3, we assume the local $\hat{x}$ axis to be directed from node 1 to node 4. We need the following expressions to evaluate the element stiffness matrix:

$$
C = \frac{x_4 - x_1}{L^{(3)}} = \frac{20 - 20}{10} = 0
$$

$$
S = \frac{z_4 - z_1}{L^{(3)}} = \frac{0 - 10}{10} = -1
$$

$$
\frac{12EI}{L^3} = \frac{12(30 \times 10^3)(400)}{(10 \times 12)^3} = 83.3
$$

$$
\frac{6EI}{L^2} = \frac{6(30 \times 10^3)(400)}{(10 \times 12)^2} = 5000 \tag{6.4.25}
$$

$$
\frac{GJ}{L} = \frac{(12 \times 10^3)(110)}{10 \times 12} = 11,000
$$

$$
\frac{4EI}{L} = \frac{4(30 \times 10^3)(400)}{10 \times 12} = 400,000
$$

Using Eqs. (6.4.25), we can obtain the upper part of the global stiffness matrix for element 3 as

$$
\underline{k}^{(3)} = \begin{bmatrix} 83.3 & 5000 & 0 \\ 5000 & 400,000 & 0 \\ 0 & 0 & 11,000 \end{bmatrix} \tag{6.4.26}
$$

Superimposing the global stiffness matrices from Eqs. (6.4.22), (6.4.24), and (6.4.26), we obtain the total stiffness matrix of the grid (with boundary conditions applied) as

$$
\underline{K}_G = \begin{bmatrix} 98.2 & 5000 & -1790 \\ 5000 & 479,000 & 0 \\ -1790 & 0 & 299,000 \end{bmatrix} \tag{6.4.27}
$$

The grid matrix equation then becomes

$$
\left\{ \begin{array}{l} F_{1y} = -100 \\ M_{1x} = 0 \\ M_{1z} = 0 \end{array} \right\} = \begin{bmatrix} 98.2 & 5000 & -1790 \\ 5000 & 479{,}000 & 0 \\ -1790 & 0 & 299{,}000 \end{bmatrix} \begin{bmatrix} d_{1y} \\ \phi_{1x} \\ \phi_{1z} \end{bmatrix} \qquad (6.4.28)
$$

The force F_{1y} is negative because the load is applied in the negative y direction. Solving for the displacement and the rotations in Eq. (6.4.28), we obtain

$$d_{1y} = -2.83 \text{ in.}$$

$$\phi_{1x} = 0.0295 \text{ rad} \qquad (6.4.29)$$

$$\phi_{1z} = -0.0169 \text{ rad}$$

Having solved for the unknown displacement and the rotations, we can obtain the local element forces on formulating the element equations in a manner similar to that for the beam and the plane frame. The local forces (which are needed in the design/analysis stage) are found by applying the equation $\hat{f} = \hat{k}_G T_G d$ for each element as follows:

Element 1

Using Eqs. (6.4.17) and (6.4.18) for $\hat{k}_G$ and T_G and Eq. (6.4.29), we obtain

$$
T_G d = \begin{bmatrix} 1 & 0 & 0 & 0 & 0 & 0 \\ 0 & -0.894 & 0.447 & 0 & 0 & 0 \\ 0 & -0.447 & -0.894 & 0 & 0 & 0 \\ 0 & 0 & 0 & 1 & 0 & 0 \\ 0 & 0 & 0 & 0 & -0.894 & 0.447 \\ 0 & 0 & 0 & 0 & -0.447 & -0.894 \end{bmatrix} \begin{Bmatrix} -2.83 \\ 0.0295 \\ -0.0169 \\ 0 \\ 0 \\ 0 \end{Bmatrix}
$$

Multiplying the matrices, we obtain

$$
T_G d = \begin{Bmatrix} -2.83 \\ -0.0339 \\ 0.00192 \\ 0 \\ 0 \\ 0 \end{Bmatrix} \qquad (6.4.30)
$$

Then $\hat{f} = \hat{k}_G T_G d$ becomes

$$
\begin{Bmatrix} \hat{f}_{1y} \\ \hat{m}_{1x} \\ \hat{m}_{1z} \\ \hat{f}_{2y} \\ \hat{m}_{2x} \\ \hat{m}_{2z} \end{Bmatrix} = \begin{bmatrix} 7.45 & 0 & 1000 & -7.45 & 0 & 1000 \\ 0 & 4920 & 0 & 0 & -4920 & 0 \\ 1000 & 0 & 179{,}000 & -1000 & 0 & 89{,}500 \\ -7.45 & 0 & -1000 & 7.45 & 0 & -1000 \\ 0 & -4920 & 0 & 0 & 4920 & 0 \\ 1000 & 0 & 89{,}500 & -1000 & 0 & 179{,}000 \end{bmatrix} \begin{Bmatrix} -2.83 \\ -0.0339 \\ 0.00192 \\ 0 \\ 0 \\ 0 \end{Bmatrix}
$$

$$(6.4.31)$$

Multiplying the matrices in Eq. (6.4.31), we obtain the local element forces as

$$
\begin{Bmatrix} \hat{f}_{1y} \\ \hat{m}_{1x} \\ \hat{m}_{1z} \\ \hat{f}_{2y} \\ \hat{m}_{2x} \\ \hat{m}_{2z} \end{Bmatrix} = \begin{Bmatrix} -19.2 \text{ k} \\ -167 \text{ k-in.} \\ -2480 \text{ k-in.} \\ 19.2 \text{ k} \\ 167 \text{ k-in.} \\ -2660 \text{ k-in.} \end{Bmatrix} \tag{6.4.32}
$$

The directions of the forces acting on element 1 are shown in the free-body diagram of element 1 in Figure 6-18.

Element 2

Similarly, using $\underline{\hat{f}} = \underline{\hat{k}}_G \underline{T}_G \underline{d}$ for element 2, with the direction cosines in Eqs. (6.4.23), we obtain

$$
\begin{Bmatrix} \hat{f}_{1y} \\ \hat{m}_{1x} \\ \hat{m}_{1z} \\ \hat{f}_{3y} \\ \hat{m}_{3x} \\ \hat{m}_{3z} \end{Bmatrix} = \begin{bmatrix} 7.45 & 0 & 1000 & -7.45 & 0 & 1000 \\ 0 & 4920 & 0 & 0 & -4920 & 0 \\ 1000 & 0 & 179{,}000 & -1000 & 0 & 89{,}500 \\ -7.45 & 0 & -1000 & 7.45 & 0 & -1000 \\ 0 & -4920 & 0 & 0 & 4920 & 0 \\ 1000 & 0 & 89{,}500 & -1000 & 0 & 179{,}000 \end{bmatrix}
$$

$$
\times \begin{bmatrix} 1 & 0 & 0 & 0 & 0 & 0 \\ 0 & -0.894 & -0.447 & 0 & 0 & 0 \\ 0 & 0.447 & -0.894 & 0 & 0 & 0 \\ 0 & 0 & 0 & 1 & 0 & 0 \\ 0 & 0 & 0 & 0 & -0.894 & -0.447 \\ 0 & 0 & 0 & 0 & 0.447 & -0.894 \end{bmatrix} \begin{Bmatrix} -2.83 \\ 0.0295 \\ -0.0169 \\ 0 \\ 0 \\ 0 \end{Bmatrix} \tag{6.4.33}
$$

Multiplying the matrices in Eq. (6.4.33), we obtain the local element forces as

$$
\hat{f}_{1y} = 7.23 \text{ k}
$$
$$
\hat{m}_{1x} = -92.5 \text{ k-in.}
$$
$$
\hat{m}_{1z} = 2240 \text{ k-in.}
$$
$$
\hat{f}_{3y} = -7.23 \text{ k} \tag{6.4.34}
$$
$$
\hat{m}_{3x} = 92.5 \text{ k-in.}
$$
$$
\hat{m}_{3z} = -295 \text{ k-in.}
$$

Element 3

Finally, using the direction cosines in Eqs. (6.4.25), we obtain the local element forces as

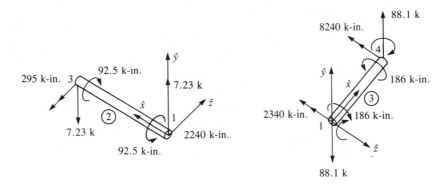

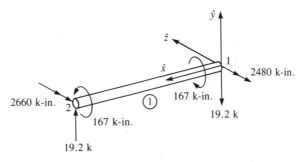

FIGURE 6-18   Free-body diagrams of each element of Figure 6-17

$$
\begin{Bmatrix} \hat{f}_{1y} \\ \hat{m}_{1x} \\ \hat{m}_{1z} \\ \hat{f}_{3y} \\ \hat{m}_{3x} \\ \hat{m}_{3z} \end{Bmatrix} =
\begin{bmatrix}
83.3 & 0 & 5000 & -83.3 & 0 & 5000 \\
0 & 11{,}000 & 0 & 0 & -11{,}000 & 0 \\
5000 & 0 & 400{,}000 & -5000 & 0 & 200{,}000 \\
-83.3 & 0 & -5000 & 83.33 & 0 & -5000 \\
0 & -11{,}000 & 0 & 0 & 11{,}000 & 0 \\
5000 & 0 & 200{,}000 & -5000 & 0 & 400{,}000
\end{bmatrix}
$$

$$
\times
\begin{bmatrix}
1 & 0 & 0 & 0 & 0 & 0 \\
0 & 0 & -1 & 0 & 0 & 0 \\
0 & 1 & 0 & 0 & 0 & 0 \\
0 & 0 & 0 & 1 & 0 & 0 \\
0 & 0 & 0 & 0 & 0 & -1 \\
0 & 0 & 0 & 0 & 1 & 0
\end{bmatrix}
\begin{Bmatrix}
-2.83 \\ 0.0295 \\ -0.0169 \\ 0 \\ 0 \\ 0
\end{Bmatrix}
\qquad (6.4.35)
$$

Multiplying the matrices in Eq. (6.4.35), we obtain the local element forces as

$$\hat{f}_{1y} = -88.1 \text{ k}$$

$$\hat{m}_{1x} = 186 \text{ k-in.}$$

$$\hat{m}_{1z} = -2340 \text{ k-in.}$$

$$\hat{f}_{4y} = 88.1 \text{ k}$$

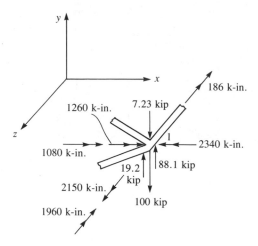

FIGURE   6-19   Free-body diagram of node 1 of Figure 6-17

$$\hat{m}_{4x} = -186 \text{ k-in.}$$

$$\hat{m}_{4z} = -8240 \text{ k-in.} \qquad (6.4.36)$$

Free-body diagrams for each element are shown in Figure 6-18. Each element is in equilibrium. For each element, the $\hat{x}$ axis is shown directed from the first node to the second node, the $\hat{y}$ axis coincides with the global y axis, and the $\hat{z}$ axis is perpendicular to the $\hat{x} - \hat{y}$ plane with its direction given by the right-hand rule.

To verify equilibrium of node 1, we draw a free-body diagram of the node showing all forces and moments transferred from node 1 of each element, as in Figure 6-19. In Figure 6-19, the local forces and moments from each element have been transformed to global components and any applied nodal forces have been included. To perform this transformation, recall that, in general, $\hat{f} = \underline{T} f$ and therefore, $f = \underline{T}^T \hat{f}$ because $\underline{T}^T = \underline{T}^{-1}$. Since we are transforming forces at node 1 of each element, only the upper 3×3 part of Eq. (6.4.18) for $\underline{T}_G$ need be applied. Therefore, by premultiplying the local element forces and moments at node 1 by the transpose of the transformation matrix for each element, we obtain the global nodal forces and moments as follows:

Element 1

$$\begin{Bmatrix} f_{1y} \\ m_{1x} \\ m_{1z} \end{Bmatrix} = \begin{bmatrix} 1 & 0 & 0 \\ 0 & -0.894 & -0.447 \\ 0 & 0.447 & -0.894 \end{bmatrix} \begin{Bmatrix} -19.2 \\ -167 \\ -2480 \end{Bmatrix}$$

Simplifying, we obtain the global-coordinate force and moments as

$$f_{1y} = -19.2 \text{ k} \qquad m_{1x} = 1260 \text{ k-in.} \qquad m_{1z} = 2150 \text{ k-in.} \qquad (6.4.37)$$

where $f_{1y} = \hat{f}_{1y}$ because $y = \hat{y}$.

Element 2

$$\begin{Bmatrix} f_{1y} \\ m_{1x} \\ m_{1z} \end{Bmatrix} = \begin{bmatrix} 1 & 0 & 0 \\ 0 & -0.894 & 0.447 \\ 0 & -0.447 & -0.894 \end{bmatrix} \begin{Bmatrix} 7.23 \\ -92.5 \\ 2240 \end{Bmatrix}$$

Simplifying, we obtain the global-coordinate force and moments as

$$f_{1y} = 7.23 \text{ k} \qquad m_{1x} = 1080 \text{ k-in.} \qquad m_{1z} = -1960 \text{ k-in.} \quad (6.4.38)$$

Element 3

$$\begin{Bmatrix} f_{1y} \\ m_{1x} \\ m_{1z} \end{Bmatrix} = \begin{bmatrix} 1 & 0 & 0 \\ 0 & 0 & 1 \\ 0 & -1 & 0 \end{bmatrix} \begin{Bmatrix} -88.1 \\ 186 \\ -2340 \end{Bmatrix}$$

Simplifying, we obtain the global-coordinate force and moments as

$$f_{1y} = -88.1 \text{ k} \qquad m_{1x} = -2340 \text{ k-in.} \qquad m_{1z} = -186 \text{ k-in.} \quad (6.4.39)$$

Then forces and moments from each element that are equal in magnitude but opposite in sign will be applied to node 1. Hence, the free-body diagram of node 1 is shown in Figure 6-19. Force and moment equilibrium are verified as follows:

$$\sum F_{1y} = -100 - 7.23 + 19.2 + 88.1 = 0.07 \text{ k} \qquad \text{(close to zero)}$$

$$\sum M_{1x} = -1260 - 1080 + 2340 = 0.0 \text{ k-in.}$$

$$\sum M_{1z} = -2150 + 1960 + 186 = -4.00 \text{ k-in.} \qquad \text{(close to zero)}$$

Thus, we have verified the solution to be correct within the accuracy associated with a longhand solution. ∎

E X A M P L E 6.6

Analyze the grid shown in Figure 6-20. The grid consists of two elements, is fixed at nodes 1 and 3, and is subjected to a downward vertical load of 22 kN. The global-coordinate axes and element lengths are shown in the figure. Let $E = 210$ GPa, $G = 84$ GPa, $I = 16.6 \times 10^{-5}$ m^4, and $J = 4.6 \times 10^{-5}$ m^4. As in Example 6.5, we use the boundary conditions and express only the

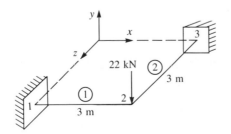

F I G U R E 6-20 Grid example

part of the stiffness matrix associated with the degrees of freedom at node 2. The boundary conditions at nodes 1 and 3 are

$$d_{1y} = \phi_{1x} = \phi_{1z} = 0 \qquad d_{3y} = \phi_{3x} = \phi_{3z} = 0 \qquad (6.4.40)$$

The global stiffness matrices for each element are obtained as follows.

Element 1

For element 1, we have the local $\hat{x}$ axis coincident with the global x axis. Therefore, we obtain

$$C = \frac{x_2 - x_1}{L^{(1)}} = \frac{3}{3} = 1 \qquad S = \frac{z_2 - z_1}{L^{(1)}} = \frac{3 - 3}{3} = 0$$

Other expressions needed to evaluate the stiffness matrix are

$$\frac{12EI}{L^3} = \frac{12(210 \times 10^6 \,\text{kN/m}^2)(16.6 \times 10^{-5} \,\text{m}^4)}{(3 \,\text{m})^3} = 1.55 \times 10^4$$

$$\frac{6EI}{L^2} = \frac{6(210 \times 10^6)(16.6 \times 10^{-5})}{(3)^2} = 2.32 \times 10^4$$

$$\frac{GJ}{L} = \frac{(84 \times 10^6)(4.6 \times 10^{-5})}{3} = 1.28 \times 10^3 \qquad (6.4.41)$$

$$\frac{4EI}{L} = \frac{4(210 \times 10^6)(16.6 \times 10^{-5})}{3} = 4.65 \times 10^4$$

Considering the boundary condition Eqs. (6.4.40), using the results of Eqs. (6.4.41) in Eq. (6.4.17) for $\hat{k}_G$ and Eq. (6.4.18) for $\underline{T}_G$, and then applying Eq. (6.4.19), we obtain the reduced part of the global stiffness matrix associated only with the degrees of freedom at node 2 as

$$\underline{k}^{(1)} = \begin{bmatrix} 1 & 0 & 0 \\ 0 & 1 & 0 \\ 0 & 0 & 1 \end{bmatrix} \begin{bmatrix} 1.55 & 0 & -2.32 \\ 0 & 0.128 & 0 \\ -2.32 & 0 & 4.65 \end{bmatrix} (10^4) \begin{bmatrix} 1 & 0 & 0 \\ 0 & 1 & 0 \\ 0 & 0 & 1 \end{bmatrix}$$

Since the local axes associated with element 1 are parallel to the global axes, we observe that $\underline{T}_G$ is merely the identity matrix; therefore, $\underline{k}_G = \hat{k}_G$. Performing the matrix multiplications, we obtain

$$\underline{k}^{(1)} = \begin{bmatrix} 1.55 & 0 & -2.32 \\ 0 & 0.128 & 0 \\ -2.32 & 0 & 4.65 \end{bmatrix} (10^4) \qquad (6.4.42)$$

Element 2

For element 2, we assume the local $\hat{x}$ axis to be directed from node 2 to node 3 for the formulation of $\underline{k}$. Therefore,

$$C = \frac{x_3 - x_2}{L^{(2)}} = \frac{0 - 0}{3} = 0 \qquad S = \frac{z_3 - z_2}{L^{(2)}} = \frac{0 - 3}{3} = -1 \qquad (6.4.43)$$

Other expressions used in Eq. (6.4.17) are identical to those obtained in Eqs. (6.4.41) for element 1. Evaluating Eq. (6.4.19) for the global stiffness matrix, we obtain

$$\underline{k}^{(2)} = \begin{bmatrix} 1 & 0 & 0 \\ 0 & 0 & 1 \\ 0 & -1 & 0 \end{bmatrix} \begin{bmatrix} 1.55 & 0 & 2.32 \\ 0 & 0.128 & 0 \\ 2.32 & 0 & 4.65 \end{bmatrix} (10^4) \begin{bmatrix} 1 & 0 & 0 \\ 0 & 0 & -1 \\ 0 & 1 & 0 \end{bmatrix}$$

where the reduced part of $\underline{k}$ is now associated with node 2 for element 2. Again performing the matrix multiplications, we have

$$\underline{k}^{(2)} = \begin{bmatrix} 1.55 & 2.32 & 0 \\ 2.32 & 4.65 & 0 \\ 0 & 0 & 0.128 \end{bmatrix} (10^4) \qquad (6.4.44)$$

Superimposing the global stiffness matrices from Eqs. (6.4.42) and (6.4.44), we obtain the total global stiffness matrix (with boundary conditions applied) as

$$\underline{K}_G = \begin{bmatrix} 3.10 & 2.32 & -2.32 \\ 2.32 & 4.78 & 0 \\ -2.32 & 0 & 4.78 \end{bmatrix} (10^4) \qquad (6.4.45)$$

The grid matrix equation becomes

$$\begin{Bmatrix} F_{2y} = -22 \\ M_{2x} = 0 \\ M_{2z} = 0 \end{Bmatrix} = \begin{bmatrix} 3.10 & 2.32 & -2.32 \\ 2.32 & 4.78 & 0 \\ -2.32 & 0 & 4.78 \end{bmatrix} \begin{Bmatrix} d_{2y} \\ \phi_{2x} \\ \phi_{2z} \end{Bmatrix} (10^4) \qquad (6.4.46)$$

Solving for the displacement and the rotations in Eq. (6.4.46), we obtain

$$d_{2y} = -0.259 \times 10^{-2} \text{ m}$$
$$\phi_{2x} = 0.126 \times 10^{-2} \text{ rad} \qquad (6.4.47)$$
$$\phi_{2z} = -0.126 \times 10^{-2} \text{ rad}$$

We determine the local element forces by applying the local equation $\hat{f} = \hat{k}_G \underline{T}_G \underline{d}$ for each element as follows.

Element 1

Using Eq. (6.4.17) for $\hat{k}_G$, Eq. (6.4.18) for $\underline{T}_G$, and Eqs. (6.4.47), we obtain

$$\underline{T}_G \underline{d} = \begin{bmatrix} 1 & 0 & 0 & 0 & 0 & 0 \\ 0 & 1 & 0 & 0 & 0 & 0 \\ 0 & 0 & 1 & 0 & 0 & 0 \\ 0 & 0 & 0 & 1 & 0 & 0 \\ 0 & 0 & 0 & 0 & 1 & 0 \\ 0 & 0 & 0 & 0 & 0 & 1 \end{bmatrix} \begin{Bmatrix} 0 \\ 0 \\ 0 \\ -0.259 \times 10^{-2} \\ 0.126 \times 10^{-2} \\ -0.126 \times 10^{-2} \end{Bmatrix}$$

Multiplying the matrices, we have

$$\underline{T_G}\underline{d} = \begin{Bmatrix} 0 \\ 0 \\ 0 \\ -0.259 \times 10^{-2} \\ 0.126 \times 10^{-2} \\ -0.126 \times 10^{-2} \end{Bmatrix} \qquad (6.4.48)$$

Using Eqs. (6.4.17), (6.4.41), and (6.4.48), we obtain the local element forces as

$$\begin{Bmatrix} \hat{f}_{1y} \\ \hat{m}_{1x} \\ \hat{m}_{1z} \\ \hat{f}_{2y} \\ \hat{m}_{2x} \\ \hat{m}_{2z} \end{Bmatrix} = (10^4) \begin{bmatrix} 1.55 & 0 & 2.32 & -1.55 & 0 & 2.32 \\ & 0.128 & 0 & 0 & -0.128 & 0 \\ & & 4.65 & -2.32 & 0 & 2.33 \\ & & & 1.55 & 0 & -2.32 \\ & & & & 0.128 & 0 \\ & \text{Symmetry} & & & & 4.65 \end{bmatrix}$$

$$\times \begin{Bmatrix} 0 \\ 0 \\ 0 \\ -0.259 \times 10^{-2} \\ 0.126 \times 10^{-2} \\ -0.126 \times 10^{-2} \end{Bmatrix} \qquad (6.4.49)$$

Multiplying the matrices in Eq. (6.4.49), we obtain

$$\hat{f}_{1y} = 11.0\,\text{kN} \qquad \hat{m}_{1x} = -1.50\,\text{kN} \cdot \text{m} \qquad \hat{m}_{1z} = 31.0\,\text{kN} \cdot \text{m}$$
$$\hat{f}_{2y} = -11.0\,\text{kN} \qquad \hat{m}_{2x} = 1.50\,\text{kN} \cdot \text{m} \qquad \hat{m}_{2z} = 1.50\,\text{kN} \cdot \text{m} \qquad (6.4.50)$$

Element 2

We can obtain the local element forces for element 2 in a similar manner. Since the procedure is the same as that used to obtain the element 1 local forces, we will not show the details, but only list the final results as follows:

$$\hat{f}_{2y} = -11.0\,\text{kN} \qquad \hat{m}_{2x} = 1.50\,\text{kN} \cdot \text{m} \qquad \hat{m}_{2z} = -1.50\,\text{kN} \cdot \text{m}$$
$$\hat{f}_{3y} = 11.0\,\text{kN} \qquad \hat{m}_{3x} = -1.50\,\text{kN} \cdot \text{m} \qquad \hat{m}_{3z} = -31.0\,\text{kN} \cdot \text{m}$$
$$(6.4.51)$$

Free-body diagrams showing the local element forces are shown in Figure 6-21. ■

6.5   Beam Element Arbitrarily Oriented in Space

In this section we develop the stiffness matrix for the beam element arbitrarily oriented in space, or three dimensions. This element can then be used to analyze frames in three-dimensional space.

First we consider bending about two axes, as shown in Figure 6-22.

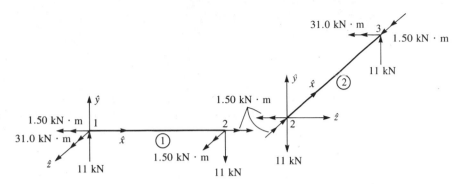

F I G U R E 6-21   Free-body diagram of each element of Figure 6-20

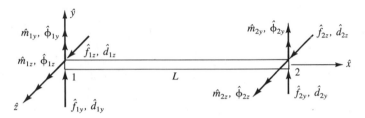

F I G U R E 6-22   Bending about two axes $\hat{y}$ and $\hat{z}$

We establish the following sign convention for the axes. Now we choose positive $\hat{x}$ from node 1 to 2. Then $\hat{y}$ is the principal axis for which the moment of inertia is minimum, I_y. By the right-hand rule we establish $\hat{z}$, and the maximum moment of inertia is I_z.

Bending in $\hat{x}$-$\hat{z}$ Plane

First consider bending in the x-z plane to $\hat{m}_y$. Then clockwise rotation $\hat{\phi}_y$ is in the same sense as before for single bending. The stiffness matrix due to bending in the x-z plane is then

$$
\underline{\hat{k}}_y = \frac{EI_y}{L^4}
\begin{bmatrix}
12L & -6L^2 & -12L & -6L^2 \\
 & 4L^3 & 6L^2 & 2L^3 \\
 & & 12L & 6L^2 \\
\text{Symmetry} & & & 4L^3
\end{bmatrix}
\tag{6.5.1}
$$

where I_y is the moment of inertia of the cross section about the principal axis $\hat{y}$, the weak axis; that is, $I_y < I_z$.

Bending in the $\hat{x}$-$\hat{y}$ Plane

Now we consider bending in the $\hat{x}$-$\hat{y}$ plane due to $\hat{m}_z$. Now positive rotation $\hat{\phi}_z$ is counterclockwise instead of clockwise. Therefore, some signs change in the stiffness matrix for bending in the $\hat{x}$-$\hat{y}$ plane. The resulting stiffness matrix is

$$\hat{k}_z = \frac{EI_z}{L^4} \begin{bmatrix} 12L & 6L^2 & -12L & 6L^2 \\ & 4L^3 & -6L^2 & 2L^3 \\ & & 12L & -6L^2 \\ \text{Symmetry} & & & 4L^3 \end{bmatrix}$$

(6.5.2)

Direct superposition of Eqs. (6.5.1) and (6.5.2) with the axial stiffness matrix Eq. (3.1.14) and the torsional stiffness matrix Eq. (6.4.14) yields the element stiffness matrix for the beam or frame element in three-dimensional space as

$$\hat{k} = \begin{bmatrix}
\dfrac{AE}{L} & 0 & 0 & 0 & 0 & 0 & -\dfrac{AE}{L} & 0 & 0 & 0 & 0 & 0 \\[2mm]
0 & \dfrac{12EI_z}{L^3} & 0 & 0 & 0 & \dfrac{6EI_z}{L^2} & 0 & -\dfrac{12EI_z}{L^3} & 0 & 0 & 0 & \dfrac{6EI_z}{L^2} \\[2mm]
0 & 0 & \dfrac{12EI_y}{L^3} & 0 & -\dfrac{6EI_y}{L^2} & 0 & 0 & 0 & -\dfrac{12EI_y}{L^3} & 0 & -\dfrac{6EI_y}{L^2} & 0 \\[2mm]
0 & 0 & 0 & \dfrac{GJ}{L} & 0 & 0 & 0 & 0 & 0 & -\dfrac{GJ}{L} & 0 & 0 \\[2mm]
0 & 0 & -\dfrac{6EI_y}{L^2} & 0 & \dfrac{4EI_y}{L} & 0 & 0 & 0 & \dfrac{6EI_y}{L^2} & 0 & \dfrac{2EI_y}{L} & 0 \\[2mm]
0 & \dfrac{6EI_z}{L^2} & 0 & 0 & 0 & \dfrac{4EI_z}{L} & 0 & -\dfrac{6EI_z}{L^2} & 0 & 0 & 0 & \dfrac{2EI_z}{L} \\[2mm]
-\dfrac{AE}{L} & 0 & 0 & 0 & 0 & 0 & \dfrac{AE}{L} & 0 & 0 & 0 & 0 & 0 \\[2mm]
0 & -\dfrac{12EI_z}{L^3} & 0 & 0 & 0 & -\dfrac{6EI_z}{L^2} & 0 & \dfrac{12EI_z}{L^3} & 0 & 0 & 0 & -\dfrac{6EI_z}{L^2} \\[2mm]
0 & 0 & -\dfrac{12EI_y}{L^3} & 0 & \dfrac{6EI_y}{L^2} & 0 & 0 & 0 & \dfrac{12EI_y}{L^3} & 0 & \dfrac{6EI_y}{L^2} & 0 \\[2mm]
0 & 0 & 0 & -\dfrac{GJ}{L} & 0 & 0 & 0 & 0 & 0 & \dfrac{GJ}{L} & 0 & 0 \\[2mm]
0 & 0 & -\dfrac{6EI_y}{L^2} & 0 & \dfrac{2EI_y}{L} & 0 & 0 & 0 & \dfrac{6EI_y}{L^2} & 0 & \dfrac{4EI_y}{L} & 0 \\[2mm]
0 & \dfrac{6EI_z}{L^2} & 0 & 0 & 0 & \dfrac{2EI_z}{L} & 0 & -\dfrac{6EI_z}{L^2} & 0 & 0 & 0 & \dfrac{4EI_z}{L}
\end{bmatrix}$$

(6.5.3)

The transformation from local to global axis system is accomplished as follows:

$$\underline{k} = \underline{T}^T \hat{\underline{k}} \underline{T}$$

(6.5.4)

where $\hat{\underline{k}}$ is given by Eq. (6.5.3) and $\underline{T}$ is given by

$$\underline{T} = \begin{bmatrix} \underline{\lambda}_{3 \times 3} & & & \\ & \underline{\lambda}_{3 \times 3} & & \\ & & \underline{\lambda}_{3 \times 3} & \\ & & & \underline{\lambda}_{3 \times 3} \end{bmatrix}$$

where

$$\underline{\lambda} = \begin{bmatrix} C_{x\hat{x}} & C_{y\hat{x}} & C_{z\hat{x}} \\ C_{x\hat{y}} & C_{y\hat{y}} & C_{z\hat{y}} \\ C_{x\hat{z}} & C_{y\hat{z}} & C_{z\hat{z}} \end{bmatrix}$$

(6.5.6)

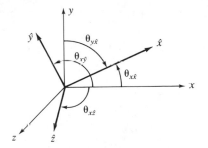

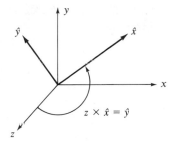

FIGURE 6-23   Direction
cosines associated with the x axis

FIGURE 6-24

Here $C_{y\hat{x}}$ and $C_{x\hat{y}}$ are not necessarily equal. The direction cosines are shown in part in Figure 6-23.

Remember that direction cosines of the $\hat{x}$ axis member are

$$\underline{\hat{x}} = \cos\theta_{x\hat{x}}\bar{i} + \cos\theta_{y\hat{x}}\bar{j} + \cos\theta_{z\hat{x}}\bar{k} \qquad (6.5.7)$$

where

$$\cos\theta_{x\hat{x}} = \frac{x_2 - x_1}{L} = l$$

$$\cos\theta_{y\hat{x}} = \frac{y_2 - y_1}{L} = m \qquad (6.5.8)$$

$$\cos\theta_{z\hat{x}} = \frac{z_2 - z_1}{L} = n$$

The $\hat{y}$ axis is selected to be perpendicular to the $\hat{x}$ and z axes in such a way that the cross product of global z with $\hat{x}$ results in the $\hat{y}$ axis, as shown in Figure 6-24. Therefore,

$$z \times \hat{x} = \hat{y} = \frac{1}{D}\begin{vmatrix} \bar{i} & \bar{j} & \bar{k} \\ 0 & 0 & 1 \\ l & m & n \end{vmatrix} \qquad (6.5.9)$$

$$\hat{y} = -\frac{m}{D}\bar{i} + \frac{l}{D}\bar{j} \qquad (6.5.10)$$

and

$$D = (l^2 + m^2)^{1/2}$$

The $\hat{z}$ axis will be determined by the orthogonality condition $\hat{z} = \hat{x} \times \hat{y}$ as follows:

$$\hat{z} = \hat{x} \times \hat{y} = \frac{1}{D}\begin{vmatrix} \bar{i} & \bar{j} & \bar{k} \\ l & m & n \\ -m & l & 0 \end{vmatrix} \qquad (6.5.11)$$

or

$$\hat{z} = -\frac{ln}{D}\bar{i} - \frac{mn}{D}\bar{j} + D\bar{k} \qquad (6.5.12)$$

Combining Eqs. (6.5.7), (6.5.10), and (6.5.12), the 3×3 transformation matrix becomes

$$
\underline{\lambda}_{3\times3} =
\begin{bmatrix}
l & m & n \\[6pt]
-\dfrac{m}{D} & \dfrac{l}{D} & 0 \\[10pt]
-\dfrac{ln}{D} & -\dfrac{mn}{D} & D
\end{bmatrix}
\tag{6.5.13}
$$

This vector $\underline{\lambda}$ rotates a vector from the local coordinate system into the global one. This is the $\underline{\lambda}$ used in the $\underline{T}$ matrix. In summary, we have

$$
\begin{aligned}
\cos\theta_{x\hat{y}} &= -\frac{m}{D} \\[10pt]
\cos\theta_{y\hat{y}} &= \frac{l}{D} \\[10pt]
\cos\theta_{z\hat{y}} &= 0 \\[10pt]
\cos\theta_{x\hat{z}} &= -\frac{ln}{D} \\[10pt]
\cos\theta_{y\hat{z}} &= -\frac{mn}{D} \\[10pt]
\cos\theta_{z\hat{z}} &= D
\end{aligned}
\tag{6.5.14}
$$

Two exceptions arise when local and global axes have special orientations with respect to each other. If the local $\hat{x}$ axis coincides with the global z axis, then the member is parallel to the global z axis and the $\hat{y}$ axis becomes uncertain, as shown in Figure 6-25(a). In this case the local $\hat{y}$ axis is selected as the global y axis. Then for the positive $\hat{x}$ axis in the same direction as the global z, $\underline{\lambda}$ becomes

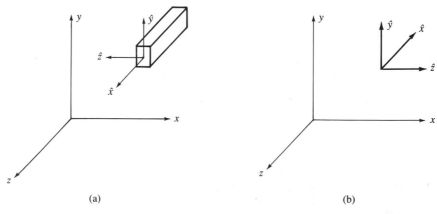

(a)   (b)

F I G U R E   6-25   Special cases of transformation matrices

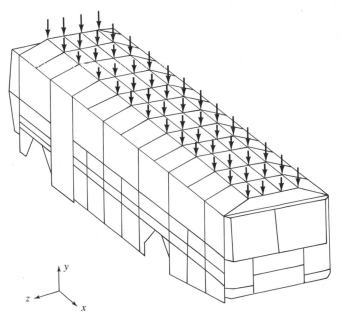

F I G U R E   6-26   Finite element model of bus frame subjected to roof load

$$\underline{\lambda} = \begin{bmatrix} 0 & 0 & 1 \\ 0 & 1 & 0 \\ -1 & 0 & 0 \end{bmatrix} \qquad\qquad (6.5.15)$$

For the positive $\hat{x}$ axis opposite the global z [Figure 6-25(b)], $\underline{\lambda}$ becomes

$$\underline{\lambda} = \begin{bmatrix} 0 & 0 & -1 \\ 0 & 1 & 0 \\ 1 & 0 & 0 \end{bmatrix} \qquad\qquad (6.5.16)$$

An example using the frame element in three-dimensional space is shown in Figure 6-26. Figure 6-26 shows a bus frame subjected to a static roof-crush analysis. In this model 599 frame elements and 357 nodes were used. A total downward load of 100 kN was uniformly spread over the 56 nodes of the roof portion of the frame. Figure 6-27 shows the rear of the frame and the displaced view of the rear frame. Other frame models with additional loads simulating rollover and front-end collisions were studied in [6].

6.6   **Concept of Substructure Analysis**

Sometimes structures are too large to be analyzed as a single system or treated as a whole; that is, the final stiffness matrix and equations for solution exceed the memory capacity of the computer. A procedure to overcome this problem is to separate the whole structure into smaller units called **substruc-**

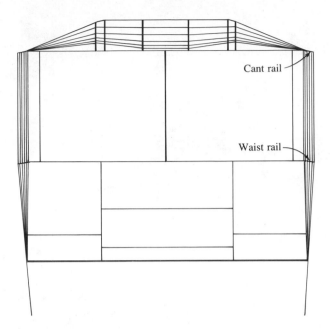

F I G U R E   6-27   Displaced view of the frame made of square section members

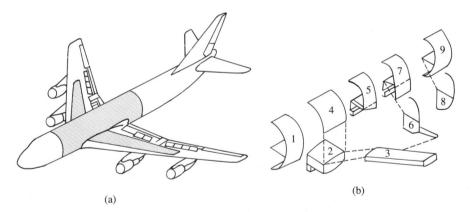

(a)   (b)

F I G U R E   6-28   Airplane frame showing substructuring (a) Boeing 747 Aircraft (shaded area indicates portion of the airframe analyzed by finite element method) and (b) substructures for finite element analysis of shaded region

tures. For example, the space frame of an airplane, as shown in Figure 6-28(a), may require thousands of nodes and elements to completely model and describe the response of the whole structure. If we separate the aircraft into substructures, such as parts of the fuselage or body, wing sections, etc., as shown in Figure 6-28(b), then we can solve the problem more readily and on computers with limited memory.

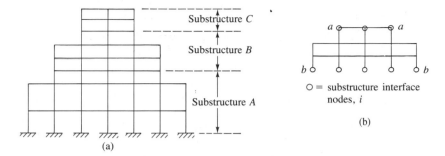

FIGURE 6-29   (a) Rigid frame for substructure analysis and
(b) substructure *B*

The analysis of the airplane frame is performed by treating each substructure separately while ensuring force and displacement compatibility at the intersections where partitioning occurs.

To describe the procedure of substructuring, consider the rigid frame shown in Figure 6-29 (even though this frame could be analyzed as a whole). First we define individual separate substructures. Normally, we make these substructures of similar size, and to reduce computations we make as few cuts as possible. We then separate the frame into three parts, *A*, *B*, and *C*.

We now analyze a typical substructure *B* shown in Figure 6-29(b). This substructure includes the beams at the top (*a-a*), but the beams at the bottom (*b-b*) are included in substructure *A*, although the beams at top could be included in substructure *C* and the beams at the bottom could be included in substructure *B*.

The force-displacement equations for substructure *B* are partitioned with the interface displacements and forces separated from the interior ones as follows:

$$\left\{ \begin{matrix} \underline{F}_i^B \\ \underline{F}_e^B \end{matrix} \right\} = \begin{bmatrix} \underline{K}_{ii}^B & \underline{K}_{ie}^B \\ \underline{K}_{ei}^B & \underline{K}_{ee}^B \end{bmatrix} \left\{ \begin{matrix} \underline{d}_i^B \\ \underline{d}_e^B \end{matrix} \right\} \qquad (6.6.1)$$

where the superscript *B* denotes the substructure *B*, subscript *i* denotes the interface nodal forces and displacements, and subscript *e* denotes the interior nodal forces and displacements to be eliminated by static condensation. Using static condensation, Eq. (6.6.1) becomes

$$\underline{F}_i^B = \underline{K}_{ii}^B \underline{d}_i^B + \underline{K}_{ie}^B \underline{d}_e^B \qquad (6.6.2)$$

$$\underline{F}_e^B = \underline{K}_{ei}^B \underline{d}_i^B + \underline{K}_{ee}^B \underline{d}_e^B \qquad (6.6.3)$$

We eliminate the interior displacements $\underline{d}_e$ by solving Eq. (6.6.3) for $\underline{d}_e^B$, as follows:

$$\underline{d}_e^B = [\underline{K}_{ee}^B]^{-1}[\underline{F}_e^B - \underline{K}_{ei}^B \underline{d}_i^B] \qquad (6.6.4)$$

Then we substitute Eq. (6.6.4) for $\underline{d}_e^B$ into Eq. (6.6.2) to obtain

$$\underline{F}_i^B - \underline{K}_{ie}^B [\underline{K}_{ee}^B]^{-1} \underline{F}_e^B = (\underline{K}_{ii}^B - \underline{K}_{ie}^B [\underline{K}_{ee}^B]^{-1} \underline{K}_{ei}^B) \underline{d}_i^B \qquad (6.6.5)$$

We define

$$\underline{\bar{F}}_i^B = \underline{K}_{ie}^B [\underline{K}_{ee}^B]^{-1} \underline{F}_e^B \quad \text{and} \quad \underline{\bar{K}}_{ii}^B = \underline{K}_{ii}^B - \underline{K}_{ie}^B [\underline{K}_{ee}^B]^{-1} \underline{K}_{ei}^B \quad (6.6.6)$$

Substituting Eq. (6.6.6) into (6.6.5), we obtain

$$\underline{F}_i^B - \underline{\bar{F}}_i^B = \underline{\bar{K}}_{ii}^B \underline{d}_i^B \tag{6.6.7}$$

Similarly, we can write force-displacement equations for substructures A and C. These equations can be partitioned in a manner similar to Eq. (6.6.1) to obtain

$$\left\{ \begin{matrix} \underline{F}_i^A \\ \underline{F}_e^A \end{matrix} \right\} = \begin{bmatrix} \underline{K}_{ii}^A & \underline{K}_{ie}^A \\ \underline{K}_{ei}^A & \underline{K}_{ee}^A \end{bmatrix} \left\{ \begin{matrix} \underline{d}_i^A \\ \underline{d}_e^A \end{matrix} \right\} \tag{6.6.8}$$

Eliminating $\underline{d}_e^A$, we obtain

$$\underline{F}_i^A - \underline{\bar{F}}_i^A = \underline{\bar{K}}_{ii}^A \underline{d}_i^A \tag{6.6.9}$$

Similarly, for substructure C, we have

$$\underline{F}_i^C - \underline{\bar{F}}_i^C = \underline{\bar{K}}_{ii}^C \underline{d}_i^C \tag{6.6.10}$$

The whole frame is now considered to be made of superelements A, B, and C connected at interface nodal points (each superelement being made up of a collection of individual smaller elements). Using compatibility, we have

$$\underline{d}_{i\,\text{top}}^A = \underline{d}_{i\,\text{bottom}}^B \quad \text{and } \underline{d}_{i\,\text{top}}^B = \underline{d}_{i\,\text{bottom}}^C \tag{6.6.11}$$

That is, the interface displacements at the common locations where cuts were made must be the same.

The response of the whole structure can now be obtained by direct superposition of Eqs. (6.6.7), (6.6.9), and (6.6.10), where now the final equations are expressed in terms of the interface displacements at the eight interface nodes only [Figure 6-29(b)] as

$$\underline{F}_i - \underline{\bar{F}}_i = \underline{\bar{K}}_{ii} \underline{d}_i \tag{6.6.12}$$

The solution of Eq. (6.6.12) gives the displacements at the interface nodes. To obtain the displacements within each substructure, we use the force-displacement Eqs. (6.6.4) for $\underline{d}_e^B$ with similar equations for substructures A and C. Example 6.7 illustrates the concept of substructure analysis. In order to solve by hand, a relatively simple structure is used.

EXAMPLE 6.7

Solve for the displacement and rotation at node 3 for the beam in Figure 6-30 by using substructuring. Let $E = 29 \times 10^3$ ksi and $I = 1000$ in.[4]
To illustrate the substructuring concept, we divide the beam into two substructures, labeled 1 and 2 in Figure 6-31. The 10-kip force has been

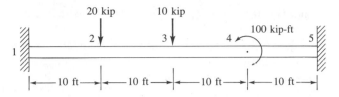

FIGURE 6-30 Beam analyzed by substructuring

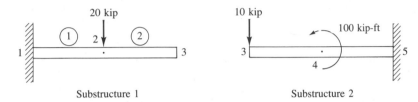

Substructure 1   Substructure 2

FIGURE 6-31 Beam of Figure 6-30 separated into substructures

assigned to node 3 of substructure 2, although it could have been assigned to either substructure or a fraction of it assigned to each substructure.

The stiffness matrix for each beam element is given by Eq. (5.1.14) as

$$
\underline{k}^{(1)} = \underline{k}^{(2)} = \underline{k}^{(3)} = \underline{k}^{(4)} = \frac{29 \times 10^6}{(120)^3}
\begin{array}{c}
\begin{matrix} 1 & 2 \\ 2 & 3 \\ 3 & 4 \\ 4 & 5 \end{matrix} \\
\begin{bmatrix}
12 & 6(120) & -12 & 6(120) \\
6(120) & 4(120)^2 & -6(120) & 2(120)^2 \\
-12 & -6(120) & 12 & -6(120) \\
6(120) & 2(120)^2 & -6(120) & 4(120)^2
\end{bmatrix}
\end{array}
$$

$$(6.6.13)$$

$$
= 16.78
\begin{bmatrix}
12 & 720 & -12 & 720 \\
720 & 57,600 & -720 & 28,800 \\
-12 & -720 & 12 & -720 \\
720 & 28,800 & -720 & 57,600
\end{bmatrix}
$$

$$(6.6.14)$$

For substructure 1, we add the stiffness matrices of elements 1 and 2 together. The equations are

$$
16.78
\begin{bmatrix}
12 + 12 & -720 + 720 & -12 & 720 \\
-720 + 720 & 57,600 + 57,600 & -720 & 28,800 \\
\hline
-12 & -720 & 12 & -720 \\
720 & 28,800 & -720 & 57,600
\end{bmatrix}
\begin{Bmatrix}
d_{2y} \\ \phi_2 \\ d_{3y} \\ \phi_3
\end{Bmatrix}
=
\begin{Bmatrix}
-20 \\ 0 \\ 0 \\ 0
\end{Bmatrix}
$$

$$(6.6.15)$$

where the boundary conditions $d_{1y} = \phi_1 = 0$ were used to reduce the equations.

Rewriting Eq. (6.6.15) with the interface displacements first allows us to use Eq. (6.6.6) to condense out, or eliminate, the interior degrees of freedom, d_{2y} and ϕ_2. These reordered equations are

$$
\begin{aligned}
16.78(12d_{3y} - 720\phi_3 - 12d_{2y} - 720\phi_2) &= 0 \\
16.78(-720d_{3y} + 57{,}600\phi_3 + 720d_{2y} + 28{,}800\phi_2) &= 0 \\
16.78(-12d_{3y} + 720\phi_3 + 24d_{2y} + \phi_2) &= -20 \\
16.78(-720d_{2y} + 28{,}800\phi_3 + 0d_{2y} + 115{,}200\phi_2) &= 0
\end{aligned}
\qquad (6.6.16)
$$

Using Eq. (6.6.6), we obtain equations for the interface degrees of freedom as

$$
16.78 \left\{ \begin{bmatrix} 12 & -720 \\ -720 & 57{,}600 \end{bmatrix} - \begin{bmatrix} -12 & -720 \\ 720 & 28{,}800 \end{bmatrix} \begin{bmatrix} 24 & 0 \\ 0 & 115{,}200 \end{bmatrix}^{-1} \right.
$$
$$
\left. \begin{bmatrix} -12 & 720 \\ -720 & 28{,}800 \end{bmatrix} \right\} \begin{Bmatrix} d_{3y} \\ \phi_3 \end{Bmatrix}
$$
$$
= \begin{Bmatrix} 0 \\ 0 \end{Bmatrix} - \begin{bmatrix} -12 & -720 \\ 720 & 28{,}800 \end{bmatrix} \begin{bmatrix} 24 & 0 \\ 0 & 115{,}200 \end{bmatrix}^{-1} \begin{Bmatrix} -20 \\ 0 \end{Bmatrix}
$$
$$
\qquad (6.6.17)
$$

Simplifying Eq. (6.6.17), we obtain

$$
\begin{bmatrix} 25.17 & -3020 \\ -3020 & 483{,}264 \end{bmatrix} \begin{Bmatrix} d_{3y} \\ \phi_3 \end{Bmatrix} = \begin{Bmatrix} -10 \\ 600 \end{Bmatrix}
\qquad (6.6.18)
$$

For substructure 2, we add the stiffness matrices of elements 3 and 4 together. The equations are

$$
16.78 \begin{bmatrix} 12 & 720 & -12 & 720 \\ 720 & 57{,}600 & -720 & 28{,}800 \\ -12 & -720 & 12+12 & -720+720 \\ 720 & 28{,}800 & -720+720 & 57{,}600+57{,}600 \end{bmatrix} \begin{Bmatrix} d_{3y} \\ \phi_3 \\ d_{4y} \\ \phi_4 \end{Bmatrix} = \begin{Bmatrix} -10 \\ 0 \\ 0 \\ 1200 \end{Bmatrix}
$$
$$
\qquad (6.6.19)
$$

where boundary conditions $d_{5y} = \phi_5 = 0$ were used to reduce the equations.

Using static condensation, Eq. (6.6.6), we obtain equations with only the interface displacements d_{3y} and ϕ_3. These equations are

$$
16.78 \left\{ \begin{bmatrix} 12 & 720 \\ 720 & 57{,}600 \end{bmatrix} - \begin{bmatrix} -12 & 720 \\ -720 & 28{,}800 \end{bmatrix} \begin{bmatrix} 24 & 0 \\ 0 & 115{,}200 \end{bmatrix}^{-1} \right.
$$
$$
\left. \begin{bmatrix} -12 & -720 \\ 720 & 28{,}800 \end{bmatrix} \right\} \begin{Bmatrix} d_{3y} \\ \phi_3 \end{Bmatrix}
$$
$$
= \begin{Bmatrix} -10 \\ 0 \end{Bmatrix} - \begin{bmatrix} -12 & 720 \\ -720 & 28{,}800 \end{bmatrix} \begin{bmatrix} 24 & 0 \\ 0 & 115{,}200 \end{bmatrix}^{-1} \begin{Bmatrix} 0 \\ 1200 \end{Bmatrix}
$$
$$
\qquad (6.6.20)
$$

Simplifying Eq. (6.6.20), we obtain

$$\begin{bmatrix} 25.17 & 3020 \\ 3020 & 483{,}264 \end{bmatrix} \begin{Bmatrix} d_{3y} \\ \phi_3 \end{Bmatrix} = \begin{Bmatrix} -17.5 \\ -300 \end{Bmatrix} \qquad (6.6.21)$$

Adding Eqs. (6.6.18) and (6.6.21), we obtain the final nodal equilibrium equations at the interface degrees of freedom as

$$\begin{bmatrix} 50.34 & 0 \\ 0 & 966{,}528 \end{bmatrix} \begin{Bmatrix} d_{3y} \\ \phi_3 \end{Bmatrix} = \begin{Bmatrix} -27.5 \\ 300 \end{Bmatrix} \qquad (6.6.22)$$

Solving Eq. (6.6.22) for the displacement and rotation at node 3, we obtain

$$d_{3y} = -0.5463 \text{ in.}$$
$$\phi_3 = 0.0003104 \text{ rad} \qquad (6.6.23)$$

We could now return to Eq. (6.6.15) or (6.6.16) to obtain d_{2y} and ϕ_2 and to Eq. (6.6.19) to obtain d_{4y} and ϕ_4.   ∎

We emphasize that this example is used as a simple illustration of substructuring and is not typical of the size of problems where substructuring is normally performed. Generally, substructuring is used when the number of degrees of freedom is very large, as might occur, for instance, for very large structures such as the airframe in Figure 6-28.

6.7   Description of a Computer Program for Plane Frame and Grid Analysis

We will now describe a computer program called PFRAME (listed on the disk accompanying the text) that can be used to solve large two-dimensional rigid plane frame and grid problems. The program is based on the flowchart of Section 4.3 (with a frame or grid replacing the truss).

The general concepts necessary for modeling and entering data into the program are dissimilar enough from that of the truss to warrant discussion as follows:

1. Establish the global-coordinate axes (x and y for a plane frame; x and z for a grid). At each node, the degrees of freedom will then be an x and a y displacement, and a bending rotation about the z axis for a plane frame; and a y displacement, and rotations about the x and z axes for a grid.

2. Number the element and the nodes. Identify the total numbers of elements and nodes.

3. Specify the nodal coordinates. Then x_j and y_j represent the coordinates of each node j for a plane frame, and x_j and z_j represent the coordinates of each node j for a grid.

4. Specify the support or boundary conditions; that is, specify the degrees of freedom that are to be zero for the plane frame or the grid.

5. Specify the appropriate loads in global-coordinate components at each node. Here F_{1j}, F_{2j}, and M_{zj} are the x and y components of force and the bending moment about the z axis acting at node j for a plane frame, and F_{2j}, M_{xj}, and M_{zj} are the y component of force and the moments about the global x and z axes for a grid. (Again, this notation is different from that previously used in this text because the program was written with the first subscript denoting the global direction, and the second subscript denoting the node at which the forces or moments act.)

6. Specify the connectivity or topology. Identify which nodes are connected to which elements.

7. Describe the element properties. For each beam element of a plane frame, the modulus of elasticity, the cross-sectional area, and the principal moment of inertia are needed; for each grid element, the modulus of elasticity, the shear modulus, the principal moment of inertia, and the torsional constant are needed.

To use computer program PFRAME, a description of the input data is necessary. Hence, we will now describe the argument list in SUBROUTINE DATA, which reads data into the program, and the variables in COMMON/FRAM/ as follows:

SUBROUTINE DATA (NNODE,IFIX,Q,E,A,XI,MUD,G,XJ)

COMMON/FRAM/NELE,NODE(2,50),XC(50),YC(50),ZC(50),IND

This subroutine reads in the data needed for the plane frame and grid analysis program. The argument list and the variables in the COMMON statement are transferred back to the calling program. As in the truss program, the number of nonzero upper codiagonals needed in the stiffness matrix is computed and stored in variable MUD. All data are input in free format, separated by commas.

Description of Variables

I is a dummy variable representing the x and y directions and the rotation about the z axis for a plane frame, and representing the y direction and the rotations about the x and z axes for a grid (in the order I = 1, 2, 3; that is, I = 1 represents the x direction, I = 2 the y direction, and I = 3 the rotation about the z axis for a plane frame).

J is a dummy variable that indicates the node number.

K is a dummy variable that indicates the element number.

NELE is the number of elements in the finte element model.

NNODE is the number of nodes in the finite element model.

IFIX(I,J)   For a plane frame, IFIX(I,J) specifies whether a displacement component in the global x or y direction or a rotation about the z axis is fixed; that is, IFIX(1,J) specifies whether the x component of displacement is fixed or free and IFIX(3,J) specifies whether the rotation about the z axis is fixed or free. For a grid, IFIX(I,J) specifies whether the vertical displacement (y direction displacement) or a rotation about the global x or z axis is fixed; that is, IFIX(1,J) specifies whether the vertical, y directed displacement is fixed or free, IFIX(2,J) specifies whether the rotation about the x axis is fixed or free, and IFIX(3,J) specifies whether the rotation about the z axis is fixed or free.

If IFIX(I,J) = 1, the Ith displacement or rotation component at node J is fixed.

If IFIX(I,J) = 0, the Ith displacement or rotation component at node J is free.

XC(J), YC(J), and ZC(J) are the x, y, and z coordinates of node J. For plane frames, the z coordinate must be zero for all nodes because all the elements are assumed to lie in the $x - y$ plane; for grids, the y coordinate must be zero for all nodes because all the elements are assumed to lie in the $x - z$ plane.

Q(I,J) are the components of the external forces and moments, with respect to global coordinates, applied at node J. For plane frames, the three components are the x and y forces and a bending moment about the z axis; for grids, the three components are a y force and moments about the x and z axes.

NODE(J,K) are the two nodes of element K.

E(K), G(K), A(K), XI(K), and XJ(K) are the modulus of elasticity, shear modulus, cross-sectional area, moment of inertia about the z axis, and torsional constant for an element. G(K) and XJ(K) can be set to zero for plane frame analysis because they do not enter the plane frame equations.

IND is an indicator used to denote either plane frame or grid analysis. For plane frame analysis, set IND = 1; for grid analysis, set IND = 0.

Input Data

The order of input data, by sequence of parameters needed by PFRAME, is as follows:

1. Indicator for plane frame or grid analysis

   IND

2. Title (in columns 1–80)

   Title of problem; any alphanumeric data to identify the problem. (This will be printed as a heading on the first page of output.)

3. Basic parameter line

   NELE,NNODE

4. Node data (one line for each node)

   J,IFIX(1,J)IFIX(2,J),IFIX(3,J),XC(J),
   YC(J),ZC(J),Q(1,J),Q(2,J)Q(3,J)

5. Element data (one line for each element)

    K,NODE(1,K),NODE(2,K),E(K),G(K),A(K),XI(K),XJ(K)

E X A M P L E   6.8

We now consider the plane frame solved longhand in Example 6.1, to illustrate the input data for PFRAME. The frame model is again shown in Figure 6-32.

Table 6-2 shows the data file used in PFRAME to solve the plane frame of Figure 6-32. The order of input is based on the general description of the input data. In Table 6-2, line 1 is the indicator of the type of problem being solved. Line 2 is the identifying title of the problem. Line 3 gives the numbers of elements and nodes, lines 4–7 describe node information, and lines 8–10 describe element information.

The complete longhand solution of this problem given in Section 6.2 can be compared with the computer program results given in Table 6-3.   ∎

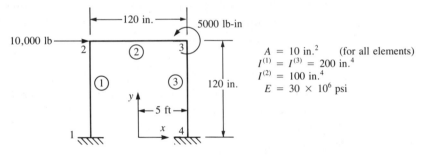

FIGURE   6-32   Plane frame used to illustrate the input data for computer program PFRAME

T A B L E   6-2   Input for the plane frame of Figure 6-32

Line	Data on the Line (Beginning in Column 1)
1	1
2	PLANE FRAME EXAMPLE 6.8
3	3,4
4	1,1,1,1,$-$60.,0.,0.,0.,0.,0.
5	2,0,0,0,$-$60.,120.,0.,10000.,0.,0.
6	3,0,0,0,60.,120.,0.,0.,0.,5000.
7	4,1,1,1,60.,0.,0.,0.,0.,0.
8	1,1,2,30.E+6,12.E+6,10.,200.,0.
9	2,2,3,30.E+6,12.E+6,10.,100.,0.
10	3,3,4,30.E+6,12.E+6,10.,200.,0.

T A B L E   6-3   Output of computer program PFRAME for the plane frame
of Figure 6-32

PLANE FRAME EXAMPLE 6.8
NUMBER OF ELEMENTS = 3
NUMBER OF NODES   = 4

NODE POINTS

K	IFIX	XC(K)	YC(K)	ZC(K)
1	1 1 1	− 60.000000	0.000000	0.000000
2	0 0 0	− 60.000000	120.000000	0.000000
3	0 0 0	60.000000	120.000000	0.000000
4	1 1 1	60.000000	0.000000	0.000000

	FORCE(1,K)	FORCE(2,K)	FORCE(3,K)
	0.000000	0.000000	0.000000
	10000.000000	0.000000	0.000000
	0.000000	0.000000	5000.000000
	0.000000	0.000000	0.000000

ELEMENTS

K	NODE(I,K)		E(K)	G(K)	A(K)
1	1	2	3.0000000E + 07	1.2000000E + 07	1.0000000E + 01
2	2	3	3.0000000E + 07	1.2000000E + 07	1.0000000E + 01
3	3	4	3.0000000E + 07	1.2000000E + 07	1.0000000E + 01

	XI(K)	XJ(K)
	2.0000000E + 02	0.0000000E + 00
	1.0000000E + 02	0.0000000E + 00
	2.0000000E + 02	0.0000000E + 00

NODE	DISPLACEMENTS		Z-ROTATION
	X	Y	THETA
1	0.00000E + 00	0.00000E + 00	0.00000E + 00
2	0.21136E + 00	0.14813E − 02	−0.15260E − 02
3	0.20936E + 00	−0.14813E − 02	−0.14860E − 02
4	0.00000E + 00	0.00000E + 00	0.00000E + 00

ELEMENT

K	NODE(I,K)		X-FORCE	Y-FORCE	Z-MOMENT
1	1	2	−0.3703E + 04	0.4992E + 04	0.3758E + 06
2	2	3	0.5008E + 04	−0.3703E + 04	−0.2232E + 06
3	3	4	0.3703E + 04	0.5008E + 04	0.2262E + 06

			X-FORCE	Y-FORCE	Z-MOMENT
			0.3703E + 04	−0.4992E + 04	0.2232E + 06
			−0.5008E + 04	0.3703E + 04	−0.2212E + 06
			−0.3703E + 04	−0.5008E + 04	0.3748E + 06

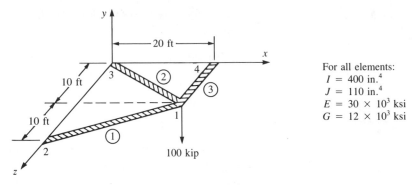

FIGURE 6-33   Grid used to illustrate the input data for computer program PFRAME

TABLE 6-4   Input for the grid of Figure 6-33 (all data have been converted to pounds and inches)

Line	Data on the Line (Beginning in Column 1)
1	0
2	GRID EXAMPLE 6.9
3	3,4
4	1,0,0,0,240.,0.,120., $-$ 100000.,0.,0.
5	2,1,1,1,0.,0.,240.,0.,0.,0.
6	3,1,1,1,0.,0.,0.,0.,0.,0.
7	4,1,1,1,240.,0.,0.,0.,0.,0.
8	1,1,2,30.E + 6,12.E + 6,1.,400.,110.
9	2,1,3,30.E + 6,12.E + 6,1.,400.,110.
10	3,1,4,30.E + 6,12.E + 6,1.,400.,110.

E X A M P L E   6.9

We now consider the grid solved longhand in Example 6.5, to illustrate the input data for PFRAME for a grid analysis. The grid is again shown in Figure 6-33.

Table 6-4 shows the data file used in PFRAME to solve the grid of Figure 6-33. Again, the order of input is based on the general description of the input data, as provided at the beginning of Section 6-7.

In Table 6-4, line 1 is the indicator that a grid analysis is to be performed. Line 2 is the identifying title of the problem. Line 3 gives the numbers of elements and nodes, lines 4–7 describe node information, and lines 8–10 describe element information. Specifically, in line 4, the load of 100 kips in the negative y direction is included. The complete longhand solution to this problem given in Section 6.4 can be compared with the computer program solution provided in Table 6-5.   ∎

T A B L E 6-5   Output of computer program PFRAME for the grid of Figure 6-33

GRID EXAMPLE 6.9
NUMBER OF ELEMENTS = 3
NUMBER OF NODES   = 4

NODE POINTS

K	IFIX	XC(K)	YC(K)	ZC(K)
1	0 0 0	240.000000	0.000000	120.000000
2	1 1 1	0.000000	0.000000	240.000000
3	1 1 1	0.000000	0.000000	0.000000
4	1 1 1	240.000000	0.000000	0.000000

FORCE(1,K)	FORCE(2,K)	FORCE(3,K)
− 100000.000000	0.000000	0.000000
0.000000	0.000000	0.000000
0.000000	0.000000	0.000000
0.000000	0.000000	0.000000

ELEMENTS

K	NODE(I,K)		E(K)	G(K)	A(K)
1	1	2	3.0000000E + 07	1.2000000E + 7	0.0000000E + 00
2	1	3	3.0000000E + 07	1.2000000E + 7	0.0000000E + 00
3	1	4	3.0000000E + 07	1.2000000E + 7	0.0000000E + 00

XI(K)	XJ(K)
4.0000000E + 02	1.1000000E + 02
4.0000000E + 02	1.1000000E + 02
4.0000000E + 02	1.1000000E + 02

Node	DISPLACEMENT	THETA-X	THETA-Z
1	−0.28327E + 01	0.29728E − 01	−0.17120E − 01
2	0.00000E + 00	0.00000E + 00	0.00000E + 00
3	0.00000E + 00	0.00000E + 00	0.00000E + 00
4	0.00000E + 00	0.00000E + 00	0.00000E + 00

ELEMENTS

K	NODE(I,K)		Y-FORCE	X-MOMENT	Z-MOMENT
1	1	2	−0.1917E + 05	−0.1674E + 06	−0.2482E + 07
2	1	3	0.7420E + 04	−0.9252E + 05	0.2240E + 07
3	1	4	−0.8825E + 05	0.1869E + 06	−0.2342E + 07

Y-FORCE	X-MOMENT	Z-MOMENT
0.1917E + 05	0.1674E + 06	−0.2662E + 07
−0.7420E + 04	0.9252E + 05	−0.2959E + 06
0.8825E + 05	−0.1869E + 06	−0.8248E + 07

REFERENCES

[1] SYSTEM/360, Scientific Subroutine Package, IBM.

[2] Budynas, R. G., *Advanced Strength and Applied Stress Analysis*, McGraw-Hill, New York, 1977.

[3] Allen, H. G., and Bulson, P. S., *Background to Buckling*, McGraw-Hill, London, 1980.

[4] Roark, R. J., and Young, W. C., *Formulas for Stress and Strain*, 5th ed., McGraw-Hill, New York, 1975.

[5] Logan, D. L., *Mechanics of Materials*, Harper Collins, New York, 1991.

[6] Parakh, Zal K., *Finite Element Analysis of Bus Frames Under Simulated Crash Loadings*, M. S. Thesis, Rose-Hulman Institute of Technology, Terre Haute, Indiana, May 1989.

[7] Martin, H. C., *Introduction to Matrix Methods of Structural Analysis*, McGraw-Hill, New York, 1966.

PROBLEMS

Solve all problems using the finite element stiffness method.

6.1   For the rigid frame shown in Figure P6-1, determine (1) the displacement components and the rotation at node 2, (2) the support reactions, and (3) the forces in each element. Then check equilibrium at node 2. Let $E = 30 \times 10^6$ psi, $A = 10$ in.^{2}, and $I = 500$ in.^{4} for both elements.

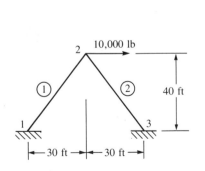

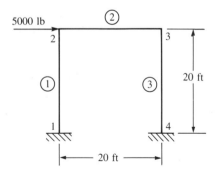

FIGURE   P6-1

FIGURE   P6-2

6.2   For the rigid frame shown in Figure P6-2, determine (1) the nodal displacement components and rotations, (2) the support reactions, and (3) the forces in each element. Let $E = 30 \times 10^6$ psi, $A = 10$ in.^{2}, and $I = 200$ in.^{4} for all elements.

6.3   For the rigid stairway frame shown in Figure P6-3, determine (1) the displacements at node 2, (2) the support reactions, and (3) the local nodal forces acting on each element. Draw the bending moment diagram for the whole frame. Remember that the angle between elements 1 and 2 is preserved as deformation takes place, similarly for the angle between elements 2 and 3. Furthermore, owing to symmetry, $d_{2x} = -d_{3x}$, $d_{2y} = d_{3y}$, and $\phi_2 = -\phi_3$. What size A36 steel channel section would be needed to keep the allowable bending stress less than two thirds of the yield stress? (For A36 steel, the yield stress is 36,000 psi.)

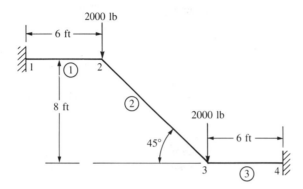

FIGURE P6-3

6.4   For the rigid frame shown in Figure P6-4, determine (1) the nodal displacements and rotation at node 4, (2) the reactions, and (3) the forces in each element. Then check equilibrium at node 4. Finally, draw the shear force and bending moment diagrams for each element. Let $E = 30 \times 10^3$ ksi, $A = 8$ in.^{2}, and $I = 800$ in.^{4} for all elements.

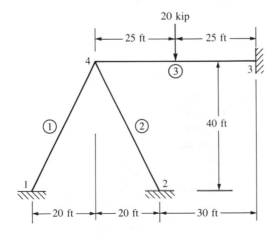

FIGURE P6-4

6.5–6.15   For the grid frames shown in Figures P6-5–P6-15, determine the displacements and rotations of the nodes, the element forces, and the reactions. The values of E, A, and I to be used are listed next to each figure.

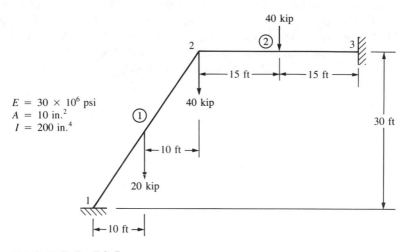

$E = 30 \times 10^6$ psi
$A = 10$ in.^{2}
$I = 200$ in.^{4}

F I G U R E   P6-5

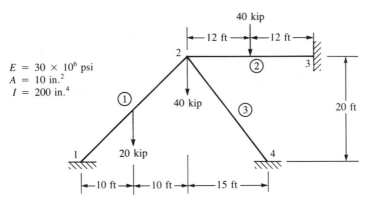

$E = 30 \times 10^6$ psi
$A = 10$ in.^{2}
$I = 200$ in.^{4}

F I G U R E   P6-6

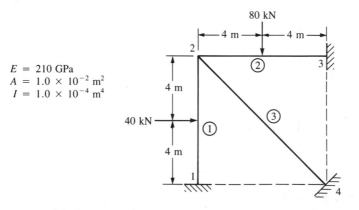

$E = 210$ GPa
$A = 1.0 \times 10^{-2}$ m^2
$I = 1.0 \times 10^{-4}$ m^4

F I G U R E   P6-7

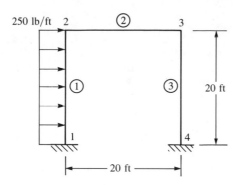

$E = 30 \times 10^6$ psi
$A = 15$ in.^{2}
$I = 250$ in.^{4}

F I G U R E   P6-8

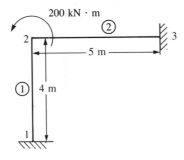

$E = 210$ GPa
$A = 2 \times 10^{-2}$ m^2
$I = 2 \times 10^{-4}$ m^4

F I G U R E   P6-9

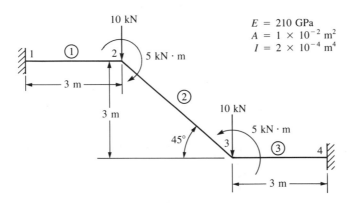

$E = 210$ GPa
$A = 1 \times 10^{-2}$ m^2
$I = 2 \times 10^{-4}$ m^4

F I G U R E   P6-10

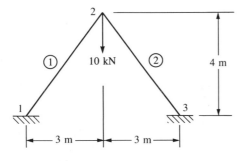

$E = 70$ GPa
$A = 3 \times 10^{-2}$ m^2
$I = 3 \times 10^{-4}$ m^4

F I G U R E   P6-11

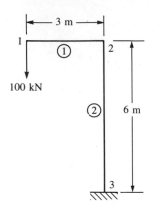

$E = 210$ GPa
$A = 8 \times 10^{-2}$ m^2
$I = 1.2 \times 10^{-4}$ m^4

FIGURE P6-12

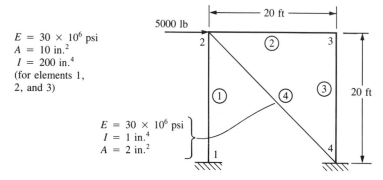

$E = 30 \times 10^6$ psi
$A = 10$ in.^{2}
$I = 200$ in.^{4}
(for elements 1, 2, and 3)

$E = 30 \times 10^6$ psi
$I = 1$ in.^{4}
$A = 2$ in.^{2}

FIGURE P6-13

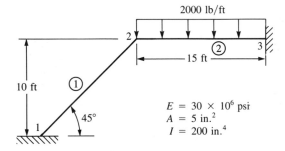

$E = 30 \times 10^6$ psi
$A = 5$ in.^{2}
$I = 200$ in.^{4}

FIGURE P6-14

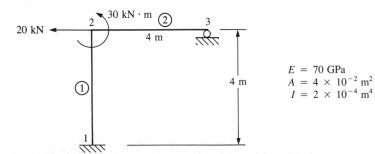

$E = 70$ GPa
$A = 4 \times 10^{-2}$ m^2
$I = 2 \times 10^{-4}$ m^4

FIGURE P6-15

6.16–6.18 Solve the structures in Figures P6-16–P6-18 by using substructuring.

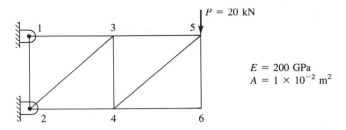

FIGURE P6-16 (Substructure the truss at nodes 3 and 4)

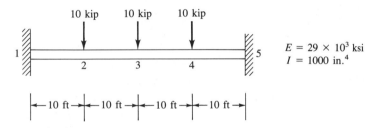

FIGURE P6-17 (Substructure the beam at node 3)

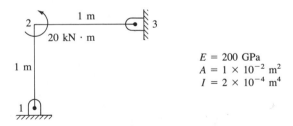

FIGURE P6-18 (Substructure the frame at node 2)

Solve Problems 6.19–6.39 by using computer program PFRAME accompanying the text or any other suitable program.

6.19 For the rigid frame shown in Figure P6-19, determine (1) the nodal displacement components and (2) the support reactions. (3) Draw the shear force and bending moment diagrams. For all elements, let $E = 30 \times 10^6$ psi, $I = 200$ in.^{4}, and $A = 10$ in.^{2}

6.20 For the rigid frame shown in Figure P6-20, determine (1) the nodal displacement components and (2) the support reactions. (3) Draw the shear force and bending moment diagrams. Let $E = 30 \times 10^6$ psi, $I = 200$ in.^{4}, and $A = 10$ in.^{2} for all elements, except as noted in the figure.

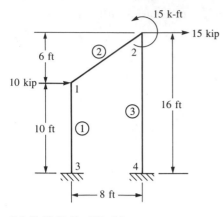

FIGURE P6-19

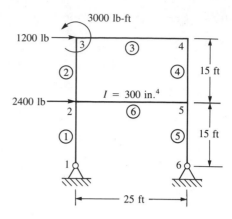

FIGURE P6-20

6.21 For the slant-legged rigid frame shown in Figure P6-21, size the structure for minimum weight based on a maximum bending stress of 20 ksi in the horizontal beam elements and a maximum compressive stress (due to bending and direct axial load) of 15 ksi in the slant-legged elements. Use the same element size for the two slant-legged elements and the same element size for the two 10-foot sections of the horizontal element. Assume A36 steel is used.

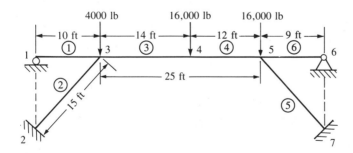

FIGURE P6-21

6.22 For the rigid building frame shown in Figure P6-22, determine the forces in each element and calculate the bending stresses. Assume all the vertical elements have $A = 10$ in.^{2} and $I = 100$ in.^{4} and all horizontal elements have $A = 15$ in.^{2} and $I = 150$ in.^{4} Let $E = 29 \times 10^6$ psi for all elements. Let $c = 5$ in. for the vertical elements and $c = 6$ in. for the horizontal elements, where c denotes the distance from the neutral axis to the top or bottom of the beam cross section, as used in the bending stress formula $\sigma = (Mc/I)$.

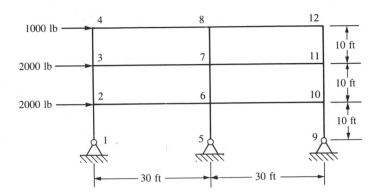

FIGURE P6-22

6.23–6.38 For the rigid frames or beams shown in Figures P6-23–P6-38, determine the displacements and rotations at the nodes, the element forces, and the reactions.

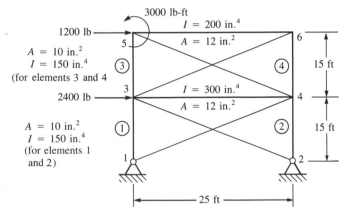

FIGURE P6-23

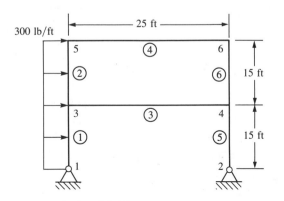

FIGURE P6-24

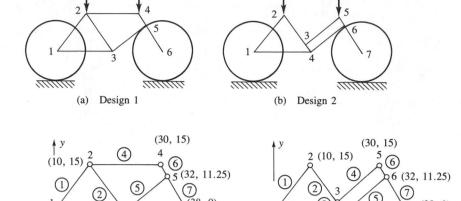

(a) Design 1   (b) Design 2

(a) Design 1   (b) Design 2

Case 1 $A_1 = 0.1$ in.^{2}
$E = 30 \times 10^6$ psi $A_2 = A_3 = A_4 = A_5 = 0.15$ in.^{2}
Case 2 $A_6 = A_7 = A_8 = 0.3$ in.^{2}
$E = 10 \times 10^6$ psi $I_1 = 0.01$ in.^{4}
                        $I_2 = I_3 = I_4 = I_5 = 0.02$ in.^{4}
                        $I_6 = I_7 = I_8 = 0.1$ in.^{4}

FIGURE P6-25   Two bicycle frame models (coordinates shown in inches)

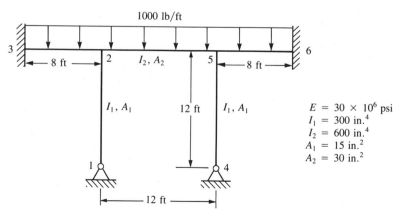

FIGURE P6-26

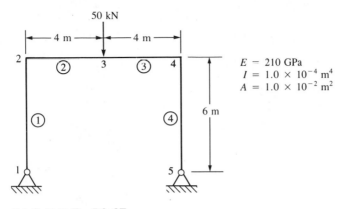

FIGURE P6-27

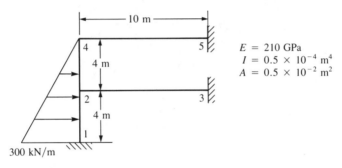

FIGURE P6-28

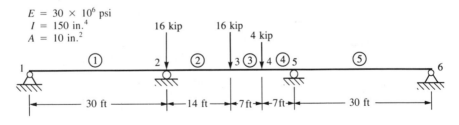

FIGURE P6-29

$E = 30 \times 10^6$ psi
$I = 200$ in.^{4}
$A = 12$ in.^{2}

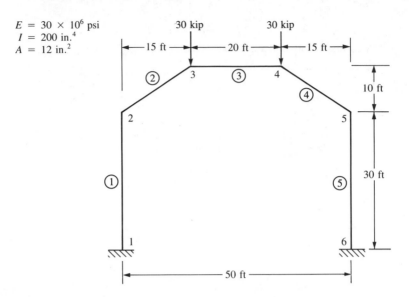

FIGURE P6-30

$E = 30 \times 10^6$ psi
$I = 100$ in.^{4}
$A = 8$ in.^{2}

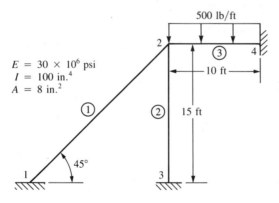

FIGURE P6-31

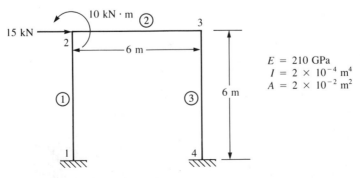

$E = 210$ GPa
$I = 2 \times 10^{-4}$ m^4
$A = 2 \times 10^{-2}$ m^2

FIGURE P6-32

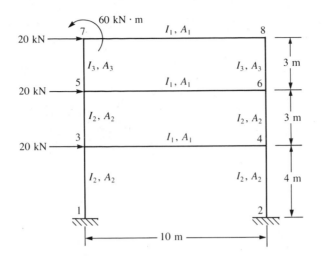

FIGURE P6-33

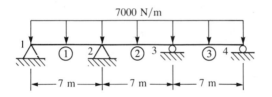

FIGURE P6-34

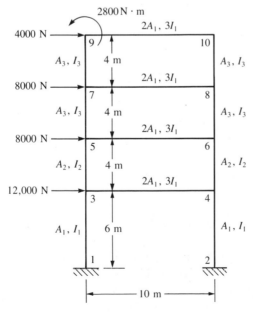

FIGURE P6-35

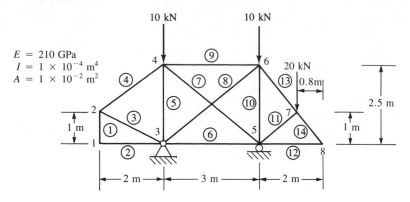

$E = 210$ GPa
$I = 1 \times 10^{-4}$ m^4
$A = 1 \times 10^{-2}$ m^2

10 kN   10 kN

20 kN

0.8m

2.5 m

1 m

2 m   3 m   2 m

FIGURE   P6-36

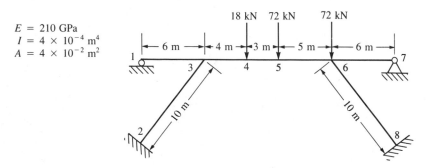

$E = 210$ GPa
$I = 4 \times 10^{-4}$ m^4
$A = 4 \times 10^{-2}$ m^2

18 kN   72 kN   72 kN

6 m   4 m   3 m   5 m   6 m

10 m   10 m

FIGURE   P6-37

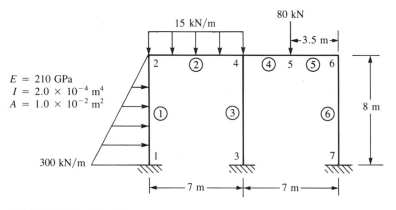

80 kN

15 kN/m

3.5 m

$E = 210$ GPa
$I = 2.0 \times 10^{-4}$ m^4
$A = 1.0 \times 10^{-2}$ m^2

8 m

300 kN/m

7 m   7 m

FIGURE   P6-38

6.39   Consider the plane structure shown in Figure P6-39. First assume the structure to be a plane frame with rigid joints and analyze using program **PFRAME**.
Then assume the structure to be pin-jointed and analyze as a plane truss using
program **TRUSS**. If the structure is actually a truss, is it appropriate to model
it as a rigid frame? How can you model the truss using the frame element? In
other words, what idealization could you make in your model to use the
PFRAME program to approximate a truss?

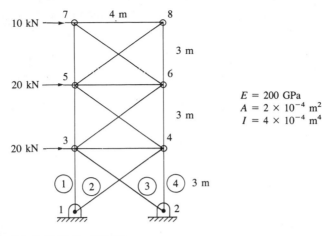

10 kN → 7   4 m   8

3 m

5   6
20 kN →

$E = 200$ GPa
$A = 2 \times 10^{-4}$ m^2
$I = 4 \times 10^{-4}$ m^4

3 m

3   4
20 kN →

1   2   3   4   3 m

1   2

FIGURE P6-39

6.40 For the tapered beam shown in Figure P6-40, determine the maximum deflection using one, two, four, and eight elements. Calculate the moment of inertia at the midlength station for each element. Let $E = 30 \times 10^6$ psi, $I_0 = 100$ in.^{4}, and $L = 100$ in. Run cases where $n = 1$, 3, and 7. Use the computer program PFRAME. The analytical solution for $n = 7$ is given by [7]

$$v_1 = \frac{PL^3}{49EI_0}(1/7 \ln 8 + 2.5) = \frac{1}{17.55}\frac{PL^3}{EI_0}$$

$$\theta_1 = \frac{PL^2}{49EI_0}(\ln 8 - 7) = -\frac{1}{9.95}\frac{PL^2}{EI_0}$$

$$I(x) = I_0\left(1 + n\frac{x}{L}\right)$$

where n = arbitrary numerical factor
    I_0 = moment of inertia of section at $x = 0$

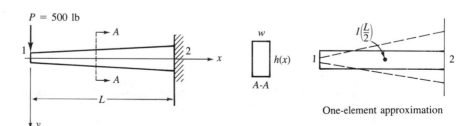

FIGURE P6-40   Tapered cantilever beam

6.41   Derive the stiffness matrix for the nonprismatic torsion bar shown in Figure P6-41. The radius of the shaft is given by

$$r = r_0 + (x/L)r_0, \text{ where } r_0 \text{ is the radius at } x = 0.$$

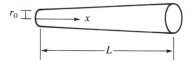

FIGURE P6-41

6.42   Derive the total potential energy for the prismatic circular cross-section torsion bar shown in Figure P6-42. Also determine the equivalent nodal torques for the bar subjected to uniform torque per unit length (lb-in./in.). Let G be the shear modulus and J be the polar moment of inertia of the bar.

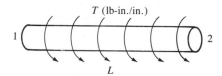

FIGURE P6-42

6.43   For the grid shown in Figure P6-43, determine the nodal displacements and the local element forces. Let $E = 30 \times 10^6$ psi, $G = 12 \times 10^6$ psi, $I = 200$ in.^{4}, and $J = 100$ in.^{4} for both elements.

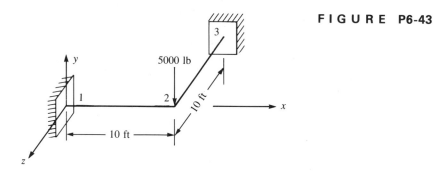

FIGURE P6-43

6.44   Resolve Problem 6-43 with an additional nodal moment of 1000 k-in. applied about the x axis at node 2.

6.45–6.46   For the grids shown in Figures P6-45 and P6-46, determine the nodal displacements and the local element forces. Let $E = 210$ GPa, $G = 84$ GPa, $I = 2 \times 10^{-4}$ m^4, $J = 1 \times 10^{-4}$ m^4, and $A = 1 \times 10^{-2}$ m^2.

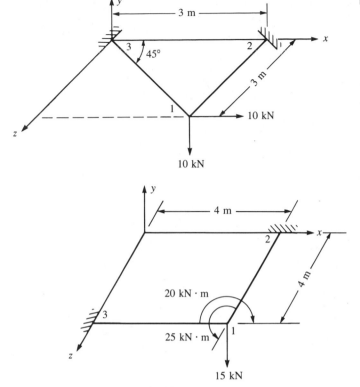

FIGURE P6-46

6.47–6.52 Solve the grid structures shown in Figures P6-47–P6-52 using computer program PFRAME (from the disk that accompanies this text) or any other suitable program. For grids P6-47–P6-49, let $E = 30 \times 10^6$ psi, $G = 12 \times 10^6$ psi, $I = 200$ in.^{4}, and $J = 100$ in.^{4}, except as noted in the figures. In Figure P6-49, let the cross elements have $I = 50$ in.^{4} and $J = 20$ in.^{4}, with dimensions and loads as in Figure P6-48. For grids P6-50–P6-52, let $E = 210$ GPa, $G = 84$ GPa, $I = 2 \times 10^{-4}$ m^4, $J = 1 \times 10^{-4}$ m^4, and $A = 1 \times 10^{-2}$ m^2.

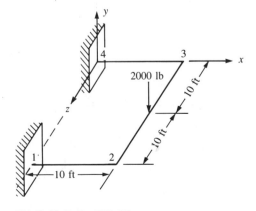

FIGURE P6-47

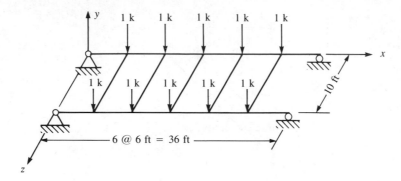

F I G U R E   P6-48

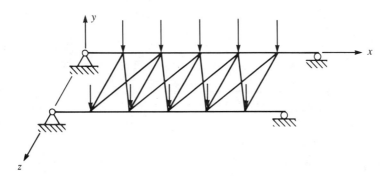

F I G U R E   P6-49

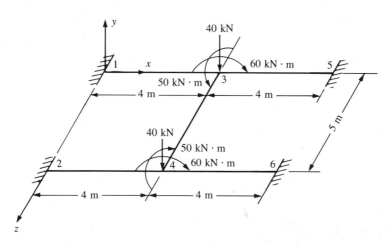

F I G U R E   P6-50

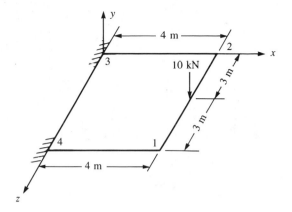

FIGURE P6-51

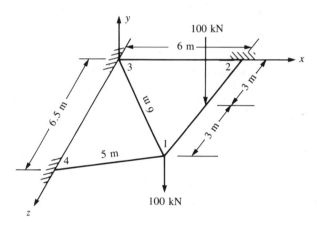

FIGURE P6-52

Development of the Plane Stress and Plane Strain Stiffness Equations

INTRODUCTION

In Chapters 2 through 6, we considered only line elements. Line elements are connected only at common nodes, forming framed or articulated structures such as trusses, frames, and grids. Line elements have geometric properties such as cross-sectional area and moment of inertia associated with their cross sections. However, only one local coordinate $\hat{x}$ along the length of the element is required to describe a position along the element (hence, they are called *line elements*). Nodal compatibility is then enforced during the formulation of the nodal equilibrium equations for a line element.

This chapter considers the two-dimensional finite element. Two-dimensional (planar) elements are thin-plate elements such that two coordinates define a position on the element surface. The elements are connected at common nodes and/or along common edges to form continuous structures such as those shown in Figures 1-3, 1-4, 1-6, and 7-1. Nodal compatibility is then enforced during the formulation of the nodal equilibrium equations for two-dimensional elements. If proper displacement functions are chosen, compatibility along common edges is also obtained. The two-dimensional element is extremely important for (1) plane stress analysis, which includes problems such as plates with holes, fillets, or other changes in geometry that are loaded in their plane resulting in local stress concentrations, such as illustrated in Figure 7-1; and (2) plane strain analysis, which includes problems such as a long underground box culvert subjected to a uniform load acting constantly over its length, as illustrated in Figure 1-3, or a long cylindrical control rod subjected to a load that remains constant over the rod length (or depth), as illustrated in Figure 1-4.

We begin this chapter with the development of the stiffness matrix for a basic two-dimensional or plane finite element, called the *constant-strain trian-*

gular element. We consider the constant-strain triangle (CST) stiffness matrix because its derivation is the simplest among the available two-dimensional elements.

We will derive the CST stiffness matrix by using the principle of minimum potential energy because the energy formulation is the most feasible for the development of the equations for both two- and three-dimensional finite elements.

We will then present a simple, thin-plate plane stress example problem to illustrate the assemblage of the plane element stiffness matrices using the direct stiffness method as presented in Chapter 2. We will present the total solution, including the stresses within the plate.

7.1 General Steps in the Formulation of the Plane Triangular Element Equations

We will now follow the steps described in Chapter 1 to formulate the governing equations for a plane stress/plane strain triangular element. First, we will describe the concepts of plane stress and plane strain. Then we will provide a brief description of the steps and basic equations pertaining to a plane triangular element. In Section 7.2, we derive the explicit element stiffness matrix and equations.

The concepts of plane stress and plane strain are important because the developments in this chapter are directly applicable only to systems assumed to behave in a plane stress or plane strain manner. Therefore, we will now describe these concepts in detail.

Plane Stress

Plane stress *is defined to be a state of stress in which the normal stress and the shear stresses directed perpendicular to the plane are assumed to be zero.* For instance, in Figures 7-1(a) and 7-1(b), the plates in the x-y plane shown subjected to surface tractions T in the plane are under a state of plane stress; that is, the normal stress σ_z and the shear stresses τ_{xz} and τ_{yz} are assumed to be zero. Generally, members that are thin (those with a small z dimension

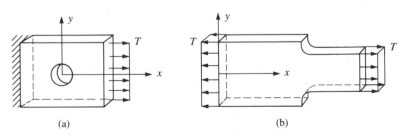

(a)   (b)

FIGURE 7-1   Plane stress problems: (a) plate with hole; (b) plate with fillet

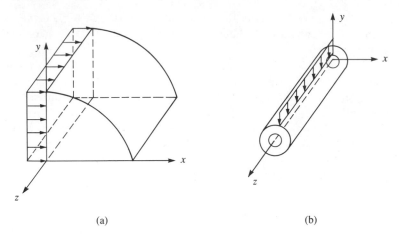

(a)   (b)

F I G U R E   7-2   Plane strain problems: (a) dam subjected to horizontal loading; (b) pipe subjected to a vertical load

compared to the in-plane x and y dimensions) and whose loads act only in the x-y plane can be considered to be under plane stress.

Plane Strain

Plane strain *is defined to be a state of strain in which the strain normal to the x-y plane ε_z and the shear strains γ_{xz} and γ_{yz} are assumed to be zero.* The assumptions of plane strain are realistic for long bodies (say, in the z direction) with constant cross-sectional area subjected to loads that act only in the x and/or y directions and do not vary in the z direction. Some plane strain examples are shown in Figure 7-2 (and in Figures 1-3 and 1-4). In these examples, only a unit thickness (1 in. or 1 ft) of the structure is considered because each unit thickness behaves identically (except near the ends). The finite element models of the structures in Figure 7-2 consist of appropriately discretized cross sections in the x-y plane with the loads acting over unit thicknesses in the x and/or y directions only.

Two-Dimensional State of Stress and Strain

The concept of two-dimensional state of stress and strain and the stress/strain relationships for plane stress and plane strain are necessary to understand fully the development and applicability of the stiffness matrix for the plane stress/plane strain triangular element. Therefore, we briefly outline the essential concepts of two-dimensional stress and strain (see References [1] and [2] and Appendix C for more details on this subject).

First, we illustrate the two-dimensional state of stress using Figure 7-3. The infinitesimal element with sides dx and dy has normal stresses σ_x and σ_y acting in the x and y directions (here on the vertical and horizontal faces), respectively. The shear stress τ_{xy} acts on the x edge (vertical face) in the y direction. The shear stress τ_{yx} acts on the y edge (horizontal face) in the x

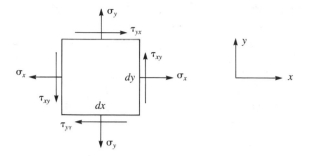

FIGURE 7-3   Two-dimensional state of stress

direction. Moment equilibrium of the element results in τ_{xy} being equal in magnitude to τ_{yx}. Hence, three independent stresses exist and are represented by the vector column matrix

$$\{\sigma\} = \begin{Bmatrix} \sigma_x \\ \sigma_y \\ \tau_{xy} \end{Bmatrix} \tag{7.1.1}$$

The stresses given by Eq. (7.1.1) will be expressed in terms of the nodal displacement degrees of freedom. Hence, once the nodal displacements are determined, these stresses can be evaluated directly.

Recall from strength of materials (see Reference [2]) that the **principal stresses**, which are the maximum and minimum normal stresses in the two-dimensional plane, can be obtained from the following expressions:

$$\sigma_1 = \frac{\sigma_x + \sigma_y}{2} + \sqrt{\left(\frac{\sigma_x - \sigma_y}{2}\right)^2 + \tau_{xy}^2} = \sigma_{max}$$

$$\sigma_2 = \frac{\sigma_x + \sigma_y}{2} - \sqrt{\left(\frac{\sigma_x - \sigma_y}{2}\right)^2 + \tau_{xy}^2} = \sigma_{min} \tag{7.1.2}$$

Also, the **principal angle** θ_p, which defines the normal whose direction is perpendicular to the plane on which the maximum or minimum principal stress acts, is defined by

$$\tan 2\theta_p = \frac{2\tau_{xy}}{\sigma_x - \sigma_y} \tag{7.1.3}$$

Figure 7-4 shows the principal stresses σ_1 and σ_2 and the angle θ_p. Recall (as Figure 7-4 indicates) that the shear stress is zero on the planes having principal (maximum and minimum) normal stresses.

In Figure 7-5, we show an infinitesimal element used to represent the general two-dimensional state of strain at some point in a structure. The element is shown to be displaced by amounts u and v in the x and y directions at point A, and to displace or extend an additional (incremental) amount $(\partial u/\partial x)\,dx$ along line AB, and $(\partial v/\partial y)\,dy$ along line AC in the x and y directions, respectively. Furthermore, observing lines AB and AC, we see that

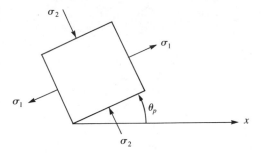

FIGURE 7-4   Principal stresses and their directions

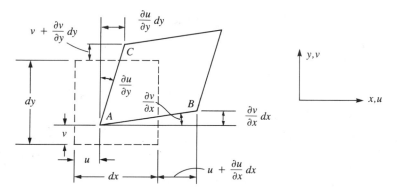

FIGURE 7-5   Displacements and rotations of lines of an element in the *x-y* plane

point B moves upward an amount $(\partial v/\partial x)\, dx$ with respect to A, and point C moves to the right an amount $(\partial u/\partial y)\, dy$ with respect to A.

Based on the general definitions of normal and shear strains and use of Figure 7-5, we obtain

$$\varepsilon_x = \frac{\partial u}{\partial x} \qquad \varepsilon_y = \frac{\partial v}{\partial y} \qquad \gamma_{xy} = \frac{\partial u}{\partial y} + \frac{\partial v}{\partial x} \qquad (7.1.4)$$

Hence, recall that the strains ε_x and ε_y are the changes in length per unit length of material fibers originally parallel to the x and y axes, respectively, when the element undergoes deformation. These strains are then called *normal* (or *extensional* or *longitudinal*) *strains*. The strain γ_{xy} is the change in the original right angle made between dx and dy when the element undergoes deformation. The strain γ_{xy} is then called a *shear strain*.

The strains given by Eqs. (7.1.4) are generally represented by the vector column matrix

$$\{\varepsilon\} = \left\{ \begin{array}{c} \varepsilon_x \\ \varepsilon_y \\ \gamma_{xy} \end{array} \right\} \qquad (7.1.5)$$

The relationships between strains and displacements referred to the x and y directions given by Eqs. (7.1.4) are sufficient for your understanding of subsequent material in this chapter.

We now present the stress/strain relationships for isotropic materials for both plane stress and plane strain. For plane stress, we assume the following stresses to be zero:

$$\sigma_z = \tau_{xz} = \tau_{yz} = 0 \tag{7.1.6}$$

Applying Eq. (7.1.6) to the three-dimensional stress/strain relationship (see Appendix C), the shear strains $\gamma_{xz} = \gamma_{yz} = 0$, but $\varepsilon_z \neq 0$. For plane stress conditions, we then have

$$\{\sigma\} = [D]\{\varepsilon\} \tag{7.1.7}$$

where

$$[D] = \frac{E}{1-v^2}\begin{bmatrix} 1 & v & 0 \\ v & 1 & 0 \\ 0 & 0 & \dfrac{1-v}{2} \end{bmatrix} \tag{7.1.8}$$

is called the *stress/strain matrix* (or *constitutive matrix*), E is the modulus of elasticity, and v is Poisson's ratio. In Eq. (7.1.7), $\{\sigma\}$ and $\{\varepsilon\}$ are defined by Eqs. (7.1.1) and (7.1.5).

For plane strain, we assume the following strains to be zero:

$$\varepsilon_z = \gamma_{xz} = \gamma_{yz} = 0 \tag{7.1.9}$$

Applying Eq. (7.1.9) to the three-dimensional stress/strain relationship, the shear stresses $\tau_{xz} = \tau_{yz} = 0$, but $\sigma_z \neq 0$. The stress/strain matrix then becomes

$$[D] = \frac{E}{(1+v)(1-2v)}\begin{bmatrix} 1-v & v & 0 \\ v & 1-v & 0 \\ 0 & 0 & \dfrac{1-2v}{2} \end{bmatrix} \tag{7.1.10}$$

The basic partial differential equations for plane stress, as derived in Reference [1], are

$$\frac{\partial^2 u}{\partial x^2} + \frac{\partial^2 u}{\partial y^2} = \frac{1+v}{2}\left(\frac{\partial^2 u}{\partial y^2} - \frac{\partial^2 v}{\partial x\,\partial y}\right)$$
$$\frac{\partial^2 v}{\partial x^2} + \frac{\partial^2 v}{\partial y^2} = \frac{1+v}{2}\left(\frac{\partial^2 v}{\partial x^2} - \frac{\partial^2 u}{\partial x\,\partial y}\right) \tag{7.1.11}$$

Steps in the Formulation of the Element Stiffness Equations

To illustrate the steps and introduce the basic equations necessary for the plane triangular element, consider the thin plate subjected to tensile loads T in Figure 7-6.

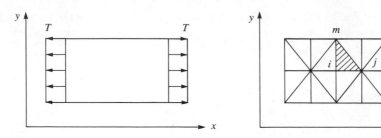

FIGURE 7-6   Thin plate in tension

FIGURE 7-7   Discretized plate of Figure 7-6 using triangular elements

Step 1   Select Element Type

To analyze the plate, we first discretize the plate as shown in Figure 7-7. The discretized plate has been divided into triangular elements, each with nodes such as i, j, and m. We use triangular elements because boundaries of irregularly shaped bodies can be closely approximated, and because the expressions related to the triangular element are comparatively simple. This discretization is called a *coarse-mesh generation* if few large elements are used. Each node has two degrees of freedom—namely, an x and a y displacement. We will let u_i and v_i represent the node i displacement components in the x and y directions, respectively.

The nodal displacements for an element with nodes i, j, and m are given by

$$\{d\} = \begin{Bmatrix} \underline{d}_i \\ \underline{d}_j \\ \underline{d}_m \end{Bmatrix} \qquad (7.1.12)$$

where the nodes are ordered counterclockwise and where

$$\{d_i\} = \begin{Bmatrix} u_i \\ v_i \end{Bmatrix} \qquad (7.1.13)$$

represents the nodal displacement components at node i. Using Eq. (7.1.13), we express Eq. (7.1.12) in expanded form as

$$\{d\} = \begin{Bmatrix} u_i \\ v_i \\ u_j \\ v_j \\ u_m \\ v_m \end{Bmatrix} \qquad (7.1.14)$$

Step 2   Select Displacement Functions

We define the general displacement functions in terms of the nodal displacements $\{d\}$ as

$$\{\psi\} = \begin{Bmatrix} u(x, y) \\ v(x, y) \end{Bmatrix} \qquad (7.1.15)$$

where $u(x, y)$ and $v(x, y)$ describe displacements at interior points of the element. The functions $u(x, y)$ and $v(x, y)$ must be compatible for the type of element used. Thus, for a constant-strain triangular element, the functions u and v must be linear (as will be indicated in Section 7.2). This will ensure that no tearing occurs at nodes and along boundary lines common to elements.

Step 3   Define the Strain/Displacement and Stress/Strain Relationships

We introduce the strain/displacement and stress/strain relationships because the element strains and stresses are used in the strain energy expression in the principle of minimum potential energy. These relationships are given by Eqs. (7.1.4) and (7.1.7), with the appropriate $[D]$ matrix given by Eq. (7.1.8) or Eq. (7.1.10).

Step 4   Derive the Element Stiffness Matrix and Equations

Using the principle of minimum potential energy, we derive the basic element stiffness matrix and equations as

$$\{f\} = [k]\{d\} \qquad (7.1.16)$$

This principle facilitates the derivation of the equations better than the direct approach that was initially used for one-dimensional elements.

Step 5   Assemble the Element Equations to Obtain the Global Equations and Introduce Boundary Conditions

We assemble the global or total structure stiffness matrix and equations by using the direct stiffness method as

$$\{F\} = [K]\{d\} \qquad (7.1.17)$$

where $\{F\}$ is the equivalent global nodal load matrix obtained by lumping distributed edge loads and element body force loads at the nodes (as well as including concentrated nodal loads), and $[K]$ is the global structure stiffness matrix.

Step 6   Solve for the Nodal Displacements

We determine the unknown global nodal displacements by solving the system of algebraic equations given by Eq. (7.1.17). In solving Eq. (7.1.17) remember that the boundary conditions on displacements must be taken into account.

Step 7   Solve for the Element Forces (Stresses)

Finally, we determine the element strains after expressing the strains in terms of the nodal displacements $\{d\}$. This expression will be developed in Section 7.2. The stresses are then obtained from Eq. (7.1.7).

7.2   Derivation of the Constant-Strain Triangular Element Stiffness Matrix and Equations

Step 1   Select Element Type

Consider the basic triangular element (taken from a discretized body) shown in Figure 7-8, with nodes i, j, and m labeled in a counterclockwise manner. Here all formulations are based on this counterclockwise system of labeling of nodes, although a formulation based on a clockwise system of labeling could be used. Remember that a consistent labeling procedure for the whole body is necessary to avoid problems in the calculations such as negative element areas. Here (x_i, y_i), (x_j, y_j), and (x_m, y_m) are the known nodal coordinates of nodes i, j, and m, respectively.

The nodal displacement matrix is given by

$$\{d\} = \begin{Bmatrix} \underline{d}_i \\ \underline{d}_j \\ \underline{d}_m \end{Bmatrix} = \begin{Bmatrix} u_i \\ v_i \\ u_j \\ v_j \\ u_m \\ v_m \end{Bmatrix} \qquad (7.2.1)$$

Step 2   Select Displacement Functions

We select a linear displacement function for each element as

$$u(x, y) = a_1 + a_2 x + a_3 y$$
$$v(x, y) = a_4 + a_5 x + a_6 y \qquad (7.2.2)$$

The linear function ensures that compatibility will be satisfied. A linear function with specified endpoints has only one path through which to pass—that is, through the two points. Hence, the linear function ensures that the displacements along the edge and at the nodes shared by adjacent elements are equal. Using Eqs. (7.2.2) in Eq. (7.1.15), the general displacement function $\{\psi\}$, which stores the functions u and v, can be expressed as

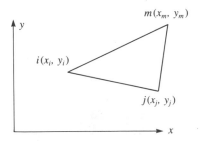

F I G U R E   7-8   Basic triangular element

$$\{\psi\} = \begin{Bmatrix} a_1 + a_2x + a_3y \\ a_4 + a_5x + a_6y \end{Bmatrix} = \begin{bmatrix} 1 & x & y & 0 & 0 & 0 \\ 0 & 0 & 0 & 1 & x & y \end{bmatrix} \begin{Bmatrix} a_1 \\ a_2 \\ a_3 \\ a_4 \\ a_5 \\ a_6 \end{Bmatrix} \qquad (7.2.3)$$

To obtain the a's in Eqs. (7.2.2), we begin by substituting the coordinates of the nodal points into Eqs. (7.2.2) to yield

$$u_i = a_1 + a_2x_i + a_3y_i$$

$$u_j = a_1 + a_2x_j + a_3y_j$$

$$u_m = a_1 + a_2x_m + a_3y_m$$

$$v_i = a_4 + a_5x_i + a_6y_i \qquad (7.2.4)$$

$$v_j = a_4 + a_5x_j + a_6y_j$$

$$v_m = a_4 + a_5x_m + a_6y_m$$

where $u_i = u(x_i, y_i)$, $u_j = u(x_j, y_j)$, etc. We can solve for the a's beginning with the first three of Eqs. (7.2.4) expressed in matrix form as

$$\begin{Bmatrix} u_i \\ u_j \\ u_m \end{Bmatrix} = \begin{bmatrix} 1 & x_i & y_i \\ 1 & x_j & y_j \\ 1 & x_m & y_m \end{bmatrix} \begin{Bmatrix} a_1 \\ a_2 \\ a_3 \end{Bmatrix} \qquad (7.2.5)$$

or, solving for the a's, we have

$$\{a\} = [x]^{-1}\{u\} \qquad (7.2.6)$$

where $[x]$ is the 3×3 matrix on the right side of Eq. (7.2.5). The method of cofactors (see Appendix A) is one possible method for finding the inverse of $[x]$. Thus,

$$[x]^{-1} = \frac{1}{2A} \begin{bmatrix} \alpha_i & \alpha_j & \alpha_m \\ \beta_i & \beta_j & \beta_m \\ \gamma_i & \gamma_j & \gamma_m \end{bmatrix} \qquad (7.2.7)$$

where

$$2A = \begin{vmatrix} 1 & x_i & y_i \\ 1 & x_j & y_j \\ 1 & x_m & y_m \end{vmatrix} \qquad (7.2.8)$$

is the determinant of $[x]$, which on evaluation is

$$2A = x_i(y_j - y_m) + x_j(y_m - y_i) + x_m(y_i - y_j) \qquad (7.2.9)$$

Here A is the area of the triangle, and

$$
\begin{aligned}
\alpha_i &= x_j y_m - y_j x_m & \alpha_j &= y_i x_m - x_i y_m & \alpha_m &= x_i y_j - y_i x_j \\
\beta_i &= y_j - y_m & \beta_j &= y_m - y_i & \beta_m &= y_i - y_j \\
\gamma_i &= x_m - x_j & \gamma_j &= x_i - x_m & \gamma_m &= x_j - x_i
\end{aligned}
\qquad (7.2.10)
$$

Having determined $[x]^{-1}$, we can now express Eq. (7.2.6) in expanded matrix form as

$$
\begin{Bmatrix} a_1 \\ a_2 \\ a_3 \end{Bmatrix} = \frac{1}{2A} \begin{bmatrix} \alpha_i & \alpha_j & \alpha_m \\ \beta_i & \beta_j & \beta_m \\ \gamma_i & \gamma_j & \gamma_m \end{bmatrix} \begin{Bmatrix} u_i \\ u_j \\ u_m \end{Bmatrix}
\qquad (7.2.11)
$$

Similarly, using the last three of Eqs. (7.2.4), we can obtain

$$
\begin{Bmatrix} a_4 \\ a_5 \\ a_6 \end{Bmatrix} = \frac{1}{2A} \begin{bmatrix} \alpha_i & \alpha_j & \alpha_m \\ \beta_i & \beta_j & \beta_m \\ \gamma_i & \gamma_j & \gamma_m \end{bmatrix} \begin{Bmatrix} v_i \\ v_j \\ v_m \end{Bmatrix}
\qquad (7.2.12)
$$

We will derive the general x displacement function $u(x, y)$ of $\{\psi\}$ (v will follow analogously) in terms of the coordinate variables x and y, known coordinate variables α_i, α_j, ..., γ_m, and unknown nodal displacements u_i, u_j, and u_m. Beginning with Eqs. (7.2.2) expressed in matrix form, we have

$$
\{u\} = \begin{bmatrix} 1 & x & y \end{bmatrix} \begin{Bmatrix} a_1 \\ a_2 \\ a_3 \end{Bmatrix}
\qquad (7.2.13)
$$

Substituting Eq. (7.2.11) into Eq. (7.2.13), we obtain

$$
\{u\} = \frac{1}{2A} \begin{bmatrix} 1 & x & y \end{bmatrix} \begin{bmatrix} \alpha_i & \alpha_j & \alpha_m \\ \beta_i & \beta_j & \beta_m \\ \gamma_i & \gamma_j & \gamma_m \end{bmatrix} \begin{Bmatrix} u_i \\ u_j \\ u_m \end{Bmatrix}
\qquad (7.2.14)
$$

Expanding Eq. (7.2.14), we have

$$
\{u\} = \frac{1}{2A} \begin{bmatrix} 1 & x & y \end{bmatrix} \begin{Bmatrix} \alpha_i u_i + \alpha_j u_j + \alpha_m u_m \\ \beta_i u_i + \beta_j u_j + \beta_m u_m \\ \gamma_i u_i + \gamma_j u_j + \gamma_m u_m \end{Bmatrix}
\qquad (7.2.15)
$$

Multiplying the two matrices in Eq. (7.2.15) and rearranging, we obtain

$$
u(x, y) = \frac{1}{2A} \{ (\alpha_i + \beta_i x + \gamma_i y) u_i + (\alpha_j + \beta_j x + \gamma_j y) u_j
$$
$$
+ (\alpha_m + \beta_m x + \gamma_m y) u_m \}
\qquad (7.2.16)
$$

Similarly, replacing u_i by v_i, u_j by v_j, and u_m by v_m in Eq. (7.2.16), we have the y displacement given by

$$
v(x, y) = \frac{1}{2A} \{ (\alpha_i + \beta_i x + \gamma_i y) v_i + (\alpha_j + \beta_j x + \gamma_j y) v_j
$$
$$
+ (\alpha_m + \beta_m x + \gamma_m y) v_m \}
\qquad (7.2.17)
$$

To express Eqs. (7.2.16) and (7.2.17) for u and v in simpler form, we define

$$N_i = \frac{1}{2A}(\alpha_i + \beta_i x + \gamma_i y)$$

$$N_j = \frac{1}{2A}(\alpha_j + \beta_j x + \gamma_j y) \qquad (7.2.18)$$

$$N_m = \frac{1}{2A}(\alpha_m + \beta_m x + \gamma_m y)$$

Thus, using Eqs. (7.2.18), we can rewrite Eqs. (7.2.16) and (7.2.17) as

$$u(x, y) = N_i u_i + N_j u_j + N_m u_m$$
$$v(x, y) = N_i v_i + N_j v_j + N_m v_m \qquad (7.2.19)$$

Expressing Eqs. (7.2.19) in matrix form, we obtain

$$\{\psi\} = \begin{Bmatrix} u(x, y) \\ v(x, y) \end{Bmatrix} = \begin{Bmatrix} N_i u_i + N_j u_j + N_m u_m \\ N_i v_i + N_j v_j + N_m v_m \end{Bmatrix}$$

or

$$\{\psi\} = \begin{bmatrix} N_i & 0 & N_j & 0 & N_m & 0 \\ 0 & N_i & 0 & N_j & 0 & N_m \end{bmatrix} \begin{Bmatrix} u_i \\ v_i \\ u_j \\ v_j \\ u_m \\ v_m \end{Bmatrix} \qquad (7.2.20)$$

Finally, expressing Eq. (7.2.20) in abbreviated matrix form, we have

$$\{\psi\} = [N]\{d\} \qquad (7.2.21)$$

where $[N]$ is given by

$$[N] = \begin{bmatrix} N_i & 0 & N_j & 0 & N_m & 0 \\ 0 & N_i & 0 & N_j & 0 & N_m \end{bmatrix} \qquad (7.2.22)$$

We have now expressed the general displacements as functions of $\{d\}$, in terms of the shape functions N_i, N_j, and N_m. The shape functions represent the shape of $\{\psi\}$ when plotted over the surface of a typical element. For instance, N_i represents the shape of the variable u when plotted over the surface of the element for $u_i = 1$ and all other degrees of freedom equal to zero; that is, $u_j = u_m = v_i = v_j = v_m = 0$. In addition, $u(x_i, y_i)$ must be equal to u_i. Therefore, we must have $N_i = 1$, $N_j = 0$, and $N_m = 0$ at (x_i, y_i). Similarly, $u(x_j, y_j) = u_j$. Therefore, $N_i = 0$, $N_j = 1$, and $N_m = 0$ at (x_j, y_j). Figure 7-9 shows the shape variation of N_i plotted over the surface of a typical element. Note that N_i does not equal zero except along a line connecting and including nodes j and m.

Finally, $N_i + N_j + N_m = 1$ for all x and y locations on the surface of the element. The proof of this relationship follows that given for the bar element in Section 3.2.

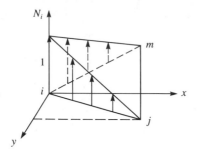

FIGURE 7-9   Variation of N_i over the x-y surface of a typical element

Step 3   Define the Strain/Displacement and Stress/Strain Relationships

We express the element strains and stresses in terms of the unknown nodal displacements.

Element Strains

The strains associated with the two-dimensional element are given by

$$\{\varepsilon\} = \left\{ \begin{array}{c} \varepsilon_x \\ \varepsilon_y \\ \gamma_{xy} \end{array} \right\} = \left\{ \begin{array}{c} \dfrac{\partial u}{\partial x} \\[2mm] \dfrac{\partial v}{\partial y} \\[2mm] \dfrac{\partial u}{\partial y} + \dfrac{\partial v}{\partial x} \end{array} \right\} \qquad (7.2.23)$$

Using Eqs. (7.2.19) for the displacements, we have

$$\frac{\partial u}{\partial x} = u_{,x} = \frac{\partial}{\partial x}(N_i u_i + N_j u_j + N_m u_m) \qquad (7.2.24)$$

or $$u_{,x} = N_{i,x} u_i + N_{j,x} u_j + N_{m,x} u_m \qquad (7.2.25)$$

where the comma followed by a variable indicates differentiation with respect to that variable. We have used $u_{i,x} = 0$ because $u_i = u(x_i, y_i)$ is a constant value; similarly, $u_{j,x} = 0$ and $u_{m,x} = 0$.

Using Eqs. (7.2.18), we can evaluate the expressions for the derivatives of the shape functions in Eq. (7.2.25) as follows:

$$N_{i,x} = \frac{1}{2A} \frac{\partial}{\partial x}(\alpha_i + \beta_i x + \gamma_i y) = \frac{\beta_i}{2A} \qquad (7.2.26)$$

Similarly, $$N_{j,x} = \frac{\beta_j}{2A} \quad \text{and} \quad N_{m,x} = \frac{\beta_m}{2A} \qquad (7.2.27)$$

Therefore, using Eqs. (7.2.26) and (7.2.27) in Eq. (7.2.25), we have

$$\frac{\partial u}{\partial x} = \frac{1}{2A}(\beta_i u_i + \beta_j u_j + \beta_m u_m) \qquad (7.2.28)$$

Similarly, we can obtain

$$\frac{\partial v}{\partial y} = \frac{1}{2A}(\gamma_i v_i + \gamma_j v_j + \gamma_m v_m)$$

$$\frac{\partial u}{\partial y} + \frac{\partial v}{\partial x} = \frac{1}{2A}(\gamma_i u_i + \beta_i v_i + \gamma_j u_j + \beta_j v_j + \gamma_m u_m + \beta_m v_m)$$

(7.2.29)

Using Eqs. (7.2.28) and (7.2.29) in Eq. (7.2.23), we obtain

$$\{\varepsilon\} = \frac{1}{2A}\begin{bmatrix} \beta_i & 0 & \beta_j & 0 & \beta_m & 0 \\ 0 & \gamma_i & 0 & \gamma_j & 0 & \gamma_m \\ \gamma_i & \beta_i & \gamma_j & \beta_j & \gamma_m & \beta_m \end{bmatrix}\begin{Bmatrix} u_i \\ v_i \\ u_j \\ v_j \\ u_m \\ v_m \end{Bmatrix}$$

(7.2.30)

or

$$\{\varepsilon\} = [\underline{B}_i \quad \underline{B}_j \quad \underline{B}_m]\begin{Bmatrix} \underline{d}_i \\ \underline{d}_j \\ \underline{d}_m \end{Bmatrix}$$

(7.2.31)

where

$$[B_i] = \frac{1}{2A}\begin{bmatrix} \beta_i & 0 \\ 0 & \gamma_i \\ \gamma_i & \beta_i \end{bmatrix} \quad [B_j] = \frac{1}{2A}\begin{bmatrix} \beta_j & 0 \\ 0 & \gamma_j \\ \gamma_j & \beta_j \end{bmatrix} \quad [B_m] = \frac{1}{2A}\begin{bmatrix} \beta_m & 0 \\ 0 & \gamma_m \\ \gamma_m & \beta_m \end{bmatrix}$$

(7.2.32)

Finally, in simplified matrix form, Eq. (7.2.31) can be written as

$$\{\varepsilon\} = [B]\{d\}$$

(7.2.33)

where

$$[B] = [\underline{B}_i \quad \underline{B}_j \quad \underline{B}_m]$$

(7.2.34)

The $\underline{B}$ matrix is independent of the x and y coordinates. It depends solely on the element nodal coordinates, as seen from Eqs. (7.2.32) and (7.2.10). The strains in Eq. (7.2.33) will be constant; hence, the element is called a *constant-strain triangle* (CST).

Stress/Strain Relationship

In general, the in-plane stress/strain relationship is given by

$$\begin{Bmatrix} \sigma_x \\ \sigma_y \\ \tau_{xy} \end{Bmatrix} = [D]\begin{Bmatrix} \varepsilon_x \\ \varepsilon_y \\ \gamma_{xy} \end{Bmatrix}$$

(7.2.35)

where $[D]$ is given by Eq. (7.1.8) for plane stress problems, and by Eq. (7.1.10) for plane strain problems. Using Eq. (7.2.33) in Eq. (7.2.35), we obtain the in-plane stresses in terms of the unknown nodal degrees of freedom as

$$\{\sigma\} = [D][B]\{d\}$$

(7.2.36)

Step 4   Derive the Element Stiffness Matrix and Equations

Using the principle of minimum potential energy, we can generate the equations for a typical constant-strain triangular element. Keep in mind that for the basic plane stress element, the total potential energy is now a function of the nodal displacements $u_i, v_i, u_j, \ldots, v_m$ (that is, $\{d\}$) such that

$$\pi_p = \pi_p(u_i, v_i, u_j, \ldots, v_m) \tag{7.2.37}$$

Here the total potential energy is given by

$$\pi_p = U + \Omega_b + \Omega_p + \Omega_s \tag{7.2.38}$$

where the strain energy is given by

$$U = \frac{1}{2} \iiint_V \{\varepsilon\}^T \{\sigma\}\, dV \tag{7.2.39}$$

or, using Eq. (7.2.35), we have

$$U = \frac{1}{2} \iiint_V \{\varepsilon\}^T [D]\{\varepsilon\}\, dV \tag{7.2.40}$$

where we have used $[D]^T = [D]$ in Eq. (7.2.40).

The potential energy of the body forces is given by

$$\Omega_b = -\iiint_V \{\psi\}^T \{X\}\, dV \tag{7.2.41}$$

where $\{\psi\}$ is again the general displacement function, and $\{X\}$ is the body weight/unit volume or weight density (typically, in units of pounds per cubic inch or kilonewtons per cubic meter).

The potential energy of concentrated loads is given by

$$\Omega_p = -\{d\}^T \{P\} \tag{7.2.42}$$

where $\{d\}$ represents the usual nodal displacements, and $\{P\}$ now represents the concentrated external loads.

The potential energy of distributed loads (or surface tractions) is given by

$$\Omega_s = -\iint_S \{\psi\}^T \{T\}\, dS \tag{7.2.43}$$

where $\{T\}$ represents the surface tractions (typically, in units of pounds per square inch or kilonewtons per square meter), and S represents the surfaces over which the tractions $\{T\}$ act.

Using Eq. (7.2.21) for $\{\psi\}$ and Eq. (7.2.33) for the strains in Eqs. (7.2.40)–(7.2.43), we have

$$\pi_p = \frac{1}{2} \iiint_V \{d\}^T [B]^T [D][B]\{d\}\, dV - \iiint_V \{d\}^T [N]^T \{X\}\, dV$$

$$- \{d\}^T \{P\} - \iint_S \{d\}^T [N]^T \{T\}\, dS \tag{7.2.44}$$

The nodal displacements $\{d\}$ are independent of the general x-y coordinates, so $\{d\}$ can be taken out of the integrals of Eq. (7.2.44). Therefore,

$$\pi_p = \frac{1}{2}\{d\}^T \iiint_V [B]^T[D][B]\, dV\{d\} - \{d\}^T \iiint_V [N]^T\{X\}\, dV$$

$$- \{d\}^T\{P\} - \{d\}^T \iint_S [N]^T\{T\}\, dS \qquad (7.2.45)$$

From Eqs. (7.2.41)–(7.2.43) we can see that the last three terms of Eq. (7.2.45) represent the total load system $\{f\}$ on an element; that is,

$$\{f\} = \iiint_V [N]^T\{X\}\, dV + \{P\} + \iint_S [N]^T\{T\}\, dS \qquad (7.2.46)$$

where the first, second, and third terms on the right side of Eq. (7.2.46) represent the body forces, the concentrated nodal forces, and the surface tractions, respectively. Using Eq. (7.2.46) in Eq. (7.2.45), we obtain

$$\pi_p = \frac{1}{2}\{d\}^T \iiint_V [B]^T[D][B]\, dV\{d\} - \{d\}^T\{f\} \qquad (7.2.47)$$

Taking the first variation, or equivalently as shown in Chapter 3, the partial derivative of π_p with respect to the nodal displacements since $\pi_p = \pi_p(\underline{d})$ (as was previously done for the bar and beam elements in Chapters 3 and 5, respectively), we obtain

$$\frac{\partial \pi_p}{\partial \{d\}} = \left[\iiint_V [B]^T[D][B]\, dV \right]\{d\} - \{f\} = 0 \qquad (7.2.48)$$

Rewriting Eq. (7.2.48), we have

$$\iiint_V [B]^T[D][B]\, dV\{d\} = \{f\} \qquad (7.2.49)$$

where the partial derivative with respect to matrix $\{d\}$ was previously defined by Eq. (3.8.9). From Eq. (7.2.49) we can see that

$$[k] = \iiint_V [B]^T[D][B]\, dV \qquad (7.2.50)$$

For an element with constant thickness t, Eq. (7.2.50) becomes

$$[k] = t \iint_A [B]^T[D][B]\, dx\, dy \qquad (7.2.51)$$

where the integrand is not a function of x or y for the constant-strain triangular element, and thus, can be taken out of the integral to yield

$$[k] = tA[B]^T[D][B] \qquad (7.2.52)$$

where A is given by Eq. (7.2.9), $[B]$ is given by Eq. (7.2.34), and $[D]$ is given by Eq. (7.1.8) or Eq. (7.1.10). We will assume elements of constant thickness. (This assumption is convergent to the actual situation as the element size is decreased.)

From Eq. (7.2.52) we see that $[k]$ is a function of the nodal coordinates (because $[B]$ and A are defined in terms of them) and of the mechanical properties E and v (of which $[D]$ is a function). The expansion of Eq. (7.2.52) for an element is

$$[k] = \begin{bmatrix} [k_{ii}] & [k_{ij}] & [k_{im}] \\ [k_{ji}] & [k_{jj}] & [k_{jm}] \\ [k_{mi}] & [k_{mj}] & [k_{mm}] \end{bmatrix} \qquad (7.2.53)$$

where the 2×2 submatrices are given by

$$[k_{ii}] = [B_i]^T[D][B_i]tA$$
$$[k_{ij}] = [B_i]^T[D][B_j]tA \qquad (7.2.54)$$
$$[k_{im}] = [B_i]^T[D][B_m]tA$$

and so forth. In Eqs. (7.2.54), $[B_i]$, $[B_j]$, and $[B_m]$ are defined by Eqs. (7.2.32). The $[k]$ matrix is seen to be a 6×6 matrix (equal in order to the number of degrees of freedom per node, two, times the total number of nodes per element, three).

In general, Eq. (7.2.46) must be used to evaluate the surface and body forces. When Eq. (7.2.46) is used to evaluate the surface and body forces, these forces are called *consistent loads* because they are derived from the consistent (energy) approach. For higher-order elements, typically with quadratic or cubic displacement functions, Eq. (7.2.46) should be used. However, for the CST element, the body and surface forces can be lumped at the nodes with equivalent results (this is illustrated in Section 7.3) and added to any concentrated nodal forces to obtain the element force matrix. The element equations are then given by

$$\begin{Bmatrix} f_{1x} \\ f_{1y} \\ f_{2x} \\ f_{2y} \\ f_{3x} \\ f_{3y} \end{Bmatrix} = \begin{bmatrix} k_{11} & k_{12} & \cdots & k_{16} \\ k_{21} & k_{22} & \cdots & k_{26} \\ \vdots & \vdots & & \vdots \\ k_{61} & k_{62} & \cdots & k_{66} \end{bmatrix} \begin{Bmatrix} u_1 \\ v_1 \\ u_2 \\ v_2 \\ u_3 \\ v_3 \end{Bmatrix} \qquad (7.2.55)$$

Step 5   **Assemble the Element Equations to Obtain the Global Equations and Introduce Boundary Conditions**

We obtain the global structure stiffness matrix and equations by using the direct stiffness method as

$$[K] = \sum_{e=1}^{N} [k^{(e)}] \qquad (7.2.56)$$

and
$$\{F\} = [K]\{d\} \tag{7.2.57}$$

where, in Eq. (7.2.56), all element stiffness matrices are defined in terms of the global x-y coordinate system, $\{d\}$ is now the total structure displacement matrix, and

$$\{F\} = \sum_{e=1}^{N} \{f^{(e)}\} \tag{7.2.58}$$

is the column of equivalent global nodal loads obtained by lumping body forces and distributed loads at the proper nodes (as well as including concentrated nodal loads) or by consistently using Eq. (7.2.46). (Further details regarding the treatment of body forces and surface tractions will be given in Section 7.3.)

In the formulation of the element stiffness matrix Eq. (7.2.52), the matrix has been derived for a general orientation in global coordinates. Equation (7.2.52) then applies for all elements. All element matrices are expressed in the global-coordinate orientation. Therefore, no transformation from local to global equations is necessary. However, for completeness, we will now describe the method to use if the local axes for the constant-strain triangular element are not parallel to the global axes for the whole structure.

If the local axes for the constant-strain triangular element are not parallel to the global axes for the whole structure, we must apply rotation-of-axes transformations similar to those introduced in Chapter 3 by Eq. (3.3.16) to the element stiffness matrix, as well as to the element nodal force and displacement matrices. We illustrate the transformation of axes for the triangular element shown in Figure 7-10, considering the element to have local axes $\hat{x}$-$\hat{y}$ not parallel to global axes x-y. Local nodal forces are shown in the figure. The transformation from local to global equations follows the procedure outlined in Section 3.4. We have the same general expressions, Eqs. (3.4.14), (3.4.16), and (3.4.22), to relate local to global displacements, forces, and stiffness matrices, respectively; that is,

$$\underline{\hat{d}} = \underline{T}\underline{d} \qquad \underline{\hat{f}} = \underline{T}\underline{f} \qquad \underline{k} = \underline{T}^T\underline{\hat{k}}\underline{T} \tag{7.2.59}$$

where Eq. (3.4.15) for the transformation matrix $\underline{T}$ used in Eqs. (7.2.59) must

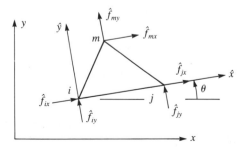

FIGURE 7-10 Triangular element with local axes not parallel to global axes

be expanded because two additional degrees of freedom are present in the constant-strain triangular element. Thus, Eq. (3.4.15) is expanded to

$$\underline{T} = \begin{bmatrix} C & S & 0 & 0 & 0 & 0 \\ -S & C & 0 & 0 & 0 & 0 \\ 0 & 0 & C & S & 0 & 0 \\ 0 & 0 & -S & C & 0 & 0 \\ 0 & 0 & 0 & 0 & C & S \\ 0 & 0 & 0 & 0 & -S & C \end{bmatrix} \qquad (7.2.60)$$

where $C = \cos \theta$, $S = \sin \theta$, and θ is shown in Figure 7-10.

Step 6   Solve for the Nodal Displacements

We determine the unknown global structure nodal displacements by solving the system of algebraic equations given by Eq. (7.2.57).

Step 7   Solve for the Element Forces (Stresses)

Having solved for the nodal displacements, we obtain the strains and stresses in the global x and y directions in the elements by using Eqs. (7.2.33) and (7.2.36). Finally, we determine the maximum and minimum in-plane principal stresses σ_1 and σ_2 by using the transformation Eqs. (7.1.2), where these stresses are usually assumed to act at the centroid of the element. The angle that one of the principal stresses makes with the x axis is given by Eq. (7.1.3).

E X A M P L E   7.1

Evaluate the stiffness matrix for the element shown in Figure 7-11. The coordinates are shown in units of inches. Assume plane stress conditions. Let $E = 30 \times 10^6$ psi, $v = 0.25$, and thickness $t = 1$ in. Assume the element nodal displacements have been determined to be $u_1 = 0.0$, $v_1 = 0.0025$ in., $u_2 = 0.0012$ in., $v_2 = 0.0$, $u_3 = 0.0$, and $v_3 = 0.0025$ in. Determine the element stresses.

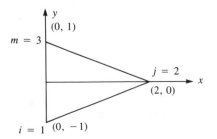

FIGURE 7-11   Plane stress element for stiffness matrix evaluation

We use Eq. (7.2.52) to obtain the element stiffness matrix. To evaluate $\underline{k}$, we first use Eqs. (7.2.10) to obtain the β's and γ's as follows:

$$\beta_i = y_j - y_m = 0 - 1 = -1 \qquad \gamma_i = x_m - x_j = 0 - 2 = -2$$
$$\beta_j = y_m - y_i = 1 - (-1) = 2 \qquad \gamma_j = x_i - x_m = 0 - 0 = 0 \qquad (7.2.61)$$
$$\beta_m = y_i - y_j = -1 - 0 = -1 \qquad \gamma_m = x_j - x_i = 2 - 0 = 2$$

Using Eqs. (7.2.32) and (7.2.34), matrix $\underline{B}$ is obtained as

$$\underline{B} = \frac{1}{2(2)} \begin{bmatrix} -1 & 0 & 2 & 0 & -1 & 0 \\ 0 & -2 & 0 & 0 & 0 & 2 \\ -2 & -1 & 0 & 2 & 2 & -1 \end{bmatrix} \qquad (7.2.62)$$

where we have used $A = 2$ in.^{2} in Eq. (7.2.62).

Using Eq. (7.1.8) for plane stress conditions,

$$\underline{D} = \frac{30 \times 10^6}{1 - (0.25)^2} \begin{bmatrix} 1 & 0.25 & 0 \\ 0.25 & 1 & 0 \\ 0 & 0 & \dfrac{1 - 0.25}{2} \end{bmatrix} \qquad (7.2.63)$$

Substituting Eqs. (7.2.62) and (7.2.63) into Eq. (7.2.52), we obtain

$$\underline{k} = \frac{(2)30 \times 10^6}{4(0.9375)} \begin{bmatrix} -1 & 0 & -2 \\ 0 & -2 & -1 \\ 2 & 0 & 0 \\ 0 & 0 & 2 \\ -1 & 0 & 2 \\ 0 & 2 & -1 \end{bmatrix}$$

$$\times \begin{bmatrix} 1 & 0.25 & 0 \\ 0.25 & 1 & 0 \\ 0 & 0 & 0.375 \end{bmatrix} \frac{1}{2(2)} \begin{bmatrix} -1 & 0 & 2 & 0 & -1 & 0 \\ 0 & -2 & 0 & 0 & 0 & 2 \\ -2 & -1 & 0 & 2 & 2 & -1 \end{bmatrix}$$

Performing the matrix triple product, we have

$$\underline{k} = 4.0 \times 10^6 \begin{bmatrix} 2.5 & 1.25 & -2 & -1.5 & -0.5 & 0.25 \\ 1.25 & 4.375 & -1 & -0.75 & -0.25 & -3.625 \\ -2 & -1 & 4 & 0 & -2 & 1 \\ -1.5 & -0.75 & 0 & 1.5 & 1.5 & -0.75 \\ -0.5 & -0.25 & -2 & 1.5 & 2.5 & -1.25 \\ 0.25 & -3.625 & 1 & -0.75 & -1.25 & 4.375 \end{bmatrix}$$

$$(7.2.64)$$

To evaluate the stresses, we use Eq. (7.2.36). Substituting Eqs. (7.2.62) and (7.2.63), along with the given nodal displacements, into Eq. (7.2.36), we obtain

$$\begin{Bmatrix} \sigma_x \\ \sigma_y \\ \tau_{xy} \end{Bmatrix} = \frac{30 \times 10^6}{1 - (0.25)^2} \begin{bmatrix} 1 & 0.25 & 0 \\ 0.25 & 1 & 0 \\ 0 & 0 & 0.375 \end{bmatrix}$$

$$\times \frac{1}{2(2)} \begin{bmatrix} -1 & 0 & 2 & 0 & -1 & 0 \\ 0 & -2 & 0 & 0 & 0 & 2 \\ -2 & -1 & 0 & 2 & 2 & -1 \end{bmatrix} \begin{Bmatrix} 0.0 \\ 0.0025 \\ 0.0012 \\ 0.0 \\ 0.0 \\ 0.0025 \end{Bmatrix} \qquad (7.2.65)$$

Performing the matrix triple product in Eq. (7.2.65), we have

$$\sigma_x = 19{,}200 \text{ psi} \qquad \sigma_y = 4800 \text{ psi} \qquad \tau_{xy} = -15{,}000 \text{ psi} \qquad (7.2.66)$$

Finally, the principal stresses and principal angle are obtained by substituting the results from Eqs. (7.2.66) into Eqs. (7.1.2) and (7.1.3) as follows:

$$\sigma_1 = \frac{19{,}200 + 4800}{2} + \left[\left(\frac{19{,}200 - 4800}{2} \right)^2 + (-15{,}000)^2 \right]^{1/2}$$

$$= 28{,}639 \text{ psi}$$

$$\sigma_2 = \frac{19{,}200 + 4800}{2} - \left[\left(\frac{19{,}200 - 4800}{2} \right)^2 + (-15{,}000)^2 \right]^{1/2}$$

$$= -4639 \text{ psi} \qquad (7.2.67)$$

$$\theta_p = \frac{1}{2} \tan^{-1} \left[\frac{2(-15{,}000)}{19{,}200 - 4800} \right] = -32.2° \qquad \blacksquare$$

7.3   Treatment of Body and Surface Forces

Body Forces

Using the first term on the right side of Eq. (7.2.46), we can evaluate the body forces at the nodes as

$$\{f_b\} = \iiint_V [N]^T \{X\} \, dV \qquad (7.3.1)$$

where

$$\{X\} = \begin{Bmatrix} X_b \\ Y_b \end{Bmatrix} \qquad (7.3.2)$$

and X_b and Y_b are the weight densities in the x and y directions, respectively. These forces may arise, for instance, because of actual body weight (gravitational forces), angular velocity (called *centrifugal body forces*, as described in Chapter 10), or inertial forces in dynamics.

In Eq. (7.3.1), $[N]$ is not constant; therefore, the integration must be

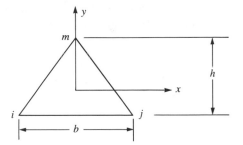

FIGURE 7-12   Element with centroidal coordinate axes

carried out. Without lack of generality, the integration is simplified if the origin of the coordinates is chosen at the centroid of the element. For example, consider the element with coordinates shown in Figure 7-12. With the origin of the coordinate placed at the centroid of the element, we have, from the definition of the centroid, $\iint x \, dA = \iint y \, dA = 0$ and therefore,

$$\iint \beta_i x \, dA = \iint \gamma_i y \, dA = 0 \qquad (7.3.3)$$

and

$$\alpha_i = \alpha_j = \alpha_m = \frac{2A}{3} \qquad (7.3.4)$$

Using Eqs. (7.3.2), (7.3.3), and (7.3.4) in Eq. (7.3.1), the body force at node i is then represented by

$$\{f_{bi}\} = \begin{Bmatrix} X_b \\ Y_b \end{Bmatrix} \frac{tA}{3} \qquad (7.3.5)$$

Similarly, considering the j and m node body forces, we obtain the same results as in Eq. (7.3.5). In matrix form, the element body forces are

$$\{f_b\} = \begin{Bmatrix} f_{bix} \\ f_{biy} \\ f_{bjx} \\ f_{bjy} \\ f_{bmx} \\ f_{bmy} \end{Bmatrix} = \begin{Bmatrix} X_b \\ Y_b \\ X_b \\ Y_b \\ X_b \\ Y_b \end{Bmatrix} \frac{At}{3} \qquad (7.3.6)$$

From the results of Eq. (7.3.6), we can conclude that the body forces are distributed to the nodes in three equal parts. The signs depend on the directions of X_b and Y_b with respect to the positive x and y global coordinates. For the case of body weight only, due to the gravitational force associated with the y direction, we have only Y_b ($X_b = 0$).

Surface Forces

Using the third term on the right side of Eq. (7.2.46), we can evaluate the surface forces at the nodes as

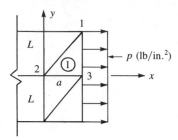

FIGURE 7-13   Element with
uniform load acting on one edge

$$\{f_s\} = \int\int_S [N]^T\{T\}\, dS \qquad (7.3.7)$$

We will now illustrate the use of Eq. (7.3.7) by considering the example of a uniform stress p (say, in pounds per square inch) acting between nodes 1 and 3 on the edge of element 1 in Figure 7-13. In Eq. (7.3.7), the surface traction now becomes

$$\{T\} = \begin{Bmatrix} p_x \\ p_y \end{Bmatrix} = \begin{Bmatrix} p \\ 0 \end{Bmatrix} \qquad (7.3.8)$$

and

$$[N]^T = \begin{bmatrix} N_1 & 0 \\ 0 & N_1 \\ N_2 & 0 \\ 0 & N_2 \\ N_3 & 0 \\ 0 & N_3 \end{bmatrix} \qquad (7.3.9)$$

Using Eqs. (7.3.8) and (7.3.9), we express Eq. (7.3.7) as

$$\{f_s\} = \int_0^t \int_0^L \begin{bmatrix} N_1 & 0 \\ 0 & N_1 \\ N_2 & 0 \\ 0 & N_2 \\ N_3 & 0 \\ 0 & N_3 \end{bmatrix} \begin{Bmatrix} p \\ 0 \end{Bmatrix} dz\, dy \qquad (7.3.10)$$

evaluated at $x = a$, $y = y$

where the N's are evaluated at $x = a$ and $y = y$ because the load acts on that edge. Simplifying Eq. (7.3.10), we obtain

$$\{f_s\} = t \int_0^L \begin{bmatrix} N_1 p \\ 0 \\ N_2 p \\ 0 \\ N_3 p \\ 0 \end{bmatrix} dy \qquad (7.3.11)$$

evaluated at $x = a$, $y = y$

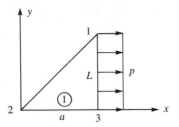

FIGURE 7-14 Representative element subjected to edge loading

Now, by Eqs. (7.2.18) (with $i = 1$), we have

$$N_1 = \frac{1}{2A}(\alpha_1 + \beta_1 x + \gamma_1 y) \qquad (7.3.12)$$

For convenience, we choose the coordinate system for the element as shown in Figure 7-14. Using the definition Eqs. (7.2.10), we obtain

$$\alpha_i = x_j y_m - y_j x_m$$

or, with $i = 1$, $j = 2$, and $m = 3$,

$$\alpha_1 = x_2 y_3 - y_2 x_3 \qquad (7.3.13)$$

Substituting the coordinates into Eq. (7.3.13), we obtain

$$\alpha_1 = 0 \qquad (7.3.14)$$

Similarly, again using Eqs. (7.2.10), we obtain

$$\beta_1 = 0 \qquad \gamma_1 = a \qquad (7.3.15)$$

Therefore, substituting Eqs. (7.3.14) and (7.3.15) into Eq. (7.3.12), we obtain

$$N_1 = \frac{ay}{2A} \qquad (7.3.16)$$

Similarly, using Eqs. (7.2.18), we can show that

$$N_2 = \frac{L(a - x)}{2A} \quad \text{and} \quad N_3 = \frac{Lx - ay}{2A} \qquad (7.3.17)$$

On substituting Eqs. (7.3.16) and (7.3.17) for N_1, N_2, and N_3 into Eq. (7.3.11), evaluating N_1, N_2, and N_3 at $x = a$ and $y = y$ (the coordinates corresponding to the location of the surface load p), and then integrating with respect to y, we obtain

$$\{f_s\} = \frac{t}{2(aL/2)} \begin{Bmatrix} a\left(\dfrac{L^2}{2}\right)p \\ 0 \\ 0 \\ 0 \\ \left(L^2 - \dfrac{L^2}{2}\right)ap \\ 0 \end{Bmatrix} \qquad (7.3.18)$$

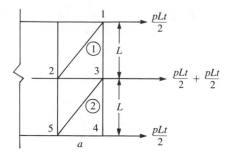

FIGURE 7-15   Surface load equivalent nodal forces

where the shape function $N_2 = 0$ between nodes 1 and 3, as should be the case according to the definitions of the shape functions. Simplifying Eq. (7.3.18), we finally obtain

$$\{f_s\} = \begin{Bmatrix} f_{s1x} \\ f_{s1y} \\ f_{s2x} \\ f_{s2y} \\ f_{s3x} \\ f_{s3y} \end{Bmatrix} = \begin{Bmatrix} pLt/2 \\ 0 \\ 0 \\ 0 \\ pLt/2 \\ 0 \end{Bmatrix} \qquad (7.3.19)$$

Figure 7-15 illustrates the results for the surface load equivalent nodal forces for both elements 1 and 2.

We can conclude that for a constant-strain triangle, a distributed load on an element edge can be treated as concentrated loads acting at the nodes associated with the loaded edge by making the two kinds of load statically equivalent [which is equivalent to applying Eq. (7.3.7)]. However, for higher-order elements such as the linear-strain triangle (discussed in Chapter 9), the load replacement should be made by using Eq. (7.3.7), which was derived by the principle of minimum potential energy. For higher-order elements, this load replacement by use of Eq. (7.3.7) is generally not equal to the apparent statically equivalent one; however, it is consistent in that this replacement results directly from the energy approach.

We now recognize the force matrix $\{f_s\}$ defined by Eq. (7.3.7), and based on the principle of minimum potential energy, to be equivalent to that based on work equivalence, which was previously used in Chapter 5 when discussing distributed loads acting on beams.

7.4   Explicit Expression for the Constant-Strain Triangle Stiffness Matrix

Although the stiffness matrix is generally formulated internally in most computer programs by performing the matrix triple product indicated by Eq. (7.4.1), it is still a valuable learning experience to explicitly evaluate the stiffness matrix for the constant-strain triangular element. Hence, we will consider the plane strain case specifically in this development.

First, recall that the stiffness matrix is given by

$$[k] = tA[B]^T[D][B] \qquad (7.4.1)$$

where, for the plane strain case, $[D]$ is given by Eq. (7.1.10) and $[B]$ is given by Eq. (7.2.34). On substituting the matrices $[D]$ and $[B]$ into Eq. (7.4.1), we obtain

$$[k] = \frac{tE}{4A(1+v)(1-2v)}
\begin{bmatrix}
\beta_i & 0 & \gamma_i \\
0 & \gamma_i & \beta_i \\
\beta_j & 0 & \gamma_j \\
0 & \gamma_j & \beta_j \\
\beta_m & 0 & \gamma_m \\
0 & \gamma_m & \beta_m
\end{bmatrix}
\begin{bmatrix}
1-v & v & 0 \\
v & 1-v & 0 \\
0 & 0 & \dfrac{1-2v}{2}
\end{bmatrix}
\begin{bmatrix}
\beta_i & 0 & \beta_j & 0 & \beta_m & 0 \\
0 & \gamma_i & 0 & \gamma_j & 0 & \gamma_m \\
\gamma_i & \beta_i & \gamma_j & \beta_j & \gamma_m & \beta_m
\end{bmatrix}$$

$$(7.4.2)$$

On multiplying the matrices in Eq. (7.4.2), we obtain Eq. (7.4.3), the explicit constant-strain triangle stiffness matrix for the plane strain case. Note that $[k]$ is a function of the difference in the x and y nodal coordinates, as indicated by the γ's and β's, of the material properties E and v, and of the thickness t and surface area A of the element.

$$\underline{k} = \frac{tE}{4A(1+v)(1-2v)}$$

$$\times
\begin{bmatrix}
\beta_i^2(1-v)+\gamma_i^2\left(\dfrac{1-2v}{2}\right) & \beta_i\gamma_i v + \beta_i\gamma_i\left(\dfrac{1-2v}{2}\right) & \beta_i\beta_j(1-v)+\gamma_i\gamma_j\left(\dfrac{1-2v}{2}\right) \\[3mm]
 & \gamma_i^2(1-v)+\beta_i^2\left(\dfrac{1-2v}{2}\right) & \beta_j\gamma_i v + \beta_i\gamma_j\left(\dfrac{1-2v}{2}\right) \\[3mm]
 & & \beta_j^2(1-v)+\gamma_j^2\left(\dfrac{1-2v}{2}\right) \\[3mm]
\text{Symmetry} & &
\end{bmatrix}$$

$$\begin{bmatrix}
\beta_i\gamma_j v + \beta_j\gamma_i\left(\dfrac{1-2v}{2}\right) & \beta_i\beta_m(1-v)+\gamma_i\gamma_m\left(\dfrac{1-2v}{2}\right) & \beta_i\gamma_m v + \beta_m\gamma_i\left(\dfrac{1-2v}{2}\right) \\[3mm]
\gamma_i\gamma_j(1-v)+\beta_i\beta_j\left(\dfrac{1-2v}{2}\right) & \beta_m\gamma_i v + \beta_i\gamma_m\left(\dfrac{1-2v}{2}\right) & \gamma_i\gamma_m(1-v)+\beta_i\beta_m\left(\dfrac{1-2v}{2}\right) \\[3mm]
\beta_j\gamma_j v + \beta_j\gamma_j\left(\dfrac{1-2v}{2}\right) & \beta_j\beta_m(1-v)+\gamma_j\gamma_m\left(\dfrac{1-2v}{2}\right) & \beta_j\gamma_m v + \gamma_j\beta_m\left(\dfrac{1-2v}{2}\right) \\[3mm]
\gamma_j^2(1-v)+\beta_j^2\left(\dfrac{1-2v}{2}\right) & \beta_m\gamma_j v + \beta_j\gamma_m\left(\dfrac{1-2v}{2}\right) & \gamma_j\gamma_m(1-v)+\beta_j\beta_m\left(\dfrac{1-2v}{2}\right) \\[3mm]
 & \beta_m^2(1-v)+\gamma_m^2\left(\dfrac{1-2v}{2}\right) & \gamma_m\beta_m v + \beta_m\gamma_m\left(\dfrac{1-2v}{2}\right) \\[3mm]
 & & \gamma_m^2(1-v)+\beta_m^2\left(\dfrac{1-2v}{2}\right)
\end{bmatrix}$$

$$(7.4.3)$$

For the plane stress case, we need only replace $1 - v$ by 1, $(1 - 2v)/2$ by $(1 - v)/2$, and $(1 + v)(1 - 2v)$ outside the brackets by $1 - v^2$ in Eq. (7.4.3).

Finally, it should be noted that for Poisson's ratio v approaching 0.5, as in rubberlike materials and plastic solids, for instance, a material becomes incompressible [2]. For plane strain, as v approaches 0.5, the denominator becomes zero in the material property matrix (see Eq. (7.1.10)) and hence in the stiffness matrix, Eq. (7.4.3). A value of v near 0.5—such as 0.49—can cause ill-conditioned structural equations. A special formulation (called a *penalty formulation*; see Reference [3]) has been used in this case.

7.5 Finite Element Solution of a Plane Stress Problem

To illustrate the finite element method for a plane stress problem, we now present a detailed solution.

E X A M P L E   7.2

For a thin plate subjected to the surface traction shown in Figure 7-16, determine the nodal displacements and the element stresses. The plate thickness $t = 1$ in., $E = 30 \times 10^6$ psi, and $v = 0.30$.

Discretization

To illustrate the finite element method solution for the plate, we first discretize the plate into two elements, as shown in Figure 7-17. It should be understood

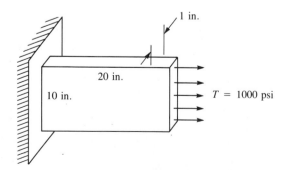

F I G U R E   7-16   Thin plate subjected to tensile stress

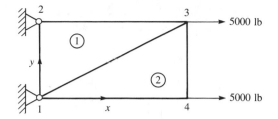

FIGURE 7-17   Discretized plate

that the coarseness of the mesh will not yield as true a predicted behavior of the plate as would a finer mesh, particularly near the fixed edge. However, since we are performing a longhand solution, we will use a coarse discretization for simplicity (but without loss of generality of the method).

In Figure 7-17, the tensile surface traction has been converted to nodal forces as follows:

$$F = \tfrac{1}{2}TA$$

$$F = \tfrac{1}{2}(1000 \text{ psi})(1 \text{ in.} \times 10 \text{ in.})$$

$$F = 5000 \text{ lb}$$

In general, for higher-order elements, Eq. (7.3.7) should be used to convert distributed surface tractions to nodal forces. However, for the CST element, we have shown in Section 7.3 that a statically equivalent force replacement can be used directly, as has been done here.

The governing global matrix equation is

$$\{F\} = [K]\{d\} \qquad (7.5.1)$$

Expanding matrices in Eq. (7.5.1), we obtain

$$
\begin{Bmatrix} F_{1x} \\ F_{1y} \\ F_{2x} \\ F_{2y} \\ F_{3x} \\ F_{3y} \\ F_{4x} \\ F_{4y} \end{Bmatrix}
=
\begin{Bmatrix} R_{1x} \\ R_{1y} \\ R_{2x} \\ R_{2y} \\ 5000 \\ 0 \\ 5000 \\ 0 \end{Bmatrix}
= [K]
\begin{Bmatrix} d_{1x} \\ d_{1y} \\ d_{2x} \\ d_{2y} \\ d_{3x} \\ d_{3y} \\ d_{4x} \\ d_{4y} \end{Bmatrix}
= [K]
\begin{Bmatrix} 0 \\ 0 \\ 0 \\ 0 \\ d_{3x} \\ d_{3y} \\ d_{4x} \\ d_{4y} \end{Bmatrix}
\qquad (7.5.2)
$$

where $[K]$ is an 8×8 matrix before deleting rows and columns to account for the fixed boundary support conditions at nodes 1 and 2.

Assemblage of the Stiffness Matrix

We assemble the global stiffness matrix by superposition of the individual element stiffness matrices. By Eq. (7.2.52), the stiffness matrix for an element is

$$[k] = tA[B]^T[D][B] \qquad (7.5.3)$$

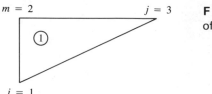

m = 2   j = 3   **FIGURE 7-18**   Element 1
                                                        of the discretized plate

i = 1

In Figure 7-18 for element 1, we have coordinates $x_i = 0$, $y_i = 0$, $x_j = 20$, $y_j = 10$, $x_m = 0$, and $y_m = 10$, since the global coordinate axes are set up at node 1, and

$$A = \tfrac{1}{2}bh$$

$$A = (\tfrac{1}{2})(20)(10) = 100 \text{ in.}^2$$

or, in general, A can be obtained equivalently by the nodal coordinate formula of Eq. (7.2.9).

We will now evaluate $[B]$, where $[B]$ is given by Eq. (7.2.34), expanded here as

$$[B] = \frac{1}{2A}\begin{bmatrix} \beta_i & 0 & \beta_j & 0 & \beta_m & 0 \\ 0 & \gamma_i & 0 & \gamma_j & 0 & \gamma_m \\ \gamma_i & \beta_i & \gamma_j & \beta_j & \gamma_m & \beta_m \end{bmatrix} \qquad (7.5.4)$$

and, from Eqs. (7.2.10),

$$\beta_i = y_j - y_m = 10 - 10 = 0$$

$$\beta_j = y_m - y_i = 10 - 0 = 10$$

$$\beta_m = y_i - y_j = 0 - 10 = -10$$

$$\gamma_i = x_m - x_j = 0 - 20 = -20 \qquad (7.5.5)$$

$$\gamma_j = x_i - x_m = 0 - 0 = 0$$

$$\gamma_m = x_j - x_i = 20 - 0 = 20$$

Therefore, substituting Eqs. (7.5.5) into Eq. (7.5.4), we obtain

$$[B] = \frac{1}{200}\begin{bmatrix} 0 & 0 & 10 & 0 & -10 & 0 \\ 0 & -20 & 0 & 0 & 0 & 20 \\ -20 & 0 & 0 & 10 & 20 & -10 \end{bmatrix} \qquad (7.5.6)$$

For plane stress, the $[D]$ matrix is conveniently expressed here as

$$[D] = \frac{E}{(1 - v^2)}\begin{bmatrix} 1 & v & 0 \\ v & 1 & 0 \\ 0 & 0 & \dfrac{1-v}{2} \end{bmatrix} \qquad (7.5.7)$$

With $v = 0.3$ and $E = 30 \times 10^6$ psi, we obtain

$$[D] = \frac{30(10^6)}{0.91} \begin{bmatrix} 1 & 0.3 & 0 \\ 0.3 & 1 & 0 \\ 0 & 0 & 0.35 \end{bmatrix} \qquad (7.5.8)$$

Then

$$[B]^T[D] = \frac{30(10^6)}{200(0.91)} \begin{bmatrix} 0 & 0 & -20 \\ 0 & -20 & 0 \\ 10 & 0 & 0 \\ 0 & 0 & 10 \\ -10 & 0 & 20 \\ 0 & 20 & -10 \end{bmatrix} \begin{bmatrix} 1 & 0.3 & 0 \\ 0.3 & 1 & 0 \\ 0 & 0 & 0.35 \end{bmatrix} \qquad (7.5.9)$$

Simplifying Eq. (7.5.9) yields

$$[B]^T[D] = \frac{(0.15)(10^6)}{0.91} \begin{bmatrix} 0 & 0 & -7 \\ -6 & -20 & 0 \\ 10 & 3 & 0 \\ 0 & 0 & 3.5 \\ -10 & -3 & 7 \\ 6 & 20 & -3.5 \end{bmatrix} \qquad (7.5.10)$$

Using Eqs. (7.5.10) and (7.5.6) in Eq. (7.5.3), we have the stiffness matrix for element 1 as

$$[k] = (1)(100)\frac{(0.15)(10^6)}{0.91} \begin{bmatrix} 0 & 0 & -7 \\ -6 & -20 & 0 \\ 10 & 3 & 0 \\ 0 & 0 & 3.5 \\ -10 & -3 & 7 \\ 6 & 20 & -3.5 \end{bmatrix}$$

$$\times \frac{1}{2(100)} \begin{bmatrix} 0 & 0 & 10 & 0 & -10 & 0 \\ 0 & -20 & 0 & 0 & 0 & 20 \\ -20 & 0 & 0 & 10 & 20 & -10 \end{bmatrix} \qquad (7.5.11)$$

Finally, simplifying Eq. (7.5.11) yields

$$[k] = \frac{75,000}{0.91} \begin{matrix} i=1 & & j=3 & & m=2 & \\ \begin{bmatrix} 140 & 0 & 0 & -70 & -140 & 70 \\ 0 & 400 & -60 & 0 & 60 & -400 \\ 0 & -60 & 100 & 0 & -100 & 60 \\ -70 & 0 & 0 & 35 & 70 & -35 \\ -140 & 60 & -100 & 70 & 240 & -130 \\ 70 & -400 & 60 & -35 & -130 & 435 \end{bmatrix} \end{matrix} \qquad (7.5.12)$$

where the numbers above the columns indicate the nodal order of the degrees of freedom in the element 1 stiffness matrix.

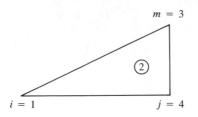

$i = 1$ $j = 4$

In Figure 7-19 for element 2, we have $x_i = 0$, $y_i = 0$, $x_j = 20$, $y_j = 0$, $x_m = 20$, and $y_m = 10$. Then, from Eqs. (7.2.10), we have

$$\beta_i = y_j - y_m = 0 - 10 = -10$$

$$\beta_j = y_m - y_i = 10 - 0 = 10$$

$$\beta_m = y_i - y_j = 0 - 0 = 0$$

$$\gamma_i = x_m - x_j = 20 - 20 = 0 \qquad (7.5.13)$$

$$\gamma_j = x_i - x_m = 0 - 20 = -20$$

$$\gamma_m = x_j - x_i = 20 - 0 = 20$$

Therefore, using Eqs. (7.5.13) in Eq. (7.5.4) yields

$$[B] = \frac{1}{200}\begin{bmatrix} -10 & 0 & 10 & 0 & 0 & 0 \\ 0 & 0 & 0 & -20 & 0 & 20 \\ 0 & -10 & -20 & 10 & 20 & 0 \end{bmatrix} \qquad (7.5.14)$$

The $[D]$ matrix is again given by

$$[D] = \frac{30(10^6)}{0.91}\begin{bmatrix} 1 & 0.3 & 0 \\ 0.3 & 1 & 0 \\ 0 & 0 & 0.35 \end{bmatrix} \qquad (7.5.15)$$

Then using Eqs. (7.5.14) and (7.5.15), we obtain

$$[B]^T[D] = \frac{30(10^6)}{200(0.91)}\begin{bmatrix} -10 & 0 & 0 \\ 0 & 0 & -10 \\ 10 & 0 & -20 \\ 0 & -20 & 10 \\ 0 & 0 & 20 \\ 0 & 20 & 0 \end{bmatrix}\begin{bmatrix} 1 & 0.3 & 0 \\ 0.3 & 1 & 0 \\ 0 & 0 & 0.35 \end{bmatrix} \qquad (7.5.16)$$

Simplifying Eq. (7.5.16) yields

$$[B]^T[D] = \frac{(0.15)(10^6)}{0.91}\begin{bmatrix} -10 & -3 & 0 \\ 0 & 0 & -3.5 \\ 10 & 3 & -7 \\ -6 & -20 & 3.5 \\ 0 & 0 & 7 \\ 6 & 20 & 0 \end{bmatrix} \qquad (7.5.17)$$

Finally, substituting Eqs. (7.5.17) and (7.5.14) into Eq. (7.5.3), we obtain the stiffness matrix for element 2 as

$$[k] = (1)(100)\frac{(0.15)(10^6)}{0.91}\begin{bmatrix} -10 & -3 & 0 \\ 0 & 0 & -3.5 \\ 10 & 3 & -7 \\ -6 & -20 & 3.5 \\ 0 & 0 & 7 \\ 6 & 20 & 0 \end{bmatrix}$$

$$\times \frac{1}{2(100)}\begin{bmatrix} -10 & 0 & 10 & 0 & 0 & 0 \\ 0 & 0 & 0 & -20 & 0 & 20 \\ 0 & -10 & -20 & 10 & 20 & 0 \end{bmatrix} \qquad (7.5.18)$$

Equation (7.5.18) simplifies to

$$[k] = \frac{75{,}000}{0.91}\begin{array}{cccccc} i=1 & & j=4 & & m=3 & \\ \begin{bmatrix} 100 & 0 & -100 & 60 & 0 & -60 \\ 0 & 35 & 70 & -35 & -70 & 0 \\ -100 & 70 & 240 & -130 & -140 & 60 \\ 60 & -35 & -130 & 435 & 70 & -400 \\ 0 & -70 & -140 & 70 & 140 & 0 \\ -60 & 0 & 60 & -400 & 0 & 400 \end{bmatrix} \end{array} \qquad (7.5.19)$$

where the degrees of freedom in the element 2 stiffness matrix are shown above the columns in Eq. (7.5.19). Rewriting the element stiffness matrices, Eqs. (7.5.12) and (7.5.19), expanded to the order of and rearranged according to increasing nodal degrees of freedom of the total $\underline{K}$ matrix (where we have factored out a constant 5), we obtain

Element 1

$$[k] = \frac{375{,}000}{0.91}\begin{array}{cccccccc} 1 & & 2 & & 3 & & 4 & \\ \begin{bmatrix} 28 & 0 & -28 & 14 & 0 & -14 & 0 & 0 \\ 0 & 80 & 12 & -80 & -12 & 0 & 0 & 0 \\ -28 & 12 & 48 & -26 & -20 & 14 & 0 & 0 \\ 14 & -80 & -26 & 87 & 12 & -7 & 0 & 0 \\ 0 & -12 & -20 & 12 & 20 & 0 & 0 & 0 \\ -14 & 0 & 14 & -7 & 0 & 7 & 0 & 0 \\ 0 & 0 & 0 & 0 & 0 & 0 & 0 & 0 \\ 0 & 0 & 0 & 0 & 0 & 0 & 0 & 0 \end{bmatrix} \end{array}$$

$$(7.5.20)$$

Element 2

$$[k] = \frac{375{,}000}{0.91}
\begin{array}{c}
1234 \\
\begin{bmatrix}
20 & 0 & 0 & 0 & 0 & -12 & -20 & 12 \\
0 & 7 & 0 & 0 & -14 & 0 & 14 & -7 \\
0 & 0 & 0 & 0 & 0 & 0 & 0 & 0 \\
0 & 0 & 0 & 0 & 0 & 0 & 0 & 0 \\
0 & -14 & 0 & 0 & 28 & 0 & -28 & 14 \\
-12 & 0 & 0 & 0 & 0 & 80 & 12 & -80 \\
-20 & 14 & 0 & 0 & -28 & 12 & 48 & -26 \\
12 & -7 & 0 & 0 & 14 & -80 & -26 & 87
\end{bmatrix}
\end{array}$$

$$(7.5.21)$$

Using superposition of the element stiffness matrices, Eqs. (7.5.20) and (7.5.21), now that the order of the degrees of freedom are the same, we obtain the total global stiffness matrix as

$$[K] = \frac{375{,}000}{0.91}
\begin{array}{c}
1234 \\
\begin{bmatrix}
48 & 0 & -28 & 14 & 0 & -26 & -20 & 12 \\
0 & 87 & 12 & -80 & -26 & 0 & 14 & -7 \\
-28 & 12 & 48 & -26 & -20 & 14 & 0 & 0 \\
14 & -80 & -26 & 87 & 12 & -7 & 0 & 0 \\
0 & -26 & -20 & 12 & 48 & 0 & -28 & 14 \\
-26 & 0 & 14 & -7 & 0 & 87 & 12 & -80 \\
-20 & 14 & 0 & 0 & -28 & 12 & 48 & -26 \\
12 & -7 & 0 & 0 & 14 & -80 & -26 & 87
\end{bmatrix}
\end{array}$$

$$(7.5.22)$$

[Alternatively, we could have applied the direct stiffness method to Eqs. (7.5.12) and (7.5.19) to obtain Eq. (7.5.22).] Substituting $[K]$ into $\{F\} = [K]\{d\}$ of Eq. (7.5.2), we have

$$
\begin{Bmatrix}
R_{1x} \\ R_{1y} \\ R_{2x} \\ R_{2y} \\ 5000 \\ 0 \\ 5000 \\ 0
\end{Bmatrix}
= \frac{375{,}000}{0.91}
\begin{bmatrix}
48 & 0 & -28 & 14 & 0 & -26 & -20 & 12 \\
0 & 87 & 12 & -80 & -26 & 0 & 14 & -7 \\
-28 & 12 & 48 & -26 & -20 & 14 & 0 & 0 \\
14 & -80 & -26 & 87 & 12 & -7 & 0 & 0 \\
0 & -26 & -20 & 12 & 48 & 0 & -28 & 14 \\
-26 & 0 & 14 & -7 & 0 & 87 & 12 & -80 \\
-20 & 14 & 0 & 0 & -28 & 12 & 48 & -26 \\
12 & -7 & 0 & 0 & 14 & -80 & -26 & 87
\end{bmatrix}
\begin{Bmatrix}
0 \\ 0 \\ 0 \\ 0 \\ d_{3x} \\ d_{3y} \\ d_{4x} \\ d_{4y}
\end{Bmatrix}
$$

$$(7.5.23)$$

Applying the support or boundary conditions by eliminating rows and columns corresponding to displacement matrix rows and columns equal to zero [namely, rows and columns 1–4 in Eq. (7.5.23)], we obtain

$$
\begin{Bmatrix} 5000 \\ 0 \\ 5000 \\ 0 \end{Bmatrix} = \frac{375{,}000}{0.91} \begin{bmatrix} 48 & 0 & -28 & 14 \\ 0 & 87 & 12 & -80 \\ -28 & 12 & 48 & -26 \\ 14 & -80 & -26 & 87 \end{bmatrix} \begin{Bmatrix} d_{3x} \\ d_{3y} \\ d_{4x} \\ d_{4y} \end{Bmatrix} \qquad (7.5.24)
$$

Transposing the displacement matrix to the left side, we have

$$
\begin{Bmatrix} d_{3x} \\ d_{3y} \\ d_{4x} \\ d_{4y} \end{Bmatrix} = \frac{0.91}{375{,}000} \begin{bmatrix} 48 & 0 & -28 & 14 \\ 0 & 87 & 12 & -80 \\ -28 & 12 & 48 & -26 \\ 14 & -80 & -26 & 87 \end{bmatrix}^{-1} \begin{Bmatrix} 5000 \\ 0 \\ 5000 \\ 0 \end{Bmatrix} \qquad (7.5.25)
$$

Solving for the displacements in Eq. (7.5.25), we obtain

$$
\begin{Bmatrix} d_{3x} \\ d_{3y} \\ d_{4x} \\ d_{4y} \end{Bmatrix} = \frac{0.91}{75} \begin{Bmatrix} 0.05024 \\ 0.00034 \\ 0.05470 \\ 0.00878 \end{Bmatrix} \qquad (7.5.26)
$$

Simplifying Eq. (7.5.26), the final displacements are given by

$$
\begin{Bmatrix} d_{3x} \\ d_{3y} \\ d_{4x} \\ d_{4y} \end{Bmatrix} = \begin{Bmatrix} 609.6 \\ 4.2 \\ 663.7 \\ 104.1 \end{Bmatrix} \times 10^{-6} \text{ in.} \qquad (7.5.27)
$$

Comparing the finite element solution to an analytical solution, as a first approximation, we have

$$
\delta = \frac{PL}{AE} = \frac{(10{,}000)20}{10(30 \times 10^6)} = 670 \times 10^{-6} \text{ in.}
$$

for a one-dimensional bar subjected to tensile force. Hence, the nodal x displacement components of Eq. (7.5.27) for the two-dimensional plate appear to be reasonably correct, considering the coarseness of the mesh.

We now determine the stresses in each element by using Eq. (7.2.36):

$$
\{\sigma\} = [D][B]\{d\} \qquad (7.5.28)
$$

In general, for element 1, we then have

$$
\{\sigma\} = \frac{E}{(1 - v^2)} \begin{bmatrix} 1 & v & 0 \\ v & 1 & 0 \\ 0 & 0 & \dfrac{1-v}{2} \end{bmatrix} \times \left(\frac{1}{2A}\right) \begin{bmatrix} \beta_1 & 0 & \beta_3 & 0 & \beta_2 & 0 \\ 0 & \gamma_1 & 0 & \gamma_3 & 0 & \gamma_2 \\ \gamma_1 & \beta_1 & \gamma_3 & \beta_3 & \gamma_2 & \beta_2 \end{bmatrix} \begin{Bmatrix} d_{1x} \\ d_{1y} \\ d_{3x} \\ d_{3y} \\ d_{2x} \\ d_{2y} \end{Bmatrix}
$$
$$(7.5.29)$$

Substituting numerical values for $[B]$, given by Eq. (7.5.6); for $[D]$, given by

Eq. (7.5.8); and the appropriate part of $\{d\}$, given by Eq. (7.5.27); we obtain

$$
\{\sigma\} = \frac{30(10^6)(10^{-6})}{0.91(200)} \begin{bmatrix} 1 & 0.3 & 0 \\ 0.3 & 1 & 0 \\ 0 & 0 & 0.35 \end{bmatrix}
$$

$$
\times \begin{bmatrix} 0 & 0 & 10 & 0 & -10 & 0 \\ 0 & -20 & 0 & 0 & 0 & 20 \\ -20 & 0 & 0 & 10 & 20 & -10 \end{bmatrix} \begin{Bmatrix} 0 \\ 0 \\ 609.6 \\ 4.2 \\ 0 \\ 0 \end{Bmatrix} \qquad (7.5.30)
$$

Simplifying Eq. (7.5.30), we obtain

$$
\begin{Bmatrix} \sigma_x \\ \sigma_y \\ \tau_{xy} \end{Bmatrix} = \begin{Bmatrix} 1005 \\ 301 \\ 2.4 \end{Bmatrix} \text{ psi} \qquad (7.5.31)
$$

In general, for element 2, we have

$$
\{\sigma\} = \frac{E}{(1-v^2)}\left(\frac{1}{2A}\right) \begin{bmatrix} 1 & v & 0 \\ v & 1 & 0 \\ 0 & 0 & \dfrac{1-v}{2} \end{bmatrix} \times \begin{bmatrix} \beta_1 & 0 & \beta_4 & 0 & \beta_3 & 0 \\ 0 & \gamma_1 & 0 & \gamma_4 & 0 & \gamma_3 \\ \gamma_1 & \beta_1 & \gamma_4 & \beta_4 & \gamma_3 & \beta_3 \end{bmatrix} \begin{Bmatrix} d_{1x} \\ d_{1y} \\ d_{4x} \\ d_{4y} \\ d_{3x} \\ d_{3y} \end{Bmatrix}
$$

$$(7.5.32)$$

Substituting numerical values into Eq. (7.5.32), we obtain

$$
\{\sigma\} = \frac{30(10^6)(10^{-6})}{0.91(200)} \begin{bmatrix} 1 & 0.3 & 0 \\ 0.3 & 1 & 0 \\ 0 & 0 & 0.35 \end{bmatrix}
$$

$$
\times \begin{bmatrix} -10 & 0 & 10 & 0 & 0 & 0 \\ 0 & 0 & 0 & -20 & 0 & 20 \\ 0 & -10 & -20 & 10 & 20 & 0 \end{bmatrix} \begin{Bmatrix} 0 \\ 0 \\ 663.7 \\ 104.1 \\ 609.6 \\ 4.2 \end{Bmatrix} \qquad (7.5.33)
$$

Simplifying Eq. (7.5.33), we obtain

$$
\begin{Bmatrix} \sigma_x \\ \sigma_y \\ \tau_{xy} \end{Bmatrix} = \begin{Bmatrix} 995 \\ -1.2 \\ -2.4 \end{Bmatrix} \text{ psi} \qquad (7.5.34)
$$

The principal stresses can now be determined from Eq. (7.1.2), and the principal angle made by one of the principal stresses can be determined from Eq. (7.1.3). (The other principal stress will be directed $90°$ from the first.) We determine these principal stresses for element 2 as

$$\sigma_1 = \frac{\sigma_x + \sigma_y}{2} + \left[\left(\frac{\sigma_x - \sigma_y}{2}\right)^2 + \tau_{xy}^2\right]^{1/2}$$

$$\sigma_1 = \frac{995 + (-1.2)}{2} + \left[\left(\frac{995 - (-1.2)}{2}\right)^2 + (-2.4)^2\right]^{1/2}$$

$$\sigma_1 = 497 + 498 = 995 \text{ psi}$$

$$\sigma_2 = \frac{995 + (-1.2)}{2} - 498 = -1.1 \text{ psi}$$

The principal angle is then

$$\theta_p = \tfrac{1}{2} \tan^{-1}\left[\frac{2\tau_{xy}}{\sigma_x - \sigma_y}\right]$$

or

$$\theta_p = \tfrac{1}{2} \tan^{-1}\left[\frac{2(-2.4)}{995 - (-1.2)}\right] = 0°$$

Owing to the uniform stress of 1000 psi acting only in the x direction on the edge of the plate, we would expect the stress $\sigma_x (= \sigma_1)$ to be near 1000 psi in each element. Thus, the results from Eqs. (7.5.31) and (7.5.34) for σ_x are quite good. We would expect the stress σ_y to be very small (at least near the free edge). The restraint of element 1 at nodes 1 and 2 causes a relatively large element stress σ_y, whereas the restraint of element 2 at only one node causes a very small stress σ_y. The shear stresses τ_{xy} remain close to zero, as expected. Had the number of elements been increased, with smaller ones used near the support edge, even more realistic results would have been obtained. However, a finer discretization would result in a cumbersome longhand solution, and thus was not used here. Use of the computer program CSFEP described in Chapter 8 is recommended for a more detailed solution to this plate problem and certainly for solving more-complex plane stress/strain problems. ∎

REFERENCES

[1] Timoshenko, S., and Goodier, J., *Theory of Elasticity*, 3rd ed., McGraw-Hill, New York, 1970.

[2] Logan, D. L., *Mechanics of Materials*, Harper Collins, New York, 1991.

[3] Cook, R. D., Malkus, D. S. and Plesha, M. E., *Concepts and Applications of Finite Element Analysis*, 3rd ed., Wiley, New York, 1989.

PROBLEMS

7.1   Sketch the variations of the shape functions N_j and N_m, given by Eqs. (7.2.18), over the surface of the triangular element with nodes i, j, and m. Check that $N_i + N_j + N_m = 1$ anywhere on the element.

7.2   For a simple three-noded triangular element, show explicitly that Eq. (7.2.47) indeed results in Eq. (7.2.48); that is, substitute the expression for $[B]$ and the plane stress condition for $[D]$ into Eq. (7.2.47), and then differentiate π_p with respect to each nodal degree of freedom in Eq. (7.2.47) to obtain Eq. (7.2.48).

7.3   Evaluate the stiffness matrix for the elements shown in Figure P7-3. The coordinates are in units of inches. Assume plane stress conditions. Let $E = 30 \times 10^6$ psi, $v = 0.25$, and thickness $t = 1$ in.

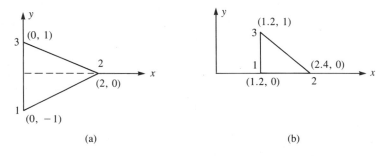

(a)   (b)

F I G U R E   P7-3

7.4   For the elements given in Problem 7.3, the nodal displacements are given as

$$u_1 = 0.0 \qquad v_1 = 0.0025 \text{ in.} \qquad u_2 = 0.0012 \text{ in.}$$

$$v_2 = 0.0 \qquad u_3 = 0.0 \qquad v_3 = 0.0025 \text{ in.}$$

Determine the element stresses σ_x, σ_y, τ_{xy}, σ_1, and σ_2, and the principal angle θ_p. Use the values of E, v, and t given in Problem 7.3.

7.5   Evaluate the stiffness matrix for the elements shown in Figure P7-5. The coordinates are given in units of millimeters. Assume plane stress conditions. Let $E = 210$ GPa, $v = 0.25$, and $t = 10$ mm.

7.6   For the elements given in Problem 7.5, the nodal displacements are given as

$$u_1 = 2.0 \text{ mm} \qquad v_1 = 1.0 \text{ mm} \qquad u_2 = 0.5 \text{ mm}$$

$$v_2 = 0.0 \text{ mm} \qquad u_3 = 3.0 \text{ mm} \qquad v_3 = 1.0 \text{ mm}$$

Determine the element stresses σ_x, σ_y, τ_{xy}, σ_1, and σ_2, and the principal angle θ_p. Use the values of E, v, and t given in Problem 7.5.

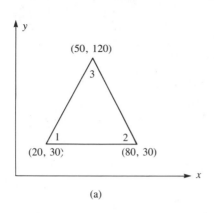

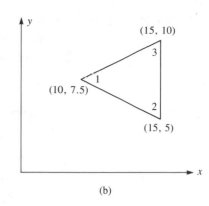

FIGURE P7-5

7.7 For the plane strain elements shown in Figure P7-7, the nodal displacements
are given as•

$$u_1 = 0.001 \text{ in.} \qquad v_1 = 0.005 \text{ in.} \qquad u_2 = 0.001 \text{ in.}$$

$$v_2 = 0.0025 \text{ in.} \qquad u_3 = 0.0 \text{ in.} \qquad v_3 = 0.0 \text{ in.}$$

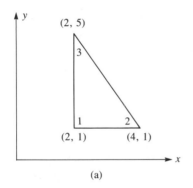

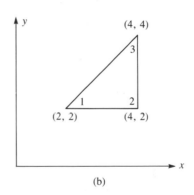

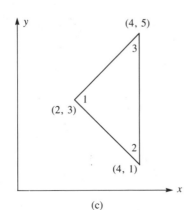

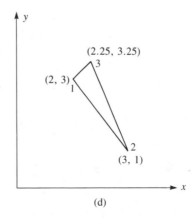

FIGURE P7-7

Determine the element stresses σ_x, σ_y, τ_{xy}, σ_1, and σ_2, and the principal angle θ_p. Let $E = 30 \times 10^6$ psi and $v = 0.25$, and use unit thickness for plane strain. All coordinates are in inches.

7.8 For the plane strain elements shown in Figure P7-8, the nodal displacements are given as

$$u_1 = 0.005 \, \text{mm} \qquad v_1 = 0.002 \, \text{mm} \qquad u_2 = 0.0 \, \text{mm}$$

$$v_2 = 0.0 \, \text{mm} \qquad u_3 = 0.005 \, \text{mm} \qquad v_3 = 0.0 \, \text{mm}$$

Determine the element stresses σ_x, σ_y, τ_{xy}, σ_1, and σ_2, and the principal angle θ_p. Let $E = 70$ GPa and $v = 0.3$, and use unit thickness for plane strain. All coordinates are in millimeters.

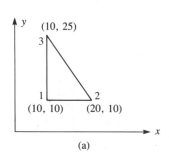

(a)

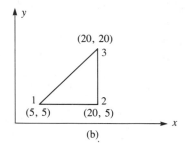

(b)

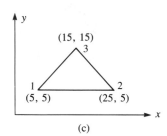

(c)

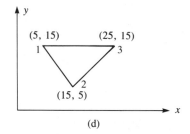

(d)

FIGURE P7-8

7.9 Determine the nodal forces for (a) a linearly varying pressure p_x on the edge of the triangular element shown in Figure P7-9(a) and (b) the quadratic varying pressure shown in Figure P7-9(b) by evaluating the surface integral given by Eq. (7.3.7). Assume the element thickness is equal to t.

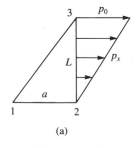

(a)

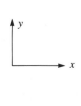

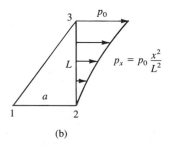

(b)

FIGURE P7-9

7.10   Determine the nodal displacements and the element stresses, including principal stresses, for the thin plate of Section 7.5 with a uniform shear load (instead of a tensile load) acting on the right edge, as shown in Figure P7-10. Use $E = 30 \times 10^6$ psi, $v = 0.30$, and $t = 1$ in. (*Hint:* The $[K]$ matrix derived in Section 7.5 and given by Eq. (7.5.22) can be used to solve the problem.)

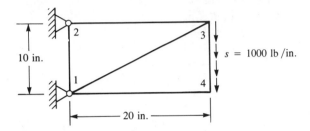

FIGURE P7-10

7.11   Determine the nodal displacements and the element stresses, including principal stresses, due to the loads shown for the thin plates in Figure P7-11. Use $E = 210$ GPa, $v = 0.30$, and $t = 5$ mm. Assume plane stress conditions apply. The recommended discretized plates are shown in the figures.

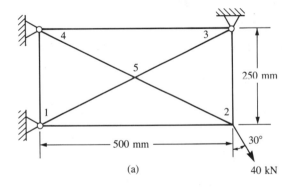

(a)

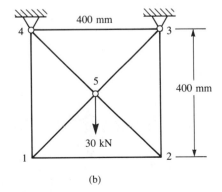

(b)

FIGURE P7-11

7.12 Evaluate the body force matrix for the plate shown in Figure P7-11(b). Assume the weight density to be 77.1 kN/m³.

7.13 For the plane structures modeled by triangular elements shown in Figure P7-13, show that numbering in the direction that has fewer nodes, as in Figure P7-13(a) (as opposed to numbering in the direction that has more nodes), results in a reduced bandwidth. Illustrate this fact by filling in, with *X*'s, the occupied elements in $\underline{K}$ for each mesh, as was done in Chapter 4. Compare the bandwidths for each case.

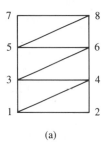 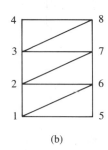

    (a)   (b)

FIGURE P7-13

7.14 Go through the detailed steps to evaluate Eq. (7.3.6).

7.15 How would you treat a linearly varying thickness for a three-noded triangle?

7.16 Compute the stiffness matrix of element 1 of the two-triangle element model of the rectangular plate in plane stress shown in Figure P7-16. Then use it to compute the stiffness matrix of element 2.

FIGURE P7-16

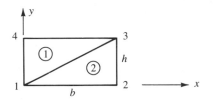

Practical Considerations

in Modeling;

Interpreting Results;

and a Computer Program

for Plane

Stress/Strain Analysis

INTRODUCTION

In this chapter, we will describe some modeling guidelines, including generally recommended mesh size, natural subdivisions, and more on use of symmetry and associated boundary conditions. This is followed by discussion of equilibrium, compatibility, and convergence of solution. We will then consider interpretation of stress results.

Next, we introduce the concept of static condensation, which enables us to apply the concept of the basic constant-strain triangle stiffness matrix to a quadrilateral element. Thus, both three-sided and four-sided two-dimensional elements can be used in the finite element models of actual bodies. This concept is incorporated in the computer program called CSFEP.

We then describe the use of the program CSFEP (listed on the disk at the back of this text). This program facilitates the solution of complex, large-number-of-degree-of-freedom plane stress/plane strain problems that generally cannot be solved longhand because of the larger number of equations involved. Also, problems for which longhand solutions do not exist (such as those involving complex geometries and complex loads or where unrealistic, often gross, assumptions were previously made to simplify the problem to allow it to be described by a classical differential equation solution approach) can now be solved with a higher degree of confidence in the results using the finite element approach (with its resulting system of algebraic equations).

A general description of the input data and a typical data file are provided to help in using the program.

8.1   Finite Element Modeling

We will now discuss various concepts that should be considered when modeling any problem for solution by the finite element method.

General Considerations

Finite element modeling is partly an art guided by visualizing physical interactions taking place within the body. One appears to acquire good modeling techniques through experience and by working with experienced people. General-purpose programs provide some guidelines for specific types of problems [12]. In subsequent parts of this section some significant concepts that should be considered are described.

In modeling, the user is first confronted with the sometimes difficult task of understanding the physical behavior taking place and understanding the physical behavior of the various elements available for use. Choosing the proper type of element or elements to match as closely as possible the physical behavior of the problem is one of the numerous decisions that must be made by the user. Understanding the boundary conditions imposed on the problem can, at times, be a difficult task. Also, it is often difficult to determine the kinds of loads and magnitudes and location of the loads that must be applied to a body. Again, working with more experienced users and searching the literature can help overcome these difficulties.

Aspect Ratio and Element Shapes

The **aspect ratio** *is defined as the ratio of the longest dimension to the shortest dimension of a quadrilateral element.* In many cases, as the aspect ratio increases, the inaccuracy of the solution increases. To illustrate this point, Figure 8-1(a) shows five different finite element models used to analyze a beam subjected to bending. Figure 8-1(b) is a plot of the resulting error in the displacement at point *A* of the beam versus the aspect ratio, Table 8-1 reports a comparison of results for the displacements at points *A* and *B* for the five models, and the exact solution (see Reference [2]).

There are exceptions for which aspect ratios approaching 50 still produce satisfactory results; for example, if the stress gradient is close to zero at some location of the actual problem, then large aspect ratios at that location still produce reasonable results.

In general, an element yields best results if its shape is compact and regular. Although different elements have different sensitivities to shape dis-

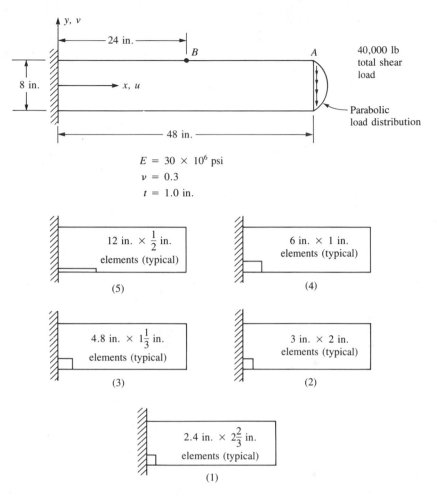

FIGURE 8-1   (a) Beam with loading; effects of the aspect ratio illustrated by five cases with different aspect ratios

tortions, try to maintain (1) aspect ratios low as in Figure 8-1, cases 1 and 2, and (2) corner angles of quadrilaterals near 90°. Figure 8-2 shows elements with poor shapes that tend to promote poor results. If few of these poor element shapes exist in a model, then usually only results near these elements are poor.

Use of Symmetry

The appropriate use of symmetry* will often expedite the modeling of a problem. Use of symmetry allows us to consider a reduced problem instead of the actual problem. Thus, we can use a finer subdivision of elements with

---

*Again, *symmetry* means correspondence in size, shape, and position of loads; material properties; and boundary conditions that are on opposite sides of a dividing line or plane.

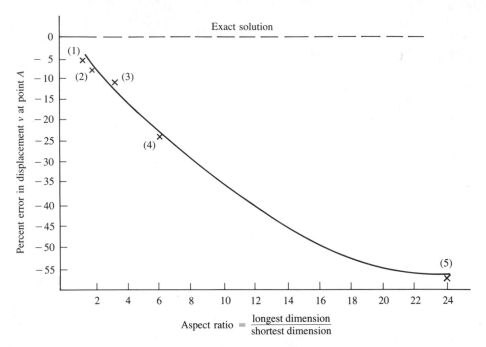

FIGURE 8-1   (b) Inaccuracy of solution as a function of the aspect ratio (numbers in parentheses correspond to the cases listed in Table 8-1)

TABLE 8-1   Comparison of results for various aspect ratios

Case	Aspect Ratio	Number of Nodes	Number of Elements	Vertical Displacement, v, (inches) Point A v	Point B v	Percent Error in Displacement at A
1	1.1	84	60	−1.093	−0.346	5.2
2	1.5	85	64	−1.078	−0.339	6.4
3	3.6	77	60	−1.014	−0.328	11.9
4	6.0	81	64	−0.886	−0.280	23.0
5	24.0	85	64	−0.500	−0.158	56.0
Exact solution (see Reference [2])				−1.152	−0.360	

less labor and computer costs. For other discussion on use of symmetry, see Reference [3].)

Figures 8-3–8-5 illustrate the use of symmetry in modeling (1) a soil mass subjected to foundation loading, (2) a uniaxially loaded member with a fillet, and (3) a plate with a hole subjected to internal pressure. Note that at the plane of symmetry the displacement in the direction perpendicular to the plane must be equal to zero. This is modeled by the rollers at nodes 2, 3, 4, 5,

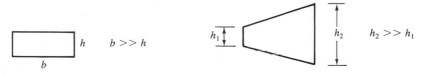

(a)   Large aspect ratio   (b)   Approaching a triangular shape

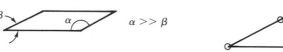

(c)   Very large and small   (d)   Triangular quadrilateral
      corner angles

F I G U R E   8-2   Elements with poor shapes

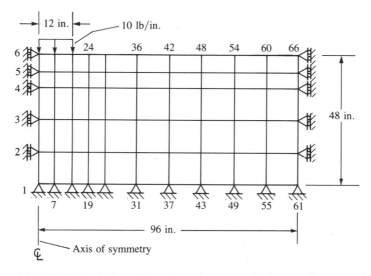

F I G U R E   8-3   Use of symmetry applied to a soil mass subjected to
                           foundation loading (number of nodes = 66, number of
                           elements = 50) (2.54 cm = 1 in., 4.445 N = 1 lb)

and 6 in Figure 8-3, where the plane of symmetry is the vertical plane passing
through nodes 1–6, perpendicular to the plane of the model. In Figures 8-4(a)
and 8-5(a), there are two planes of symmetry. Thus, we need model only one
fourth of the actual members, as shown in Figures 8-4(b) and 8-5(b). Therefore, rollers are used at nodes along both the vertical and horizontal planes
of symmetry.

   In vibration and buckling problems, symmetry must be used with caution, since symmetry in geometry does not imply symmetry in all vibration or
buckling modes.

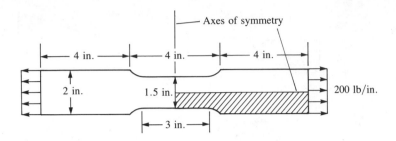

(a) Plane stress uniaxially loaded member with fillet

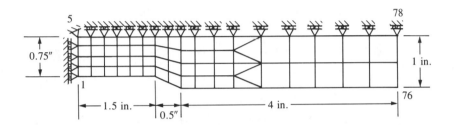

(b) Enlarged finite element model of the cross-hatched quarter of the member
(number of nodes = 78, number of elements = 60) (2.54 cm = 1 in.)

F I G U R E   8-4   Use of symmetry applied to a uniaxially loaded member
with a fillet

Natural Subdivisions at Discontinuities

Figure 8-6 illustrates various natural subdivisions for finite element discretization. For instance, nodes are required at locations of concentrated loads or discontinuity in loads, as shown in Figure 8-6(a) and (b). Nodal lines are defined by abrupt changes of plate thickness, as in Figure 8-6(c), and by abrupt changes of material properties, as in Figure 8-6(d) and (e). Other natural subdivisions occur at re-entrant corners, as in Figure 8-6(f), and along holes in members, as in Figure 8-5.

Sizing of Elements and Mesh Refinement

The discretization depends on the geometry of the structure, the loading pattern, and the boundary conditions. For instance, regions of stress concentration or high stress gradient due to fillets, holes, or re-entrant corners require a finer mesh near those regions, as indicated in Figures 8-4, 8-5, and 8-6(f).

Finally, Figure 8-4 illustrates the use of triangular elements for transitions from smaller quadrilaterals to larger quadrilaterals. This transition is necessary because for simple CST elements, intermediate nodes along element edges are inconsistent with the energy formulation of the CST equations. If

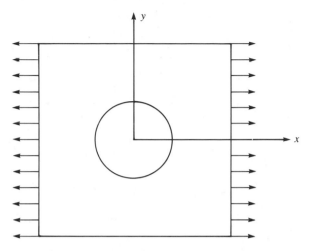

(a) Plate with hole under plane stress

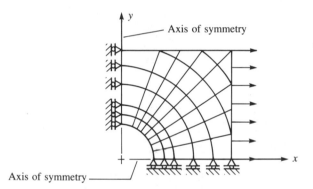

(b) Finite element model of one quarter of the plate

F I G U R E 8-5   Problem reduction using axes of symmetry applied to a plate with a hole subjected to tensile force

intermediate nodes were used, no assurance of compatibility would be possible, and resulting holes could occur in the deformed model. Using higher-order elements, such as the linear-strain triangle described in Chapter 9, allows us to use intermediate nodes along element edges and maintain compatibility.

Infinite Medium

Figure 8-3 shows a typical model used to represent an infinite medium (a soil mass subjected to a foundation load). The guideline for the finite element model is that enough material must be included such that the displacements at nodes and stresses within the elements become negligibly small at locations far from the foundation load. Just how much of the medium should be

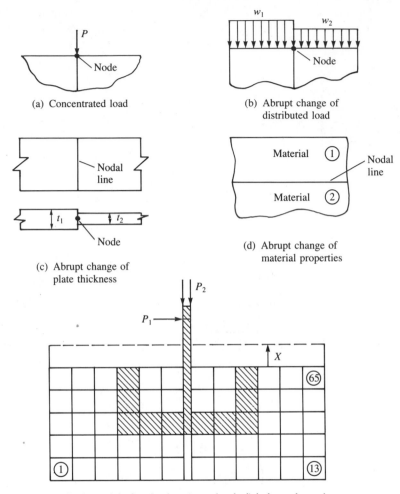

(a) Concentrated load

(b) Abrupt change of
distributed load

(c) Abrupt change of
plate thickness

(d) Abrupt change of
material properties

(e) Basic model of an implant (cross-hatched) in bone, located
at various depths X beneath the bony surface, using
rectangular elements

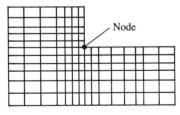

(f) Re-entrant corner

F I G U R E   8-6   Natural subdivisions at discontinuities

modeled can be determined by a trial-and-error procedure in which the horizontal and vertical distances from the load are varied and the resulting effects on the displacements and stresses are observed. Alternatively, the experiences of other investigators working on similar problems may prove helpful. For a homogeneous soil mass, experience has shown that the influence of the footing becomes insignificant if the horizontal distance of the model is taken as approximately four to six times the width of the footing and the vertical distance is taken as approximately four to ten times the width of the footing (see References [4]–[6]). Also, the use of infinite elements is described in [13].

After choosing the horizontal and vertical dimensions of the model, we must idealize the boundary conditions. Usually, the horizontal displacement becomes negligible far from the load, and we restrain the horizontal movement of all the nodal points on that boundary (the right-side boundary in Figure 8-3). Hence, rollers are used to restrain the horizontal motion along the right side. The bottom boundary can be completely fixed, as is modeled in Figure 8-3 by using pin supports at each nodal point along the bottom edge. Alternatively, the bottom can be constrained only against vertical movement. The choice depends on the soil conditions at the bottom of the model. Usually, complete fixity is assumed if the lower boundary is taken as bedrock.

In Figure 8-3, the left-side vertical boundary is taken to be directly under the center of the load because symmetry has been assumed. As previously stated when discussing symmetry, all nodal points along the line of symmetry are restrained against horizontal displacement.

Finally, Reference [11] is recommended for additional discussion regarding guidelines in modeling with different element types, such as beams, plane stress/plane strain, and three-dimensional solids.

Checking the Model

The discretized finite element model should be checked carefully before results are computed. Ideally, a model should be checked by an analyst not involved in the preparation of the model, who is then more likely to be objective.

Preprocessors with their detailed graphical display capabilities (Figure 8-7) now make it comparatively easy to find errors, particularly the more obvious ones involved with a misplaced node or missing element or a misplaced load or boundary support. Preprocessors include such niceties as color, shrink plots, rotated views, sectioning, exploded views, and removal of hidden lines to aid in error detection.

Most commercial codes also include warnings regarding overly distorted element shapes and checking for sufficient supports. However, the user must still select the proper element types, place supports and forces in proper locations, use consistent units, etc., to obtain a successful analysis.

FIGURE 8-7   Plate with hole discretized using preprocessor program

Checking the Results and Typical Postprocessor Results

The results should be checked for consistency by making sure that intended support nodes have zero displacement, as required. If symmetry exists, then stresses and displacements should exhibit this symmetry. Computed results from the finite element program should be compared with results from other available techniques, even if these techniques may be cruder than the finite element results. For instance, approximate mechanics of material formulas, experimental data, and numerical analysis of simpler but similar problems may be used for comparison, particularly if you have no real idea of the magnitude of the answers. Remember to use all results with some degree of caution, as errors can crop up in such sources as textbook or handbook comparison solutions and experimental results.

In the end, the analyst should probably spend as much time processing, checking, and analyzing results as spent in data preparation.

Finally, we present some typical postprocessor results for some plane stress problems (see Figures 8-8 and 8-9).

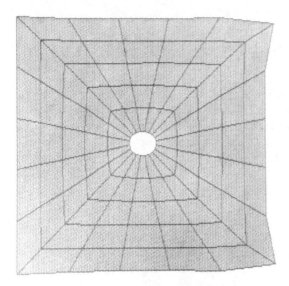

FIGURE 8-8   Displaced plate with hole subjected to right-edge tensile load

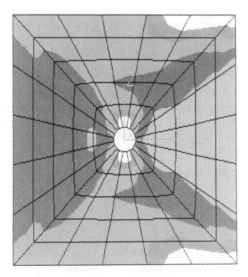

FIGURE 8-9   Maximum principal stress contour for plate with hole subjected to tensile load

8.2 Equilibrium and Compatibility of Finite Element Results

An approximate solution for a stress analysis problem using the finite element method based on assumed displacement fields does not generally satisfy all the requirements for equilibrium and compatibility that an exact theory-of-elasticity solution satisfies. However, remember that relatively few exact solutions exist. Hence, the finite element method is a very practical one for obtaining reasonable, but approximate, numerical solutions. Recall the advantages of the finite element method as described in Chapter 1, and as illustrated numerous times throughout this text.

We now describe some of the approximations generally inherent with finite element solutions.

1. Equilibrium of nodal forces and moments is satisfied. This is true because the global equation $\underline{F} = \underline{K}\underline{d}$ is a nodal equilibrium equation whose solution for $\underline{d}$ is such that the sums of all forces and moments applied to each node are zero. Equilibrium of the whole structure is also satisfied because the structure reactions are included in the global forces, and hence, in the nodal equilibrium equations. Numerous example problems, particularly involving truss and frame analysis in Chapters 3 and 6, respectively, have illustrated the equilibrium of nodes and of total structures.

2. Equilibrium within an element is not always satisfied. However, for the constant-strain bar of Chapter 3 and the constant-strain triangle of Chapter 7, element equilibrium is satisfied. Also the cubic displacement function is shown to satisfy the basic beam equilibrium differential equation in Chapter 5, and hence, to satisfy element force and moment equilibrium. However, elements such as the linear-strain triangle of Chapter 9, the axisymmetric element of Chapter 10, and the rectangular element of Chapter 11 usually only approximately satisfy the element equilibrium equations.

3. Equilibrium is not usually satisfied between elements. A differential element including parts of two adjacent finite elements is usually not in equilibrium (see Figure 8-10). For line elements, such as used for truss and frame analysis, interelement equilibrium is satisfied as shown in example problems in Chapters 3–6. However, for two- and three-dimensional elements, interelement equilibrium is not usually satisfied. For instance, the results of Example 7.2 indicate that the normal stress along the diagonal edge between the two elements is different in the two elements. Also, the coarseness of the mesh causes this lack of interelement equilibrium to be even more pronounced. The normal and shear stresses at a free edge usually are not zero even though theory predicts them to be. Again, Example 7.2 illustrates this, with free-edge stresses σ_y and τ_{xy} not equal to zero. However, as more elements are used (refined mesh) the σ_y and τ_{xy} stresses on the stress-free edges will approach zero.

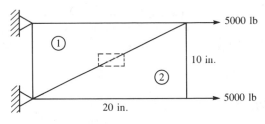

Example 7.2

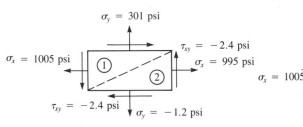

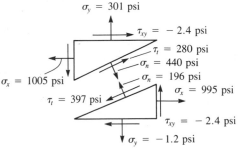

Stresses on a differential
element common to both finite
elements illustrating violation
of equilibrium

Stress along the diagonal between elements
showing normal and shear stresses
σ_n and τ_t (*note:* σ_n and τ_t are not
equal in magnitude but opposite in
sign for the two elements, hence,
interelement equilibrium is not satisfied)

F I G U R E 8-10   Example 7.2 illustrating violation of equilibrium of a differential
element and along the diagonal edge between two elements (the
coarseness of the mesh amplifies the violation of equilibrium)

4. Compatibility is satisfied within an element as long as the element
   displacement field is continuous. Hence, individual elements do not
   tear apart.

5. In the formulation of the element equations, compatibility is
   invoked at the nodes. Hence, elements remain connected at their
   common nodes. Similarly, the structure remains connected to its
   support nodes because boundary conditions are invoked at these
   nodes.

6. Compatibility may or may not be satisfied along interelement
   boundaries. For line elements such as bars and beams, interelement
   boundaries are merely nodes. Therefore, the preceding Statement 5
   applies for these line elements. The constant-strain traingle of
   Chapter 7 and the rectangular element of Chapter 11 remain
   straight sided when deformed and therefore, interelement
   compatibility exists for these elements; that is, these plane elements
   deform along common lines without openings, overlaps, or
   discontinuities. Incompatible elements, those that allow gaps or
   overlaps between elements, can be acceptable and even desirable.

Incompatible element formulations, in some cases, have been shown to converge more rapidly to the exact solution (see Reference [1]). (For more on this special topic, consult References [7] and [8].)

8.3   Convergence of Solution

In Section 3.2, we presented guidelines for the selection of so-called compatible and complete displacement functions as they related to the bar element. Those four guidelines are generally applicable, and satisfaction of them has been shown to ensure monotonic convergence of the solution of a particular problem (see Reference [9]). Furthermore, it has been shown (see Reference [10]) that these compatible and complete displacement functions used in the displacement formulation of the finite element method yield an upper bound on the true stiffness, and hence, a lower bound on the displacement of the problem, as shown in Figure 8-11.

Hence, as the mesh size is reduced—that is, as the number of elements is increased—we are ensured of monotonic convergence of the solution when compatible and complete displacement functions are used. Examples of this convergence are given in References [1] and [11], and in Table 8-2 for the beam with loading shown in Figure 8-1(a). All elements in the table are

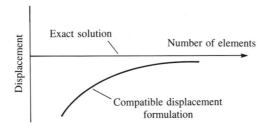

F I G U R E 8-11   Convergence of a finite element solution based on the compatible displacement formulation

T A B L E 8-2   Comparison of results for different numbers of elements

Case	Number of Nodes	Number of Elements	Aspect Ratio	Vertical Displacement, v (inches) Point A
1	21	12	2	−0.740
2	39	24	1	−0.980
3	45	32	3	−0.875
4	85	64	1.5	−1.078
5	105	80	1.2	−1.100
Exact solution (see Reference [2])				−1.152

rectangular. The results in Table 8-2 indicate the influence of the number of elements (or the number of degrees of freedom as measured by the number of nodes) on the convergence toward a common solution, in this case the exact one. We again observe the influence of the aspect ratio. The higher the aspect ratio, even with a larger number of degrees of freedom, the worse the answer, as indicated by comparing cases 2 and 3.

8.4 Interpretation of Stresses

In the stiffness or displacement formulation of the finite element method used throughout this text, the primary quantities determined are the interelement nodal displacements of the assemblage. The secondary quantities, such as strain and stress in an element, are then obtained through use of $\{\varepsilon\} = [B]\{d\}$ and $\{\sigma\} = [D][B]\{d\}$. For elements using linear-displacement models, such as the bar and the constant-strain triangle, $[B]$ is constant, and since we assume $[D]$ to be constant, the stresses are constant over the element. In this case, it is common practice to assign the stress to the centroid of the element with acceptable results.

However, as illustrated in Section 3.10 for the axial member, stresses are not predicted as accurately as the displacements. For example, remember the constant-strain or constant-stress element has been used in modeling the beam in Figure 8-1. Therefore, the stress in each element is assumed constant. Figure 8-12 compares the exact beam theory solution for bending stress through the beam depth at the centroidal location of the elements next to the wall with the finite element solution of case 4 in Table 8-2. This finite element model consists of four elements through the beam depth. Therefore, only four stress values are obtained through the depth. Again, the best approximation of the stress appears to occur at the midpoint of each element, since the

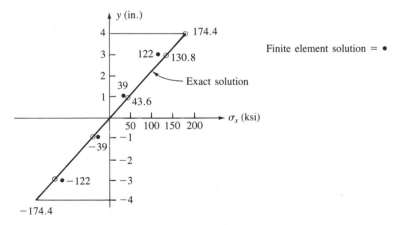

F I G U R E 8-12   Comparison of the finite element and exact solutions of bending stress through beam cross section

derivative of displacement is better predicted between the nodes than at the nodes.

For higher-order elements, such as the linear-strain triangle of Chapter 9, [B], and hence the stresses, are functions of the coordinates. The common practice is then to directly evaluate the stresses at the centroid of the element. An alternative procedure sometimes is to use an average (possibly weighted) value of the stresses evaluated at each node of the element. Finally, another procedure is to evaluate the stresses in all elements at a shared node and then use an average of these element nodal stresses to represent the stress at the node.

8.5   Static Condensation

We will now consider the concept of static condensation because this concept is used in developing the stiffness matrix of a quadrilateral element in the computer program CSFEP.

Consider the basic quadrilateral element with external nodes 1, 2, 3, and 4 shown in Figure 8-13. An imaginary node 5 is temporarily introduced at the intersection of the diagonals of the quadrilateral to create four triangles. We then superimpose the stiffness matrices of the four triangles to create the stiffness matrix of the quadrilateral element, where the internal imaginary node 5 degrees of freedom are said to be *condensed out* so as to never enter the final equations. Hence, only the degrees of freedom associated with the four *actual* external corner nodes enter the equations.

We begin the static condensation procedure by partitioning the equilibrium equations as

$$
\begin{bmatrix} k_{11} & k_{12} \\ k_{21} & k_{22} \end{bmatrix} \begin{Bmatrix} d_a \\ d_i \end{Bmatrix} = \begin{Bmatrix} F_a \\ F_i \end{Bmatrix}
\tag{8.5.1}
$$

where d_i is the vector of internal displacements corresponding to the imaginary internal node (node 5 in Figure 8-13), F_i is the vector of loads at the internal node, and d_a and F_a are the actual nodal degrees of freedom and loads, respectively, at the actual nodes. Rewriting Eq. (8.5.1), we have

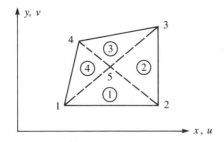

FIGURE 8-13   Quadrilateral element with an internal node

$$[k_{11}]\{d_a\} + [k_{12}]\{d_i\} = \{F_a\} \qquad (8.5.2)$$

$$[k_{21}]\{d_a\} + [k_{22}]\{d_i\} = \{F_i\} \qquad (8.5.3)$$

Solving for $\{d_i\}$ in Eq. (8.5.3), we obtain

$$\{d_i\} = -[k_{22}]^{-1}[k_{21}]\{d_a\} + [k_{22}]^{-1}\{F_i\} \qquad (8.5.4)$$

Substituting Eq. (8.5.4) into Eq. (8.5.2), we obtain the condensed equilibrium equation

$$[k_c]\{d_a\} = \{F_c\} \qquad (8.5.5)$$

where

$$[k_c] = [k_{11}] - [k_{12}][k_{22}]^{-1}[k_{21}] \qquad (8.5.6)$$

$$\{F_c\} = \{F_a\} - [k_{12}][k_{22}]^{-1}\{F_i\} \qquad (8.5.7)$$

and $[k_c]$ and $\{F_c\}$ are called the *condensed stiffness matrix* and the *condensed load vector*, respectively. Equation (8.5.5) can now be solved for the actual corner node displacements in the usual manner of solving simultaneous linear equations.

Both constant-strain triangular (CST) and constant-strain quadrilateral elements are used to analyze plane stress/plane strain problems. The quadrilateral element has the stiffness of four CST elements. An advantage of the four-CST quadrilateral is that the solution becomes less dependent on the *skew* of the subdivision mesh, as shown in Figure 8-14. Here **skew** means the *directional stiffness bias* that can be built into a model through certain discretization patterns, since the stiffness matrix of an element is a function of its nodal coordinates, as indicated by Eq. (7.2.52). The four-CST mesh of Figure 8-14(c) represents a reduction in the skew effect over the meshes of Figure 8-14(a) and (b). Figure 8-14(b) is generally worse than Figure 8-14(a) because the use of long, narrow triangles results in an element stiffness matrix that is stiffer along the narrow direction of the triangle.

The resulting stiffness matrix of the quadrilateral element will be an 8×8 matrix consisting of the stiffnesses of four triangles, as was shown in Figure 8-13. The stiffness matrix is first assembled according to the usual direct stiffness method. Then we apply static condensation as outlined in Eqs. (8.5.1)–(8.5.7) to remove the internal node 5 degrees of freedom.

The stiffness matrix of a typical triangular element (labeled element 1 in Figure 8-13) with nodes 1, 2, and 5 is given in general form by

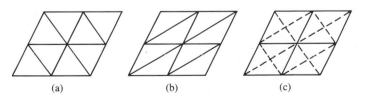

(a) (b) (c)

F I G U R E 8-14 Skew effects in finite element modeling

$$[k^{(1)}] = \begin{bmatrix} \underline{k}_{11}^{(1)} & \underline{k}_{12}^{(1)} & \underline{k}_{15}^{(1)} \\ \underline{k}_{21}^{(1)} & \underline{k}_{22}^{(1)} & \underline{k}_{25}^{(1)} \\ \underline{k}_{51}^{(1)} & \underline{k}_{52}^{(1)} & \underline{k}_{55}^{(1)} \end{bmatrix} \tag{8.5.8}$$

where the superscript in parentheses again refers to the element number, and each submatrix $[k_{ij}^{(1)}]$ is of order 2×2. The stiffness matrix of the quadrilateral, assembled using Eq. (8.5.8) along with similar stiffness matrices for elements 2, 3, and 4 of Figure 8-12 is given by the following (before static condensation is used):

$$[k] = \begin{matrix} (u_1, v_1) & (u_2, v_2) & (u_3, v_3) & (u_4, v_4) & (u_5, v_5) \end{matrix}$$

$$[k] = \left[\begin{array}{ccccc} [k_{11}^{(1)}] + [k_{11}^{(4)}] & [k_{12}^{(1)}] & [0] & [k_{14}^{(4)}] & [k_{15}^{(1)}] + [k_{15}^{(4)}] \\[2ex] [k_{21}^{(1)}] & [k_{22}^{(1)}] + [k_{22}^{(2)}] & [k_{23}^{(2)}] & [0] & [k_{25}^{(1)}] + [k_{25}^{(2)}] \\[2ex] [0] & [k_{32}^{(2)}] & [k_{33}^{(2)}] + [k_{33}^{(3)}] & [k_{34}^{(3)}] & [k_{35}^{(2)}] + [k_{35}^{(3)}] \\[2ex] [k_{41}^{(4)}] & [0] & [k_{43}^{(3)}] & [k_{44}^{(3)}] + [k_{44}^{(4)}] & [k_{45}^{(3)}] + [k_{45}^{(4)}] \\[2ex] \hline [k_{51}^{(1)}] + [k_{51}^{(4)}] & [k_{52}^{(1)}] + [k_{52}^{(2)}] & [k_{53}^{(2)}] + [k_{53}^{(3)}] & [k_{54}^{(3)}] + [k_{54}^{(4)}] & ([k_{55}^{(1)}] + [k_{55}^{(2)}]) + ([k_{55}^{(3)}] + [k_{55}^{(4)}]) \end{array} \right] \tag{8.5.9}$$

where the order of the degrees of freedom are shown above the columns of the stiffness matrix and the partitioning scheme used in static condensation is indicated by the dotted lines. Before static condensation is applied, the stiffness matrix is of order 10×10.

E X A M P L E   8.1

Consider the quadrilateral with internal node 5 and dimensions as shown in Figure 8-15 to illustrate the application of static condensation.

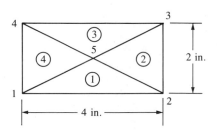

FIGURE 8-15   Quadrilateral with an internal node

Recall that the original stiffness matrix of the quadrilateral is 10×10, but static condensation will result in an 8×8 stiffness matrix after removal of the degrees of freedom (u_5, v_5) at node 5.

Using the CST stiffness matrix of Eq. (7.4.3) for plane strain, we have

$$[k^{(1)}] = [k^{(3)}] = \frac{E}{4.16} \begin{bmatrix} \overset{\overset{3}{1}}{1.5} & \overset{\overset{}{}}{1.0} & 0.1 & \overset{\overset{4}{2}}{0.2} & -1.6 & \overset{\overset{5}{5}}{-1.2} \\ & 3.0 & -0.2 & 2.6 & -0.8 & -5.6 \\ & & 1.5 & -1.0 & -1.6 & 1.2 \\ & & & 3.0 & 0.8 & -5.6 \\ & & & & 3.2 & 0.0 \\ \text{Symmetry} & & & & & 11.2 \end{bmatrix} \quad (8.5.10)$$

Similarly, from Figure 8-15, we can show that

$$[k^{(2)}] = [k^{(4)}] = \frac{E}{4.16} \begin{bmatrix} \overset{\overset{2}{4}}{1.5} & \overset{}{-1.0} & -0.1 & \overset{\overset{3}{1}}{0.2} & -1.4 & \overset{\overset{5}{5}}{0.8} \\ & 3.0 & -0.2 & -2.6 & 1.2 & -0.4 \\ & & 1.5 & 1.0 & -1.4 & -0.8 \\ & & & 3.0 & -1.2 & -0.4 \\ & & & & 2.8 & 0.0 \\ \text{Symmetry} & & & & & 0.8 \end{bmatrix}$$

$$(8.5.11)$$

where the numbers above the columns in Eqs. (8.5.10) and (8.5.11) indicate the orders of the degrees of freedom associated with each stiffness matrix. Here the quantity in the denominator of Eq. (7.4.3), $4A(1 + v)(1 - 2v)$, is equal to 4.16 in Eqs. (8.5.10) and (8.5.11) because $A = 2$ in.^{2} and v is taken to be 0.3. Also, the thickness t of the element has been taken as 1 in. Now we can superimpose the stiffness terms as indicated by Eq. (8.5.9) to obtain the general expression for a four-CST element. The resulting assembled total stiffness matrix before static condensation is applied is given by

$$[k] = \frac{E}{4.16} \begin{bmatrix} 3.0 & 2.0 & 0.1 & 0.2 & 0.0 & 0.0 & -0.1 & -0.2 & \vdots & -3.0 & -2.0 \\ & 6.0 & -0.2 & 2.6 & 0.0 & 0.0 & 0.2 & -2.6 & \vdots & -2.0 & -6.0 \\ & & 3.0 & -2.0 & -0.1 & 0.2 & 0.0 & 0.0 & \vdots & -3.0 & 2.0 \\ & & & 6.0 & -0.2 & -2.6 & 0.0 & 0.0 & \vdots & 2.0 & -6.0 \\ & & & & 3.0 & 2.0 & 0.1 & 0.2 & \vdots & -3.0 & -2.0 \\ & & & & & 6.0 & -0.2 & 2.6 & \vdots & -2.0 & -6.0 \\ & & & & & & 3.0 & -2.0 & \vdots & -3.0 & 2.0 \\ & & & & & & & 6.0 & \vdots & 2.0 & -6.0 \\ \hline & & & & & & & & \vdots & 12.0 & 0.0 \\ \text{Symmetry} & & & & & & & & \vdots & & 24.0 \end{bmatrix}$$

$$(8.5.12)$$

After partitioning Eq. (8.5.12) and using Eq. (8.5.6), the condensed stiffness matrix is given by

$$[k_c] = \frac{E}{4.16} \begin{bmatrix} \overset{u_1}{2.08} & \overset{v_1}{1.00} & \overset{u_2}{-0.48} & \overset{v_2}{0.20} & \overset{u_3}{-0.92} & \overset{v_3}{-1.00} & \overset{u_4}{-0.68} & \overset{v_4}{-0.20} \\ & 4.17 & -0.20 & 1.43 & -1.00 & -1.83 & 0.20 & -3.77 \\ & & 2.08 & -1.00 & -0.68 & 0.20 & -0.92 & 1.00 \\ & & & 4.17 & -0.20 & -3.77 & 1.00 & -1.83 \\ & & & & 2.08 & 1.00 & -0.48 & 0.20 \\ & & & & & 4.17 & -0.20 & 1.43 \\ & & & & & & 2.08 & -1.00 \\ \text{Symmetry} & & & & & & & 4.17 \end{bmatrix}$$

$$(8.5.13)$$

■

8.6 Flowchart of a CST Stiffness Program

Figure 8-16 is a flowchart of a finite element program (called CSFEP) used for the analysis of plane stress/strain problems.

8.7 Description of a Computer Program for Plane Stress/Strain Analysis

We will now describe a computer program called CSFEP* (listed on the disk at the back of the text) that can be used to solve large plane stress/strain problems. The program is based on the flowchart of Section 8.6, and is a modified version of a program found in Reference [1].

To use program CSFEP, we first describe the input data. Hence, the argument list in SUBROUTINE DATAIN and the associated COMMON blocks used to read in and transfer data through different parts of the program are described as follows:

```
SUBROUTINE DATAIN(MAXEL,MAXNP,MAXMAT,MAXSLC,
ISTOP)COMMON NNP,NEL,NMAT,NSLC,NOPT,NBODY,MTYP,
E(10),PR(10),RO(10),TH(10),IE(120,5),X(120),Y(120),ULX(120),
VLY(120),KODE(120), ISC(20),JSC(20),SURTRX(20,2),SURTRY(20,2)
COMMON/THRE/IMOUT,MOUT
```

SUBROUTINE DATAIN reads the data needed by program CSFEP and transfers the variables back to the calling program and other appropriate

---

*Desai, C. S., and Abel, J. F. *Introduction to the Finite Element Method.* New York: Van Nostrand Reinhold, 1972.

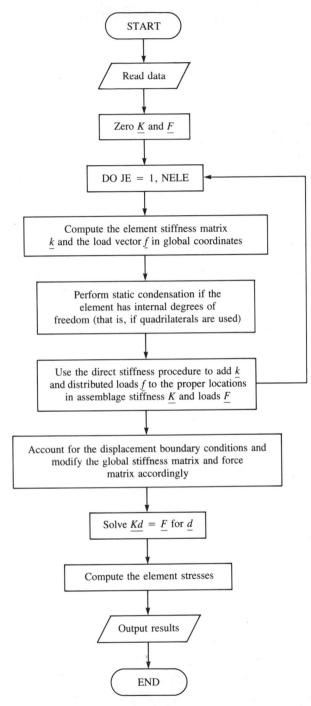

FIGURE 8-16 Flowchart of plane stress/strain finite element program CSFEP

subroutines through the arguments in the COMMON statements. Again, as in the other programs, all data is input in free format, separated by commas.

Description of Variables

E(NMAT) is the modulus of elasticity of a material type.

IE(M,5): IE(M,1) through IE(M,4) store the four corner nodes I, J, K, and L of quadrilateral element M, and IE(M,5) denotes the type of material in the element.

I, J, K, and L are the indices of the four nodes of the quadrilateral in SUBROUTINE QUAD.

I, J, and K are the indices of the three nodes of the triangular element in SUBROUTINE CST.

IMOUT: Any number except zero will cause output of the element stiffness matrix.

ISC(NSLC) is the I node of side I–J of an element having a surface traction on side I–J.

JSC(NSLC) is the J node of side I–J of an element having a surface traction on side I–J.

KODE(NNP) is an index of displacement and concentrated load conditions at one of the nodes NNP.

M is an index denoting the node number or the element number.

MAXBW is the maximum semibandwidth allowed by the storage allocation declarations.

MAXEL is the maximum number of elements allowed by the storage allocation declarations.

MAXMAT is the maximum number of material types allowed by the storage allocation declarations.

MAXNP is the maximum number of node points allowed by the storage allocation declarations.

MAXSLC is the maximum number of surface load cases allowed by the storage allocation declarations.

MOUT: Any number except zero will cause output of the global stiffness matrix.

MTYP is an index that denotes the type of material in the element.

NBODY is an option for body force.

NBODY = 0 for no weight.

NBODY = 1 for a weight force in the negative y direction.

NEL is the number of elements (NEL ≤ MAXEL).

NMAT is the number of different material types $(1 \leq$ NMAT $\leq$ MAXMAT).

NNP is the number of nodal points (NNP ≤ MAXNP).

NOPT is an option for plane strain/stress.

NOPT = 1 for plane strain.

NOPT = 2 for plane stress.

NSLC is the number of surface tractions (NSLC ≤ MAXSLC).

PR(NMAT) is Poisson's ratio for a material type.

RO(NMAT) is the weight density of a material type.

SURTRX(NSLC,I) and SURTRY(NSLC,I) are the x and y components of the prescribed distributed loads at node I along side I–J of an element.

SURTRX(NSLC,J) and SURTRY(NSLC,J) are the x and y components of the prescribed distributed loads at node J along side I–J of an element.

TH(NMAT) is the thickness of a material type.

TITLE is an array for the title of the problem (up to eighty alphanumeric characters).

ULX(NNP) and VLY(NNP) are the concentrated loads or displacements in the x and y directions at one of the nodes NNP.

X(NNP) and Y(NNP) are the x and y coordinates of one of the nodes NNP.

Input Data

Since program CSFEP has the capability for node and element generation, the actual line numbers of data may vary. Hence, we list the input data by sequence of parameters needed by CSFEP as follows:

1. Title (columns 1–80)
   Title of problem; any alphanumeric data to identify the problem. This will be printed as a heading on the first page of output.

2. Basic parameters line

   NNP,NEL,NMAT,NSLC,NOPT,NBODY

3. Option for printing out stiffness matrices

   IMOUT,MOUT

4. Material properties (one line for each NMAT)

   E(NMAT),PR(NMAT),RO(NMAT),TH(NMAT)

5. Nodal point data (see the following Notes 3 and 4)

   M,KODE(M),X(M),Y(M),ULX(M),VLY(M)

6. Element data (see Notes 5 and 6)

   M,I,J,K,L,MTYP

7. Surface tractions (as many lines as the value of NSLC) (see Note 7); skip this line if NSLC = 0 in line 2

   ISC(NSLC),JSC(NSLC),SURTRX(NSLC,1),SURTRX(NSLC,2),
   SURTRY(NSLC,1),SURTRY(NSLC,2)

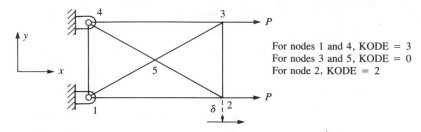

FIGURE 8-17   Model illustrating KODE

Notes on Input Data for CSFEP

The following notes provide further explanation of the preceding description of input data:

1. Data lines must be in proper sequence.

2. Units must be consistent.

3. Usually, one line is needed for each node. However, if some nodes fall on a straight line and are equidistant, data for only the first and the last points of this group are needed. Intermediate nodal point data are automatically generated by linear interpolation. Nodes must be numbered successively to generate data (that is, 1, 2, 3, 4, ... along a line and not, say, 1, 4 , 7, and 10 along a line).

4. Forces and/or displacements prescribed at a node are identified by KODE as follows (also see Figure 8-17):

   KODE   Force/Displacement Boundary Conditions

   0 ULX = Prescribed load in the x direction

     VLY = Prescribed load in the y direction

   1 ULX = Prescribed displacement in the x direction

     VLY = Prescribed load in the y direction

   2 ULX = Prescribed load in the x direction

     VLY = Prescribed displacement in the y direction

   3 ULX = Prescribed displacement in the x direction

     VLY = Prescribed displacement in the y direction

   The sign of an applied force or displacement follows the sign of the coordinate directions. For instance, a force in the positive x direction is positive, and so on. For the nodes automatically generated as in Note 3, KODE = 0, ULX = 0, and VLY = 0 are assigned for the generated nodes.

5. IE(M,1), IE(M,2), IE(M,3), and IE(M,4) denote the four corner nodes I, J, K, and L of quadrilateral element M. The program also

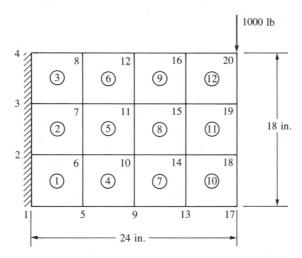

FIGURE 8-18   Model illustrating element data generation

permits use of triangular elements, which are indicated by repeating the third node; that is, IE(M,3) = IE(M,4) (or K = L). For a right-handed coordinate system, the nodes must be input counterclockwise around the element, or a negative area will result. IE(M,5) (MTYP) denotes the type of material in the element. The maximum difference between numbers of any two nodes for a given element must be less than MAXBW/2.

6. Usually, one line is needed for each element. However, if some elements are on a line in such a way that their corner-node indices each increase by one compared to the previous element, only the data for the first element on the line need be input. However, note that data for the last element of the assemblage must be input. For example, in Figure 8-18, only data for elements 1, 4, 7, 10, and 12 are needed. The omitted element data are generated internally by the computer. All generated elements are assigned the same material type as the previously input element.

7. Surface tractions must be specified between two adjacent nodes only. The three possible cases are shown in Figure 8-19. For case (a), only SURTRX(NSLC,I) and SURTRX(NSLC,J) are nonzero, and the others are zero. For case (b), only SURTRY(NSLC,I) and SURTRY(NSLC,J) are nonzero, and the others are zero. For both tractions, all columns are input. For case (c), you will need to compute the components of the tractions manually before entering them into the computer. In addition, you must multiply all surface intensities by the thickness of the element before the intensities are input into the computer. Units of surface traction will then be force/length. The signs of tractions follow the directions of the coordinate axes. A traction in the negative y direction is negative, and so on. Each side

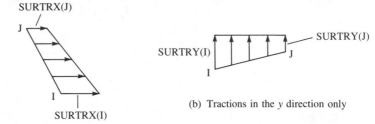

(a) Tractions in the *x* direction only

(b) Tractions in the *y* direction only

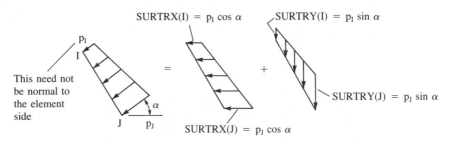

(c) Tractions in both the *x* and *y* directions

F I G U R E   8-19   Three possible cases of surface tractions on element
side I-J

I-J of an element having a traction is counted in the total number of
surface tractions, NSLC.

8. Any data that are zeros should be input as zeros; do not assume
   blanks will be treated as zeros.

9. The following upper limits have been set in this program (these parameters can be modified to fit the computer system available to you).

   Maximum number of node points, MAXNP = 120
   Maximum number of elements, MAXEL = 120
   Maximum number of different materials, MAXMAT = 10
   Maximum bandwidth, MAXBW = 50
   Maximum number of surface tractions, MAXSLC = 20

E X A M P L E   8.2

We now consider the thin plate solved longhand in Section 7.5 to illustrate
the input data for CSFEP. Here $E = 30 \times 10^6$ psi, $v = 0.30$, and $t = 1$ in.
  The data file for the plate shown in Figure 8-20 is given in Table 8-3. Line
1 of Table 8-3 is the identifying title of the problem being solved. On line 2
are the numbers of node points, elements, material types, and surface load
cases, the option for plane stress, and the option not to include body forces

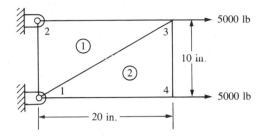

FIGURE 8-20 Example problem used to illustrate the data file

TABLE 8-3 Input for the thin plate of Figure 8-20

Line	Data on the Line (Beginning in Column 1)
1	PLATE IN TENSION OF SECTION 7.5
2	4,2,1,0,2,0
3	0,0
4	30.E + 6,0.30,0.283,1.
5	1,3,0.,0.,0.,0.
6	2,3,0.,10.,0.,0.
7	3,0,20.,10.,5000.,0.
8	4,0,20.,0.,5000.,0.
9	1,1,3,2,2,1
10	2,1,4,3,3,1

(self-weight). On line 3 are the options for element and global stiffness matrix printout (here zeros indicate they will not be printed). Line 4 lists the modulus of elasticity, Poisson's ratio, density, and thickness of the material type used. Lines 5–8 describe node information, and lines 9 and 10 describe element information. The solution for this problem given in Section 7.5 can be compared to the computer program answers given in Table 8-4. ■

TABLE 8-4 Output of computer program CSFEP for the thin plate of Figure 8-20

PLATE IN TENSION OF SECTION 7.5

INPUT TABLE 1. . BASIC PARAMETERS
NUMBER OF NODAL POINTS.   4
NUMBER OF ELEMENTS   2
NUMBER OF DIFFERENT MATERIALS 1
NUMBER OF SURFACE LOAD CARDS 0
1 = PLANE STRAIN, 2 = PLANE STRESS   2
BODY FORCES (1 = IN − Y DIREC., 0 = NONE)   0

T A B L E   8-4 (*Continued*)

INPUT TABLE 2. . MATERIAL PROPERTIES

MATERIAL, NUMBER	MODULUS OF ELASTICITY	POISSON'S RATIO,	MATERIAL DENSITY	MATERIAL THICKNESS
1	0.3000E + 08	0.3000E + 00	0.2830E + 00	0.1000E + 01

INPUT TABLE 3. . NODAL POINT DATA

NODAL POINT	TYPE	X	Y	X-DISP. OR LOAD	Y-DISP. OR LOAD
1	3	0.0000E + 00	0.0000E + 00	0.0000E + 00	0.0000E + 00
2	3	0.0000E + 00	0.1000E + 02	0.0000E + 00	0.0000E + 00
3	0	0.2000E + 02	0.1000E + 02	0.5000E + 04	0.0000E + 00
4	0	0.2000E + 02	0.0000E + 00	0.5000E + 04	0.0000E + 00

INPUT TABLE 4. . ELEMENT DATA

	GLOBAL INDICES OF ELEMENT NODES				
ELEMENT	1	2	3	4	MATERIAL
1	1	3	2	2	1
2	1	4	3	3	1

OUTPUT TABLE 1. . NODAL DISPLACEMENTS

NODE	U = X-DISP.	V = Y-DISP.
1	0.00000000E + 00	0.00000000E + 00
2	0.00000000E + 00	0.00000000E + 00
3	0.62499999E − 03	0.00000000E + 00
4	0.66666672E − 03	0.83333332E − 04

OUTPUT TABLE 2. . STRESSES AT ELEMENT CENTROIDS

ELEMENT	X	Y	SIGMA(X)	SIGMA(Y)	TAU(X,Y)
1	6.67	6.67	1.0000E + 03	2.5000E + 02	0.0000E + 00
2	13.33	3.33	1.0000E + 03	3.0518E − 05	−8.7311E − 05

ELEMENT			SIGMA(1)	SIGMA(2)	ANGLE
1			1.0000E + 03	2.5000E + 02	0.0000E + 00
2			1.0000E + 03	3.0518E − 05	−5.0026E − 06

REFERENCES

[1] Desai, C. S., and Abel, J. F., *Introduction to the Finite Element Method*, Van Nostrand Reinhold, New York, 1972.

[2] Timoshenko, S., and Goodier, J., *Theory of Elasticity*, 3rd ed., McGraw-Hill, New York, 1970.

[3] Glockner, P. G., "Symmetry in Structural Mechanics," *Journal of the Structural Division*, American Society of Civil Engineers, Vol. 99, No. ST1, pp. 71–89, 1973.

[4] Yamada, Y., "Dynamic Analysis of Civil Engineering Structures," *Recent Advances in Matrix Methods of Structural Analysis and Design*, R. H. Gallagher, Y. Yamada, and J. T. Oden, Eds., University of Alabama Press, Alabama, pp. 487–512, 1970.

[5] Koswara, H., *A Finite Element Analysis of Underground Shelter Subjected to Ground Shock Load*, M. S. Thesis, Rose-Hulman Institute of Technology, Terre Haute, Indiana, 1983.

[6] Dunlop, P., Duncan, J. M., and Seed, H. B., "Finite Element Analyses of Slopes in Soil," *Journal of the Soil Mechanics and Foundations Division*, Proceedings of the American Society of Civil Engineers, Vol. 96, No. SM2, March 1970.

[7] Cook, R. D., Malkus, D. S., and Plesha, M. E., *Concepts and Applications of Finite Element Analysis*, 3rd ed., Wiley, New York, 1989.

[8] Taylor, R. L., Beresford, P. J., and Wilson, E. L., "A Nonconforming Element for Stress Analysis," *International Journal for Numerical Methods in Engineering*, Vol. 10, No. 6, pp. 1211–1219, 1976.

[9] Melosh, R. J., "Basis for Derivation of Matrices for the Direct Stiffness Method," *Journal of the American Institute of Aeronautics and Astronautics*, Vol. 1, No. 7, pp. 1631–1637, July 1963.

[10] Fraeijes de Veubeke, B., "Upper and Lower Bounds in Matrix Structural Analysis," *Matrix Methods of Structural Analysis*, AGARDograph 72, B. Fraeijes de Veubeke, Ed., Macmillan, New York, 1964.

[11] Dunder, V., and Ridlon, S., "Practical Applications of Finite Element Method," *Journal of the Structural Division*, American Society of Civil Engineers, No. ST1, pp. 9–21, 1978.

[12] Swanson, J. A., *ANSYS—Engineering Systems User's Manual*, Swanson Analysis Systems, Inc., Elizabeth, Pennsylvania.

[13] Bettess, P., "More on Infinite Elements," *International Journal for Numerical Methods in Engineering*, Vol. 15, pp. 1613–1626, 1980.

PROBLEMS

8.1 For the finite element mesh shown in Figure P8-1, comment on the goodness of the mesh. Indicate the mistakes in the model. Explain and show how to correct them.

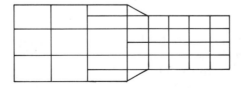

F I G U R E P8-1

8.2   Comment on the mesh sizing in Figure P8-2. Is it reasonable? If not, explain why not.

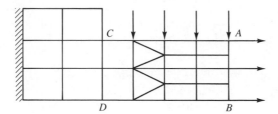

FIGURE   P8-2

8.3   What happens if the material property $v = 0.5$ in the plane strain case? Is this possible? Explain.

8.4   Under what conditions is the structure in Figure P8-4 a plane strain problem? Under what conditions is the structure a plane stress problem?

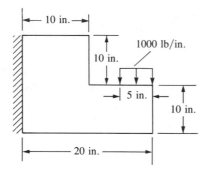

FIGURE   P8-4

8.5   Show that Eq. (8.5.13) is obtained by static condensation of Eq. (8.5.12).

Solve the following problems using computer program CSFEP on the disk at the back of the text (or any other suitable program). In some of these problems, we suggest that students be assigned separate parts (or models) to facilitate parametric studies.

8.6   Determine the free-end displacements and the element stresses for the plate discretized into four triangular elements and subjected to the tensile forces shown in Figure P8-6. Compare your results to the solution given in Section 7.5. Why are these results different? Let $E = 30 \times 10^6$ psi, $v = 0.30$, and $t = 1$ in.

8.7   Determine the stresses in the plate with the hole subjected to the tensile stress shown in Figure P8-7. Graph the stress variation σ_x versus the distance, y, from the hole. Let $E = 30 \times 10^6$ psi, $v = 0.25$, and $t = 1$ in. (Use approximately 25, 50, 75, 100, and then 120 nodes in your finite element model.) Use symmetry as appropriate.

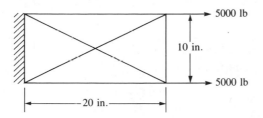

FIGURE   P8-6

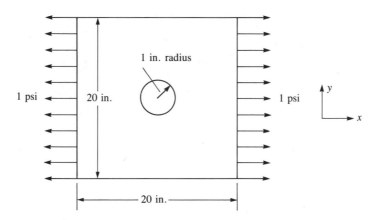

FIGURE   P8-7

8.8 Solve the following problem of a tensile plate with a concentrated load applied at the top, as shown in Figure P8-8. Determine at what depth the effect of the load dies out. Plot stress σ_y versus distance from the load. At distances of 1 in., 2 in., 4 in., 6 in., 10 in., 15 in., 20 in., and 30 in. from the load, list σ_y versus these distances. Let the width of the plate be $b = 4$ in., thickness of the plate be $t = 0.25$ in., and length be $L = 40$ in. Look up the concept of St. Venant's principle to see how it explains the stress behavior in this problem.

8.9 For the connecting rod shown in Figure P8-9, determine the maximum principal stresses and their location. Let $E = 30 \times 10^6$ psi, $v = 0.25$, $t = 1$ in., and $P = 1000$ lb.

8.10 Determine the maximum principal stresses and their location for the member with fillet subjected to tensile forces shown in Figure P8-10. Let $E = 30 \times 10^6$ psi and $v = 0.25$. Then let $E = 10 \times 10^6$ psi and $v = 0.30$. Let $t = 1$ in. for both cases. Compare your answers for the two cases.

8.11 Determine the stresses in the member with a re-entrant corner as shown in Figure P8-11. At what location are the principal stresses largest? Let $E = 30 \times 10^6$ psi and $v = 0.25$. Use plane strain conditions.

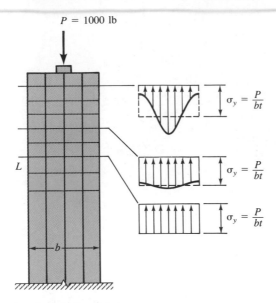

$P = 1000$ lb

$\sigma_y = \dfrac{P}{bt}$

$\sigma_y = \dfrac{P}{bt}$

$\sigma_y = \dfrac{P}{bt}$

L

b

FIGURE P8-8

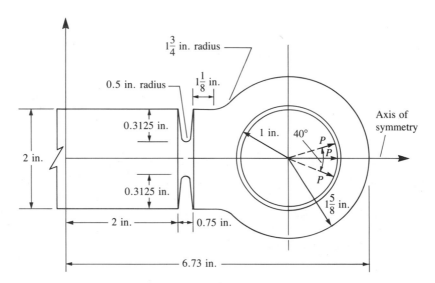

$1\frac{3}{4}$ in. radius

0.5 in. radius

$1\frac{1}{8}$ in.

0.3125 in.

1 in. 40°

P

P

P

Axis of symmetry

2 in.

0.3125 in.

$1\frac{5}{8}$ in.

2 in.

0.75 in.

6.73 in.

FIGURE P8-9

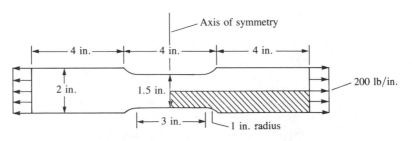

Axis of symmetry

4 in.

4 in.

4 in.

200 lb/in.

2 in.

1.5 in.

3 in.

1 in. radius

FIGURE P8-10

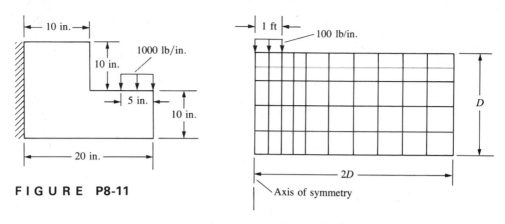

FIGURE P8-11

FIGURE P8-12

8.12 Determine the stresses in the soil mass subjected to the strip footing load shown in Figure P8-12. Use a width of $2D$ and depth of D, where D is 3 ft, 4 ft, 6 ft, 8 ft, and 10 ft. Plot the maximum stress contours on your finite element model for each case. Compare your results. Comment regarding your observations on modeling infinite media. Let $E = 30,000$ psi and $v = 0.30$. Use plane strain conditions.

8.13 For the tooth implant subjected to loads shown in Figure P8-13, determine the maximum principal stresses. Let $E = 1.6 \times 10^6$ psi and $v = 0.3$ for the dental restorative implant material (cross-hatched), and let $E = 1 \times 10^6$ psi and $v = 0.35$ for the bony material. Let $X = 0.05$ in., 0.1 in., 0.2 in., 0.3 in., and 0.5 in., where X represents the various depths of the implant beneath the bony surface. Rectangular elements are used in the finite element model shown in Figure P8-13. Assume the thickness of each element to be $t = 0.25$ in.

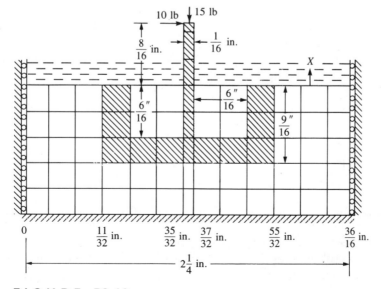

FIGURE P8-13

8.14 Determine the mid-depth deflection at the free end and the maximum principal stresses and their location for the beam subjected to the shear load variation shown in Figure P8-14. Do this using 64 rectangular elements all of size 12 in. × $\frac{1}{2}$ in.; then all of size 6 in. × 1 in.; then all of size 3 in. × 2 in. Then use 60 rectangular elements all of size 2.4 in. × $2\frac{2}{3}$ in.; then all of size 4.8 in. × $1\frac{1}{3}$ in. Compare the free-end deflections and the maximum principal stresses in each case and to the exact solution. Let $E = 30 \times 10^6$ psi, $v = 0.3$, and $t = 1$ in. Comment on the accuracy of both displacements and stresses.

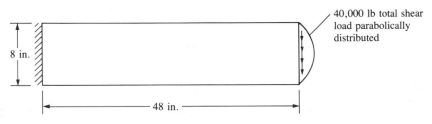

FIGURE P8-14

8.15 Determine the stresses in the shear wall shown in Figure P8-15. At what location are the principal stresses largest? Let $E = 21$ GPa, $v = 0.25$, $t_{wall} = 0.10$ m, and $t_{beam} = 0.20$ m.

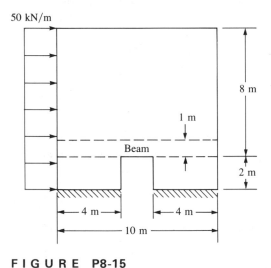

FIGURE P8-15

8.16 Determine the stresses in the plates with the round and square holes subjected to the tensile stresses shown in Figure P8-16. Compare the largest principal stresses for each plate. Let $E = 210$ GPa, $v = 0.25$, and $t = 5$ mm.

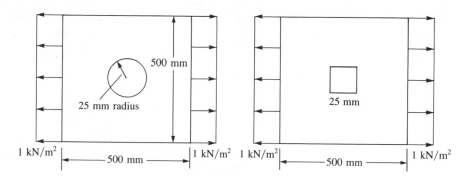

FIGURE   P8-16

8.17 For the concrete overpass structure shown in Figure P8-17, determine the maximum principal stresses and their location. Assume plane strain conditions. Let $E = 3.0 \times 10^6$ psi and $v = 0.30$.

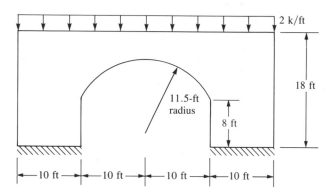

FIGURE   P8-17

8.18 For the steel culvert shown in Figure P8-18, determine the maximum principal stresses and their location and the largest displacement and its location. Let $E_{steel} = 210$ GPa and let $v = 0.30$.

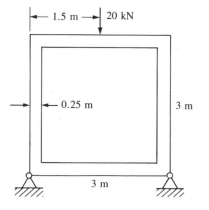

FIGURE   P8-18

8.19   For the tensile member shown in Figure P8-19 with two holes, determine the maximum principal stresses and their location. Let $E = 210$ GPa, $v = 0.25$, and $t = 10$ mm. Then let $E = 70$ GPa and $v = 0.30$. Compare your results.

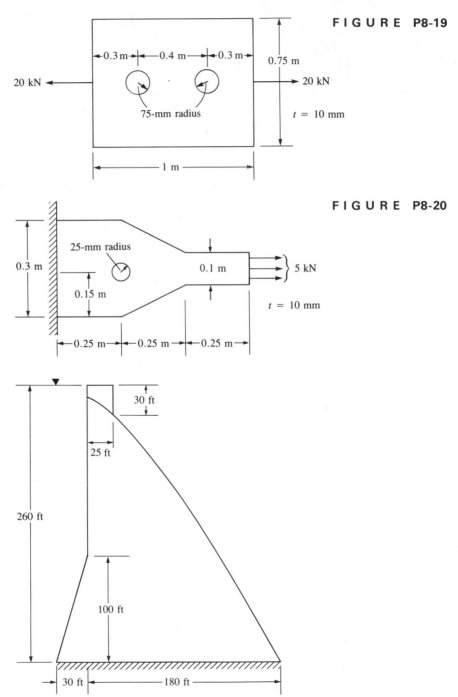

FIGURE   P8-19

FIGURE   P8-20

FIGURE   P8-21

8.20 For the plate shown in Figure P8-20, determine the maximum principal stresses and their location. Let $E = 210$ GPa and $v = 0.25$.

8.21 For the concrete dam shown subjected to water pressure in Figure P8-21, determine the principal stresses. Let $E = 3.5 \times 10^6$ psi and $v = 0.30$. Assume plane strain conditions. Perform the analysis for self-weight and then for hydrostatic (water) pressure against the dam vertical face as shown.

8.22 Determine the stresses in the wrench shown in Figure P8-22. Let $E = 200$ GPa and $v = 0.25$, and assume uniform thickness $t = 10$ mm.

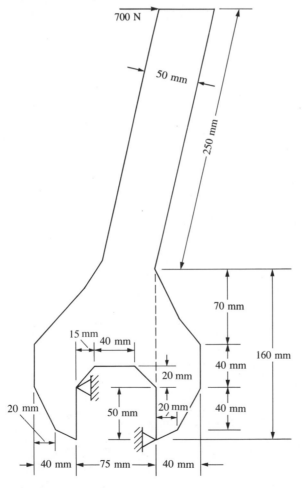

F I G U R E   P8-22

8.23 Determine the principal stresses in the blade implant and the bony material shown in Figure P8-23. Let $E_{blade} = 20$ GPa, $v_{blade} = 0.30$, $E_{bone} = 12$ GPa, and $v_{bone} = 0.35$. Assume plane stress conditions with $t = 5$ mm.

8.24 Determine the stresses in the plate shown in Figure P8-24. Let $E = 210$ GPa and $v = 0.25$. The element thickness is 10 mm.

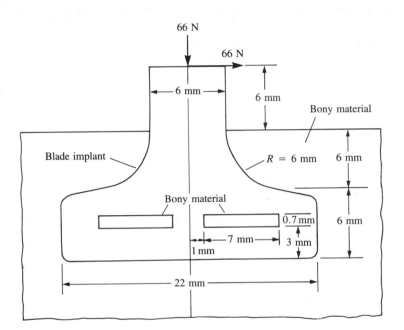

F I G U R E   P8-23

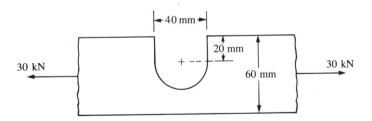

F I G U R E   P8-24

CHAPTER 9

Development of the

Linear-Strain Triangle

Equations

INTRODUCTION

In this chapter, we consider the development of the stiffness matrix and equations for a higher-order triangular element, called the *linear-strain triangle* (LST). This element is available in many commercial computer programs and has some advantages over the constant-strain triangle described in Chapter 7.

The LST element has six nodes and twelve unknown displacement degrees of freedom. The displacement functions for the element are quadratic instead of linear (as in the CST).

The procedures for development of the equations for the LST element follow the same steps as those used in Chapter 7 for the CST element. However, the number of equations now becomes twelve instead of six, making a longhand solution extremely cumbersome. Hence, we will use a computer to perform many of the mathematical operations.

After deriving the element equations, we will compare results from problems solved using the LST element with those solved using the CST element. The introduction of the higher-order LST element will illustrate the possible advantages of higher-order elements, and should enhance your general understanding of the concepts involved with finite element procedures.

9.1 Derivation of the Linear-Strain Triangular Element Stiffness Matrix and Equations

We will now derive the LST stiffness matrix and element equations. The steps used here are identical to those used for the CST element, and much of the notation is the same.

349

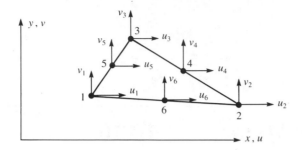

FIGURE 9-1   Basic six-node triangular element

Step 1   Select Element Type

Consider the triangular element shown in Figure 9-1 with the usual end nodes and three additional nodes conveniently located at the midpoints of the sides. Thus, a computer program can automatically compute the midpoint coordinates once the coordinates of the corner nodes are given as input.

The unknown nodal displacements are now given by

$$\{d\} = \begin{Bmatrix} \underline{d}_1 \\ \underline{d}_2 \\ \underline{d}_3 \\ \underline{d}_4 \\ \underline{d}_5 \\ \underline{d}_6 \end{Bmatrix} = \begin{Bmatrix} u_1 \\ v_1 \\ u_2 \\ v_2 \\ u_3 \\ v_3 \\ u_4 \\ v_4 \\ u_5 \\ v_5 \\ u_6 \\ v_6 \end{Bmatrix} \qquad (9.1.1)$$

Step 2   Select a Displacement Function

We now select a quadratic displacement function in each element as

$$u(x, y) = a_1 + a_2 x + a_3 y + a_4 x^2 + a_5 xy + a_6 y^2$$
$$v(x, y) = a_7 + a_8 x + a_9 y + a_{10} x^2 + a_{11} xy + a_{12} y^2$$
$$(9.1.2)$$

Again, the number of coefficients a_i (12) equals the total number of degrees of freedom for the element. The displacement compatibility among adjoining elements is satisfied because three nodes are located along each side and a parabola is defined by three points on its path. Since adjacent elements are connected at common nodes, their displacement compatibility across the boundaries will be maintained.

In general, when considering triangular elements, a complete polynomial in Cartesian coordinates can be used to describe the displacement field within an element. Using internal nodes as necessary for the higher-order cubic and

Terms in Pascal Triangle	Polynomial Degree	Number of Terms	Triangle
1	0 (constant)	1	
$x \quad y$	1 (linear)	3	CST (Chap. 7)
$x^2 \quad xy \quad y^2$	2 (quadratic)	6	LST (Chap. 9)
$x^3 \quad x^2y \quad xy^2 \quad y^3$	3 (cubic)	10	QST

F I G U R E   9-2   Relation between type of plane triangular element and polynomial coefficients based on Pascal

quartic elements, all terms of a truncated Pascal triangle are used in the displacement field or, equivalently, the shape functions, as shown by Figure 9-2; that is, a complete linear function is used for the CST element considered previously in Chapter 7. The complete quadratic function is used for the LST of this chapter. The complete quadratic function is also used for the quadratic-strain triangle (QST), with an internal node necessary as the tenth node.

The general displacement functions, Eqs. (9.1.2), expressed in matrix form are now

$$\{\psi\} = \begin{Bmatrix} u \\ v \end{Bmatrix} = \begin{bmatrix} 1 & x & y & x^2 & xy & y^2 & 0 & 0 & 0 & 0 & 0 & 0 \\ 0 & 0 & 0 & 0 & 0 & 0 & 1 & x & y & x^2 & xy & y^2 \end{bmatrix} \begin{Bmatrix} a_1 \\ a_2 \\ \vdots \\ a_{12} \end{Bmatrix}$$

$$(9.1.3)$$

Alternatively, we can express Eq. (9.1.3) as

$$\{\psi\} = [M^*]\{a\} \qquad (9.1.4)$$

where $[M^*]$ is defined to be the first matrix on the right side of Eq. (9.1.3). The coefficients a_1 through a_{12} can be obtained by substituting the coordinates into u and v as follows:

$$\begin{Bmatrix} u_1 \\ u_2 \\ \vdots \\ u_6 \\ v_1 \\ \vdots \\ v_5 \\ v_6 \end{Bmatrix} = \begin{bmatrix} 1 & x_1 & y_1 & x_1^2 & x_1y_1 & y_1^2 & 0 & 0 & 0 & 0 & 0 & 0 \\ 1 & x_2 & y_2 & x_2^2 & x_2y_2 & y_2^2 & 0 & 0 & 0 & 0 & 0 & 0 \\ \vdots & \vdots & \vdots & \vdots & \vdots & \vdots & \vdots & \vdots & \vdots & \vdots & \vdots & \vdots \\ 1 & x_6 & y_6 & x_6^2 & x_6y_6 & y_6^2 & 0 & 0 & 0 & 0 & 0 & 0 \\ 0 & 0 & 0 & 0 & 0 & 0 & 1 & x_1 & y_1 & x_1^2 & x_1y_1 & y_1^2 \\ \vdots & \vdots & \vdots & \vdots & \vdots & \vdots & \vdots & \vdots & \vdots & \vdots & \vdots & \vdots \\ 0 & 0 & 0 & 0 & 0 & 0 & 1 & x_5 & y_5 & x_5^2 & x_5y_5 & y_5^2 \\ 0 & 0 & 0 & 0 & 0 & 0 & 1 & x_6 & y_6 & x_6^2 & x_6y_6 & y_6^2 \end{bmatrix} \begin{Bmatrix} a_1 \\ a_2 \\ \vdots \\ a_6 \\ a_7 \\ \vdots \\ a_{11} \\ a_{12} \end{Bmatrix}$$

$$(9.1.5)$$

or solving for the a_i's, we have

$$
\left\{\begin{array}{c} a_1 \\ \vdots \\ a_6 \\ a_7 \\ \vdots \\ a_{12} \end{array}\right\} = \left[\begin{array}{cccccccccccc} 1 & x_1 & y_1 & x_1^2 & x_1 y_1 & y_1^2 & 0 & 0 & 0 & 0 & 0 & 0 \\ \vdots & \vdots & \vdots & \vdots & \vdots & \vdots & \vdots & \vdots & \vdots & \vdots & \vdots & \vdots \\ 1 & x_6 & y_6 & x_6^2 & x_6 y_6 & y_6^2 & 0 & 0 & 0 & 0 & 0 & 0 \\ 0 & 0 & 0 & 0 & 0 & 0 & 1 & x_1 & y_1 & x_1^2 & x_1 y_1 & y_1^2 \\ \vdots & \vdots & \vdots & \vdots & \vdots & \vdots & \vdots & \vdots & \vdots & \vdots & \vdots & \vdots \\ 0 & 0 & 0 & 0 & 0 & 0 & 1 & x_6 & y_6 & x_6^2 & x_6 y_6 & y_6^2 \end{array}\right]^{-1}
$$

$$
\times \left\{\begin{array}{c} u_1 \\ \vdots \\ u_6 \\ v_1 \\ \vdots \\ v_6 \end{array}\right\} \tag{9.1.6}
$$

or, alternatively, we can express Eq. (9.1.6) as

$$\{a\} = [X]^{-1}\{d\} \tag{9.1.7}$$

where $[X]$ is the 12×12 matrix on the right side of Eq. (9.1.6). It is best to invert the $[X]$ matrix by using a digital computer. Then the a_i's, in terms of nodal displacements, are substituted into Eq. (9.1.4). Note that only the 6×6 part of $[X]$ in Eq. (9.1.6) really must be inverted. Finally, using Eq. (9.1.7) in Eq. (9.1.4), we can obtain the general displacement expressions in terms of the shape functions and the nodal degrees of freedom as

$$\{\psi\} = [N]\{d\} \tag{9.1.8}$$

where

$$[N] = [M^*][X]^{-1} \tag{9.1.9}$$

Step 3   Define the Strain/Displacement and Stress/Strain Relationships

The element strains are again given by

$$
\{\varepsilon\} = \left\{\begin{array}{c} \varepsilon_x \\ \varepsilon_y \\ \gamma_{xy} \end{array}\right\} = \left\{\begin{array}{c} \dfrac{\partial u}{\partial x} \\[2mm] \dfrac{\partial v}{\partial y} \\[2mm] \dfrac{\partial v}{\partial x} + \dfrac{\partial u}{\partial y} \end{array}\right\} \tag{9.1.10}
$$

or, using Eq. (9.1.3),

$$
\{\varepsilon\} = \left[\begin{array}{cccccccccccc} 0 & 1 & 0 & 2x & y & 0 & 0 & 0 & 0 & 0 & 0 & 0 \\ 0 & 0 & 0 & 0 & 0 & 0 & 0 & 0 & 1 & 0 & x & 2y \\ 0 & 0 & 1 & 0 & x & 2y & 0 & 1 & 0 & 2x & y & 0 \end{array}\right] \left\{\begin{array}{c} a_1 \\ a_2 \\ \vdots \\ a_{12} \end{array}\right\} \tag{9.1.11}
$$

We observe that Eq. (9.1.11) yields a linear strain variation in the element. Therefore, the element is called a *linear-strain triangle* (LST). Rewriting Eq. (9.1.11), we have

$$\{\varepsilon\} = [M']\{a\} \qquad (9.1.12)$$

where $[M']$ is the first matrix on the right side of Eq. (9.1.11). Substituting Eq. (9.1.6) for the a_i's into Eq. (9.1.12), we have $\{\varepsilon\}$ in terms of the nodal displacements as

$$\{\varepsilon\} = [B]\{d\} \qquad (9.1.13)$$

where $[B]$ is a function of the variables x and y and the coordinates (x_1, y_1) through (x_6, y_6) given by

$$[B] = [M'][X]^{-1} \qquad (9.1.14)$$

where Eq. (9.1.7) has been used in expressing Eq. (9.1.14).

The stresses are again given by

$$\begin{Bmatrix} \sigma_x \\ \sigma_y \\ \tau_{xy} \end{Bmatrix} = [D] \begin{Bmatrix} \varepsilon_x \\ \varepsilon_y \\ \gamma_{xy} \end{Bmatrix} \qquad (9.1.15)$$

where $[D]$ is given by Eq. (7.1.8) for plane stress or by Eq. (7.1.10) for plane strain.

Step 4   **Derive the Element Stiffness Matrix and Equations**

We determine the stiffness matrix in a manner similar to that used in Section 7.2, by using

$$[k] = \iint\limits_{V}\int [B]^T [D][B]\, dV \qquad (9.1.16)$$

However, the $[B]$ matrix is now a function of x and y as given by Eq. (9.1.14). Therefore, we must perform the integration in Eq. (9.1.16). Finally, the $[B]$ matrix is of the form

$$[B] = \frac{1}{2A} \begin{bmatrix} \beta_1 & 0 & \beta_2 & 0 & \beta_3 & 0 & \beta_4 & 0 & \beta_5 & 0 & \beta_6 & 0 \\ 0 & \gamma_1 & 0 & \gamma_2 & 0 & \gamma_3 & 0 & \gamma_4 & 0 & \gamma_5 & 0 & \gamma_6 \\ \gamma_1 & \beta_1 & \gamma_2 & \beta_2 & \gamma_3 & \beta_3 & \gamma_4 & \beta_4 & \gamma_5 & \beta_5 & \gamma_6 & \beta_6 \end{bmatrix} \qquad (9.1.17)$$

where the β's and γ's are now functions of x and y as well as of the nodal coordinates, as is illustrated in Section 9.2 by Eq. (9.2.8). The stiffness matrix is then seen to be a 12×12 matrix on multiplying the matrices in Eq. (9.1.16). The stiffness matrix, Eq. (9.1.16), is very cumbersome to obtain in explicit form, so it will not be given here. However, if the origin of the coordinates is considered to be at the centroid of the element, the integrations become amenable (see Reference [9]). Alternatively, area coordinates (see References [3], [8], and [9]) can be used to obtain an explicit form of the stiffness matrix.

However, even the use of area coordinates usually involves tedious calculations. Therefore, the integration is best carried out numerically. (Numerical integration is described in Section 11.4.)

The element body forces and surface forces should not be automatically lumped at the nodes, but for a consistent formulation (one that is formulated from the same shape functions used to formulate the stiffness matrix), Eqs. (7.3.1) and (7.3.7), respectively, should be used. (Problems 9.3 and 9.4 illustrate this concept.) These forces can be added to any concentrated nodal forces to obtain the element force matrix. Here the element force matrix is of order 12 × 1 because in general, there could be an x and a y component of force at each of the six nodes associated with the element. The element equations are then given by

$$
\begin{Bmatrix} f_{1x} \\ f_{1y} \\ \vdots \\ f_{6y} \end{Bmatrix} = \begin{bmatrix} k_{11} & \cdots & k_{1,12} \\ k_{21} & & k_{2,12} \\ \vdots & & \vdots \\ k_{12,1} & \cdots & k_{12,12} \end{bmatrix} \begin{Bmatrix} u_1 \\ v_1 \\ \vdots \\ v_6 \end{Bmatrix} \qquad (9.1.18)
$$
$$
(12 \times 1) \qquad\qquad (12 \times 12) \qquad\quad (12 \times 1)
$$

Steps 5, 6, and 7

Steps 5, 6, and 7, involving assembling the global stiffness matrix and equations, determining the unknown global nodal displacements, and calculating the stresses, are identical to those in Section 7.2 for the CST. However, instead of constant stresses in each element, we now have a linear variation of the stresses in each element. Common practice is to use the centroidal element stresses, or the average of the nodal element stresses.

9.2   Example LST Stiffness Determination

To illustrate some of the procedures outlined in Section 9.1 for deriving an LST stiffness matrix, consider the following example. Figure 9-3 shows a

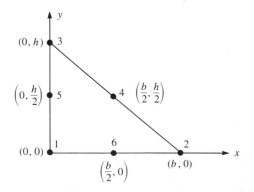

FIGURE   9-3   LST triangle for evaluation of stiffness matrix

specific LST and its coordinates. The triangle is of base dimension b and height h, with midside nodes.

Using the first six equations of Eq. (9.1.5), we calculate the coefficients a_1 through a_6 by evaluating the displacement u at each of the six known coordinates of each node as follows:

$$u_1 = u(0, 0) = a_1$$

$$u_2 = u(b, 0) = a_1 + a_2 b + a_4 b^2$$

$$u_3 = u(0, h) = a_1 + a_3 h + a_6 h^2$$

$$u_4 = u\left(\frac{b}{2}, \frac{h}{2}\right) = a_1 + a_2\frac{b}{2} + a_3\frac{h}{2} + a_4\left(\frac{b}{2}\right)^2 + a_5\frac{bh}{4} + a_6\left(\frac{h}{2}\right)^2 \qquad (9.2.1)$$

$$u_5 = u\left(0, \frac{h}{2}\right) = a_1 + a_3\frac{h}{2} + a_6\left(\frac{h}{2}\right)^2$$

$$u_6 = u\left(\frac{b}{2}, 0\right) = a_1 + a_2\frac{b}{2} + a_4\left(\frac{b}{2}\right)^2$$

Solving Eqs. (9.2.1) simultaneously for the a_i's, we obtain

$$a_1 = u_1 \qquad a_2 = \frac{4u_6 - 3u_1 - u_2}{b} \qquad a_3 = \frac{4u_5 - 3u_1 - u_3}{h}$$

$$a_4 = \frac{2(u_2 - 2u_6 + u_1)}{b^2} \qquad a_5 = \frac{4(u_1 + u_4 - u_5 - u_6)}{bh} \qquad (9.2.2)$$

$$a_6 = \frac{2(u_3 - 2u_5 + u_1)}{h^2}$$

Substituting Eqs. (9.2.2) into the displacement expression for u from Eqs. (9.1.2), we have

$$u = u_1 + \left[\frac{4u_6 - 3u_1 - u_2}{b}\right]x + \left[\frac{4u_5 - 3u_1 - u_3}{h}\right]y$$

$$+ \left[\frac{2(u_2 - 2u_6 + u_1)}{b^2}\right]x^2 + \left[\frac{4(u_1 + u_4 - u_5 - u_6)}{bh}\right]xy$$

$$+ \left[\frac{2(u_3 - 2u_5 + u_1)}{h^2}\right]y^2 \qquad (9.2.3)$$

Similarly, solving for a_7 through a_{12} by evaluating the displacement v at each of the six nodes and then substituting the results into the expression for v from Eqs. (9.1.2), we obtain

$$v = v_1 + \left[\frac{4v_6 - 3v_1 - v_2}{b}\right]x + \left[\frac{4v_5 - 3v_1 - v_3}{h}\right]y$$

$$+ \left[\frac{2(v_2 - 2v_6 + v_1)}{b^2}\right]x^2 + \left[\frac{4(v_1 + v_4 - v_5 - v_6)}{bh}\right]xy$$

$$+ \left[\frac{2(v_3 - 2v_5 + v_1)}{h^2}\right]y^2 \qquad (9.2.4)$$

Using Eqs. (9.2.3) and (9.2.4), we can express the general displacement expressions in terms of the shape functions as

$$
\left\{ \begin{matrix} u \\ v \end{matrix} \right\} = \begin{bmatrix} N_1 & 0 & N_2 & 0 & N_3 & 0 & N_4 & 0 & N_5 & 0 & N_6 & 0 \\ 0 & N_1 & 0 & N_2 & 0 & N_3 & 0 & N_4 & 0 & N_5 & 0 & N_6 \end{bmatrix} \left\{ \begin{matrix} u_1 \\ v_1 \\ \vdots \\ v_6 \end{matrix} \right\}
$$

$$(9.2.5)$$

where the shape functions are given by

$$
N_1 = 1 - \frac{3x}{b} - \frac{3y}{h} + \frac{2x^2}{b^2} + \frac{4xy}{bh} + \frac{2y^2}{h^2} \qquad N_2 = \frac{-x}{b} + \frac{2x^2}{b^2}
$$

$$
N_3 = \frac{-y}{h} + \frac{2y^2}{h^2} \qquad N_4 = \frac{4xy}{bh} \qquad N_5 = \frac{4y}{h} - \frac{4xy}{bh} - \frac{4y^2}{h^2} \qquad (9.2.6)
$$

$$
N_6 = \frac{4x}{b} - \frac{4x^2}{b^2} - \frac{4xy}{bh}
$$

Using Eq. (9.2.5) in Eq. (9.1.10), and performing the differentiations indicated on u and v, we obtain

$$
\varepsilon = \underline{B}\underline{d} \tag{9.2.7}
$$

where $\underline{B}$ is of the form of Eq. (9.1.17), with the resulting β's and γ's in Eq. (9.1.17) given by

$$
\beta_1 = -3h + \frac{4hx}{b} + 4y \qquad \beta_2 = -h + \frac{4hx}{b} \qquad \beta_3 = 0
$$

$$
\beta_4 = 4y \qquad \beta_5 = -4y \qquad \beta_6 = 4h - \frac{8hx}{b} - 4y
$$

$$(9.2.8)$$

$$
\gamma_1 = -3b + 4x + \frac{4by}{h} \qquad \gamma_2 = 0 \qquad \gamma_3 = -b + \frac{4by}{h}
$$

$$
\gamma_4 = 4x \qquad \gamma_5 = 4b - 4x - \frac{8by}{h} \qquad \gamma_6 = -4x
$$

The stiffness matrix for a constant thickness element can now be obtained on substituting Eqs. (9.2.8) into Eq. (9.1.17) to obtain $\underline{B}$, then substituting $\underline{B}$ into Eq. (9.1.16) and using calculus to set up the appropriate integration. The explicit expression for the 12 × 12 stiffness matrix, being extremely cumbersome to obtain, is not given here. Stiffness matrix expressions for higher-order elements are found in References [1] and [2].

9.3   Comparison of Elements

For a given number of nodes, a better representation of true stress and displacement is generally obtained using the LST element than is obtained

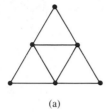
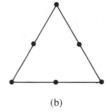

(a)   (b)

FIGURE 9-4   Basic triangular element: (a) four-CST and (b) one-LST

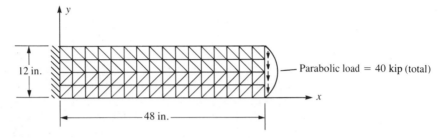

FIGURE 9-5   Cantilever beam used to compare the CST and LST elements with a 4 × 16 mesh

TABLE 9-1   Models used to compare CST and LST results for the cantilever beam of Figure 9-5

Series of Tests Run	Number of Nodes	Number of Degrees of Freedom, n_d	Number of Triangular Elements
A-1 4 × 16 mesh	85	160	128 CST
A-2 8 × 32	297	576	512 CST
B-1 2 × 8	85	160	32 LST
B-2 4 × 16	297	576	128 LST

with the same number of nodes using a much finer subdivision into simple CST elements. For example, using one LST yields better results than using four CST elements with the same number of nodes (see Figure 9-4), and hence, the same number of degrees of freedom (except for the case when constant stress exists).

We now present results to compare the CST of Chapter 7 with the LST of Chapter 9. Consider the cantilever beam subjected to a parabolic load variation acting as shown in Figure 9-5. Let $E = 30 \times 10^6$ psi, $v = 0.25$, and $t = 1.0$ in.

Table 9-1 lists the series of tests run to compare results using the CST and LST elements. Table 9-2 shows comparisons of free-end (tip) deflection and stress σ_x for each element type used to model the cantilever beam. From Table 9-2, we can observe that the larger the number of degrees of freedom for a given type of triangular element, the closer the solution converges to the exact

T A B L E   9-2   Comparison of CST and LST results for the cantilever beam of Figure 9-5

Run	n_d	Bandwidth n_b	Tip Deflection (in.)	σ_x (ksi)	Location x (in.), y (in.)
A-1	160	14	−0.29555	67.236	2.250, 11.250
A-2	576	22	−0.33850	81.302	1.125, 11.630
B-1	160	18	−0.33470	58.885	4.500, 10.500
B-2	576	22	−0.35159	69.956	2.250, 11.250
Exact solution			−0.36133	80.000	0, 12

one (compare run A-1 to run A-2, and B-1 to B-2). For a given number of nodes, the LST analysis yields somewhat better results than the CST analysis (compare run A-1 to run B-1). Although the CST element is rather poor in modeling bending, we observe from Table 9-2 that the element can be used to model a beam in bending if sufficient number of elements are used through the depth of the beam. In general, both the LST and CST analyses yield sufficient results for most plane stress/strain problems, provided a sufficient number of elements is used. In fact, most commercial programs incorporate the use of CST and/or LST elements for plane stress/strain problems although these elements are used primarily as transition elements (usually during mesh generation). The four-sided isoparametric plane stress/strain element is most frequently used in commercial programs and is described in Chapter 11.

Also, recall that finite element displacements will always be less than the exact ones, because finite element models are always predicted to be stiffer than the actual structures when using the displacement formulation of the finite element method. (The reason for the stiffer model was discussed in Sections 3.8 and 8.3. Proof of this assertion can be found in References [4], [5], [6], and [7].)

Finally, Figure 9-6 (from Reference [8]) illustrates a comparison of CST and LST models of a plate subjected to parabolically distributed edge loads. Figure 9-6 shows that the LST model converges to the exact solution for horizontal displacement at point A faster than does the CST model. However, the CST model is quite acceptable even for modest numbers of degrees of freedom. For example, a CST model with 100 nodes (200 degrees of freedom) often yields nearly as accurate a solution as does an LST model with the same number of degrees of freedom.

In conclusion, the results of Table 9-2 and Figure 9-6 indicate that the LST model might be preferred over the CST model for plane stress applications when a relatively small number of nodes is used. However, the use of triangular elements of higher order, such as the LST, is not visibly more advantageous when large numbers of nodes are used, particularly when the cost of formation of the element stiffnesses, equation bandwidth, and overall complexities involved in the computer modeling are considered.

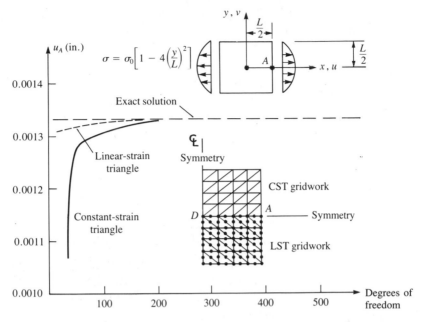

F I G U R E   9-6   Plates subjected to parabolically distributed edge loads;
comparison of results for triangular elements (from
Reference [8]). (Gallagher, R. H. *Finite Element Analysis*:
Fundamentals, © 1975, pp. 269, 270. Reprinted by
permission of Prentice Hall, Inc., Englewood Cliffs, N.J.)

REFERENCES

[1] Pederson, P., "Some Properties of Linear Strain Triangles and Optimal Finite
Element Models," *International Journal for Numerical Methods in Engineering*,
Vol. 7, pp. 415–430, 1973.

[2] Tocher, J. L., and Hartz, B. J., "Higher-Order Finite Element for Plane Stress,"
Journal of the Engineering Mechanics Division, Proceedings of the American
Society of Civil Engineers, Vol. 93, No. EM4, pp. 149–174, Aug. 1967.

[3] Bowes, W. H., and Russell, L. T., *Stress Analysis by the Finite Element Method for
Practicing Engineers*, Lexington Books, Toronto, 1975.

[4] Fraeijes de Veubeke, B., "Upper and Lower Bounds in Matrix Structural Analysis," *Matrix Methods of Structural Analysis*, AGAR-Dograph 72, B. Fraeijes de
Veubeke, Ed., Macmillan, New York, 1964.

[5] McLay, R. W., *Completeness and Convergence Properties of Finite Element Displacement Functions—A General Treatment*, American Institute of Aeronautics
and Astronautics Paper No. 67-143, AIAA 5th Aerospace Meeting, New York,
1967.

[6] Tong, P., and Pian, T. H. H., "The Convergence of Finite Element Method in
Solving Linear Elastic Problems," *International Journal of Solids and Structures*,
Vol. 3, pp. 865–879, 1967.

[7] Cowper, G. R., "Variational Procedures and Convergence of Finite-Element Methods," *Numerical and Computer Methods in Structural Mechanics*, S. J. Fenves, N. Perrone, A. R. Robinson, and W. C. Schnobrich, Eds., Academic Press, New York, 1973.

[8] Gallagher, R., *Finite Element Analysis Fundamentals*, Prentice Hall, Englewood Cliffs, New Jersey, 1975.

[9] Zienkiewicz, O. C., *The Finite Element Method*, 3rd ed., McGraw-Hill, New York, 1977.

PROBLEMS

9.1 Evaluate the shape functions given by Eq. (9.2.6). Sketch the variation of each function over the surface of the triangular element shown in Figure 9-3.

9.2 Express the strains ε_x, ε_y, and γ_{xy} for the element of Figure 9-3 by using the results given in Section 9.2. Evaluate these strains at the centroid of the element, then evaluate the stresses at the centroid in terms of E and v. Assume plane stress conditions apply.

9.3 For the element of Figure 9-3 (shown again as Figure P9-3) subjected to the uniform pressure shown acting over the vertical side, determine the nodal force replacement system using Eq. (7.3.7). Assume an element thickness of t.

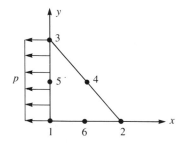

FIGURE P9-3

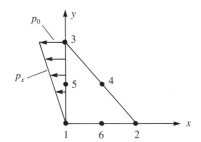

FIGURE P9-4

9.4 For the element of Figure 9-3 (shown as Figure P9-4) subjected to the linearly varying line load shown acting over the vertical side, determine the nodal force replacement system using Eq. (7.3.7). Compare this result to that of Problem 7.9. Are these results expected? Explain.

9.5 For the linear-strain elements shown in Figure P9-5, determine the strains ε_x, ε_y, and γ_{xy}. Evaluate the stresses σ_x, σ_y, and τ_{xy} at the centroids. The coordinates of the nodes are shown in units of inches. Let $E = 30 \times 10^6$ psi, $v = 0.25$, and $t = 0.25$ in. for both elements. Assume plane stress conditions apply. The nodal displacements are given as

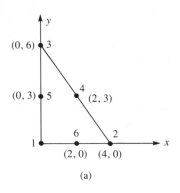
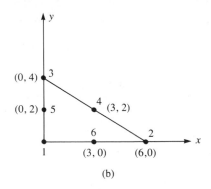

(a)   (b)

FIGURE   P9-5

$$u_1 = 0.0 \qquad\qquad v_1 = 0.0$$

$$u_2 = 0.001 \text{ in.} \qquad v_2 = 0.002 \text{ in.}$$

$$u_3 = 0.0005 \text{ in.} \qquad v_3 = 0.0002 \text{ in.}$$

$$u_4 = 0.0002 \text{ in.} \qquad v_4 = 0.0001 \text{ in.}$$

$$u_5 = 0.0 \qquad\qquad v_5 = 0.0001 \text{ in.}$$

$$u_6 = 0.0005 \text{ in.} \qquad v_6 = 0.001 \text{ in.}$$

(**Hint:** Use the results of Section 9.2.)

9.6   For the linear-strain element shown in Figure P9-6, determine the strains ε_x, ε_y, and γ_{xy}. Evaluate these strains at the centroid of the element, then evaluate the stresses σ_x, σ_y, and τ_{xy} at the centroid. The coordinates of the nodes are shown in units of millimeters. Let $E = 210$ GPa, $v = 0.25$, and $t = 10$ millimeters. Assume plane stress conditions apply. Use the nodal displacements given in Problem 9.5 (converted to millimeters).

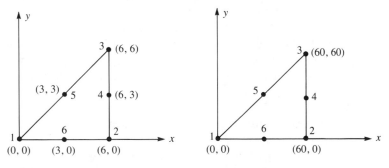

FIGURE   P9-6   FIGURE   P9-7

9.7   Evaluate the shape functions for the linear-strain triangle shown in Figure P9-7. Then evaluate the $\underline{B}$ matrix. Units are millimeters.

10

Axisymmetric Elements

INTRODUCTION

In previous chapters, we have been concerned with line elements (Chapters 2–6) and two-dimensional elements (Chapters 7–9). In this chapter, we consider a special two-dimensional element called the *axisymmetric element*. This element is quite useful when symmetry with respect to geometry and loading exists about an axis of the body being analyzed. Problems such as soil masses subjected to circular footing loads, and thick-walled pressure vessels can often be analyzed using the element developed in this chapter.

We begin with the development of the stiffness matrix for the simplest axisymmetric element, the triangular torus, whose vertical cross section is a plane triangle.

We then present the longhand solution of a thick-walled pressure vessel to illustrate the use of the axisymmetric element equations. This is followed by a description of some typical large-scale problems that have been modeled using the axisymmetric element considered in this chapter.

10.1 Derivation of the Stiffness Matrix

In this section, we will derive the stiffness matrix and the body and surface force matrices for the axisymmetric element. However, before the development, we will first present some fundamental concepts prerequisite to the understanding of the derivation. Axisymmetric elements are triangular tori such that each element is symmetric with respect to geometry and loading about an axis such as the z axis in Figure 10-1. Hence, the z axis is called the *axis of symmetry* or the *axis of revolution*. Each vertical cross section of the element is a plane triangle. The nodal points of an axisymmetric triangular element describe circumferential lines, as indicated in Figure 10-1.

In plane stress problems, stresses exist only in the x-y plane. In axisymmetric problems, the radial displacements develop circumferential strains that induce stresses σ_r, σ_θ, σ_z, and τ_{rz}, where r, θ, and z indicate the radial, circumferential, and longitudinal directions, respectively. Triangular torus

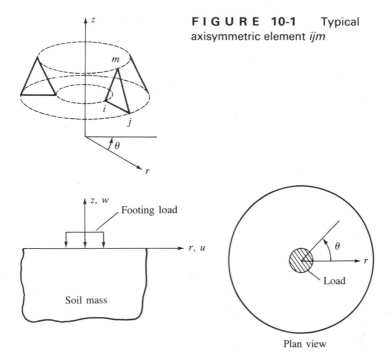

FIGURE 10-1   Typical axisymmetric element *ijm*

FIGURE 10-2   Semi-infinite half-space modeled by axisymmetric elements

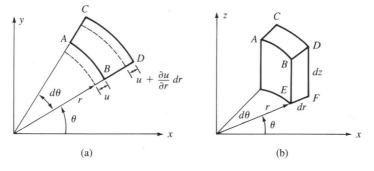

(a)   (b)

FIGURE 10-3   (a) Plane cross-section of (b) axisymmetric element

elements are often used to idealize the axisymmetric system because they can be used to simulate complex surfaces and are simple to work with. For instance, the axisymmetric problem of a semi-infinite half-space loaded by a circular area (circular footing) shown in Figure 10-2 can be solved using the axisymmetric element developed in this chapter.

Because of symmetry about the z axis, the stresses are independent of the θ coordinate. Therefore, all derivatives with respect to θ vanish and the displacement component v (tangent to the θ direction), the shear strains $\gamma_{r\theta}$ and $\gamma_{\theta z}$, and the shear stresses $\tau_{r\theta}$ and $\tau_{\theta z}$ are all zero.

Figure 10-3 shows an axisymmetric ring element and its cross-section to

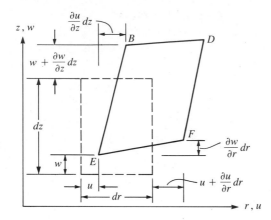

FIGURE 10-4 Displacement and rotations of lines of element in the *r-z* plane

represent the general state of strain for an axisymmetric problem. It is most convenient to express the displacements of an element *ABCD* in the plane of a cross section in cylindrical coordinates. We then let u and w denote the displacements in the radial and longitudinal directions, respectively. The side *AB* of the element is displaced an amount u and side *CD* is then displaced an amount $u + (\partial u/\partial r)\, dr$ in the radial direction. The normal strain in the radial direction is then given by

$$\varepsilon_r = \frac{\partial u}{\partial r} \qquad (10.1.1a)$$

In general, the strain in the tangential direction depends on the tangential displacement v and on the radial displacement u. However, for axisymmetric deformation behavior, recall that the tangential displacement v is equal to zero. Hence, the tangential strain is due only to the radial displacement. Having only radial displacement u, the new length of the arc $\widehat{AB}$ is $(r + u)\, d\theta$, and the tangential strain is then given by

$$\varepsilon_\theta = \frac{(r + u)\, d\theta - r\, d\theta}{r\, d\theta} = \frac{u}{r} \qquad (10.1.1b)$$

Next, we consider the longitudinal element *BDEF* to obtain the longitudinal strain and the shear strain. In Figure 10-4, the element is shown to displace by amounts u and w in the radial and longitudinal directions at point E, and to displace additional amounts $(\partial w/\partial z)\, dz$ along line *BE* and $(\partial u/\partial r)\, dr$ along line *EF*. Furthermore, observing lines *EF* and *BE*, we see that point F moves upward an amount $(\partial w/\partial r)\, dr$ with respect to point E and point B moves to the right an amount $(\partial u/\partial z)\, dz$ with respect to point E. Again, from the basic definitions of normal and shear strain, we have the longitudinal normal strain given by

$$\varepsilon_z = \frac{\partial w}{\partial z} \qquad (10.1.1c)$$

and the shear strain in the *r-z* plane given by

$$\gamma_{rz} = \frac{\partial u}{\partial z} + \frac{\partial w}{\partial r} \qquad (10.1.1d)$$

Summarizing the strain/displacement relationships (10.1.1(a)–(d)) in one equation for easier reference, we have

$$\varepsilon_r = \frac{\partial u}{\partial r} \qquad \varepsilon_\theta = \frac{u}{r} \qquad \varepsilon_z = \frac{\partial w}{\partial z} \qquad \gamma_{rz} = \frac{\partial u}{\partial z} + \frac{\partial w}{\partial r} \qquad (10.1.1e)$$

The isotropic stress/strain relationship, obtained by simplifying the general stress/strain relationships given in Appendix C, is

$$\begin{Bmatrix} \sigma_r \\ \sigma_z \\ \sigma_\theta \\ \tau_{rz} \end{Bmatrix} = \frac{E}{(1 + v)(1 - 2v)} \begin{bmatrix} 1 - v & v & v & 0 \\ v & 1 - v & v & 0 \\ v & v & 1 - v & 0 \\ 0 & 0 & 0 & \frac{1 - 2v}{2} \end{bmatrix} \begin{Bmatrix} \varepsilon_r \\ \varepsilon_z \\ \varepsilon_\theta \\ \gamma_{rz} \end{Bmatrix}$$

$$(10.1.2)$$

The theoretical development follows that of the plane stress/strain problem given in Chapter 7.

Step 1   Select Element Type

An axisymmetric solid is shown discretized in Figure 10-5(a), along with a typical triangular element. The element has three nodes with two degrees of freedom per node. The stresses in the axisymmetric problem are shown in Figure 10-5(b).

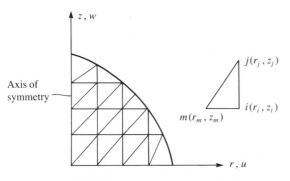

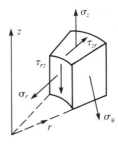

(a) Typical slice through an axisymmetric
    solid discretized into triangular
    elements

(b) Stresses in the axisymmetric problem

F I G U R E   10-5   Discretized axisymmetric solid

Step 2   **Select Displacement Functions**

The element displacement functions are taken to be

$$u(r, z) = a_1 + a_2 r + a_3 z$$
$$w(r, z) = a_4 + a_5 r + a_6 z \tag{10.1.3}$$

so that we have the same linear displacement functions as used in the plane stress, constant-strain triangle. Again, the total number of a_i's (six) introduced in the displacement functions is the same as the total number of degrees of freedom for the element. The nodal displacements are

$$\{d\} = \begin{Bmatrix} \underline{d}_i \\ \underline{d}_j \\ \underline{d}_m \end{Bmatrix} = \begin{Bmatrix} u_i \\ w_i \\ u_j \\ w_j \\ u_m \\ w_m \end{Bmatrix} \tag{10.1.4}$$

and u evaluated at node i is

$$u(r_i, z_i) = u_i = a_1 + a_2 r_i + a_3 z_i \tag{10.1.5}$$

The general displacement function is then expressed in matrix form as

$$\{\psi\} = \begin{Bmatrix} u \\ w \end{Bmatrix} = \begin{Bmatrix} a_1 + a_2 r + a_3 z \\ a_4 + a_5 r + a_6 z \end{Bmatrix} = \begin{bmatrix} 1 & r & z & 0 & 0 & 0 \\ 0 & 0 & 0 & 1 & r & z \end{bmatrix} \begin{Bmatrix} a_1 \\ a_2 \\ a_3 \\ a_4 \\ a_5 \\ a_6 \end{Bmatrix} \tag{10.1.6}$$

Substituting the coordinates of the nodal points into Eq. (10.1.6), we can solve for the a_i's in a manner similar to that of Chapter 7. The resulting expressions are

$$\begin{Bmatrix} a_1 \\ a_2 \\ a_3 \end{Bmatrix} = \begin{bmatrix} 1 & r_i & z_i \\ 1 & r_j & z_j \\ 1 & r_m & z_m \end{bmatrix}^{-1} \begin{Bmatrix} u_i \\ u_j \\ u_m \end{Bmatrix} \tag{10.1.7}$$

and

$$\begin{Bmatrix} a_4 \\ a_5 \\ a_6 \end{Bmatrix} = \begin{bmatrix} 1 & r_i & z_i \\ 1 & r_j & z_j \\ 1 & r_m & z_m \end{bmatrix}^{-1} \begin{Bmatrix} w_i \\ w_j \\ w_m \end{Bmatrix} \tag{10.1.8}$$

Performing the inversion operations in Eqs. (10.1.7) and (10.1.8), we have

$$\begin{Bmatrix} a_1 \\ a_2 \\ a_3 \end{Bmatrix} = \frac{1}{2A} \begin{bmatrix} \alpha_i & \alpha_j & \alpha_m \\ \beta_i & \beta_j & \beta_m \\ \gamma_i & \gamma_j & \gamma_m \end{bmatrix} \begin{Bmatrix} u_i \\ u_j \\ u_m \end{Bmatrix} \qquad (10.1.9)$$

and

$$\begin{Bmatrix} a_4 \\ a_5 \\ a_6 \end{Bmatrix} = \frac{1}{2A} \begin{bmatrix} \alpha_i & \alpha_j & \alpha_m \\ \beta_i & \beta_j & \beta_m \\ \gamma_i & \gamma_j & \gamma_m \end{bmatrix} \begin{Bmatrix} w_i \\ w_j \\ w_m \end{Bmatrix} \qquad (10.1.10)$$

where

$$\begin{array}{lll} \alpha_i = r_j z_m - z_j r_m & \alpha_j = r_m z_i - z_m r_i & \alpha_m = r_i z_j - z_i r_j \\ \beta_i = z_j - z_m & \beta_j = z_m - z_i & \beta_m = z_i - z_j \\ \gamma_i = r_m - r_j & \gamma_j = r_i - r_m & \gamma_m = r_j - r_i \end{array} \qquad (10.1.11)$$

We define the shape functions, similar to Eqs. (7.2.18), as

$$N_i = \frac{1}{2A}(\alpha_i + \beta_i r + \gamma_i z)$$

$$N_j = \frac{1}{2A}(\alpha_j + \beta_j r + \gamma_j z) \qquad (10.1.12)$$

$$N_m = \frac{1}{2A}(\alpha_m + \beta_m r + \gamma_m z)$$

Substituting Eqs. (10.1.7) and (10.1.8) into Eq. (10.1.6), along with the shape function Eqs. (10.1.12), the general displacement function is

$$\{\psi\} = \begin{Bmatrix} u(r, z) \\ w(r, z) \end{Bmatrix} = \begin{bmatrix} N_i & 0 & N_j & 0 & N_m & 0 \\ 0 & N_i & 0 & N_j & 0 & N_m \end{bmatrix} \begin{Bmatrix} u_i \\ w_i \\ u_j \\ w_j \\ u_m \\ w_m \end{Bmatrix} \qquad (10.1.13)$$

or

$$\{\psi\} = [N]\{d\} \qquad (10.1.14)$$

Step 3 Define the Strain/Displacement and Stress/Strain Relationships

Using Eqs. (10.1.1) and (10.1.3), the strains become

$$\{\varepsilon\} = \begin{Bmatrix} a_2 \\ a_6 \\ \dfrac{a_1}{r} + a_2 + \dfrac{a_3 z}{r} \\ a_3 + a_5 \end{Bmatrix} \qquad (10.1.15)$$

Rewriting Eq. (10.1.15) with the a_i's as a separate column matrix, we have

$$
\begin{Bmatrix} \varepsilon_r \\ \varepsilon_z \\ \varepsilon_\theta \\ \gamma_{rz} \end{Bmatrix} =
\begin{bmatrix}
0 & 1 & 0 & 0 & 0 & 0 \\
0 & 0 & 0 & 0 & 0 & 1 \\
\dfrac{1}{r} & 1 & \dfrac{z}{r} & 0 & 0 & 0 \\
0 & 0 & 1 & 0 & 1 & 0
\end{bmatrix}
\begin{Bmatrix} a_1 \\ a_2 \\ a_3 \\ a_4 \\ a_5 \\ a_6 \end{Bmatrix}
\tag{10.1.16}
$$

Substituting Eqs. (10.1.7) and (10.1.8) into Eq. (10.1.16) and making use of Eq. (10.1.11), we obtain

$$
\{\varepsilon\} = \frac{1}{2A}
\begin{bmatrix}
\beta_i & 0 & \beta_j & 0 & \beta_m & 0 \\
0 & \gamma_i & 0 & \gamma_j & 0 & \gamma_m \\
\dfrac{\alpha_i}{r}+\beta_i+\dfrac{\gamma_i z}{r} & 0 & \dfrac{\alpha_j}{r}+\beta_j+\dfrac{\gamma_j z}{r} & 0 & \dfrac{\alpha_m}{r}+\beta_m+\dfrac{\gamma_m z}{r} & 0 \\
\gamma_i & \beta_i & \gamma_j & \beta_j & \gamma_m & \beta_m
\end{bmatrix}
\begin{Bmatrix} u_i \\ w_i \\ u_j \\ w_j \\ u_m \\ w_m \end{Bmatrix}
\tag{10.1.17}
$$

or, rewriting Eq. (10.1.17) in simplified matrix form,

$$
\{\varepsilon\} = [\underline{B}_i \quad \underline{B}_j \quad \underline{B}_m]
\begin{Bmatrix} u_i \\ w_i \\ u_j \\ w_j \\ u_m \\ w_m \end{Bmatrix}
\tag{10.1.18}
$$

where

$$
[B_i] = \frac{1}{2A}
\begin{bmatrix}
\beta_i & 0 \\
0 & \gamma_i \\
\dfrac{\alpha_i}{r}+\beta_i+\dfrac{\gamma_i z}{r} & 0 \\
\gamma_i & \beta_i
\end{bmatrix}
\tag{10.1.19}
$$

Similarly, we obtain submatrices $\underline{B}_j$ and $\underline{B}_m$ by replacing the subscript i with j and then with m in Eq. (10.1.19). Rewriting Eq. (10.1.18) in compact matrix form, we have

$$
\{\varepsilon\} = [B]\{d\}
\tag{10.1.20}
$$

where

$$
[B] = [\underline{B}_i \quad \underline{B}_j \quad \underline{B}_m]
\tag{10.1.21}
$$

Note that $[B]$ is a function of the r and z coordinates. Therefore, in general, the strain ε_θ will not be constant.

The stresses are given by

$$
\{\sigma\} = [D][B]\{d\}
\tag{10.1.22}
$$

where $[D]$ is given by the first matrix on the right side of Eq. (10.1.2). (As mentioned in Chapter 7, for $v = 0.5$, a special formula must be used [9].)

Step 4   **Derive the Element Stiffness Matrix and Equations**

The stiffness matrix is

$$[k] = \int\int\int_V [B]^T[D][B] \, dV \qquad (10.1.23)$$

or

$$[k] = 2\pi \int\int_A [B]^T[D][B]r \, dr \, dz \qquad (10.1.24)$$

after integrating along the circumferential boundary. The $[B]$ matrix, Eq. (10.1.21), is a function of r and z. Therefore, $[k]$ is a function of r and z and is of order 6×6.

We can evaluate Eq. (10.1.24) for $[k]$ by one of three methods:

1. Numerical integration (Gaussian quadrature) as discussed in Chapter 11.

2. Explicit multiplication and term-by-term integration (see Reference [1]).

3. Evaluate $[B]$ for a centroidal point $(\bar{r}, \bar{z})$ of the element

$$r = \bar{r} = \frac{r_i + r_j + r_m}{3} \qquad z = \bar{z} = \frac{z_i + z_j + z_m}{3} \qquad (10.1.25)$$

and define $[B(\bar{r}, \bar{z})] = [\bar{B}]$. Therefore, as a first approximation,

$$[k] = 2\pi\bar{r}A[\bar{B}]^T[D][\bar{B}] \qquad (10.1.26)$$

If the triangular subdivisions are consistent with the final stress distribution (that is, small elements in regions of high stress gradients), then acceptable results can be obtained by Method 3.

Distributed Body Forces

Loads such as gravity (in the direction of the z axis) or centrifugal forces in rotating machine parts (in the direction of the r axis) are considered to be body forces (as shown in Figure 10-6). The body forces can be found by

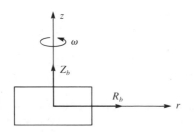

F I G U R E   10-6   Axisymmetric element with body forces per unit volume

$$\{f_b\} = 2\pi \iint_A [N]^T \begin{Bmatrix} R_b \\ Z_b \end{Bmatrix} r \, dr \, dz \qquad (10.1.27)$$

where $R_b = \omega^2 \rho r$ for a machine part moving with a constant angular velocity ω about the z axis, with material mass density ρ, and radial coordinate r, and Z_b is the body force per unit volume due to the force of gravity.

Considering the body force at node i, we have

$$\{f_{bi}\} = 2\pi \iint_A [N_i]^T \begin{Bmatrix} R_b \\ Z_b \end{Bmatrix} r \, dr \, dz \qquad (10.1.28)$$

where

$$[N_i]^T = \begin{bmatrix} N_i & 0 \\ 0 & N_i \end{bmatrix} \qquad (10.1.29)$$

Multiplying and integrating in Eq. (10.1.28), we obtain

$$\{f_{bi}\} = \frac{2\pi}{3} \begin{Bmatrix} \bar{R}_b \\ Z_b \end{Bmatrix} A\bar{r} \qquad (10.1.30)$$

where the origin of the coordinates has been taken as the centroid of the element, and $\bar{R}_b$ is the radially directed body force per unit volume evaluated at the centroid of the element. The body forces at nodes j and m are identical to those given by Eq. (10.1.30) for node i. Hence, for an element, we have

$$\{f_b\} = \frac{2\pi\bar{r}A}{3} \begin{Bmatrix} \bar{R}_b \\ Z_b \\ \bar{R}_b \\ Z_b \\ \bar{R}_b \\ Z_b \end{Bmatrix} \qquad (10.1.31)$$

where

$$\bar{R}_b = \omega^2 \rho \bar{r} \qquad (10.1.32)$$

Equation (10.1.31) is a first approximation to the radially directed body force distribution.

Surface Forces

Surface forces can be found by

$$\{f_s\} = \iint_S [N]^T \{T\} \, dS \qquad (10.1.33)$$

For radial and axial pressures p_r and p_z, respectively, we have

$$\{f_s\} = \iint_S [N]^T \begin{Bmatrix} p_r \\ p_z \end{Bmatrix} dS \qquad (10.1.34)$$

For example, along the vertical face jm of an element, let uniform loads p_r and p_z be applied, as shown in Figure 10-7. We can use Eq. (10.1.34) written for

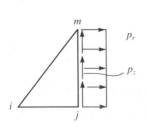

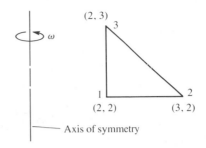

FIGURE 10-7 Axisymmetric element with surface forces

FIGURE 10-8 Axisymmetric element subjected to angular velocity

each node separately. For instance, for node j, substituting N_j from Eqs. (10.1.12) into Eq. (10.1.34), we have

$$\{f_{sj}\} = \int_{z_i}^{z_m} \frac{1}{2A} \begin{bmatrix} \alpha_j + \beta_j r + \gamma_j z & 0 \\ 0 & \alpha_j + \beta_j r + \gamma_j z \end{bmatrix} \begin{Bmatrix} p_r \\ p_z \end{Bmatrix} 2\pi r_j \, dz \qquad (10.1.35)$$

Performing the integration of Eq. (10.1.35) explicitly, along with similar evaluations for f_{si} and f_{sm}, we obtain the total distribution of surface force to nodes i, j, and m as

$$\{f_s\} = \frac{2\pi r_j(z_m - z_j)}{2} \begin{Bmatrix} 0 \\ 0 \\ p_r \\ p_z \\ p_r \\ p_z \end{Bmatrix} \qquad (10.1.36)$$

Steps 5, 6, and 7

Steps 5, 6, and 7, which involve assembling the total stiffness matrix, total force matrix, and total set of equations; solving for the nodal degrees of freedom; and calculating the element stresses, are analogous to those of Chapter 7 for the CST element, except the stresses are not constant in each element. They are usually determined by one of two methods that we use to determine the LST element stresses. Either we determine the centroidal element stresses, or we determine the nodal stresses for the element and then average them. The latter method has been shown to be more accurate in some cases (see Reference [2]).

EXAMPLE 10.1

For the element of an axisymmetric body rotating with a constant angular velocity $\omega = 100$ revolutions per minute as shown in Figure 10-8, evaluate the approximate body force matrix. Include the weight of the material, where the weight density ρ_w is 0.283 lb/in.[3] The coordinates of the element (in inches) are shown in the figure.

We need to evaluate Eq. (10.1.31) to obtain the approximate body force matrix. Therefore,

$Z_b = 0.283$ lb/in.[3]

$$\bar{R}_b = \omega^2 \rho \bar{r} = \left[\left(100 \frac{\text{rev}}{\text{min}} \right) \left(2\pi \frac{\text{rad}}{\text{rev}} \right) \left(\frac{1 \text{ min}}{60 \text{ s}} \right) \right]^2 \frac{(0.283 \text{ lb/in.}^3)}{(32.2 \times 12) \text{ in./s}^2} (2.333 \text{ in.})$$

$\bar{R}_b = 0.187$ lb/in.[3]

$$\frac{2\pi \bar{r} A}{3} = \frac{2\pi (2.333)(0.5)}{3} = 2.44 \text{ in.}^3$$

$f_{b1r} = (2.44)(0.187) = 0.457$ lb

$f_{b1z} = -(2.44)(0.283) = -0.691$ lb   (downward)

Since we are using the first approximation Eq. (10.1.31), all r directed nodal body forces are equal, and all z directed body forces are equal. Therefore,

$$f_{b2r} = 0.457 \text{ lb} \qquad f_{b2z} = -0.691 \text{ lb}$$

$$f_{b3r} = 0.457 \text{ lb} \qquad f_{b3z} = -0.691 \text{ lb} \qquad \blacksquare$$

10.2   Solution of an Axisymmetric Pressure Vessel

To illustrate the use of the equations developed in Section 10.1, we will now solve an axisymmetric stress problem.

E X A M P L E   10.2

For the long, thick-walled cylinder under internal pressure p equal to 1 psi shown in Figure 10-9, determine the displacements and stresses.

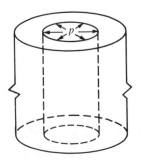

F I G U R E   10-9   Thick-walled cylinder subjected to internal pressure

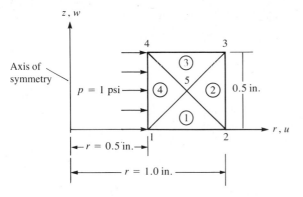

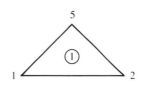

FIGURE 10-11   Element
1 of the discretized cylinder

FIGURE 10-10   Discretized cylinder slice

Discretization

To illustrate the finite element solution for the cylinder, we first discretize the cylinder into four triangular elements, as shown in Figure 10-10. A horizontal slice of the cylinder represents the total cylinder behavior. Since we are performing a longhand solution, a coarse mesh of elements is used for simplicity's sake (but without loss of generality of the method). The governing global matrix equation is

$$\begin{Bmatrix} F_{1r} \\ F_{1z} \\ F_{2r} \\ F_{2z} \\ F_{3r} \\ F_{3z} \\ F_{4r} \\ F_{4z} \\ F_{5r} \\ F_{5z} \end{Bmatrix} = [K] \begin{Bmatrix} u_1 \\ w_1 \\ u_2 \\ w_2 \\ u_3 \\ w_3 \\ u_4 \\ w_4 \\ u_5 \\ w_5 \end{Bmatrix} \qquad (10.2.1)$$

where the $[K]$ matrix is of order 10×10.

Assemblage of the Stiffness Matrix

We assemble the $[K]$ matrix in the usual manner by superposition of the individual element stiffness matrices. For simplicity's sake, we will use the first approximation method given by Eq. (10.1.26) to evaluate the element matrices. Therefore,

$$[k] = 2\pi \bar{r} A [\bar{B}]^T [D][\bar{B}] \qquad (10.2.2)$$

For element 1 (Figure 10-11), the coordinates are $r_i = 0.5$, $z_i = 0$, $r_j = 1.0$, $z_j = 0$, $r_m = 0.75$, and $z_m = 0.25$ ($i = 1$, $j = 2$, and $m = 5$ for element 1) for the global-coordinate axes as set up in Figure 10-10.

We now evaluate $[\bar{B}]$, where $[\bar{B}]$ is given by Eq. (10.1.19) evaluated at the centroid of the element $r = \bar{r}$, $z = \bar{z}$, and expanded here as

$$
[\bar{B}] = \frac{1}{2A}
\begin{bmatrix}
\beta_i & 0 & \beta_j & 0 & \beta_m & 0 \\
0 & \gamma_i & 0 & \gamma_j & 0 & \gamma_m \\
\dfrac{\alpha_i}{\bar{r}} + \beta_i + \dfrac{\gamma_i \bar{z}}{\bar{r}} & 0 & \dfrac{\alpha_j}{\bar{r}} + \beta_j + \dfrac{\gamma_j \bar{z}}{\bar{r}} & 0 & \dfrac{\alpha_m}{\bar{r}} + \beta_m + \dfrac{\gamma_m \bar{z}}{\bar{r}} & 0 \\
\gamma_i & \beta_i & \gamma_j & \beta_j & \gamma_m & \beta_m
\end{bmatrix}
\tag{10.2.3}
$$

where, using element coordinates in Eqs. (10.1.11), we have

$$\alpha_i = r_j z_m - z_j r_m = (1.0)(0.25) - (0.0)(0.75) = 0.25$$

$$\alpha_j = r_m z_i - z_m r_i = (0.75)(0) - (0.25)(0.5) = -0.125$$

$$\alpha_m = r_i z_j - z_i r_j = (0.5)(0.0) - (0)(1.0) = 0.0$$

$$\beta_i = z_j - z_m = 0.0 - 0.25 = -0.25$$

$$\beta_j = z_m - z_i = 0.25 - 0 = 0.25 \tag{10.2.4}$$

$$\beta_m = z_i - z_j = 0.0 - 0.0 = 0.0$$

$$\gamma_i = r_m - r_j = 0.75 - 1.0 = -0.25$$

$$\gamma_j = r_i - r_m = 0.5 - 0.75 = -0.25$$

$$\gamma_m = r_j - r_i = 1.0 - 0.5 = 0.5$$

and

$$\bar{r} = 0.5 + \tfrac{1}{2}(0.5) = 0.75 \text{ in.} \qquad \bar{z} = \tfrac{1}{3}(0.25) = 0.0833 \text{ in.}$$

$$A = \tfrac{1}{2}(0.5)(0.25) = 0.0625 \text{ in.}^2$$

Substituting the results from Eqs. (10.2.4) into Eq. (10.2.3), we obtain

$$
[\bar{B}] = \frac{1}{0.125}
\begin{bmatrix}
-0.25 & 0 & 0.25 & 0 & 0 & 0 \\
0 & -0.25 & 0 & -0.25 & 0 & 0.5 \\
0.0556 & 0 & 0.0556 & 0 & 0.0556 & 0 \\
-0.25 & -0.25 & -0.25 & 0.25 & 0.5 & 0
\end{bmatrix}
\tag{10.2.5}
$$

For the axisymmetric stress case, the matrix $[D]$ is given in Eq. (10.1.2) as

$$
[D] = \frac{E}{(1+v)(1-2v)}
\begin{bmatrix}
1-v & v & v & 0 \\
v & 1-v & v & 0 \\
v & v & 1-v & 0 \\
0 & 0 & 0 & \dfrac{1-2v}{2}
\end{bmatrix}
\tag{10.2.6}
$$

With $v = 0.3$ and $E = 30 \times 10^6$ psi, we obtain

$$[D] = \frac{30(10^6)}{(1 + 0.3)[1 - 2(0.3)]} \begin{bmatrix} 1 - 0.3 & 0.3 & 0.3 & 0 \\ 0.3 & 1 - 0.3 & 0.3 & 0 \\ 0.3 & 0.3 & 1 - 0.3 & 0 \\ 0 & 0 & 0 & \dfrac{1 - 2(0.3)}{2} \end{bmatrix}$$

(10.2.7)

or, simplifying Eq. (10.2.7),

$$[D] = 57.7(10^6) \begin{bmatrix} 0.7 & 0.3 & 0.3 & 0 \\ 0.3 & 0.7 & 0.3 & 0 \\ 0.3 & 0.3 & 0.7 & 0 \\ 0 & 0 & 0 & 0.2 \end{bmatrix}$$

(10.2.8)

Using Eqs. (10.2.5) and (10.2.8), we obtain

$$[\bar{B}]^T[D] = \frac{57.7(10^6)}{0.125} \begin{bmatrix} -0.158 & -0.0583 & -0.0361 & -0.05 \\ -0.075 & -0.175 & -0.075 & -0.05 \\ 0.192 & 0.0917 & 0.114 & -0.05 \\ -0.075 & -0.175 & -0.075 & 0.05 \\ 0.0167 & 0.0166 & 0.0388 & 0.1 \\ 0.15 & 0.35 & 0.15 & 0 \end{bmatrix}$$

(10.2.9)

Substituting Eqs. (10.2.5) and (10.2.9) into Eq. (10.2.2), we obtain the stiffness matrix for element 1 as

$$[k^{(1)}] = (10^6) \begin{matrix} i = 1 \qquad\qquad j = 2 \qquad\qquad m = 5 \\ \begin{bmatrix} 54.46 & 29.45 & -31.63 & 2.26 & -29.37 & -31.71 \\ 29.45 & 61.17 & -11.33 & 33.98 & -31.72 & -95.15 \\ -31.63 & -11.33 & 72.59 & -38.52 & -20.31 & 49.84 \\ 2.26 & 33.98 & -38.52 & 61.17 & 22.66 & -95.15 \\ -29.37 & -31.72 & -20.31 & 22.66 & 56.72 & 9.06 \\ -31.71 & -95.15 & 49.84 & -95.15 & 9.06 & 190.31 \end{bmatrix} \end{matrix}$$

(10.2.10)

where the numbers above the columns indicate the nodal orders of degrees of freedom in the element 1 stiffness matrix.

For element 2 (Figure 10-12), the coordinates are $r_i = 1.0$, $z_i = 0.0$, $r_j = 1.0$, $z_j = 0.5$, $r_m = 0.75$, and $z_m = 0.25$ ($i = 2$, $j = 3$, and $m = 5$ for element 2). Therefore,

$$\alpha_i = (1.0)(0.25) - (0.5)(0.75) = -0.125$$

$$\alpha_j = (0.75)(0.0) - (0.25)(1.0) = -0.25$$

$$\alpha_m = (1.0)(0.5) - (0.0)(1.0) = 0.5$$

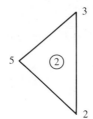

FIGURE 10-12 Element 2 of the discretized cylinder

$$\beta_i = 0.5 - 0.25 = 0.25 \qquad \beta_j = 0.25 - 0.0 = 0.25$$

$$\beta_m = 0.0 - 0.5 = -0.5 \qquad \gamma_i = 0.75 - 1.0 = -0.25$$

$$\gamma_j = 1.0 - 0.75 = 0.25 \qquad \gamma_m = 1.0 - 1.0 = 0.0 \qquad (10.2.11)$$

and

$$\bar{r} = 0.9167 \text{ in.} \qquad \bar{z} = 0.25 \text{ in.} \qquad A = 0.0625 \text{ in.}^2$$

Using Eqs. (10.2.11) in Eq. (10.2.2) and proceeding as for element 1, we obtain the stiffness matrix for element 2 as

$$
[k^{(2)}] = (10^6)
\begin{array}{c}
\begin{array}{ccc}
\qquad\quad i = 2 & \qquad\qquad j = 3 & \qquad\qquad m = 5
\end{array} \\
\begin{bmatrix}
85.75 & -46.07 & 52.52 & 12.84 & -118.92 & 33.23 \\
-46.07 & 74.77 & -12.84 & -41.54 & 45.32 & -33.23 \\
52.52 & -12.84 & 85.74 & 46.07 & -118.92 & -33.23 \\
12.84 & -41.54 & 46.07 & 74.77 & -45.32 & -33.23 \\
-118.92 & 45.32 & -118.92 & -45.32 & 216.41 & 0 \\
33.23 & -33.23 & -33.23 & -33.23 & 0 & 66.46
\end{bmatrix}
\end{array}
$$

$$(10.2.12)$$

We obtain the stiffness matrices for elements 3 and 4 in a manner similar to that used to obtain the stiffness matrices for elements 1 and 2. Thus,

$$
[k^{(3)}] = (10^6)
\begin{array}{c}
\begin{array}{ccc}
\qquad\quad i = 3 & \qquad\qquad j = 4 & \qquad\qquad m = 5
\end{array} \\
\begin{bmatrix}
72.58 & 38.52 & -31.63 & 11.33 & -20.31 & -49.84 \\
38.52 & 61.17 & -2.26 & 33.98 & -22.66 & -95.15 \\
-31.63 & -2.26 & 54.46 & -29.45 & -29.37 & 31.72 \\
11.33 & 33.98 & -29.45 & 61.17 & 31.72 & -95.15 \\
-20.31 & -22.66 & -29.37 & 31.72 & 56.72 & -9.06 \\
-49.84 & -95.15 & 31.72 & -95.15 & -9.06 & 190.31
\end{bmatrix}
\end{array}
$$

$$(10.2.13)$$

and

$$
[k^{(4)}] = (10^6)
\begin{matrix}
i = 4 \qquad\qquad\qquad j = 1 \qquad\qquad m = 5 \\
\begin{bmatrix}
41.53 & -21.90 & 20.39 & 0.75 & -66.45 & 21.14 \\
-21.90 & 47.57 & -0.75 & -26.43 & 36.24 & -21.14 \\
20.39 & -0.75 & 41.53 & 21.90 & -66.45 & -21.14 \\
0.75 & -26.43 & 21.90 & 47.57 & -36.24 & -21.14 \\
-66.45 & 36.24 & -66.45 & -36.24 & 169.14 & 0 \\
21.14 & -21.14 & -21.14 & -21.14 & 0 & 42.28
\end{bmatrix}
\end{matrix}
$$

$$(10.2.14)$$

Using superposition of the element stiffness matrices, Eqs. (10.2.10), (10.2.12), (10.2.13), and (10.2.14), where we rearrange the elements of each stiffness matrix in order of increasing nodal degrees of freedom, we obtain the global stiffness matrix as

$$
[K] = (10^6)
\begin{bmatrix}
95.99 & 51.35 & -31.63 & 2.26 & 0 \\
51.35 & 108.74 & -11.33 & 33.98 & 0 \\
-31.63 & -11.33 & 158.34 & -84.59 & 52.52 \\
2.26 & 33.98 & -84.59 & 135.94 & -12.84 \\
0 & 0 & 52.52 & -12.84 & 158.33 \\
0 & 0 & 12.84 & -41.54 & 84.59 \\
20.39 & 0.75 & 0 & 0 & -31.63 \\
-0.75 & -26.43 & 0 & 0 & 11.33 \\
-95.82 & -67.96 & -139.2 & 67.98 & -139.2 \\
-52.86 & -116.3 & 83.07 & -128.4 & -83.07
\end{bmatrix}
$$

$$
\begin{bmatrix}
0 & 20.39 & -0.75 & -95.82 & -52.86 \\
0 & 0.75 & 26.43 & -67.96 & -116.3 \\
12.84 & 0 & 0 & -139.2 & 83.07 \\
-41.54 & 0 & 0 & 67.98 & -128.4 \\
84.59 & -31.63 & 11.33 & -139.2 & -83.07 \\
135.94 & -2.26 & 33.98 & -67.98 & -128.4 \\
-2.26 & 95.99 & -51.35 & -95.82 & 52.86 \\
33.98 & -51.35 & 108.74 & 67.96 & -116.3 \\
-67.98 & -95.82 & 67.96 & 498.99 & 0 \\
-128.4 & 52.86 & -116.3 & 0 & 489.36
\end{bmatrix}
$$

$$(10.2.15)$$

The applied nodal forces are given by Eq. (10.1.36) as

$$
F_{1r} = F_{4r} = \frac{2\pi(0.5)(0.5)}{2}(1) = 0.785 \text{ lb} \qquad (10.2.16)
$$

All other nodal forces are zero. Using Eq. (10.2.15) for $[K]$ and Eq. (10.2.16) for the nodal forces in Eq. (10.2.1), and solving for the nodal displacements, we obtain

$$u_1 = 0.0322 \times 10^{-6} \text{ in.} \qquad w_1 = 0.00115 \times 10^{-6} \text{ in.}$$

$$u_2 = 0.0219 \times 10^{-6} \text{ in.} \qquad w_2 = 0.00206 \times 10^{-6} \text{ in.}$$

$$u_3 = 0.0219 \times 10^{-6} \text{ in.} \qquad w_3 = -0.00211 \times 10^{-6} \text{ in.}$$

$$u_4 = 0.0322 \times 10^{-6} \text{ in.} \qquad w_4 = -0.00115 \times 10^{-6} \text{ in.}$$

$$u_5 = 0.0244 \times 10^{-6} \text{ in.} \qquad w_5 = 0 \qquad\qquad (10.2.17)$$

We now determine the stresses in each element by using Eq. (10.1.22) as

$$\{\sigma\} = [D][\bar{B}]\{d\} \qquad\qquad (10.2.18)$$

For element 1, we use Eq. (10.2.5) for $[\bar{B}]$, Eq. (10.2.8) for $[D]$, and Eq. (10.2.17) for $\{d\}$ in Eq. (10.2.18) to obtain

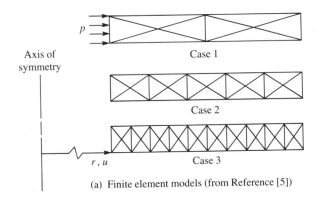

(a) Finite element models (from Reference [5])

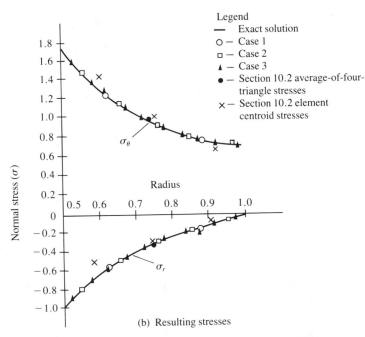

(b) Resulting stresses

FIGURE 10-13   Finite element analysis of a thick-walled cylinder under internal pressure

$$\sigma_r = -0.338 \text{ psi} \qquad \sigma_z = -0.0126 \text{ psi}$$

$$\sigma_\theta = 0.942 \text{ psi} \qquad \tau_{rz} = -0.1037 \text{ psi}$$

Similarly, for element 2, we obtain

$$\sigma_r = -0.105 \text{ psi} \qquad \sigma_z = -0.0747 \text{ psi}$$

$$\sigma_\theta = 0.690 \text{ psi} \qquad \tau_{rz} = 0.000 \text{ psi}$$

For element 3, the stresses are

$$\sigma_r = -0.337 \text{ psi} \qquad \sigma_z = -0.0125 \text{ psi}$$

$$\sigma_\theta = 0.942 \text{ psi} \qquad \tau_{rz} = 0.1037 \text{ psi}$$

For element 4, the stresses are

$$\sigma_r = -0.470 \text{ psi} \qquad \sigma_z = 0.1493 \text{ psi}$$

$$\sigma_\theta = 1.426 \text{ psi} \qquad \tau_{rz} = 0.000 \text{ psi}$$

Figure 10-13 shows the exact solution, along with the results determined here and the results from Reference [5]. Observe that agreement with the exact solution is quite good except for the limited results due to the very coarse mesh used in the longhand example, and in case 1 of Reference [5]. In Reference [5], stresses have been plotted at the center of the quadrilaterals and were obtained by averaging the stresses in the four connecting triangles.

■

10.3 Applications of Axisymmetric Elements

Numerous structural (and nonstructural) systems can be classified as axisymmetric. Some typical structural systems whose behavior is modeled accurately using the axisymmetric element developed in this chapter are represented in Figures 10-14 and 10-15.

Figure 10-14 illustrates the finite element model of a steel-reinforced concrete pressure vessel. The vessel is a thick-walled cylinder with flat heads. An axis of symmetry (the *z* axis) exists such that only one half of the *r-z* plane passing through the middle of the structure need be modeled. The concrete was modeled by using the axisymmetric triangular element developed in this chapter. The steel elements were laid out along the boundaries of the concrete elements so as to maintain continuity (or perfect bond assumption) between the concrete and the steel. The vessel was then subjected to an internal pressure as shown in the figure. Note that the nodes along the axis of symmetry should be supported by rollers preventing motion perpendicular to the axis of symmetry.

Figure 10-15 shows a finite element model of a high-strength steel die used in a thin-plastic-film-making process [7]. The die is an irregularly shaped disk. An axis of symmetry with respect to geometry and loading exists as shown. The die was modeled by using simple quadrilateral axisymmetric

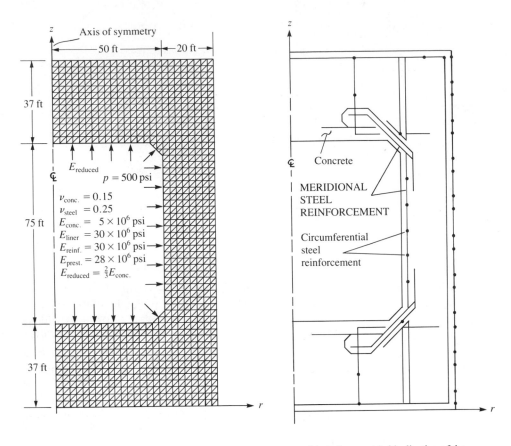

(a) Two-dimensional view of a
finite element idealization
for a prestressed concrete reactor
vessel (PCRV)

(b) Axisymmetric idealization of the
steel reinforcement

F I G U R E **10-14** Model of steel-reinforced concrete pressure vessel (from
Reference [4]) (North Holland Physics Publishing, Amsterdam)

elements. The locations of high stress were of primary concern. Figure 10-16
shows a plot of the Von Mises stress contours for the die of Figure 10-15. The
Von Mises stress (see Reference [8]) is often used as a failure criterion in
design. The Von Mises stress σ_{VM} may be related to the principal stresses by
the expression

$$\sigma_{VM} = \frac{1}{\sqrt{2}}\sqrt{(\sigma_1 - \sigma_2)^2 + (\sigma_2 - \sigma_3)^2 + (\sigma_3 - \sigma_1)^2}$$

where the principal stresses are given by σ_1, σ_2, and σ_3. These results were
obtained from the commercial computer code ANSYS.

Other dies with modifications in geometry were also studied to evaluate
the most suitable die before the construction of an expensive prototype.
Confidence in the acceptability of the prototype was enhanced by doing these

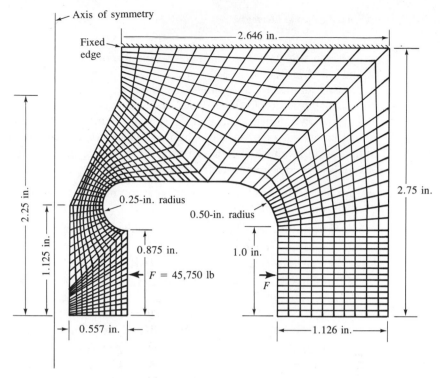

FIGURE 10-15   Model of a high-strength steel die (924 nodes and 830 elements)

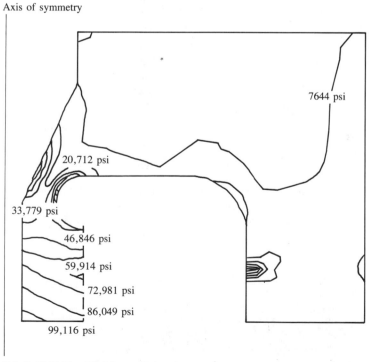

FIGURE 10-16   Von Mises stress contour plot of axisymmetric model of Figure 10-15 (also producing a radial inward deflection of about 0.015 in.)

comparison studies. Other examples of the use of the axisymmetric element can be found in References [2]–[6].

In this chapter, we have shown the finite element analysis of axisymmetric systems using a simple three-noded triangular element to be analogous to that of the two-dimensional plane stress problem using three-noded triangular elements as developed in Chapter 7. Therefore, the plane stress/strain computer program CSFEP described in Chapter 8 could easily be adapted to the analysis of axisymmetric structures by addition of a few steps as follows:

1.   The addition of the 4×4 stress/strain matrix $[D]$ given in Eq. (10.1.2) would be necessary.

2.   The addition of the 4×6 strain/displacement matrix $[B]$ given in Eq. (10.1.17) would be necessary.

3.   A subroutine for numerical integration of the stiffness matrix given by Eq. (10.1.24) would be necessary, or the explicit expressions for the triangular torus element stiffness matrix given in Reference [1] could be used. Finally, the simple, but generally less accurate, third method outlined by Eqs. (10.1.25) and (10.1.26) could be used.

4.   The body forces from Eq. (10.1.27) would have to be evaluated by one of the same three methods used for evaluating the stiffness matrix, and the surface forces would have to be evaluated by equations such as Eq. (10.1.36).

Finally, note that other axisymmetric elements, such as a simple quadrilateral (one with four corner nodes and two degrees of freedom per node, as used in the steel die analysis of Figure 10-15) or higher-order triangular elements, such as in Reference [6], in which a cubic polynomial involving ten terms (ten a's) for both u and w, could be used for axisymmetric analysis. The three-noded triangular element was described here because of its simplicity and ability to describe geometric boundaries rather easily.

REFERENCES

[1]   Utku, S., "Explicit Expressions for Triangular Torus Element Stiffness Matrix," *Journal of the American Institute of Aeronautics and Astronautics*, Vol. 6, No. 6, pp. 1174–1176, June 1968.

[2]   Zienkiewicz, O. C., *The Finite Element Method*, 3rd ed., McGraw-Hill, London, 1977.

[3]   Clough, R., and Rashid, Y., "Finite Element Analysis of Axisymmetric Solids," *Journal of the Engineering Mechanics Division*, American Society of Civil Engineers, Vol. 91, pp. 71–85, Feb. 1965.

[4]   Rashid, Y., "Analysis of Axisymmetric Composite Structures by the Finite Element Method," *Nuclear Engineering and Design*, Vol. 3, pp. 163–182, 1966.

[5]   Wilson, E., "Structural Analysis of Axisymmetric Solids," *Journal of the American Institute of Aeronautics and Astronautics*, Vol. 3, No. 12, pp. 2269–2274, Dec. 1965.

[6] Chacour, S., "A High Precision Axisymmetric Triangular Element Used in the Analysis of Hydraulic Turbine Components," Transactions of the American Society of Mechanical Engineers, *Journal of Basic Engineering*, Vol. 92, pp. 819–826, 1973.

[7] Greer, R. D., *The Analysis of a Film Tower Die Utilizing the ANSYS Finite Element Package*, M. S. Thesis, Rose-Hulman Institute of Technology, Terre Haute, Indiana, May 1989.

[8] Logan, D. L., *Mechanics of Materials*, Harper Collins, New York, 1991.

[9] Cook, R. D., Malkus, D. S., and Plesha, M. E., *Concepts and Applications of Finite Element Analysis*, 3rd ed., Wiley, New York, 1989.

PROBLEMS

10.1 For the elements shown in Figure P10-1, evaluate the stiffness matrices using Eq. (10.2.2). The coordinates are shown in the figures. Let $E = 30 \times 10^6$ psi and $v = 0.25$ for each element.

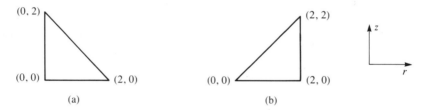

FIGURE P10-1

10.2 Evaluate the nodal forces used to replace the linearly varying surface traction shown in Figure P10-2.

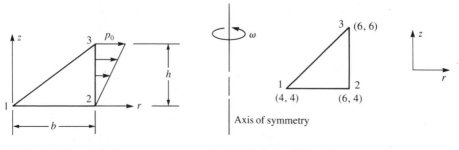

FIGURE P10-2   **FIGURE P10-3**

10.3 For an element of an axisymmetric body rotating with a constant angular velocity $\omega = 20$ rpm as shown in Figure P10-3, evaluate the body-force matrix. The coordinates of the element are shown in the figure. Let the weight density ρ_w be 0.283 lb/in.^{3}

10.4   For the axisymmetric elements shown in Figure P10-4, determine the element stresses. Let $E = 30 \times 10^6$ psi and $v = 0.25$. The coordinates (in inches) are shown in the figures and the nodal displacements for each element are $u_1 = 0.0001$ in., $w_1 = 0.0002$ in., $u_2 = 0.0005$ in., $w_2 = 0.0006$ in., $u_3 = 0$, and $w_3 = 0$.

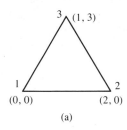
(a)

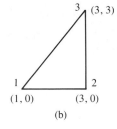
(b)

FIGURE   P10-4

10.5   Explicitly show that the integration of Eq. (10.1.35) yields the j surface forces given by Eq. (10.1.36).

10.6   For the elements shown in Figure P10-6, evaluate the stiffness matrices using Eq. (10.2.2). The coordinates (in mm) are shown in the figures. Let $E = 210$ GPa and $v = 0.25$ for each element.

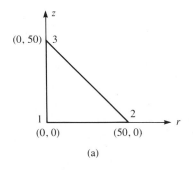
(a)

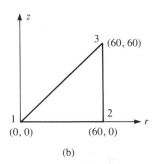
(b)

FIGURE   P10-6

10.7   For the axisymmetric elements shown in Figure P10-7, determine the element stresses. Let $E = 210$ GPa and $v = 0.25$. The coordinates (in millimeters) are shown in the figures and the nodal displacements for each element are

$$u_1 = 0.05 \text{ mm} \qquad w_1 = 0.03 \text{ mm}$$

$$u_2 = 0.02 \text{ mm} \qquad w_2 = 0.02 \text{ mm}$$

$$u_3 = 0.0 \text{ mm} \qquad w_3 = 0.0 \text{ mm}$$

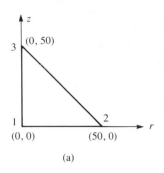

FIGURE P10-7

Solve the following axisymmetric problems using the computer program CSFEP modified for the axisymmetric case (or any other suitable program).

10.8 The soil mass in Figure P10-8 is loaded by a force transmitted through a circular footing as shown. Determine the stresses in the soil. Compare the values of σ_r using an axisymmetric model with the σ_y values using a plane stress model. Let $E = 3000$ psi and $v = 0.45$ for the soil mass.

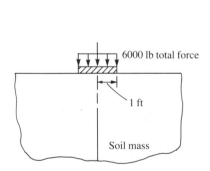

FIGURE P10-8

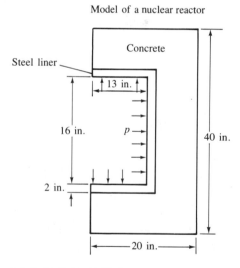

Model of a nuclear reactor

FIGURE P10-9

10.9 Perform a stress analysis of the pressure vessel shown in Figure P10-9. Let $E = 5 \times 10^6$ psi and $v = 0.15$ for the concrete, and let $E = 29 \times 10^6$ psi and $v = 0.25$ for the steel liner. The steel liner is 2 in. thick. Let the pressure p equal 500 psi.

10.10 Perform a stress analysis of the disk shown in Figure P10-10 if it rotates with constant angular velocity of $\omega = 50$ rpm. Let $E = 30 \times 10^6$ psi, $v = 0.25$, and the weight density $\rho_w = 0.283$ lb/in.^{3}

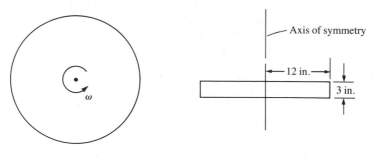

F I G U R E   P10-10

10.11   For the die casting shown in Figure P10-11, determine the maximum stresses and their location. Let $E = 30 \times 10^6$ psi and $v = 0.25$. The dimensions are shown in the figure.

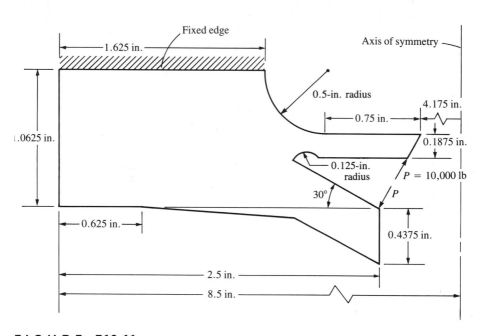

F I G U R E   P10-11

10.12   For the axisymmetric connecting rod shown in Figure P10-12, determine the stresses σ_z, σ_r, σ_θ, and τ_{rz}. Plot stress contours (lines of constant stress) for each of the normal stresses. Let $E = 30 \times 10^6$ psi and $v = 0.25$. The applied loading and boundary conditions are shown in the figure. A typical discretized rod is shown in the figure for illustrative purposes only.

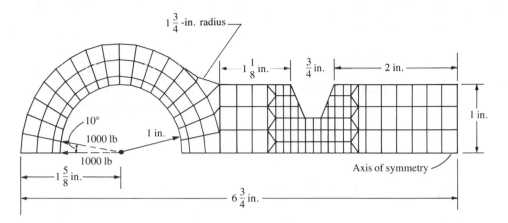

FIGURE P10-12

10.13 For the thick-walled open-ended cylindrical pipe subjected to internal pressure, use five layers of elements to obtain the circumferential stress, σ_θ, and principal stresses and maximum radial displacement. Compare these results to the exact solution. Let $E = 30 \times 10^6$ psi and $v = 0.3$.

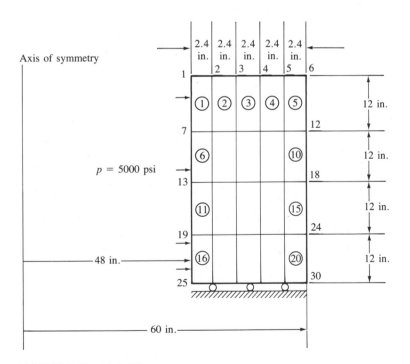

FIGURE P10-13

Isoparametric Formulation

INTRODUCTION

In this chapter, we introduce the isoparametric formulation of the element stiffness matrices. After considering the linear-strain triangular element in Chapter 9, we can see that the development of element matrices and equations expressed in terms of a global coordinate system becomes an enormously difficult task (if at all possible) except for the simplest of elements such as the constant-strain triangle of Chapter 7. Hence, the isoparametric formulation was developed (see Reference [1]). The isoparametric method may appear somewhat tedious (and confusing initially), but it will lead to a simple computer program formulation, and it is generally applicable for two- and three-dimensional stress analysis and for nonstructural problems. The isoparametric formulation allows elements to be created that are nonrectangular and have curved sides. Furthermore, numerous commercial computer programs (as described in Chapter 1) have adapted this formulation for their various libraries of elements.

We first illustrate the isoparametric formulation to develop the simple bar element stiffness matrix. Use of the bar element affords a relatively easy understanding of the method because simple expressions result.

We then consider the development of the rectangular plane stress element stiffness matrix in terms of a global coordinate system that will be convenient for use with the element. These concepts will be useful in understanding some of the procedures used with the isoparametric formulation of the simple quadrilateral element stiffness matrix, which we will develop subsequently.

Next, we will introduce a numerical integration method for evaluating the quadrilateral element stiffness matrix, and illustrate the adaptability of the isoparametric formulation to common numerical integration methods.

Finally, we will consider some higher-order elements and their associated shape functions.

11.1 Isoparametric Formulation of the Bar Element Stiffness Matrix

The term *isoparametric* is derived from the use of the same shape functions (or interpolation functions) $[N]$ to define the element's geometric shape as are used to define the displacements within the element. Thus, when the shape function is $u = a_1 + a_2 s$ for the displacement, we use $x = a_1 + a_2 s$ for the description of the nodal coordinate of a point on the bar element, and hence, the physical shape of the element.

Isoparametric element equations are formulated using a **natural (or intrinsic) coordinate system** s that is defined by element geometry and not by the element orientation in the global coordinate system. In other words, axial coordinate s is attached to the bar and remains directed along the axial length of the bar, regardless of how the bar is oriented in space. There is a relationship (called a *transformation mapping*) between the natural coordinate system s and the global coordinate system x for each element of a specific structure, and this relationship must be used in the element equation formulations.

We will now develop the isoparametric formulation of the simple linear bar element stiffness matrix.

Step 1 Select Element Type

First, the natural coordinate s is attached to the element, with the origin located at the center of the element, as shown in Figure 11-1(a). The s axis need not be parallel to the x axis—this is only for convenience.

We consider the bar element to have two degrees of freedom—axial displacements u_1 and u_2 at each node associated with the global x axis.

For the special case when the s and x axes are parallel to each other, the s and x coordinates can be related by

$$x = x_c + \frac{L}{2}s \qquad (11.1.1)$$

where x_c is the global coordinate of the element centroid.

The shape functions used to define a position within the bar are found in a manner similar to that used in Chapter 3 to define displacement within a bar (see Section 3.1). We begin by relating the natural coordinate to the global coordinate by

$$x = a_1 + a_2 s \qquad (11.1.2)$$

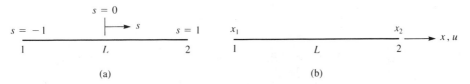

(a)

(b)

FIGURE 11-1 Linear bar element in (a) natural coordinate system s and in (b) global coordinate system x

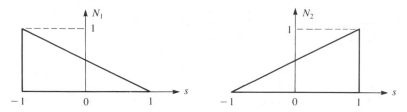

FIGURE 11-2   Shape function variations with natural coordinates

where we note that s is such that $-1 \leq s \leq 1$. Solving for the a_i's in terms of x_1 and x_2, we obtain

$$x = \tfrac{1}{2}[(1-s)x_1 + (1+s)x_2] \qquad (11.1.3)$$

or, in matrix form, we can express Eq. (11.1.3) as

$$\{x\} = [N_1 \quad N_2]\begin{Bmatrix} x_1 \\ x_2 \end{Bmatrix} \qquad (11.1.4)$$

where the shape functions in Eq. (11.1.4) are

$$N_1 = \frac{1-s}{2} \qquad N_2 = \frac{1+s}{2} \qquad (11.1.5)$$

The linear shape functions in Eqs. (11.1.5) map the s coordinate of any point in the element to the x coordinate. For instance, when we substitute $s = -1$ into Eq. (11.1.3), we obtain $x = x_1$. These shape functions are shown in Figure 11-2, where we can see that they have the same properties as defined for the interpolation functions of Section 3.1. Hence, N_1 represents the physical shape of the coordinate x when plotted over the length of the element for $x_1 = 1$ and $x_2 = 0$, and N_2 represents the coordinate x when plotted over the length of the element for $x_2 = 1$ and $x_1 = 0$. Again, we must have $N_1 + N_2 = 1$.

Step 2   Select a Displacement Function

The displacement function within the bar is now defined by the same shape functions, Eqs. (11.1.5), as used to define the element shape; that is,

$$\{u\} = [N_1 \quad N_2]\begin{Bmatrix} u_1 \\ u_2 \end{Bmatrix} \qquad (11.1.6)$$

When a particular coordinate s of the point of interest is substituted into $[N]$, Eq. (11.1.6) yields the displacement of a point on the bar in terms of the nodal degrees of freedom u_1 and u_2. Since u and x are defined by the same shape functions at the same nodes, the element is called *isoparametric*.

Step 3   Define the Strain/Displacement and Stress/Strain Relationships

We now want to formulate element matrix $[B]$ to evaluate $[k]$. We use the isoparametric formulation to illustrate its manipulations. For a simple bar element, no real advantage may appear evident. However, for higher-order elements, the advantage will become clear because relatively simple computer program formulations will result.

To construct the element stiffness matrix, we must determine the strain, which is defined in terms of the derivative of the displacement with respect to x. The displacement u, however, is now a function of s as given by Eq. (11.1.6). Therefore, we must apply the chain rule of differentiation to the function u as follows:

$$\frac{du}{ds} = \frac{du}{dx}\frac{dx}{ds} \qquad (11.1.7)$$

We can evaluate (du/ds) and (dx/ds) using Eqs. (11.1.6) and (11.1.3). We seek $(du/dx) = \varepsilon_x$. Therefore, we solve Eq. (11.1.7) for (du/dx) as

$$\frac{du}{dx} = \frac{\left(\dfrac{du}{ds}\right)}{\left(\dfrac{dx}{ds}\right)} \qquad (11.1.8)$$

Using Eqs. (11.1.6) and (11.1.3) in Eq. (11.1.8), we obtain

$$\{\varepsilon_x\} = \begin{bmatrix} -\dfrac{1}{L} & \dfrac{1}{L} \end{bmatrix} \begin{Bmatrix} u_1 \\ u_2 \end{Bmatrix} \qquad (11.1.9)$$

where we have used

$$\frac{dx}{ds} = \frac{x_2 - x_1}{2} = \frac{L}{2} \qquad (11.1.10)$$

in obtaining Eq. (11.1.9). Since $\{\varepsilon\} = [B]\{d\}$, the strain/displacement matrix $[B]$ is then given in Eq. (11.1.9) as

$$[B] = \begin{bmatrix} -\dfrac{1}{L} & \dfrac{1}{L} \end{bmatrix} \qquad (11.1.11)$$

For higher-order elements, such as the quadratic bar with three nodes, $[B]$ becomes a function of natural coordinate s (see Problem 11.3).

Step 4   **Derive the Element Stiffness Matrix and Equations**

The stiffness matrix is

$$[k] = \int_0^L [B]^T[D][B]A\,dx \qquad (11.1.12)$$

However, in general, we must transform the coordinate x to s because $[B]$ is, in general, a function of s. This general type of transformation is given by (see References [4] and [5])

$$\int_0^L f(x)\,dx = \int_{-1}^1 f(s)|\underline{J}|\,ds \qquad (11.1.13)$$

where $\underline{J}$ is called the *Jacobian*. In the one-dimensional case, we have $|\underline{J}| = \underline{J}$. For the simple bar element, from Eq. (11.1.3), we have

$$|\underline{J}| = \frac{dx}{ds} = \frac{L}{2} \tag{11.1.14}$$

because $x_2 - x_1 = L$. Observe that in Eq. (11.1.14), the Jacobian relates an element length in the global coordinate system to an element length in the natural coordinate system. In general, $|\underline{J}|$ is a function of s and depends on the numerical values of the nodal coordinates. This can be seen by working Problems 11.1–11.3 and by looking at Eq. (11.3.21) for the quadrilateral element. (Section 11.3 further discusses the Jacobian.) Using Eqs. (11.1.13) and (11.1.14) in Eq. (11.1.12), we obtain the stiffness matrix in natural coordinates as

$$[k] = \frac{L}{2} \int_{-1}^{1} [B]^T E[B] A \, ds \tag{11.1.15}$$

where, for the one-dimensional case, we have used the modulus of elasticity $E = [D]$ in Eq. (11.1.15). Substituting Eq. (11.1.11) in Eq. (11.1.15) and performing the simple integration, we obtain

$$[k] = \frac{AE}{L} \begin{bmatrix} 1 & -1 \\ -1 & 1 \end{bmatrix} \tag{11.1.16}$$

which is the same as Eq. (3.1.14). For higher-order one-dimensional elements, the integration in closed form becomes difficult if not impossible (see Problem 11.4). Even the simple rectangular element stiffness matrix is difficult to evaluate in closed form (see Section 11.3). However, the use of numerical integration, as described in Section 11.4, illustrates the distinct advantage of the isoparametric formulation of the equations.

11.2   Rectangular Plane Stress Element

We will now develop the rectangular plane stress element stiffness matrix. We will later refer to this element in the isoparametric formulation of a general quadrilateral element.

Two advantages of the rectangular element over the triangular element are ease of data input and simpler interpretation of output stresses. A disadvantage of the rectangular element is that the simple linear-displacement rectangle with its associated straight sides poorly approximates the real boundary condition edges.

The usual steps outlined in Chapter 1 will be followed to obtain the element stiffness matrix and related equations.

Step 1   Select Element Type

Consider the rectangular element shown in Figure 11-3 (all interior angles are 90°) with corner nodes 1, 2, 3, and 4 (again labeled counterclockwise) and base and height dimensions $2b$ and $2h$, respectively.

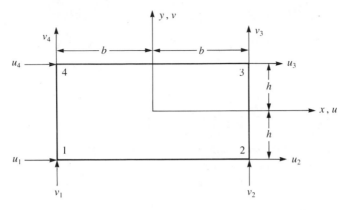

FIGURE   11-3   Basic four-node rectangular element

The unknown nodal displacements are now given by

$$\{d\} = \begin{Bmatrix} u_1 \\ v_1 \\ u_2 \\ v_2 \\ u_3 \\ v_3 \\ u_4 \\ v_4 \end{Bmatrix} \qquad (11.2.1)$$

Step 2 **Select Displacement Functions**

For a compatible displacement field, the element displacement functions u and v must be linear along each edge because only two points (the corner nodes) exist along each edge. We then select the linear displacement functions as

$$u(x, y) = a_1 + a_2 x + a_3 y + a_4 xy$$
$$v(x, y) = a_5 + a_6 x + a_7 y + a_8 xy \qquad (11.2.2)$$

We can proceed in the usual manner to eliminate the a_i's from Eqs. (11.2.2) to obtain

$$u(x, y) = \frac{1}{4bh}[(b - x)(h - y)u_1 + (b + x)(h - y)u_2$$

$$+ (b + x)(h + y)u_3 + (b - x)(h + y)u_4] \qquad (11.2.3)$$

$$v(x, y) = \frac{1}{4bh}[(b - x)(h - y)v_1 + (b + x)(h - y)v_2$$

$$+ (b + x)(h + y)v_3 + (b - x)(h + y)v_4]$$

These displacement expressions, Eqs. (11.2.3), can be expressed equivalently in terms of the shape functions and unknown nodal displacements as

$$\{\psi\} = [N]\{d\} \qquad (11.2.4)$$

where the shape functions are given by

$$N_1 = \frac{(b - x)(h - y)}{4bh} \qquad N_2 = \frac{(b + x)(h - y)}{4bh}$$

$$N_3 = \frac{(b + x)(h + y)}{4bh} \qquad N_4 = \frac{(b - x)(h + y)}{4bh} \qquad (11.2.5)$$

and the N_i's are again such that $N_1 = 1$ at node 1 and $N_1 = 0$ at all the other nodes, with similar requirements for the other shape functions. In expanded form, Eq. (11.2.4) becomes

$$\begin{Bmatrix} u \\ v \end{Bmatrix} = \begin{bmatrix} N_1 & 0 & N_2 & 0 & N_3 & 0 & N_4 & 0 \\ 0 & N_1 & 0 & N_2 & 0 & N_3 & 0 & N_4 \end{bmatrix} \begin{Bmatrix} u_1 \\ v_1 \\ u_2 \\ v_2 \\ u_3 \\ v_3 \\ u_4 \\ v_4 \end{Bmatrix} \qquad (11.2.6)$$

Step 3   Define the Strain/Displacement and Stress/Strain Relationships

Again the element strains for the two-dimensional stress state are given by

$$\begin{Bmatrix} \varepsilon_x \\ \varepsilon_y \\ \gamma_{xy} \end{Bmatrix} = \begin{Bmatrix} \dfrac{\partial u}{\partial x} \\[2mm] \dfrac{\partial v}{\partial y} \\[2mm] \dfrac{\partial u}{\partial y} + \dfrac{\partial v}{\partial x} \end{Bmatrix} \qquad (11.2.7)$$

Using Eq. (11.2.6) in Eq. (11.2.7) and taking the derivatives of u and v as indicated, we can express the strains in terms of the unknown nodal displacements as

$$\{\varepsilon\} = [B]\{d\} \qquad (11.2.8)$$

where

$$[B] = \frac{1}{4bh} \begin{bmatrix} -(h - y) & 0 & (h - y) & 0 \\ 0 & -(b - x) & 0 & -(b + x) \\ -(b - x) & -(h - y) & -(b + x) & (h - y) \end{bmatrix}$$

$$\begin{matrix} (h + y) & 0 & -(h + y) & 0 \\ 0 & (b + x) & 0 & (b - x) \\ (b + x) & (h + y) & (b - x) & -(h + y) \end{matrix} \qquad (11.2.9)$$

From Eqs. (11.2.8) and (11.2.9), we observe that ε_x is a function of y, ε_y is a function of x, and γ_{xy} is a function of both x and y. The stresses are again given by the formulas in Eq. (9.1.15) where now $[B]$ is that of Eq. (11.2.9) and $\{d\}$ is that of Eq. (11.2.1).

Step 4 Derive the Element Stiffness Matrix and Equations

The stiffness matrix is determined by

$$[k] = \int_{-h}^{h} \int_{-b}^{b} [B]^T [D][B] t \, dx \, dy \qquad (11.2.10)$$

with $[D]$ again given by the usual plane stress or plane strain conditions, Eq. (7.1.8) or Eq. (7.1.10). Since the $[B]$ matrix is a function of x and y, integration of Eq. (11.2.10) must be performed. The $[k]$ matrix for the rectangular element is now of order 8×8.

The element force matrix is determined by Eq. (7.2.46) as

$$\{f\} = \int\int\int_V [N]^T \{X\} \, dV + \{P\} + \int\int_S [N]^T \{T\} \, dS \qquad (11.2.11)$$

where $[N]$ is the rectangular matrix in Eq. (11.2.6), and N_1 through N_4 are given by Eqs. (11.2.5). The element equations are then given by

$$\{f\} = [k]\{d\} \qquad (11.2.12)$$

Steps 5, 6, and 7

Steps 5, 6, and 7, involving assembling the global stiffness matrix and equations, determining the unknown nodal displacements, and calculating the stress, are identical to those in Section 7.2 for the CST. However, the stresses within each element now vary in both the x and y directions.

11.3 Isoparametric Formulation of the Plane Element Stiffness Matrix

Recall that the term *isoparametric* is derived from the use of the same shape functions to define the element shape as are used to define the displacements within the element. Thus, when the shape function is $u = a_1 + a_2 s + a_3 t + a_4 st$ for the displacement, we use $x = a_1 + a_2 s + a_3 t + a_4 st$ for the description of a coordinate point in the plane element.

The natural coordinate system s-t is defined by element geometry and not by the element orientation in the global coordinate system x-y. Similar to the bar element example, there is a transformation mapping between the two coordinate systems for each element of a specific structure, and this relationship must be used in the element formulation.

We will now discuss the isoparametric formulation of the simple linear

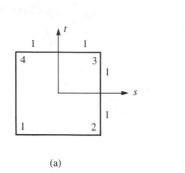
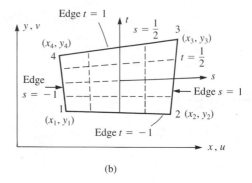

(a)   (b)

FIGURE 11-4   (a) Linear square element in s-t coordinates and
(b) square element mapped into quadrilateral in x-y
coordinates whose size and shape are determined by
the eight nodal coordinates

plane element stiffness matrix. This formulation is general enough to be applied to more-complicated (higher-order) elements such as a quadratic plane element with three nodes along an edge, which can have straight or quadratic curved sides. Higher-order elements have additional nodes and use different shape functions as compared to the linear element, but the steps in the development of the stiffness matrices are the same. We will briefly discuss these elements following the linear plane element formulation.

Step 1   Select Element Type

First, the natural s-t coordinates are attached to the element, with the origin at the center of the element, as shown in Figure 11-4(a). The s and t axes need not be orthogonal, and neither has to be parallel to the x or y axes. The orientation of s-t coordinates is such that the four corner nodes and the edges of the quadrilateral are bounded by $+1$ or -1. This orientation will later allow us to more fully take advantage of common numerical integration schemes.

We consider the quadrilateral to have eight degrees of freedom, $u_1, v_1,$ $\ldots, u_4,$ and v_4 associated with the global x and y directions. The element then has straight sides but is otherwise of arbitrary shape, as shown in Figure 11-4(b).

For the special case when the distorted element becomes a rectangular element with sides parallel to the global x-y coordinates (see Figure 11-3), the s-t coordinates can be related to the global element coordinates x and y by

$$x = x_c + bs \qquad y = y_c + ht \qquad (11.3.1)$$

where x_c and y_c are the global coordinates of the element centroid.

As we have shown for a rectangular element, the shape functions that define the displacements within the element are given by Eqs. (11.2.5). These same shape functions will now be used to map the square of Figure 11-4(a) in isoparametric coordinates s and t to the quadrilateral of Figure 11-4(b) in x-y

coordinates whose size and shape are determined by the eight nodal coordinates $x_1, y_1, x_2, y_2, \ldots, x_4$, and y_4; that is, letting

$$x = a_1 + a_2 s + a_3 t + a_4 st$$
$$y = a_5 + a_6 s + a_7 t + a_8 st$$

(11.3.2)

and solving for the a_i's in terms of $x_1, x_2, x_3, x_4, y_1, y_2, y_3$, and y_4, we establish a form similar to Eqs. (11.2.3) such that

$$x = \tfrac{1}{4}[(1 - s)(1 - t)x_1 + (1 + s)(1 - t)x_2$$
$$+ (1 + s)(1 + t)x_3 + (1 - s)(1 + t)x_4]$$
$$y = \tfrac{1}{4}[(1 - s)(1 - t)y_1 + (1 + s)(1 - t)y_2$$
$$+ (1 + s)(1 + t)y_3 + (1 - s)(1 + t)y_4]$$

(11.3.3)

or, in matrix form, we can express Eqs. (11.3.3) as

$$\begin{Bmatrix} x \\ y \end{Bmatrix} = \begin{bmatrix} N_1 & 0 & N_2 & 0 & N_3 & 0 & N_4 & 0 \\ 0 & N_1 & 0 & N_2 & 0 & N_3 & 0 & N_4 \end{bmatrix} \begin{Bmatrix} x_1 \\ y_1 \\ x_2 \\ y_2 \\ x_3 \\ y_3 \\ x_4 \\ y_4 \end{Bmatrix}$$

(11.3.4)

where the shape functions of Eq. (11.3.4) are now

$$N_1 = \frac{(1 - s)(1 - t)}{4} \qquad N_2 = \frac{(1 + s)(1 - t)}{4}$$
$$N_3 = \frac{(1 + s)(1 + t)}{4} \qquad N_4 = \frac{(1 - s)(1 + t)}{4}$$

(11.3.5)

The shape functions of Eqs. (11.3.5) are linear. These shape functions are seen to map the s and t coordinates of any point in the square element of Figure 11-4(a) to those x and y coordinates in the quadrilateral element of Figure 11-4(b). For instance, consider square element node 1 coordinates, where $s = -1$ and $t = -1$. Using Eqs. (11.3.4) and (11.3.5), the left side of Eq. (11.3.4) becomes

$$x = x_1 \qquad y = y_1$$

(11.3.6)

Similarly, we can map the other local nodal coordinates at nodes 2, 3, and 4 such that the square element in s-t isoparametric coordinates is mapped into a quadrilateral element in global coordinates. Also observe the property that $N_1 + N_2 + N_3 + N_4 = 1$ for all values of s and t.

We further observe that the shape functions in Eq. (11.3.5) are again such that N_1 through N_4 have the properties that N_i ($i = 1, 2, 3, 4$) is equal to one at node i and equal to zero at all other nodes. The physical shapes of N_i as

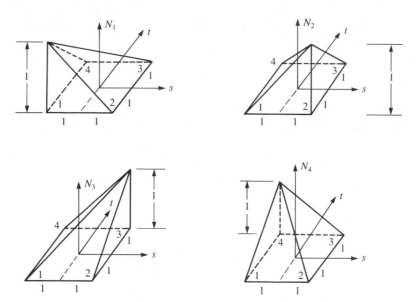

F I G U R E   11-5   Variations of the shape functions over a linear square
element

they vary over the element with natural coordinates are shown in Figure 11-5.
For instance, N_1 represents the geometric shape for $x_1 = 1$, $y_1 = 1$, and x_2, y_2,
x_3, y_3, x_4, and y_4 all equal to zero.

   Until this point in the discussion, we have always developed the element
shape functions either by assuming some relationship between the natural
and global coordinates in terms of the generalized coordinates (a_i's) as in Eqs.
(11.3.2), or similarly, by assuming a displacement function in terms of the a_i's.
However, physical intuition can often guide us in directly expressing shape
functions based on the following two criteria set forth in Section 3.2 and used
on numerous occasions; that is,

$$\sum_{i=1}^{n} N_i = 1 \qquad (i = 1, 2, \ldots, n)$$

where n = the number of shape functions corresponding to displacement
shape functions N_i, and $N_i = 1$ at node i and $N_i = 0$ at all nodes other than i.
In addition, a third criterion is based on Lagrangian interpolation when
displacement continuity is to be satisfied, or Hermitian interpolation if additional slope continuity needs to be satisfied, as in the beam element of Chapter 5. (For a description of the use of Lagrangian and Hermitian interpolation
to develop shape functions, consult References [4] and [6].)

Step 2   Select Displacement Functions

The displacement functions within an element are now similarly defined by
the same shape functions as used to define the element shape; that is,

$$\left\{ \begin{matrix} u \\ v \end{matrix} \right\} = \begin{bmatrix} N_1 & 0 & N_2 & 0 & N_3 & 0 & N_4 & 0 \\ 0 & N_1 & 0 & N_2 & 0 & N_3 & 0 & N_4 \end{bmatrix} \left\{ \begin{matrix} u_1 \\ v_1 \\ u_2 \\ v_2 \\ u_3 \\ v_3 \\ u_4 \\ v_4 \end{matrix} \right\} \qquad (11.3.7)$$

where u and v are displacements parallel to the global x and y coordinates, and the shape functions are given by Eqs. (11.3.5).

Comparing Eqs. (11.2.6) and (11.3.7), we see similarities between the rectangular element with sides of lengths $2b$ and $2h$ (Figure 11-3) and the square element with sides of length 2. If we let $b = 1$ and $h = 1$, the two sets of shape functions, Eqs. (11.2.5) and (11.3.5), are identical.

Step 3 **Define the Strain/Displacement and Stress/Strain Relationships**

We now want to formulate element matrix $\underline{B}$ to evaluate $\underline{k}$. However, because it becomes tedious and difficult (if not impossible) to write the shape functions in terms of the x and y coordinates, as seen in Chapter 9, we will carry out the formulation in terms of the isoparametric coordinates s and t. This may appear tedious, but it is easier to use the s and t coordinate expressions than to attempt to use the x and y coordinate expressions. This approach also leads to a simple computer program formulation.

To construct an element stiffness matrix, we must determine the strains, which are defined in terms of the derivatives of the displacements with respect to the x and y coordinates. The displacements, however, are now functions of the s and t coordinates, as given by Eq. (11.3.7), with the shape functions given by Eqs. (11.3.5). Before, we could determine $(\partial f/\partial x)$ and $(\partial f/\partial y)$, where, in general, f is a function representing the displacement functions u or v. However, u and v are now expressed in terms of s and t. Therefore, we need to apply the chain rule of differentiation because it will not be possible to express s and t as functions of x and y directly. For f as a function of x and y, the chain rule yields

$$\frac{\partial f}{\partial s} = \frac{\partial f}{\partial x}\frac{\partial x}{\partial s} + \frac{\partial f}{\partial y}\frac{\partial y}{\partial s}$$

$$\frac{\partial f}{\partial t} = \frac{\partial f}{\partial x}\frac{\partial x}{\partial t} + \frac{\partial f}{\partial y}\frac{\partial y}{\partial t}$$

$$(11.3.8)$$

In Eq. (11.3.8), $(\partial f/\partial s)$, $(\partial f/\partial t)$, $(\partial x/\partial s)$, $(\partial y/\partial s)$, $(\partial x/\partial t)$, and $(\partial y/\partial t)$ are all known using Eqs. (11.3.7) and (11.3.4). We still seek $(\partial f/\partial x)$ and $(\partial f/\partial y)$. The strains can then be found; for example, $\varepsilon_x = (\partial u/\partial x)$. Therefore, we solve Eqs. (11.3.8) for $(\partial f/\partial x)$ and $(\partial f/\partial y)$ using Cramer's rule, which involves evaluation of determinants (see Appendix B), as

$$\frac{\partial f}{\partial x} = \frac{\begin{vmatrix} \dfrac{\partial f}{\partial s} & \dfrac{\partial y}{\partial s} \\[2ex] \dfrac{\partial f}{\partial t} & \dfrac{\partial y}{\partial t} \end{vmatrix}}{\begin{vmatrix} \dfrac{\partial x}{\partial s} & \dfrac{\partial y}{\partial s} \\[2ex] \dfrac{\partial x}{\partial t} & \dfrac{\partial y}{\partial t} \end{vmatrix}} \qquad \frac{\partial f}{\partial y} = \frac{\begin{vmatrix} \dfrac{\partial x}{\partial s} & \dfrac{\partial f}{\partial s} \\[2ex] \dfrac{\partial x}{\partial t} & \dfrac{\partial f}{\partial t} \end{vmatrix}}{\begin{vmatrix} \dfrac{\partial x}{\partial s} & \dfrac{\partial y}{\partial s} \\[2ex] \dfrac{\partial x}{\partial t} & \dfrac{\partial y}{\partial t} \end{vmatrix}} \qquad (11.3.9)$$

where the determinant in the denominator is the determinant of the *Jacobian* matrix $\underline{J}$. Hence, the Jacobian matrix is given by

$$[J] = \begin{bmatrix} \dfrac{\partial x}{\partial s} & \dfrac{\partial y}{\partial s} \\[2ex] \dfrac{\partial x}{\partial t} & \dfrac{\partial y}{\partial t} \end{bmatrix} \qquad (11.3.10)$$

We now want to express the element strains as

$$\underline{\varepsilon} = \underline{Bd} \qquad (11.3.11)$$

where $\underline{B}$ must now be expressed as a function of s and t. We start with the usual relationship between strains and displacements given in matrix form as

$$\left\{ \begin{array}{c} \varepsilon_x \\ \varepsilon_y \\ \gamma_{xy} \end{array} \right\} = \begin{bmatrix} \dfrac{\partial(\ )}{\partial x} & 0 \\[2ex] 0 & \dfrac{\partial(\ )}{\partial y} \\[2ex] \dfrac{\partial(\ )}{\partial y} & \dfrac{\partial(\ )}{\partial x} \end{bmatrix} \left\{ \begin{array}{c} u \\ v \end{array} \right\} \qquad (11.3.12)$$

where the rectangular matrix on the right side of Eq. (11.3.12) is an *operator matrix*; that is, $\partial(\ )/\partial x$ and $\partial(\ )/\partial y$ represent the partial derivatives of any variable we put inside the parentheses.

Using Eqs. (11.3.9), we have

$$\begin{aligned} \frac{\partial(\ )}{\partial x} &= \frac{1}{|\underline{J}|}\left[\frac{\partial y}{\partial t}\frac{\partial(\ )}{\partial s} - \frac{\partial y}{\partial s}\frac{\partial(\ )}{\partial t} \right] \\[2ex] \frac{\partial(\ )}{\partial y} &= \frac{1}{|\underline{J}|}\left[\frac{\partial x}{\partial s}\frac{\partial(\ )}{\partial t} - \frac{\partial x}{\partial t}\frac{\partial(\ )}{\partial s} \right] \end{aligned} \qquad (11.3.13)$$

where $|\underline{J}|$ is the determinant of $\underline{J}$. Using Eq. (11.3.13) in Eq. (11.3.12) we obtain the strains expressed in terms of the natural coordinates (s. as

$$\begin{Bmatrix} \varepsilon_x \\ \varepsilon_y \\ \gamma_{xy} \end{Bmatrix} = \frac{1}{|J|} \begin{bmatrix} \dfrac{\partial y}{\partial t}\dfrac{\partial(\ )}{\partial s} - \dfrac{\partial y}{\partial s}\dfrac{\partial(\ )}{\partial t} & 0 \\[4mm] 0 & \dfrac{\partial x}{\partial s}\dfrac{\partial(\ )}{\partial t} - \dfrac{\partial x}{\partial t}\dfrac{\partial(\ )}{\partial s} \\[4mm] \dfrac{\partial x}{\partial s}\dfrac{\partial(\ )}{\partial t} - \dfrac{\partial x}{\partial t}\dfrac{\partial(\ )}{\partial s} & \dfrac{\partial y}{\partial t}\dfrac{\partial(\ )}{\partial s} - \dfrac{\partial y}{\partial s}\dfrac{\partial(\ )}{\partial t} \end{bmatrix} \begin{Bmatrix} u \\ v \end{Bmatrix} \qquad (11.3.14)$$

Using Eq. (11.3.7), we can express Eq. (11.3.14) in terms of the shape functions and global coordinates in compact matrix form as

$$\underline{\varepsilon} = \underline{D'}\underline{N}\underline{d} \qquad (11.3.15)$$

where $\underline{D'}$ is an operator matrix given by

$$\underline{D'} = \frac{1}{|\underline{J}|} \begin{bmatrix} \dfrac{\partial y}{\partial t}\dfrac{\partial(\ )}{\partial s} - \dfrac{\partial y}{\partial s}\dfrac{\partial(\ )}{\partial t} & 0 \\[4mm] 0 & \dfrac{\partial x}{\partial s}\dfrac{\partial(\ )}{\partial t} - \dfrac{\partial x}{\partial t}\dfrac{\partial(\ )}{\partial s} \\[4mm] \dfrac{\partial x}{\partial s}\dfrac{\partial(\ )}{\partial t} - \dfrac{\partial x}{\partial t}\dfrac{\partial(\ )}{\partial s} & \dfrac{\partial y}{\partial t}\dfrac{\partial(\ )}{\partial s} - \dfrac{\partial y}{\partial s}\dfrac{\partial(\ )}{\partial t} \end{bmatrix} \qquad (11.3.16)$$

and $\underline{N}$ is the 2 × 8 shape function matrix given as the first matrix on the right side of Eq. (11.3.7) and $\underline{d}$ is the column matrix on the right side of Eq. (11.3.7). Defining $\underline{B}$ as

$$\begin{matrix} \underline{B} & = & \underline{D'} & \underline{N} \\ (3 \times 8) & & (3 \times 2) & (2 \times 8) \end{matrix} \qquad (11.3.17)$$

we have $\underline{B}$ expressed as a function of s and t and thus have the strains in terms of s and t. Here $\underline{B}$ is of order 3 × 8 as indicated in Eq. (11.3.17).

Step 4   Derive the Element Stiffness Matrix and Equations

We now want to express the stiffness matrix in terms of s-t coordinates. For a constant thickness element, we have

$$[k] = \iint_A [B]^T [D] [B] t \, dx \, dy \qquad (11.3.18)$$

However, $\underline{B}$ is now a function of s and t, so we must integrate with respect to s and t. Once again, to transform the variables and the region from x and y to s and t, we must have a standard procedure that involves the determinant of $\underline{J}$. This general type of transformation (see References [4] and [5]) is given by

$$\iint_A f(x, y) \, dx \, dy = \iint_A f(s, t)|\underline{J}| \, ds \, dt \qquad (11.3.19)$$

where the inclusion of $|\underline{J}|$ into the integrand on the right side of Eq. (11.3.19) results from a theorem of integral calculus (see Reference [5] for the complete proof of this theorem). Using Eq. (11.3.19) in Eq. (11.3.18), we obtain

$$[k] = \int_{-1}^{1} \int_{-1}^{1} [B]^T [D][B] t |\underline{J}| \, ds \, dt \qquad (11.3.20)$$

The determinant $|\underline{J}|$ is a polynomial in s and t, and is tedious to evaluate even for the simplest case of the linear plane element. However, using Eq. (11.3.10) for $[J]$ and Eqs. (11.3.3) for x and y, we can evaluate $|\underline{J}|$ as

$$|\underline{J}| = \tfrac{1}{8}\{X_c\}^T \begin{bmatrix} 0 & 1-t & t-s & s-1 \\ t-1 & 0 & s+1 & -s-t \\ s-t & -s-1 & 0 & t+1 \\ 1-s & s+t & -t-1 & 0 \end{bmatrix} \{Y_c\} \qquad (11.3.21)$$

where

$$\{X_c\}^T = [x_1 \quad x_2 \quad x_3 \quad x_4] \qquad (11.3.22)$$

and

$$\{Y_c\} = \begin{Bmatrix} y_1 \\ y_2 \\ y_3 \\ y_4 \end{Bmatrix} \qquad (11.3.23)$$

We observe that $|\underline{J}|$ is a function of s and t and the known global coordinates $x_1, x_2, \ldots, y_4$.

The explicit form of $\underline{B}$ can be obtained by substituting Eq. (11.3.16) for $\underline{D}'$ and Eqs. (11.3.5) for the shape functions into Eq. (11.3.17). The matrix multiplications yield

$$\underline{B}(s, t) = \frac{1}{|\underline{J}|} [\underline{B}_1 \quad \underline{B}_2 \quad \underline{B}_3 \quad \underline{B}_4] \qquad (11.3.24)$$

where the submatrices of $\underline{B}$ are given by

$$\underline{B}_i = \begin{bmatrix} a(N_{i,s}) - b(N_{i,t}) & 0 \\ 0 & c(N_{i,t}) - d(N_{i,s}) \\ c(N_{i,t}) - d(N_{i,s}) & a(N_{i,s}) - b(N_{i,t}) \end{bmatrix} \qquad (11.3.25)$$

Here i is a dummy variable equal to 1, 2, 3, and 4, and

$$a = \tfrac{1}{4}[y_1(s-1) + y_2(-1-s) + y_3(1+s) + y_4(1-s)]$$
$$b = \tfrac{1}{4}[y_1(t-1) + y_2(1-t) + y_3(1+t) + y_4(-1-t)]$$
$$c = \tfrac{1}{4}[x_1(t-1) + x_2(1-t) + x_3(1+t) + x_4(-1-t)] \qquad (11.3.26)$$
$$d = \tfrac{1}{4}[x_1(s-1) + x_2(-1-s) + x_3(1+s) + x_4(1-s)]$$

Using the shape functions defined by Eqs. (11.3.5), we have

$$N_{1,s} = \tfrac{1}{4}(t-1) \qquad N_{1,t} = \tfrac{1}{4}(s-1) \qquad \text{(etc.)} \qquad (11.3.27)$$

Hence, $\underline{B}$ is a function of s and t in both the numerator and denominator [because of $|\underline{J}|$ given by Eq. (11.3.21)] and of the known global coordinates x_1 through y_4. Since $|\underline{J}|$ and $\underline{B}$ are such as to result in complicated expressions within the integral of Eq. (11.3.20), the integration to determine the element stiffness matrix is usually done numerically. A method for numerically integrating Eq. (11.3.20) is given in Section 11.4.

The element body force matrix will now be determined from

$$\{f_b\} = \int_{-1}^{1} \int_{-1}^{1} [N]^T \quad \{X\} \ t|\underline{J}| \ ds \ dt \qquad (11.3.28)$$
$$(8 \times 1) \qquad\qquad (8 \times 2)(2 \times 1)$$

and the surface force matrix, say, along edge $t = 1$ (see Figure 11-4(b)] with overall length L, is

$$\{f_s\} = \int_{-1}^{1} [N]^T \quad \{T\} \ t\frac{L}{2} \ ds \qquad (11.3.29)$$
$$(4 \times 1) \qquad (4 \times 2)(2 \times 1)$$

or

$$\begin{Bmatrix} f_{s3s} \\ f_{s3t} \\ f_{s4s} \\ f_{s4t} \end{Bmatrix} = \int_{-1}^{1} \begin{bmatrix} N_3 & 0 & N_4 & 0 \\ 0 & N_3 & 0 & N_4 \end{bmatrix}^T \begin{Bmatrix} p_s \\ p_t \end{Bmatrix} t\frac{L}{2} \ ds \qquad (11.3.30)$$

since $N_1 = 0$ and $N_2 = 0$ along edge $t = 1$, and hence, no nodal forces exist at nodes 1 and 2.

11.4 Gaussian Quadrature (Numerical Integration)

In this section, we will describe Gauss's method, one of the many schemes for numerical evaluation of definite integrals, because it has proven most useful for finite element work.

To evaluate the integral

$$I = \int_{-1}^{1} y \ dx \qquad (11.4.1)$$

where $y = y(x)$, we might choose (sample or evaluate) y at the midpoint and multiply by the length of the interval, as shown in Figure 11-6, to arrive at $I = 2y_1$, a result that is exact if the curve happens to be a straight line. This is an example of what is called **one-point Gaussian quadrature** because only one sampling point was used.

Generalization of this formula leads to

$$I = \int_{-1}^{1} y \ dx = \sum_{i=1}^{n} W_i y_i \qquad (11.4.2)$$

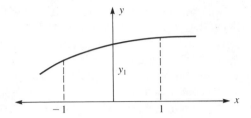

F I G U R E   11-6   Gaussian quadrature using one sampling point

T A B L E   11-1   Table for Gauss points for integration from minus one to one

Number of Points	Locations, x_i	Associated Weights, W_i
1	$x_1 = 0.000\ldots$	2.000
2	$x_1, x_2 = \pm 0.57735026918962$	1.000
3	$x_1, x_3 = \pm 0.77459666924148$	$\frac{5}{9} = 0.555\ldots$
	$x_2 = 0.000\ldots$	$\frac{8}{9} = 0.888\ldots$

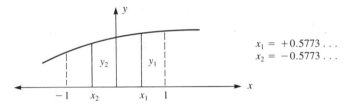

$$x_1 = +0.5773\ldots$$
$$x_2 = -0.5773\ldots$$

F I G U R E   11-7   Gaussian quadrature using two sampling points

—that is, to approximate the integral, we evaluate the function at several sampling points n, multiply each value y_i by the appropriate weight W_i, and add the terms. Gauss's method chooses the sampling points so that for a given number of points, the best possible accuracy is obtained. Sampling points are located symmetrically with respect to the center of the interval. Symmetrically paired points are given the same weight W_i. Table 11-1 gives appropriate sampling points and weighting coefficients for the first three orders—that is, one, two, or three sampling points (see Reference [2] for more complete tables). For example, using two points (see Figure 11-7), we simply have $I = y_1 + y_2$ because $W_1 = W_2 = 1.000$. This is the exact result if $y = f(x)$ is a polynomial containing terms up to and including x^3. In general, Gaussian quadrature using n points (Gauss points) is exact if the integrand is a polynomial of degree $2n - 1$ or less. In using n points, we effectively replace the given function $y = f(x)$ by a polynomial of degree $2n - 1$. The accuracy of the numerical integration depends on how well the polynomial fits the given curve.

If the function $f(x)$ is not a polynomial, Gaussian quadrature is inexact, but it becomes more accurate as more Gauss points are used. Also, it is important to understand that the ratio of two polynomials is, in general, not a polynomial; therefore, Gaussian quadrature will not yield exact integration of the ratio.

To illustrate the derivation of a Gauss formula, consider, for example, $y = C_0 + C_1 x + C_2 x^2 + C_3 x^3$. If integrated between -1 and 1, the area under this curve is $A = 2C_0 + 2C_2/3$. Using two (symmetrically located) Gauss points $x = \pm a$, we propose to calculate the area as $A_G = Wy(-a) + Wy(+a) = 2W(C_0 + C_2 a^2)$. If the error, $e = A - A_G$, is to vanish for any C_0 and C_2, we must have $(\partial e/\partial C_0) = (\partial e/\partial C_2) = 0$, from which we find $W = 1$ and $a = 1/\sqrt{3} = 0.5773\ldots$, which are the W_i's and a_i's (x_i's) for two-point Gaussian quadrature given in Table 11-1.

In two dimensions, we obtain the quadrature formula by integrating first with respect to one coordinate and then with respect to the other as

$$I = \int_{-1}^{1}\int_{-1}^{1} f(s, t)\, ds\, dt = \int_{-1}^{1}\left[\sum_i W_i f(s_i, t)\right] dt$$

$$= \sum_j W_j\left[\sum_i W_i f(s_i, t_j)\right] = \sum_i \sum_j W_i W_j f(s_i, t_j) \qquad (11.4.3)$$

In Eq. (11.4.3), we need not use the same number of Gauss points in each direction (that is, i does not have to equal j), but this is usually done. Thus, for example, a four-point Gauss rule (often described as a 2×2 rule) is shown in Figure 11-8. Equation (11.4.3) with $i = 1, 2$ and $j = 1, 2$ yields

$$I = W_1 W_1 f(s_1, t_1) + W_1 W_2 f(s_1, t_2) + W_2 W_1 f(s_2, t_1) + W_2 W_2 f(s_2, t_2) \qquad (11.4.4)$$

where the four sampling points are at $s_i, t_i = \pm 0.5773\ldots = \pm 1/\sqrt{3}$, and the weights are all 1.000. Hence, the double summation in Eq. (11.4.3) can really be interpreted as a single summation over the four points for the rectangle.

In general, in three dimensions, we have

$$I = \int_{-1}^{1}\int_{-1}^{1}\int_{-1}^{1} f(s, t, z)\, ds\, dt\, dz = \sum_i \sum_j \sum_k W_i W_j W_k f(s_i, t_j, z_k)$$

Now if the limits are $\int_0^1 f(x)\, dx = \sum_{i=1}^{n} W_i f(x_i)$, then the weights W_i and

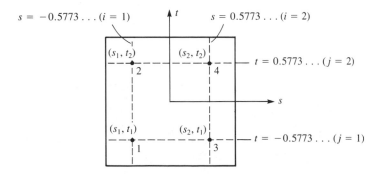

FIGURE 11-8 Four-point Gaussian quadrature in two dimensions

TABLE 11-2   Table of Gauss points for four-point Gaussian quadrature for integration from 0 to 1

Locations, x_i	Associated Weights, W_i
0.06943185	0.1739274
0.3300095	0.3260725
0.6699905	0.3260725
0.9305682	0.1739274

locations x_i are different than for limits from -1 to 1 (as listed in Table 11-2), but the procedures are the same. In Table 11-2, the weights are $W_1 = W_4$ and $W_2 = W_3$. As an example of the use of Table 11-2, we evaluate the function in Example 11.1.

E X A M P L E   11.1

Evaluate the integral of $\sin \pi x$ using numerical integration:

$$I = \int_0^1 \sin \pi x \, dx$$

Using Table 11-2, we have

$$I = \sum_{i=1}^{4} W_i \sin \pi x_i$$

$$= W_1 \sin \pi x_1 + W_2 \sin \pi x_2 + W_3 \sin \pi x_3 + W_4 \sin \pi x_4$$

$$= 0.1739 \sin \pi(0.0694) + 0.3261 \sin \pi(0.3300)$$

$$+ 0.3261 \sin \pi(0.6700) + 0.1739 \sin \pi(0.9306)$$

$$= 0.6366$$

where four significant figures have been used. For comparison, the exact answer by direct integration is also 0.6366. Note that if we wanted to use a three-point Gaussian quadrature, the locations x_i and weights W_i would be different from those listed in Table 11-2.   ∎

11.5 **Evaluation of the Stiffness Matrix by Gaussian Quadrature**

For the two-dimensional element, we have shown in previous chapters that

$$\underline{k} = \int\int_A \underline{B}^T(x, y)\underline{D}\,\underline{B}(x, y)t \, dx \, dy \qquad (11.5.1)$$

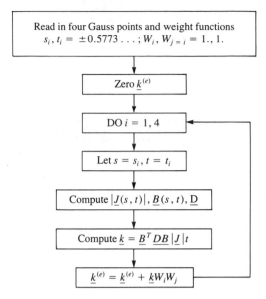

Read in four Gauss points and weight functions
$s_i, t_i = \pm 0.5773 \ldots; W_i, W_{j=i} = 1., 1.$

Zero $\underline{k}^{(e)}$

DO $i = 1, 4$

Let $s = s_i, t = t_i$

Compute $|\underline{J}(s, t)|, \underline{B}(s, t), \underline{D}$

Compute $\underline{k} = \underline{B}^T \underline{DB} |\underline{J}| t$

$\underline{k}^{(e)} = \underline{k}^{(e)} + \underline{k} W_i W_j$

FIGURE 11-9   Flowchart to evaluate $\underline{k}^{(e)}$ by four-point Gaussian quadrature

where, in general, the integrand is a function of x and y and nodal coordinate values.

We have shown in Section 11.3 that $\underline{k}$ for a quadrilateral element can be evaluated in terms of a local set of coordinates s-t with limits from minus one to one within the element, and in terms of global nodal coordinates as given by Eq. (11.3.20). We repeat Eq. (11.3.20) here for convenience as

$$\underline{k} = \int_{-1}^{1} \int_{-1}^{1} \underline{B}^T(s, t)\underline{DB}(s, t)|\underline{J}|t \; ds \; dt \qquad (11.5.2)$$

where $|\underline{J}|$ is defined by Eq. (11.3.21) and $\underline{B}$ is defined by Eq. (11.3.24). In Eq. (11.5.2), each coefficient of the integrand $\underline{B}^T\underline{DB}|\underline{J}|$ must be evaluated by numerical integration in the same manner as $f(s, t)$ was integrated in Eq. (11.4.4).

A flowchart to evaluate $\underline{k}$ of Eq. (11.5.2) for an element using four-point Gaussian quadrature is given in Figure 11-9. The four-point Gaussian quadrature rule is relatively easy to use. Also it has been shown to yield good results [7]. In Figure 11-9, in explicit form for four-point Gaussian quadrature (now using the single summation notation with $i = 1$ to 4), we have

$$\underline{k} = \underline{B}^T(s_1, t_1)\underline{DB}(s_1, t_1)|\underline{J}(s_1, t_1)|tW_1 W_1$$
$$+ \underline{B}^T(s_2, t_2)\underline{DB}(s_2, t_2)|\underline{J}(s_2, t_2)|tW_2 W_2$$
$$+ \underline{B}^T(s_3, t_3)\underline{DB}(s_3, t_3)|\underline{J}(s_3, t_3)|tW_3 W_3$$
$$+ \underline{B}^T(s_4, t_4)\underline{DB}(s_4, t_4)|\underline{J}(s_4, t_4)|tW_4 W_4 \qquad (11.5.3)$$

where $s_1 = t_1 = -0.5773, \quad s_2 = -0.5773, \quad t_2 = 0.5773, \quad s_3 = 0.5773, \quad t_3 =$

-0.5773, and $s_4 = t_4 = 0.5773$ as shown in Figure 11-8, and $W_1 = W_2 = W_3 = W_4 = 1.000$.

E X A M P L E   11.2

Evaluate the stiffness matrix for the quadrilateral element shown in Figure 11-10 using the four-point Gaussian quadrature rule. Let $E = 30 \times 10^6$ psi and $v = 0.25$. The global coordinates are shown in inches. Assume $t = 1$ in.

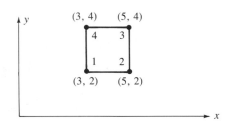

F I G U R E   11-10   Quadrilateral element for stiffness evaluation

Using Eq. (11.5.3), we evaluate the k matrix. Using the four-point rule, the four points are

$$(s_1, t_1) = (-0.5773, -0.5773)$$

$$(s_2, t_2) = (-0.5773, 0.5773)$$

$$(s_3, t_3) = (0.5773, -0.5773)$$

$$(s_4, t_4) = (0.5773, 0.5773)$$

(11.5.4a)

with weights $W_1 = W_2 = W_3 = W_4 = 1.000$.

Therefore, by Eq. (11.5.3), we have

$$k = \underline{B}^T(-0.5773, -0.5773)\underline{DB}(-0.5773, -0.5773)$$

$$\times |\underline{J}(-0.5773, -0.5773)|(1)(1.000)(1.000)$$

$$+ \underline{B}^T(-0.5773, 0.5773)\underline{DB}(-0.5773, 0.5773)$$

$$\times |\underline{J}(-0.5773, 0.5773)|(1)(1.000)(1.000)$$

$$+ \underline{B}^T(0.5773, -0.5773)\underline{DB}(0.5773, -0.5773)$$

$$\times |\underline{J}(0.5773, -0.5773)|(1)(1.000)(1.000)$$

$$+ \underline{B}^T(0.5773, 0.5773)\underline{DB}(0.5773, 0.5773)$$

$$\times |\underline{J}(0.5773, 0.5773)|(1)(1.000)(1.000)$$

(11.5.4b)

To evaluate k, we first evaluate $|\underline{J}|$ at each Gauss point by using Eq. (11.3.21). For instance, one part of $|\underline{J}|$ is given by

$$|\underline{J}(-0.5773, -0.5773)| = \tfrac{1}{8}[3 \quad 5 \quad 5 \quad 3]$$

$$\times \begin{bmatrix} 0 & 1-(-0.5773) & -0.5773-(-0.5773) & -0.5773-1 \\ -0.5773-1 & 0 & -0.5773+1 & -0.5773-(-0.5773) \\ -0.5773-(-0.5773) & -0.5773-1 & 0 & -0.5773+1 \\ 1-(-0.5773) & -0.5773+(-0.5773) & -0.5773-1 & 0 \end{bmatrix}$$

$$\times \begin{Bmatrix} 2 \\ 2 \\ 4 \\ 4 \end{Bmatrix} = 1.000 \qquad \qquad (11.5.4c)$$

Similarly,

$$|\underline{J}(-0.5773, 0.5773)| = 1.000$$

$$|\underline{J}(0.5773, -0.5773)| = 1.000 \qquad (11.5.4d)$$

$$|\underline{J}(0.5773, 0.5773)| = 1.000$$

Even though $|\underline{J}| = 1$ in this example, in general, $|\underline{J}| \neq 1$ and varies in space.

Then using Eqs. (11.3.24) and (11.3.25), we evaluate $\underline{B}$. For instance, one part of $\underline{B}$ is

$$\underline{B}(-0.5773, -0.5773) = \frac{1}{|\underline{J}(-0.5773, -0.5773)|}[\underline{B}_1 \quad \underline{B}_2 \quad \underline{B}_3 \quad \underline{B}_4]$$

where by Eq. (11.3.25)

$$\underline{B}_1 = \begin{bmatrix} aN_{1,s} - bN_{1,t} & 0 \\ 0 & cN_{1,t} - dN_{1,s} \\ cN_{1,t} - dN_{1,s} & aN_{1,s} - bN_{1,t} \end{bmatrix} \qquad (11.5.4e)$$

and by Eqs. (11.3.26) and (11.3.27), a, b, c, d, $N_{1,s}$, and $N_{1,t}$ are evaluated. For instance,

$$a = \tfrac{1}{4}[y_1(s-1) + y_2(-1-s) + y_3(1+s) + y_4(1-s)]$$
$$= \tfrac{1}{4}[2(-0.5773-1) + 2(-1-(0.5773))$$
$$+ 4(1+(-0.5773)) + 4(1-(-0.5773))]$$
$$= 1.00 \qquad (11.5.4f)$$

with similar computations used to obtain b, c, and d. Also

$$N_{1,s} = \tfrac{1}{4}(t-1) = \tfrac{1}{4}(-0.5773-1) = -0.3943$$
$$N_{1,t} = \tfrac{1}{4}(s-1) = \tfrac{1}{4}(-0.5773-1) = -0.3943 \qquad (11.5.4g)$$

Similarly, $\underline{B}_2$, $\underline{B}_3$, and $\underline{B}_4$ must be evaluated like $\underline{B}_1$, at $(-0.5773, -0.5773)$. We then repeat the calculations to evaluate $\underline{B}$ at the other Gauss points [Eq. (11.5.4(a))].

Using a computer program written specifically to evaluate $\underline{B}$ and then $\underline{k}$, we obtain the final form of $\underline{B}$ as

$$B = \begin{bmatrix} -0.1057 & 0 & 0.1057 & 0 & 0 & -0.1057 & 0 & -0.3943 \\ -0.1057 & -0.1057 & -0.3943 & 0.1057 & 0.3943 & 0 & -0.3943 & 0 \\ 0 & 0.3943 & 0 & 0.1057 & 0.3943 & 0.3943 & 0.1057 & -0.3943 \end{bmatrix}$$
$$(11.5.4h)$$

From Eq. (7.1.8), the matrix $\underline{D}$ is

$$\underline{D} = \frac{E}{1-v^2} \begin{bmatrix} 1 & v & 0 \\ v & 1 & 0 \\ 0 & 0 & \frac{1-v}{2} \end{bmatrix} = \begin{bmatrix} 32 & 8 & 0 \\ 8 & 32 & 0 \\ 0 & 0 & 12 \end{bmatrix} \times 10^6 \text{ psi} \qquad (11.5.4i)$$

Finally, the matrix $\underline{k}$ becomes

$$\underline{k} = 10^4 \begin{bmatrix} 1466 & 500 & -866 & -99 & -733 & -500 & 133 & 99 \\ 500 & 1466 & 99 & 133 & -500 & -733 & -99 & -866 \\ -866 & 99 & 1466 & -500 & 133 & -99 & -733 & 500 \\ -99 & 133 & -500 & 1466 & 99 & -866 & 500 & -733 \\ -733 & -500 & 133 & 99 & 1466 & 500 & -866 & -99 \\ -500 & -733 & -99 & -866 & 500 & 1466 & 99 & 133 \\ 133 & -99 & -733 & 500 & -866 & 99 & 1466 & -500 \\ 99 & -866 & 500 & -733 & -99 & 133 & -500 & 1466 \end{bmatrix}$$
$$(11.5.4j) \quad \blacksquare$$

11.6   Higher-Order Shape Functions

In general, higher-order element shape functions can be developed by adding additional nodes to the sides of the linear element. These elements result in higher-order strain variations within each element, and convergence to the exact solution thus occurs at a faster rate using fewer elements. (However, a trade-off exists because a more complicated element takes up so much computation time that even with few elements in the model, the computation time can become larger than for the simple linear element.) Another advantage of the use of higher-order elements is that curved boundaries of irregularly shaped bodies can be approximated more closely than by the use of simple straight-sided linear elements.

To illustrate the concept of higher-order elements, we will consider the quadratic and cubic element shape functions as described in Reference [3]. Figure 11-11 shows a quadratic isoparametric element with four corner nodes and four additional midside nodes. The shape functions of the quadratic element are based on the incomplete cubic polynomial such that coordinates x and y are

$$x = a_1 + a_2 s + a_3 t + a_4 st + a_5 s^2 + a_6 t^2 + a_7 s^2 t + a_8 st^2$$

$$y = a_9 + a_{10} s + a_{11} t + a_{12} st + a_{13} s^2 + a_{14} t^2 + a_{15} s^2 t + a_{16} st^2$$
$$(11.6.1)$$

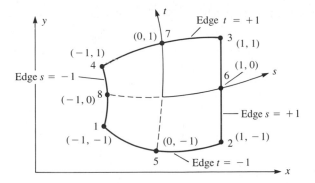

F I G U R E 11-11   Quadratic isoparametric element

Again, we have the same number of degrees of freedom (2 per node times 8 nodes equals 16) as we have a_i's. To describe the shape functions, two forms are required—one for corner nodes and one for midside nodes, as given in Reference [3]. For the corner nodes ($i = 1, 2, 3, 4$),

$$N_1 = \tfrac{1}{4}(1 - s)(1 - t)(-s - t - 1)$$

$$N_2 = \tfrac{1}{4}(1 + s)(1 - t)(s - t - 1)$$

$$N_3 = \tfrac{1}{4}(1 + s)(1 + t)(s + t - 1)$$

$$N_4 = \tfrac{1}{4}(1 - s)(1 + t)(-s + t - 1)$$

(11.6.2)

or, in compact index notation, we express Eqs. (11.6.2) as

$$N_i = \tfrac{1}{4}(1 + ss_i)(1 + tt_i)(ss_i + tt_i - 1) \qquad (11.6.3)$$

where i is the number of the shape function and

$$s_i = -1, 1, 1, -1 \quad \text{for } i = 1, 2, 3, 4$$

$$t_i = -1, -1, 1, 1 \quad \text{for } i = 1, 2, 3, 4$$

(11.6.4)

For the midside nodes ($i = 5, 6, 7, 8$),

$$N_5 = \tfrac{1}{2}(1 - t)(1 + s)(1 - s)$$

$$N_6 = \tfrac{1}{2}(1 + s)(1 + t)(1 - t)$$

$$N_7 = \tfrac{1}{2}(1 + t)(1 + s)(1 - s)$$

$$N_8 = \tfrac{1}{2}(1 - s)(1 + t)(1 - t)$$

(11.6.5)

or, in index notation,

$$N_i = \tfrac{1}{2}(1 - s^2)(1 + tt_i) \qquad t_i = -1, 1 \quad \text{for } i = 5, 7$$

$$N_i = \tfrac{1}{2}(1 + ss_i)(1 - t^2) \qquad s_i - 1, -1 \quad \text{for } i = 6, 8$$

(11.6.6)

We can observe from Eqs. (11.6.2) and (11.6.5) than an edge (and displacement) can vary with s^2 (along t constant) or with t^2 (along s constant). Furthermore, $N_i = 1$ at node i and $N_i = 0$ at the other nodes, as it must be according to our usual definition of shape functions.

The displacement functions are given by

$$\begin{Bmatrix} u \\ v \end{Bmatrix} = \begin{bmatrix} N_1 & 0 & N_2 & 0 & N_3 & 0 & N_4 & 0 & N_5 & 0 & N_6 & 0 & N_7 & 0 & N_8 & 0 \\ 0 & N_1 & 0 & N_2 & 0 & N_3 & 0 & N_4 & 0 & N_5 & 0 & N_6 & 0 & N_7 & 0 & N_8 \end{bmatrix}$$

$$\times \begin{Bmatrix} u_1 \\ v_1 \\ u_2 \\ v_2 \\ \vdots \\ v_8 \end{Bmatrix} \tag{11.6.7}$$

and the strain matrix is now

$$\varepsilon = \underline{D'Nd}$$

with

$$\underline{B} = \underline{D'N}$$

We can develop the matrix $\underline{B}$ using Eq. (11.3.17) with $\underline{D'}$ from Eq. (11.3.16) and with $\underline{N}$ now the 2 × 16 matrix given in Eq. (11.6.7), where the N's are defined in explicit form by Eq. (11.6.2) and (11.6.5).

To evaluate the matrix $\underline{B}$ and the matrix $\underline{k}$ for the eight-noded quadratic isoparametric element, we now use the nine-point Gauss rule (often described as a 3 × 3 rule). Results using 2 × 2 and 3 × 3 rules have shown significant differences, and the 3 × 3 rule is recommended by Bathe and Wilson [7]. Table 11-1 indicates the locations of points and the associated weights. The 3 × 3 rule is shown in Figure 11-12.

The cubic element in Figure 11-13 has four corner nodes and additional nodes taken to be at one-third and two-thirds of the length along each side. The shape functions of the cubic element (as derived in Reference [3]) are based on the incomplete quartic polynomial such that

$$x = a_1 + a_2 s + a_3 t + a_4 s^2 + a_5 st + a_6 t^2 + a_7 s^2 t + a_8 st^2$$

$$+ a_9 s^3 + a_{10} t^3 + a_{11} s^3 t + a_{12} st^3 \tag{11.6.8}$$

with a similar polynomial for y. For the corner nodes ($i = 1, 2, 3, 4$),

$$N_i = \tfrac{1}{32}(1 + ss_i)(1 + tt_i)[9(s^2 + t^2) - 10] \tag{11.6.9}$$

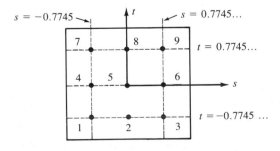

FIGURE 11-12 3 × 3 rule in two dimensions

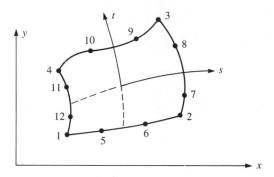

FIGURE 11-13   Cubic isoparametric element

with s_i and t_i given by Eqs. (11.6.4). For the nodes on sides $s = \pm 1$ ($i = 7, 8$, 11, 12),

$$N_i = \tfrac{9}{32}(1 + ss_i)(1 + 9tt_i)(1 - t^2) \qquad (11.6.10)$$

with $s_i = \pm 1$ and $t_i = \pm\tfrac{1}{3}$. For the nodes on sides $t = \pm 1$ ($i = 5, 6, 9, 10$),

$$N_i = \tfrac{9}{32}(1 + tt_i)(1 + 9ss_i)(1 - s^2) \qquad (11.6.11)$$

with $t_i = \pm 1$ and $s_i = \pm\tfrac{1}{3}$.

Having the shape functions for the quadratic element given by Eqs. (11.6.2) and (11.6.5) or for the cubic element given by Eqs. (11.6.9), (11.6.10), and (11.6.11), we can again use Eq. (11.3.17) to obtain $\underline{B}$ and then Eq. (11.3.20) to set up $\underline{k}$ for numerical integration for the plane element. The cubic element requires a 3×3 rule (nine points) to evaluate the matrix $\underline{k}$ exactly. We then conclude that what is really desired is a library of shape functions that can be used in the general equations developed for stiffness matrices, distributed load, and body force, and applied not only to stress analysis but to non-structural problems as well.

Since in this discussion the element shape functions N_i relating x and y to nodal coordinates x_i and y_i are of the same form as the shape functions relating u and v to nodal displacements u_i and v_i, this is said to be an *isoparametric formulation*. For instance, for the linear element $x = \sum_{i=1}^{4} N_i x_i$ and the displacement function $u = \sum_{i=1}^{4} N_i u_i$, use the same shape functions N_i given by Eq. (11.3.5). If instead the shape functions for the coordinates are of lower order (say, linear for x) than the shape functions used for displacements (say, quadratic for u), this is called a *subparametric formulation*.

Finally, referring to Figure 11-13, note that an element can have a linear shape along, say, one edge (1-2), a quadratic along, say, two edges (2-3 and 1-4), and a cubic along the other edge (3-4). Hence, the simple linear element can be mixed with different higher-order elements in regions of a model where rapid stress variation is expected. The advantage of the use of higher-order elements is further illustrated in Reference [3].

REFERENCES

[1] Irons, B. M., "Engineering Applications of Numerical Integration in Stiffness Methods," *Journal of the American Institute of Aeronautics and Astronautics*, Vol. 4, No. 11, pp. 2035–2037, 1966.

[2] Stroud, A. H., and Secrest, D., *Gaussian Quadrature Formulas*, Prentice Hall, Englewood Cliffs, New Jersey, 1966.

[3] Ergatoudis, I., Irons, B. M., and Zienkiewicz, O. C., "Curved Isoparametric, Quadrilateral Elements for Finite Element Analysis," *International Journal of Solids and Structures*, Vol. 4, pp. 31–42, 1968.

[4] Zienkiewicz, O. C., *The Finite Element Method*, 3rd ed., McGraw-Hill, London, 1977.

[5] Thomas, B. G., and Finney, R. L., *Calculus and Analytic Geometry*, Addison-Wesley, Reading, Massachusetts, 1984.

[6] Gallagher, R., *Finite Element Analysis Fundamentals*, Prentice Hall, Englewood Cliffs, New Jersey, 1975.

[7] Bathe, K. J., and E. L. Wilson, *Numerical Methods in Finite Element Analysis*, Prentice Hall, Englewood Cliffs, New Jersey, 1976.

PROBLEMS

11.1   For the three-noded linear strain bar with three coordinates of nodes x_1, x_2, and x_3 shown in Figure P11-1, obtain an expression for J showing J to be a function of s.

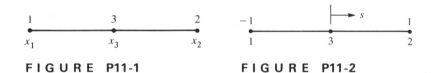

FIGURE P11-1   **FIGURE P11-2**

11.2   Show that if node 3 in Figure P11-2 is at the middle of the bar of length L, then $J = L/2$. Let the coordinates of the nodes be x_1, x_2, and x_3. Assume the bar is a three-noded linear-strain isoparametric element.

11.3   For the three-noded linear-strain one-dimensional isoparametric element shown in Figure P11-2, determine (a) the shape functions N_1, N_2, and N_3, and (b) the strain/displacement matrix $[B]$. Assume $u = a_1 + a_2 s + a_3 s^2$.

11.4   For the four-noded quadratic-strain one-dimensional isoparametric element shown in Figure P11-4, determine (a) the shape functions N_1, N_2, N_3, and N_4, and (b) the strain/displacement matrix $[B]$. Assume $u = a_1 + a_2 s + a_3 s^2 + a_4 s^3$.

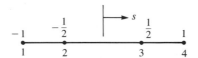

11.5 Show that the sum $N_1 + N_2 + N_3 + N_4$ is equal to 1 anywhere on a rectangular element, where N_1 through N_4 are defined by Eqs. (11.2.5).

11.6 For the rectangular element of Figure 11-3, the nodal displacements are given by

$$u_1 = 0 \qquad\qquad v_1 = 0 \qquad\qquad u_2 = 0.005 \text{ in.}$$

$$v_2 = 0.0025 \text{ in.} \qquad u_3 = 0.0025 \text{ in.} \qquad v_3 = -0.0025 \text{ in.}$$

$$u_4 = 0 \qquad\qquad v_4 = 0$$

For $b = 2$ in., $h = 1$ in., $E = 30 \times 10^6$ psi, and $v = 0.3$, determine the element strains and stresses at the centroid of the element and at the corner nodes.

11.7 Derive $|\underline{J}|$ given by Eq. (11.3.21) for a four-noded isoparametric element.

11.8 Show that for the quadrilateral element described in Section 11.3, $[J]$ can be expressed as

$$[J] = \begin{bmatrix} N_{1,s} & N_{2,s} & N_{3,s} & N_{4,s} \\ N_{1,t} & N_{2,t} & N_{3,t} & N_{4,t} \end{bmatrix} \begin{bmatrix} x_1 & y_1 \\ x_2 & y_2 \\ x_3 & y_3 \\ x_4 & y_4 \end{bmatrix}$$

11.9 Derive Eq. (11.3.24) by substituting Eq. (11.3.16) for $\underline{D}'$ and Eqs. (11.3.5) for the shape functions into Eq. (11.3.17).

11.10 Use Eq. (11.3.30) with $p_s = 0$ and $p_t = p$ (a constant) alongside 3-4 of the element shown in Figure 11-4(b) to obtain the nodal forces.

11.11 For the element shown in Figure P11-11, replace the distributed load with the energy equivalent nodal forces by evaluating Eq. (11.3.29). Let $t = 0.1$ in.

11.12 Use Gaussian quadrature with three or four Gauss points and Tables 11-1 and 11-2, as appropriate, to evaluate the following integrals:

$$\text{(a)} \int_{-1}^{1} \cos\frac{s}{2}\, ds \qquad \text{(b)} \int_{-1}^{1} s^2\, ds \qquad \text{(c)} \int_{-1}^{1} s^4\, ds$$

$$\text{(d)} \int_{-1}^{1} \frac{\cos s}{1 - s^2}\, ds \qquad \text{(e)} \int_{0}^{1} s^3\, ds \qquad \text{(f)} \int_{0}^{1} s \cos s\, ds$$

11.13 For the quadrilateral elements shown in Figure P11-13, write a computer program to evaluate the stiffness matrices using four-point Gaussian quadrature as outlined in Section 11.5. Let $E = 30 \times 10^6$ psi and $v = 0.25$. The global coordinates (in inches) are shown in the figures.

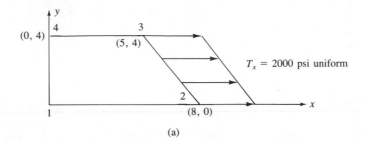

(a)

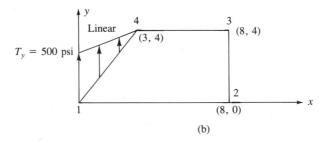

(b)

FIGURE P11-11

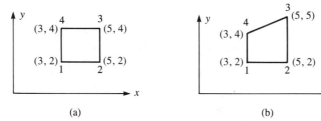

(a)

(b)

FIGURE P11-13

11.14   For the quadrilateral elements shown in Figure P11-14, evaluate the stiffness matrices using four-point Gaussian quadrature. Let $E = 210$ GPa and $v = 0.25$. The global coordinates (in millimeters) are shown in the figures.

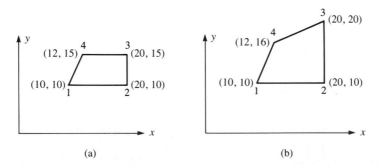

(a)

(b)

FIGURE P11-14

11.15 Evaluate the matrix $\underline{B}$ for the quadratic quadrilateral element shown in Figure 11.11 (see Section 11.6).

11.16 Evaluate the stiffness matrix for the three-noded bar of Problem 11.1 using two-point Gaussian quadrature.

Three-Dimensional

Stress Analysis

INTRODUCTION

In this chapter, we consider the three-dimensional, or solid, element. This element is useful for the stress analysis of general three-dimensional bodies that require more precise analysis than is possible through two-dimensional and/or axisymmetric analysis. Examples of three-dimensional problems are arch dams, thick-walled pressure vessels, and solid forging parts as used, for instance, in the heavy equipment and automotive industries.

The tetrahedron is the basic three-dimensional element, and it is used in the development of the shape functions, stiffness matrix, and force matrices in terms of a global coordinate system. We follow this development with the isoparametric formulation of the stiffness matrix for the hexahedron, or brick element. Finally, we will provide some typical three-dimensional applications.

12.1 Three-Dimensional Stress and Strain

We begin by considering the three-dimensional infinitesimal element in Cartesian coordinates with dimensions dx, dy, and dz, and normal and shear stresses as shown in Figure 12-1. This element conveniently represents the state of stress on three mutually perpendicular planes of a body in a state of three-dimensional stress. As usual, normal stresses are perpendicular to the faces of the element, and are represented by σ_x, σ_y, and σ_z. Shear stresses act in the faces (planes) of the element, and are represented by τ_{xy}, τ_{yz}, τ_{zx}, and so on.

From moment equilibrium of the element, we show in Appendix C that

$$\tau_{xy} = \tau_{yx} \qquad \tau_{yz} = \tau_{zy} \qquad \tau_{zx} = \tau_{xz}$$

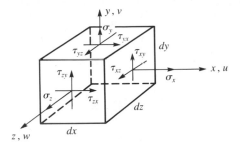

F I G U R E   12-1   Three-dimensional stresses on an element

Hence, there are only three independent shear stresses, along with the three normal stresses.

The element strain/displacement relationships are obtained in Appendix C. They are repeated here, for convenience, as

$$\varepsilon_x = \frac{\partial u}{\partial x} \qquad \varepsilon_y = \frac{\partial v}{\partial y} \qquad \varepsilon_z = \frac{\partial w}{\partial z} \qquad (12.1.1)$$

where u, v, and w are the displacements associated with the x, y, and z directions. The shear strains γ are now given by

$$\gamma_{xy} = \frac{\partial u}{\partial y} + \frac{\partial v}{\partial x} = \gamma_{yx}$$

$$\gamma_{yz} = \frac{\partial v}{\partial z} + \frac{\partial w}{\partial y} = \gamma_{zy} \qquad (12.1.2)$$

$$\gamma_{zx} = \frac{\partial w}{\partial x} + \frac{\partial u}{\partial z} = \gamma_{xz}$$

where, similar to shear stresses, only three independent shear strains exist.

We again represent the stresses and strains by column matrices as

$$\{\sigma\} = \begin{Bmatrix} \sigma_x \\ \sigma_y \\ \sigma_z \\ \tau_{xy} \\ \tau_{yz} \\ \tau_{zx} \end{Bmatrix} \qquad \{\varepsilon\} = \begin{Bmatrix} \varepsilon_x \\ \varepsilon_y \\ \varepsilon_z \\ \gamma_{xy} \\ \gamma_{yz} \\ \gamma_{zx} \end{Bmatrix} \qquad (12.1.3)$$

The stress/strain relationships for an isotropic material are again given by

$$\{\sigma\} = [D]\{\varepsilon\} \qquad (12.1.4)$$

where $\{\sigma\}$ and $\{\varepsilon\}$ are defined by Eqs. (12.1.3), and the constitutive matrix $[D]$ (see also Appendix C) is now given by

$$[D] = \frac{E}{(1+v)(1-2v)} \begin{bmatrix} 1-v & v & v & 0 & 0 & 0 \\ & 1-v & v & 0 & 0 & 0 \\ & & 1-v & 0 & 0 & 0 \\ & & & \dfrac{1-2v}{2} & 0 & 0 \\ & & & & \dfrac{1-2v}{2} & 0 \\ & \text{Symmetry} & & & & \dfrac{1-2v}{2} \end{bmatrix}$$

$$(12.1.5)$$

12.2   Tetrahedral Element

We now develop the tetrahedral stress element stiffness matrix by again using the usual steps outlined in Chapter 1. The development is seen to be an extension of the plane element previously described in Chapter 7. This extension was suggested in References [1] and [2].

Step 1   Select Element Type

Consider the tetrahedral element shown in Figure 12-2 with corner nodes 1, 2, 3, and 4. This element is a four-noded solid. The nodes of the element must be numbered such that when viewed from the last node (say node 4), the first three nodes are numbered in a counterclockwise manner, such as 1, 2, 3, 4 or 2, 3, 1, 4. This ordering of nodes avoids the calculation of negative volumes and is consistent with the counterclockwise node numbering associated with the CST element in Chapter 7. The unknown nodal displacements are now given by

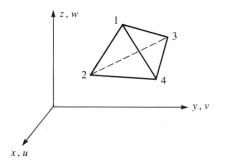

FIGURE 12-2   Tetrahedral solid element

$$\{d\} = \begin{Bmatrix} u_1 \\ v_1 \\ w_1 \\ \vdots \\ u_4 \\ v_4 \\ w_4 \end{Bmatrix} \qquad (12.2.1)$$

Hence, there are 3 degrees of freedom per node, or 12 total degrees of freedom per element.

Step 2 **Select Displacement Functions**

For a compatible displacement field, the element displacement functions u, v, and w must be linear along each edge because only two points (the corner nodes) exist along each edge, and the functions must be linear in each plane side of the tetrahedron. We then select the linear displacement functions as

$$u(x, y, z) = a_1 + a_2 x + a_3 y + a_4 z$$
$$v(x, y, z) = a_5 + a_6 x + a_7 y + a_8 z \qquad (12.2.2)$$
$$w(x, y, z) = a_9 + a_{10} x + a_{11} y + a_{12} z$$

In the same manner as in Chapter 7, we can express the a_i's in terms of the known nodal coordinates $(x_1, y_1, z_1, \dots, z_4)$ and the unknown nodal displacements $(u_1, v_1, w_1, \dots, w_4)$ of the element. Skipping the straightforward but tedious details, we obtain

$$u(x, y, z) = \frac{1}{6V} \{(\alpha_1 + \beta_1 x + \gamma_1 y + \delta_1 z)u_1$$
$$+ (\alpha_2 + \beta_2 x + \gamma_2 y + \delta_2 z)u_2$$
$$+ (\alpha_3 + \beta_3 x + \gamma_3 y + \delta_3 z)u_3$$
$$+ (\alpha_4 + \beta_4 x + \gamma_4 y + \delta_4 z)u_4\} \qquad (12.2.3)$$

where $6V$ is obtained by evaluating the determinant

$$6V = \begin{vmatrix} 1 & x_1 & y_1 & z_1 \\ 1 & x_2 & y_2 & z_2 \\ 1 & x_3 & y_3 & z_3 \\ 1 & x_4 & y_4 & z_4 \end{vmatrix} \qquad (12.2.4)$$

and V represents the volume of the tetrahedron. The coefficients α_i, β_i, γ_i, and δ_i ($i = 1, 2, 3, 4$) in Eq. (12.2.3) are given by

$$
\alpha_1 = \begin{vmatrix} x_2 & y_2 & z_2 \\ x_3 & y_3 & z_3 \\ x_4 & y_4 & z_4 \end{vmatrix} \qquad \beta_1 = -\begin{vmatrix} 1 & y_2 & z_2 \\ 1 & y_3 & z_3 \\ 1 & y_4 & z_4 \end{vmatrix}
$$

$$(12.2.5)$$

$$
\gamma_1 = \begin{vmatrix} 1 & x_2 & z_2 \\ 1 & x_3 & z_3 \\ 1 & x_4 & z_4 \end{vmatrix} \qquad \delta_1 = -\begin{vmatrix} 1 & x_2 & y_2 \\ 1 & x_3 & y_3 \\ 1 & x_4 & y_4 \end{vmatrix}
$$

and

$$
\alpha_2 = -\begin{vmatrix} x_1 & y_1 & z_1 \\ x_3 & y_3 & z_3 \\ x_4 & y_4 & z_4 \end{vmatrix} \qquad \beta_2 = \begin{vmatrix} 1 & y_1 & z_1 \\ 1 & y_3 & z_3 \\ 1 & y_4 & z_4 \end{vmatrix}
$$

$$(12.2.6)$$

$$
\gamma_2 = -\begin{vmatrix} 1 & x_1 & z_1 \\ 1 & x_3 & z_3 \\ 1 & x_4 & z_4 \end{vmatrix} \qquad \delta_2 = \begin{vmatrix} 1 & x_1 & y_1 \\ 1 & x_3 & y_3 \\ 1 & x_4 & y_4 \end{vmatrix}
$$

and

$$
\alpha_3 = \begin{vmatrix} x_1 & y_1 & z_1 \\ x_2 & y_2 & z_2 \\ x_4 & y_4 & z_4 \end{vmatrix} \qquad \beta_3 = -\begin{vmatrix} 1 & y_1 & z_1 \\ 1 & y_2 & z_2 \\ 1 & y_4 & z_4 \end{vmatrix}
$$

$$(12.2.7)$$

$$
\gamma_3 = \begin{vmatrix} 1 & x_1 & z_1 \\ 1 & x_2 & z_2 \\ 1 & x_4 & z_4 \end{vmatrix} \qquad \delta_3 = -\begin{vmatrix} 1 & x_1 & y_1 \\ 1 & x_2 & y_2 \\ 1 & x_4 & y_4 \end{vmatrix}
$$

and

$$
\alpha_4 = -\begin{vmatrix} x_1 & y_1 & z_1 \\ x_2 & y_2 & z_2 \\ x_3 & y_3 & z_3 \end{vmatrix} \qquad \beta_4 = \begin{vmatrix} 1 & y_1 & z_1 \\ 1 & y_2 & z_2 \\ 1 & y_3 & z_3 \end{vmatrix}
$$

$$(12.2.8)$$

$$
\gamma_4 = -\begin{vmatrix} 1 & x_1 & z_1 \\ 1 & x_2 & z_2 \\ 1 & x_3 & z_3 \end{vmatrix} \qquad \delta_4 = \begin{vmatrix} 1 & x_1 & y_1 \\ 1 & x_2 & y_2 \\ 1 & x_3 & y_3 \end{vmatrix}
$$

Expressions for v and w are obtained by simply substituting v_i's for all u_i's and then w_i's for all u_i's in Eq. (12.2.3).

The displacement expression for u given by Eq. (12.2.3), with similar expressions for v and w, can be written equivalently in expanded form in terms of the shape functions and unknown nodal displacements as

$$
\begin{Bmatrix} u \\ v \\ w \end{Bmatrix} = \begin{bmatrix} N_1 & 0 & 0 & N_2 & 0 & 0 & N_3 & 0 & 0 & N_4 & 0 & 0 \\ 0 & N_1 & 0 & 0 & N_2 & 0 & 0 & N_3 & 0 & 0 & N_4 & 0 \\ 0 & 0 & N_1 & 0 & 0 & N_2 & 0 & 0 & N_3 & 0 & 0 & N_4 \end{bmatrix} \begin{Bmatrix} u_1 \\ v_1 \\ w_1 \\ \vdots \\ u_4 \\ v_4 \\ w_4 \end{Bmatrix}
$$

$$(12.2.9)$$

where the shape functions are given by

$$N_1 = \frac{(\alpha_1 + \beta_1 x + \gamma_1 y + \delta_1 z)}{6V} \qquad N_2 = \frac{(\alpha_2 + \beta_2 x + \gamma_2 y + \delta_2 z)}{6V}$$

$$N_3 = \frac{(\alpha_3 + \beta_3 x + \gamma_3 y + \delta_3 z)}{6V} \qquad N_4 = \frac{(\alpha_4 + \beta_4 x + \gamma_4 y + \delta_4 z)}{6V} \qquad (12.2.10)$$

and the rectangular matrix on the right side of Eq. (12.2.9) is the shape function matrix $[N]$.

Step 3 **Define the Strain/Displacement and Stress/Strain Relationships**

The element strains for the three-dimensional stress state are given by

$$\{\varepsilon\} = \begin{Bmatrix} \varepsilon_x \\ \varepsilon_y \\ \varepsilon_z \\ \gamma_{xy} \\ \gamma_{yz} \\ \gamma_{zx} \end{Bmatrix} = \begin{Bmatrix} \dfrac{\partial u}{\partial x} \\[6pt] \dfrac{\partial v}{\partial y} \\[6pt] \dfrac{\partial w}{\partial z} \\[6pt] \dfrac{\partial u}{\partial y} + \dfrac{\partial v}{\partial x} \\[6pt] \dfrac{\partial v}{\partial z} + \dfrac{\partial w}{\partial y} \\[6pt] \dfrac{\partial w}{\partial x} + \dfrac{\partial u}{\partial z} \end{Bmatrix} \qquad (12.2.11)$$

Using Eq. (12.2.9) in Eq. (12.2.11), we obtain

$$\{\varepsilon\} = [B]\{d\} \qquad (12.2.12)$$

where

$$[B] = [\underline{B}_1 \quad \underline{B}_2 \quad \underline{B}_3 \quad \underline{B}_4] \qquad (12.2.13)$$

The submatrix $\underline{B}_1$ in Eq. (12.2.13) is defined by

$$\underline{B}_1 = \begin{bmatrix} N_{1,x} & 0 & 0 \\ 0 & N_{1,y} & 0 \\ 0 & 0 & N_{1,z} \\ N_{1,y} & N_{1,x} & 0 \\ 0 & N_{1,z} & N_{1,y} \\ N_{1,z} & 0 & N_{1,x} \end{bmatrix} \qquad (12.2.14)$$

where, again, the comma after the subscript indicates differentation with respect to the variable that follows. Submatrices $\underline{B}_2$, $\underline{B}_3$, and $\underline{B}_4$ are defined by simply indexing the subscript in Eq. (12.2.14) from 1 to 2, 3, and then 4, respectively. Substituting the shape functions from Eqs. (12.2.10) into Eq. (12.2.14), $\underline{B}_1$ is expressed as

$$\underline{B}_1 = \frac{1}{6V} \begin{bmatrix} \beta_1 & 0 & 0 \\ 0 & \gamma_1 & 0 \\ 0 & 0 & \delta_1 \\ \gamma_1 & \beta_1 & 0 \\ 0 & \delta_1 & \gamma_1 \\ \delta_1 & 0 & \beta_1 \end{bmatrix} \qquad (12.2.15)$$

with similar expressions for $\underline{B}_2$, $\underline{B}_3$, and $\underline{B}_4$.

The element stresses are related to the element strains by

$$\{\sigma\} = [D]\{\varepsilon\} \qquad (12.2.16)$$

where the constitutive matrix for an elastic material is now given by Eq. (12.1.5).

Step 4   Derive the Element Stiffness Matrix and Equations

The element stiffness matrix is given by

$$[k] = \iiint_V [B]^T[D][B]\, dV \qquad (12.2.17)$$

Since both matrices $[B]$ and $[D]$ are constant for the simple tetrahedral element, Eq. (12.2.17) can be simplified to

$$[k] = [B]^T[D][B]V \qquad (12.2.18)$$

where, again, V is the volume of the element. The element stiffness matrix is now of order 12×12.

The element body force matrix is given by

$$\{f_b\} = \iiint_V [N]^T\{X\}\, dV \qquad (12.2.19)$$

where $[N]$ is given by the 3×12 matrix in Eq. (12.2.9), and

$$\{X\} = \begin{Bmatrix} X_b \\ Y_b \\ Z_b \end{Bmatrix} \qquad (12.2.20)$$

For constant body forces, the nodal components of the total resultant body forces can be shown to be distributed to the nodes in four equal parts.

Again, the surface forces are given by

$$\{f_s\} = \iint_S [N]^T\{T\}\, dS \qquad (12.2.21)$$

For example, consider the case of uniform pressure p acting on the face with nodes 1, 2, and 3. The resulting nodal forces become

$$\{f_s\} = \int\int_S [N]^T|_{\substack{\text{evaluated on}\\\text{surface } 1,2,3}} \begin{Bmatrix} p_x \\ p_y \\ p_z \end{Bmatrix} dS \qquad (12.2.22)$$

where p_x, p_y, and p_z are the x, y, and z components, respectively, of p. Simplifying and integrating Eq. (12.2.22), we can show that

$$\{f_s\} = \frac{S_{123}}{3} \begin{Bmatrix} p_x \\ p_y \\ p_z \\ p_x \\ p_y \\ p_z \\ p_x \\ p_y \\ p_z \\ 0 \\ 0 \\ 0 \end{Bmatrix} \qquad (12.2.23)$$

where S_{123} is the area of the surface associated with nodes 1, 2, and 3. The use of volume coordinates, as explained in Reference [8], facilitates the integration of Eq. (12.2.22).

EXAMPLE 12.1

Evaluate the matrices necessary to determine the stiffness matrix for the tetrahedral element shown in Figure 12-3. Let $E = 30 \times 10^6$ psi and $v = 0.30$. The coordinates are shown in the figure in units of inches.

To evaluate the element stiffness matrix, we first determine the element

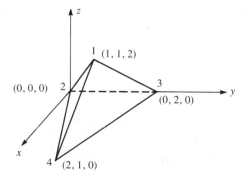

FIGURE 12-3   Tetrahedral element

volume V, and all α's, β's, γ's, and δ's from Eqs. (12.2.4)–(12.2.8). From Eq. (12.2.4), we have

$$6V = \begin{vmatrix} 1 & 1 & 1 & 2 \\ 1 & 0 & 0 & 0 \\ 1 & 0 & 2 & 0 \\ 1 & 2 & 1 & 0 \end{vmatrix} = 8 \text{ in.}^3 \qquad (12.2.24)$$

From Eqs. (12.2.5), we obtain

$$\alpha_1 = \begin{vmatrix} 0 & 0 & 0 \\ 0 & 2 & 0 \\ 2 & 1 & 0 \end{vmatrix} = 0 \qquad \beta_1 = -\begin{vmatrix} 1 & 0 & 0 \\ 1 & 2 & 0 \\ 1 & 1 & 0 \end{vmatrix} = 0 \qquad (12.2.25)$$

and similarly,

$$\gamma_1 = 0 \qquad \delta_1 = 4$$

From Eqs. (12.2.6)–(12.2.8), we obtain

$$\begin{array}{cccc} \alpha_2 = 8 & \beta_2 = -2 & \gamma_2 = -4 & \delta_2 = -1 \\ \alpha_3 = 0 & \beta_3 = -2 & \gamma_3 = 4 & \delta_3 = -1 \\ \alpha_4 = 0 & \beta_4 = 4 & \gamma_4 = 0 & \delta_4 = -2 \end{array} \qquad (12.2.26)$$

Note that α's have typical units of cubic inches or cubic meters, whereas β's, γ's, and δ's have units of square inches or square meters.

Next, the shape functions are determined using Eqs. (12.2.10) and the results from Eqs. (12.2.25) and (12.2.26) as

$$N_1 = \frac{4z}{8} \qquad\qquad N_2 = \frac{8 - 2x - 4y - z}{8}$$

$$N_3 = \frac{-2x + 4y - z}{8} \qquad N_4 = \frac{4x - 2z}{8} \qquad (12.2.27)$$

Note that $N_1 + N_2 + N_3 + N_4 = 1$ is again satisfied.

The 6×3 submatrices of the matrix $\underline{B}$, Eq. (12.2.13), are now evaluated using Eqs. (12.2.14) and (12.2.27) as

$$\underline{B}_1 = \begin{bmatrix} 0 & 0 & 0 \\ 0 & 0 & 0 \\ 0 & 0 & \frac{1}{2} \\ 0 & 0 & 0 \\ 0 & \frac{1}{2} & 0 \\ \frac{1}{2} & 0 & 0 \end{bmatrix} \qquad \underline{B}_2 = \begin{bmatrix} -\frac{1}{4} & 0 & 0 \\ 0 & -\frac{1}{2} & 0 \\ 0 & 0 & -\frac{1}{8} \\ -\frac{1}{2} & -\frac{1}{4} & 0 \\ 0 & -\frac{1}{8} & -\frac{1}{2} \\ -\frac{1}{8} & 0 & -\frac{1}{4} \end{bmatrix}$$

$$\underline{B}_3 = \begin{bmatrix} -\frac{1}{4} & 0 & 0 \\ 0 & \frac{1}{2} & 0 \\ 0 & 0 & -\frac{1}{8} \\ \frac{1}{2} & -\frac{1}{4} & 0 \\ 0 & -\frac{1}{8} & \frac{1}{2} \\ -\frac{1}{8} & 0 & -\frac{1}{4} \end{bmatrix} \qquad \underline{B}_4 = \begin{bmatrix} \frac{1}{2} & 0 & 0 \\ 0 & 0 & 0 \\ 0 & 0 & -\frac{1}{4} \\ 0 & \frac{1}{2} & 0 \\ 0 & -\frac{1}{4} & 0 \\ -\frac{1}{4} & 0 & \frac{1}{2} \end{bmatrix} \qquad (12.2.28)$$

Next, the matrix $\underline{D}$ is evaluated using Eq. (12.1.5) as

$$[D] = \frac{30 \times 10^6}{(1 + 0.3)(1 - 0.6)} \begin{bmatrix} 0.7 & 0.3 & 0.3 & 0 & 0 & 0 \\ & 0.7 & 0.3 & 0 & 0 & 0 \\ & & 0.7 & 0 & 0 & 0 \\ & & & 0.2 & 0 & 0 \\ & & & & 0.2 & 0 \\ \text{Symmetry} & & & & & 0.2 \end{bmatrix} \quad (12.2.29)$$

Finally, substituting the results from Eqs. (12.2.24), (12.2.28), and (12.2.29) into Eq. (12.2.18), we obtain the element stiffness matrix. The resulting 12×12 matrix, being cumbersome to obtain by longhand calculations, is best left for the computer to evaluate. ∎

12.3 Isoparametric Formulation

We now describe the isoparametric formulation of the stiffness matrix for some three-dimensional hexahedral elements. The basic (linear) hexahedral element now has eight corner nodes with isoparametric coordinates given by s, t, and z' as shown in Figure 12-4. The element faces are now defined by s, t, $z' = \pm 1$. (We use s, t, and z' for the coordinate axes because they are probably simpler to use than Greek letters ξ, η, and ζ).

The formulation of the stiffness matrix follows steps analogous to the isoparametric formulation of the stiffness matrix for the plane element in Chapter 11.

First, we expand Eq. (11.3.4) to include the z coordinate as follows:

$$\begin{Bmatrix} x \\ y \\ z \end{Bmatrix} = \sum_{i=1}^{8} \left(\begin{bmatrix} N_i & 0 & 0 \\ 0 & N_i & 0 \\ 0 & 0 & N_i \end{bmatrix} \begin{Bmatrix} x_i \\ y_i \\ z_i \end{Bmatrix} \right) \quad (12.3.1)$$

where the shape functions are now given by

$$N_i = \frac{(1 + ss_i)(1 + tt_i)(1 + z'z_i')}{8} \quad (12.3.2)$$

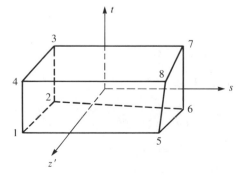

F I G U R E 12-4   Linear hexahedral element

with $s_i, t_i, z'_i = \pm 1$ and $i = 1, 2, \ldots, 8$. For instance,

$$N_1 = \frac{(1 + ss_1)(1 + tt_1)(1 + z'z'_1)}{8} \tag{12.3.3}$$

and when, from Figure 12-4, $s_1 = -1, t_1 = -1$, and $z'_1 = +1$ are used in Eq. (12.3.3), we obtain

$$N_1 = \frac{(1 - s)(1 - t)(1 + z')}{8} \tag{12.3.4}$$

Explicit forms of the other shape functions follow similarly.

The displacement functions now include w such that

$$\begin{Bmatrix} u \\ v \\ w \end{Bmatrix} = \sum_{i=1}^{8} \left(\begin{bmatrix} N_i & 0 & 0 \\ 0 & N_i & 0 \\ 0 & 0 & N_i \end{bmatrix} \begin{Bmatrix} u_i \\ v_i \\ w_i \end{Bmatrix} \right) \tag{12.3.5}$$

with the same shape functions as defined by Eq. (12.3.2), and the size of the shape function matrix now 3×24.

The Jacobian matrix, Eq. (11.3.10), is now expanded to

$$[J] = \begin{bmatrix} \dfrac{\partial x}{\partial s} & \dfrac{\partial y}{\partial s} & \dfrac{\partial z}{\partial s} \\[2mm] \dfrac{\partial x}{\partial t} & \dfrac{\partial y}{\partial t} & \dfrac{\partial z}{\partial t} \\[2mm] \dfrac{\partial x}{\partial z'} & \dfrac{\partial y}{\partial z'} & \dfrac{\partial z}{\partial z'} \end{bmatrix} \tag{12.3.6}$$

Since the strain/displacement relationships, given by Eq. (12.2.11) in terms of global coordinates, include differentiation with respect to z, we expand Eq. (11.3.9) as follows:

$$\frac{\partial f}{\partial x} = \frac{\begin{vmatrix} \dfrac{\partial f}{\partial s} & \dfrac{\partial y}{\partial s} & \dfrac{\partial z}{\partial s} \\[2mm] \dfrac{\partial f}{\partial t} & \dfrac{\partial y}{\partial t} & \dfrac{\partial z}{\partial t} \\[2mm] \dfrac{\partial f}{\partial z'} & \dfrac{\partial y}{\partial z'} & \dfrac{\partial z}{\partial z'} \end{vmatrix}}{|\underline{J}|} \qquad \frac{\partial f}{\partial y} = \frac{\begin{vmatrix} \dfrac{\partial x}{\partial s} & \dfrac{\partial f}{\partial s} & \dfrac{\partial z}{\partial s} \\[2mm] \dfrac{\partial x}{\partial t} & \dfrac{\partial f}{\partial t} & \dfrac{\partial z}{\partial t} \\[2mm] \dfrac{\partial x}{\partial z'} & \dfrac{\partial f}{\partial z'} & \dfrac{\partial z}{\partial z'} \end{vmatrix}}{|\underline{J}|}$$

$$\frac{\partial f}{\partial z} = \frac{\begin{vmatrix} \dfrac{\partial x}{\partial s} & \dfrac{\partial y}{\partial s} & \dfrac{\partial f}{\partial s} \\[2mm] \dfrac{\partial x}{\partial t} & \dfrac{\partial y}{\partial t} & \dfrac{\partial f}{\partial t} \\[2mm] \dfrac{\partial x}{\partial z'} & \dfrac{\partial y}{\partial z'} & \dfrac{\partial f}{\partial z'} \end{vmatrix}}{|\underline{J}|} \tag{12.3.7}$$

TABLE 12-1 Table of Gauss points for linear hexahedral element with associated weights

Points, i	s_i	t_i	z_i'	Weight, W_i
1	$-1/\sqrt{3}$	$-1/\sqrt{3}$	$1/\sqrt{3}$	1
2	$1/\sqrt{3}$	$-1/\sqrt{3}$	$1/\sqrt{3}$	1
3	$1/\sqrt{3}$	$1/\sqrt{3}$	$1/\sqrt{3}$	1
4	$-1/\sqrt{3}$	$1/\sqrt{3}$	$1/\sqrt{3}$	1
5	$-1/\sqrt{3}$	$-1/\sqrt{3}$	$-1/\sqrt{3}$	1
6	$1/\sqrt{3}$	$-1/\sqrt{3}$	$-1/\sqrt{3}$	1
7	$1/\sqrt{3}$	$1/\sqrt{3}$	$-1/\sqrt{3}$	1
8	$-1/\sqrt{3}$	$1/\sqrt{3}$	$-1/\sqrt{3}$	1

Note: $1/\sqrt{3} = 0.57735$.

Using Eqs. (12.3.7) by substituting u, v, and then w for f and using the definitions of the strains, we can express the strains in terms of natural coordinates (s, t, z') to obtain an equation similar to Eq. (11.3.14). In compact form, we can again express the strains in terms of the shape functions and global nodal coordinates similar to Eq. (11.3.15). The matrix $\underline{B}$, given by a form similar to Eq. (11.3.17), is now a function of s, t, and z', and of order 6×24.

The 24×24 stiffness matrix is now given by

$$[k] = \int_{-1}^{1} \int_{-1}^{1} \int_{-1}^{1} [B]^T [D] [B] |\underline{J}| \, ds \, dt \, dz' \tag{12.3.8}$$

Again, it is best to evaluate $[k]$ by numerical integration (also see Section 11.4); that is, we evaluate (integrate) the eight-noded hexahedral element stiffness matrix using a $2 \times 2 \times 2$ rule (or two-point rule). Actually eight points defined in Table 12-1 are used to evaluate $\underline{k}$ as

$$\underline{k} = \sum_{i=1}^{8} \underline{B}^T(s_i, t_i, z_i') \underline{D} \underline{B}(s_i, t_i, z_i') \underline{J}(s_i, t_i, z_i') W_i W_j W_k$$

where $W_i = W_j = W_k$ for the two-point rule.

For the quadratic hexahedral element shown in Figure 12-5, we have a total of 20 nodes with the inclusion of a total of 12 midside nodes. The development of the stiffness matrix follows the same steps previously outlined for the linear hexahedral element, where the shape functions now take on new forms. Again, letting s_i, t_i, $z_i' = \pm 1$, we have for the corner nodes ($i = 1, 2, \ldots, 8$),

$$N_i = \frac{(1 + ss_i)(1 + tt_i)(1 + z'z_i')}{8} (ss_i + tt_i + z'z_i' - 2) \tag{12.3.9}$$

For the midside nodes at $s_i = 0, t_i = \pm 1, z_i' = \pm 1$ ($i = 17, 18, 19, 20$), we have

$$N_i = \frac{(1 - s^2)(1 + tt_i)(1 + z'z_i')}{4} \tag{12.3.10}$$

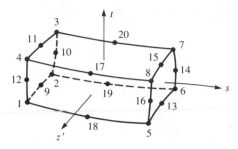

FIGURE 12-5 Quadratic hexahedral isoparametric element

For the midside nodes at $s_i = \pm 1, t_i = 0, z_i' = \pm 1$ ($i = 10, 12, 14, 16$), we have

$$N_i = \frac{(1 + ss_i)(1 - t^2)(1 + z'z_i')}{4} \qquad (12.3.11)$$

Finally, for the midside nodes at $s_i = \pm 1$, $t_i = \pm 1$, $z_i' = 0$ ($i = 9, 11, 13, 15$), we have

$$N_i = \frac{(1 + ss_i)(1 + tt_i)(1 - z'^2)}{4} \qquad (12.3.12)$$

Note that the stiffness matrix for the quadratic solid element is of order 60×60 because it has 20 nodes and 3 degrees of freedom per node.

The stiffness matrix for this 20-noded quadratic solid element can be evaluated using a $3 \times 3 \times 3$ rule (27 points). However, a special 14-point rule may be a better choice (see References [9] and [10]).

Figures 1-7 and 12-6 show applications of the use of linear and quadratic (curved sides) solid elements to model three-dimensional solids.

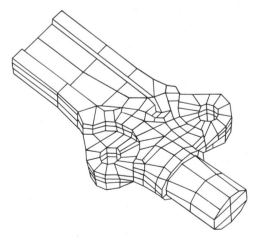

FIGURE 12-6 Finite element model of a forging using linear and quadratic solid elements

It has been shown in Reference [3] that use of the simple eight-node hexahedral element yields better results than use of the tetrahedral element discussed in Section 12.1. In any case, the use of three-dimensional elements results in a large number of equations to be solved simultaneously. Large bandwidths are usually created, causing large computer storage requirements. For instance, a model using a simple cube with, say, 20 by 20 by 20 nodes (= 8000 total nodes) for a region requires 8000 times 3 degrees of freedom per node (= 24,000) simultaneous equations. Furthermore, the bandwidth now involves an interconnection of some 20 by 20 (= 400) nodes or 400 times 3 degrees of freedom per node (= 1200) variables.

Finally, References [4], [5], [6], and [7] report on three-dimensional programs and analysis procedures using solid elements such as a family of subparametric curvilinear elements, linear tetrahedral elements, and eight-node linear and twenty-node quadratic isoparametric elements.

REFERENCES

[1] Martin, H. C., "Plane Elasticity Problems and the Direct Stiffness Method." *The Trend in Engineering*, Vol. 13, pp. 5–19, Jan. 1961.

[2] Gallagher, R. H., Padlog, J., and Bijlaard, P. P., "Stress Analysis of Heated Complex Shapes," *Journal of the American Rocket Society*, pp. 700–707, May 1962.

[3] Melosh, R. J., "Structural Analysis of Solids," *Journal of the Structural Division*, American Society of Civil Engineers, pp. 205–223, Aug. 1963.

[4] Chacour, S., "DANUTA, a Three Dimensional Finite Element Program Used in the Analysis of Turbo-Machinery," Transactions of the American Society of Mechanical Engineers, *Journal of Basic Engineering*, March 1972.

[5] Rashid, Y. R., "Three-Dimensional Analysis of Elastic Solids-I: Analysis Procedure," *International Journal of Solids and Structures*, Vol. 5, pp. 1311–1331, 1969.

[6] Rashid, Y. R., "Three-Dimensional Analysis of Elastic Solids-II: The Computational Problem," *International Journal of Solids and Structures*, Vol. 6, pp. 195–207, 1970.

[7] *Three-Dimensional Continuum Computer Programs for Structural Analysis*, Cruse, T. A., and Griffin, D. S., Eds., American Society of Mechanical Engineers, 1972.

[8] Zienkiewicz, O. C., *The Finite Element Method*, 3rd ed., McGraw-Hill, London, 1977.

[9] Irons, B. M., "Quadrature Rules for Brick Based Finite Elements," *International Journal for Numerical Methods in Engineering*, Vol. 3, No. 2, pp. 293–294, 1971.

[10] Hellen, T. K., "Effective Quadrature Rules for Quadratic Solid Isoparametric Finite Elements," *International Journal for Numerical Methods in Engineering*, Vol. 4, No. 4, pp. 597–599, 1972.

PROBLEMS

12.1   Evaluate the matrix $\underline{B}$ for the tetrahedral solid element shown in Figure P12-1.

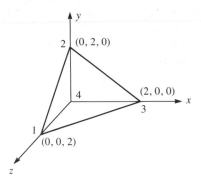

FIGURE P12-1

12.2   For the element shown in Figure P12-1, assume the nodal displacements have been determined to be

$$u_1 = 0.005 \text{ in.} \qquad v_1 = 0.0 \qquad w_1 = 0.0$$

$$u_2 = 0.001 \text{ in.} \qquad v_2 = 0.0 \qquad w_2 = 0.001 \text{ in.}$$

$$u_3 = 0.005 \text{ in.} \qquad v_3 = 0.0 \qquad w_3 = 0.0$$

$$u_4 = -0.001 \text{ in.} \qquad v_4 = 0.0 \qquad w_4 = 0.005 \text{ in.}$$

Determine the strains and then the stresses in the element. Let $E = 30 \times 10^6$ psi and $v = 0.3$.

12.3   Show that for constant body force Z_b acting on an element ($X_b = 0$ and $Y_b = 0$),

$$\{f_{bi}\} = \frac{V}{4} \left\{ \begin{array}{c} 0 \\ 0 \\ Z_b \end{array} \right\}$$

where $\{f_{bi}\}$ represents the body forces at node i of the element with volume V.

12.4   Evaluate the $\underline{B}$ matrix for the tetrahedral solid element shown in Figure P12-4. The coordinates are in units of millimeters.

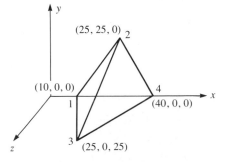

FIGURE P12-4

12.5   For the element shown in Figure P12-4, assume the nodal displacements have been determined to be

$$u_1 = 0.0 \qquad\qquad v_1 = 0.0 \qquad\qquad w_1 = 0.0$$

$$u_2 = 0.01 \text{ mm} \qquad v_2 = 0.02 \text{ mm} \qquad w_2 = 0.01 \text{ mm}$$

$$u_3 = 0.02 \text{ mm} \qquad v_3 = 0.01 \text{ mm} \qquad w_3 = 0.005 \text{ mm}$$

$$u_4 = 0.0 \qquad\qquad v_4 = 0.01 \text{ mm} \qquad w_4 = 0.01 \text{ mm}$$

Determine the strains and then the stresses in the element. Let $E = 210$ GPa and $v = 0.3$.

12.6   Express the explicit shape functions N_2 through N_8, similar to N_1 given by Eq. (12.3.4), for the linear hexahedral element shown in Figure 12-4.

12.7   Express the explicit shape functions for the corner nodes of the quadratic hexahedral element shown in Figure 12-5.

12.8   Write a computer program to evaluate $\underline{k}$ of Eq. (12.3.8) using a $2 \times 2 \times 2$ Gaussian Quadrature rule.

Heat Transfer and

Mass Transport

INTRODUCTION

In this chapter, we present the first use in this text of the finite element method for solution of nonstructural problems. We first consider the heat-transfer problem, although many similar problems, such as seepage through porous media, torsion of shafts, and magnetostatics (as listed in Reference [3]) can also be treated by the same form of equations (but with different physical characteristics) as that for heat transfer.

A study of the heat-transfer problem will allow the determination of the temperature distribution within a body. We can then determine the amount of heat moving into or out of the body, and the thermal stresses.

We begin with a derivation of the basic differential equation for heat conduction in one dimension, and then extend this derivation to the two-dimensional case. We will then review the units used for the physical quantities involved in heat transfer.

In preceding chapters dealing with stress analysis, we used the principle of minimum potential energy to derive the element equations, where an assumed displacement function within each element was used as a starting point in the derivation. We will now use a similar procedure for the nonstructural heat-transfer problem. We define an assumed temperature function within each element. Instead of minimizing a potential energy functional, a similar functional is minimized to obtain the element equations. In doing so, matrices analogous to the stiffness and force matrices of the structural problem result.

We will consider both one- and two-dimensional finite element formulations of the heat-transfer problem and provide illustrative examples of the determination of the temperature distribution along the length of a rod, and within a two-dimensional body.

Next, we will consider the contribution of fluid mass transport. The one-dimensional mass-transport phenomenon is included in the basic heat-transfer differential equation. Since it is not readily apparent that a variational formulation is possible for this problem, we will apply Galerkin's residual method directly to the differential equation to obtain the finite element equations. (You should note that the mass transport stiffness matrix is asymmetric.) We will compare an analytical solution to the finite element solution for a heat exchanger design/analysis problem to show the excellent agreement.

Finally, we will describe a computer program for two-dimensional heat transfer only and present an example of its use.

13.1 Derivation of the Basic Differential Equation

One-Dimensional Heat Conduction (Without Convection)

We now consider the derivation of the basic differential equation for the one-dimensional problem of heat conduction without convection. The purpose of this derivation is to present a physical insight into the heat-transfer phenomena, which must be understood so that the finite element formulation of the problem can be fully understood. (For additional information on heat transfer, consult texts such as References [1] and [2].) We begin with the control volume shown in Figure 13-1. By conservation of energy, we have

$$E_{in} + E_{generated} = \Delta U + E_{out} \qquad (13.1.1)$$

or
$$q_x A\, dt + QA\, dx\, dt = \Delta U + q_{x+dx} A\, dt \qquad (13.1.2)$$

where

E_{in} is the energy entering the control volume, in units of joules (J) or $kW \cdot h$ or Btu.

ΔU is the change in stored energy, in units of $kW \cdot h$ (kWh) or Btu.

q_x is the heat conducted (heat flux) into the control volume at surface edge x, in units of kW/m^2 or $Btu/(h\text{-}ft^2)$.

q_{x+dx} is the heat conducted out of control volume at the surface edge $x + dx$.

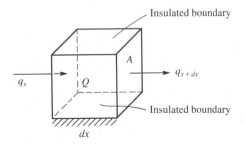

FIGURE 13-1   Control volume for one-dimensional heat conduction

t is time, in h or s (in U.S. customary units) or *s* (in SI units).

Q is the internal heat source (heat generated per unit time per unit volume), in kW/m^3 or $Btu/(h\text{-}ft^3)$ (a heat sink is negative).

A is the cross-sectional area perpendicular to heat flow *q*, in m^2 or ft^2.

By Fourier's law of heat conduction,

$$q_x = -K_{xx} \frac{dT}{dx} \qquad (13.1.3)$$

where

K_{xx} is the thermal conductivity in the *x* direction, in $kW/(m \cdot {}^\circ C)$ or $Btu/(h\text{-}ft\text{-}{}^\circ F)$.

T is the temperature, in °C or °F.

$\frac{dT}{dx}$ is the temperature gradient, in °C/m or °F/ft.

Equation (13.1.3) states that the heat flux in the *x* direction is proportional to the gradient of temperature in the *x* direction. The minus sign in Eq. (13.1.3) implies that, by convention, heat flow is positive in the direction opposite the direction of temperature increase. Equation (13.1.3) is analogous to the one-dimensional stress/strain law for the stress analysis problem—that is, to $\sigma_x = E(du/dx)$. Similarly,

$$q_{x+dx} = -K_{xx} \frac{dT}{dx}\bigg|_{x+dx} \qquad (13.1.4)$$

where the gradient in Eq. (13.1.4) is evaluated at $x + dx$. By Taylor series expansion, for any general function $f(x)$, we have

$$f_{x+dx} = f_x + \frac{df}{dx} dx + \frac{d^2 f}{dx^2} \frac{dx^2}{2} + \cdots$$

Therefore, using a two-term Taylor series, Eq. (13.1.4) becomes

$$q_{x+dx} = -\left[K_{xx} \frac{dT}{dx} + \frac{d}{dx}\left(K_{xx} \frac{dT}{dx} \right) dx \right] \qquad (13.1.5)$$

The change in stored energy can be expressed by

$$\Delta U = (\text{specific heat}) \times (\text{mass}) \times (\text{change in temperature})$$

$$= c(\rho A\, dx)\, dT \qquad (13.1.6)$$

where *c* is the specific heat in $kW \cdot h/(kg \cdot {}^\circ C)$ or $Btu/(slug\text{-}{}^\circ F)$, and ρ is the mass density in kg/m^3 or $slug/ft^3$. On substituting Eqs. (13.1.3), (13.1.5), and (13.1.6) into Eq. (13.1.2), dividing Eq. (13.1.2) by $A\, dx\, dt$, and simplifying, we have the one-dimensional heat conduction equation as

$$\frac{\partial}{\partial x}\left(K_{xx} \frac{\partial T}{\partial x} \right) + Q = \rho c \frac{\partial T}{\partial t} \qquad (13.1.7)$$

For steady state, any differentiation with respect to time is equal to zero, so

Eq. (13.1.7) becomes

$$\frac{d}{dx}\left(K_{xx}\frac{dT}{dx}\right) + Q = 0 \qquad (13.1.8)$$

For constant thermal conductivity and steady state, Eq. (13.1.7) becomes

$$K_{xx}\frac{d^2T}{dx^2} + Q = 0 \qquad (13.1.9)$$

The boundary conditions are of the form

$$T = T_B \qquad \text{on } S_1 \qquad (13.1.10)$$

where T_B represents a known boundary temperature and S_1 is a surface where the temperature is known, and

$$q_x^* = -K_{xx}\frac{dT}{dx} = \text{constant} \qquad \text{on } S_2 \qquad (13.1.11)$$

where S_2 is a surface where the prescribed heat flux q_x^* or temperature gradient is known. On an insulated boundary, $q_x^* = 0$.

Two-Dimensional Heat Conduction (Without Convection)

Consider the two-dimensional heat conduction problem in Figure 13-2. Similar to the one-dimensional case, for steady-state conditions, we can show that for material properties coinciding with the global x and y directions,

$$\frac{\partial}{\partial x}\left(K_{xx}\frac{\partial T}{\partial x}\right) + \frac{\partial}{\partial y}\left(K_{yy}\frac{\partial T}{\partial y}\right) + Q = 0 \qquad (13.1.12)$$

with boundary conditions

$$T = T_B \qquad \text{on } S_1 \qquad (13.1.13)$$

$$K_{xx}\frac{\partial T}{\partial x}C_x + K_{yy}\frac{\partial T}{\partial y}C_y = \text{constant} \qquad \text{on } S_2 \qquad (13.1.14)$$

where C_x and C_y are the direction cosines of the unit vector $\mathbf{n}$ normal to the surface S_2 shown in Figure 13-3.

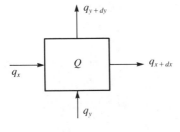

FIGURE 13-2   Control volume for two-dimensional heat conduction

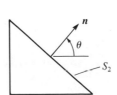

FIGURE 13-3   Unit vector normal to surface S_2

13.2   Heat Transfer with Convection

For a conducting solid in contact with a fluid, if heat transfer is taking place, the fluid will be in motion either through external pumping action (**forced convection**) or through the buoyancy forces created within the fluid by the temperature differences within it (**natural** or **free convection**).

We will now consider the derivation of the basic differential equation for one-dimensional heat conduction with convection. Figure 13-4 shows the control volume used in the derivation. Again, by Eq. (13.1.1) for conservation of energy, we have

$$q_x A \, dt + QA \, dx \, dt = c(\rho A \, dx) \, dT + q_{x+dx} A \, dt + q_h P \, dx \, dt \qquad (13.2.1)$$

In Eq. (13.2.1), all terms have the same meaning as in Section 13.1, except the heat flow by convective heat transfer is given by (see Reference [1] or [2])

$$q_h = h(T - T_\infty) \qquad (13.2.2)$$

where

h is the heat-transfer or convection coefficient, in $kW/(m^2 \cdot {}^\circ C)$ or $Btu/(h\text{-}ft^2\text{-}{}^\circ F)$.

T is the temperature of the solid surface, at the solid/fluid interface.

T_∞ is the temperature of the fluid (here the free-stream fluid temperature).

P in Eq. (13.2.1) denotes the perimeter around the constant cross-sectional area A.

Again, using Eqs. (13.1.3)–(13.1.6) and (13.2.2) in Eq. (13.2.1), dividing by $A \, dx \, dt$, and simplifying, we obtain the equation for one-dimensional heat conduction with convection as

$$\frac{\partial}{\partial x}\left(K_{xx}\frac{\partial T}{\partial x}\right) + Q = \rho c \frac{\partial T}{\partial t} + \frac{hP}{A}(T - T_\infty) \qquad (13.2.3)$$

with possible boundary conditions on (1) temperature, given by Eq. (13.1.10), and/or (2) temperature gradient, given by Eq. (13.1.11), and/or (3) loss of heat by convection from the ends of the one-dimensional body, as shown in Figure 13-5. Equating the heat flow in the solid wall to the heat flow in the fluid at the solid/fluid interface, we have

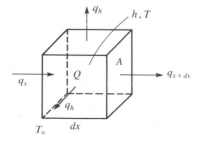

FIGURE 13-4   Control volume for one-dimensional heat conduction with convection

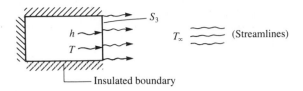

F I G U R E   13-5   Model illustrating convective heat transfer (arrows on surface S_3 indicate heat transfer by convection)

$$-K_{xx}\frac{dT}{dx} = h(T - T_\infty) \qquad \text{on } S_3 \qquad (13.2.4)$$

as a boundary condition for the problem of heat conduction with convection.

13.3 Typical Units; Thermal Conductivities, *K*; and Heat-Transfer Coefficients, *h*

Table 13-1 lists some typical units used for the heat-transfer problem.

T A B L E   13-1   Typical units for heat transfer

Variable	SI	U.S. Customary
Thermal conductivity, K	$kW/(m \cdot {}^\circ C)$	$Btu/(h\text{-}ft\text{-}{}^\circ F)$
Temperature, T	${}^\circ C$ or K	${}^\circ F$ or ${}^\circ R$
Internal heat source, Q	kW/m^3	$Btu/(h\text{-}ft^3)$
Heat flux, q	kW/m^2	$Btu/(h\text{-}ft^2)$
Convection coefficient, h	$kW/(m^2 \cdot {}^\circ C)$	$Btu/(h\text{-}ft^2\text{-}{}^\circ F)$
Energy, E	$kW \cdot h$	Btu
Specific heat, c	$(kW \cdot h)/(kg \cdot {}^\circ C)$	$Btu/(slug\text{-}{}^\circ F)$
Mass density, ρ	kg/m^3	$slug/ft^3$

Table 13-2 lists some typical thermal conductivities of various solids and liquids. The thermal conductivity K, in Btu/(h-ft-°F) or W/(m · °C), measures the amount of heat energy (Btu or W · h) that will flow through a unit length (ft or m) of a given substance in a unit time (h) to raise the temperature one degree (°F or °C).

Table 13-3 lists approximate ranges of values of convection coefficients for various conditions of convection. The heat transfer coefficient h, in Btu/(h-ft²-°F) or W/(m² · °C), measures the amount of heat energy (Btu or W · h) that will flow across a unit area (ft² or m²) of a given substance in a unit time (h) to raise the temperature 1° (°F or °C).

Natural or **free convection** occurs when, for instance, a heated plate is exposed to ambient room air without an external source of motion. This

T A B L E   13-2   Typical thermal conductivities of some solids and liquids

| | Solids | |
Material	K, Btu/(h-ft-°F)	K, W/(m·°C)
Aluminum, 0°C (32°F)	117	202
Steel (1% carbon), 0°C	20	35
Fiberglass, 20°C (68°F)	0.020	0.035
Concrete, 0°C	0.468–0.81	0.81–1.40
Earth, coarse gravelly, 20°C	0.300	0.520
Wood, oak, radial direction, 20°C	0.098	0.17
	Liquids	
Engine Oil, 20°C	0.084	0.145
Dry air, atmospheric pressure, 20°C	0.014	0.0243

T A B L E   13-3   Approximate values of convection heat-transfer coefficients (from Reference [1])

Mode	h, Btu/(h-ft²-°F)	h, W/(m²·°C)
Free convection, air	1–5	5–25
Forced convection, air	2–100	10–500
Forced convection, water	20–3,000	100–15,000
Boiling water	500–5,000	2,500–25,000
Condensation of water vapor	1,000–20,000	5,000–100,000

movement of the air, experienced as a result of the density gradients near the plate, is called *natural* or *free convection*. **Forced convection** is experienced, for instance, in the case of a fan blowing air over a plate.

13.4   Finite Element Formulation Using a Variational Method

The temperature distribution influences the amount of heat moving into or out of a body and also influences the stresses in a body. Thermal stresses occur in all bodies that experience a temperature gradient from some equilibrium state but are not free to expand in all directions. To evaluate thermal stresses, we need to know the temperature distribution in the body. The finite element method is a realistic method for predicting quantities such as temperature distribution and thermal stresses in a body. In this section, we formulate the two-dimensional heat-transfer equations in general terms using a variational method.

Equations (13.1.12)–(13.1.14) and (13.2.2) can be shown to be derivable (as shown, for instance, in References [4]–[6]) by the minimization of the following functional (analogous to the potential energy functional π_p):

$$\pi_h = U + \Omega_Q + \Omega_q + \Omega_h \qquad (13.4.1)$$

where

$$U = \frac{1}{2} \iiint_V \left[K_{xx} \left(\frac{\partial T}{\partial x} \right)^2 + K_{yy} \left(\frac{\partial T}{\partial y} \right)^2 \right] dV$$

$$\Omega_Q = - \iiint_V QT \, dV \qquad \Omega_q = - \iint_{S_2} q^*T \, dS \qquad \Omega_h = \frac{1}{2} \iint_{S_3} h(T - T_\infty)^2 \, dS$$

$$(13.4.2)$$

and where S_2 and S_3 are separate surface areas over which heat flow (flux) q^* and convection loss $h(T - T_\infty)$ are specified. We cannot specify q^* and h on the same surface because they cannot occur simultaneously on the same surface, as indicated by Eqs. (13.4.2).

Step 1 **Select Element Type**

We choose the two-dimensional triangular element similar to that used in Chapter 7 and shown in Figure 13-15.

Step 2 **Choose a Temperature Function**

We define the temperature functions T within each element in terms of the shape functions N and the nodal temperatures t as

$$\{T\} = [N]\{t\} \qquad (13.4.3)$$

where

$$\{t\} = \begin{Bmatrix} t_1 \\ t_2 \\ t_3 \\ \vdots \\ t_i \end{Bmatrix} \qquad (13.4.4)$$

(For the simple three-node triangular element of constant temperature gradient, $i = 3$.)

Step 3 **Define the Temperature Gradient/Temperature and Heat Flux/Temperature Gradient Relationships**

We define the gradient matrix analogous to the strain matrix used in the stress analysis problem as

$$\{g\} = \begin{Bmatrix} \dfrac{\partial T}{\partial x} \\[2mm] \dfrac{\partial T}{\partial y} \end{Bmatrix} \qquad (13.4.5)$$

The heat flux is given by

$$\begin{Bmatrix} q_x \\ q_y \end{Bmatrix} = -[D]\{g\}$$

where the material property matrix is

$$[D] = \begin{bmatrix} K_{xx} & 0 \\ 0 & K_{yy} \end{bmatrix} \tag{13.4.6}$$

Step 4   Derive the Element Conduction Matrix and Equations

Using Eqs. (13.4.3)–(13.4.6) in Eq. (13.4.2) and then using Eq. (13.4.1), π_h can be written in matrix form as

$$\pi_h = \frac{1}{2} \iiint_V [\{g\}^T[D]\{g\}] \, dV - \iiint_V \{t\}^T[N]^T Q \, dV$$

$$- \iint_{S_2} \{t\}^T[N]^T q^* \, dS + \frac{1}{2} \iint_{S_3} h[(\{t\}^T[N]^T - T_\infty)^2] \, dS \tag{13.4.7}$$

Using Eq. (13.4.3) in Eq. (13.4.5), we have

$$\{g\} = \begin{bmatrix} \dfrac{\partial N_1}{\partial x} & \dfrac{\partial N_2}{\partial x} & \cdots & \dfrac{\partial N_i}{\partial x} \\[2mm] \dfrac{\partial N_1}{\partial y} & \dfrac{\partial N_2}{\partial y} & \cdots & \dfrac{\partial N_i}{\partial y} \end{bmatrix} \begin{Bmatrix} t_1 \\ t_2 \\ t_3 \\ \vdots \\ t_i \end{Bmatrix} \tag{13.4.8}$$

Rewriting Eq. (13.4.8) in compact matrix form, we have

$$\{g\} = [B]\{t\} \tag{13.4.9}$$

where $[B]$ is the rectangular matrix on the right side of Eq. (13.4.8). On substituting Eq. (13.4.9) into Eq. (13.4.7) and using the fact that the nodal temperatures $\{t\}$ are independent of the general coordinates x and y and can therefore be taken outside of the integrals, we have

$$\pi_h = \frac{1}{2}\{t\}^T \iiint_V [B]^T[D][B] \, dV \{t\} - \{t\}^T \iiint_V [N]^T Q \, dV$$

$$- \{t\}^T \iint_{S_2} [N]^T q^* \, dS + \frac{1}{2} \iint_{S_3} h[\{t\}^T[N]^T[N]\{t\}$$

$$- (\{t\}^T[N]^T + [N]\{t\})T_\infty + T_\infty^2] \, dS \tag{13.4.10}$$

In Eq. (13.4.10), the minimization is most easily accomplished by explicitly writing the surface integral S_3 with $\{t\}$ left inside the integral as shown. On minimizing Eq. (13.4.10) with respect to $\{t\}$, we obtain

$$\frac{\partial \pi_h}{\partial \{t\}} = \int\int\int_V [B]^T [D][B] \, dV \{t\} - \int\int\int_V [N]^T Q \, dV$$

$$- \int\int_{S_2} [N]^T q^* \, dS + \int\int_{S_3} h[N]^T [N] \, dS \, \{t\}$$

$$- \int\int_{S_3} [N]^T h T_\infty \, dS = 0 \qquad (13.4.11)$$

Simplifying Eq. (13.4.11), we obtain

$$\left[\int\int\int_V [B]^T [D][B] \, dV + \int\int_{S_3} h[N]^T [N] \, ds \right] \{t\} = \{f_Q\} + \{f_q\} + \{f_h\}$$

$$(13.4.12)$$

where the force matrices have been defined by

$$\{f_Q\} = \int\int\int_V [N]^T Q \, dV \qquad \{f_q\} = \int\int_{S_2} [N]^T q^* \, dS$$

$$(13.4.13)$$

$$\{f_h\} = \int\int_{S_3} [N]^T h T_\infty \, dS$$

In Eq. (13.4.13), the first term $\{f_Q\}$ (heat source) is of the same form as the body force term, and the second term $\{f_q\}$ (heat flux) and third term $\{f_h\}$ (heat transfer or convection) are similar to surface tractions (distributed loading) in the stress analysis problem. You can observe this fact by comparing Eq. (13.4.13) with Eq. (7.2.46). Since we are formulating element equations of the form $f = \underline{kt}$, we have the element conduction matrix* for the heat-transfer problem given in Eq. (13.4.12) by

$$[k] = \int\int\int_V [B]^T [D][B] \, dV + \int\int_{S_3} h[N]^T [N] \, dS \qquad (13.4.14)$$

where the first and second integrals in Eq. (13.4.14) are the contributions of conduction and convection, respectively. Using Eq. (13.4.14) in Eq. (13.4.12), for each element, we have

$$\{f\} = [k]\{t\} \qquad (13.4.15)$$

---

*The element conduction matrix is often called the *stiffness matrix* because *stiffness matrix* is becoming a generally accepted term used to describe the matrix of known coefficients multiplied by the unknown degrees of freedom, such as temperatures, displacements, and so on.

Step 5   **Assemble the Element Equations to Obtain the Global Equations and Introduce Boundary Conditions**

We obtain the global or total structure conduction matrix using the same procedure as for the structural problem (called the *direct stiffness method* as described in Section 2.4); that is,

$$[K] = \sum_{e=1}^{N} [k^{(e)}] \qquad (13.4.16)$$

typically in units of kW/°C or Btu/(h-°F). The global force matrix is the sum of all element heat sources and is given by

$$\{F\} = \sum_{e=1}^{N} \{f^{(e)}\} \qquad (13.4.17)$$

typically in units of kW or Btu/h. The global equations are then

$$\{F\} = [K]\{t\} \qquad (13.4.18)$$

with the prescribed nodal temperature boundary conditions given by Eq. (13.1.13). Note that the boundary conditions on heat flux, Eq. (13.1.11), and convection, Eq. (13.2.4), are actually accounted for in the same manner as distributed loading was accounted for in the stress analysis problem; that is, they are included in the column of force matrices through a consistent approach (using the same shape functions used to derive $[k]$), as given by Eqs. (13.4.13).

The heat-transfer problem is now amenable to solution by the finite element method. The procedure used for solution is similar to that for the stress analysis problem. In Section 13.6, we will derive the specific equations used to solve the two-dimensional heat-transfer problem, after we first derive the equations used for the one-dimensional problem in Section 13.5.

13.5   One-Dimensional Finite Element Formulation

We first consider the one-dimensional finite element formulation of the heat-transfer problem, along with examples, to illustrate the solution of this type of problem.

Step 1   **Select Element Type**

The basic element with nodes 1 and 2 is shown in Figure 13-6.

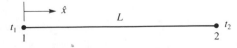

F I G U R E   13-6   Basic one-dimensional temperature element

Step 2 **Choose a Temperature Function**

We choose the temperature function, similar to the displacement function of Chapter 3, as

$$T = N_1 t_1 + N_2 t_2 \qquad (13.5.1)$$

where t_1 and t_2 are the nodal temperatures to be determined, and

$$N_1 = 1 - \frac{\hat{x}}{L} \qquad N_2 = \frac{\hat{x}}{L} \qquad (13.5.2)$$

are again the same shape functions as used for the bar element. The $[N]$ matrix is then given by

$$[N] = \left[1 - \frac{\hat{x}}{L} \quad \frac{\hat{x}}{L} \right] \qquad (13.5.3)$$

Step 3 **Define the Temperature Gradient/Temperature and Heat Flux/Temperature Gradient Relationships**

The temperature gradient matrix $\{g\}$, analogous to the strain matrix $\{\varepsilon\}$, is given by

$$\{g\} = \left\{ \frac{dT}{d\hat{x}} \right\} = [B]\{t\} \qquad (13.5.4)$$

where $[B]$ is obtained from its definition given in Eq. (13.4.8); that is,

$$[B] = \left[\frac{dN_1}{d\hat{x}} \quad \frac{dN_2}{d\hat{x}} \right]$$

Using Eqs. (13.5.2) in the definition for $[B]$, we have

$$[B] = \left[-\frac{1}{L} \quad \frac{1}{L} \right] \qquad (13.5.5)$$

The heat flux/temperature gradient relationship is given by

$$q_x = -[D]\{g\}$$

where the material property matrix is now given by

$$[D] = [K_{xx}] \qquad (13.5.6)$$

Step 4 **Derive the Element Conduction Matrix and Equations**

Using Eqs. (13.4.14), (13.5.5), and (13.5.6), the conduction part of the $[k]$ matrix for the one-dimensional element becomes

$$[k_c] = \iiint_V [B]^T [D][B] \, dV = \int_0^L \left\{ \begin{array}{c} -\dfrac{1}{L} \\ \dfrac{1}{L} \end{array} \right\} [K_{xx}] \left[-\frac{1}{L} \quad \frac{1}{L} \right] A \, dx$$

$$= \frac{AK_{xx}}{L^2} \int_0^L \left[\begin{array}{cc} 1 & -1 \\ -1 & 1 \end{array} \right] dx \qquad (13.5.7)$$

or, finally,

$$[k_c] = \frac{AK_{xx}}{L}\begin{bmatrix} 1 & -1 \\ -1 & 1 \end{bmatrix} \qquad (13.5.8)$$

The convection part of the $[k]$ matrix becomes

$$[k_h] = \int\!\!\int_{S_3} h[N]^T[N]\,dS = hP \int_0^L \left\{ \begin{array}{c} 1 - \dfrac{\hat{x}}{L} \\ \dfrac{\hat{x}}{L} \end{array} \right\} \left[1 - \dfrac{\hat{x}}{L} \quad \dfrac{\hat{x}}{L} \right] d\hat{x}$$

or, on integrating,

$$[k_h] = \frac{hPL}{6}\begin{bmatrix} 2 & 1 \\ 1 & 2 \end{bmatrix} \qquad (13.5.9)$$

where $dS = P\,d\hat{x}$

and P is the perimeter of the element (assumed to be constant). Therefore, adding Eqs. (13.5.8) and (13.5.9), the $[k]$ matrix is

$$[k] = \frac{AK_{xx}}{L}\begin{bmatrix} 1 & -1 \\ -1 & 1 \end{bmatrix} + \frac{hPL}{6}\begin{bmatrix} 2 & 1 \\ 1 & 2 \end{bmatrix} \qquad (13.5.10)$$

When h is zero on the boundary of an element, the second term on the right side of Eq. (13.5.10) (convection portion of $[k]$) is zero. This corresponds, for instance, to an insulated boundary.

The force matrix terms are

$$\{f_Q\} = \int\!\!\int\!\!\int_V [N]^T Q\,dV = QA \int_0^L \left\{ \begin{array}{c} 1 - \dfrac{\hat{x}}{L} \\ \dfrac{\hat{x}}{L} \end{array} \right\} d\hat{x} = \frac{QAL}{2}\left\{ \begin{array}{c} 1 \\ 1 \end{array} \right\} \qquad (13.5.11)$$

and

$$\{f_q\} = \int\!\!\int_{S_2} q^*[N]^T\,dS = q^*P \int_0^L \left\{ \begin{array}{c} 1 - \dfrac{\hat{x}}{L} \\ \dfrac{\hat{x}}{L} \end{array} \right\} d\hat{x} = \frac{q^*PL}{2}\left\{ \begin{array}{c} 1 \\ 1 \end{array} \right\} \qquad (13.5.12)$$

and

$$\{f_h\} = \int\!\!\int_{S_3} hT_\infty[N]^T\,dS = \frac{hT_\infty PL}{2}\left\{ \begin{array}{c} 1 \\ 1 \end{array} \right\} \qquad (13.5.13)$$

Therefore, adding Eqs. (13.5.11)–(13.5.13), we obtain

$$\{f\} = \frac{QAL + q^*PL + hT_\infty PL}{2}\left\{ \begin{array}{c} 1 \\ 1 \end{array} \right\} \qquad (13.5.14)$$

Equation (13.5.14) indicates that one-half of the assumed uniform heat source Q goes to each node, one-half of the prescribed uniform heat flux q^* (positive

FIGURE 13-7 Convection force from the end of an element

q^* enters the body) goes to each node, and one-half of the convection from the perimeter surface hT_∞ goes to each node of an element.

Finally, we must consider the convection from the free end of an element. For simplicity's sake, we will assume convection occurs only from the right end of the element, as shown in Figure 13-7. The additional convection term contribution to the stiffness matrix is given by

$$[k_h]_{\text{end}} = \int\int_{S_{\text{end}}} h[N]^T[N] \, dS \qquad (13.5.15)$$

Now $N_1 = 0$ and $N_2 = 1$ at the right end of the element. Substituting the N's into Eq. (13.5.15), we obtain

$$[k_h]_{\text{end}} = \int\int_{S_{\text{end}}} h\begin{Bmatrix}0\\1\end{Bmatrix}[0 \quad 1] \, dS = hA\begin{bmatrix}0 & 0\\0 & 1\end{bmatrix} \qquad (13.5.16)$$

The convection force from the free end of the element is obtained from the application of Eq. (13.5.13) with the shape functions now evaluated at the right end (where convection occurs) and with S_3 (the surface over which convection occurs) now equal to the cross-sectional area A of the rod. Hence,

$$\{f_h\}_{\text{end}} = hT_\infty A\begin{Bmatrix}N_1(\hat{x} = L)\\N_2(\hat{x} = L)\end{Bmatrix} = hT_\infty A\begin{Bmatrix}0\\1\end{Bmatrix} \qquad (13.5.17)$$

represents the convective force from the right end of an element where $N_1(\hat{x} = L)$ represents N_1 *evaluated at* $\hat{x} = L$, etc.

Step 5 **Assemble the Element Equations to Obtain the Global Equations and Introduce Boundary Conditions**

We now assemble the total stiffness matrix $[K]$, total force matrix $\{F\}$, and total set of equations as

$$[K] = \sum_{e=1}^{N} [k^{(e)}] \qquad \{F\} = \sum_{e=1}^{N} \{f^{(e)}\}$$

and

$$\{F\} = [K]\{t\}$$

This assemblage procedure is similar to the direct stiffness approach but is now based on the requirement that the temperatures at a common node between two elements are equal. This requirement is analogous to inter-element compatibility for displacements as described in Chapter 2. The boundary conditions on nodal temperatures are given by Eq. (13.1.13).

Step 6   Solve for the Nodal Temperatures

We now solve for the global nodal temperature, $\{t\}$, where the appropriate nodal temperature boundary conditions, Eq. (13.1.13), are specified.

Step 7   Solve for the Element Temperature Gradients and Heat Fluxes

Finally, we calculate the element temperature gradients from Eq. (13.5.4), and heat fluxes, typically from Eq. (13.1.3).

To illustrate the use of the equations developed in this section, we will now solve some one-dimensional heat-transfer problems.

E X A M P L E   13.1

Determine the temperature distribution along the length of the rod shown in Figure 13-8 with an insulated perimeter. The temperature at the left end is 100°F and the free-stream temperature is 10°F. Let $h = 10$ Btu/(h-ft²-°F) and $K_{xx} = 20$ Btu/(h-ft-°F). The value of h is typical for forced air convection and the value of K_{xx} is a typical conductivity for carbon steel (see Tables 13-2 and 13-3).

The finite element discretization is shown in Figure 13-9. For simplicity's sake, we will use four elements, each 10 in. long. There will be convective heat loss only over the right end of the rod because we consider the left end and the perimeter to be insulated. We calculate the stiffness matrices for each element as follows:

$$\frac{AK_{xx}}{L} = \frac{\pi(1 \text{ in.})^2[20 \text{ Btu/(h-ft-°F)}](1 \text{ ft}^2)}{\left(\dfrac{10 \text{ in.}}{12 \text{ in./ft}}\right)(144 \text{ in.}^2)}$$

$$= 0.5236 \text{ Btu/(h-°F)}$$

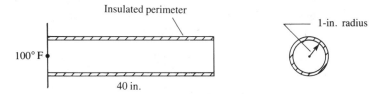

F I G U R E   13-8   One-dimensional rod subjected to temperature variation

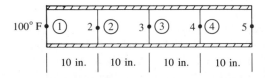

F I G U R E   13-9   Finite element discretized rod

$$\frac{hPL}{6} = \frac{[10 \text{ Btu/(h-ft}^2\text{-°F)}](2\pi)}{6}\left(\frac{1 \text{ in.}}{12 \text{ in./ft}}\right)\left(\frac{10 \text{ in.}}{12 \text{ in./ft}}\right)$$

$$= 0.7272 \text{ Btu/(h-°F)}$$

$$hT_\infty PL = [10 \text{ Btu/(h-ft}^2\text{-°F)}](10\text{°F})(2\pi)\left(\frac{1 \text{ in.}}{12 \text{ in./ft}}\right)\left(\frac{10 \text{ in.}}{12 \text{ in./ft}}\right)$$

$$= 43.63 \text{ Btu/h} \qquad (13.5.18)$$

In general, from Eqs. (13.5.10) and (13.5.15), we have

$$[k] = \frac{AK_{xx}}{L}\begin{bmatrix} 1 & -1 \\ -1 & 1 \end{bmatrix} + \frac{hPL}{6}\begin{bmatrix} 2 & 1 \\ 1 & 2 \end{bmatrix} + \int\int_{S_{end}} h[N]^T[N]\, dS \qquad (13.5.19)$$

Substituting Eqs. (13.5.18) into Eq. (13.5.19) for element 1, we have

$$[k^{(1)}] = 0.5236\begin{bmatrix} 1 & -1 \\ -1 & 1 \end{bmatrix} \text{Btu/(h-°F)} \qquad (13.5.20)$$

where the second and third terms on the right side of Eq. (13.5.19) are zero because there are no convection terms associated with element 1. Similarly, for elements 2 and 3, we have

$$[k^{(2)}] = [k^{(3)}] = [k^{(1)}] \qquad (13.5.21)$$

However, element 4 has an additional (convection) term owing to heat loss from the flat surface at its right end. Hence, using Eq. (13.5.16), we have

$$[k^{(4)}] = [k^{(1)}] + hA\begin{bmatrix} 0 & 0 \\ 0 & 1 \end{bmatrix}$$

$$= 0.5236\begin{bmatrix} 1 & -1 \\ -1 & 1 \end{bmatrix} + [10 \text{ Btu/(h-ft}^2\text{-°F)}]\pi\left(\frac{1 \text{ in.}}{12 \text{ in./ft}}\right)^2\begin{bmatrix} 0 & 0 \\ 0 & 1 \end{bmatrix}$$

$$= \begin{bmatrix} 0.5236 & -0.5236 \\ -0.5236 & 0.7418 \end{bmatrix} \text{Btu/(h-°F)} \qquad (13.5.22)$$

In general, we would use Eqs. (13.5.11), (13.5.12), (13.5.13), and (13.5.17) to obtain the element force matrices. However, in this example, $Q = 0$ (no heat source), $q^* = 0$ (no heat flux), and there is no convection except from the right end. Therefore,

$$\{f^{(1)}\} = \{f^{(2)}\} = \{f^{(3)}\} = 0 \qquad (13.5.23)$$

and

$$\{f^{(4)}\} = hT_\infty A\begin{Bmatrix} 0 \\ 1 \end{Bmatrix}$$

$$= [10 \text{ Btu/(h-ft}^2\text{-°F)}](10\text{°F})\pi\left(\frac{1 \text{ in.}}{12 \text{ in./ft}}\right)^2\begin{Bmatrix} 0 \\ 1 \end{Bmatrix}$$

$$= 2.182\begin{Bmatrix} 0 \\ 1 \end{Bmatrix} \text{Btu/h} \qquad (13.5.24)$$

The assembly of the element stiffness matrices, Eqs. (13.5.20)–(13.5.22), and the element force matrices, Eqs. (13.5.23) and (13.5.24), using the direct stiffness method, produces the following system of equations:

$$
\begin{bmatrix}
0.5236 & -0.5236 & 0 & 0 & 0 \\
-0.5236 & 1.0472 & -0.5236 & 0 & 0 \\
0 & -0.5236 & 1.0472 & -0.5236 & 0 \\
0 & 0 & -0.5236 & 1.0472 & -0.5236 \\
0 & 0 & 0 & -0.5236 & 0.7418
\end{bmatrix}
\begin{Bmatrix}
t_1 \\ t_2 \\ t_3 \\ t_4 \\ t_5
\end{Bmatrix}
=
\begin{Bmatrix}
F_1 \\ 0 \\ 0 \\ 0 \\ 2.182
\end{Bmatrix}
$$

$$(13.5.25)$$

where F_1 corresponds to an unknown rate of heat flow at node 1 (analogous to an unknown support force in the stress analysis problem). We have a known nodal temperature boundary condition of $t_1 = 100°F$. This non-homogeneous boundary condition must be treated in the same manner as was described for the stress analysis problem (see Chapter 4). We modify the stiffness (conduction) matrix and force matrix as follows:

$$
\begin{bmatrix}
1 & 0 & 0 & 0 & 0 \\
0 & 1.0472 & -0.5236 & 0 & 0 \\
0 & -0.5236 & 1.0472 & -0.5236 & 0 \\
0 & 0 & -0.5236 & 1.0472 & -0.5236 \\
0 & 0 & 0 & -0.5236 & 0.7418
\end{bmatrix}
\begin{Bmatrix}
t_1 \\ t_2 \\ t_3 \\ t_4 \\ t_5
\end{Bmatrix}
=
\begin{Bmatrix}
100 \\ 52.36 \\ 0 \\ 0 \\ 2.182
\end{Bmatrix}
$$

$$(13.5.26)$$

where the terms in the first row and column of the stiffness matrix corresponding to the known temperature condition, $t_1 = 100°F$, have been set equal to 0 except for the main diagonal, which has been set equal to 1, and the first row of the force matrix has been set equal to the known nodal temperature at node 1. Also, the term $(-0.5236) \times (100°F) = -52.36$ on the left side of the second equation of Eq. (13.5.25) has been transposed to the right side in the second row (as $+52.36$) of Eq. (13.5.26). The second through fifth equations of Eq. (13.5.26) corresponding to the rows of unknown nodal temperatures can now be solved (typically by Gaussian elimination). The resulting solution is given by

$$t_2 = 85.93°F \qquad t_3 = 71.87°F \qquad t_4 = 57.81°F \qquad t_5 = 43.75°F$$

$$(13.5.27)$$

For this elementary problem, the closed-form solution of the differential equation for conduction, Eq. (13.1.9), with the left-end boundary condition given by Eq. (13.1.10) and the right-end boundary condition given by Eq. (13.2.4) yields a linear temperature distribution through the length of the rod. The evaluation of this linear temperature function at 10-in. intervals (corresponding to the nodal points used in the finite element model) yields the same temperatures as obtained in this example by the finite element method. Since the temperature function was assumed to be linear in each finite element, this comparison is as expected. Note F_1 could be determined by the first of Eqs. (13.5.25). ∎

E X A M P L E 13.2

To illustrate more fully the use of the equations developed in Section 13.5, we will now solve the heat-transfer problem shown in Figure 13-10. For the one-dimensional rod insulated only at the left end, determine the temperatures at 3-in. increments along the length of the rod, and the rate of heat flow through element 1. Let $K_{xx} = 3$ Btu/(h-in.-°F), $h = 1.0$ Btu/(h-in.²-°F), and $T_\infty = 0°F$. The temperature at the left end of the rod is 200°F.

The finite element discretization is shown in Figure 13-11. Three elements are sufficient to enable the determination of temperatures at the four points along the rod, although more elements would yield answers more closely approximating the analytical solution obtained by solving the differential equation such as Eq. (13.2.3) with the partial derivative with respect to time equal to zero. There will be convective heat loss over the perimeter and the right end of the rod. The insulated left end will not have convective heat loss. Using Eqs. (13.5.10) and (13.5.16), we calculate the stiffness matrices for each element as follows:

$$\frac{AK_{xx}}{L} = \frac{(4\pi)(3)}{3} = 4\pi \ \text{Btu/(h-°F)}$$

$$\frac{hPL}{6} = \frac{(1)(4\pi)(3)}{6} = 2\pi \ \text{Btu/(h-°F)} \tag{13.5.28}$$

$$hA = (1)(4\pi) = 4\pi \ \text{Btu/(h-°F)}$$

Substituting the results of Eqs. (13.5.28) into Eq. (13.5.10), we obtain the stiffness matrix for element 1 as

$$[k^{(1)}] = 4\pi \begin{bmatrix} 1 & -1 \\ -1 & 1 \end{bmatrix} + 2\pi \begin{bmatrix} 2 & 1 \\ 1 & 2 \end{bmatrix}$$

$$= 4\pi \begin{bmatrix} 2 & -\frac{1}{2} \\ -\frac{1}{2} & 2 \end{bmatrix} \text{Btu/(h-°F)} \tag{13.5.29}$$

Since there is no convection across the ends of element 1 (its left end is insulated and its right end is inside the whole rod and thus not exposed to fluid motion), the contribution to the stiffness matrix owing to convection

200°F

9 in.

F I G U R E 13-10 One-dimensional rod subjected to temperature variation

200°F

3 in. 3 in. 3 in.

F I G U R E 13-11 Finite element discretized rod

from an end of the element, such as given by Eq. (13.5.16), is zero. Similarly,

$$[k^{(2)}] = [k^{(1)}] = 4\pi \begin{bmatrix} 2 & -\frac{1}{2} \\ -\frac{1}{2} & 2 \end{bmatrix} \text{Btu/(h-°F)} \qquad (13.5.30)$$

However, element 3 has an additional (convection) term owing to heat loss from the exposed surface at its right end. Therefore, Eq. (13.5.16) yields a contribution to the element 3 stiffness matrix, which is then given by

$$[k^{(3)}] = [k^{(1)}] + hA \begin{bmatrix} 0 & 0 \\ 0 & 1 \end{bmatrix}$$

$$= 4\pi \begin{bmatrix} 2 & -\frac{1}{2} \\ -\frac{1}{2} & 2 \end{bmatrix} + 4\pi \begin{bmatrix} 0 & 0 \\ 0 & 1 \end{bmatrix}$$

$$= 4\pi \begin{bmatrix} 2 & -\frac{1}{2} \\ -\frac{1}{2} & 3 \end{bmatrix} \text{Btu/(h-°F)} \qquad (13.5.31)$$

In general, we calculate the force matrices by using Eqs. (13.5.14) and (13.5.17). Since $Q = 0$, $q^* = 0$, and $T_\infty = 0$°F, all force terms are equal to zero.

The assembly of the element matrices, Eqs. (13.5.29)–(13.5.31), using the direct stiffness method, produces the following system of equations:

$$4\pi \begin{bmatrix} 2 & -\frac{1}{2} & 0 & 0 \\ -\frac{1}{2} & 4 & -\frac{1}{2} & 0 \\ 0 & -\frac{1}{2} & 4 & -\frac{1}{2} \\ 0 & 0 & -\frac{1}{2} & 3 \end{bmatrix} \begin{Bmatrix} t_1 \\ t_2 \\ t_3 \\ t_4 \end{Bmatrix} = \begin{Bmatrix} F_1 \\ 0 \\ 0 \\ 0 \end{Bmatrix} \qquad (13.5.32)$$

We have a known nodal temperature boundary condition of $t_1 = 200$°F. As in Example 13.1, we modify the conduction matrix and force matrix as follows:

$$4\pi \begin{bmatrix} 1 & 0 & 0 & 0 \\ 0 & 4 & -\frac{1}{2} & 0 \\ 0 & -\frac{1}{2} & 4 & -\frac{1}{2} \\ 0 & 0 & -\frac{1}{2} & 3 \end{bmatrix} \begin{Bmatrix} t_1 \\ t_2 \\ t_3 \\ t_4 \end{Bmatrix} = \begin{Bmatrix} 200 \\ 400\pi \\ 0 \\ 0 \end{Bmatrix} \qquad (13.5.33)$$

where the terms in the first row and column of the conduction matrix corresponding to the known temperature condition, $t_1 = 200$°F, have been set equal to zero except for the main diagonal, which has been set to equal one, and the row of the force matrix has been set equal to the known nodal temperature at node 1. Also, the term $(-1/2)(200)(4\pi) = -400\pi$ on the left side of the second equation of Eq. (13.5.32) has been transposed to the right side in the second row (as $+400\pi$) of Eq. (13.5.33). The second through fourth equations of Eq. (13.5.33), corresponding to the rows of unknown nodal temperatures, can now be solved. The resulting solution is given by

$$t_2 = 25.4°F \qquad t_3 = 3.24°F \qquad t_4 = 0.54°F \qquad (13.5.34)$$

Next, we determine the heat flux for element 1 by using Eqs. (13.1.3) and (13.5.4) as

$$q^{(1)} = -K_{xx}[B]\{t\} \qquad (13.5.35)$$

Using Eq. (13.5.5) in Eq. (13.5.35), we have

$$q^{(1)} = -K_{xx} \begin{bmatrix} -\dfrac{1}{L} & \dfrac{1}{L} \end{bmatrix} \begin{Bmatrix} t_1 \\ t_2 \end{Bmatrix} \qquad (13.5.36)$$

Substituting the numerical values into Eq. (13.5.36), we obtain

$$q^{(1)} = -3 \begin{bmatrix} -\dfrac{1}{3} & \dfrac{1}{3} \end{bmatrix} \begin{Bmatrix} 200 \\ 25.4 \end{Bmatrix}$$

or
$$q^{(1)} = 174.6 \text{ Btu/(h-in.}^2) \qquad (13.5.37)$$

We then determine the rate of heat flow $\bar{q}$ by multiplying Eq. (13.5.37) by the cross-sectional area over which q acts. Therefore,

$$\bar{q}^{(1)} = 174.6(4\pi) = 2194 \text{ Btu/h} \qquad (13.5.38) \quad \blacksquare$$

EXAMPLE 13.3

The plane wall shown in Figure 13-12 is 1 m thick. The left surface of the wall ($x = 0$) is maintained at a constant temperature of 200°C, and the right surface ($x = L = 1$ m) is insulated. The thermal conductivity is $K_{xx} = 25$ W/(m · °C) and there is a uniform generation of heat inside the wall of $Q = 400$ W/m³. Determine the temperature distribution through the wall thickness.

This problem is assumed to be approximated as a one-dimensional heat-transfer problem. The discretized model of the wall is shown in Figure 13-13. For simplicity, we use four equal-length elements all with unit cross-sectional area ($A = 1$ m²). The unit area represents a typical cross section of the wall. The perimeter of the wall model is then insulated to obtain the correct conditions.

Using Eqs. (13.5.10) and (13.5.16), we calculate the element stiffness matrices as follows:

$$\frac{AK_{xx}}{L} = \frac{(1 \text{ m}^2)[25 \text{ W/(m · °C)}]}{0.25 \text{ m}} = 100 \text{ W/°C}$$

For each identical element, we have

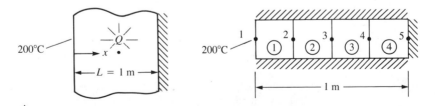

▲
FIGURE 13-12 Conduction in a plane wall subjected to uniform heat generation

FIGURE 13-13 Discretized model of Figure 13-12

$$[k] = 100 \begin{bmatrix} 1 & -1 \\ -1 & 1 \end{bmatrix} \text{W/°C} \qquad (13.5.39)$$

Since no convection occurs, h is equal to zero; therefore, there is no convection contribution to $\underline{k}$.

The element force matrices are given by Eq. (13.5.14). With $Q = 400$ W/m^3, $q = 0$, and $h = 0$, Eq. (13.5.14) becomes

$$\{f\} = \frac{QAL}{2} \begin{Bmatrix} 1 \\ 1 \end{Bmatrix} \qquad (13.5.40)$$

Evaluating Eq. (13.5.40) for a typical element, such as element 1, we obtain

$$\begin{Bmatrix} f_{1x} \\ f_{2x} \end{Bmatrix} = \frac{(400 \text{ W/m}^3)(1 \text{ m}^2)(0.25 \text{ m})}{2} \begin{Bmatrix} 1 \\ 1 \end{Bmatrix} = \begin{Bmatrix} 50 \\ 50 \end{Bmatrix} \text{W} \qquad (13.5.41)$$

The force matrices for all other elements are equal to Eq. (13.5.41).

The assemblage of the element matrices, Eqs. (13.5.39) and (13.5.41) and the other force matrices similar to Eq. (13.5.41), yields

$$100 \begin{bmatrix} 1 & -1 & 0 & 0 & 0 \\ -1 & 2 & -1 & 0 & 0 \\ 0 & -1 & 2 & -1 & 0 \\ 0 & 0 & -1 & 2 & -1 \\ 0 & 0 & 0 & -1 & 1 \end{bmatrix} \begin{Bmatrix} t_1 \\ t_2 \\ t_3 \\ t_4 \\ t_5 \end{Bmatrix} = \begin{Bmatrix} F_1 + 50 \\ 100 \\ 100 \\ 100 \\ 50 \end{Bmatrix} \qquad (13.5.42)$$

Substituting the known temperature $t_1 = 200°C$ into Eq. (13.5.42), dividing both sides of Eq. (13.5.42) by 100, and transposing known terms to the right side, we have

$$\begin{bmatrix} 1 & 0 & 0 & 0 & 0 \\ 0 & 2 & -1 & 0 & 0 \\ 0 & -1 & 2 & -1 & 0 \\ 0 & 0 & -1 & 2 & -1 \\ 0 & 0 & 0 & -1 & 1 \end{bmatrix} \begin{Bmatrix} t_1 \\ t_2 \\ t_3 \\ t_4 \\ t_5 \end{Bmatrix} = \begin{Bmatrix} 200°C \\ 201 \\ 1 \\ 1 \\ 0.5 \end{Bmatrix} \qquad (13.5.43)$$

The second through fifth equations of Eq. (13.5.43) can now be solved simultaneously to yield

$$t_2 = 203.5°C \qquad t_3 = 206°C \qquad t_4 = 207.5°C \qquad t_5 = 208°C \qquad (13.5.44)$$

Using the first of Eqs. (13.5.42), the rate of heat flow out the left end is

$$F_1 = 100(t_1 - t_2) - 50$$
$$F_1 = 100(200 - 203.5) - 50$$
$$F_1 = -400 \text{ W}$$

The closed-form solution of the differential equation for conduction, Eq. (13.1.9), with the left-end boundary condition given by Eq. (13.1.10) and the right-end boundary condition given by Eq. (13.1.11), and with $q_x^* = 0$, is shown in Reference [2] to yield a parabolic temperature distribution through the wall. Evaluating the expression for the temperature function given in

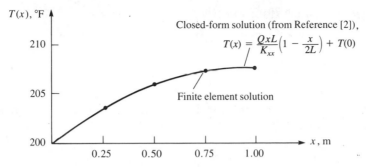

FIGURE 13-14   Comparison of the finite element and closed-form
solutions for Example 13.3

Reference [2] for values of x corresponding to the node points of the finite element model, we obtain

$$t_2 = 203.5°C \qquad t_3 = 206°C \qquad t_4 = 207.5°C \qquad t_5 = 208°C \qquad (13.5.45)$$

Figure 13-14 is a plot of the closed-form solution and the finite element solution for the temperature variation through the wall. The finite element nodal values and the closed-form values are equal, because the consistent equivalent force matrix has been used. (This was also discussed in Section 3.9 and 3.10 for the axial bar subjected to distributed loading). ∎

Finally, remember that the most important advantage of the finite element method comes in being able to approximate, with high confidence, more-complicated problems, such as those with more than one thermal conductivity, for which closed-form solutions are difficult (if not impossible) to obtain. The automation of the finite element method through general computer programs makes the method extremely powerful.

13.6 Two-Dimensional Finite Element Formulation

Since many bodies can be modeled as two-dimensional heat-transfer problems, we now develop the equations for an element appropriate for these problems. Examples using this element then follow.

Step 1 Select Element Type

The three-noded triangular element in Figure 13-15 is the basic element for the solution of the two-dimensional heat-transfer problem.

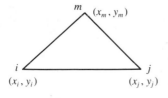

FIGURE 13-15   Basic triangular element

Step 2   **Select a Temperature Function**

The temperature function is given by

$$\{T\} = [N_i \quad N_j \quad N_m] \begin{Bmatrix} t_i \\ t_j \\ t_m \end{Bmatrix} \qquad (13.6.1)$$

where t_i, t_j, and t_m are the nodal temperatures, and the shape functions are again given by Eqs. (7.2.18); that is,

$$N_i = \frac{1}{2A}(\alpha_i + \beta_i x + \gamma_i y) \qquad (13.6.2)$$

with similar expressions for N_j and N_m. Here the α's, β's, and γ's are defined by Eqs. (7.2.10).

Step 3   **Define the Temperature Gradient/Temperature and Heat Flux/Temperature Gradient Relationships**

The gradient matrix $\{g\}$, analogous to the strain matrix $\{\varepsilon\}$ of the stress analysis problem, is given by

$$\{g\} = [B]\{t\} \qquad (13.6.3)$$

where the $[B]$ matrix is obtained by substituting the three equations suggested by Eq. (13.6.2) in Eq. (13.4.8) as

$$[B] = \frac{1}{2A}\begin{bmatrix} \beta_i & \beta_j & \beta_m \\ \gamma_i & \gamma_j & \gamma_m \end{bmatrix} \qquad (13.6.4)$$

The heat flux/temperature gradient relationship is now

$$\begin{Bmatrix} q_x \\ q_y \end{Bmatrix} = -[D]\{g\}$$

where the material property matrix is

$$[D] = \begin{bmatrix} K_{xx} & 0 \\ 0 & K_{yy} \end{bmatrix} \qquad (13.6.5)$$

Step 4   **Derive the Element Conduction Matrix and Equations**

The element stiffness matrix from Eq. (13.4.14) is

$$[k] = \iiint_V [B]^T [D][B] \, dV + \iint_{S_3} h[N]^T [N] \, dS \qquad (13.6.6)$$

where

$$[k_c] = \iiint_V [B]^T [D][B] \, dV$$

$$= \iiint_V \frac{1}{4A^2} \begin{bmatrix} \beta_i & \gamma_i \\ \beta_j & \gamma_j \\ \beta_m & \gamma_m \end{bmatrix} \begin{bmatrix} K_{xx} & 0 \\ 0 & K_{yy} \end{bmatrix} \begin{bmatrix} \beta_i & \beta_j & \beta_m \\ \gamma_i & \gamma_j & \gamma_m \end{bmatrix} dV \qquad (13.6.7)$$

Assuming constant thickness in the element and noting that all terms of the integrand of Eq. (13.6.7) are constant, we have

$$[k_c] = \iiint_V [B]^T[D][B] \, dV = tA[B]^T[D][B] \qquad (13.6.8)$$

Equation (13.6.8) is the true conduction portion of the total stiffness matrix Eq. (13.6.6). The second integral of Eq. (13.6.6) (the convection portion of the total stiffness matrix) is defined by

$$[k_h] = \iint_{S_3} h[N]^T[N] \, dS \qquad (13.6.9)$$

We can explicitly multiply the matrices in Eq. (13.6.9) to obtain

$$[k_h] = h \iint_{S_3} \begin{bmatrix} N_iN_i & N_iN_j & N_iN_m \\ N_jN_i & N_jN_j & N_jN_m \\ N_mN_i & N_mN_j & N_mN_m \end{bmatrix} dS \qquad (13.6.10)$$

To illustrate the use of Eq. (13.6.10), consider the side between nodes i and j of the triangular element to be subjected to convection (see Figure 13-16). Then $N_m = 0$ along side i-j, and we obtain

$$[k_h] = \frac{hL_{i\text{-}j}t}{6} \begin{bmatrix} 2 & 1 & 0 \\ 1 & 2 & 0 \\ 0 & 0 & 0 \end{bmatrix} \qquad (13.6.11)$$

where $L_{i\text{-}j}$ is the length of side i-j.

The evaluation of the force matrix integrals in Eq. (13.4.13) is as follows:

$$\{f_Q\} = \iiint_V Q[N]^T \, dV = Q \iiint_V [N]^T \, dV \qquad (13.6.12)$$

for constant heat source Q. Thus it can be shown (left to your discretion) that this integral is equal to

$$\{f_Q\} = \frac{QV}{3} \begin{Bmatrix} 1 \\ 1 \\ 1 \end{Bmatrix} \qquad (13.6.13)$$

where $V = At$ is the volume of the element. Equation (13.6.13) indicates that heat is generated by the body in three equal parts to the nodes (like body forces in the elasticity problem). The second force matrix in Eq. (13.4.13) is

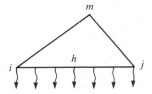

FIGURE 13-16   Heat loss by convection from side i-j

$$\{f_q\} = \int\int_{S_2} q*[N]^T \, dS = \int\int_{S_2} q* \left\{ \begin{array}{c} N_i \\ N_j \\ N_m \end{array} \right\} dS \qquad (13.6.14)$$

This reduces to

$$\frac{q*L_{i\text{-}j}t}{2} \left\{ \begin{array}{c} 1 \\ 1 \\ 0 \end{array} \right\} \qquad \text{on side } i\text{-}j \qquad (13.6.15)$$

$$\frac{q*L_{j\text{-}m}t}{2} \left\{ \begin{array}{c} 0 \\ 1 \\ 1 \end{array} \right\} \qquad \text{on side } j\text{-}m \qquad (13.6.16)$$

$$\frac{q*L_{m\text{-}i}t}{2} \left\{ \begin{array}{c} 1 \\ 0 \\ 1 \end{array} \right\} \qquad \text{on side } m\text{-}i \qquad (13.6.17)$$

where $L_{i\text{-}j}$, $L_{j\text{-}m}$, $L_{m\text{-}i}$ are the lengths of the sides of the element, and $q*$ is assumed constant over each edge. The integral $\int\int_{S_3} hT_\infty[N]^T \, dS$ can be found similar to Eq. (13.6.14) by simply replacing $q*$ with hT_∞ in Eqs. (13.6.15)–(13.6.17).

Steps 5, 6, and 7

Steps 5, 6, and 7 are identical to those described in Section 13.5.

To illustrate the use of the equations presented in Section 13.6, we will now solve some two-dimensional heat-transfer problems.

E X A M P L E   13.4

For the two-dimensional body shown in Figure 13-17, determine the temperature distribution. The temperature at the left side of the body is maintained at 100°F. The edges on the top and bottom of the body are insulated. There is heat convection from the right side with convection coefficient $h = 20$ Btu/(h-ft²-°F). The free-stream temperature is $T_\infty = 50$°F. The coefficients of thermal conductivity are $K_{xx} = K_{yy} = 25$ Btu/(h-ft-°F). The dimensions are shown in the figure. Assume the thickness to be 1 ft.

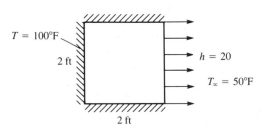

FIGURE 13-17 Two-dimensional body subjected to temperature variation and convection

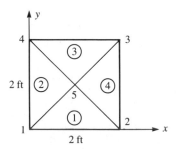

FIGURE 13-18 Discretized two-dimensional body

The finite element discretization is shown in Figure 13-18. We will use four triangular elements of equal size for simplicity of the longhand solution. There will be convective heat loss only over the right side of the body because the other faces are insulated. We now calculate the element stiffness matrices using Eq. (13.6.8) applied for all elements, and Eq. (13.6.11) applied for element 4 only, because convection is occurring only across one edge of element 4.

Element 1

The coordinates of the element 1 nodes are $x_1 = 0$, $y_1 = 0$, $x_2 = 2$, $y_2 = 0$, $x_5 = 1$, and $y_5 = 1$. Using these coordinates and Eqs. (7.2.10), we obtain

$$\beta_1 = 0 - 1 = -1 \qquad \beta_2 = 1 - 0 = 1 \qquad \beta_5 = 0 - 0 = 0$$
$$\gamma_1 = 1 - 2 = -1 \qquad \gamma_2 = 0 - 1 = -1 \qquad \gamma_5 = 2 - 0 = 2 \tag{13.6.18}$$

Using Eqs. (13.6.18) in Eq. (13.6.8), we obtain

$$[k_c^{(1)}] = \frac{1(1)}{2(2)} \begin{bmatrix} -1 & -1 \\ 1 & -1 \\ 0 & 2 \end{bmatrix} \begin{bmatrix} 25 & 0 \\ 0 & 25 \end{bmatrix} \begin{bmatrix} -1 & 1 & 0 \\ -1 & -1 & 2 \end{bmatrix} \tag{13.6.19}$$

Simplifying Eq. (13.6.19), we obtain

$$[k_c^{(1)}] = \begin{matrix} 1 & 2 & 5 \\ \begin{bmatrix} 12.5 & 0 & -12.5 \\ 0 & 12.5 & -12.5 \\ -12.5 & -12.5 & 25 \end{bmatrix} \end{matrix} \text{Btu/(h-°F)} \tag{13.6.20}$$

where the numbers above the columns indicate the node numbers associated with the matrix.

Element 2

The coordinates of the element 2 nodes are $x_1 = 0$, $y_1 = 0$, $x_5 = 1$, $y_5 = 1$, $x_4 = 0$, and $y_4 = 2$. Using these coordinates, we obtain

$$\beta_1 = 1 - 2 = -1 \qquad \beta_5 = 2 - 0 = 2 \qquad \beta_4 = 0 - 1 = -1$$
$$\gamma_1 = 0 - 1 = -1 \qquad \gamma_5 = 0 - 0 = 0 \qquad \gamma_4 = 1 - 0 = 1 \tag{13.6.21}$$

Using Eqs. (13.6.21) in Eq. (13.6.8), we obtain

$$[k_c^{(2)}] = \frac{1}{4} \begin{bmatrix} -1 & -1 \\ 2 & 0 \\ -1 & 1 \end{bmatrix} \begin{bmatrix} 25 & 0 \\ 0 & 25 \end{bmatrix} \begin{bmatrix} -1 & 2 & -1 \\ -1 & 0 & 1 \end{bmatrix} \tag{13.6.22}$$

Simplifying Eq. (13.6.22), we obtain

$$[k_c^{(2)}] = \begin{matrix} 1 & 5 & 4 \\ \begin{bmatrix} 12.5 & -12.5 & 0 \\ -12.5 & 25 & -12.5 \\ 0 & -12.5 & 12.5 \end{bmatrix} \end{matrix} \text{Btu/(h-°F)} \tag{13.6.23}$$

Element 3

The coordinates of the element 3 nodes are $x_4 = 0$, $y_4 = 2$, $x_5 = 1$, $y_5 = 1$, $x_3 = 2$, and $y_3 = 2$. Using these coordinates, we obtain

$$\beta_4 = 1 - 2 = -1 \qquad \beta_5 = 2 - 2 = 0 \qquad \beta_3 = 2 - 1 = 1$$

$$\gamma_4 = 2 - 1 = 1 \qquad \gamma_5 = 0 - 2 = -2 \qquad \gamma_3 = 1 - 0 = 1 \qquad (13.6.24)$$

Using Eqs. (13.6.24) in Eq. (13.6.8), we obtain

$$[k_c^{(3)}] = \begin{array}{c} \\ \begin{array}{ccc} 4 & 5 & 3 \end{array} \\ \left[\begin{array}{ccc} 12.5 & -12.5 & 0 \\ -12.5 & 25 & -12.5 \\ 0 & -12.5 & 12.5 \end{array} \right] \end{array} \text{Btu/(h-}^\circ\text{F)} \qquad (13.6.25)$$

Element 4

The coordinates of the element 4 nodes are $x_2 = 2$, $y_2 = 0$, $x_3 = 2$, $y_3 = 2$, $x_5 = 1$, and $y_5 = 1$. Using these coordinates, we obtain

$$\beta_2 = 2 - 1 = 1 \qquad \beta_3 = 1 - 0 = 1 \qquad \beta_5 = 0 - 2 = -2$$

$$\gamma_2 = 1 - 2 = -1 \qquad \gamma_3 = 2 - 1 = 1 \qquad \gamma_5 = 2 - 2 = 0 \qquad (13.6.26)$$

Using Eqs. (13.6.26) in Eq. (13.6.8), we obtain

$$[k_c^{(4)}] = \begin{array}{c} \\ \begin{array}{ccc} 2 & 3 & 5 \end{array} \\ \left[\begin{array}{ccc} 12.5 & 0 & -12.5 \\ 0 & 12.5 & -12.5 \\ -12.5 & -12.5 & 25 \end{array} \right] \end{array} \text{Btu/(h-}^\circ\text{F)} \qquad (13.6.27)$$

For element 4, we have a convection contribution to the total stiffness matrix because side $2 - 3$ is exposed to the free-stream temperature. Using Eq. (13.6.11) with $i = 2$ and $j = 3$, we obtain

$$[k_h^{(4)}] = \frac{(20)(2)(1)}{6} \left[\begin{array}{ccc} 2 & 1 & 0 \\ 1 & 2 & 0 \\ 0 & 0 & 0 \end{array} \right] \qquad (13.6.28)$$

Simplifying Eq. (13.6.28) yields

$$[k_h^{(4)}] = \begin{array}{c} \\ \begin{array}{ccc} 2 & 3 & 5 \end{array} \\ \left[\begin{array}{ccc} 13.3 & 6.67 & 0 \\ 6.67 & 13.3 & 0 \\ 0 & 0 & 0 \end{array} \right] \end{array} \text{Btu/(h-}^\circ\text{F)} \qquad (13.6.29)$$

Adding Eqs. (13.6.27) and (13.6.29), we obtain the element 4 total stiffness matrix as

$$[k^{(4)}] = \begin{array}{c} \\ \begin{array}{ccc} 2 & 3 & 5 \end{array} \\ \left[\begin{array}{ccc} 25.83 & 6.67 & -12.5 \\ 6.67 & 25.83 & -12.5 \\ -12.5 & -12.5 & 25 \end{array} \right] \end{array} \text{Btu/(h-}^\circ\text{F)} \qquad (13.6.30)$$

Superimposing the stiffness matrices given by Eqs. (13.6.20), (13.6.23), (13.6.25), and (13.6.30), we obtain the total stiffness matrix for the body as

$$\underline{K} = \begin{bmatrix} 25 & 0 & 0 & 0 & -25 \\ 0 & 38.33 & 6.67 & 0 & -25 \\ 0 & 6.67 & 38.33 & 0 & -25 \\ 0 & 0 & 0 & 25 & -25 \\ -25 & -25 & -25 & -25 & 100 \end{bmatrix} \text{Btu/(h-°F)} \qquad (13.6.31)$$

Next, we determine the element force matrices by using Eqs. (13.6.15)–(13.6.17) with q^* replaced by hT_∞. Since $Q = 0$, $q^* = 0$, and we have convective heat transfer only from side 2–3, element 4 is the only one that contributes nodal forces. Hence,

$$\{f^{(4)}\} = \begin{Bmatrix} f_2 \\ f_3 \\ f_5 \end{Bmatrix} = \frac{hT_\infty L_{2-3} t}{2} \begin{Bmatrix} 1 \\ 1 \\ 0 \end{Bmatrix} \qquad (13.6.32)$$

Substituting the appropriate numerical values into Eq. (13.6.32) yields

$$\{f^{(4)}\} = \frac{(20)(50)(2)(1)}{2} \begin{Bmatrix} 1 \\ 1 \\ 0 \end{Bmatrix} = \begin{Bmatrix} 1000 \\ 1000 \\ 0 \end{Bmatrix} \frac{\text{Btu}}{\text{h}} \qquad (13.6.33)$$

Using Eqs. (13.6.31) and (13.6.33), the total assembled system of equations is

$$\begin{bmatrix} 25 & 0 & 0 & 0 & -25 \\ 0 & 38.33 & 6.67 & 0 & -25 \\ 0 & 6.67 & 38.33 & 0 & -25 \\ 0 & 0 & 0 & 25 & -25 \\ -25 & -25 & -25 & -25 & 100 \end{bmatrix} \begin{Bmatrix} t_1 \\ t_2 \\ t_3 \\ t_4 \\ t_5 \end{Bmatrix} = \begin{Bmatrix} F_1 \\ 1000 \\ 1000 \\ F_4 \\ 0 \end{Bmatrix} \qquad (13.6.34)$$

We have known nodal temperature boundary conditions of $t_1 = 100°$F and $t_4 = 100°$F. We again modify the stiffness and force matrices as follows:

$$\begin{bmatrix} 1 & 0 & 0 & 0 & 0 \\ 0 & 38.33 & 6.67 & 0 & -25 \\ 0 & 6.67 & 38.33 & 0 & -25 \\ 0 & 0 & 0 & 1 & 0 \\ 0 & -25 & -25 & 0 & 100 \end{bmatrix} \begin{Bmatrix} t_1 \\ t_2 \\ t_3 \\ t_4 \\ t_5 \end{Bmatrix} = \begin{Bmatrix} 100 \\ 1000 \\ 1000 \\ 100 \\ 5000 \end{Bmatrix} \qquad (13.6.35)$$

The terms in the first and fourth rows and columns corresponding to the known temperature conditions $t_1 = 100°$F and $t_4 = 100°$F have been set equal to zero except for the main diagonal, which has been set equal to one, and the first and fourth rows of the force matrix have been set equal to the known nodal temperatures. Also, the term $(-25)(100°\text{F}) + (-25) \times (100°\text{F}) = -5000$ on the left side of the fifth equation of Eq. (13.6.34) has been transposed to the right side in the fifth row (as $+5000$) of Eq. (13.6.35). The second, third, and fifth equations of Eq. (13.6.35), corresponding to the rows of unknown nodal temperatures, can now be solved in the usual manner. The resulting solution is given by

$$t_2 = 69.33°F \qquad t_3 = 69.33°F \qquad t_5 = 84.62°F \quad (13.6.36) \quad ■$$

E X A M P L E   13.5

For the two-dimensional body shown in Figure 13-19, determine the temperature distribution. The temperature of the top side of the body is maintained at 100°C. The body is insulated on the other edges. A uniform heat source of $Q = 1000$ W/m³ acts over the whole plate, as shown in the figure. Assume a constant thickness of 1 m. Let $K_{xx} = K_{yy} = 25$ W/(m·°C).

We need only consider the left half of the body as we have a vertical plane of symmetry passing through the body 2 m from both the left and right edges. This vertical plane can be considered to be an insulated boundary. The finite element model is shown in Figure 13-20.

We will now calculate the element stiffness matrices. Since the magnitude of the coordinates are the same as in Example 13.4, the element stiffness matrices are the same as Eqs. (13.6.20), (13.6.23), (13.6.25), and (13.6.27). Remember that there is no convection from any side of an element, so the convection contribution $[k_h]$ to the stiffness matrix is zero. Superimposing the element stiffness matrices, we obtain the total stiffness matrix as

$$\underline{K} = \begin{bmatrix} 25 & 0 & 0 & 0 & -25 \\ 0 & 25 & 0 & 0 & -25 \\ 0 & 0 & 25 & 0 & -25 \\ 0 & 0 & 0 & 25 & -25 \\ -25 & -25 & -25 & -25 & 100 \end{bmatrix} \text{W/°C} \qquad (13.6.37)$$

Since the heat source Q is acting uniformly over each element, we use Eq. (13.6.13) to evaluate the nodal forces for each element as

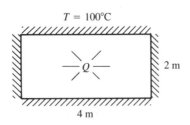

F I G U R E   13-19   Two-dimensional body subjected to a heat source

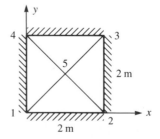

F I G U R E   13-20   Discretized body

$$\{f^{(e)}\} = \frac{QV}{3}\begin{Bmatrix} 1 \\ 1 \\ 1 \end{Bmatrix} = \frac{1000(1 \text{ m}^3)}{3}\begin{Bmatrix} 1 \\ 1 \\ 1 \end{Bmatrix} = \begin{Bmatrix} 333 \\ 333 \\ 333 \end{Bmatrix} \text{ W} \qquad (13.6.38)$$

We then use Eq. (13.6.37), and Eq. (13.6.38) applied to each element, to assemble the total system of equations as

$$\begin{bmatrix} 25 & 0 & 0 & 0 & -25 \\ 0 & 25 & 0 & 0 & -25 \\ 0 & 0 & 25 & 0 & -25 \\ 0 & 0 & 0 & 25 & -25 \\ -25 & -25 & -25 & -25 & 100 \end{bmatrix}\begin{Bmatrix} t_1 \\ t_2 \\ t_3 \\ t_4 \\ t_5 \end{Bmatrix} = \begin{Bmatrix} 666 \\ 666 \\ F_3 \\ F_4 \\ 1333 \end{Bmatrix} \qquad (13.6.39)$$

We have known nodal temperature boundary conditions of $t_3 = 100°C$ and $t_4 = 100°C$. In the usual manner, as was shown in Example 13.4, we modify the stiffness and force matrices of Eq. (13.6.39) to obtain

$$\begin{bmatrix} 25 & 0 & 0 & 0 & -25 \\ 0 & 25 & 0 & 0 & -25 \\ 0 & 0 & 1 & 0 & 0 \\ 0 & 0 & 0 & 1 & 0 \\ -25 & -25 & 0 & 0 & 100 \end{bmatrix}\begin{Bmatrix} t_1 \\ t_2 \\ t_3 \\ t_4 \\ t_5 \end{Bmatrix} = \begin{Bmatrix} 666 \\ 666 \\ 100 \\ 100 \\ 6333 \end{Bmatrix} \qquad (13.6.40)$$

Equation (13.6.40) satisfies the boundary temperature conditions and is equivalent to Eq. (13.6.39); that is, the first, second, and fifth equations of Eq. (13.6.40) are the same as the first, second, and fifth equations of Eq. (13.6.39), and the third and fourth equations of Eq. (13.6.40) identically satisfy the boundary temperature conditions at nodes 3 and 4. The first, second, and fifth equations of Eq. (13.6.40), corresponding to the rows of unknown nodal temperatures, can now be solved simultaneously. The resulting solution is given by

$$t_1 = 180°C \qquad t_2 = 180°C \qquad t_5 = 153°C \qquad (13.6.41) \quad \blacksquare$$

13.7 Line or Point Sources

A common practical heat-transfer problem is that of a source of heat generation present within a very small volume or area of some larger media. When such heat sources exist within small volumes or areas, they may be idealized as **line** or **point sources**. Practical examples that can be modeled as line sources include hot-water pipes embedded within a medium such as concrete or earth, and conducting electrical wires embedded within a material.

A line or point source can be considered by simply including a node at the location of the source when the discretized finite element model is created. The value of the line source can then be added to the row of the global force matrix corresponding to the global degree of freedom assigned to the node. However, another procedure can be used to treat the line source when it is more convenient to leave the source within an element.

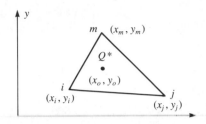

We now consider the line source of magnitude Q^*, with typical units of Btu/(h-ft), located at (x_o, y_o) within the two-dimensional element shown in Figure 13-21. The heat source Q is no longer constant over the element volume.

Using Eq. (13.4.13), we can express the heat source matrix as

$$\{f_Q\} = \iiint_V \left\{\begin{array}{c} N_i \\ N_j \\ N_m \end{array}\right\}\Bigg|_{x=x_o, y=y_o} \frac{Q^*}{A^*}\, dV \tag{13.7.1}$$

where A^* is the cross-sectional area over which Q^* acts, and the N's are evaluated at $x = x_o$ and $y = y_o$. Equation (13.7.1) can be rewritten as

$$\{f_Q\} = \iint_{A^*} \int_0^t \left\{\begin{array}{c} N_i \\ N_j \\ N_m \end{array}\right\}\Bigg|_{x=x_o, y=y_o} \frac{Q^*}{A^*}\, dA\, dz \tag{13.7.2}$$

Because the N's are evaluated at $x = x_o$ and $y = y_o$, they are no longer functions of x and y. Thus, we can simplify Eq. (13.7.2) to

$$\{f_Q\} = \left\{\begin{array}{c} N_i \\ N_j \\ N_m \end{array}\right\}\Bigg|_{x=x_o, y=y_o} Q^*t \qquad \text{Btu/h} \tag{13.7.3}$$

From Eq. (13.7.3), we can see that the portion of the line source Q^* distributed to each node is based on the values of N_i, N_j, and N_m, which are evaluated using the coordinates (x_o, y_o) of the line source. Recalling that the sum of the N's at any point within an element is equal to one [that is, $N_i(x_o, y_o) + N_j(x_o, y_o) + N_m(x_o, y_o) = 1$], we see that no more than the total amount of Q^* is distributed, and that

$$Q_i^* + Q_j^* + Q_m^* = Q^* \tag{13.7.4}$$

E X A M P L E   13.6

A line source $Q^* = 65$ Btu/(h-in.) is located at coordinates (5, 2) in the element shown in Figure 13-22. Determine the amount of Q^* allocated to each node. All nodal coordinates are in units of inches. Assume an element thickness of $t = 1$ in.

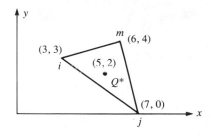

FIGURE 13-22   Line source located within a triangular element

We first evaluate the α's, β's, and γ's, defined by Eqs. (7.2.10), associated with each shape function as follows:

$$\alpha_i = x_j y_m - x_m y_j = 7(4) - 6(0) = 28$$

$$\alpha_j = x_m y_i - x_i y_m = 6(3) - 3(4) = 6$$

$$\alpha_m = x_i y_j - x_j y_i = 3(0) - 7(3) = -21$$

$$\beta_i = y_j - y_m = 0 - 4 = -4$$

$$\beta_j = y_m - y_i = 4 - 3 = 1 \qquad (13.7.5)$$

$$\beta_m = y_i - y_j = 3 - 0 = 3$$

$$\gamma_i = x_m - x_j = 6 - 7 = -1$$

$$\gamma_j = x_i - x_m = 3 - 6 = -3$$

$$\gamma_m = x_j - x_i = 7 - 3 = 4$$

Also,
$$2A = \begin{vmatrix} 1 & x_i & y_i \\ 1 & x_j & y_j \\ 1 & x_m & y_m \end{vmatrix} = \begin{vmatrix} 1 & 3 & 3 \\ 1 & 7 & 0 \\ 1 & 6 & 4 \end{vmatrix} = 13 \qquad (13.7.6)$$

Substituting the results of Eqs. (13.7.5) and (13.7.6) into Eq. (13.6.2) yields

$$N_i = \frac{1}{13}[28 - 4x - 1y]$$

$$N_j = \frac{1}{13}[6 + x - 3y] \qquad (13.7.7)$$

$$N_m = \frac{1}{13}[-21 + 3x + 4y]$$

Equations (13.7.7) for N_i, N_j, and N_m evaluated at $x = 5$ and $y = 2$ are

$$N_i = \frac{1}{13}[28 - 4(5) - 1(2)] = \frac{6}{13}$$

$$N_j = \frac{1}{13}[6 + 5 - 3(2)] = \frac{5}{13} \qquad (13.7.8)$$

$$N_m = \frac{1}{13}[-21 + 3(5) + 4(2)] = \frac{2}{13}$$

Therefore, using Eq. (13.7.3), we obtain

$$\left\{\begin{matrix} f_{Qi} \\ f_{Qj} \\ f_{Qm} \end{matrix}\right\} = Q^*t \left\{\begin{matrix} N_i \\ N_j \\ N_m \end{matrix}\right\}_{\substack{x=x_0=5 \\ y=y_0=2}} = \frac{65(1)}{13} \left\{\begin{matrix} 6 \\ 5 \\ 2 \end{matrix}\right\} = \left\{\begin{matrix} 30 \\ 25 \\ 10 \end{matrix}\right\} \text{ Btu/h} \qquad (13.7.9)$$

■

13.8   One-Dimensional Heat Transfer with Mass Transport

We now consider the derivation of the basic differential equation for one-dimensional heat flow where the flow is due to conduction, convection, and **mass transport** (or **transfer**) of the fluid. The purpose of this derivation including mass transport is to show how Galerkin's residual method can be directly applied to a problem for which the variational method is difficult to apply.

The control volume used in the derivation is shown in Figure 13-23. Again, from Eq. (13.1.1) for conservation of energy, we obtain

$$q_x A \, dt + QA \, dx \, dt = c\rho A \, dx \, dT + q_{x+dx}A \, dt + q_h P \, dx \, dt + q_m \, dt \qquad (13.8.1)$$

All of the terms in Eq. (13.8.1) have the same meaning as in Sections 13.1 and 13.2, except the additional mass-transport term is given by (see Reference [1])

$$q_m = \dot{m}cT \qquad (13.8.2)$$

where the additional variable $\dot{m}$ is the *mass flow rate* in typical units of kg/h or slug/h.

Again, using Eqs. (13.1.3)–(13.1.6), (13.2.2), and (13.8.2) in Eq. (13.8.1) and differentiating with respect to x and t, we obtain

$$\frac{\partial}{\partial x}\left(K_{xx}\frac{\partial T}{\partial x}\right) + Q = \frac{\dot{m}c}{A}\frac{\partial T}{\partial x} + \frac{hP}{A}(T - T_\infty) + \rho c \frac{\partial T}{\partial t} \qquad (13.8.3)$$

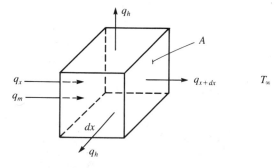

FIGURE 13-23   Control volume for one-dimensional heat conduction with convection and mass transport

Equation (13.8.3) is the basic one-dimensional differential equation for heat transfer with mass transport.

13.9 Finite Element Formulation of Heat Transfer with Mass Transport by Galerkin's Method

Having obtained the differential equation for heat transfer with mass transport, Eq. (13.8.3), we now derive the finite element equations by applying Galerkin's residual method, as outlined in Section 3.11, directly to the differential equation. We assume here that $Q = 0$ and that we have steady-state conditions so that differentiation with respect to time is zero.

The residual R is now given by

$$R(T) = -\frac{d}{dx}\left(K_{xx}\frac{dT}{dx}\right) + \frac{\dot{m}c}{A}\frac{dT}{dx} + \frac{hP}{A}(T - T_\infty) \qquad (13.9.1)$$

Applying Galerkin's criterion, Eq. (3.9.3), to Eq. (13.9.1), we have

$$\int_0^L \left[-\frac{d}{dx}\left(K_{xx}\frac{dT}{dx}\right) + \frac{\dot{m}c}{A}\frac{dT}{dx} + \frac{hP}{A}(T - T_\infty)\right] N_i \, dx = 0 \qquad (i = 1, 2)$$
$$(13.9.2)$$

where the shape functions are given by Eqs. (13.5.2). Applying integration by parts to the first term of Eq. (13.9.2), we obtain

$$u = N_i \qquad du = \frac{dN_i}{dx}\, dx$$
$$(13.9.3)$$
$$dv = -\frac{d}{dx}\left(K_{xx}\frac{dT}{dx}\right) dx \qquad v = -K_{xx}\frac{dT}{dx}$$

Using Eqs. (13.9.3) in the general formula for integration by parts [see Eq. (3.9.6)], we obtain

$$\int_0^L \left[-\frac{d}{dx}\left(K_{xx}\frac{dT}{dx}\right)\right] N_i \, dx = -K_{xx}\frac{dT}{dx}N_i\Big|_0^L + \int_0^L K_{xx}\frac{dT}{dx}\frac{dN_i}{dx}\,dx$$
$$(13.9.4)$$

Substituting Eq. (13.9.4) into Eq. (13.9.2), we obtain

$$\int_0^L \left(K_{xx}\frac{dT}{dx}\frac{dN_i}{dx}\right) dx + \int_0^L \left[\frac{\dot{m}c}{A}\frac{dT}{dx} + \frac{hP}{A}(T - T_\infty)\right] N_i \, dx = K_{xx}\frac{dT}{dx}N_i\Big|_0^L$$
$$(13.9.5)$$

Using Eq. (13.5.2) in (13.5.1) for T, we obtain

$$\frac{dT}{dx} = -\frac{t_1}{L} + \frac{t_2}{L} \qquad (13.9.6)$$

From Eq. (13.5.2), we obtain

$$\frac{dN_1}{dx} = -\frac{1}{L} \qquad \frac{dN_2}{dx} = \frac{1}{L} \qquad (13.9.7)$$

By letting $N_i = N_1 = 1 - (x/L)$ and substituting Eqs. (13.9.6) and (13.9.7) into Eq. (13.9.5), along with Eq. (13.5.1) for T, we obtain the first finite element equation

$$\int_0^L K_{xx}\left(-\frac{t_1}{L} + \frac{t_2}{L}\right)\left(-\frac{1}{L}\right)dx + \int_0^L \frac{\dot{m}c}{A}\left(-\frac{t_1}{L} + \frac{t_2}{L}\right)\left(1 - \frac{x}{L}\right)dx$$

$$+ \int_0^L \frac{hP}{A}\left[\left(1 - \frac{x}{L}\right)t_1 + \left(\frac{x}{L}\right)t_2 - T_\infty\right]\left(\frac{1-x}{L}\right)dx = q_{x1}^* \qquad (13.9.8)$$

where the definition for q_x given by Eq. (13.1.3) has been used in Eq. (13.9.8). Equation (13.9.8) has a boundary condition q_{x1}^* at $x = 0$ only since $N_1 = 1$ at $x = 0$ and $N_1 = 0$ at $x = L$. Integrating Eq. (13.9.8), we obtain

$$\left(\frac{K_{xx}A}{L} - \frac{\dot{m}c}{2} + \frac{hPL}{3}\right)t_1 + \left(-\frac{K_{xx}A}{L} + \frac{\dot{m}c}{2} + \frac{hPL}{6}\right)t_2 = q_{x1}^* + \frac{hPL}{2}T_\infty$$

$$(13.9.9)$$

where q_{x1}^* is defined to be q_x evaluated at node 1.

To obtain the second finite element equation, we let $N_i = N_2 = x/L$ in Eq. (13.9.5) and again use Eqs. (13.9.6), (13.9.7), and (13.5.1) in Eq. (13.9.5) to obtain

$$\left(-\frac{K_{xx}A}{L} - \frac{\dot{m}c}{2} + \frac{hPL}{6}\right)t_1 + \left(\frac{K_{xx}A}{L} + \frac{\dot{m}c}{2} + \frac{hPL}{3}\right)t_2 = q_{x2}^* + \frac{hPL}{2}T_\infty$$

$$(13.9.10)$$

where q_{x2}^* is defined to be q_x evaluated at node 2. Rewriting Eqs. (13.9.9) and (13.9.10) in matrix form yields

$$\left[\frac{K_{xx}A}{L}\begin{bmatrix} 1 & -1 \\ -1 & 1 \end{bmatrix} + \frac{\dot{m}c}{2}\begin{bmatrix} -1 & 1 \\ -1 & 1 \end{bmatrix} + \frac{hPL}{6}\begin{bmatrix} 2 & 1 \\ 1 & 2 \end{bmatrix}\right]\begin{Bmatrix} t_1 \\ t_2 \end{Bmatrix}$$

$$= \frac{hPLT_\infty}{2}\begin{Bmatrix} 1 \\ 1 \end{Bmatrix} + \begin{Bmatrix} q_{x1}^* \\ q_{x2}^* \end{Bmatrix} \qquad (13.9.11)$$

Applying the element equation $\{f\} = [k]\{t\}$ to Eq. (13.9.11), we see that the element stiffness (conduction) matrix is now composed of three parts:

$$[k] = [k_c] + [k_h] + [k_m] \qquad (13.9.12)$$

where

$$[k_c] = \frac{K_{xx}A}{L}\begin{bmatrix} 1 & -1 \\ -1 & 1 \end{bmatrix} \qquad [k_h] = \frac{hPL}{6}\begin{bmatrix} 2 & 1 \\ 1 & 2 \end{bmatrix} \qquad [k_m] = \frac{\dot{m}c}{2}\begin{bmatrix} -1 & 1 \\ -1 & 1 \end{bmatrix}$$

$$(13.9.13)$$

and the element nodal force and unknown nodal temperature matrices are

$$\{f\} = \frac{hPLT_\infty}{2}\begin{Bmatrix} 1 \\ 1 \end{Bmatrix} + \begin{Bmatrix} q_{x1}^* \\ q_{x2}^* \end{Bmatrix} \qquad \{t\} = \begin{Bmatrix} t_1 \\ t_2 \end{Bmatrix} \qquad (13.9.14)$$

We observe from Eq. (13.9.13) that the mass transport stiffness matrix $[k_m]$ is asymmetric, and hence, $[k]$ is asymmetric. Also, if heat flux exists, it usually occurs across the free ends of a system. Therefore, q_{x1} and q_{x2} usually occur only at the free ends of a system modeled by this element. When the elements are assembled, the heat fluxes q_{x1} and q_{x2} are usually equal but opposite at the node common to two elements, unless there is an internal concentrated heat flux in the system. Furthermore, for insulated ends, the q_x^*'s also go to zero.

To illustrate the use of the finite element equations developed in this section for heat transfer with mass transport, we will now solve the following problem.

E X A M P L E 13.7

Air is flowing at a rate of 4.72 lb/h inside a round tube with a diameter of 1 in. and length of 5 in., as shown in Figure 13-24. The initial temperature of the air entering the tube is 100°F. The wall of the tube has a uniform constant temperature of 200°F. The specific heat of the air is 0.24 Btu/(lb-°F), the convection coefficient between the air and the inner wall of the tube is 2.7 Btu/(h-ft²-°F), and the thermal conductivity is 0.017 Btu/(h-ft-°F). Determine the temperature of the air along the length of the tube and the heat flow at the inlet and outlet of the tube. Here the flow rate and specific heat are given in force units (lb) instead of mass units (slug). This is not a problem because the units cancel in the $\dot{m}c$ product in the formulation of the equations.

We first determine the element stiffness and force matrices using Eqs. (13.9.13) and (13.9.14). To do this, we evaluate the following factors:

$$\frac{K_{xx}A}{L} = \frac{(0.017)\left[\dfrac{\pi(1)}{4(144)}\right]}{1.25/12} = 0.891 \times 10^{-3} \text{ Btu/(h-°F)}$$

$$\dot{m}c = (4.72)(0.24) = 1.133 \text{ Btu/(h-°F)} \tag{13.9.15}$$

$$\frac{hPL}{6} = \frac{(2.7)(0.262)(0.104)}{6} = 0.0123 \text{ Btu/(h-°F)}$$

$$hPLT_\infty = (2.7)(0.262)(0.104)(200) = 14.71 \text{ Btu/h}$$

We can see from Eqs. (13.9.15) that the conduction portion of the stiffness matrix is negligible. Therefore, we neglect this contribution to the total stiffness matrix and obtain

$$\underline{k}^{(1)} = \frac{1.133}{2}\begin{bmatrix} -1 & 1 \\ -1 & 1 \end{bmatrix} + 0.0123\begin{bmatrix} 2 & 1 \\ 1 & 2 \end{bmatrix} = \begin{bmatrix} -0.542 & 0.579 \\ -0.554 & 0.591 \end{bmatrix} \tag{13.9.16}$$

Similarly, since all elements have the same properties,

$$\underline{k}^{(2)} = \underline{k}^{(3)} = \underline{k}^{(4)} = \underline{k}^{(1)} \tag{13.9.17}$$

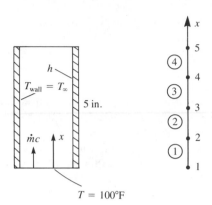

FIGURE 13-24   Air flowing through a tube, and the finite element model

Using Eqs. (13.9.14) and (13.9.15), we obtain the element force matrices as

$$\underline{f}^{(1)} = \underline{f}^{(2)} = \underline{f}^{(3)} = \underline{f}^{(4)} = \begin{Bmatrix} 7.35 \\ 7.35 \end{Bmatrix} \qquad (13.9.18)$$

Assembling the global stiffness matrix using Eqs. (13.9.16) and (13.9.17) and the global force matrix using Eq. (13.9.18), we obtain the global equations as

$$\begin{bmatrix} -0.542 & 0.579 & 0 & 0 & 0 \\ -0.554 & 0.591 - 0.542 & 0.579 & 0 & 0 \\ 0 & -0.554 & 0.591 - 0.542 & 0.579 & 0 \\ 0 & 0 & -0.554 & 0.591 - 0.542 & 0.579 \\ 0 & 0 & 0 & -0.554 & 0.591 \end{bmatrix}$$

$$\times \begin{Bmatrix} t_1 \\ t_2 \\ t_3 \\ t_4 \\ t_5 \end{Bmatrix} = \begin{Bmatrix} F_1 \\ 14.7 \\ 14.7 \\ 14.7 \\ 7.35 \end{Bmatrix} \qquad (13.9.19)$$

Applying the boundary condition $t_1 = 100°\text{F}$, we rewrite Eq. (13.9.19) using the usual procedures outlined in Section 4.2 to obtain

$$\begin{bmatrix} 1 & 0 & 0 & 0 & 0 \\ 0 & 0.049 & 0.579 & 0 & 0 \\ 0 & -0.554 & 0.049 & 0.579 & 0 \\ 0 & 0 & -0.554 & 0.049 & 0.579 \\ 0 & 0 & 0 & -0.554 & 0.591 \end{bmatrix} \begin{Bmatrix} t_1 \\ t_2 \\ t_3 \\ t_4 \\ t_5 \end{Bmatrix} = \begin{Bmatrix} 100 \\ 14.7 + 55.4 \\ 14.7 \\ 14.7 \\ 7.35 \end{Bmatrix}$$

$$(13.9.20)$$

Solving the second through fifth equations of Eq. (13.9.20) for the unknown temperatures, we obtain

$$t_2 = 106.1°\text{F} \qquad t_3 = 112.1°\text{F} \qquad t_4 = 117.6°\text{F} \qquad t_5 = 122.6°\text{F}$$

$$(13.9.21)$$

Using Eq. (13.8.2), we obtain the heat flow into and out of the tube as

$$q_{in} = \dot{m}ct_1 = (4.72)(0.24)(100) = 113.28 \text{ Btu/h}$$
$$q_{out} = \dot{m}ct_5 = (4.72)(0.24)(122.6) = 138.9 \text{ Btu/h}$$

(13.9.22)

where, again, the conduction contribution to q is negligible; that is, $-kA\,\Delta T$ is negligible. The analytical solution in Reference [7] yields

$$t_5 = 123.0°\text{F} \qquad q_{out} = 139.33 \text{ Btu/h} \qquad (13.9.23)$$

The finite element solution is then seen to compare quite favorably with the analytical solution. ∎

The element with the stiffness matrix given by Eq. (13.9.13) has been used in Reference [8] to analyze heat exchangers. Both double-pipe and shell-and-tube heat exchangers were modeled to predict the length of tube needed to perform the task of proper heat exchange between two counterflowing fluids. Excellent agreement was found between the finite element solution and the analytical solutions described in Reference [9].

Finally, remember that when the variational formulation of a problem is difficult to obtain but the differential equation describing the problem is available, a residual method such as Galerkin's method can be used to solve the problem.

13.10 Flowchart of a Heat-Transfer Program

Figure 13-25 is a flowchart of a finite element program (called HEAT) used for the analysis of two-dimensional heat-transfer problems.

13.11 Description of a Computer Program for Heat Transfer

We will now describe a computer program called HEAT (the source code for HEAT is included on the disk enclosed in the back of the text) that can be used to solve two-dimensional heat-transfer (without mass transport) problems. The program is based on the flowchart of Section 13.10 and is a modified version of that found in Reference [10].

To use the computer program HEAT, we first describe the input data. As in the CSFEP program, the node numbers associated with an element must be entered counterclockwise. Only triangular elements can be used in this program. Again, all data is input in free format, separated by commas. Consistent units should be used. The thickness t, being common to all $\underline{k}$ and $\underline{f}$ matrices, cancels out of the system of equations and hence is not needed as input.

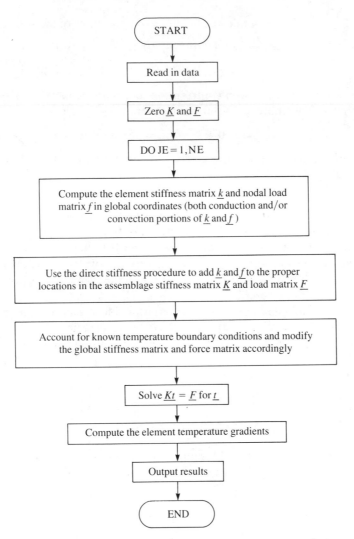

Flowchart of two-dimensional heat-transfer program HEAT

Description of Variables

H is the heat transfer or surface convection coefficient.

ISIDE(K,1) and ISIDE(K,2) are the sides of the element K, if any, that are subjected to convection heat transfer—that is, exposed to free-stream fluid. A maximum of two sides of an element may be subjected to convection heat transfer. The sides are numbered as 1, 2, and 3, as determined by the order of the node numbers attached to the element. For instance, ISIDE(K,1) = 1 and ISIDE(K,2) = 3 indicate that sides 1 and 3 of element K are exposed to free-stream fluid as shown in Figure 13-26. If ISIDE(K,1) and ISIDE(K,2) are zero, no convective heat transfer occurs.

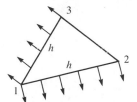

For node 1 taken to be the first node of the element:
Side 1 is side 1-2
Side 2 is side 2-3
Side 3 is side 3-1

F I G U R E 13-26 Element subjected to convection heat transfer from two sides

J is a dummy variable that indicates the node number of an element (J = 1, 2, 3).

K is a dummy variable that indicates the element number.

KXX and KYY are the thermal conductivity coefficients in the global x and y directions, respectively.

NE is the number of triangular elements.

NHS is the total number of nodal heat sources in the finite element model.

NNP is the number of node points.

NPQ is the number of the node with an applied heat source.

NPT is the number of the node with a known temperature.

NST is the number of specified nodal temperatures.

NS(K,J) represents the three nodes of element K numbered in a counter-clockwise manner.

Q is the magnitude of the nodal heat source at node NPQ.

TEMP is the known nodal temperature of node NPT.

TINF is the free-stream fluid temperature.

XC(J) and YC(J) are the x and y coordinates of node J.

Input Data

The program HEAT does not have the capability for node and element generation. Therefore, all node and element information must be entered separately. The order of input data by sequence of parameters needed by HEAT is as follows:

1. Title (Columns 1–80)

Title of problem; any alphanumeric data to identify the problem. This will be printed as a heading on the output.

2. Basic parameters line

NNP,NE,NHS,NST,KXX,KYY,H,TINF

3. Nodal point data (one line for each node)

   J,XC(J),YC(J)

4. Element data (one line for each element)

   NEL(K),NS(K,1),NS(K,2),NS(K,3),ISIDE(K,1),ISIDE(K,2)

5. Heat source data (as many lines as NHS)

   NPQ,Q

6. Specified nodal temperature data (as many lines as NST)

   NPT,TEMP

E X A M P L E   13.8

We now consider the problem shown discretized in Figure 13-27 to illustrate how to input data into program HEAT. The top and bottom are insulated, whereas the right side is subjected to convection heat transfer. The left side has a constant temperature of 100°F. The heat transfer coefficient at the right surface is $h = 20$ Btu/(h-ft²-°F).

Table 13-4 shows the data file used by HEAT to solve the problem in Figure 13-27. The order of input is based on the general description of the input data. In Table 13-4, line 1 is the identifying title of the problem being solved. Line 2 provides the number of nodes, number of elements, number of heat sources, number of specified temperatures, material thermal conductivities in the global x and y directions, the heat transfer or convection coefficient, and the fluid free-stream temperature. Lines 3–10 are the node numbers and their x and y coordinates. Lines 11–18 are the element numbers, their associated nodes, and whether there is convection heat transfer occurring from any sides of an element. For instance, in line 17, the 2 indicates that side 2 (between nodes 8 and 7) of element 7 is subjected to convection heat transfer. Lines 19 and 20 indicate the node numbers and temperatures of

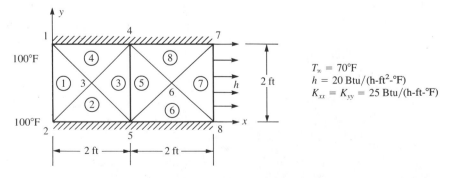

F I G U R E   13-27   Two-dimensional heat-transfer problem

TABLE 13-4   Input for the heat-transfer problem of Figure
              13-27

Line	Data on the line (beginning in column 1)
1	2-D HEAT TRANSFER PROBLEM OF FIGURE 13-27
2	8,8,0,2,25.,25.,20.,70.
3	1,0.,2.
4	2,0.,0.
5	3,1.,1.
6	4,2.,2.
7	5,2.,0.
8	6,3.,1.
9	7,4.,2.
10	8,4.,0.
11	1,1,2,3,0,0
12	2,2,5,3,0,0
13	3,3,5,4,0,0
14	4,1,3,4,0,0
15	5,4,5,6,0,0
16	6,5,8,6,0,0
17	7,6,8,7,2,0
18	8,4,6,7,0,0
19	1,100.
20	2,100.

nodes with specified temperatures. The computer program solution to this problem is given in Table 13-5 on p. 476. ■

REFERENCES

[1] Holman, J. P., *Heat Transfer*, 3rd ed., McGraw-Hill, New York, 1972.

[2] Kreith, F., and Black, W. Z., *Basic Heat Transfer*, Harper & Row, New York, 1980.

[3] Lyness, J. F., Owen, D. R. J., and Zienkiewicz, O. C., "The Finite Element Analysis of Engineering Systems Governed by a Non-Linear Quasi-Harmonic Equation," *Computers and Structures*, Vol. 5, pp. 65–79, 1975.

[4] Zienkiewicz, O. C., and Cheung, Y. K., "Finite Elements in the Solution of Field Problems," *The Engineer*, pp. 507–510, Sept. 24, 1965.

[5] Wilson, E. L., and Nickell, R. E., "Application of the Finite Element Method to Heat Conduction Analysis," *Nuclear Engineering and Design*, Vol. 4, pp. 276–286, 1966.

[6] Emery, A. F., and Carson, W. W., "An Evaluation of the Use of the Finite Element Method in the Computation of Temperature," *Journal of Heat Transfer*, American Society of Mechanical Engineers, pp. 136–145, May 1971.

T A B L E 13-5   Output of computer program HEAT for two-dimensional heat-transfer
problem of Figure 13-27

2-D HEAT TRANSFER PROBLEM OF FIGURE 13-27

KXX = 25.0   KYY = 25.0
CONVECTION COEFF = 20.0
FLUID TEMP = 70.0
SEMI-BANDWIDTH = 4

NEL	NODE	NUMBER		X(1)	Y(1)	X(2)	Y(2)	X(3)	Y(3)
1	1	2	3	0.0000	2.0000	0.0000	0.0000	1.0000	1.0000
2	2	5	3	0.0000	0.0000	2.0000	0.0000	1.0000	1.0000
3	3	5	4	1.0000	1.0000	2.0000	0.0000	2.0000	2.0000
4	1	3	4	0.0000	2.0000	1.0000	1.0000	2.0000	2.0000
5	4	5	6	2.0000	2.0000	2.0000	0.0000	3.0000	1.0000
6	5	8	6	2.0000	0.0000	4.0000	0.0000	3.0000	1.0000
7	6	8	7	3.0000	1.0000	4.0000	0.0000	4.0000	2.0000
8	4	6	7	2.0000	2.0000	3.0000	1.0000	4.0000	2.0000

CONVECTION FROM SIDE 2 OF ELEMENT   7

PRESCRIBED NODAL TEMPERATURE VALUES
1   0.1000E+03
2   0.1000E+03
RESULTING NODAL TEMPERATURE VALUES
1   0.10000E+03   2   0.10000E+03   3   0.94286E+02   4   0.88571E+02
5   0.88571E+02   6   0.82857E+02   7   0.77143E+02   8   0.77143E+02

ELEMENT RESULTANTS

ELEMENT

	GRAD(X)	GRAD(Y)	AVE TEMP
1	−0.5714E+01	0.0000E+00	0.9810E+02
2	−0.5714E+01	0.0000E+00	0.9429E+02
3	−0.5714E+01	0.0000E+00	0.9048E+02
4	−0.5714E+01	0.0000E+00	0.9429E+02
5	−0.5714E+01	0.0000E+00	0.8667E+02
6	−0.5714E+01	−0.7629E−05	0.8286E+02
7	−0.5714E+01	0.0000E+00	0.7905E+02
8	−0.5714E+01	0.3815E−05	0.8286E+02

[7] Rohsenow, W. M., and Choi, H. Y., *Heat, Mass, and Momentum Transfer*, Prentice Hall, Englewood Cliffs, New Jersey, 2nd Printing, 1963.

[8] Goncalves, L., *Finite Element Analysis of Heat Exchangers*, M.S. Thesis, Rose-Hulman Institute of Technology, Terre Haute, Indiana, 1984.

[9] Kern, D. Q., and Kraus, A. D., *Extended & Surface Heat Transfer*, McGraw-Hill, New York, 1972.

[10] Segerlind, L. J., *Applied Finite Element Analysis*, Wiley, New York, 1976.

PROBLEMS

13.1 For the one-dimensional composite bar shown in Figure P13-1, determine the interface temperatures. For element 1, let $K_{xx} = 200$ W/(m·°C), for element 2, let $K_{xx} = 100$ W/(m·°C), and for element 3, let $K_{xx} = 50$ W/(m·°C). Let $A = 0.1$ m². The left end has a constant temperature of 100°C and the right end has a constant temperature of 300°C.

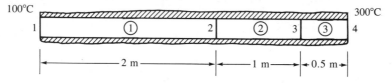

FIGURE P13-1

13.2 For the one-dimensional rod shown in Figure P13-2 (insulated except at the ends), determine the temperatures at $L/3$, $2L/3$, and L. Let $K_{xx} = 3$ Btu/(h.-in.-°F), $h = 1.0$ Btu/(h-in.²-°F), and $T_\infty = 0$°F. Also, the temperature at the left end is 200°F.

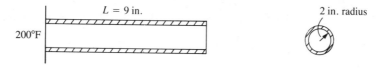

FIGURE P13-2

13.3 A rod with uniform cross-sectional area of 2 in.² and thermal conductivity of 3 Btu/(h-in.-°F) has heat flow in the x direction only (Figure P13-3). The right end is insulated. The left end is maintained at 50°F, and the system has the linearly distributed heat flux shown.

Use a two-element model and estimate the temperature at the node points and the heat flow at the left boundary.

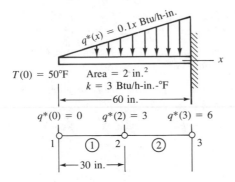

FIGURE P13-3

13.4   The rod of 1-in. radius shown in Figure P13-4 generates heat internally at the rate of uniform $Q = 10{,}000$ Btu/(h-ft^3) throughout the rod. The left edge and perimeter of the rod are insulated, and the right edge is exposed to an environment of $T_\infty = 100°$F. The convection heat-transfer coefficient between the wall and the environment is $h = 100$ Btu/(h-ft^2-°F). The thermal conductivity of the rod is $K_{xx} = 12$ Btu/(h-ft-°F). The length of the rod is 3 in. Calculate the temperature distribution in the rod. Use at least three elements in your finite element model.

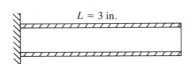

FIGURE P13-4

13.5   The fin shown in Figure P13-5 is insulated on the perimeter. The left end has a constant temperature of 100°C. A positive heat flux of $q^* = 5000$ W/m^2 acts on the right end. Let $K_{xx} = 6$ W/(m·°C) and cross-sectional area $A = 0.1$ m^2. Determine the temperatures at $L/4$, $L/2$, $3L/4$, and L, where $L = 0.4$ m.

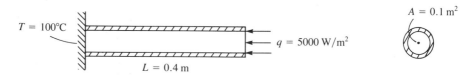

FIGURE P13-5

13.6   For the composite wall shown in Figure P13-6, determine the interface temperatures. What is the heat flux through the 8-cm portion? Use the finite element method. Use three elements with the nodes shown. 1 cm = 0.01 m.

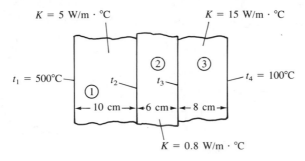

$K = 5$ W/m · °C $K = 15$ W/m · °C

$t_1 = 500°C$ t_2 t_3 $t_4 = 100°C$

← 10 cm →|← 6 cm →|← 8 cm →

$K = 0.8$ W/m · °C

FIGURE P13-6

13.7 For the composite wall, idealized by the one-dimensional model shown in Figure P13-7, determine the interface temperatures. For element 1, let $K_{xx} = 5$ W/(m · °C), for element 2, $K_{xx} = 10$ W/(m · °C), and for element 3, $K_{xx} = 15$ W/(m · °C). The left end has a constant temperature of 200°C and the right end has a constant temperature of 600°C.

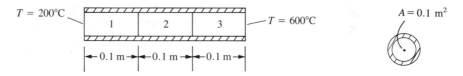

$T = 200°C$ 1   2   3 $T = 600°C$ $A = 0.1$ m^2

← 0.1 m →|← 0.1 m →|← 0.1 m →

FIGURE P13-7

13.8 Use the direct method to derive the element equations for the one-dimensional steady-state conduction heat-transfer problem shown in Figure P13-8. The bar is insulated all around, has cross-sectional area A, length L, and thermal conductivity K_{xx}. Determine the relationship between nodal temperatures t_1 and t_2 (°F) and the thermal inputs F_1 and F_2 (in Btu). Use Fourier's law of heat conduction for this case.

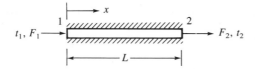

x

1   2

t_1, F_1 F_2, t_2

|← L →|

FIGURE P13-8

13.9 For the element shown in Figure P13-9, determine the $\underline{k}$ and $\underline{f}$ matrices. The conductivities are $K_{xx} = K_{yy} = 15$ Btu/(h-ft-°F) and the convection coefficient is $h = 20$ Btu/(h-ft^2-°F). Convection occurs across the i-j surface. The free-stream temperature is $T_\infty = 70°F$. The coordinates are expressed in units of feet. Let the line source be $Q^* = 150$ Btu/(h-ft) as located in the figure. Take the thickness of the element to be 1 ft.

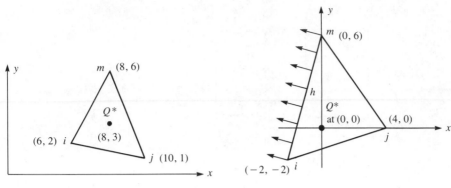

FIGURE   P13-9   FIGURE   P13-10

13.10   Calculate the $\underline{k}$ and $\underline{f}$ matrices for the element shown in Figure P13-10. The conductivities are $K_{xx} = K_{yy} = 15$ W/(m·°C) and the convection coefficient is $h = 20$ W/(m²·°C). Convection occurs across the *i-m* surface. The free-stream temperature is $T_\infty = 15$°C. The coordinates are shown expressed in units of meters. Let the line source be $Q^* = 100$ W/m located in the figure. Take the thickness of the element to be 1 m.

13.11   For the square two-dimensional body shown in Figure P13-11, determine the temperature distribution. Let $K_{xx} = K_{yy} = 25$ Btu/(h-ft-°F) and $h = 10$ Btu/(h-ft²-°F). Convection occurs across side 4-5. The free-stream temperature is $T_\infty = 50$°F. The temperatures at nodes 1 and 2 are 100°F. The dimensions of the body are shown in the figure. Take the thickness of the body to be 1 ft.

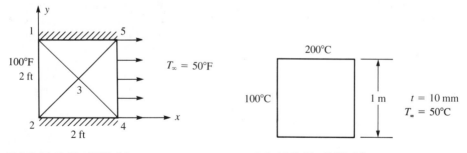

FIGURE   P13-11   FIGURE   P13-12

13.12   For the square plate shown in Figure P13-12, determine the temperature distribution. Let $K_{xx} = K_{yy} = 10$ W/(m·°C) and $h = 20$ W/(m²·°C). The temperature along the left side is maintained at 100°C and that along the top side is maintained at 200°C.

Use a computer program such as HEAT (included on the accompanying disk) to calculate the temperature distribution in the following two-dimensional bodies.

13.13 For the body shown in Figure P13-13, determine the temperature distribution. Surface temperatures are shown in the figure. The body is insulated along the top and bottom edges, and $K_{xx} = K_{yy} = 1.0$ Btu/(h-in.-°F). No internal heat generation is present.

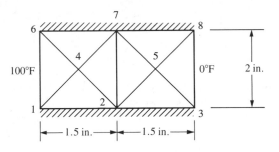

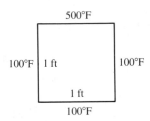

FIGURE P13-13   **FIGURE P13-14**

13.14 For the square two-dimensional body shown in Figure P13-14, determine the temperature distribution. Let $K_{xx} = K_{yy} = 10$ Btu/(h-ft-°F). The top surface is maintained at 500°F and the other three sides are maintained at 100°F. Also, plot the temperature contours on the body.

13.15 For the square two-dimensional body shown in Figure P13-15, determine the temperature distribution. Let $K_{xx} = K_{yy} = 10$ Btu/(h-ft-°F) and $h = 10$ Btu/(h-ft²-°F). The top face is maintained at 500°F, the left face is maintained at 100°F, and the other two faces are exposed to an environmental (free-stream) temperature of 100°F. Also, plot the temperature contours on the body.

FIGURE P13-15

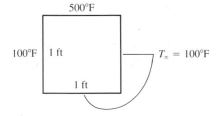

13.16 Hot water pipes are located on 2.0-ft centers in a concrete slab with $K_{xx} = K_{yy} = 0.80$ Btu/(h-ft-°F), as shown in Figure P13-16. If the outside surfaces of the concrete are at 85°F and the water has an average temperature of 200°F, determine the temperature distribution in the concrete slab. Plot the temperature contours through the concrete. Use symmetry in your finite element model.

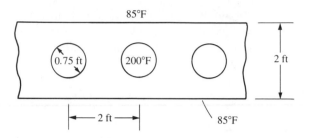

FIGURE   P13-16

13.17   The cross-section of a tall chimney shown in Figure P13-17 has an inside surface temperature of 330°F and an exterior temperature of 130°F. The thermal conductivity is $K = 0.5$ Btu/(h-ft-°F). Determine the temperature distribution within the chimney per unit length.

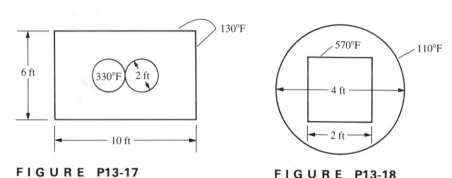

FIGURE   P13-17   **FIGURE   P13-18**

13.18   A square duct shown in Figure P13-18 carries hot gases such that its surface temperature is 570°F. The duct is insulated by a layer of circular fiberglass that has a thermal conductivity of $K = 0.020$ Btu/(h-ft-°F). The outside surface temperature of the fiberglass is maintained at 110°F. Determine the temperature distribution within the fiberglass.

13.19   The buried pipeline in Figure P13-19 transports oil with an average temperature of 60°F. The pipe is located 15 ft below the surface of the earth. The thermal conductivity of the earth is 0.6 Btu/(h-ft-°F). The surface of the earth is 50°F. Determine the temperature distribution in the earth.

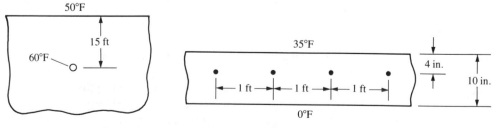

FIGURE   P13-19   **FIGURE   P13-20**

13.20 A 10-in.-thick concrete bridge deck is embedded with heating cables, as shown in Figure P13-20. If the lower surface is at 0°F, the rate of heat generation (assumed to be the same in each cable) is 100 Btu/(h-in.) and the top surface of the concrete is at 35°F. The thermal conductivity of the concrete is 0.500 Btu/(h-ft-°F). What is the temperature distribution in the slab? Use symmetry in your model.

13.21 For the circular body with holes shown in Figure P13-21, determine the temperature distribution. The inside surfaces of the holes have temperatures of 150°C. The outside of the circular body is insulated and has a temperature of 30°C. Let $K_{xx} = K_{yy} = 10$ W/(m·°C).

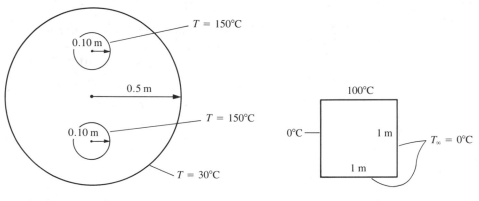

F I G U R E   P13-21   **F I G U R E   P13-22**

13.22 For the square two-dimensional body shown in Figure P13-22, determine the temperature distribution. Let $K_{xx} = K_{yy} = 10$ W/(m·°C) and $h = 10$ W/(m²·°C). The top face is maintained at 100°C, the left face is maintained at 0°C, and the other two faces are exposed to a free-stream temperature of 0°C. Also, plot the temperature contours on the body.

13.23 A 200-mm-thick concrete bridge deck is embedded with heating cables as shown in Figure P13-23. If the lower surface is at −10°C and the upper surface is at 5°C, what is the temperature distribution in the slab? The heating cables are line sources generating heat of $Q^* = 50$ W/m. The thermal conductivity of the concrete is 1.2 W/(m·°C). Use symmetry in your model.

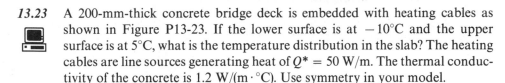

F I G U R E   P13-23

13.24   For the two-dimensional body shown in Figure P13-24, determine the temperature distribution. Let the left and right ends have constant temperatures of 200°C and 100°C, respectively. Let $K_{xx} = K_{yy} = 5$ W/(m · °C). The body is insulated all around.

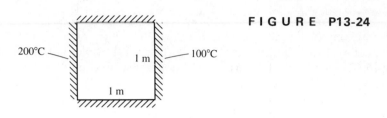

F I G U R E   P13-24

13.25   For the two-dimensional body shown in Figure P13-25, determine the temperature distribution. The top and bottom sides are insulated. The right side is subjected to heat transfer by convection. Let $K_{xx} = K_{yy} = 10$ W/(m · °C).

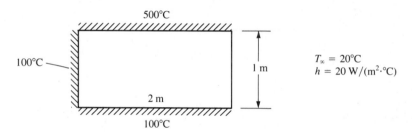

F I G U R E   P13-25

13.26   For the two-dimensional body shown in Figure P13-26, determine the temperature distribution. The left and right sides are insulated. The top surface is subjected to heat transfer by convection. The bottom and internal portion surfaces are maintained at 300°C.

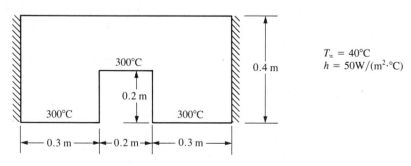

F I G U R E   P13-26

Fluid Flow

INTRODUCTION

In this chapter we consider the flow of fluid through porous media, such as the flow of water through an earthen dam, through pipes, and around solid bodies. We will observe that the form of the equations is the same as that for heat transfer described in Chapter 13.

We begin with a derivation of the basic differential equation in one dimension for an ideal fluid in a steady state, not rotating (that is, the fluid particles are translating only), incompressible (constant mass density), and inviscid (having no viscosity). We then extend this derivation to the two-dimensional case. We also consider the units used for the physical quantities involved in fluid flow. For more advanced topics, such as viscous flow, compressible flow, and three-dimensional problems, consult Reference [1].

We will use the same procedure to develop the element equations as in the heat-transfer problem; that is, we define an assumed fluid head for the flow through porous media (seepage) problem or velocity potential for flow of fluid through pipes and around solid bodies within each element. Then both a direct approach similar to that used in Chapters 2, 3, and 5 to develop the element equations and the minimization of a functional as used in Chapter 13 are used to obtain the element equations. These equations result in matrices analogous to the stiffness and force matrices of the stress analysis problem or the conduction and associated force matrices of the heat-transfer problem.

Next, we consider both one- and two-dimensional finite element formulations of the fluid-flow problem and provide examples of one-dimensional fluid flow through porous media and through pipes, and flow within a two-dimensional region.

Finally, we describe a computer program for two-dimensional fluid flow only and present an example of its use.

14.1 Derivation of the Basic Differential Equations

Fluid Flow Through a Porous Media

Let us first consider the derivation of the basic differential equation for the one-dimensional problem of fluid flow through a porous medium. The purpose of this derivation is to present a physical insight into the fluid-flow phenomena, which must be understood so that the finite element formulation of the problem can be fully comprehended. (For additional information on fluid flow, consult References [2] and [3]). We begin by considering the control volume shown in Figure 14-1. By conservation of mass, we have

$$M_{in} + M_{generated} = M_{out} \qquad (14.1.1)$$

or
$$\rho v_x A \, dt + \rho Q \, dt = \rho v_{x+dx} A \, dt \qquad (14.1.2)$$

where

M_{in} is the mass entering the control volume, in units of kilograms or slugs.

$M_{generated}$ is the mass generated within the body.

M_{out} is the mass leaving the control volume.

v_x is the velocity of the fluid flow at surface edge x, in units of m/s or in./s.

v_{x+dx} is the velocity of the fluid leaving the control volume at surface edge $x + dx$.

t is time, in s.

Q is an internal fluid source (an internal volumetric flow rate), in m³/s or in.³/s.

ρ is the mass density of the fluid, in kg/m³ or slugs/in.³

A is the cross-sectional area perpendicular to the fluid flow, in m² or in.²

By Darcy's law, we relate the velocity of fluid flow to the hydraulic gradient (the change in fluid head with respect to x) as

$$v_x = -K_{xx} \frac{d\phi}{dx} = -K_{xx} g_x \qquad (14.1.3)$$

where

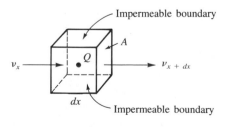

Impermeable boundary

A

v_x

Q

$v_{x + dx}$

dx

Impermeable boundary

FIGURE 14-1   Control volume for one-dimensional fluid flow

K_{xx} is the permeability coefficient of the porous media in the x direction, in m/s or in./s.

Φ is the fluid head, in m or in.

$d\phi/dx = g_x$ is the fluid head gradient or hydraulic gradient, which is a unitless quantity in the seepage problem.

Equation (14.1.3) states that the velocity in the x direction is proportional to the gradient of the fluid head in the x direction. The minus sign in Eq. (14.1.3) implies that fluid flow is positive in the direction opposite the direction of fluid head increase, or that the fluid flows in the direction of lower fluid head. Equation (14.1.3) is analogous to Fourier's law of heat conduction, Eq. (13.1.3).

Similarly,

$$v_{x+dx} = -K_{xx}\frac{d\phi}{dx}\bigg|_{x+dx} \qquad (14.1.4)$$

where the gradient is now evaluated at $x + dx$. By Taylor series expansion, similar to that used in obtaining Eq. (13.1.5), we have

$$v_{x+dx} = -\left[K_{xx}\frac{d\phi}{dx} + \frac{d}{dx}\left(K_{xx}\frac{d\phi}{dx}\right)dx\right] \qquad (14.1.5)$$

where a two-term Taylor series has been used in Eq. (14.1.5). On substituting Eqs. (14.1.3) and (14.1.5) into Eq. (14.1.2), dividing Eq. (14.1.2) by $\rho A\, dx\, dt$, and simplifying, we have the one-dimensional fluid flow through a porous media equation as

$$\frac{d}{dx}\left(K_{xx}\frac{d\phi}{dx}\right) + \bar{Q} = 0 \qquad (14.1.6)$$

where $\bar{Q} = Q/A\, dx$ is the volume flow rate per unit volume in units 1/s. For a constant permeability coefficient, Eq. (14.1.6) becomes

$$K_{xx}\frac{d^2\phi}{dx^2} + \bar{Q} = 0 \qquad (14.1.7)$$

The boundary conditions are of the form

$$\Phi = \Phi_B \qquad \text{on } S_1 \qquad (14.1.8)$$

where Φ_B represents a known boundary fluid head and S_1 is a surface where this head is known and

$$v_x^* = -K_{xx}\frac{d\phi}{dx} = \text{constant} \qquad \text{on } S_2 \qquad (14.1.9)$$

where S_2 is a surface where the prescribed velocity v_x^* or gradient is known. On an impermeable boundary, $v_x^* = 0$.

Comparing this derivation to that for the one-dimensional heat conduction problem in Section 13.1, we observe numerous analogies among the variables; that is, ϕ is analogous to the temperature function T, v_x is analogous to heat flux, and K_{xx} is analogous to thermal conductivity.

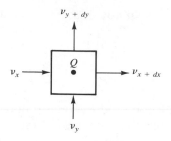

FIGURE 14-2   Control volume for two-dimensional fluid flow

Now consider the two-dimensional fluid flow through a porous medium, as shown in Figure 14-2. As in the one-dimensional case, we can show that for material properties coinciding with the global x and y directions,

$$\frac{\partial}{\partial x}\left(K_{xx}\frac{\partial \phi}{\partial x}\right) + \frac{\partial}{\partial y}\left(K_{yy}\frac{\partial \phi}{\partial y}\right) + \bar{Q} = 0 \qquad (14.1.10)$$

with boundary conditions

$$\phi = \phi_B \qquad \text{on } S_1 \qquad (14.1.11)$$

and

$$K_{xx}\frac{\partial \phi}{\partial x}C_x + K_{yy}\frac{\partial \phi}{\partial y}C_y = \text{constant} \qquad \text{on } S_2 \qquad (14.1.12)$$

where C_x and C_y are direction cosines of the unit vector normal to the surface S_2, as previously shown in Figure 13-3.

Fluid Flow in Pipes and Around Solid Bodies

We now consider the steady-state irrotational flow of an incompressible and inviscid fluid. For the ideal fluid, the fluid particles do not rotate; they only translate, and the friction between the fluid and the surfaces is ignored. Also, the fluid does not penetrate into the surrounding body or separate from the surface of the body, which could create voids.

The equations for this fluid motion can be expressed in terms of the stream function or the velocity potential function. We will use the velocity potential analogous to the fluid head that was used for the derivation of the differential equation for flow through a porous media in the preceding subsection.

The velocity v of the fluid is related to the velocity potential function ϕ by

$$v_x = \frac{\partial \phi}{\partial x} \quad \text{and} \quad v_y = \frac{\partial \phi}{\partial y} \qquad (14.1.13)$$

where v_x and v_y are the velocities in the x and y directions, respectively. In the absence of sources or sinks Q, conservation of mass in two dimensions yields the two-dimensional differential equation as

$$\frac{\partial^2 \phi}{\partial x^2} + \frac{\partial^2 \phi}{\partial y^2} = 0 \qquad (14.1.14)$$

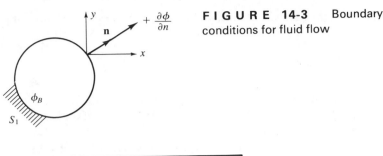

FIGURE 14-3   Boundary conditions for fluid flow

FIGURE 14-4   Known velocities at left and right edges of a pipe

Equation (14.1.14) is analogous to Eq. (14.1.10) when we set $K_{xx} = K_{yy} = 1$ and $Q = 0$. Hence, Eq. (14.1.14) is just a special form of Eq. (14.1.10). The boundary conditions are

$$\phi = \phi_B \qquad \text{on } S_1 \qquad\qquad (14.1.15)$$

and

$$\frac{\partial \phi}{\partial x} C_x + \frac{\partial \phi}{\partial y} C_y = \text{constant} \quad \text{on } S_2 \qquad (14.1.16)$$

where C_x and C_y are again direction cosines of unit vector **n** normal to surface S_2. Also see Figure 14-3. That is, Eq. (14.1.15) states that the velocity potential Φ_B is known on a boundary surface S_1, whereas Eq. (14.1.16) states that the potential gradient or velocity is known normal to a surface S_2, as indicated for flow out of the pipe shown in Figure 14-3.

 To clarify the sign convention on the S_2 boundary condition, consider the case of fluid flowing through a pipe in the positive x direction, as shown in Figure 14-4. Assume we know the velocities at the left edge (1) and the right edge (2). By Eq. (14.1.13) the velocity of the fluid is related to the velocity potential by

$$v_x = -\frac{\partial \phi}{\partial x}$$

At the left edge (1) assume we know $v_x = v_{x1}$. Then

$$v_{x1} = -\frac{\partial \phi}{\partial x}$$

But the normal is always positive away, or outward, from the surface. Therefore, positive n_1 is directed to the left, whereas positive x is to the right, resulting in

$$\frac{\partial \phi}{\partial n_1} = -\frac{\partial \phi}{\partial x} = v_{x1} = v_{n1}$$

At the right edge (2) assume we know $v_x = v_{x2}$. Now the normal n_2 is in the same direction as x. Therefore,

$$\frac{\partial \phi}{\partial n_2} = \frac{\partial \phi}{\partial x} = -v_{x2} = -v_{n2}$$

We conclude that the boundary flow velocity is positive if directed into the surface (region), as at the left edge, and is negative if directed away from the surface, as at the right edge.

At an impermeable boundary, the flow velocity and thus the derivative of the velocity potential normal to the boundary must be zero. At a boundary of uniform or constant velocity, any convenient magnitude of velocity potential ϕ may be specified as the gradient of the potential function; see, for instance, Eq. (14.1.13). This idea is also illustrated by Example 14.3.

14.2   One-Dimensional Finite Element Formulation

We can proceed directly to the one-dimensional finite element formulation of the fluid-flow problem by now realizing that the fluid-flow problem is analogous to the heat-conduction problem of Chapter 13. We merely substitute the fluid velocity potential function ϕ for the temperature function T, the vector of nodal potentials denoted by $\{p\}$ for the nodal temperature vector $\{t\}$, fluid velocity v for heat flux q, and permeability coefficient K for flow through a porous media instead of the conduction coefficient K. If fluid flow through a pipe or around a solid body is considered, then K is taken as unity. The steps are as follows.

Step 1   Select Element Type

The basic two-noded element is again used, as shown in Figure 14-5, with nodal fluid heads, or potentials, denoted by p_1 and p_2.

Step 2   Choose A Potential Function

We choose the potential function ϕ similar to the temperature function of Section 13.5 as

$$\Phi = N_1 p_1 + N_2 p_2 \qquad (14.2.1)$$

where p_1 and p_2 are the nodal potentials (or fluid heads in the case of the seepage problem) to be determined, and

$$N_1 = 1 - \frac{\hat{x}}{L} \qquad N_2 = \frac{\hat{x}}{L} \qquad (14.2.2)$$

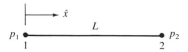

FIGURE 14-5   Basic one-dimensional fluid-flow element

are again the same shape functions used for the temperature element. The matrix $[N]$ is then

$$[N] = \left[1 - \frac{\hat{x}}{L} \quad \frac{\hat{x}}{L} \right] \tag{14.2.3}$$

Step 3 Define the Gradient/Potential and Velocity/Gradient Relationships

The hydraulic gradient matrix $\{g\}$ is given by

$$\{g\} = \left\{ \frac{d\phi}{d\hat{x}} \right\} = [B]\{p\} \tag{14.2.4}$$

where $[B]$ is identical to Eq. (13.5.5), given by

$$[B] = \left[-\frac{1}{L} \quad \frac{1}{L} \right] \tag{14.2.5}$$

and

$$\{p\} = \left\{ \begin{matrix} p_1 \\ p_2 \end{matrix} \right\} \tag{14.2.6}$$

The velocity/gradient relationship based on Darcy's law is given by

$$v_x = -[D]\{g\} \tag{14.2.7}$$

where the material property matrix is now given by

$$[D] = [K_{xx}] \tag{14.2.8}$$

with K_{xx} the permeability of the porous media in the x direction. Typical permeabilities of some granular materials are listed in Table 14-1. High permeabilities occur when $K > 10^{-1}$ cm/s, and when $K < 10^{-7}$ the material is considered to be nearly impermeable. For ideal flow through a pipe or over a solid body, we arbitrarily—but conveniently—let $K = 1$.

Step 4 Derive the Element Stiffness Matrix and Equations

The fluid-flow problem has a stiffness matrix that can be found using the first term on the right side of Eq. (13.4.14). That is, the fluid-flow stiffness matrix is analogous to the conduction part of the stiffness matrix in the heat-transfer

T A B L E 14-1   Permeabilities of granular materials

Material	K (cm/s)
Clay	1×10^{-8}
Sandy clay	1×10^{-3}
Ottawa sand	$2-3 \times 10^{-2}$
Coarse gravel	1

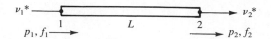

FIGURE 14-6   Fluid element subjected to nodal velocities

problem. There is no comparable convection matrix to be added to the stiffness matrix. However, we will choose to use a direct approach similar to that used initially to develop the stiffness matrix for the bar element in Chapter 3.

Consider the fluid element shown in Figure 14-6 with length L and uniform cross-sectional area A. Recall that the stiffness matrix is defined in the structure problem to relate nodal forces to nodal displacements or in the temperature problem to relate nodal rates of heat flow to nodal temperatures. In the fluid-flow problem, we define the stiffness matrix to relate nodal volumetric fluid-flow rates to nodal potentials or fluid heads as $\underline{f} = \underline{k}\underline{p}$. Therefore,

$$f = v^*A \tag{14.2.9}$$

defines the volumetric flow rate f in units of cubic meters or cubic inches per second. Now using Eqs. (14.2.7) and (14.2.8) in (14.2.9), we obtain

$$f = -K_{xx}Ag \tag{14.2.10}$$

in scalar form; based on Eqs. (14.2.4) and (14.2.5) g is given in explicit form by

$$g = \frac{p_2 - p_1}{L} \tag{14.2.11}$$

Applying Eqs. (14.2.10) and (14.2.11) at nodes 1 and 2, we obtain

$$f_1 = -K_{xx}A\frac{p_2 - p_1}{L} \tag{14.2.12}$$

and

$$f_2 = K_{xx}A\frac{p_2 - p_1}{L} \tag{14.2.13}$$

where f_1 is directed into the element, indicating fluid flowing into the element (p_1 must be greater than p_2 to push the fluid through the element, actually resulting in positive f_1), whereas f_2 is directed away from the element, indicating fluid flowing out of the element; hence the negative sign changes to a positive one in Eq. (14.2.13). Expressing Eqs. (14.2.12) and (14.2.13) together in matrix form, we have

$$\begin{Bmatrix} f_1 \\ f_2 \end{Bmatrix} = \frac{AK_{xx}}{L} \begin{bmatrix} 1 & -1 \\ -1 & 1 \end{bmatrix} \begin{Bmatrix} p_1 \\ p_2 \end{Bmatrix} \tag{14.2.14}$$

The stiffness matrix is then

$$\underline{k} = \frac{AK_{xx}}{L} \begin{bmatrix} 1 & -1 \\ -1 & 1 \end{bmatrix} \tag{14.2.15}$$

Equation (14.2.15) is analogous to Eq. (13.5.7) for the heat-conduction ele-

F I G U R E 14-7   Additional sources of volumetric fluid flow rates

ment or Eq. (3.1.14) for the one-dimensional (axial stress) bar element. The permeability or stiffness matrix will have units of square meters or square inches per second.

In general, the basic element may be subjected to internal sources or sinks, such as due to a pump, or to surface-edge flow rates, such as from a river or stream. To include these or similar effects, consider the element of Figure 14-6 to now include a uniform internal source Q acting over the whole element and a uniform surface flow rate source q^* acting over the surface, as shown in Figure 14-7. The force matrix terms are

$$\{f_Q\} = \iiint_V [N]^T Q \, dV = \frac{QAL}{2} \begin{Bmatrix} 1 \\ 1 \end{Bmatrix} \qquad (14.2.16)$$

where Q will have units of $m^3/(m^3 \cdot s)$, or $1/s$, and

$$\{f_q\} = \iint_{S_2} q^* [N]^T \, dS = \frac{q^* L t}{2} \begin{Bmatrix} 1 \\ 1 \end{Bmatrix} \qquad (14.2.17)$$

where q^* will have units of m/s or in./s. Equations (14.2.16) and (14.2.17) indicate that one-half of the uniform volumetric flow rate per unit volume Q (a source being positive and a sink being negative) is allocated to each node and one-half the surface flow rate (again a source is positive) is allocated to each node.

Step 5 **Assemble the Element Equations to Obtain the Global Equations and Introduce Boundary Conditions**

We assemble the total stiffness matrix $[K]$, total force matrix $\{F\}$, and total set of equations as

$$[K] = \sum [k^{(e)}] \qquad \{F\} = \sum \{f^{(e)}\} \qquad (14.2.18)$$

and $$\{F\} = [K]\{p\} \qquad (14.2.19)$$

The assemblage procedure is similar to the direct stiffness approach, but it is now based on the requirement that the potentials at a common node between two elements be equal. The boundary conditions on nodal potentials are given by Eq. (14.1.15).

Step 6 **Solve for the Nodal Potentials**

We now solve for the global nodal potentials, $\{p\}$, where the appropriate nodal potential boundary conditions, Eq. (14.1.15), are specified.

Step 7 Solve for the Element Velocities and Volumetric Flow Rates

Finally, we calculate the element velocities from Eq. (14.2.7) and the volumetric flow rate Q_f as

$$Q_f = (v)(A) \qquad (14.2.20)$$

E X A M P L E   14.1

Determine (a) the fluid head distribution along the length of the coarse gravelly medium shown in Figure 14-8, (b) the velocity in the upper part, and (c) the volumetric flow rate in the upper part. The fluid head at the top is 10 in. and that at the bottom is 1 in. Let the permeability coefficient be $K_{xx} = 0.5$ in./s. Assume a cross-sectional area of $A = 1$ in.^{2}

The finite element discretization is shown in Figure 14-9. For simplicity, we will use three elements, each 10 in. long.

We calculate the stiffness matrices for each element as follows:

$$\frac{AK_{xx}}{L} = \frac{(1 \text{ in.}^2)(0.5 \text{ in./s})}{10 \text{ in.}} = 0.05 \text{ in.}^2/s$$

Using Eq. (14.2.15) for elements 1, 2, and 3, we have

$$[k^{(1)}] = [k^{(2)}] = [k^{(3)}] = 0.05 \begin{bmatrix} 1 & -1 \\ -1 & 1 \end{bmatrix} \text{in.}^2/s \qquad (14.2.21)$$

In general, we would use Eqs. (14.2.16) and (14.2.17) to obtain element forces. However, in this example $Q = 0$ (no sources or sinks) and $q^* = 0$ (no applied surface flow rates). Therefore,

$$\{f^{(1)}\} = \{f^{(2)}\} = \{f^{(3)}\} = 0 \qquad (14.2.22)$$

The assembly of the element stiffness matrices from Eq. (14.2.21), using

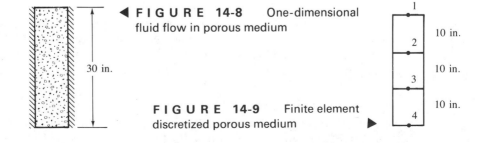

◀ **F I G U R E   14-8**   One-dimensional fluid flow in porous medium

30 in.

F I G U R E   14-9   Finite element discretized porous medium ▶

1

10 in.

2

10 in.

3

10 in.

4

the direct stiffness method, produces the following system of equations:

$$0.05 \begin{bmatrix} 1 & -1 & 0 & 0 \\ -1 & 2 & -1 & 0 \\ 0 & -1 & 2 & -1 \\ 0 & 0 & -1 & 1 \end{bmatrix} \begin{Bmatrix} p_1 \\ p_2 \\ p_3 \\ p_4 \end{Bmatrix} = \begin{Bmatrix} 0 \\ 0 \\ 0 \\ 0 \end{Bmatrix} \qquad (14.2.23)$$

Known nodal fluid head boundary conditions are $p_1 = 10$ in. and $p_4 = 1$ in. These nonhomogeneous boundary conditions are treated as described in Chapter 4 for the stress analysis problem. We modify the stiffness (permeability) matrix and force matrix as follows:

$$0.05 \begin{bmatrix} 1 & -1 & 0 & 0 \\ 0 & 2 & -1 & 0 \\ 0 & -1 & 2 & 0 \\ 0 & 0 & 0 & 1 \end{bmatrix} \begin{Bmatrix} p_1 \\ p_2 \\ p_3 \\ p_4 \end{Bmatrix} = \begin{Bmatrix} 10 \\ 0.5 \\ 0.05 \\ 1 \end{Bmatrix} \qquad (14.2.24)$$

where the terms in the first and fourth rows and columns of the stiffness matrix corresponding to the known fluid heads $p_1 = 10$ in. and $p_4 = 1$ in. have been set equal to 0 except for the main diagonal, which has been set equal to 1, and the first and fourth rows of the force matrix have been set equal to the known nodal fluid heads at nodes 1 and 4. Also the terms $(-0.05) \times (10 \text{ in.}) = -0.5$ in. on the left side of the second equation of Eq. (14.2.24) and $(-0.05) \times (1 \text{ in.}) = -0.05$ in. on the left side of the third equation of Eq. (14.2.24) have been transposed to the right side in the second and third rows (as $+0.5$ and $+0.05$). The second and third equations of Eq. (14.2.24) can now be solved. The resulting solution is given by

$$p_2 = 7 \text{ in.} \qquad p_3 = 4 \text{ in.} \qquad (14.2.25)$$

Next we use Eq. (14.2.7) to determine the fluid velocity in element 1 as

$$v_x^{(1)} = -K_{xx}[B]\{p^{(1)}\} \qquad (14.2.26)$$

$$= -K_{xx}\left[-\frac{1}{L} \quad \frac{1}{L}\right]\begin{Bmatrix} p_1 \\ p_2 \end{Bmatrix} \qquad (14.2.27)$$

or
$$v_x^{(1)} = 0.15 \text{ in./s} \qquad (14.2.28)$$

You can verify that the velocities in the other elements are also 0.15 in./s because the cross section is constant and the material properties are uniform. We then determine the volumetric flow rate Q_f in element 1 using Eq. (14.2.20) as

$$Q_f = (0.15 \text{ in./s})(1 \text{ in.}^2) = 0.15 \text{ in.}^3/\text{s} \qquad (14.2.29)$$

This volumetric flow rate is constant throughout the length of the medium.

∎

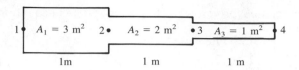

FIGURE 14-10 Variable cross-section pipe subjected to fluid flow

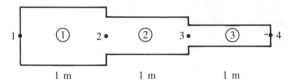

FIGURE 14-11 Discretized pipe

EXAMPLE 14.2

For the smooth pipe of variable cross section shown in Figure 14-10, determine the potential at the junctions, determine the velocities in each section of pipe and the volumetric flow rate. The potential at the left end is $p_1 = 10 \text{ m}^2/\text{s}$ and that at the right end is $p_4 = 1 \text{ m}^2/\text{s}$.

For the fluid flow through a smooth pipe, $K_{xx} = 1$. The pipe has been discretized into three elements and four nodes, as shown in Figure 14-11. Using Eq. (14.2.15), the element stiffness matrices are

$$\underline{k}^{(1)} = \frac{3}{1}\begin{bmatrix} 1 & -1 \\ -1 & 1 \end{bmatrix}, \quad \underline{k}^{(2)} = \frac{2}{1}\begin{bmatrix} 1 & -1 \\ -1 & 1 \end{bmatrix}, \quad \underline{k}^{(3)} = \frac{1}{1}\begin{bmatrix} 1 & -1 \\ -1 & 1 \end{bmatrix}$$

$$(14.2.30)$$

There are no applied fluid sources. Therefore $f^{(1)} = f^{(2)} = f^{(3)} = 0$. The assembly of the element stiffness matrices produces the following system of equations:

$$\begin{bmatrix} 3 & -3 & 0 & 0 \\ -3 & 5 & -2 & 0 \\ 0 & -2 & 3 & -1 \\ 0 & 0 & -1 & 1 \end{bmatrix}\begin{Bmatrix} 10 \\ p_2 \\ p_3 \\ 1 \end{Bmatrix} = \begin{Bmatrix} 0 \\ 0 \\ 0 \\ 0 \end{Bmatrix} \qquad (14.2.31)$$

Solving the second and third of Eqs. (14.2.31) for p_2 and p_3 in the usual manner, we obtain

$$p_2 = 8.365 \text{ m}^2/\text{s} \quad \text{and} \quad p_3 = 5.91 \text{ m}^2/\text{s} \qquad (14.2.32)$$

Using Eqs. (14.2.7) and (14.2.20), the velocities and volumetric flow rates in each element are

$$v_x^{(1)} = -[B]\{p^{(1)}\}$$

$$= -\begin{bmatrix} -\dfrac{1}{L} & \dfrac{1}{L} \end{bmatrix} \begin{Bmatrix} 10 \\ 8.365 \end{Bmatrix}$$

$$= 1.635 \text{ m/s}$$

$$Q_f^{(1)} = Av_x^{(1)} = 3(1.635) = 4.91 \text{ m}^3/\text{s}$$

$$v_x^{(2)} = -(-8.365 + 5.91) = 2.455 \text{ m/s}$$

$$Q_f^{(2)} = 2.455(2) = 4.91 \text{ m}^3/\text{s}$$

$$v_x^{(3)} = -(-5.91 + 1) = 4.91 \text{ m/s}$$

$$Q_f^{(3)} = 4.91(1) = 4.91 \text{ m}^3/\text{s}$$

The potential, being higher at the left and decreasing to the right, indicates that the velocities are to the right. The volumetric flow rate is constant throughout the pipe, as conservation of mass would indicate. ∎

We now illustrate how you can solve a fluid-flow problem where the boundary condition is a known fluid velocity, but none of the p's are initially known.

E X A M P L E 14.3

For the smooth pipe shown discretized in Figure 14-12 with uniform cross section of 1 in.², determine the flow velocities at the center and right end knowing the velocity at the left end is $v_x = 2$ in./s.
   Using Eq. (14.2.15), the element stiffness matrices are

$$\underline{k}^{(1)} = \frac{1}{10}\begin{bmatrix} 1 & -1 \\ -1 & 1 \end{bmatrix} \quad \text{and} \quad \underline{k}^{(2)} = \frac{1}{10}\begin{bmatrix} 1 & -1 \\ -1 & 1 \end{bmatrix} \qquad (14.2.33)$$

Assembling the element stiffness matrices produces the following equations:

$$\frac{1}{10}\begin{bmatrix} 1 & -1 & 0 \\ -1 & 2 & -1 \\ 0 & -1 & 1 \end{bmatrix}\begin{Bmatrix} p_1 \\ p_2 \\ p_3 \end{Bmatrix} = \begin{Bmatrix} f_1 \\ f_2 \\ f_3 \end{Bmatrix} \qquad (14.2.34)$$

The specified boundary condition is $v_x = 2$ in./s, so that by Eq. (14.2.9), we

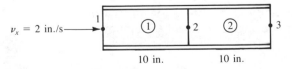

FIGURE 14-12   Discretized pipe for fluid-flow problem

have

$$f_1 = v_1 A = (2 \text{ in./s})(1 \text{ in.}^2) = 2 \text{ in.}^3/\text{s} \qquad (14.2.35)$$

Because p_1, p_2, and p_3 in Eq. (14.2.34) are not known, we cannot determine these potentials directly. The problem is similar to that occurring if we try to solve the structural problem without prescribing displacements sufficient to prevent rigid body motion of the structure. This was discussed in Chapter 2. Since the p's correspond to displacements in the structural problem, it appears that we must specify at least one value of p in order to obtain a solution. We then proceed as follows. Select a convenient value for p_3, (for instance set $p_3 = 0$). (Since the velocities are functions of the derivatives or differences in p's, a value of $p_3 = 0$ is acceptable.) Then p_1 and p_2 are the unknowns. The solution will yield p_1 and p_2 relative to $p_3 = 0$. Therefore, from the first two of Eqs. (14.2.34), we have

$$\frac{1}{10} \begin{bmatrix} 1 & -1 \\ -1 & 2 \end{bmatrix} \begin{Bmatrix} p_1 \\ p_2 \end{Bmatrix} = \begin{Bmatrix} 2 \\ 0 \end{Bmatrix} \qquad (14.2.36)$$

where $f_1 = 2 \text{ in.}^3/\text{s}$ from Eq. (14.2.35) and $f_2 = 0$, because there is no applied fluid force at node 2.

Solving Eq. (14.2.36), we obtain

$$p_1 = 40 \quad \text{and} \quad p_2 = 20 \qquad (14.2.37)$$

These are not absolute values for p_1 and p_2, rather, they are relative to p_3. The fluid velocities in each element are absolute values, since velocities depend on the differences in p's. These differences are the same no matter what value for p_3 was chosen. You can verify this by choosing $p_3 = 10$, for instance, and resolving for the velocities. [You would find $p_1 = 50$ and $p_2 = 30$ and the same v's as in Eq. (14.2.38).]

$$v_x^{(1)} = -\begin{bmatrix} \dfrac{1}{L} & \dfrac{1}{L} \end{bmatrix} \begin{Bmatrix} 40 \\ 20 \end{Bmatrix} = 2 \text{ in./s}$$

and $(14.2.38)$

$$v_x^{(2)} = -\begin{bmatrix} -\dfrac{1}{L} & \dfrac{1}{L} \end{bmatrix} \begin{Bmatrix} 20 \\ 0 \end{Bmatrix} = 2 \text{ in./s} \qquad \blacksquare$$

14.3 Two-Dimensional Finite Element Formulation

Since many fluid flow problems can be modeled as two-dimensional problems, we now develop the equations for an element appropriate for these problems. Examples using this element then follow.

Step 1

The three-noded triangular element in Figure 14-13 is the basic element for the solution of the two-dimensional fluid-flow problem.

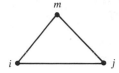

FIGURE 14-13   Basic triangular element

Step 2

The potential function is

$$[\phi] = [N_i \quad N_j \quad N_m] \begin{Bmatrix} p_i \\ p_j \\ p_m \end{Bmatrix} \qquad (14.3.1)$$

where p_i, p_j, and p_m are the nodal potentials (for groundwater flow, ϕ is the piezometric fluid head function, and the p's are the nodal heads), and the shape functions are again given by Eqs. (7.2.18) or (13.6.2) as

$$N_i = \frac{1}{2A}(\alpha_i + \beta_i x + \gamma_i y) \qquad (14.3.2)$$

with similar expressions for N_j and N_m. The α's, β's, and γ's are defined by Eqs. (7.2.10).

Step 3

The gradient matrix, $\{g\}$, is given by

$$\{g\} = [B]\{p\} \qquad (14.3.3)$$

where the matrix $[B]$ is again given by

$$[B] = \frac{1}{2A} \begin{bmatrix} \beta_i & \beta_j & \beta_m \\ \gamma_i & \gamma_j & \gamma_m \end{bmatrix} \qquad (14.3.4)$$

and

$$\{g\} = \begin{Bmatrix} g_x \\ g_y \end{Bmatrix} \qquad (14.3.5)$$

with

$$g_x = \frac{\partial \phi}{\partial x} \quad \text{and} \quad g_y = \frac{\partial \phi}{\partial y} \qquad (14.3.6)$$

The velocity/gradient matrix relationship is now

$$\begin{Bmatrix} v_x \\ v_y \end{Bmatrix} = -[D]\{g\} \qquad (14.3.7)$$

where the material property matrix is

$$[D] = \begin{bmatrix} K_{xx} & 0 \\ 0 & K_{yy} \end{bmatrix} \qquad (14.3.8)$$

and the K's are permeabilities (for the seepage problem) of the porous media in the x and y directions. For fluid flow around a solid object or through a smooth pipe, $K_{xx} = K_{yy} = 1$.

Step 4

The element stiffness matrix is given by

$$[k] = \int\int\int_V [B]^T [D][B]\, dV \tag{14.3.9}$$

Assuming constant-thickness triangular elements and noting that the integrand terms are constant, we have

$$[k] = tA[B]^T [D][B] \tag{14.3.10}$$

which can be simplified to

$$[k] = \frac{tK_{xx}}{4A} \begin{bmatrix} \beta_i^2 & \beta_i\beta_j & \beta_i\beta_m \\ \beta_i\beta_j & \beta_j^2 & \beta_j\beta_m \\ \beta_i\beta_m & \beta_j\beta_m & \beta_m^2 \end{bmatrix} + \frac{tK_{yy}}{4A} \begin{bmatrix} \gamma_i^2 & \gamma_i\gamma_j & \gamma_i\gamma_m \\ \gamma_i\gamma_j & \gamma_j^2 & \gamma_j\gamma_m \\ \gamma_i\gamma_m & \gamma_j\gamma_m & \gamma_m^2 \end{bmatrix} \tag{14.3.11}$$

The force matrices are

$$\{f_Q\} = \int\int\int_V Q[N]^T\, dV = Q \int\int\int_V [N]^T\, dV \tag{14.3.12}$$

for constant volumetric flow rate per unit volume over the whole element. On evaluating

$$\{f_Q\} = \frac{QV}{3} \begin{Bmatrix} 1 \\ 1 \\ 1 \end{Bmatrix} \tag{14.3.13}$$

the second force matrix is

$$\{f_q\} = \int\int_{S_2} q^*[N]^T\, dS = \int\int_{S_2} q^* \begin{Bmatrix} N_i \\ N_j \\ N_m \end{Bmatrix} dS \tag{14.3.14}$$

This reduces to

$$\{f_q\} = \frac{q^* L_{i-j} t}{2} \begin{Bmatrix} 1 \\ 1 \\ 0 \end{Bmatrix} \text{ on side } i\text{-}j \tag{14.3.15}$$

with similar terms on sides *j-m* and *m-i* [see Eqs. (13.6.16) and (13.6.17). Here L_{i-j} is the length of side *i-j* of the element and q^* is the assumed constant surface flow rate. Both Q and q^* are positive quantities if fluid is being added to the element. The units on Q and q^* are m³/(m³ · s) and m/s. The total force matrix is then the sum of $\{f_Q\}$ and $\{f_q\}$.

E X A M P L E   14.4

For the two-dimensional sandy soil region shown in Figure 14-14, determine the potential distribution. The potential (fluid head) on the left side is a

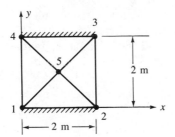

FIGURE 14-14 Two-dimensional porous media

constant 10.0 m and that on the right side is 0.0. The upper and lower edges are impermeable. The permeabilities are $K_{xx} = K_{yy} = 25 \times 10^{-5}$ m/s. Assume unit thickness.

The finite element model is shown in Figure 14-14. We use only the four triangular elements of equal size for simplicity of the longhand solution. For increased accuracy in results, we would need to refine the mesh. This body has the same magnitude of coordinates as Figure 13-20. Therefore, the total stiffness matrix is given by Eq. (13.6.37) as

$$\underline{K} = \begin{bmatrix} 25 & 0 & 0 & 0 & -25 \\ 0 & 25 & 0 & 0 & -25 \\ 0 & 0 & 25 & 0 & -25 \\ 0 & 0 & 0 & 25 & -25 \\ -25 & -25 & -25 & -25 & 100 \end{bmatrix} \times 10^{-5} \qquad (14.3.16)$$

The force matrices are zero, since $Q = 0$ and $q^* = 0$. Applying the boundary conditions, we have

$$p_1 = p_4 = 10.0 \quad \text{and} \quad p_2 = p_3 = 0$$

The assembled total system of equations is then

$$10^{-5} \begin{bmatrix} 25 & 0 & 0 & 0 & -25 \\ 0 & 25 & 0 & 0 & -25 \\ 0 & 0 & 25 & 0 & -25 \\ 0 & 0 & 0 & 25 & -25 \\ -25 & -25 & -25 & -25 & 100 \end{bmatrix} \begin{Bmatrix} 10 \\ 0 \\ 0 \\ 10 \\ p_5 \end{Bmatrix} = \begin{Bmatrix} 0 \\ 0 \\ 0 \\ 0 \\ 0 \end{Bmatrix} \qquad (14.3.17)$$

Solving the fifth of Eqs. (14.3.17) for p_5, we obtain

$$p_5 = 5 \text{ m}$$

By using Eq. (14.2.45), the velocity in element 2 is obtained as

$$\begin{Bmatrix} v_x^{(2)} \\ v_y^{(2)} \end{Bmatrix} = - \begin{bmatrix} -25 & 0 \\ 0 & -25 \end{bmatrix} \times 10^{-5} \frac{1}{2A} \begin{bmatrix} -1 & 2 & -1 \\ -1 & 0 & 1 \end{bmatrix} \begin{Bmatrix} p_1 \\ p_5 \\ p_4 \end{Bmatrix} \qquad (14.3.18)$$

where $\beta_1 = -1, \beta_5 = 2, \beta_4 = -1, \gamma_1 = -1, \gamma_5 = 0$, and $\gamma_4 = 1$ were obtained from Eq. (13.6.21). Simplifying Eq. (14.3.18), we obtain

$$v_x^{(2)} = 125 \times 10^{-5} \text{ m/s} \quad \text{and} \quad v_y^{(2)} = 0 \qquad \blacksquare$$

A line or point fluid source from a pump, for instance, can be handled in the same manner as described in Section 13.7 for heat sources. If the source is at a node when the discretized finite element model is created, then the source can be added to the row of the global force matrix corresponding to the global degree of freedom assigned to the node. If the source is within an element, we can use Section 13.7 to allocate the source to the proper nodes, as illustrated by the following example.

EXAMPLE 14.5

A pump, pumping fluid at $Q^* = 6500$ m^2/h, is located at coordinates (5, 2) in the element shown in Figure 14-15. Determine the amount of Q^* allocated to each node. All nodal coordinates are in units of meters. Assume unit thickness of $t = 1$ mm.

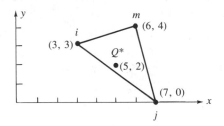

FIGURE 14-15   Triangular element with pump located within element

The magnitudes of the numbers are the same as in Example 13.6. Therefore, the shape functions are identical to Eq. (13.7.7); when evaluated at the source $x = 5$ m, $y = 2$ m, they are equal to Eq. (13.7.8). Using Eq. (13.7.3), we obtain the amount of Q^* allocated to each node or the force matrix as

$$
\left\{ \begin{array}{c} f_{Qi} \\ f_{Qj} \\ f_{Qm} \end{array} \right\} = Q^* t \left\{ \begin{array}{c} N_i \\ N_j \\ N_m \end{array} \right\} \Bigg|_{\substack{x=x_0=5\text{ m} \\ y=y_0=2\text{ m}}}
$$

$$
= \frac{6500\,(1\text{ mm})}{(13)\left(\dfrac{1000\text{ mm}}{1\text{ m}}\right)} \left\{ \begin{array}{c} 6 \\ 5 \\ 2 \end{array} \right\} = \left\{ \begin{array}{c} 3.0 \\ 2.5 \\ 1.0 \end{array} \right\} \text{ m}^3/\text{h} \qquad \blacksquare
$$

14.4 Flowchart of a Fluid-Flow Program

Figure 14-16 is a flowchart of a finite element program (called FLUID) used for the analysis of two-dimensional irrotational fluid flow. The program can be used to solve seepage or groundwater problems and potential flow of fluids over and around solid bodies.

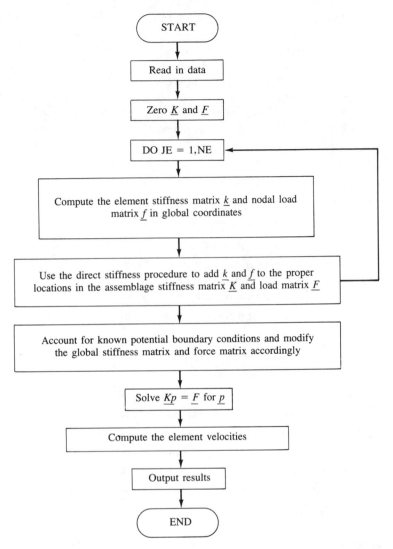

FIGURE   14-16   Flowchart of two-dimensional fluid-flow program FLUID

14.5   Description of a Computer Program for Fluid Flow

We will now describe a computer program called FLUID (included on the disk enclosed in back of the text) that can be used to solve two-dimensional fluid-flow problems. The fluid flow must be irrotational and the fluid incompressible. Only steady-state conditions can occur. The program is based on the flowchart of Section 14.4 and is a modified version of that found in Reference [4].

To use the computer program FLUID, we first describe the input data. As in CSFEP and HEAT programs, the node numbers associated with an element must be entered counterclockwise. Only triangular elements can be used in the program. Consistent units should be used. The thickness, t, being common to all $\underline{K}$ and $\underline{f}$ matrices, cancels out of the system of equations and hence is not needed as input.

The velocities calculated in the program are those that occur in groundwater flow and streamline flow. Velocities for potential flow are not calculated using program FLUID.

Description of Variables

FLOW is the magnitude of flow at node NPF.

J is a dummy variable representing the node of an element ($J = 1, 2, 3$).

K is a dummy variable representing the element number.

OPT is an option for the output of the individual loading cases.
 OPT $= 0$ for no output of loading cases.
 OPT $= 1$ for output of all loading cases.

NE is the number of triangular elements (the maximum number is 100).

NNP is the number of nodes (the maximum number is 100).

NS(K, J) represents the three nodes of the element K numbered in a counterclockwise manner.

NHPSI is the number of the node with a specified head.

NPF is the number of the node with a specified flow.

NSH is the number of specified heads.

NSP is the number of specified flows.

PSI is the magnitude of the head at node NHPSI.

PXX and PYY are the permeabilities in the global x and y directions, respectively.

Input Data

The program FLUID is not capable of node or element generation. All data must be entered individually. The sequence for entering data for the program FLUID is as follows:

 1. Title (maximum of 80 characters)

  This is the title of the problem. It may be any alphanumeric characters. It will be printed as a heading on the output.

 2. Basic parameters line

  OPT,NNP,NE,NSP,NSH,PXX,PYY

 3. Nodal point data (one line for each node)

  J,XC(J),YC(J)

4. Element data (one line for each element)

NEL(K),NS(K,1),NS(K,2),NS(K,3)

5. Specified nodal flow data (one for each specified flow)

NPF,FLOW

6. Specified nodal head data (one for each specified head)

NHPSI,PSI

E X A M P L E   14.6

We now consider the problem shown discretized in Figure 14-17 to illustrate
how to input data into program FLUID. The top and bottom sides are
impervious, whereas the right side has a constant head of 3 cm and the left
side has a constant head of 4 cm.

Table 14-2 shows the data file used by FLUID to solve the problem in
Figure 14-17. The order of input is based on the general description of the
input data.

The computer program solution to this problem is given in Table 14-3
on p. 506.

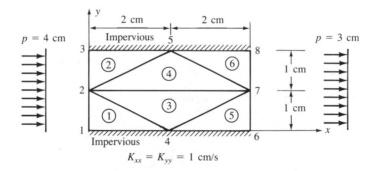

$$K_{xx} = K_{yy} = 1 \text{ cm/s}$$

FIGURE 14-17   Two-dimensional fluid-flow problem

T A B L E 14-2   Input for the fluid-flow problem of Figure 14-17

Line	Data on the line (beginning in column 1)
1	2D FLUID MECHANICS EXAMPLE
2	1, 8, 6, 0, 6, 1., 1.
3	1, 0., 0.
4	2, 0., 1.
5	3, 0., 2.
6	4, 2., 0.
7	5, 2., 2.
8	6, 4., 0.
9	7, 4., 1.
10	8, 4., 2.
11	1, 1, 4, 2
12	2, 2, 5, 3
13	3, 2, 4, 7
14	4, 2, 7, 5
15	5, 4, 6, 7
16	6, 5, 7, 8
17	1, 4.
18	2, 4.
19	3, 4.
20	6, 3.
21	7, 3.
22	8, 3.

T A B L E 14-3   Output of computer program FLUID for two-dimensional fluid flow problem of Figure 14-17

2D FLUID MECHANICS EXAMPLE

PXX = 1.0
PYY = 1.0

NEL	NODE	NUMBER		X(1)	Y(1)	X(2)	Y(2)	X(3)	Y(3)
1	1	4	2	.00	.00	2.00	.00	.00	1.00
2	2	5	3	.00	1.00	2.00	2.00	.00	2.00
3	2	4	7	.00	1.00	2.00	.00	4.00	1.00
4	2	7	5	.00	1.00	4.00	1.00	2.00	2.00
5	4	6	7	2.00	.00	4.00	.00	4.00	1.00
6	5	7	8	2.00	2.00	4.00	1.00	4.00	2.00

1
2D FLUID MECHANICS EXAMPLE

T A B L E 14-3 (*Continued*)

BOUNDARY VALUES

NODAL FORCES
LOADING CASE 1
    0.   .00000E + 00

PRESCRIBED NODAL VALUES
    1   .40000E + 01
    2   .40000E + 01
    3   .40000E + 01
    6   .30000E + 01
    7   .30000E + 01
    8   .30000E + 01
  1

2D FLUID MECHANICS EXAMPLE

NODAL VALUES, LOADING CASE 1

1 0.40000E + 01	2 0.40000E + 01	3 0.40000E + 01	4 0.35000E + 01
5 0.35000E + 01	6 0.30000E + 01	7 0.30000E + 01	8 0.30000E + 01

RESULTANT NODAL VALUES
    1   .40000E + 01
    2   .40000E + 01
    3   .40000E + 01
    4   .35000E + 01
    5   .35000E + 01
    6   .30000E + 01
    7   .30000E + 01
    8   .30000E + 01
  1

2D FLUID MECHANICS EXAMPLE

ELEMENT VELOCITY COMPONENTS

ELEMENT	VEL(X)	VEL(Y)
1	.25000E + 00	.00000E + 00
2	.25000E + 00	.00000E + 00
3	.25000E + 00	.00000E + 00
4	.25000E + 00	.00000E + 00
5	.25000E + 00	.00000E + 00
6	.25000E + 00	.00000E + 00

REFERENCES

[1] Chung, T. J., *Finite Element Analysis in Fluid Dynamics*, McGraw-Hill, New York, 1978.

[2] John, J. E. A., and Haberman, W. L., *Introduction to Fluid Mechanics*, Prentice Hall, Englewood Cliffs, New Jersey, 1988.

[3] Harr, M. E., *Ground Water and Seepage*, McGraw-Hill, New York, 1962.

[4] Segerlind, L. G., *Applied Finite Element Analysis*, Wiley, New York, 1976.

PROBLEMS

14.1   For the one-dimensional flow through the porous media shown in Figure P14-1, determine the potentials at one-third and two-thirds of the length. Also determine the velocities in each element. Let $A = 2 \ m^2$.

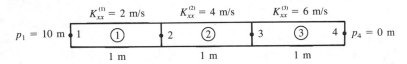

F I G U R E   P14-1

14.2   For the one-dimensional flow through the porous media shown in Figure P14-2 with fluid flux at the right end, determine the potentials at the third points. Also determine the velocities in each element. Let $A = 2 \ m^2$.

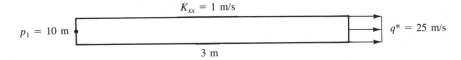

F I G U R E   P14-2

14.3   For the one-dimensional fluid flow through the stepped porous media shown in Figure P14-3, determine the potentials at the junction of each area. Also determine the velocities in each element. Let $K_{xx} = 1$ in./s.

F I G U R E   P14-3

14.4 For the one-dimensional fluid flow problem (Figure P14-4) with velocity known at the right end, determine the velocities and the volumetric flow rates at nodes 1 and 2. Let $K_{xx} = 2$ cm/s.

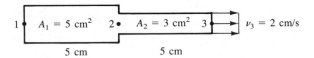

FIGURE P14-4

14.5 Derive the stiffness matrix, Eq. (14.2.15), using the first term on the right side of Eq. (13.4.14).

14.6 For the one-dimensional fluid-flow problem in Figure P14-6, determine the velocities and volumetric flow rates at nodes two and three. Let $K_{xx} = 10^{-1}$ in./s.

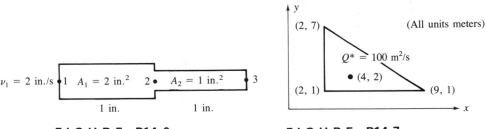

FIGURE P14-6   **FIGURE P14-7**

14.7 For the triangular element subjected to a fluid source shown in Figure P14-7, determine the amount of Q^* allocated to each node.

14.8 For the triangular element subjected to the surface fluid source shown in Figure P14-8, determine the amount of fluid force at each node.

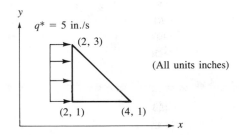

FIGURE P14-8

14.9   For the two-dimensional fluid flow shown in Figure P14-9, determine the potentials at the center and right edge.

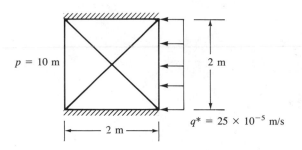

$q^* = 25 \times 10^{-5}$ m/s

FIGURE   P14-9

14.10–14.15   Using a computer program such as FLUID (on the accompanying disk) for fluid flow, determine the potential distribution in the two-dimensional bodies shown in Figures P14-10–P14-15.

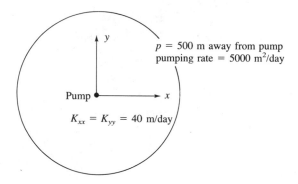

$p = 500$ m away from pump
pumping rate $= 5000$ m^2/day

Pump

$K_{xx} = K_{yy} = 40$ m/day

FIGURE   P14-10

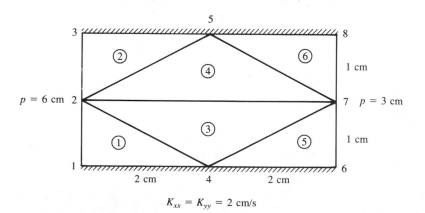

$K_{xx} = K_{yy} = 2$ cm/s

FIGURE   P14-11

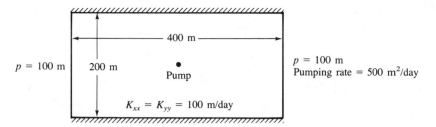

FIGURE P14-12

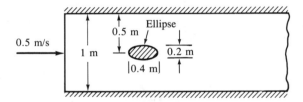

FIGURE P14-13

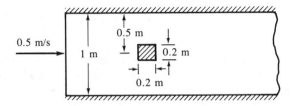

FIGURE P14-14

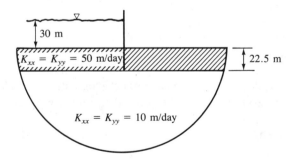

FIGURE P14-15

Thermal Stress

INTRODUCTION

In this chapter, we consider the problem of thermal stresses within a body. First, we will discuss the strain energy due to thermal stresses (stresses resulting from the constrained motion of a body or part of a body during a temperature change in the body).

The minimization of the thermal strain energy equation is shown to result in the thermal force matrix. We will then develop this thermal force matrix for the one-dimensional bar element and the two-dimensional plane stress and plane strain elements.

We will outline the procedures for solution of both one- and two-dimensional problems and then provide solutions of specific problems.

15.1 Formulation of the Thermal Stress Problem and Examples

In addition to the strains associated with the displacement functions, there may be other strains within a body due to temperature variations, swelling (moisture differential), or other causes. We will concern ourselves only with the strains due to temperature variation, ε_T, and will consider both one- and two-dimensional problems.

We will first consider the one-dimensional thermal stress problem. The linear stress/strain diagram with initial (thermal) strain ($\varepsilon_0 = \varepsilon_T$) is shown in Figure 15-1.

For the one-dimensional problem, we have, from Figure 15-1,

$$\varepsilon_x = \frac{\sigma_x}{E} + \varepsilon_T \qquad (15.1.1)$$

If, in general, we let $1/E = \underline{D}^{-1}$, then in general matrix form Eq. (15.1.1) can be written as

$$\underline{\varepsilon} = [D]^{-1}\underline{\sigma} + \underline{\varepsilon}_T \qquad (15.1.2)$$

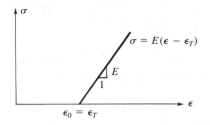

FIGURE 15-1   Linear stress/strain law with initial thermal strain

From Eq. (15.1.2), we solve for $\underline{\sigma}$ as

$$\underline{\sigma} = \underline{D}(\underline{\varepsilon} - \underline{\varepsilon}_T) \qquad (15.1.3)$$

The strain energy per unit volume (called strain energy density) is the area under the $\sigma - \varepsilon$ diagram in Figure 15-1 and is given by

$$u_0 = \tfrac{1}{2}\underline{\sigma}(\underline{\varepsilon} - \underline{\varepsilon}_T) \qquad (15.1.4)$$

Using Eq. (15.1.3) in (15.1.4), we have

$$u_0 = \tfrac{1}{2}(\underline{\varepsilon} - \underline{\varepsilon}_T)^T \underline{D}(\underline{\varepsilon} - \underline{\varepsilon}_T) \qquad (15.1.5)$$

where, in general, the transpose is needed on the strain matrix to multiply the matrices properly.

The total strain energy is then

$$U = \int_V u_0 \, dV \qquad (15.1.6)$$

Substituting Eq. (15.1.5) into (15.1.6), we obtain

$$U = \int_V \frac{1}{2}(\underline{\varepsilon} - \underline{\varepsilon}_T)^T \underline{D}(\underline{\varepsilon} - \underline{\varepsilon}_T) \, dV \qquad (15.1.7)$$

Now using $\underline{\varepsilon} = \underline{B}\underline{d}$ in Eq. (15.1.7), we obtain

$$U = \frac{1}{2}\int_V (\underline{B}\underline{d} - \underline{\varepsilon}_T)^T \underline{D}(\underline{B}\underline{d} - \underline{\varepsilon}_T) \, dV \qquad (15.1.8)$$

Simplifying Eq. (15.1.8) yields

$$U = \frac{1}{2}\int_V (\underline{d}^T\underline{B}^T\underline{D}\underline{B}\underline{d} - \underline{d}^T\underline{B}^T\underline{D}\underline{\varepsilon}_T$$

$$- \underline{\varepsilon}_T^T\underline{D}\underline{B}\underline{d} + \underline{\varepsilon}_T^T\underline{D}\underline{\varepsilon}_T) \, dV \qquad (15.1.9)$$

The first term in Eq. (15.1.9) is the usual strain energy due to stress produced from mechanical loading—that is,

$$U_L = \frac{1}{2}\int_V \underline{d}^T\underline{B}^T\underline{D}\underline{B}\underline{d} \, dV \qquad (15.1.10)$$

Terms 2 and 3 in Eq. (15.1.9) are identical and can be written together as

$$U_T = \int_V \underline{d}^T \underline{B}^T \underline{D}\varepsilon_T \, dV \qquad (15.1.11)$$

The last (fourth) term in Eq. (15.1.9) is a constant and drops out when we apply the principle of minimum potential energy by setting

$$\frac{\partial U}{\partial \underline{d}} = 0 \qquad (15.1.12)$$

Therefore, letting $U = U_L + U_T$ and substituting Eqs. (15.1.10) and (15.1.11) into Eq. (15.1.12), we obtain two contributions as

$$\frac{\partial U_L}{\partial \underline{d}} = \int_V \underline{B}^T \underline{D}\underline{B} \, dV \underline{d} \qquad (15.1.13)$$

and

$$\frac{\partial U_T}{\partial \underline{d}} = \int_V \underline{B}^T \underline{D}\varepsilon_T \, dV = \{f_T\} \qquad (15.1.14)$$

We recognize the integral term in Eq. (15.1.13) as the general form of the element stiffness matrix $\underline{k}$, whereas Eq. (15.1.14) is the load or force vector due to temperature change in the element.

We will now consider the one-dimensional thermal stress problem. We define the **thermal strain matrix** for the one-dimensional bar made of isotropic material with coefficient of thermal expansion α, and subjected to a uniform temperature rise T, as

$$\{\varepsilon_T\} = \{\varepsilon_{xT}\} = \{\alpha T\} \qquad (15.1.15)$$

where the units on α are typically (in./in.)/°F or (mm/mm)/°C.

For the simple one-dimensional bar (with a node at each end), we substitute Eq. (15.1.15) into Eq. (15.1.14) to obtain the thermal force matrix as

$$\{f_T\} = A \int_0^L [B]^T [D]\{\alpha T\} \, dx \qquad (15.1.16)$$

Recall that for the one-dimensional case, from Eqs. (3.9.15) and (3.9.13), we have

$$[D] = [E] \quad \text{and} \quad [B] = \left[-\frac{1}{L} \quad \frac{1}{L} \right] \qquad (15.1.17)$$

Substituting Eqs. (15.1.17) into Eq. (15.1.16) and simplifying, we obtain the thermal force matrix as

$$\{f_T\} = \begin{Bmatrix} f_{T1} \\ f_{T2} \end{Bmatrix} = \begin{Bmatrix} -E\alpha TA \\ E\alpha TA \end{Bmatrix} \qquad (15.1.18)$$

For the two-dimensional thermal stress problem, the thermal strain matrix for an anisotropic material is

$$\{\varepsilon_T\} = \begin{Bmatrix} \varepsilon_{xT} \\ \varepsilon_{yT} \\ \gamma_{xyT} \end{Bmatrix} \qquad (15.1.19)$$

For the case of plane stress in an isotropic material with coefficient of thermal expansion α subjected to a temperature rise T, the thermal strain matrix is

$$\{\varepsilon_T\} = \begin{Bmatrix} \alpha T \\ \alpha T \\ 0 \end{Bmatrix} \qquad (15.1.20)$$

No shear strains are caused by a change in temperature of isotropic materials, only expansion or contraction.

For the case of plane strain in an isotropic material, the thermal strain matrix is

$$\{\varepsilon_T\} = (1 + v) \begin{Bmatrix} \alpha T \\ \alpha T \\ 0 \end{Bmatrix} \qquad (15.1.21)$$

For a constant-thickness (t), constant-strain triangular element, Eq. (15.1.14) can be simplified to

$$\{f_T\} = [B]^T[D]\{\varepsilon_T\}tA \qquad (15.1.22)$$

The forces in Eq. (15.1.22) are contributed to the nodes of an element in an unequal manner and require precise evaluation. It can be shown that on substituting Eq. (7.1.8) for $[D]$, Eq. (7.2.34) for $[B]$, and Eq. (15.1.20) for $\{\varepsilon_T\}$ for a plane stress condition into Eq. (15.1.22), the constant-strain triangular element thermal force matrix is

$$\{f_T\} = \begin{Bmatrix} f_{Tix} \\ f_{Tiy} \\ \vdots \\ f_{Tmy} \end{Bmatrix} = \frac{\alpha E t T}{2(1 - v)} \begin{Bmatrix} \beta_i \\ \gamma_i \\ \beta_j \\ \gamma_j \\ \beta_m \\ \gamma_m \end{Bmatrix} \qquad (15.1.23)$$

where the β's and γ's are defined by Eqs. (7.2.10).

We will now describe the solution procedure for both one- and two-dimensional thermal stress problems.

Step 1

Evaluate the thermal force matrix, such as Eq. (15.1.18) or Eq. (15.1.23). Then treat this force matrix as an equivalent (or initial) force matrix $\underline{F}_0$ analogous to that obtained when replacing a distributed load acting on an element by equivalent nodal forces (see Chapters 5 and 6 and Appendix D).

Step 2

Apply $\underline{F} = \underline{K}\underline{d} - \underline{F}_0$, where if only thermal loading is considered, we solve $\underline{F}_0 = \underline{K}\underline{d}$ for the nodal displacements. Recall that in formulating the set of simultaneous equations, $\underline{F}$ represents the applied nodal forces, which here are assumed to be zero.

Step 3

Back-substitute the now known $\underline{d}$ into Step 2 to obtain the actual nodal forces, $\underline{F}(= \underline{K}\underline{d} - \underline{F}_0)$.

Hence, the thermal stress problem is solved in a manner similar to the distributed load problem discussed for beams and frames in Chapters 5 and 6. We will now solve the following examples to illustrate the general procedure.

E X A M P L E   15.1

For the one-dimensional bar fixed at both ends and subjected to a uniform temperature rise $T = 50°F$ as shown in Figure 15-2, determine the reactions at the fixed ends, and the axial stress in the bar. Let $E = 30 \times 10^6$ psi, $A = 4$ in.^{2}, $L = 4$ ft, and $\alpha = 7.0 \times 10^{-6}$ (in./in.)/°F.

Two elements will be sufficient to represent the bar because internal nodal displacements are not of importance here. To solve $\underline{F}_0 = \underline{K}\underline{d}$, we must determine the global stiffness matrix for the bar. Hence, for each element, we have

$$\underline{k}^{(1)} = \frac{AE}{L/2} \overset{\displaystyle 1 \qquad 2}{\begin{bmatrix} 1 & -1 \\ -1 & 1 \end{bmatrix}} \qquad \underline{k}^{(2)} = \frac{AE}{L/2} \overset{\displaystyle 2 \qquad 3}{\begin{bmatrix} 1 & -1 \\ -1 & 1 \end{bmatrix}} \qquad (15.1.24)$$

Step 1

Using Eq. (15.1.18), the thermal force matrix for each element is given by

$$\underline{f}^{(1)} = \begin{Bmatrix} -E\alpha TA \\ E\alpha TA \end{Bmatrix} \qquad \underline{f}^{(2)} = \begin{Bmatrix} -E\alpha TA \\ E\alpha TA \end{Bmatrix} \qquad (15.1.25)$$

where these forces are considered to be equivalent nodal forces.

Step 2

Applying the direct stiffness method to Eqs. (15.1.24) and (15.1.25), we assemble the global equations as

$$\begin{Bmatrix} -E\alpha TA \\ 0 \\ E\alpha TA \end{Bmatrix} = \frac{AE}{L/2} \begin{bmatrix} 1 & -1 & 0 \\ -1 & 1+1 & -1 \\ 0 & -1 & 1 \end{bmatrix} \begin{Bmatrix} d_{1x} \\ d_{2x} \\ d_{3x} \end{Bmatrix} \qquad (15.1.26)$$

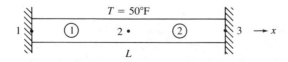

F I G U R E   15-2   Bar subjected to a uniform temperature rise

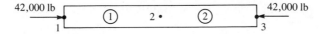

FIGURE 15-3 Free-body diagram of the bar of Figure 15-2

Applying the boundary conditions $d_{1x} = 0$ and $d_{3x} = 0$ and solving the second of Eq. (15.1.26), we obtain

$$d_{2x} = 0 \qquad (15.1.27)$$

Step 3

Back-substituting Eq. (15.1.27) into the global equation (Step 2) for the nodal forces, we obtain

$$\begin{Bmatrix} F_{1x} \\ F_{2x} \\ F_{3x} \end{Bmatrix} = \begin{Bmatrix} 0 \\ 0 \\ 0 \end{Bmatrix} - \begin{Bmatrix} -E\alpha TA \\ 0 \\ E\alpha TA \end{Bmatrix} = \begin{Bmatrix} E\alpha TA \\ 0 \\ -E\alpha TA \end{Bmatrix} \qquad (15.1.28)$$

Using the numerical quantities for E, α, T, and A in Eq. (15.1.28), we obtain

$$F_{1x} = 42,000 \text{ lb} \qquad F_{2x} = 0 \qquad F_{3x} = -42,000 \text{ lb}$$

as shown in Figure 15-3. The stress in the bar is then

$$\sigma = \frac{42,000}{4} = 10,500 \text{ psi} \qquad \text{(compressive)} \qquad (15.1.29) \quad \blacksquare$$

EXAMPLE 15.2

For the bar assemblage shown in Figure 15-4 determine the reactions at the fixed ends and the axial stress in each bar. Bar 1 is subjected to a temperature drop of 10°C. Let bar 1 be aluminum with $E = 70$ GPa, $\alpha = 23 \times 10^{-6}$ (mm/mm)/°C, $A = 12 \times 10^{-4} \text{ m}^2$, and $L = 2$ m. Let bars 2 and 3 be brass with $E = 100$ GPa, $\alpha = 20 \times 10^{-6}$ (mm/mm)/°C, $A = 6 \times 10^{-4} \text{ m}^2$, and $L = 2$ m.

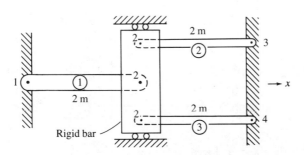

FIGURE 15-4 Bar assemblage for thermal stress analysis

We begin the solution by determining the stiffness matrices for each element.

Element 1

$$k^{(1)} = \frac{(12 \times 10^{-4})(70 \times 10^6)}{2} \begin{bmatrix} 1 & -1 \\ -1 & 1 \end{bmatrix} = 42{,}000 \begin{matrix} 1 & 2 \\ \begin{bmatrix} 1 & -1 \\ -1 & 1 \end{bmatrix} \end{matrix}$$

$$(15.1.30a)$$

Elements 2 and 3

$$k^{(2)} = k^{(3)} = \frac{(6 \times 10^{-4})(100 \times 10^6)}{2} \begin{bmatrix} 1 & -1 \\ -1 & 1 \end{bmatrix} = 30{,}000 \begin{matrix} 2 & 3 \\ 2 & 4 \\ \begin{bmatrix} 1 & -1 \\ -1 & 1 \end{bmatrix} \end{matrix}$$

$$(15.1.30b)$$

Step 1

We obtain the element thermal force matrices by evaluating Eq. (15.1.18). First, evaluating $-E\alpha TA$ for element 1, we have

$$-E\alpha TA = -(70 \times 10^6)(23 \times 10^{-6})(-10)(12 \times 10^{-4}) = 19.32 \text{ kN}$$

$$(15.1.31)$$

where the -10 term in Eq. (15.1.31) is due to the temperture drop in element 1. Using the result of Eq. (15.1.31) in Eq. (15.1.18), we obtain

$$f^{(1)} = \begin{Bmatrix} f_{1x} \\ f_{2x} \end{Bmatrix} = \begin{Bmatrix} 19.32 \\ -19.32 \end{Bmatrix}$$

$$(15.1.32)$$

There is no temperature change in elements 2 and 3, so

$$f^{(2)} = f^{(3)} = \begin{Bmatrix} 0 \\ 0 \end{Bmatrix}$$

$$(15.1.33)$$

Step 2

Assembling the global equations using Eqs. (15.1.30), (15.1.32), and (15.1.33), we obtain

$$1000 \begin{matrix} & 1 & 2 & 3 & 4 \\ & \begin{bmatrix} 42 & -42 & 0 & 0 \\ -42 & 42+30+30 & -30 & -30 \\ 0 & -30 & 30 & 0 \\ 0 & -30 & 0 & 30 \end{bmatrix} \end{matrix} \begin{Bmatrix} d_{1x} \\ d_{2x} \\ d_{3x} \\ d_{4x} \end{Bmatrix} = \begin{Bmatrix} 19.32 \\ -19.32 \\ 0 \\ 0 \end{Bmatrix}$$

$$(15.1.34)$$

where the right side thermal forces are considered to be equivalent nodal forces. Using the boundary conditions

$$d_{1x} = 0 \qquad d_{3x} = 0 \qquad d_{4x} = 0 \qquad (15.1.35)$$

we obtain from the second equation of Eq. (15.1.34),

$$1000(102)d_{2x} = -19.32$$

Solving for d_{2x}, we obtain

$$d_{2x} = -1.89 \times 10^{-4} \text{ m} \qquad (15.1.36)$$

Step 3

Back-substituting Eq. (15.1.36) into the global equation for the nodal forces, we have

$$\begin{Bmatrix} F_{1x} \\ F_{2x} \\ F_{3x} \\ F_{4x} \end{Bmatrix} = 1000 \begin{bmatrix} 42 & -42 & 0 & 0 \\ -42 & 102 & -30 & -30 \\ 0 & -30 & 30 & 0 \\ 0 & -30 & 0 & 30 \end{bmatrix} \begin{Bmatrix} 0 \\ -1.89 \times 10^{-4} \\ 0 \\ 0 \end{Bmatrix} - \begin{Bmatrix} 19.32 \\ -19.32 \\ 0 \\ 0 \end{Bmatrix}$$

$$(15.1.37)$$

Simplifying Eq. (15.1.37), we obtain

$$F_{1x} = -11.38 \text{ kN}$$

$$F_{2x} = 0.0 \text{ kN}$$

$$F_{3x} = 5.69 \text{ kN} \qquad (15.1.38)$$

$$F_{4x} = 5.69 \text{ kN}$$

A free-body diagram of the bar assemblage is shown in Figure 15-5. The stresses in each bar are then

$$\sigma^{(1)} = \frac{11.38}{12 \times 10^{-4}} = 9.48 \times 10^3 \text{ kN/m}^2 \quad (9.48 \text{ MPa})$$

$$(15.1.39)$$

$$\sigma^{(2)} = \sigma^{(3)} = \frac{5.69}{6 \times 10^{-4}} = 9.48 \times 10^3 \text{ kN/m}^2 \quad (9.48 \text{ MPa})$$

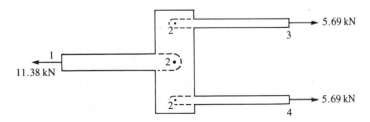

F I G U R E   15-5   Free-body diagram of the bar assemblage of Figure 15-4

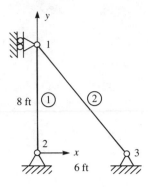

FIGURE 15-6   Plane truss for thermal stress analysis

EXAMPLE 15.3

For the plane truss shown in Figure 15-6, determine the displacements at node 1 and the axial stresses in each bar. Bar 1 is subjected to a temperature rise of 75°F. Let $E = 30 \times 10^6$ psi, $\alpha = 7 \times 10^{-6}$ (in./in.)/°F, and $A = 2$ in.² for both bar elements.

First, using Eq. (3.4.23), we determine the stiffness matrices for each element.

Element 1

Choosing $\hat{x}$ from node 2 to node 1, $\theta = 90°$, so $\cos \theta = 0$, $\sin \theta = 1$, and

$$\underline{k}^{(1)} = \frac{(2)(30 \times 10^6)}{(8 \times 12)} \begin{array}{c} \begin{array}{cccc} & 2 & & 1 \end{array} \\ \begin{bmatrix} 0 & 0 & 0 & 0 \\ & 1 & 0 & -1 \\ & & 0 & 0 \\ \text{Symmetry} & & & 1 \end{bmatrix} \end{array} \qquad (15.1.40)$$

Element 2

Choosing $\hat{x}$ from node 3 to node 1, $\theta = 180° - 53.13° = 126.87°$, so $\cos \theta = -0.6$, $\sin \theta = 0.8$, and

$$\underline{k}^{(2)} = \frac{(2)(30 \times 10^6)}{(10 \times 12)} \begin{array}{c} \begin{array}{cccc} 3 & & 1 & \end{array} \\ \begin{bmatrix} 0.36 & -0.48 & -0.36 & 0.48 \\ & 0.64 & 0.48 & -0.64 \\ & & 0.36 & -0.48 \\ \text{Symmetry} & & & 0.64 \end{bmatrix} \end{array} \qquad (15.1.41)$$

Step 1

We obtain the element thermal force matrices by evaluating Eq. (15.1.18) as follows:

$$-E\alpha TA = -(30 \times 10^6)(7 \times 10^{-6})(75)(2) = -31,500 \text{ lb} \qquad (15.1.42)$$

Using the result of Eq. (15.1.42) for element 1, we then have the local thermal force matrix as

$$\hat{f}^{(1)} = \begin{Bmatrix} \hat{f}_{2x} \\ \hat{f}_{1x} \end{Bmatrix} = \begin{Bmatrix} -31,500 \\ 31,500 \end{Bmatrix} \qquad (15.1.43)$$

There is no temperature change in element 2, so

$$\hat{f}^{(2)} = \begin{Bmatrix} \hat{f}_{3x} \\ \hat{f}_{1x} \end{Bmatrix} = \begin{Bmatrix} 0 \\ 0 \end{Bmatrix} \qquad (15.1.44)$$

Recall that by Eq. (3.4.16), $\hat{f} = \underline{T}f$. Since we have shown that $\underline{T}^{-1} = \underline{T}^T$, we can obtain the global forces by premultiplying Eq. (3.4.16) by $\underline{T}^T$ to obtain the element nodal forces in the global reference frame as

$$f = \underline{T}^T \hat{f} \qquad (15.1.45)$$

Using Eq. (15.1.45), the element 1 global nodal forces are then

$$\begin{Bmatrix} f_{2x} \\ f_{2y} \\ f_{1x} \\ f_{1y} \end{Bmatrix} = \begin{bmatrix} C & -S & 0 & 0 \\ S & C & 0 & 0 \\ 0 & 0 & C & -S \\ 0 & 0 & S & C \end{bmatrix} \begin{Bmatrix} \hat{f}_{2x} \\ \hat{f}_{2y} \\ \hat{f}_{1x} \\ \hat{f}_{1y} \end{Bmatrix} \qquad (15.1.46)$$

where the order of terms in Eq. (15.1.46) is due to the choice of the $\hat{x}$ axis from node 2 to node 1 and $\underline{T}$, given by Eq. (3.4.15), has been used.

Substituting the numerical quantities $C = 0$ and $S = 1$ (consistent with $\hat{x}$ for element 1), and $\hat{f}_{1x} = 31,500$, $\hat{f}_{1y} = 0$, $\hat{f}_{2x} = -31,500$, and $\hat{f}_{2y} = 0$ into Eq. (15.1.46), we obtain

$$f_{2x} = 0 \qquad f_{2y} = -31,500 \text{ lb} \qquad f_{1x} = 0 \qquad f_{1y} = 31,500 \text{ lb} \qquad (15.1.47)$$

These element forces are now the only equivalent global nodal ones, since element 2 is not subjected to a change in temperature.

Step 2

Assembling the global equations using Eqs. (15.1.40), (15.1.41), and (15.1.47), we obtain

$$0.50 \times 10^6 \begin{bmatrix} 0.36 & -0.48 & 0 & 0 & 0 & 0 \\ & 1.89 & 0 & -1.25 & 0 & 0 \\ & & 0 & 0 & 0 & 0 \\ & & & 1.25 & 0 & 0 \\ & & & & 0.36 & -0.48 \\ \text{Symmetry} & & & & & 0.64 \end{bmatrix} \begin{Bmatrix} d_{1x} \\ d_{1y} \\ d_{2x} \\ d_{2y} \\ d_{3x} \\ d_{3y} \end{Bmatrix} = \begin{Bmatrix} F_{1x} + 0 \\ 31,500 \\ F_{2x} + 0 \\ -31,500 + F_{2y} \\ F_{3x} + 0 \\ F_{3y} + 0 \end{Bmatrix} \qquad (15.1.48)$$

The boundary conditions are given by

$$d_{1x} = 0 \qquad d_{2x} = 0 \qquad d_{2y} = 0 \qquad d_{3x} = 0 \qquad d_{3y} = 0 \qquad (15.1.49)$$

Using the boundary condition Eqs. (15.1.49), and the second equation of Eq. (15.1.48), we obtain

$$(0.945 \times 10^6)d_{1y} = 31,500$$

or $$d_{1y} = 0.0333 \text{ in.} \tag{15.1.50}$$

Step 3

We now illustrate the procedure used to obtain the local element forces in local coordinates; that is, the local element forces are

$$\hat{f} = \hat{\underline{k}}\hat{\underline{d}} - \hat{f}_0 \tag{15.1.51}$$

We determine the actual local element nodal forces by using the relationship $\hat{\underline{d}} = \underline{T}^*\underline{d}$; the usual bar element $\underline{k}$ matrix, Eq. (3.1.14); the transformation matrix $\underline{T}^*$, Eq. (3.4.8); and the calculated displacements and initial thermal forces applicable for the element under consideration. Substituting the numerical quantities for element 1, from Eq. (15.1.51), we have

$$\begin{Bmatrix} \hat{f}_{2x} \\ \hat{f}_{1x} \end{Bmatrix} = \frac{2(30 \times 10^6)}{8 \times 12} \begin{bmatrix} 1 & -1 \\ -1 & 1 \end{bmatrix} \begin{bmatrix} 0 & 1 & 0 & 0 \\ 0 & 0 & 0 & 1 \end{bmatrix}$$

$$\times \begin{Bmatrix} d_{2x} = 0 \\ d_{2y} = 0 \\ d_{1x} = 0 \\ d_{1y} = 0.0333 \end{Bmatrix} - \begin{Bmatrix} -31,500 \\ 31,500 \end{Bmatrix} \tag{15.1.52}$$

Simplifying Eq. (15.1.52), we obtain

$$\hat{f}_{2x} = 10,700 \text{ lb} \qquad \hat{f}_{1x} = -10,700 \text{ lb} \tag{15.1.53}$$

Dividing the local element force $\hat{f}_{1x}$ (which is the far-end force consistent with the convention used in Section 3.5) by the cross-sectional area, we obtain the stress as

$$\sigma^{(1)} = \frac{-10,700}{2} = -5350 \text{ psi} \tag{15.1.54}$$

Similarly, for element 2, we have

$$\begin{Bmatrix} \hat{f}_{3x} \\ \hat{f}_{1x} \end{Bmatrix} = \frac{2(30 \times 10^6)}{10 \times 12} \begin{bmatrix} 1 & -1 \\ -1 & 1 \end{bmatrix} \begin{bmatrix} -0.6 & 0.8 & 0 & 0 \\ 0 & 0 & -0.6 & 0.8 \end{bmatrix} \begin{Bmatrix} 0 \\ 0 \\ 0 \\ 0.0333 \end{Bmatrix}$$

$$\tag{15.1.55}$$

Simplifying, Eq. (15.1.55), we obtain

$$\hat{f}_{3x} = -13,310 \text{ lb} \qquad \hat{f}_{1x} = 13,310 \text{ lb} \tag{15.1.56}$$

where no initial thermal forces were present for element 2 because the element was not subjected to a temperature change. Dividing the far-end force $\hat{f}_{1x}$ by the cross-sectional area results in

$$\sigma^{(2)} = 6660 \text{ psi} \tag{15.1.57}$$

For two- and three-dimensional stress problems, this direct division of force by cross-sectional area is not permissible. Hence, the total stress due to both applied loading and temperature change must be determined by

$$\underline{\sigma} = \underline{\sigma}_L - \underline{\sigma}_T \qquad (15.1.58)$$

We now illustrate Eq. (15.1.58) for bar element 1 of the truss of Example 15.4. For the bar, σ_L can be obtained using Eq. (3.5.6), and σ_T is obtained from

$$\sigma_T = \underline{D}\varepsilon_T = E\alpha T \qquad (15.1.59)$$

since $\underline{D} = E$ and $\varepsilon_T = \alpha T$ for the bar element. The stress in bar element 1 is then determined to be

$$\sigma^{(1)} = \frac{E}{L}[-C \quad -S \quad C \quad S]\begin{Bmatrix} d_{2x} \\ d_{2y} \\ d_{1x} \\ d_{1y} \end{Bmatrix} - E\alpha T \qquad (15.1.60)$$

Substituting the numerical quantities for element 1 into Eq. (15.1.60), we obtain

$$\sigma^{(1)} = \frac{30 \times 10^6}{8 \times 12}[0 \quad -1 \quad 0 \quad 1]\begin{Bmatrix} 0 \\ 0 \\ 0 \\ 0.0333 \end{Bmatrix} - (30 \times 10^6)(7 \times 10^{-6})(75)$$

$$(15.1.61)$$

or

$$\sigma^{(1)} = -5350 \text{ psi} \qquad (15.1.62) \quad \blacksquare$$

We will now illustrate the solutions of two plane thermal stress problems.

E X A M P L E   15.4

For the plane stress element shown in Figure 15-7, determine the element equations. The element has a 2000-lb/in.^{2} pressure acting perpendicular to side j-m and is subjected to a 30°F temperature rise.

Recall that the stiffness matrix is given by

$$[k] = [B]^T[D][B]tA \qquad (15.1.63)$$

and

$$\beta_i = y_j - y_m = -3 \qquad \gamma_i = x_m - x_j = -1$$
$$\beta_j = y_m - y_i = 3 \qquad \gamma_j = x_i - x_m = -1 \qquad (15.1.64)$$
$$\beta_m = y_i - y_j = 0 \qquad \gamma_m = x_j - x_i = 2$$

and

$$A = \frac{(3)(2)}{2} = 3 \text{ in.}^2$$

Therefore, substituting the results of Eqs. (15.1.64) into Eq. (7.2.34) for $[B]$, we obtain

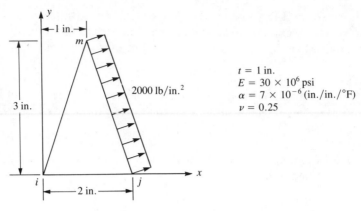

FIGURE 15-7   Plane stress element subjected to mechanical loading and a temperature change

$$[B] = \frac{1}{6}\begin{bmatrix} -3 & 0 & 3 & 0 & 0 & 0 \\ 0 & -1 & 0 & -1 & 0 & 2 \\ -1 & -3 & -1 & 3 & 2 & 0 \end{bmatrix} \qquad (15.1.65)$$

Assuming plane stress conditions to be valid, we have

$$[D] = \frac{E}{1 - v^2}\begin{bmatrix} 1 & v & 0 \\ v & 1 & 0 \\ 0 & 0 & \dfrac{1-v}{2} \end{bmatrix} = \frac{30 \times 10^6}{1 - (0.25)^2}\begin{bmatrix} 1 & 0.25 & 0 \\ 0.25 & 1 & 0 \\ 0 & 0 & 0.375 \end{bmatrix}$$

$$= (4 \times 10^6)\begin{bmatrix} 8 & 2 & 0 \\ 2 & 8 & 0 \\ 0 & 0 & 3 \end{bmatrix} \qquad (15.1.66)$$

Also,

$$[B]^T[D] = \frac{1}{6}\begin{bmatrix} -3 & 0 & -1 \\ 0 & -1 & -3 \\ 3 & 0 & -1 \\ 0 & -1 & 3 \\ 0 & 0 & 2 \\ 0 & 2 & 0 \end{bmatrix}(4 \times 10^6)\begin{bmatrix} 8 & 2 & 0 \\ 2 & 8 & 0 \\ 0 & 0 & 3 \end{bmatrix} \qquad (15.1.67)$$

Simplifying Eq. (15.1.67), we obtain

$$[B]^T[D] = \frac{4 \times 10^6}{6}\begin{bmatrix} -24 & -6 & -3 \\ -2 & -8 & -9 \\ 24 & 6 & -3 \\ -2 & -8 & 9 \\ 0 & 0 & 6 \\ 4 & 16 & 0 \end{bmatrix} \qquad (15.1.68)$$

Therefore, substituting the results of Eqs. (15.1.65) and (15.1.68) into Eq. (15.1.63) yields the element stiffness matrix as

$$[k] = (1 \text{ in.}) \frac{(3 \text{ in.}^2) \, 4 \times 10^6}{6 \quad 6} \begin{bmatrix} -24 & -6 & -3 \\ -2 & -8 & -9 \\ 24 & 6 & -3 \\ -2 & -8 & 9 \\ 0 & 0 & .6 \\ 4 & 16 & 0 \end{bmatrix} \begin{bmatrix} -3 & 0 & 3 & 0 & 0 & 0 \\ 0 & -1 & 0 & -1 & 0 & 2 \\ -1 & -3 & -1 & 3 & 2 & 0 \end{bmatrix}$$

$$(15.1.69)$$

Simplifying Eq. (15.1.69), we have the element stiffness matrix as

$$[k] = \frac{1 \times 10^6}{3} \begin{bmatrix} 75 & 15 & -69 & -3 & -6 & -12 \\ 15 & 35 & 3 & -19 & -18 & -16 \\ -69 & 3 & 75 & -15 & -6 & 12 \\ -3 & -19 & -15 & 35 & 18 & -16 \\ -6 & -18 & -6 & 18 & 12 & 0 \\ -12 & -16 & 12 & -16 & 0 & 32 \end{bmatrix} \qquad (15.1.70)$$

Using Eq. (15.1.23), the thermal force matrix is given by

$$\{f_T\} = \frac{\alpha E t T}{2(1-v)} \begin{Bmatrix} \beta_i \\ \gamma_i \\ \beta_j \\ \gamma_j \\ \beta_m \\ \gamma_m \end{Bmatrix} = \frac{(7 \times 10^{-6})(30 \times 10^6)(1)(30)}{2(1-0.25)} \begin{Bmatrix} -3 \\ -1 \\ 3 \\ -1 \\ 0 \\ 2 \end{Bmatrix}$$

$$= 4200 \begin{Bmatrix} -3 \\ -1 \\ 3 \\ -1 \\ 0 \\ 2 \end{Bmatrix}$$

or

$$\{f_T\} = \begin{Bmatrix} -12{,}600 \\ -4200 \\ 12{,}600 \\ -4200 \\ 0 \\ 8400 \end{Bmatrix} \qquad (15.1.71)$$

The force matrix due to the pressure applied alongside j-m is determined as follows:

$$L_{j\text{-}m} = [(2-1)^2 + (3-0)^2]^{1/2} = 3.163 \text{ in.}$$

$$p_x = p \cos \theta = 2000 \left(\frac{3}{3.163} \right) = 1896 \text{ lb/in.}^2 \qquad (15.1.72)$$

$$p_y = p \sin \theta = 2000 \left(\frac{1}{3.163} \right) = 632 \text{ lb/in.}^2$$

where θ is the angle measured from the x axis to the normal to surface $j\text{-}m$. Using Eq. (7.3.7) to evaluate the surface forces, we have

$$\{f_L\} = \int\!\!\int_{S_{j\text{-}m}} [N]^T \begin{Bmatrix} p_x \\ p_y \end{Bmatrix} dS$$

$$= \int\!\!\int_{S_{j\text{-}m}} \begin{bmatrix} N_i & 0 \\ 0 & N_i \\ N_j & 0 \\ 0 & N_j \\ N_m & 0 \\ 0 & N_m \end{bmatrix}_{\substack{\text{evaluated} \\ \text{alongside } j\text{-}m}} \begin{Bmatrix} p_x \\ p_y \end{Bmatrix} dS = \frac{t L_{j\text{-}m}}{2} \begin{bmatrix} 0 & 0 \\ 0 & 0 \\ 1 & 0 \\ 0 & 1 \\ 1 & 0 \\ 0 & 1 \end{bmatrix} \begin{Bmatrix} p_x \\ p_y \end{Bmatrix} \qquad (15.1.73)$$

Evaluating Eq. (15.1.73), we obtain

$$\{f_L\} = \frac{(1 \text{ in.})(3.163 \text{ in.})}{2} \begin{bmatrix} 0 & 0 \\ 0 & 0 \\ 1 & 0 \\ 0 & 1 \\ 1 & 0 \\ 0 & 1 \end{bmatrix} \begin{Bmatrix} 1896 \\ 632 \end{Bmatrix} = \begin{Bmatrix} 0 \\ 0 \\ 3000 \\ 1000 \\ 3000 \\ 1000 \end{Bmatrix} \qquad (15.1.74)$$

Using Eqs. (15.1.70), (15.1.71), and (15.1.74), the complete set of element equations is

$$\frac{1 \times 10^6}{3} \begin{bmatrix} 75 & 15 & -69 & -3 & -6 & -12 \\ & 35 & 3 & -19 & -18 & -16 \\ & & 75 & -15 & -6 & 12 \\ & & & 35 & 18 & -16 \\ & & & & 12 & 0 \\ \text{Symmetry} & & & & & 32 \end{bmatrix} \begin{Bmatrix} u_i \\ v_i \\ u_j \\ v_j \\ u_m \\ v_m \end{Bmatrix} = \begin{Bmatrix} -12,600 \\ -4200 \\ 15,600 \\ -3200 \\ 3000 \\ 9400 \end{Bmatrix}$$

$$(15.1.75)$$

where the force matrix is $\{f_T\} + \{f_L\}$, obtained by adding Eqs. (15.1.71) and (15.1.74). ∎

E X A M P L E   15.5

For the plane stress plate fixed along one edge and subjected to a uniform temperature rise of 50°C as shown in Figure 15-8, determine the nodal dis-

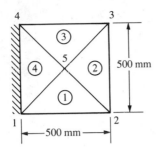

FIGURE 15-8 Discretized plate subjected to a temperature change

placements, and the stresses in each element. Let $E = 210$ GPa, $v = 0.30$, $t = 5$ mm, and $\alpha = 12 \times 10^{-6}$ (mm/mm)/°C.

The discretized plate is shown in Figure 15-8. We begin by evaluating the stiffness matrix of each element using Eq. (7.2.52).

Element 1

Element 1 has coordinates $x_1 = 0$, $y_1 = 0$, $x_2 = 0.5$, $y_2 = 0$, $x_5 = 0.25$, and $y_5 = 0.25$. From Eqs. (7.2.10), we obtain

$$\beta_1 = y_2 - y_5 = -0.25 \qquad \beta_2 = y_5 - y_1 = 0.25 \qquad \beta_5 = y_1 - y_2 = 0$$

$$\gamma_1 = x_5 - x_2 = -0.25 \qquad \gamma_2 = x_1 - x_5 = -0.25 \qquad \gamma_5 = x_2 - x_1 = 0.5$$
$$(15.1.76)$$

Using Eqs. (7.2.32) in Eq. (7.2.34), we have

$$[B] = \frac{1}{2A} \begin{bmatrix} \beta_1 & 0 & \beta_2 & 0 & \beta_5 & 0 \\ 0 & \gamma_1 & 0 & \gamma_2 & 0 & \gamma_5 \\ \gamma_1 & \beta_1 & \gamma_2 & \beta_2 & \gamma_5 & \beta_5 \end{bmatrix}$$

$$= \frac{1}{0.125} \begin{bmatrix} -0.25 & 0 & 0.25 & 0 & 0 & 0 \\ 0 & -0.25 & 0 & -0.25 & 0 & 0.5 \\ -0.25 & -0.25 & -0.25 & 0.25 & 0.5 & 0 \end{bmatrix} \qquad (15.1.77)$$

For plane stress, $[D]$ is given by

$$D = \frac{E}{(1 - v^2)} \begin{bmatrix} 1 & v & 0 \\ v & 1 & 0 \\ 0 & 0 & \frac{1-v}{2} \end{bmatrix} = \frac{210 \times 10^9}{0.91} \begin{bmatrix} 1 & 0.3 & 0 \\ 0.3 & 1 & 0 \\ 0 & 0 & 0.35 \end{bmatrix} \qquad (15.1.78)$$

We obtain the element stiffness matrix using

$$[k] = tA[B]^T[D][B] \qquad (15.1.79)$$

Substituting the results of Eqs. (15.1.77) and (15.1.78) into Eq. (15.1.79) and carrying out the multiplications, we have

$$
\underline{k} = 4.615 \times 10^7
\begin{array}{c}
\begin{array}{cccccc}
d_{1x} & d_{1y} & d_{2x} & d_{2y} & d_{5x} & d_{5y}
\end{array} \\
\begin{bmatrix}
8.4375 & 4.0625 & -4.0625 & -0.3125 & -4.375 & -3.75 \\
4.0625 & 8.4375 & 0.3125 & 4.0625 & -4.375 & -12.5 \\
-4.0625 & 0.3125 & 8.4375 & -4.0625 & -4.375 & 3.75 \\
-0.3125 & 4.0625 & -4.0625 & 8.4375 & 4.375 & -12.5 \\
-4.375 & -4.375 & -4.375 & 4.375 & 8.75 & 0 \\
-3.75 & -12.5 & 3.75 & -12.5 & 0 & 25
\end{bmatrix}
\end{array}
$$

$$(15.1.80)$$

Element 2

For element 2, the coordinates are $x_2 = 0.5$, $y_2 = 0$, $x_3 = 0.5$, $y_3 = 0.5$, $x_5 = 0.25$, and $y_5 = 0.25$. Proceeding as for element 1, we obtain

$$
\beta_2 = 0.25 \qquad \beta_3 = 0.25 \qquad \beta_5 = -0.5
$$

$$
\gamma_2 = -0.25 \qquad \gamma_3 = 0.25 \qquad \gamma_5 = 0
$$

The element stiffness matrix then becomes

$$
\underline{k} = 4.615 \times 10^7
\begin{array}{c}
\begin{array}{cccccc}
d_{2x} & d_{2y} & d_{3x} & d_{3y} & d_{5x} & d_{5y}
\end{array} \\
\begin{bmatrix}
8.4375 & -4.0625 & 4.0625 & -0.3125 & -12.5 & 4.375 \\
-4.0625 & 8.4375 & 0.3125 & -4.0625 & 3.75 & -4.375 \\
4.0625 & 0.3125 & 8.437 & 4.0625 & -12.5 & -4.375 \\
-0.3125 & -4.0625 & 4.0625 & 8.4375 & -3.75 & -4.375 \\
-12.5 & 3.75 & -12.5 & -3.75 & 25 & 0 \\
4.375 & -4.375 & -4.375 & -4.375 & 0 & 8.75
\end{bmatrix}
\end{array}
$$

$$(15.1.81)$$

Element 3

For element 3, using the same steps as for element 1, we obtain the stiffness matrix as

$$
\underline{k} = 4.615 \times 10^7
\begin{array}{c}
\begin{array}{cccccc}
d_{3x} & d_{3y} & d_{4x} & d_{4y} & d_{5x} & d_{5y}
\end{array} \\
\begin{bmatrix}
8.437 & 4.0625 & -4.0625 & -0.3125 & -4.375 & -3.75 \\
4.0625 & 8.437 & 0.3125 & 4.0625 & -4.375 & -12.5 \\
-4.0625 & 0.3125 & 8.437 & -4.0625 & -4.375 & 3.75 \\
-0.3125 & 4.0625 & -4.0625 & 8.4375 & 4.375 & -12.5 \\
-4.375 & -4.375 & -4.375 & 4.375 & 8.75 & 0 \\
-3.75 & -12.5 & 3.75 & -12.5 & 0 & 25
\end{bmatrix}
\end{array}
$$

$$(15.1.82)$$

Element 4

Finally, for element 4, we obtain

$$
\underline{k} = 4.615 \times 10^7
\begin{array}{c}
\begin{array}{cccccc}
d_{4x} & d_{4y} & d_{1x} & d_{1y} & d_{5x} & d_{5y}
\end{array} \\
\begin{bmatrix}
8.437 & -4.0625 & 4.0625 & -0.3125 & -12.5 & 4.375 \\
-4.0625 & 8.4375 & 0.3125 & -4.0625 & 3.75 & -4.375 \\
4.0625 & 0.3125 & 8.437 & 4.0625 & -12.5 & -4.375 \\
-0.3125 & -4.0625 & 4.0625 & 8.431 & -3.75 & -4.375 \\
-12.5 & 3.75 & -12.5 & -3.75 & 25 & 0 \\
4.375 & -4.375 & -4.375 & -4.375 & 0 & 8.75
\end{bmatrix}
\end{array}
$$

$$(15.1.83)$$

Using the direct stiffness method, we assemble the element stiffness matrices, Eqs. (15.1.80)–(15.1.83), to obtain the global stiffness matrix as

$$
\underline{K} = 4.615 \times 10^7
$$

d_{1x}	d_{1y}	d_{2x}	d_{2y}
16.874	8.125	−4.0625	−0.3125
8.125	16.874	0.3125	4.0625
−4.0625	0.3125	16.874	−8.125
−0.3125	4.0625	−8.125	16.875
0	0	4.0625	0.3125
0	0	−0.3125	−4.0625
4.0625	−0.3125	0	0
0.3125	−4.0625	0	0
−16.875	−8.125	−16.875	8.125
−8.125	−16.875	8.125	−16.875

d_{3x}	d_{3y}	d_{4x}	d_{4y}	d_{5x}	d_{5y}
0	0	4.0625	0.3125	−16.875	−8.125
0	0	−0.3125	−4.0625	−8.125	−16.875
4.0625	−0.3125	0	0	−16.875	8.125
0.3125	−4.0625	0	0	8.125	−16.875
16.875	8.125	−4.0625	−0.3125	−16.875	−8.125
8.125	16.875	0.3125	4.0625	−8.125	−16.875
−4.0625	0.3125	16.875	−8.125	−16.875	8.125
−0.3125	4.0625	−8.125	16.875	8.125	−16.875
−16.875	−8.125	−16.875	8.125	67.5	0
−8.125	−16.875	8.125	−16.875	0	67.5

$$(15.1.84)$$

Next, we determine the thermal force matrices for each element by using Eq. (15.1.23) as follows:

Element 1

$$\{f_T\} = \frac{\alpha E t\, T}{2(1-v)} \begin{Bmatrix} \beta_1 \\ \gamma_1 \\ \beta_2 \\ \gamma_2 \\ \beta_5 \\ \gamma_5 \end{Bmatrix} = \frac{(12 \times 10^{-6})(210 \times 10^9)(0.005\ \text{m})(50)}{2(1-0.3)} \begin{Bmatrix} -0.25 \\ -0.25 \\ 0.25 \\ -0.25 \\ 0 \\ 0.5 \end{Bmatrix}$$

$$= 450{,}000 \begin{Bmatrix} -0.25 \\ -0.25 \\ 0.25 \\ -0.25 \\ 0 \\ 0.5 \end{Bmatrix} = \begin{Bmatrix} f_{T1x} \\ f_{T1y} \\ f_{T2x} \\ f_{T2y} \\ f_{T5x} \\ f_{T5y} \end{Bmatrix} = \begin{Bmatrix} -112{,}500 \\ -112{,}500 \\ 112{,}500 \\ -112{,}500 \\ 0 \\ 225{,}000 \end{Bmatrix} \text{N} \qquad (15.1.85)$$

Element 2

$$\{f_T\} = 450{,}000 \begin{Bmatrix} 0.25 \\ -0.25 \\ 0.25 \\ 0.25 \\ -0.5 \\ 0 \end{Bmatrix} = \begin{Bmatrix} f_{T2x} \\ f_{T2y} \\ f_{T3x} \\ f_{T3y} \\ f_{T5x} \\ f_{T5y} \end{Bmatrix} = \begin{Bmatrix} 112{,}500 \\ -112{,}500 \\ 112{,}500 \\ 112{,}500 \\ -225{,}000 \\ 0 \end{Bmatrix} \text{N} \qquad (15.1.86)$$

Element 3

$$\{f_T\} = 450{,}000 \begin{Bmatrix} 0.25 \\ 0.25 \\ -0.25 \\ 0.25 \\ 0 \\ -0.5 \end{Bmatrix} = \begin{Bmatrix} f_{T3x} \\ f_{T3y} \\ f_{T4x} \\ f_{T4y} \\ f_{T5x} \\ f_{T5y} \end{Bmatrix} = \begin{Bmatrix} 112{,}500 \\ 112{,}500 \\ -112{,}500 \\ 112{,}500 \\ 0 \\ -225{,}000 \end{Bmatrix} \text{N} \qquad (15.1.87)$$

Element 4

$$\{f_T\} = 450{,}000 \begin{Bmatrix} -0.25 \\ 0.25 \\ -0.25 \\ -0.25 \\ 0.5 \\ 0 \end{Bmatrix} = \begin{Bmatrix} f_{T4x} \\ f_{T4y} \\ f_{T1x} \\ f_{T1y} \\ f_{T5x} \\ f_{T5y} \end{Bmatrix} = \begin{Bmatrix} -112{,}500 \\ 112{,}500 \\ -112{,}500 \\ -112{,}500 \\ 225{,}000 \\ 0 \end{Bmatrix} \text{N} \qquad (15.1.88)$$

We then obtain the global thermal force matrix by direct assemblage of the element force matrices. The resulting matrix is

$$
\begin{Bmatrix} f_{T1x} \\ f_{T1y} \\ f_{T2x} \\ f_{T2y} \\ f_{T3x} \\ f_{T3y} \\ f_{T4x} \\ f_{T4y} \\ f_{T5x} \\ f_{T5y} \end{Bmatrix} = \begin{Bmatrix} -225{,}000 \\ -225{,}000 \\ 225{,}000 \\ -225{,}000 \\ 225{,}000 \\ 225{,}000 \\ -225{,}000 \\ 225{,}000 \\ 0 \\ 0 \end{Bmatrix} \text{N} \qquad (15.1.89)
$$

Using Eqs. (15.1.84) and (15.1.89) and imposing the boundary conditions $d_{1x} = d_{1y} = d_{4x} = d_{4y} = 0$, we obtain the system of equations for solution as

$$
\begin{Bmatrix} f_{T2x} = 225{,}000 \\ f_{T2y} = -225{,}000 \\ f_{T3x} = 225{,}000 \\ f_{T3y} = 225{,}000 \\ f_{T5x} = 0 \\ f_{T5y} = 0 \end{Bmatrix} = 4.615 \times 10^7
$$

$$
\begin{bmatrix} 16.874 & -8.125 & 4.0625 & -0.3125 & -16.875 & 8.125 \\ -8.125 & 16.875 & 0.3125 & -4.0625 & 8.125 & -16.875 \\ 4.0625 & 0.3125 & 16.875 & 8.125 & -16.875 & -8.125 \\ -0.3125 & -4.0625 & 8.125 & 16.875 & -8.125 & -16.875 \\ -16.875 & 8.125 & -16.875 & -8.125 & 67.5 & 0 \\ 8.125 & -16.875 & -8.125 & -16.875 & 0 & 67.5 \end{bmatrix} \begin{Bmatrix} d_{2x} \\ d_{2y} \\ d_{3x} \\ d_{3y} \\ d_{5x} \\ d_{5y} \end{Bmatrix}
$$

$$(15.1.90)$$

Solving Eq. (15.1.90) for the nodal displacements, we have

$$
\begin{Bmatrix} d_{2x} \\ d_{2y} \\ d_{3x} \\ d_{3y} \\ d_{5x} \\ d_{5y} \end{Bmatrix} = \begin{Bmatrix} 3.327 \times 10^{-4} \\ -1.911 \times 10^{-4} \\ 3.327 \times 10^{-4} \\ 1.911 \times 10^{-4} \\ 2.123 \times 10^{-4} \\ 6.654 \times 10^{-9} \end{Bmatrix} \text{m} \qquad (15.1.91)
$$

We now use Eq. (15.1.58) to obtain the stresses in each element. Using Eqs. (7.2.36) and (15.1.59), we write Eq. (15.1.58) as

$$
\{\sigma\} = [D][B]\{d\} - [D]\{\varepsilon_T\} \qquad (15.1.92)
$$

Element 1

$$
\left\{ \begin{array}{c} \sigma_x \\ \sigma_y \\ \tau_{xy} \end{array} \right\} = \frac{E}{1-v^2} \begin{bmatrix} 1 & v & 0 \\ v & 1 & 0 \\ 0 & 0 & \dfrac{1-v}{2} \end{bmatrix} \frac{1}{2A} \begin{bmatrix} \beta_1 & 0 & \beta_2 & 0 & \beta_5 & 0 \\ 0 & \gamma_1 & 0 & \gamma_2 & 0 & \gamma_5 \\ \gamma_1 & \beta_1 & \gamma_2 & \beta_2 & \gamma_5 & \beta_5 \end{bmatrix} \left\{ \begin{array}{c} d_{1x} \\ d_{1y} \\ d_{2x} \\ d_{2y} \\ d_{5x} \\ d_{5y} \end{array} \right\}
$$

$$
- \frac{E}{1-v^2} \begin{bmatrix} 1 & v & 0 \\ v & 1 & 0 \\ 0 & 0 & \dfrac{1-v}{2} \end{bmatrix} \left\{ \begin{array}{c} \alpha T \\ \alpha T \\ 0 \end{array} \right\} \tag{15.1.93}
$$

Using Eqs. (15.1.76) and (15.1.91) along with the mechanical properties E, v, and α in Eq. (15.1.93), we obtain

$$
\left\{ \begin{array}{c} \sigma_x \\ \sigma_y \\ \tau_{xy} \end{array} \right\} = \frac{210 \times 10^9}{0.91} \begin{bmatrix} 1 & 0.3 & 0 \\ 0.3 & 1 & 0 \\ 0 & 0 & 0.35 \end{bmatrix}
$$

$$
\times \frac{1}{0.125} \begin{bmatrix} -0.25 & 0 & 0.25 & 0 & 0 & 0 \\ 0 & -0.25 & 0 & -0.25 & 0 & 0.5 \\ -0.25 & -0.25 & -0.25 & 0.25 & 0.5 & 0 \end{bmatrix} \left\{ \begin{array}{c} 0 \\ 0 \\ 3.327 \times 10^{-4} \\ -1.911 \times 10^{-4} \\ 2.123 \times 10^{-4} \\ 6.654 \times 10^{-9} \end{array} \right\}
$$

$$
- \frac{210 \times 10^9}{0.91} \begin{bmatrix} 1 & 0.3 & 0 \\ 0.3 & 1 & 0 \\ 0 & 0 & 0.35 \end{bmatrix} \left\{ \begin{array}{c} (12 \times 10^{-6})(50) \\ (12 \times 10^{-6})(50) \\ 0 \end{array} \right\} \tag{15.1.94}
$$

Simplifying Eq. (15.1.94) yields

$$
\left\{ \begin{array}{c} \sigma_x \\ \sigma_y \\ \tau_{xy} \end{array} \right\} = \left\{ \begin{array}{c} 1.800 \times 10^8 \\ 1.342 \times 10^8 \\ -1.600 \times 10^7 \end{array} \right\} - \left\{ \begin{array}{c} 1.8 \times 10^8 \\ 1.8 \times 10^8 \\ 0 \end{array} \right\} = \left\{ \begin{array}{c} 0 \\ -4.57 \times 10^7 \\ -1.60 \times 10^7 \end{array} \right\} \text{Pa} \tag{15.1.95}
$$

Similarly, we obtain the stresses in the other elements as follows:

Element 2

$$
\left\{ \begin{array}{c} \sigma_x \\ \sigma_y \\ \tau_{xy} \end{array} \right\} = \left\{ \begin{array}{c} 1.640 \times 10^8 \\ 2.097 \times 10^8 \\ -2150 \end{array} \right\} - \left\{ \begin{array}{c} 1.8 \times 10^8 \\ 1.8 \times 10^8 \\ 0 \end{array} \right\} = \left\{ \begin{array}{c} -1.6 \times 10^7 \\ 2.973 \times 10^7 \\ -2150 \end{array} \right\} \text{Pa} \tag{15.1.96}
$$

Element 3

$$\left\{ \begin{array}{c} \sigma_x \\ \sigma_y \\ \tau_{xy} \end{array} \right\} = \left\{ \begin{array}{c} 1.800 \times 10^8 \\ 1.342 \times 10^8 \\ 1.600 \times 10^7 \end{array} \right\} - \left\{ \begin{array}{c} 1.8 \times 10^8 \\ 1.8 \times 10^8 \\ 0 \end{array} \right\} = \left\{ \begin{array}{c} 0 \\ -4.57 \times 10^7 \\ -1.60 \times 10^7 \end{array} \right\} \text{ Pa}$$

(15.1.97)

Element 4

$$\left\{ \begin{array}{c} \sigma_x \\ \sigma_y \\ \tau_{xy} \end{array} \right\} = \left\{ \begin{array}{c} 1.960 \times 10^8 \\ 5.880 \times 10^7 \\ 2150 \end{array} \right\} - \left\{ \begin{array}{c} 1.8 \times 10^8 \\ 1.8 \times 10^8 \\ 0 \end{array} \right\} = \left\{ \begin{array}{c} 1.6 \times 10^7 \\ -1.212 \times 10^8 \\ 2150 \end{array} \right\} \text{ Pa}$$

(15.1.98) ■

PROBLEMS

15.1 For the one-dimensional bar fixed at the left end, free at the right end, and subjected to a uniform temperature rise $T = 50°F$ as shown in Figure P15-1, determine the free-end displacement, the displacement 60 in. from the fixed end, the reactions at the fixed end, and the axial stress. Let $E = 30 \times 10^6$ psi, $A = 4$ in.^{2}, and $\alpha = 7.0 \times 10^{-6}$ (in./in.)/°F.

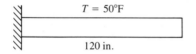

$T = 50°F$

120 in.

FIGURE P15-1

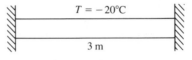

$T = -20°C$

3 m

FIGURE P15-2

15.2 For the one-dimensional bar fixed at each end and subjected to a uniform temperature drop of $T = 20°C$ as shown in Figure P15-2, determine the reactions at the fixed ends and the stress in the bar. Let $E = 210$ GPa, $A = 1 \times 10^{-2}$ m^2, and $\alpha = 12 \times 10^{-6}$ (mm/mm)/°C.

15.3 For the plane truss shown in Figure P15-3, bar element 2 is subjected to a uniform temperature rise of $T = 50°F$. Let $E = 30 \times 10^6$ psi, $A = 2$ in.^{2}, and $\alpha = 7.0 \times 10^{-6}$ (in./in.)/°F. The lengths of the truss elements are shown in the figure. Determine the stresses in each bar.
(*Hint:* See Eqs. (3.6.4) and (3.6.6) in Example 3.5 for the global and reduced K matrices.)

15.4 For the plane truss shown in Figure P15-4, bar element 1 is subjected to a uniform temperature rise of 30°F. Let $E = 30 \times 10^6$ psi, $A = 2$ in.^{2}, and $\alpha = 7.0 \times 10^{-6}$ (in./in.)/°F. The lengths of the truss elements are shown in the figure. Determine the stresses in each bar.
(*Hint:* Use Problem 3.21 for K.)

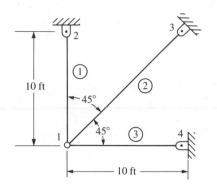

FIGURE   P15-3

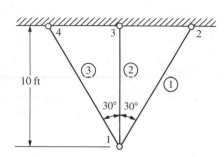

FIGURE   P15-4

15.5   For the structure shown in Figure P15-5, bar element 1 is subjected to a uniform temperature rise of $T = 20°C$. Let $E = 210$ GPa, $A = 2 \times 10^{-2}$ m^2, and $\alpha = 12 \times 10^{-6}$ (mm/mm)/°C. Determine the stresses in each bar.

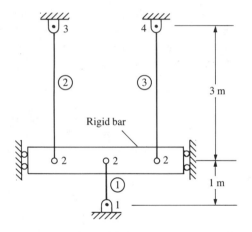

FIGURE   P15-5

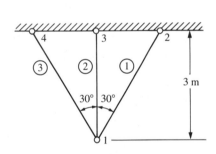

FIGURE   P15-6

15.6   For the plane truss shown in Figure P15-6, bar element 2 is subjected to a uniform temperature drop of $T = 20°C$. Let $E = 70$ GPa, $A = 4 \times 10^{-2}$ m^2, and $\alpha = 23 \times 10^{-6}$ (mm/mm)/°C. Determine the stresses in each bar and the displacement of node 1.

FIGURE   P15-7

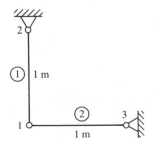

15.7 For the bar structure shown in Figure P15-7, element 1 is subjected to a temperature rise of $T = 30°C$. Let $E = 210$ GPa, $A = 3 \times 10^{-2}$ m^2, and $\alpha = 12 \times 10^{-6}$ (mm/mm)/°C. Determine the displacement of node 1 and the stresses in each bar.

15.8 A bar assemblage consists of two outer steel bars and an inner brass bar. The three-bar assemblage is then heated to raise the temperature by an amount $T = 40°F$. Let all cross-sectional areas be $A = 2$ in.^{2} and $L = 60$ in., $E_{steel} = 30 \times 10^6$ psi, $E_{brass} = 15 \times 10^6$ psi, $\alpha_{steel} = 6.5 \times 10^{-6}$/°F, and $\alpha_{brass} = 10 \times 10^{-6}$/°F. Determine (a) the displacement of node 2 and (b) the stress in the steel and brass bars. See figure P15-8.

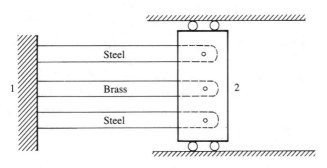

F I G U R E   P15-8

15.9 For the plane stress element shown in Figure P15-9 subjected to a uniform temperature rise of $T = 50°F$, determine the thermal force matrix $\{f_T\}$. Let $E = 10 \times 10^6$ psi, $v = 0.30$, and $\alpha = 12.5 \times 10^{-6}$ (in./in.)/°F. The coordinates (in inches) are shown in the figure. The element thickness is $t = 1$ in.

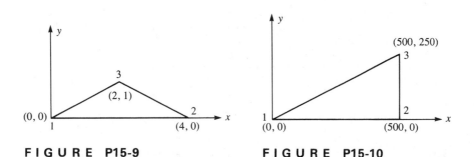

F I G U R E   P15-9   **F I G U R E   P15-10**

15.10 For the plane stress element shown in Figure P15-10 subjected to a uniform temperature rise of $T = 30°C$, determine the thermal force matrix $\{f_T\}$. Let $E = 70$ GPa, $v = 0.3$, $\alpha = 23 \times 10^{-6}$ (mm/mm)/°C, and $t = 5$ mm. The coordinates (in millimeters) are shown in the figure.

15.11 For the plane stress element shown in Figure P15-11 subjected to a uniform temperature rise of $T = 100°F$, determine the thermal force matrix $\{f_T\}$. Let $E = 30 \times 10^6$ psi, $v = 0.3$, $\alpha = 7.0 \times 10^{-6}$ (in./in.)/°F, and $t = 1$ in. The coordinates (in inches) are shown in the figure.

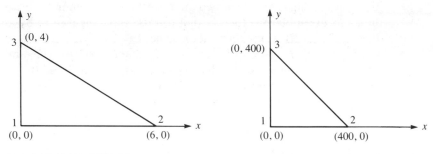

FIGURE P15-11   FIGURE P15-12

15.12 For the plane stress element shown in Figure P15-12 subjected to a uniform temperature drop of $T = 20°C$, determine the thermal force matrix $\{f_T\}$. Let $E = 210$ GPa, $v = 0.25$, and $\alpha = 12 \times 10^{-6}$ (mm/mm)/°C. The coordinates (in millimeters) are shown in the figure. The element thickness is 10 mm.

15.13 For the plane stress plate fixed along the left and right sides and subjected to a uniform temperature rise of 50°F as shown in Figure P15-13, determine the stresses in each element. Let $E = 10 \times 10^6$ psi, $v = 0.30$, $\alpha = 12.5 \times 10^{-6}$ (in./in.)/°F, and $t = \frac{1}{4}$ in. The coordinates (in inches) are shown in the figure. (*Hint:* The nodal displacements are all equal to zero. Therefore, the stresses can be determined from $\{\sigma\} = -[D]\{\varepsilon_T\}$.)

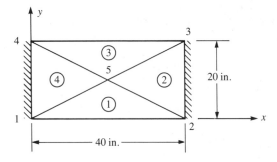

FIGURE P15-13

15.14 For the plane stress plate fixed along all edges and subjected to a uniform temperature decrease of 20°C as shown in Figure P15-14, determine the stresses in each element. Let $E = 210$ GPa, $v = 0.25$, and $\alpha = 12 \times 10^{-6}$ (mm/mm)/°C. The coordinates of the plate are shown in the figure. The plate thickness is 10 mm. (*Hint:* The nodal displacements are all equal to zero. Therefore, the stresses can be determined from $\{\sigma\} = -[D]\{\varepsilon_T\}$.)

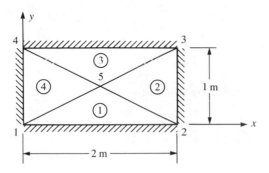

FIGURE P15-14

15.15 If the thermal expansion coefficient of a bar is given by $\alpha = \alpha_0(1 + x)$, determine the thermal force matrix. Let the bar have length L, modulus of elasticity E, and cross-sectional area A.

15.16 Assume the temperature function to vary linearly over the length of a bar as $T = t_1 + t_2 x$; that is, express the temperature function as $\{T\} = [N]\{t\}$, where $[N]$ is the shape function matrix for the two-noded bar element. In other words, $[N] = [1 - x/L \quad x/L]$. Determine the force matrix in terms of E, A, α, L, t_1, and t_2.

15.17 Derive the thermal force matrix for the axisymmetric element of Chapter 10.

Using a modified version of computer code CSFEP or another suitable code, solve the following problems.

15.18 The square plate in Figure P15-18 is subjected to uniform heating of 80°F. Determine the nodal displacements and element stresses. Let the element thickness be $t = 0.1$ in., $E = 30 \times 10^6$ psi, $v = 0.33$, and $\alpha = 10 \times 10^{-6}/°F$.

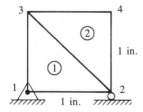   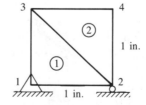

FIGURE P15-18   **FIGURE P15-19**

15.19 The square plate in Figure P15-19 has element 1 made of steel with $E = 30 \times 10^6$ psi, $v = 0.33$, and $\alpha = 10 \times 10^{-6}/°F$ and element 2 made of a material with $E = 15 \times 10^6$ psi, $v = 0.25$, and $\alpha = 50 \times 10^{-6}/°F$. Let the plate thickness be $t = 0.1$ in. Determine the nodal displacements and element stresses for element 1 subjected to 80°F temperature increase and element 2 subjected to a 50°F temperature increase.

Structural Dynamics

and Time-Dependent

Heat Transfer

INTRODUCTION

This chapter provides an elementary introduction to time-dependent problems. We will introduce the basic concepts using the single-degree-of-freedom spring-mass system. We will include discussion of the stress analysis of the one-dimensional bar, beam, truss, and plane frame. This is followed by the analysis of one-dimensional heat transfer.

We will provide the basic equations necessary for structural dynamic analysis and develop both the lumped- and the consistent-mass matrices involved in the analyses of the bar, beam, truss, and plane frame. We will describe the assembly of the global mass matrix for truss and plane frame analysis and then present numerical integration methods for handling the time derivative.

We will provide longhand solutions for the determination of the natural frequencies for bars and beams, and then illustrate the time-step integration process involved with the stress analysis of a bar subjected to a time-dependent forcing function. We will describe a computer program for the dynamic analysis of plane frames and illustrate its use with an example problem.

We will next derive the basic equations for the time-dependent one-dimensional heat-transfer problem and discuss their applications. Finally, this chapter provides the basic concepts necessary for the solution of time-dependent problems.

16.1 Dynamics of a Spring-Mass System

In this section, we discuss the motion of a single-degree-of-freedom spring-mass system to introduce the important concepts necessary for the later study of continuous systems such as bars, beams, and plane frames. In Figure 16-1, we show the single-degree-of-freedom spring-mass system subjected to a time-dependent force $F(t)$. Here k represents the spring stiffness or constant, and m represents the mass of the system.

The free-body diagram of the mass is shown in Figure 16-2. The spring force $T = kx$ and the applied force $F(t)$ act on the mass, and the mass-times-acceleration term is shown separately.

Applying Newton's second law of motion, $f = ma$, to the mass, we obtain the equation of motion in the x direction as

$$F(t) - kx = m\ddot{x} \qquad (16.1.1)$$

where a dot over a variable denotes differentiation with respect to time; that is, $(\dot{\ }) = d(\ )/dt$. Rewriting Eq. (16.1.1) in standard form, we have

$$m\ddot{x} + kx = F(t) \qquad (16.1.2)$$

Equation (16.1.2) is a linear differential equation of the second order whose standard solution for the displacement x consists of a homogeneous solution and a particular solution. Standard analytical solutions for this forced vibration can be found in texts on dynamics or vibrations such as Reference [1]. The analytical solution will not be presented here as our intent is to introduce basic concepts in vibration behavior. However, we will solve the problem defined by Eq. (16.1.2) by an approximate numerical technique in Section 16.3 (see Examples 16.1 and 16.2).

The homogeneous solution to Eq. (16.1.2) is the solution obtained when the right side is set equal to zero. A number of useful concepts regarding vibrations are obtained by considering this free vibration of the mass; that is, when $F(t) = 0$. Hence, defining

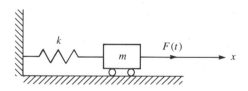

F I G U R E 16-1   Spring-mass system subjected to a time-dependent force

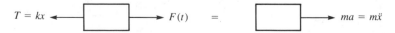

F I G U R E 16-2   Free-body diagram of the mass of Figure 16-1

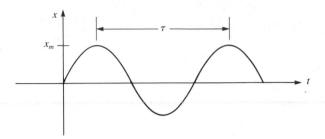

F I G U R E   16-3   Displacement/time curve for simple harmonic motion

$$\omega^2 = \frac{k}{m} \qquad (16.1.3)$$

and setting the right side of Eq. (16.1.2) equal to zero, we have

$$\ddot{x} + \omega^2 x = 0 \qquad (16.1.4)$$

where ω is called the **natural circular frequency** of the free vibration of the mass, expressed in units of radians per second or revolutions per minute (rpm). Hence, the natural circular frequency defines the number of cycles per unit time of the mass vibration. We observe from Eq. (16.1.3) that ω depends only on the spring stiffness k and the mass m of the body.

The motion defined by Eq. (16.1.4) is called **simple harmonic motion**. The displacement and acceleration are seen to be proportional but of opposite direction. Again, a standard solution to Eq. (16.1.4) can be found in Reference [1]. A typical displacement/time curve is represented by the sine curve shown in Figure 16-3, where x_m denotes the maximum displacement (called the **amplitude** of the vibration). The time interval required for the mass to complete one full cycle of motion is called the **period** of the vibration τ and is given by

$$\tau = \frac{2\pi}{\omega} \qquad (16.1.5)$$

where τ is measured in seconds.

Finally, note that all vibrations are damped to some degree by friction forces. These forces may be caused by dry or Coulomb friction between rigid bodies, by internal friction between molecules within a deformable body, or by fluid friction when a body moves in a fluid. Damping results in natural circular frequencies that are smaller than those for undamped systems; maximum displacements also are smaller when damping occurs. A basic treatment of damping can be found in Reference [1].

16.2   Direct Derivation of the Bar Element Equations

We will now derive the finite element equations for the time-dependent (dynamic) stress analysis of the one-dimensional bar. Recall that the time-

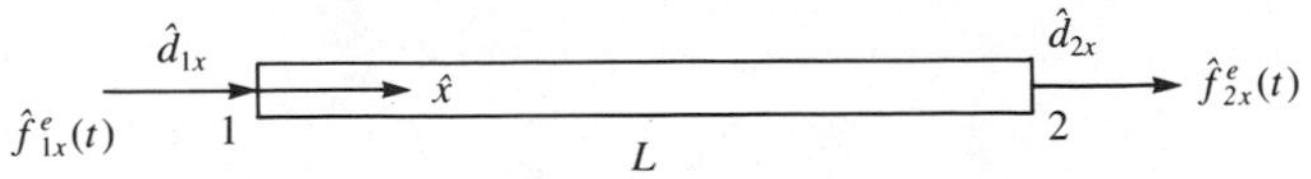

FIGURE 16-4   Bar element subjected to time-dependent loads

independent (static) stress analysis of the bar was considered in Chapter 3. The steps used in deriving the dynamic equations are the same as those used for the derivation of the static equations.

Step 1   **Select Element Type**

Figure 16-4 shows the typical bar element of length L, cross-sectional area A, and mass density ρ (with typical units of lb-s^2/in.^{4}), with nodes 1 and 2 subjected to external time-dependent loads $\hat{f}_x^e(t)$.

Step 2   **Select a Displacement Function**

Again, we assume a linear displacement function along the $\hat{x}$ axis of the bar [see Eq. (3.1.1)]; that is, we let

$$\hat{u} = a_1 + a_2 \hat{x} \qquad (16.2.1)$$

As was shown in Chapter 3, Eq. (16.2.1) can be expressed in terms of the shape functions as

$$\hat{u} = N_1 \hat{d}_{1x} + N_2 \hat{d}_{2x} \qquad (16.2.2)$$

where
$$N_1 = 1 - \frac{\hat{x}}{L} \qquad N_2 = \frac{\hat{x}}{L} \qquad (16.2.3)$$

Step 3   **Define the Strain/Displacement and Stress/Strain Relationships**

Again, the strain/displacement relationship is given by

$$\{\varepsilon_x\} = \frac{\partial \hat{u}}{\partial \hat{x}} = [B]\{\hat{d}\} \qquad (16.2.4)$$

where
$$[B] = \left[-\frac{1}{L} \quad \frac{1}{L} \right] \qquad \{\hat{d}\} = \begin{Bmatrix} \hat{d}_{1x} \\ \hat{d}_{2x} \end{Bmatrix} \qquad (16.2.5)$$

and the stress/strain relationship is given by

$$\{\sigma_x\} = [D]\{\varepsilon_x\} = [D][B]\{\hat{d}\} \qquad (16.2.6)$$

Step 4   **Derive the Element Stiffness and Mass Matrices and Equations**

The bar is generally not in equilibrium under a time-dependent force; hence, $f_{1x} \neq f_{2x}$. Therefore, we again apply Newton's second law of motion, $f = ma$, to each node. In general, the law can be written for each node as the external (applied) force f_x^e minus the internal force is equal to the nodal mass times acceleration. Equivalently, adding the internal force to the *ma* term, we have

$$\hat{f}^e_{1x} = \hat{f}_{1x} + m_1 \frac{\partial^2 \hat{d}_{1x}}{\partial t^2} \quad \text{and} \quad \hat{f}^e_{2x} = \hat{f}_{2x} + m_2 \frac{\partial^2 \hat{d}_{2x}}{\partial t^2} \qquad (16.2.7)$$

where the masses m_1 and m_2 are obtained by lumping the total mass of the bar equally at the two nodes such that

$$m_1 = \frac{\rho AL}{2} \qquad m_2 = \frac{\rho AL}{2} \qquad (16.2.8)$$

In matrix form, we express Eqs. (16.2.7) as

$$\left\{ \begin{array}{c} \hat{f}^e_{1x} \\ \hat{f}^e_{2x} \end{array} \right\} = \left\{ \begin{array}{c} \hat{f}_{1x} \\ \hat{f}_{2x} \end{array} \right\} + \left[\begin{array}{cc} m_1 & 0 \\ 0 & m_2 \end{array} \right] \left\{ \begin{array}{c} \dfrac{\partial^2 \hat{d}_{1x}}{\partial t^2} \\ \dfrac{\partial^2 \hat{d}_{2x}}{\partial t^2} \end{array} \right\} \qquad (16.2.9)$$

Using Eqs. (3.1.13) and (3.1.14), we replace $\{\hat{f}\}$ with $[\hat{k}]\{\hat{d}\}$ in Eq. (16.2.9) to obtain the element equations

$$\{\hat{f}^e(t)\} = [\hat{k}]\{\hat{d}\} + [\hat{m}]\{\ddot{\hat{d}}\} \qquad (16.2.10)$$

where

$$[\hat{k}] = \frac{AE}{L} \left[\begin{array}{cc} 1 & -1 \\ -1 & 1 \end{array} \right] \qquad (16.2.11)$$

is the bar element stiffness matrix, and

$$[\hat{m}] = \frac{\rho AL}{2} \left[\begin{array}{cc} 1 & 0 \\ 0 & 1 \end{array} \right] \qquad (16.2.12)$$

is called the **lumped-mass matrix**. Also,

$$\{\ddot{\hat{d}}\} = \frac{\partial^2 \{\hat{d}\}}{\partial t^2} \qquad (16.2.13)$$

Observe that the lumped-mass matrix has diagonal terms only. This facilitates the computation of the global equations. However, solution accuracy is usually not as good as if a consistent-mass matrix is used (see Reference [2]).

We will now develop the **consistent-mass matrix** for the bar element. Numerous methods are available to obtain the consistent-mass matrix. The generally applicable virtual work principle (which is the basis of many energy principles, such as the principle of minimum potential energy for elastic bodies previously used in this text) provides a relatively simple method for derivation of the element equations, and is included in Appendix E. However, an even simpler approach is to use D'Alembert's principle; thus, we introduce an effective body force X^e as

$$\{X^e\} = -\rho \{\ddot{u}\} \qquad (16.2.14)$$

where the minus sign is due to the fact that the acceleration produces

D'Alembert's body forces in the direction opposite the acceleration. The nodal forces associated with $\{X^e\}$ are then found by using Eq. (7.3.1), repeated here as

$$\{f_b\} = \iiint_V [N]^T\{X\}\, dV \qquad (16.2.15)$$

Substituting $\{X^e\}$ given by Eq. (16.2.14) into Eq. (16.2.15) for $\{X\}$, we obtain

$$\{f_b\} = -\iiint_V \rho[N]^T\{\ddot{u}\}\, dV \qquad (16.2.16)$$

Recalling from Eq. (16.2.2) that $\{\hat{u}\} = [N]\{\hat{d}\}$, the first and second derivatives with respect to time are

$$\{\dot{\hat{u}}\} = [N]\{\dot{\hat{d}}\} \qquad \{\ddot{\hat{u}}\} = [N]\{\ddot{\hat{d}}\} \qquad (16.2.17)$$

where $\{\dot{\hat{d}}\}$ and $\{\ddot{\hat{d}}\}$ are the nodal velocities and accelerations, respectively. Substituting Eqs. (16.2.17) into Eq. (16.2.16), we obtain

$$\{f_b\} = -\iiint_V \rho[N]^T[N]\, dV\{\ddot{\hat{d}}\} = -[\hat{m}]\{\ddot{\hat{d}}\} \qquad (16.2.18)$$

where the element mass matrix is defined as

$$[\hat{m}] = \iiint_V \rho[N]^T[N]\, dV \qquad (16.2.19)$$

This mass matrix is called the *consistent-mass matrix* because it is derived from the same shape functions $[N]$ that are used to obtain the stiffness matrix $[\hat{k}]$. In general, $[\hat{m}]$ given by Eq. (16.2.19) will be a full but symmetric matrix. Equation (16.2.19) is a general form of the consistent-mass matrix; that is, substituting the appropriate shape functions, we can generate the mass matrix for such elements as the bar, beam, and plane stress.

We will now develop the consistent-mass matrix for the bar element of Figure 16-4 by substituting the shape function Eqs. (16.2.3) into Eq. (16.2.19) as follows:

$$[\hat{m}] = \iiint_V \rho \begin{Bmatrix} 1 - \dfrac{\hat{x}}{L} \\[2mm] \dfrac{\hat{x}}{L} \end{Bmatrix} \begin{bmatrix} 1 - \dfrac{\hat{x}}{L} & \dfrac{\hat{x}}{L} \end{bmatrix} dV \qquad (16.2.20)$$

Simplifying Eq. (16.2.20), we obtain

$$[\hat{m}] = \rho A \int_0^L \begin{Bmatrix} 1 - \dfrac{\hat{x}}{L} \\[2mm] \dfrac{\hat{x}}{L} \end{Bmatrix} \begin{bmatrix} 1 - \dfrac{\hat{x}}{L} & \dfrac{\hat{x}}{L} \end{bmatrix} d\hat{x} \qquad (16.2.21)$$

or, on multiplying the matrices of Eq. (16.2.21),

$$[\dot{m}] = \rho A \int_0^L \begin{bmatrix} \left(1 - \dfrac{\hat{x}}{L}\right)^2 & \left(1 - \dfrac{\hat{x}}{L}\right)\dfrac{\hat{x}}{L} \\ \left(1 - \dfrac{\hat{x}}{L}\right)\dfrac{\hat{x}}{L} & \left(\dfrac{\hat{x}}{L}\right)^2 \end{bmatrix} d\hat{x} \qquad (16.2.22)$$

On integrating Eq. (16.2.22) term by term, we obtain the consistent-mass matrix for a bar element as

$$[\dot{m}] = \frac{\rho A L}{6} \begin{bmatrix} 2 & 1 \\ 1 & 2 \end{bmatrix} \qquad (16.2.23)$$

Step 5   **Assemble the Element Equations to Obtain the Global Equations and Introduce Boundary Conditions**

We assemble the element equations using the direct stiffness method such that interelement continuity of displacements is again satisfied at common nodes and, in addition, interelement continuity of accelerations is also satisfied; that is, we obtain the global equations

$$\{F(t)\} = [K]\{d\} + [M]\{\ddot{d}\} \qquad (16.2.24)$$

where

$$[K] = \sum_{e=1}^N [k^{(e)}] \qquad [M] = \sum_{e=1}^N [m^{(e)}] \qquad \{F\} = \sum_{e=1}^N \{f^{(e)}\} \qquad (16.2.25)$$

are the global stiffness, mass, and force matrices, respectively. Note that the global mass matrix is assembled in the same manner as the global stiffness matrix. Equation (16.2.24) represents a set of matrix equations discretized with respect to space. To obtain the solution of the equations, discretization in time is also necessary. We will describe this process in Section 16.3, and later present representative solutions illustrating these equations.

16.3   Numerical Integration in Time

We now introduce procedures for the discretization of Eq. (16.2.24) with respect to time. These procedures will allow the nodal displacements to be determined at different time increments for a given dynamic system. The general method used is called *direct integration*. There are two classifications of direct integration: explicit and implicit. We will formulate the equations for two direct integration methods. The first, and simplest, is an explicit method known as the *central difference method* (see References [3] and [4]). The second, more complicated but more versatile than the central difference method, is an implicit method known as the *Newmark–Beta* (or *Newmark's*) *method* (see Reference [5]). The versatility of Newmark's method is evidenced by its adaptation in many commercially available computer programs. Nu-

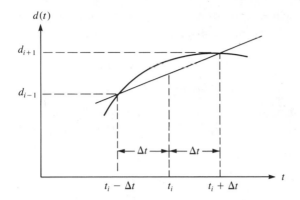

FIGURE 16-5 Numerical integration (approximation of derivative at t_i)

merous other integration methods are available in the literature. Among these methods are Wilson's method (see References [7] and [8]), which is available as an alternative to Newmark's method in some commercial computer programs, and Houboldt's method (see Reference [8]).

Central Difference Method

The central difference method is based on finite difference expressions in time for velocity and acceleration at time t given by

$$\underline{\dot{d}}_i = \frac{\underline{d}_{i+1} - \underline{d}_{i-1}}{2(\Delta t)} \tag{16.3.1}$$

$$\underline{\ddot{d}}_i = \frac{\underline{\dot{d}}_{i+1} - \underline{\dot{d}}_{i-1}}{2(\Delta t)} \tag{16.3.2}$$

where the subscripts indicate the time step; that is, for a time increment of Δt, $\underline{d}_i = \underline{d}(t)$ and $\underline{d}_{i+1} = \underline{d}(t + \Delta t)$. The procedure used in deriving Eq. (16.3.1) is illustrated by use of the displacement/time curve shown in Figure 16-5. Graphically, Eq. (16.3.1) represents the slope of the line shown in Figure 16-5; that is, given two points at increments $i - 1$ and $i + 1$ on the curve, two Δt increments apart, an approximation of the first derivative at the midpoint i of the increment is given by Eq. (16.3.1). Similarly, using a velocity/time curve, we could obtain Eq. (16.3.2), or we can see that Eq. (16.3.2) is obtained simply by differentiating Eq. (16.3.1) with respect to time.

It has been shown using, for instance, Taylor series expansions (see Reference [3]) that the acceleration can also be expressed in terms of the displacements by

$$\underline{\ddot{d}}_i = \frac{\underline{d}_{i+1} - 2\underline{d}_i + \underline{d}_{i-1}}{(\Delta t)^2} \tag{16.3.3}$$

Since we want to evaluate the nodal displacements, it is most suitable to use Eq. (16.3.3) in the form

$$\underline{d}_{i+1} = 2\underline{d}_i - \underline{d}_{i-1} + \underline{\ddot{d}}_i(\Delta t)^2 \tag{16.3.4}$$

Equation (16.3.4) will be used to determine the nodal displacements in the next time step $i + 1$ knowing the displacements at time steps i and $i - 1$ and the acceleration at time i.

From Eq. (16.2.24), we express the acceleration as

$$\ddot{\underline{d}}_i = \underline{M}^{-1}(\underline{F}_i - \underline{K}\underline{d}_i) \qquad (16.3.5)$$

To obtain an expression for $\underline{d}_{i+1}$, we first multiply Eq. (16.3.4) by the mass matrix $\underline{M}$ and then substitute Eq. (16.3.5) for $\ddot{\underline{d}}_i$ into this equation to obtain

$$\underline{M}\underline{d}_{i+1} = 2\underline{M}\underline{d}_i - \underline{M}\underline{d}_{i-1} + (\underline{F}_i - \underline{K}\underline{d}_i)(\Delta t)^2 \qquad (16.3.6)$$

Combining like terms of Eq. (16.3.6), we obtain

$$\underline{M}\underline{d}_{i+1} = (\Delta t)^2 \underline{F}_i + [2\underline{M} - (\Delta t)^2 \underline{K}]\underline{d}_i - \underline{M}\underline{d}_{i-1} \qquad (16.3.7)$$

To start the computations to determine $\underline{d}_{i+1}$, $\dot{\underline{d}}_{i+1}$, and $\ddot{\underline{d}}_{i+1}$, we need the displacement $\underline{d}_{i-1}$ initially, as indicated by Eq. (16.3.7). Using Eqs. (16.3.1) and (16.3.4), we solve for $\underline{d}_{i-1}$ as

$$\underline{d}_{i-1} = \underline{d}_i - (\Delta t)\dot{\underline{d}}_i + \frac{(\Delta t)^2}{2}\ddot{\underline{d}}_i \qquad (16.3.8)$$

The procedure for solution is then as follows:

1. Given: $\underline{d}_0$, $\dot{\underline{d}}_0$, and $\underline{F}_i(t)$.
2. If $\ddot{\underline{d}}_0$ is not initially given, solve Eq. (16.3.5) at $t = 0$ for $\ddot{\underline{d}}_0$; that is,

    $$\ddot{\underline{d}}_0 = \underline{M}^{-1}(\underline{F}_0 - \underline{K}\underline{d}_0)$$

3. Solve Eq. (16.3.8) at $t = -\Delta t$ for $\underline{d}_{-1}$; that is,

    $$\underline{d}_{-1} = \underline{d}_0 - (\Delta t)\dot{\underline{d}}_0 + \frac{(\Delta t)^2}{2}\ddot{\underline{d}}_0$$

4. Having solved for $\underline{d}_{-1}$ in Step 3, now solve for $\underline{d}_1$ using Eq. (16.3.7) as

    $$\underline{d}_1 = \underline{M}^{-1}\{(\Delta t)^2 \underline{F}_0 + [2\underline{M} - (\Delta t)^2 \underline{K}]\underline{d}_0 - \underline{M}\underline{d}_{-1}\}$$

5. With $\underline{d}_0$ initially given, and $\underline{d}_1$ determined from Step 4, use Eq. (16.3.7) to obtain

    $$\underline{d}_2 = \underline{M}^{-1}\{(\Delta t)^2 \underline{F}_1 + [2\underline{M} - (\Delta t)^2 \underline{K}]\underline{d}_1 - \underline{M}\underline{d}_0\}$$

6. Using Eq. (16.3.5), solve for $\ddot{\underline{d}}_1$ as

    $$\ddot{\underline{d}}_1 = \underline{M}^{-1}(\underline{F}_1 - \underline{K}\underline{d}_1)$$

7. Using the result of Step 5 and the boundary condition for $\underline{d}_0$ given in Step 1, determine the velocity at the first time step by Eq. (16.3.1) as

    $$\dot{\underline{d}}_1 = \frac{\underline{d}_2 - \underline{d}_0}{2(\Delta t)}$$

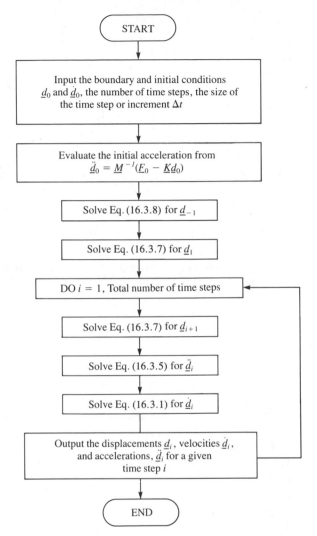

START

Input the boundary and initial conditions $\underline{d}_0$ and $\underline{\dot{d}}_0$, the number of time steps, the size of the time step or increment Δt

Evaluate the initial acceleration from
$$\underline{\ddot{d}}_0 = \underline{M}^{-1}(\underline{F}_0 - \underline{K}\underline{d}_0)$$

Solve Eq. (16.3.8) for $\underline{d}_{-1}$

Solve Eq. (16.3.7) for $\underline{d}_1$

DO $i = 1$, Total number of time steps

Solve Eq. (16.3.7) for $\underline{d}_{i+1}$

Solve Eq. (16.3.5) for $\underline{\ddot{d}}_i$

Solve Eq. (16.3.1) for $\underline{\dot{d}}_i$

Output the displacements $\underline{d}_i$, velocities $\underline{\dot{d}}_i$, and accelerations, $\underline{\ddot{d}}_i$ for a given time step i

END

FIGURE 16-6   Flowchart of the central difference method

8. Use Steps 5, 6, and 7 repeatedly to obtain the displacement, acceleration, and velocity for all other time steps.

Figure 16-6 is a flowchart of the solution procedure using the central difference equations. Note that the recurrence formulas given by equations such as Eqs. (16.3.1) and (16.3.2) are approximate but yield sufficiently accurate results provided the time step Δt is taken small in relation to the variations in acceleration. Methods for determining proper time steps for the numerical integration process are described in Section 16.5.

We will now illustrate the central difference equations as they apply to the following example problem.

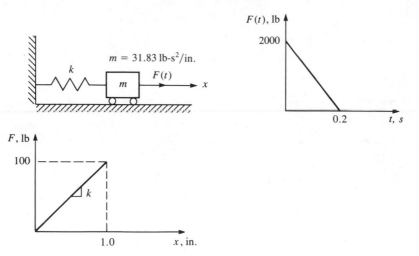

FIGURE 16-7 Spring-mass oscillator subjected to a time-dependent force

---

EXAMPLE 16.1

Determine the displacement, velocity, and acceleration at 0.05-s time intervals up to 0.2 s for the one-dimensional spring-mass oscillator subjected to the time-dependent forcing function shown in Figure 16-7. [Guidelines regarding appropriate time intervals (or time steps) are given in Section 16.5.] This forcing function is a typical one assumed for blast loads. The restoring spring force versus displacement curve is also provided. [Note that Figure 16-7 also represents a one-element bar with its left end fixed and right node subjected to $F(t)$ when a lumped mass is used.]

Since we are considering the single degree of freedom associated with the mass, the general matrix equations describing the motion reduce to single scalar equations. We will represent this single degree of freedom by d.

The solution procedure follows the steps outlined in this section and in the flowchart of Figure 16-6.

Step 1

At time $t = 0$, the initial displacement and velocity are zero; therefore,

$$d_0 = 0 \qquad \dot{d}_0 = 0$$

Step 2

The initial acceleration at $t = 0$ is obtained as

$$\ddot{d}_0 = \frac{2000 - 100(0)}{31.83} = 62.83 \text{ in./s}^2$$

where we have used $\underline{F}(0) = 2000$ lb and $\underline{K} = 100$ lb/in.

Step 3

The displacement d_{-1} is obtained as

$$d_{-1} = 0 - 0 + \frac{(0.05)^2}{2}(62.83) = 0.0785 \text{ in.}$$

Step 4

The displacement at time $t = 0.05$ s is

$$d_1 = \frac{1}{31.83}\{(0.05)^2(2000) + [2(31.83) - (0.05)^2(100)]0 - (31.83)(0.0785)\}$$

$$= 0.0785 \text{ in.}$$

Step 5

Having obtained d_1, we now determine the displacement at time $t = 0.10$ s as

$$d_2 = \frac{1}{31.83}\{(0.05)^2(1500) + [2(31.83) - (0.05)^2(100)](0.0785) - (31.83)(0)\}$$

$$= 0.274 \text{ in.}$$

Step 6

The acceleration at time $t = 0.05$ s is

$$\ddot{d}_1 = \frac{1}{31.83}[1500 - 100(0.0785)] = 46.88 \text{ in./s}^2$$

Step 7

The velocity at time $t = 0.05$ s is

$$\dot{d}_1 = \frac{0.274 - 0}{2(0.05)} = 2.74 \text{ in./s}$$

Step 8

Repeated use of Steps 5, 6, and 7 will result in the displacement, acceleration, and velocity for additional time steps as desired. We will now perform one more time-step iteration of the procedure.

Repeating Step 5 for the next time step, we have

$$d_3 = \frac{1}{31.83}\{(0.05)^2(1000) + [2(31.83) - (0.05)^2(100)](0.274)$$

$$- (31.83)(0.0785)\} = 0.546 \text{ in.}$$

Repeating Step 6 for the next time step, we have

$$\ddot{d}_2 = \frac{1}{31.83}[1000 - 100(0.274)] = 30.56 \text{ in./s}^2$$

T A B L E   16-1   Results of the analysis of Example 16.1

t, s	$F(t)$, lb	d_i, in.	Q, lb	$\ddot{d}_i$, in./s^2	$\dot{d}_i$, in./s	d_i (exact)
0	2000	0	0	62.83	0	0
0.05	1500	0.0785	7.85	46.88	2.74	0.0718
0.10	1000	0.274	27.40	30.56	4.68	0.2603
0.15	500	0.546	54.64	13.99	5.79	0.5252
0.20	0	0.854	85.35	−2.68	6.07	0.8250
0.25	0	1.154	115.4	−3.63	5.91	1.132

Finally, repeating Step 7 for the next time step, we obtain

$$\dot{d}_2 = \frac{0.546 - 0.0785}{2(0.05)} = 4.68 \text{ in./s}$$

Table 16-1 summarizes the results obtained through time $t = 0.25$ s. In Table 16-1, $Q = kd_i$ is the restoring spring force. Also, the exact analytical solution for displacement based on the equation in Reference [13] is given.   ∎

Newmark's Method

We will now outline Newmark's numerical method, which, because of its general versatility, has been adopted into numerous commercially available computer programs for purposes of structural dynamic analysis. (Complete development of the equations can be found in Reference [5].) Newmark's equations are given by

$$\dot{\underline{d}}_{i+1} = \dot{\underline{d}}_i + (\Delta t)[(1 - \gamma)\ddot{\underline{d}}_i + \gamma\ddot{\underline{d}}_{i+1}] \qquad (16.3.9)$$

$$\underline{d}_{i+1} = \underline{d}_i + (\Delta t)\dot{\underline{d}}_i + (\Delta t)^2[(\tfrac{1}{2} - \beta)\ddot{\underline{d}}_i + \beta\ddot{\underline{d}}_{i+1}] \qquad (16.3.10)$$

where β and γ are parameters chosen by the user. The parameter β is generally chosen between 0 and $\tfrac{1}{4}$, and γ is often taken to be $\tfrac{1}{2}$. For instance, choosing $\gamma = \tfrac{1}{2}$ and $\beta = 0$, it can be shown that Eqs. (16.3.9) and (16.3.10) reduce to the central difference Eqs. (16.3.1) and (16.3.2). If $\gamma = \tfrac{1}{2}$ and $\beta = \tfrac{1}{6}$ are chosen, Eqs. (16.3.9) and (16.3.10) correspond to those for which a linear acceleration assumption is valid within each time interval. For $\gamma = \tfrac{1}{2}$ and $\beta = \tfrac{1}{4}$, it has been shown that the numerical analysis is stable; that is, computed quantities such as displacement and velocities do not become unbounded regardless of the time step chosen. Furthermore, it has been found (see Reference [5]) that a time step of approximately $\tfrac{1}{10}$ of the shortest natural frequency of the structure being analyzed usually yields the best results.

To find $\underline{d}_{i+1}$, we first multiply Eq. (16.3.10) by the mass matrix $\underline{M}$ and then substitute Eq. (16.3.5) for $\ddot{\underline{d}}_{i+1}$ into this equation to obtain

$$\underline{M}\underline{d}_{i+1} = \underline{M}\underline{d}_i + (\Delta t)\underline{M}\dot{\underline{d}}_i + (\Delta t)^2\underline{M}(\tfrac{1}{2} - \beta)\ddot{\underline{d}}_i] + \beta(\Delta t)^2[\underline{F}_{i+1} - \underline{K}\underline{d}_{i+1}] \qquad (16.3.11)$$

Combining like terms of Eq. (16.3.11), we obtain

$$(\underline{M} + \beta(\Delta t)^2\underline{K})\underline{d}_{i+1} = \beta(\Delta t)^2\underline{F}_{i+1} + \underline{M}\underline{d}_i + (\Delta t)\underline{M}\underline{\dot{d}}_i + (\Delta t)^2\underline{M}(\tfrac{1}{2} - \beta)\underline{\ddot{d}}_i$$
$$(16.3.12)$$

Finally, dividing Eq. (16.3.12) by $\beta(\Delta t)^2$, we obtain

$$\underline{K}'\underline{d}_{i+1} = \underline{F}'_{i+1} \qquad\qquad (16.3.13)$$

where
$$\underline{K}' = \underline{K} + \frac{1}{\beta(\Delta t)^2}\underline{M}$$
$$(16.3.14)$$

$$\underline{F}'_{i+1} = \underline{F}_{i+1} + \frac{\underline{M}}{\beta(\Delta t)^2}\left[\underline{d}_i + (\Delta t)\underline{\dot{d}}_i + \left(\frac{1}{2} - \beta\right)(\Delta t)^2\underline{\ddot{d}}_i\right]$$

The solution procedure using Newmark's equations is as follows:

1. Starting at time $t = 0$, $\underline{d}_0$ is known from the given boundary conditions on displacement, and $\underline{\dot{d}}_0$ is known from the initial velocity conditions.

2. Solve Eq. (16.3.5) at $t = 0$ for $\underline{\ddot{d}}_0$ (unless $\underline{\ddot{d}}_0$ is known from an initial acceleration condition); that is,

$$\underline{\ddot{d}}_0 = \underline{M}^{-1}(\underline{F}_0 - \underline{K}\underline{d}_0)$$

3. Solve Eq. (16.3.13) for $\underline{d}_1$, since $\underline{F}_{i+1}$ is known for all time steps and $\underline{d}_0$, $\underline{\dot{d}}_0$, and $\underline{\ddot{d}}_0$ are now known from Steps 1 and 2.

4. Use Eq. (16.3.10) to solve for $\underline{\ddot{d}}_1$ as

$$\underline{\ddot{d}}_1 = \frac{1}{\beta(\Delta t)^2}\left[\underline{d}_1 - \underline{d}_0 - (\Delta t)\underline{\dot{d}}_0 - (\Delta t)^2\left(\frac{1}{2} - \beta\right)\underline{\ddot{d}}_0\right]$$

5. Solve Eq. (16.3.9) directly for $\underline{\dot{d}}_1$.

6. Using the results of Steps 4 and 5, go back to Step 3 to solve for $\underline{d}_2$ and then to Steps 4 and 5 to solve for $\underline{\ddot{d}}_2$ and $\underline{\dot{d}}_2$. Use Steps 3, 4, and 5 repeatedly to solve for $\underline{d}_{i+1}$, $\underline{\dot{d}}_{i+1}$, and $\underline{\ddot{d}}_{i+1}$.

Figure 16-8 is a flowchart of the solution procedure using Newmark's equations. The advantages of Newmark's method over the central difference method are that Newmark's method can be made unconditionally stable (for instance, if $\beta = \frac{1}{4}$ and $\gamma = \frac{1}{2}$), and that larger time steps can be used with better results because, in general, the difference expressions more closely approximate the true acceleration and displacement time behavior (see References [8] and [11]). Other difference formulas, such as Wilson's and Houboldt's, also yield unconditionally stable algorithms.

We will now illustrate the use of Newmark's equations as they apply to the following example problem.

E X A M P L E   16.2

Determine the displacement, velocity, and acceleration at 0.1-s time increments up to a time of 0.5 s for the one-dimensional spring-mass oscillator

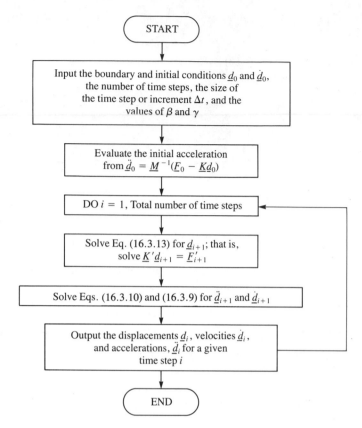

START

Input the boundary and initial conditions $\underline{d}_0$ and $\underline{\dot{d}}_0$, the number of time steps, the size of the time step or increment Δt, and the values of β and γ

Evaluate the initial acceleration from $\underline{\ddot{d}}_0 = \underline{M}^{-1}(\underline{F}_0 - \underline{K}\underline{d}_0)$

DO $i = 1$, Total number of time steps

Solve Eq. (16.3.13) for $\underline{d}_{i+1}$; that is, solve $\underline{K}'\underline{d}_{i+1} = \underline{F}'_{i+1}$

Solve Eqs. (16.3.10) and (16.3.9) for $\underline{\ddot{d}}_{i+1}$ and $\underline{\dot{d}}_{i+1}$

Output the displacements $\underline{d}_i$, velocities $\underline{\dot{d}}_i$, and accelerations, $\underline{\ddot{d}}_i$ for a given time step i

END

F I G U R E   16-8   Flowchart of numerical integration in time using Newmark's equations

subjected to the time-dependent forcing function shown in Figure 16-9, along with the restoring spring force versus displacement curve. Assume the oscillator is initially at rest. Let $\beta = \frac{1}{6}$ and $\gamma = \frac{1}{2}$, which corresponds to an assumption of linear acceleration within each time step.

Since we are again considering the single degree of freedom associated with the mass, the general matrix equations describing the motion reduce to single scalar equations. Again, we represent this single degree of freedom by d.

The solution procedure follows the steps outlined in this section and in the flowchart of Figure 16-8.

Step 1

At time $t = 0$, the initial displacement and velocity are zero; therefore,

$$d_0 = 0 \qquad \dot{d}_0 = 0$$

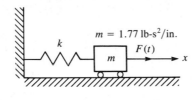

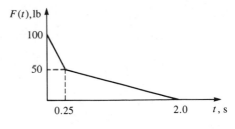

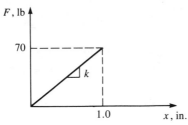

FIGURE 16-9   Spring-mass oscillator subjected to a time-dependent force

Step 2

The initial acceleration at $t = 0$ is obtained as

$$\ddot{d}_0 = \frac{100 - 70(0)}{1.77} = 56.5 \text{ in./s}^2$$

where we have used $\underline{F}_0 = 100$ lb and $\underline{K} = 70$ lb/in.

Step 3

We now solve for the displacement at time $t = 0.1$ s as

$$K' = 70 + \frac{1}{(\frac{1}{6})(0.1)^2}(1.77) = 1132 \text{ lb/in.}$$

$$F_1' = 80 + \frac{1.77}{(\frac{1}{6})(0.1)^2}\left[0 + (0.1)(0) + \left(\frac{1}{2} - \frac{1}{6}\right)(0.1)^2(56.5)\right] = 280 \text{ lb}$$

$$d_1 = \frac{280}{1132} = 0.248 \text{ in.}$$

Step 4

Solve for the acceleration at time $t = 0.1$ s as

$$\ddot{d}_1 = \frac{1}{(\frac{1}{6})(0.1)^2}\left[0.248 - 0 - (0.1)(0) - (0.1)^2\left(\frac{1}{2} - \frac{1}{6}\right)(56.5)\right]$$

$$\ddot{d}_1 = 35.4 \text{ in./s}^2$$

Step 5

Solve for the velocity at time $t = 0.1$ s as

$$\dot{d}_1 = 0 + (0.1)[(1 - \tfrac{1}{2})(56.5) + (\tfrac{1}{2})(35.4)]$$

$$\dot{d}_1 = 4.59 \text{ in./s}$$

Step 6

Repeated use of Steps 3, 4, and 5 will result in the displacement, acceleration, and velocity for additional time steps as desired. We will now perform one more time-step iteration.

Repeating Step 3 for the next time step ($t = 0.2$ s), we have

$$F_2' = 60 + \frac{1.77}{(\tfrac{1}{6})(0.1)^2}\left[0.248 + (0.1)(4.59) + \left(\frac{1}{2} - \frac{1}{6}\right)(0.1)^2(35.4)\right]$$

$$F_2' = 934 \text{ lb}$$

$$d_2 = \frac{934}{1132} = 0.825 \text{ in.}$$

Repeating Step 4 for time step $t = 0.2$ s, we obtain

$$\ddot{d}_2 = \frac{1}{(\tfrac{1}{6})(0.1)^2}\left[0.825 - 0.248 - (0.1)(4.59) - (0.1)^2\left(\frac{1}{2} - \frac{1}{6}\right)(35.4)\right]$$

$$\ddot{d}_2 = 1.27 \text{ in./s}^2$$

Finally, repeating Step 5 for time step $t = 0.2$ s, we have

$$\dot{d}_2 = 4.59 + (0.1)[(1 - \tfrac{1}{2})(35.4) + \tfrac{1}{2}(1.27)]$$

$$\dot{d}_2 = 6.42 \text{ in./s}$$

Table 16-2 summarizes the results obtained through time $t = 0.5$ s. ■

T A B L E   16-2   Results of the analysis of Example 16.2

t, s	$F(t)$, lb	d_i, in.	Q, lb	$\ddot{d}_i$, in./s^2	$\dot{d}_i$, in./s
0.	100	0	0	56.6	0
0.1	80	0.248	17.3	35.4	4.59
0.2	60	0.825	57.8	1.27	6.42
0.3	48.6	1.36	95.2	−26.2	5.17
0.4	45.7	1.72	120.4	−42.2	1.75
0.5	42.9	1.68	117.6	−42.2	−2.45

16.4   Natural Frequencies of a One-Dimensional Bar

Before solving the structural stress dynamic analysis problem, we will first describe how to determine the natural frequencies of continuous elements (specifically the bar element). The natural frequencies are necessary in a vibration analysis and also are important when choosing a proper time step for a structural dynamic analysis (as will be discussed in Section 16.5).

Natural frequencies are determined by solving Eq. (16.2.24) in the absence of a forcing function $F(t)$. Therefore, we solve the matrix equation

$$\underline{M}\underline{\ddot{d}} + \underline{K}\underline{d} = 0 \qquad (16.4.1)$$

The standard solution for $\underline{d}(t)$ is given by the harmonic equation in time

$$\underline{d}(t) = \underline{d}'e^{i\omega t} \qquad (16.4.2)$$

where $\underline{d}'$ is the part of the nodal displacement matrix called *natural modes* that is assumed to be independent of time, i is the standard imaginary number given by $i = \sqrt{-1}$, and ω is a natural frequency.

Differentiating Eq. (16.4.2) twice with respect to time, we obtain

$$\underline{\ddot{d}}(t) = \underline{d}'(-\omega^2)e^{i\omega t} \qquad (16.4.3)$$

Substitution of Eqs. (16.4.2) and (16.4.3) into Eq. (16.4.1) yields

$$-\underline{M}\omega^2\underline{d}'e^{i\omega t} + \underline{K}\underline{d}'e^{i\omega t} = 0 \qquad (16.4.4)$$

Combining terms in Eq. (16.4.4), we obtain

$$e^{i\omega t}(\underline{K} - \omega^2\underline{M})\underline{d}' = 0 \qquad (16.4.5)$$

Since $e^{i\omega t}$ is not zero, from Eq. (16.4.5), we obtain

$$(\underline{K} - \omega^2\underline{M})\underline{d}' = 0 \qquad (16.4.6)$$

Equation (16.4.6) is a set of linear homogeneous equations in terms of displacement mode $\underline{d}'$. Hence, Eq. (16.4.6) has a nontrivial solution if and only if the determinant of the coefficient matrix of $\underline{d}'$ is zero; that is, we must have

$$|\underline{K} - \omega^2\underline{M}| = 0 \qquad (16.4.7)$$

In general, Eq. (16.4.7) is a set of n algebraic equations, where n is the number of degrees of freedom associated with the problem.

To illustrate the procedure for determining the natural frequencies, we will solve the following example problem.

E X A M P L E   16.3

For the bar shown in Figure 16-10 with length $2L$, modulus of elasticity E, mass density ρ, and cross-sectional area A, determine the first two natural frequencies.

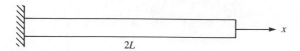

FIGURE 16-10   One-dimensional bar used for natural frequency determination

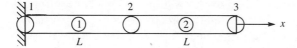

FIGURE 16-11   Discretized bar of Figure 16-10

For simplicity, the bar is discretized into two elements each of length L as shown in Figure 16-11. To solve Eq. (16.4.7), we must develop the total stiffness matrix for the bar by using Eq. (16.2.11). Either the lumped-mass matrix Eq. (16.2.12) or the consistent-mass matrix Eq. (16.2.23) can be used. In general, using the consistent-mass matrix has resulted in solutions that compare more closely to available analytical and experimental results than those found using the lumped-mass matrix. However, the longhand calculations are more tedious using the consistent-mass matrix than using the lumped-mass matrix because the consistent-mass matrix is a full symmetric matrix, whereas the lumped-mass matrix has nonzero terms only along the main diagonal. Hence, the lumped-mass matrix will be used in this analysis.

Using Eq. (16.2.11), the stiffness matrices for each element are given by

$$[\hat{k}^{(1)}] = \frac{AE}{L}\begin{matrix} 1 & 2 \\ \begin{bmatrix} 1 & -1 \\ -1 & 1 \end{bmatrix} \end{matrix} \qquad [\hat{k}^{(2)}] = \frac{AE}{L}\begin{matrix} 2 & 3 \\ \begin{bmatrix} 1 & -1 \\ -1 & 1 \end{bmatrix} \end{matrix} \qquad (16.4.8)$$

The usual direct stiffness method for assembling the element matrices, Eqs. (16.4.8), yields the global stiffness matrix for the whole bar as

$$[K] = \frac{AE}{L}\begin{bmatrix} 1 & -1 & 0 \\ -1 & 2 & -1 \\ 0 & -1 & 1 \end{bmatrix} \qquad (16.4.9)$$

Using Eq. (16.2.12), the mass matrices for each element are given by

$$[\hat{m}^{(1)}] = \frac{\rho AL}{2}\begin{matrix} 1 & 2 \\ \begin{bmatrix} 1 & 0 \\ 0 & 1 \end{bmatrix} \end{matrix} \qquad [\hat{m}^{(2)}] = \frac{\rho AL}{2}\begin{matrix} 2 & 3 \\ \begin{bmatrix} 1 & 0 \\ 0 & 1 \end{bmatrix} \end{matrix} \qquad (16.4.10)$$

The mass matrices for each element are assembled in the same manner as for the stiffness matrices. Therefore, by assembling Eqs. (16.4.10), we obtain the global mass matrix as

$$[M] = \frac{\rho AL}{2}\begin{bmatrix} 1 & 0 & 0 \\ 0 & 2 & 0 \\ 0 & 0 & 1 \end{bmatrix} \qquad (16.4.11)$$

We observe from the resulting global mass matrix that there are two mass contributions at node 2 because node 2 is common to both elements.

Substituting the global stiffness matrix Eq. (16.4.9) and the global mass matrix Eq. (16.4.11) into Eq. (16.4.6), and using the boundary condition $\hat{d}_{1x} = 0$ (or now $d_1' = 0$) to reduce the set of equations in the usual manner, we obtain

$$\left(\frac{AE}{L} \begin{bmatrix} 2 & -1 \\ -1 & 1 \end{bmatrix} - \omega^2 \frac{\rho AL}{2} \begin{bmatrix} 2 & 0 \\ 0 & 1 \end{bmatrix} \right) \begin{Bmatrix} d_2' \\ d_3' \end{Bmatrix} = \begin{Bmatrix} 0 \\ 0 \end{Bmatrix} \qquad (16.4.12)$$

To obtain a solution to the set of homogeneous equations in Eq. (16.4.12), we set the determinant of the coefficient matrix equal to zero as indicated by Eq. (16.4.7). We then have

$$\left| \frac{AE}{L} \begin{bmatrix} 2 & -1 \\ -1 & 1 \end{bmatrix} - \lambda \frac{\rho AL}{2} \begin{bmatrix} 2 & 0 \\ 0 & 1 \end{bmatrix} \right| = 0 \qquad (16.4.13)$$

where $\lambda = \omega^2$ has been used in Eq. (16.4.13). Dividing Eq. (16.4.13) by ρAL and letting $\mu = E/(\rho L^2)$, we obtain

$$\begin{vmatrix} 2\mu - \lambda & -\mu \\ -\mu & \mu - \dfrac{\lambda}{2} \end{vmatrix} = 0 \qquad (16.4.14)$$

Evaluating the determinant in Eq. (16.4.14), we obtain

$$\lambda = 2\mu \pm \mu\sqrt{2}$$

or
$$\lambda_1 = 0.60\mu \qquad \lambda_2 = 3.41\mu \qquad (16.4.15)$$

For comparison, the exact solution is given by $\lambda = 0.616\mu$, whereas the consistent-mass approach yields $\lambda = 0.648\mu$. Therefore, for bar elements, the lumped-mass approach can yield results as good as, or even better than, the results for the consistent-mass approach. However, the consistent-mass approach can be mathematically proven to yield an upper bound on the frequencies, whereas the lumped-mass approach yields results that can be below or above the exact frequencies with no mathematical proof of boundedness. From Eqs. (16.4.15), the first and second natural frequencies are given by

$$\omega_1 = \sqrt{\lambda_1} = 0.77\sqrt{\mu} \qquad \omega_2 = \sqrt{\lambda_2} = 1.85\sqrt{\mu}$$

Letting $E = 30 \times 10^6$ psi, $\rho = 0.00073$ lb-s^2/in.^{4}, and $L = 100$ in., we obtain

$$\mu = E/(\rho L^2) = (30 \times 10^6)/[(0.00073)(100)^2] = 4.12 \times 10^6 \text{ s}^{-2}$$

Therefore, we obtain the natural frequencies as

$$\omega_1 = 1.56 \times 10^3 \text{ rad/s} \qquad \omega_2 = 3.76 \times 10^3 \text{ rad/s} \qquad (16.4.16)$$

In conclusion, note that for a bar discretized such that two nodes are free to displace, there are two natural modes and two frequencies. When a system vibrates with a given natural frequency ω_i, that unique shape with arbitrary amplitude corresponding to ω_i is called the *mode*.) In general, for an n-degree-of-freedom discrete system, there are n natural modes and frequencies. A

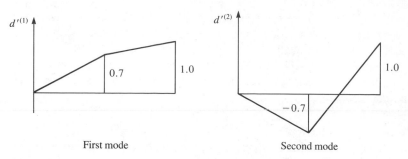

First mode   Second mode

F I G U R E   16-12   First and second modes of longitudinal vibration for
the cantilever bar of Figure 16-10

continuous system actually has an infinite number of natural modes and
frequencies. When discretized, only n degrees of freedom are created. The
lowest modes and frequencies are approximated most often; the higher frequencies are damped out more rapidly and are usually of less importance.

Substituting λ_1 from Eqs. (16.4.15) into Eq. (16.4.12) and simplifying, the
first modal equations are given by

$$1.4\mu d_2''^{(1)} - \mu d_3''^{(1)} = 0$$

$$-\mu d_2''^{(1)} + 0.7\mu d_3''^{(1)} = 0$$

(16.4.17)

It is customary to specify the value of one of the natural modes $\underline{d}'$ for a given
ω_i or λ_i. Letting $d_3''^{(1)} = 1$ and solving Eq. (16.4.17), we find $d_2''^{(1)} = 0.7$. Similarly, substituting λ_2 from Eqs. (16.4.15) into Eq. (16.4.12), we obtain the second
modal equations. For brevity's sake, these equations are not presented here.
Now letting $d_3''^{(2)} = 1$ results in $d_2''^{(2)} = -0.7$. The modal response for the first
and second natural frequencies of longitudinal vibration are plotted in Figure
16-12. The first mode means that the bar is completely in tension or compression, depending on the excitation direction. The second mode means the bar
is in compression and tension or in tension and compression.   ∎

16.5   Time-Dependent One-Dimensional Bar Analysis

E X A M P L E   16.4

To illustrate the finite element solution of a time-dependent problem, we will
solve the problem of the one-dimensional bar shown in Figure 16-13(a)
subjected to the force shown in Figure 16-13(b). We will assume the boundary
condition $d_{1x} = 0$ and the initial conditions $\underline{d}_0 = 0$ and $\underline{\dot{d}}_0 = 0$. For later
numerical computation purposes, we let parameters $\rho = 0.00073$ lb-s²/in.⁴,
$A = 1$ in.², $E = 30 \times 10^6$ psi, and $L = 100$ in. These parameters are the same
values as used in Section 16.4.

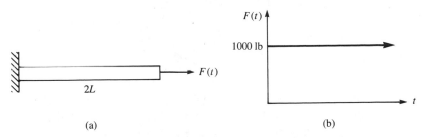

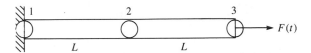

F I G U R E   16-13   (a) Bar subjected to a time-dependent force and (b) the forcing function applied to the end of the bar

F I G U R E   16-14   Discretized bar with lumped masses

Since the bar is discretized into two elements of equal length, the global stiffness and mass matrices determined in Section 16.4 and given by Eqs. (16.4.9) and (16.4.11) are applicable. We will again use the lumped-mass matrix because of its resulting computational simplicity. Figure 16-14 shows the discretized bar and the associated lumped masses.

For illustration of the numerical time integration scheme, we will use the central difference method because it is easier to apply for longhand computations (and without loss of generality).

We next select the time step to be used in the integration process. It has been mathematically shown that the time step must be less than or equal to two divided by the highest natural frequency when using the central difference method (see Reference [7]); that is, $\Delta t \le 2/\omega_{max}$. However, for practical results, we must use a time step of less than or equal to three-fourths of this value; that is,

$$\Delta t \le \frac{3}{4}\left(\frac{2}{\omega_{max}}\right) \qquad (16.5.1)$$

This time step ensures stability of the integration method. This criteria for selecting a time step demonstrates the usefulness of determining the natural frequencies of vibration, as previously described in Section 16.4, before performing the dynamic stress analysis. An alternative guide (used only for a bar) for choosing the approximate time step is

$$\Delta t = \frac{L}{c_x} \qquad (16.5.2)$$

where L is the element length, and $c_x = \sqrt{E_x/\rho}$ is called the *longitudinal wave velocity*. Evaluating the time step by using both criteria, Eqs. (16.5.1) and (16.5.2), from Eqs. (16.4.16) for ω, we obtain

$$\Delta t = \frac{3}{4}\left(\frac{2}{\omega_{max}}\right) = \frac{1.5}{3.76 \times 10^3} = 0.40 \times 10^{-3}\, \text{s} \qquad (16.5.3)$$

or

$$\Delta t = \frac{L}{c_x} = \frac{100}{\sqrt{30 \times 10^6/0.00073}} = 0.48 \times 10^{-3}\, \text{s} \qquad (16.5.4)$$

Guided by the maximum time steps calculated in Eqs. (16.5.3) and (16.5.4), we choose $\Delta t = 0.25 \times 10^{-3}$ s as a convenient time step for the computations.

Substituting the global stiffness and mass matrices, Eqs. (16.4.9) and (16.4.11), into the global dynamic Eq. (16.2.24), we obtain

$$\frac{AE}{L}\begin{bmatrix} 1 & -1 & 0 \\ -1 & 2 & -1 \\ 0 & -1 & 1 \end{bmatrix}\begin{Bmatrix} d_{1x} \\ d_{2x} \\ d_{3x} \end{Bmatrix} + \frac{\rho AL}{2}\begin{bmatrix} 1 & 0 & 0 \\ 0 & 2 & 0 \\ 0 & 0 & 1 \end{bmatrix}\begin{Bmatrix} \ddot{d}_{1x} \\ \ddot{d}_{2x} \\ \ddot{d}_{3x} \end{Bmatrix} = \begin{Bmatrix} R_1 \\ 0 \\ F_3(t) \end{Bmatrix} \qquad (16.5.5)$$

where R_1 denotes the unknown reaction at node 1. Using the procedure for solution outlined in Section 16.3 and in the flowchart of Figure 16-6, we begin as follows:

Step 1

Given: $d_{1x} = 0$ because of the fixed support at node 1, and all nodal displacements and velocities are zero at time $t = 0$; that is, $\underline{d}_0 = 0$ and $\underline{\dot{d}}_0 = 0$. Also, assume $\ddot{d}_{1x} = 0$ at all times.

Step 2

Solve for $\underline{\ddot{d}}_0$ using Eq. (16.3.5) as

$$\underline{\ddot{d}}_0 = \begin{Bmatrix} \ddot{d}_{2x} \\ \ddot{d}_{3x} \end{Bmatrix}_{t=0} = \frac{2}{\rho AL}\begin{bmatrix} \frac{1}{2} & 0 \\ 0 & 1 \end{bmatrix}\left[\begin{Bmatrix} 0 \\ 1000 \end{Bmatrix} - \frac{AE}{L}\begin{bmatrix} 2 & -1 \\ -1 & 1 \end{bmatrix}\begin{Bmatrix} 0 \\ 0 \end{Bmatrix}\right] \qquad (16.5.6)$$

where Eq. (16.5.6) accounts for the conditions $d_{1x} = 0$ and $\ddot{d}_{1x} = 0$. Simplifying Eq. (16.5.6), we obtain

$$\underline{\ddot{d}}_0 = \frac{2000}{\rho AL}\begin{Bmatrix} 0 \\ 1 \end{Bmatrix} = \begin{Bmatrix} 0 \\ 27,400 \end{Bmatrix}\, \text{in./s}^2 \qquad (16.5.7)$$

where the numerical values for ρ, A, and L have been substituted into the final numerical result in Eq. (16.5.7), and

$$\underline{M}^{-1} = \frac{2}{\rho AL}\begin{bmatrix} \frac{1}{2} & 0 \\ 0 & 1 \end{bmatrix} \qquad (16.5.8)$$

has been used in Eq. (16.5.6). The computational advantage of using the lumped-mass matrix for longhand calculations is now evident. The inverse of a diagonal matrix, such as the lumped-mass matrix, is obtained simply by inverting the diagonal elements of the matrix.

Step 3

Using Eq. (16.3.8), we solve for $\underline{d}_{-1}$ as

$$\underline{d}_{-1} = \underline{d}_0 - (\Delta t)\underline{\dot{d}}_0 + \frac{(\Delta t)^2}{2}\underline{\ddot{d}}_0 \qquad (16.5.9)$$

Substituting the initial conditions on $\underline{\dot{d}}_0$ and $\underline{d}_0$ from Step 1 and Eq. (16.5.7) for the initial acceleration $\underline{\ddot{d}}_0$ from Step 2 into Eq. (16.5.9), we obtain

$$\underline{d}_{-1} = 0 - (0.25 \times 10^{-3})(0) + \frac{(0.25 \times 10^{-3})^2}{2}(27,400)\begin{Bmatrix} 0 \\ 1 \end{Bmatrix}$$

or, on simplification,

$$\begin{Bmatrix} d_{2x} \\ d_{3x} \end{Bmatrix}_{-1} = \begin{Bmatrix} 0 \\ 0.856 \times 10^{-3} \end{Bmatrix} \text{ in.} \qquad (16.5.10)$$

Step 4

On premultiplying Eq. (16.3.7) by $\underline{M}^{-1}$, we now solve for $\underline{d}_1$ by

$$\underline{d}_1 = \underline{M}^{-1}\{(\Delta t)^2\underline{F}_0 + [2\underline{M} - (\Delta t)^2\underline{K}]\underline{d}_0 - \underline{M}\underline{d}_{-1}\} \qquad (16.5.11)$$

Substituting the numerical values for ρ, A, L, and E and the results of Eq. (16.5.10) into Eq. (16.5.11), we obtain

$$\begin{Bmatrix} d_{2x} \\ d_{3x} \end{Bmatrix}_1 = \frac{2}{0.073}\begin{bmatrix} \frac{1}{2} & 0 \\ 0 & 1 \end{bmatrix}\Bigg\{0.25 \times 10^{-3})^2\begin{Bmatrix} 0 \\ 1000 \end{Bmatrix}$$

$$+ \left[\frac{2(0.073)}{2}\begin{bmatrix} 2 & 0 \\ 0 & 1 \end{bmatrix} - (0.25 \times 10^{-3})^2(30 \times 10^4)\right.$$

$$\times \left.\begin{bmatrix} 2 & -1 \\ -1 & 1 \end{bmatrix}\right]\begin{Bmatrix} 0 \\ 0 \end{Bmatrix} - \frac{0.073}{2}\begin{bmatrix} 2 & 0 \\ 0 & 1 \end{bmatrix}\begin{Bmatrix} 0 \\ 0.856 \times 10^{-3} \end{Bmatrix}\Bigg\}$$

Simplifying, we obtain

$$\begin{Bmatrix} d_{2x} \\ d_{3x} \end{Bmatrix}_1 = \frac{2}{0.073}\begin{bmatrix} \frac{1}{2} & 0 \\ 0 & 1 \end{bmatrix}\left[\begin{Bmatrix} 0 \\ 0.0625 \times 10^{-3} \end{Bmatrix} - \begin{Bmatrix} 0 \\ 0.0312 \times 10^{-3} \end{Bmatrix}\right]$$

Finally, the nodal displacements at time $t = 0.25 \times 10^{-3}$ s become

$$\begin{Bmatrix} d_{2x} \\ d_{3x} \end{Bmatrix}_1 = \begin{Bmatrix} 0 \\ 0.858 \times 10^{-3} \end{Bmatrix} \text{ in.} \qquad (\text{at } t = 0.25 \times 10^{-3} \text{ s}) \qquad (16.5.12)$$

Step 5

With $\underline{d}_0$ initially given and $\underline{d}_1$ determined from Step 4, we use Eq. (16.3.7) to obtain

$$\underline{d}_2 = \underline{M}^{-1}\{(\Delta t)^2 \underline{F}_1 + [2\underline{M} - (\Delta t)^2 \underline{K}]\underline{d}_1 - \underline{M}\underline{d}_o\}$$

$$= \frac{2}{0.073}\begin{bmatrix} \frac{1}{2} & 0 \\ 0 & 1 \end{bmatrix}\left\{(0.25 \times 10^{-3})^2 \begin{Bmatrix} 0 \\ 1000 \end{Bmatrix} + \begin{bmatrix} \frac{2(0.073)}{2} \end{bmatrix}\begin{bmatrix} 2 & 0 \\ 0 & 1 \end{bmatrix}\right.$$

$$- (0.25 \times 10^{-3})^2 (30 \times 10^4)\begin{bmatrix} 2 & -1 \\ -1 & 1 \end{bmatrix}\right]\begin{Bmatrix} 0 \\ 0.858 \times 10^{-3} \end{Bmatrix}$$

$$- \frac{0.073}{2}\begin{bmatrix} 2 & 0 \\ 0 & 1 \end{bmatrix}\begin{Bmatrix} 0 \\ 0 \end{Bmatrix}\right\}$$

$$= \frac{2}{0.073}\begin{bmatrix} \frac{1}{2} & 0 \\ 0 & 1 \end{bmatrix}\left[\begin{Bmatrix} 0 \\ 0.0625 \times 10^{-3} \end{Bmatrix} + \begin{Bmatrix} 0.0161 \times 10^{-3} \\ 0.0466 \times 10^{-3} \end{Bmatrix}\right]$$

Simplifying, we obtain the nodal displacements at time $t = 0.50 \times 10^{-3}$ s as

$$\begin{Bmatrix} d_{2x} \\ d_{3x} \end{Bmatrix}_2 = \begin{Bmatrix} 0.221 \times 10^{-3} \\ 2.99 \times 10^{-3} \end{Bmatrix} \text{ in.} \qquad (\text{at } t = 0.50 \times 10^{-3} \text{ s}) \qquad (16.5.13)$$

Step 6

Solve for the nodal accelerations $\ddot{\underline{d}}_1$ again using Eq. (16.3.5) as

$$\ddot{\underline{d}}_1 = \frac{2}{0.073}\begin{bmatrix} \frac{1}{2} & 0 \\ 0 & 1 \end{bmatrix}\left[\begin{Bmatrix} 0 \\ 1000 \end{Bmatrix} - (30 \times 10^4)\begin{bmatrix} 2 & -1 \\ -1 & 1 \end{bmatrix}\begin{Bmatrix} 0 \\ 0.858 \times 10^{-3} \end{Bmatrix}\right]$$

Simplifying, we then obtain the nodal accelerations at time $t = 0.25 \times 10^{-3}$ s as

$$\begin{Bmatrix} \ddot{d}_{2x} \\ \ddot{d}_{3x} \end{Bmatrix}_1 = \begin{Bmatrix} 3526 \\ 20,345 \end{Bmatrix} \text{ in./s}^2 \qquad (\text{at } t = 0.25 \times 10^{-3} \text{ s}) \qquad (16.5.14)$$

Using the results of Eqs. (16.5.12) and (16.5.14) in Eq. (16.5.5), the reaction R_1 could be found.

Step 7

Using Eq. (16.5.13) from Step 5 and the boundary condition for $\underline{d}_o$ given in Step 1, we obtain $\dot{\underline{d}}_1$ as

$$\dot{\underline{d}}_1 = \frac{\left[\begin{Bmatrix} 0.221 \times 10^{-3} \\ 2.99 \times 10^{-3} \end{Bmatrix} - \begin{Bmatrix} 0 \\ 0 \end{Bmatrix}\right]}{2(0.25 \times 10^{-3})}$$

Simplifying, we obtain

$$\begin{Bmatrix} \dot{d}_{2x} \\ \dot{d}_{3x} \end{Bmatrix} = \begin{Bmatrix} 0.442 \\ 5.98 \end{Bmatrix} \text{ in./s} \qquad (\text{at } t = 0.25 \times 10^{-3} \text{ s})$$

Step 8

We now use Steps 5, 6, and 7 repeatedly to obtain the displacement, acceleration, and velocity for all other time steps. For simplicity, we calculate the acceleration only.

Repeating Step 6 with $t = 0.50 \times 10^{-3}$ s, we obtain the nodal accelerations as

$$\ddot{\underline{d}}_2 = \frac{2}{0.073}\begin{bmatrix} \frac{1}{2} & 0 \\ 0 & 1 \end{bmatrix}\left[\left\{\begin{array}{c} 0 \\ 1000 \end{array}\right\} - 30 \times 10^4 \begin{bmatrix} 2 & -1 \\ -1 & 1 \end{bmatrix}\left\{\begin{array}{c} 0.221 \times 10^{-3} \\ 2.99 \times 10^{-3} \end{array}\right\}\right]$$

On simplifying, the nodal accelerations at $t = 0.50 \times 10^{-3}$ s are

$$\left\{\begin{array}{c} \ddot{d}_{2x} \\ \ddot{d}_{3x} \end{array}\right\}_2 = \left\{\begin{array}{c} 0 \\ 27,400 \end{array}\right\} + \left\{\begin{array}{c} 10,500 \\ -22,800 \end{array}\right\} = \left\{\begin{array}{c} 10,500 \\ 4600 \end{array}\right\} \text{ in./s}^2 \text{ (at } t = 0.5 \times 10^{-3} \text{ s)}$$

$$(16.5.15) \quad \blacksquare$$

16.6 Beam Element Mass Matrices and Natural Frequencies

We now consider the lumped- and consistent-mass matrices appropriate for time-dependent beam analysis. The development of the element equations follows the same general steps as used in Section 16.2 for the bar element.

The beam element with the associated nodal degrees of freedom (transverse displacement and rotation) is shown in Figure 16-15.

The basic element equations are given by the general form, Eq. (16.2.10), with the appropriate nodal force, stiffness, and mass matrices for a beam element. The stiffness matrix for the beam element is that given by Eq. (5.1.14). A lumped-mass matrix is obtained as

$$[\hat{m}] = \frac{\rho AL}{2} \begin{array}{c} \hat{d}_{1y} \ \hat{\phi}_1 \ \hat{d}_{2y} \ \hat{\phi}_2 \\ \begin{bmatrix} 1 & 0 & 0 & 0 \\ 0 & 0 & 0 & 0 \\ 0 & 0 & 1 & 0 \\ 0 & 0 & 0 & 0 \end{bmatrix} \end{array} \qquad (16.6.1)$$

where one-half of the total beam mass has been lumped at each node, corresponding to the translational degrees of freedom. In the lumped mass approach, the inertial effect associated with possible rotational degrees of freedom has been assumed to be zero in obtaining Eq. (16.6.1), although a value may be assigned to these rotational degrees of freedom by calculating the mass moment of inertia of a fraction of the beam segment about the nodal points. For a uniform beam we could then calculate the mass moment of inertia of half of the beam segment about each end node using basic dynamics

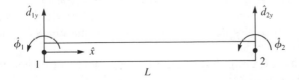

FIGURE 16-15   Beam element with nodal degrees of freedom

as
$$I = \tfrac{1}{3}(\rho AL/2)(L/2)^2$$

Again, the lumped-mass matrix given by Eq. (16.6.1) is a diagonal matrix, making matrix numerical calculations easier to perform than when using the consistent-mass matrix. The consistent-mass matrix can be obtained by applying the general Eq. (16.2.19) for the beam element, where the shape functions are now given by Eqs. (5.1.7). Therefore,

$$[\dot{m}] = \int\int\int_V \rho [N]^T [N] \, dV \tag{16.6.2}$$

$$[\dot{m}] = \int_0^L \int\int_A \rho \begin{Bmatrix} N_1 \\ N_2 \\ N_3 \\ N_4 \end{Bmatrix} [N_1 \quad N_2 \quad N_3 \quad N_4] \, dA \, d\hat{x} \tag{16.6.3}$$

with
$$N_1 = \frac{1}{L^3}(2\hat{x}^3 - 3\hat{x}^2 L + L^3)$$

$$N_2 = \frac{1}{L^3}(\hat{x}^3 L - 2\hat{x}^2 L^2 + \hat{x}L^3)$$

$$\tag{16.6.4}$$

$$N_3 = \frac{1}{L^3}(-2\hat{x}^3 + 3\hat{x}^2 L)$$

$$N_4 = \frac{1}{L^3}(\hat{x}^3 L - \hat{x}^2 L^2)$$

On substituting the shape function Eqs. (16.6.4) into Eq. (16.6.3) and performing the integration, the consistent-mass matrix becomes

$$[\dot{m}] = \frac{\rho AL}{420} \begin{bmatrix} 156 & 22L & 54 & -13L \\ 22L & 4L^2 & 13L & -3L^2 \\ 54 & 13L & 156 & -22L \\ -13L & -3L^2 & -22L & 4L^2 \end{bmatrix} \tag{16.6.5}$$

Having obtained the mass matrix for the beam element, we could proceed to formulate the global stiffness and mass matrices and equations of the form given by Eq. (16.2.24) to solve the problem of a beam subjected to a time-dependent load. We will not illustrate the procedure for solution here because it is tedious and similar to that used to solve the one-dimensional bar problem in Section 16.5. However, a computer program for the analysis of plane frames subjected to time-dependent forces is provided on the accompanying disk. Section 16.7 provides descriptions of plane frame analysis and of the computer program.

To clarify the procedure for beam analysis, we will now determine the natural frequencies of a beam.

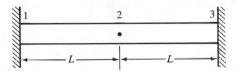

FIGURE 16-16   Beam for determination of natural frequencies

E X A M P L E   16.5

We now consider the determination of the natural frequencies of vibration for a beam fixed at both ends as shown in Figure 16-16. The beam has mass density ρ, modulus of elasticity E, cross-sectional area A, area moment of inertia I, and length $2L$. For simplicity of the longhand calculations, the beam is discretized into two elements of length L.

We can obtain the natural frequencies by using the general Eq. (16.4.7). First, we assemble the global stiffness and mass matrices (using the boundary conditions $d_{1x} = 0$, $\phi_1 = 0$, $d_{3x} = 0$, and $\phi_3 = 0$ as

$$\begin{matrix} d_{2y} & \phi_2 \end{matrix}$$

$$\underline{K} = \frac{EI}{L^3}\begin{bmatrix} 24 & 0 \\ 0 & 8L^2 \end{bmatrix} \qquad \underline{M} = \frac{\rho AL}{2}\begin{bmatrix} 2 & 0 \\ 0 & 0 \end{bmatrix} \qquad (16.6.6)$$

where Eq. (16.6.1) has been used to calculate the lumped-mass matrix. On substituting Eqs. (16.6.6) into Eq. (16.4.7), we obtain

$$\left| \frac{EI}{L^3}\begin{bmatrix} 24 & 0 \\ 0 & 8L^2 \end{bmatrix} - \omega^2 \rho AL \begin{bmatrix} 1 & 0 \\ 0 & 0 \end{bmatrix} \right| = 0 \qquad (16.6.7)$$

Dividing Eq. (16.6.7) by ρAL and simplifying, we obtain

$$\omega^2 = \frac{24EI}{\rho AL^4}$$

or

$$\omega = \frac{4.90}{L^2}\left(\frac{EI}{A\rho}\right)^{1/2}$$

The exact solution for the first natural frequency, from simple beam theory, is given by (Reference [6])

$$\omega = \frac{5.59}{L^2}\left(\frac{EI}{A\rho}\right)^{1/2}$$

This large discrepancy between the exact solution and the finite element solution is assumed to be accounted for by the coarseness of the finite element model. In Example 16.6 we show for a clamped-free beam that as the number of degrees of freedom increases, convergence to the exact solution results. ■

E X A M P L E   16.6

Determine the first natural frequency of vibration of a cantilever beam with the following data:

Length of the beam: $L = 30$ in.
Modulus of elasticity: $E = 3 \times 10^7$ psi
Moment of inertia: $I = 0.0833$ in.^{4}
Cross-sectional area: $A = 1$ in.^{2}
Mass density: $\rho = 0.00073$ lb-s^2/in.
Poisson's ratio: $v = 0.3$

The finite element longhand solution result for the first natural frequency is obtained similarly to that of Example 16.5 as

$$\omega = \frac{0.787}{L^2}\left(\frac{EI}{A\rho}\right)^2$$

The exact solution according to beam theory (Reference [1]) is

$$\omega = \frac{0.879}{L^2}\left(\frac{EI}{\rho A}\right)^2$$

According to vibration theory for a clamped-free beam (Reference [1]), we relate higher natural frequencies to the first natural frequency by

$$\frac{\omega_n}{\omega_1} = 6.2669$$

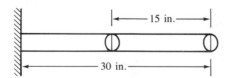

F I G U R E   16-17   Fixed-free beam (two-element model, lumped mass matrix)

T A B L E   16-3   Finite element computer solution compared to exact solution for Example 16.6

	ω_1, *rad/s*	ω_2, *rad/s*
Exact solution from beam theory	228	1434
Finite element solution		
Using 2 elements	205	1286
Using 6 elements	226	1372
Using 10 elements	227.5	1410
Using 30 elements	228.5	1430
Using 60 elements	228.5	1432

where $n = 2$ for the second natural frequency. Table 16-3 shows the computer solution as compared with the exact solution. ∎

16.7 Truss and Plane Frame Analysis

The dynamic analysis of the truss and plane frame are performed by extending the concepts presented in Sections 16.2 and 16.6 to the truss and plane frame, as has previously been done for the static analysis of trusses and frames.

The truss analysis requires the same transformation of the mass matrix from local to global coordinates as in Eq. (3.4.22) for the stiffness matrix; that is, the global mass matrix for a truss element is given by

$$\underline{m} = \underline{T}^T \underline{\hat{m}} \underline{T} \tag{16.7.1}$$

We are now dealing with motion in two or three dimensions. Therefore, we must reformulate a bar element mass matrix with both axial and transverse inertial properties because mass is included in both the global x and y directions in plane truss analysis. Considering two-dimensional motion, we express both local axial displacement $\hat{u}$ and transverse displacement $\hat{v}$ for the bar element in terms of the local axial and transverse nodal displacements as

$$\begin{Bmatrix} \hat{u} \\ \hat{v} \end{Bmatrix} = \frac{1}{L} \begin{bmatrix} L - \hat{x} & 0 & \hat{x} & 0 \\ 0 & L - \hat{x} & 0 & \hat{x} \end{bmatrix} \begin{Bmatrix} \hat{d}_{1x} \\ \hat{d}_{1y} \\ \hat{d}_{2x} \\ \hat{d}_{2y} \end{Bmatrix} \tag{16.7.2}$$

In general, $\underline{\hat{\psi}} = \underline{N}\underline{\hat{d}}$; therefore, the shape function matrix from Eq. (16.7.2) is

$$[N] = \frac{1}{L} \begin{bmatrix} L - \hat{x} & 0 & \hat{x} & 0 \\ 0 & L - \hat{x} & 0 & \hat{x} \end{bmatrix} \tag{16.7.3}$$

We can then substitute Eq. (16.7.3) into the general expression given by Eq. (16.2.19) to evaluate the bar consistent-mass matrix is

$$[\hat{m}] = \frac{\rho A L}{6} \begin{bmatrix} 2 & 0 & 1 & 0 \\ 0 & 2 & 0 & 1 \\ 1 & 0 & 2 & 0 \\ 0 & 1 & 0 & 2 \end{bmatrix} \tag{16.7.4}$$

The bar lumped-mass matrix for two-dimensional motion is obtained by simply lumping mass at each node and remembering that mass is the same in both the $\hat{x}$ and $\hat{y}$ directions. The local lumped-mass matrix is then

$$[\hat{m}] = \frac{\rho A L}{2} \begin{bmatrix} 1 & 0 & 0 & 0 \\ 0 & 1 & 0 & 0 \\ 0 & 0 & 1 & 0 \\ 0 & 0 & 0 & 1 \end{bmatrix} \tag{16.7.5}$$

The plane frame analysis first requires expanding and then combining the bar and beam mass matrices to obtain the local mass matrix. Since we recall there are six total degrees of freedom associated with a plane frame element, the bar and beam element mass matrices are expanded to order 6×6 and then superimposed. On combining the lumped-mass matrix Eqs. (16.2.12) and (16.6.1) for the bar and beam, respectively, the resulting lumped-mass matrix is

$$\underline{\hat{m}} = \frac{\rho AL}{2} \begin{matrix} \hat{d}_{1x}\ \hat{d}_{1y}\ \hat{\phi}_1\ \hat{d}_{2x}\ \hat{d}_{2y}\ \hat{\phi}_2 \\ \begin{bmatrix} 1 & 0 & 0 & 0 & 0 & 0 \\ 0 & 1 & 0 & 0 & 0 & 0 \\ 0 & 0 & 0 & 0 & 0 & 0 \\ 0 & 0 & 0 & 1 & 0 & 0 \\ 0 & 0 & 0 & 0 & 1 & 0 \\ 0 & 0 & 0 & 0 & 0 & 0 \end{bmatrix} \end{matrix} \qquad (16.7.6)$$

The mass matrix for a plane frame element arbitrarily oriented in x-y coordinates is transformed according to Eq. (16.7.1), where the transformation matrix $\underline{T}$ is now given by Eq. (6.1.10) and $\underline{\hat{m}}$ is now given by Eq. (16.7.6) if the lumped-mass matrix is used.

Since a longhand solution of the time-dependent plane frame problem is quite lengthy, only a computer program solution will be presented here. We will now describe a computer program called DFRAME (whose source code is included on the disk enclosed at the back of the text) that can be used to solve time-dependent plane frame problems.

We will apply the information presented in Chapter 6 for static analysis, as well as that presented in this chapter for dynamic analysis. We will perform the time integration using Newmark's method as explained in Section 16.3, and use the lumped-mass matrix given by Eq. (16.7.6).

Figure 16-18 is a flowchart of the finite element program DFRAME used for the dynamic analysis of plane frames. Much of the input for DFRAME is similar to that used for the static analysis of plane frames; that is, the general concepts and description of variables are the same as described in Section 6.7. Therefore, only the differences in data input are listed here.

1. Only plane frame analysis is considered, so the indicator of plane frame or grid is not needed. The first line of input is then the title line.

2. The element masses are calculated within the program and lumped at the nodes, as a translational lumped mass matrix is assumed.

3. The shear modulus G and torsional rigidity XJ are not included in the *element data* lines because a grid analysis is not available. Also, the z coordinate is not needed.

4. Additional parameter lines are included to input the time step DELT, the number of time steps NSTEP, the beta parameter BETA, and the gamma parameter GAMMA used in Newmark's

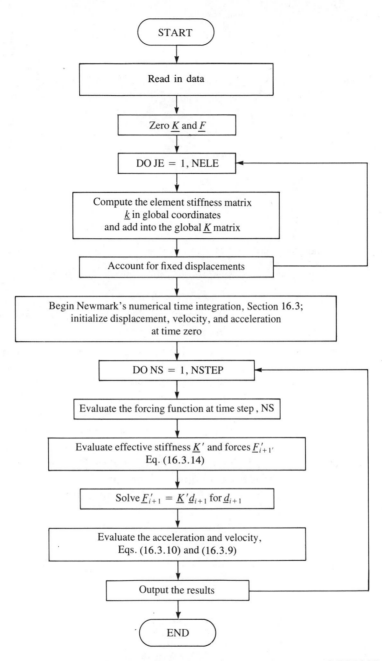

FIGURE 16-18   Flowchart of computer program DFRAME

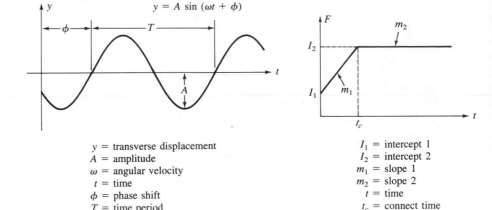

y = transverse displacement
A = amplitude
ω = angular velocity
t = time
ϕ = phase shift
T = time period

I_1 = intercept 1
I_2 = intercept 2
m_1 = slope 1
m_2 = slope 2
t = time
t_c = connect time
F = force

(a) Sinusoidal function

(b) Linear function

FIGURE 16-19   Types of forcing functions in DFRAME

numerical time integration method. These two lines of data follow all of the *element data* lines.

5. A subroutine must be added by the user to evaluate the forcing function at the various time steps chosen for numerical integration in time. For the example problem, these special subroutines in DFRAME are called FF1 and FF2. Two different forcing functions (a linear function and a sinusoidal function) are available (see Figure 16-19). Also, the fractional amplitudes of this forcing function at each node in the translational x and y directions and the rotational z direction must be input. These fractional amplitudes are input for each node of the structure on the node data lines corresponding to the nodes on which these forces act. The input data for DFRAME are now described.

Input Data

Line 1:

Format	Variable	Description
A	TITLE	Any 64 alphanumeric characters to identify the data file or the run.

Line 2:

Format	Variable	Description
I	NELE	Number of elements (maximum 50).
I	NNOD	Number of nodes in the mesh (maximum 51).

Line 3:

Format	Variable	Description
I	K	Node number
I	IFIX(1, K)	Boundary condition in x direction for node K. = 0 free (nodal load applied) = 1 fixed (zero displacement)
I	IFIX(2, K)	Boundary condition in y direction for node K. = 0 free (nodal load applied) = 1 fixed (zero displacement)
I	IFIX(3, K)	Boundary condition in z direction for node K. = 0 free (nodal moment applied) = 1 fixed (zero rotation)
G	XC(K)	The x coordinate of node K.
G	YC(K)	The y coordinate of node K.
G	FORC(1, K)	The *percentage* of the applied forcing function that acts in the x direction at node K (see line 8; expressed in decimal form)
G	FORC(2, K)	The percent of the applied forcing function that acts in the y direction at node K (expressed in decimal form).
G	FORC(3, K)	The percent of the applied couple that acts as an applied bending moment at node K (expressed in decimal form). Note: Distributed loads and couples cannot be handled directly and must be input as nodal forces and moments.

Line 4:

Format	Variable	Description
I	K	Element number.
I	NODE(1, K)	Node 1 of element K.
I	NODE(2, K)	Node 2 of element K.
G	E(K)	Young's modulus for element K.
G	A(K)	Cross-sectional area for element K.
G	XI(K)	Moment of inertia of cross-section for element K.
G	RHO(K)	Mass density of element K.

Line 5:

Format	Variable	Description
G	DELT	The time step for the analysis.
I	NSTEP	Number of time steps (maximum 150).

Line 6:

Format	Variable	Description
G	BETA	Beta is a value used in Newmark's method. (It is usually 0.25.)
G	GAMMA	Gamma is a value used in Newmark's method. (It is usually 0.50.)

Line 7:

Format	Variable	Description
I	ANS2	The type of forcing function to be input in line 8. = 1 Sinusoidal forcing function = 2 Linear forcing function

Line 8:
Sinusoidal forcing function: See Figure 16-19.

Format	Variable	Description
G	AMP	Amplitude of sinusoidal function.
G	OMEGA	Frequency of sinusoidal function.
G	ALPHA	Phase shift of sinusoidal function.

OR

Linear forcing function: See Figure 16-19.

Format	Variable	Description
G	SLOPE1	Slope of section 1 of linear function.
G	SECT1	y intercept of first section.
G	SLOPE2	Slope of section 2 of linear function.
G	SECT2	y intercept of second section.
G	TIMEC	Time at which the two sections connect. Note: Make sure that this time is correct or you might not get the results that you expected!

EXAMPLE 16.7

To illustrate the input data and solution to DFRAME, we will analyze the steel three-story rigid frame shown in Figure 16-20. The frame is one of a series of such frames spaced at 15-ft intervals in the direction into the page. The time-dependent forcing function indicated in Figure 16-20 is applied at each floor level and is assumed to be a suitable idealization of a distributed dynamic pressure applied to the walls. The loads per square foot (psf) on each floor and the weights per square foot of each wall are shown in the figure. The cross-sectional areas and moments of inertia for the elements are given in the figure, and $E = 30 \times 10^6$ psi for all elements.

We assume the translational motion in the vertical direction is to be prevented. Element mass densities ρ are needed in line 4 of the data file. These densities are calculated as follows. We first obtain the total mass at a given floor level and then divide this mass by the cross-sectional area and length of the element. For instance, for element 6 we have

$$M_6 = \frac{W_6}{g} = \frac{[(104 \text{ psf})(30 \text{ ft})(15 \text{ ft})}{386.4 \text{ in./s}^2} = 121 \text{ lb-s}^2/\text{in.}$$

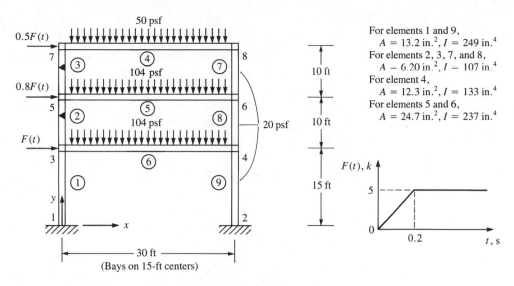

FIGURE 16-20   Plane frame subjected to a forcing function

and $\qquad \rho_6 = \dfrac{121 \text{ lb-s}^2/\text{in.}}{(24.7 \text{ in.}^2)(360 \text{ in.})} = 0.0136 \text{ lb-s}^2/\text{in.}^4$

Other element mass densities are obtained in a similar manner.

Table 16-4 shows the data file used in DFRAME to solve the plane frame of Figure 16-20. The order of input is based on the general description of input data provided on pp. 570–572.

In Table 16-4, line 1 is the identifying title of the problem. Line 2 indicates the numbers of elements and nodes. Lines 3–10 describe node information,

T A B L E 16-4   Input for the plane frame of Figure 16-20

Line	Data on the Line (Beginning in Column 1)
1	DYNAMIC ANALYSIS OF THE FRAME OF FIGURE 16-20
2	9,8
3	1,1,1,1,0.,0.,0.,0.,0.
4	2,1,1,1,360.,0.,0.,0.,0.
5	3,0,1,0,0.,180.,1.0,0.,0.
6	4,0,1,0,360.,180.,0.,0.,0.
7	5,0,1,0,0.,300.,0.8,0.,0.
8	6,0,1,0,360.,300.,0.,0.,0.
9	7,0,1,0,0.,420.,0.5,0.,0.
10	8,0,1,0,360.,420.,0.,0.,0.
11	1,1,3,30.E+6,13.2,249.,4.9E-3
12	2,3,5,30.E+6,6.2,107.,0.0104

T A B L E   16-4 (*Continued*)

Line	Data on the Line (Beginning in Column 1)
13	3,5,7,30.E+6,6.2,107.,0.0104
14	4,7,8,30.E+6,12.3,133.,0.01315
15	5,5,6,30.E+6,24.7,237.,0.0136
16	6,3,4,30.E+6,24.7,237.,0.0136
17	7,6,8,30.E+6,6.2,107.,0.0104
18	8,4,6,30.E+6,6.2,107.,0.0104
19	9,2,4,30.E+6,13.2,249.,4.9E-3
20	0.05,15
21	.25,.5
22	2
23	25000.,0.,0.,5000.,0.2

T A B L E   16-5   Solution for node 8 of the plane frame of Figure 16-20

t, s	d_{8x}, in.	$\dot{d}_{8x}$, in./s	$\ddot{d}_{8x}$, in./s^2
0	0	0	0
0.05	0.00557	0.223	8.918
0.10	0.0336	0.898	18.08
0.15	0.1057	1.986	25.46
0.20	0.2434	3.522	35.96
0.25	0.4631	5.267	33.85
0.30	0.7664	6.864	30.01
0.35	1.140	8.073	18.37
0.40	1.555	8.551	0.721
0.45	1.973	8.163	−16.23
0.50	2.353	7.015	−29.67
0.55	2.665	5.456	−32.72
0.60	2.895	3.768	−34.40
0.65	3.041	2.071	−33.06
0.70	3.102	0.3507	−35.76
0.75	3.074	−1.441	−35.91

where the last three numbers are now the percentage of the forcing function applied in the x and y directions and the applied couple at a node expressed in decimal form. Also, we assume translational motion in the y direction to be prevented, as indicated by IFIX = 1 for this direction. Lines 11–19 describe element information, where the shear modulus G and the torsional rigidity XJ are not included here because the option for grid analysis is not included in DFRAME. Line 20 indicates the time step and number of time steps. Line 21 indicates the value of BETA and GAMMA. Line 22 indicates the choice of the linear forcing function. Line 23 indicates the slope of section 1 of the linear function, the y intercept of the first section, the slope of section 2 of the linear function, the y intercept of the second section, and the time at which the two sections connect.

The computer program results are given in Table 16-5. For brevity, only the displacement, velocity, and acceleration in the x direction at node 8 are presented.   ∎

16.8 Time-Dependent Heat Transfer

In this section, we consider the time-dependent heat conduction problem in one dimension only. The basic differential equation for time-dependent heat conduction in one dimension was given previously by Eq. (13.1.7) with the boundary conditions given by Eqs. (13.1.10) and (13.1.11).

The finite element formulation of the equations can be obtained by minimization of the following functional:

$$\pi_h = \frac{1}{2} \int\int\int_V \left[K_{xx} \left(\frac{\partial T}{\partial x} \right)^2 - 2(Q - c\rho\dot{T})T \right] dV$$

$$- \int\int_{S_2} q^*T \, dS + \frac{1}{2} \int\int_{S_3} h(T - T_\infty)^2 \, ds \qquad (16.8.1)$$

Equation (16.8.1) is similar to Eq. (13.4.2) except only the one-dimensional case is considered and the Q term is now replaced with

$$Q - c\rho\dot{T} \qquad (16.8.2)$$

where, again, c is the specific heat of the material, and the dot over the variable T denotes differentiation with respect to time. Again, Eq. (13.5.10) obtained in Section 13.5 for the conductivity or stiffness matrix and Eqs. (13.5.11)–(13.5.13) for the force matrix terms are applicable here.

The term given by Eq. (16.8.2) yields an additional contribution to the basic element equations previously obtained for the time-independent problem as follows:

$$\Omega_Q = - \int\int\int_V T(Q - c\rho\dot{T}) \, dV \qquad (16.8.3)$$

Again, the temperature function is given by

$$\{T\} = [N]\{\hat{t}\} \qquad (16.8.4)$$

where $[N]$ is the shape function matrix given by Eq. (13.5.3) or Eqs. (16.2.3) for the simple one-dimensional element, and $\{\hat{t}\}$ is the nodal temperature matrix. Substituting Eq. (16.8.4) into Eq. (16.8.3) and differentiating with respect to time where indicated yields

$$\Omega_Q = - \iiint_V ([N]\{\hat{t}\}Q - c\rho[N]\{\hat{t}\}[N]\{\dot{\hat{t}}\})\,dV \qquad (16.8.5)$$

where the fact that $[N]$ is a function only of the coordinate system has been taken into account. Equation (16.8.5) must be minimized with respect to the nodal temperatures as follows:

$$\frac{\partial \Omega_Q}{\partial \{\hat{t}\}} = - \iiint_V [N]^T Q\,dV + \iiint_V c\rho[N]^T[N]\,dV\{\dot{\hat{t}}\} \qquad (16.8.6)$$

where we have assumed that $\{\dot{\hat{t}}\}$ remains constant during the differentiation with respect to $\{\hat{t}\}$. Equation (16.8.6) results in the additional time-dependent term added to Eq. (13.4.15). Hence, using previous definitions for the stiffness and force matrices, we obtain the element equations as

$$\{\hat{f}\} = [\hat{k}]\{\hat{t}\} + [\hat{m}]\{\dot{\hat{t}}\} \qquad (16.8.7)$$

where now

$$[\hat{m}] = \iiint_V c\rho[N]^T[N]\,dV \qquad (16.8.8)$$

For an element with constant cross-sectional area A, the differential volume is $dV = A\,d\hat{x}$. Substituting the one-dimensional shape function matrix Eq. (13.5.3) into Eq. (16.8.8) yields

$$[\hat{m}] = c\rho A \int_0^L \left\{ \begin{matrix} 1 - \dfrac{\hat{x}}{L} \\[2mm] \dfrac{\hat{x}}{L} \end{matrix} \right\} \left[1 - \dfrac{\hat{x}}{L} \quad \dfrac{\hat{x}}{L} \right] d\hat{x}$$

or

$$[\hat{m}] = \frac{c\rho AL}{6} \begin{bmatrix} 2 & 1 \\ 1 & 2 \end{bmatrix} \qquad (16.8.9)$$

Equation (16.8.9) is analogous to the consistent-mass matrix Eq. (16.2.23). The lumped-mass matrix for the heat conduction problem is then

$$[\hat{m}] = \frac{c\rho AL}{2} \begin{bmatrix} 1 & 0 \\ 0 & 1 \end{bmatrix} \qquad (16.8.10)$$

which is analogous to Eq. (16.2.12) for the one-dimensional stress element.

The time-dependent heat-transfer problem can now be solved in a manner analogous to that for the stress analysis problem. We present the numerical time integration scheme.

Numerical Time Integration

The numerical time integration method described here is similar to Newmark's method used for structural dynamic analysis and can be used to solve time-dependent or transient heat-transfer problems.

We begin by assuming that two temperature states $\underline{T}_i$ at time t_i and $\underline{T}_{i+1}$

at time t_{i+1} are related by

$$\underline{T}_{i+1} = \underline{T}_i + [(1 - \beta)\underline{\dot{T}}_i + \beta\underline{\dot{T}}_{i+1}](\Delta t) \qquad (16.8.11)$$

Equation (16.8.11) is known as the *generalized trapezoid rule*. Similar to Newmark's method for numerical time integration of the second-order equations of structural dynamics, Eq. (16.8.11) includes a parameter β that is chosen by the user.

Next we express Eq. (16.8.7) in global form as

$$\{F\} = [K]\{T\} + [M]\{\dot{T}\} \qquad (16.8.12)$$

We now write Eq. (16.8.12) for time t_i and then for time t_{i+1}. We then multiply the first of these two equations by $1 - \beta$ and the second by β to obtain

$$(1 - \beta)(\underline{K}\underline{T}_i + \underline{M}\underline{\dot{T}}_i) = (1 - \beta)\underline{F}_i \qquad (16.8.13a)$$

$$\beta(\underline{K}\underline{T}_{i+1} + \underline{M}\underline{\dot{T}}_{i+1}) = \beta\underline{F}_{i+1} \qquad (16.8.13b)$$

Next we add Eqs. (16.8.13a and b) together to obtain

$$\underline{M}[(1 - \beta)\underline{\dot{T}}_i + \beta\underline{\dot{T}}_{i+1}] + \underline{K}[(1 - \beta)\underline{T}_i + \beta\underline{T}_{i+1}] = (1 - \beta)\underline{F}_i + \beta\underline{F}_{i+1} \qquad (16.8.14)$$

Now using Eq. (16.8.11), we can eliminate the time derivative terms from Eq. (16.8.14) to write

$$\frac{\underline{M}(\underline{T}_{i+1} - \underline{T}_i)}{\Delta t} + \underline{K}[(1 - \beta)\underline{T}_i + \beta\underline{T}_{i+1}] = (1 - \beta)\underline{F}_i + \beta\underline{F}_{i+1} \qquad (16.8.15)$$

Rewriting Eq. (16.8.15) by grouping the $\underline{T}_{i+1}$ terms on the left side, we have

$$\left(\frac{1}{\Delta t}\underline{M} + \beta\underline{K}\right)\underline{T}_{i+1} = \left[\frac{1}{\Delta t}\underline{M} - (1 - \beta)\underline{K}\right]\underline{T}_i + (1 - \beta)\underline{F}_i + \beta\underline{F}_{i+1} \qquad (16.8.16)$$

The time integration to solve for $\underline{T}$ begins as follows. Given a known initial temperature $\underline{T}_0$ at time $t = 0$ and a time step Δt, we solve Eq. (16.8.16) for $\underline{T}_1$ at $t = \Delta t$. Then using $\underline{T}_1$, we determine $\underline{T}_2$ at $t = 2(\Delta t)$, and so on. For a constant Δt, the left-side coefficient of $\underline{T}_{i+1}$ need be evaluated only one time (assuming $\underline{M}$ and $\underline{K}$ do not vary with time). The matrix Eq. (16.8.16) can then be solved in the usual manner, such as by Gauss elimination. For a one-dimensional heat-transfer analysis, element $\underline{k}$ is given by Eqs. (13.5.10) and (13.5.16), whereas f is given by Eqs. (13.5.14) and (13.5.17).

It has been shown that depending on the value of β, the time step Δt may have an upper limit for the numerical analysis to be stable. If $\beta < \frac{1}{2}$, the largest Δt for stability as shown in Reference [12] is

$$\Delta t = \frac{2}{(1 - 2\beta)\lambda_{\max}} \qquad (16.8.17)$$

where $\lambda_{\max}$ is the largest eigenvalue of

$$(\underline{K} - \lambda\underline{M})\underline{T}' = 0 \qquad (16.8.18)$$

in which, as in Eq. (16.4.2), we have

$$\underline{T}(t) = \underline{T}'e^{i\lambda t} \qquad (16.8.19)$$

with $\underline{T}'$ representing the natural modes. If $\beta \geq \frac{1}{2}$, the numerical analysis is unconditionally stable; that is, stability of solution (but not accuracy) is guaranteed for Δt greater than that given by Eq. (16.8.17), or as Δt becomes indefinitely large. Various numerical integration methods result depending on specific values of β:

$\beta = 0$: Forward difference, or Euler [3], which is said to be conditionally stable (that is, Δt must be no greater than that given by Eq. (16.8.17) to obtain a stable solution).

$\beta = \frac{1}{2}$: Crank-Nicolson, or trapezoid, rule, which is unconditionally stable.
$\beta = \frac{2}{3}$: Galerkin, which is unconditionally stable.
$\beta = 1$: Backward difference, which is unconditionally stable.

If $\beta = 0$, the numerical integration method is called *explicit*; that is, we can solve for $\underline{T}_{i+1}$ directly at time Δt knowing only previous information at $t = \underline{T}_i$. If $\beta > 0$, the method is called implicit. If a diagonal mass–type matrix $\underline{M}$ exists and $\beta = 0$, the computational effort for each time step is small (see Example 16.4, where a lumped-mass matrix was used), but so must be Δt. The choice of $\beta > \frac{1}{2}$ is often used. However, if $\beta = \frac{1}{2}$ and sharp transients exist, the method generates spurious oscillations in the solution. Using $\beta > \frac{1}{2}$, along with smaller Δt [12], is probably better. Example 16.8 illustrates the solution of a one-dimensional time-dependent heat-transfer problem using the numerical time integration scheme [see Eq. (16.8.16)].

E X A M P L E   16.8

A circular fin is made of pure copper with a thermal conductivity of $k = 400$ W/(m · °C), $h = 150$ W/(m² · °C), mass density $\rho = 8900$ kg/m³, and specific heat $c = 375$ J/(kg · °C) (1 joule (J) = 1 watt · second). The initial temperature of the fin is 25°C. The fin length is 2 cm and the diameter is 0.4 cm. The right tip of the fin is insulated. The base of the fin is then suddenly increased to a temperature of 85°C and maintained at this temperature. Use the consistent form of the capacitance matrix, a time step of 0.1 s, and $\beta = \frac{2}{3}$. Use two elements of equal length. Determine the temperature distribution up to 3 s.

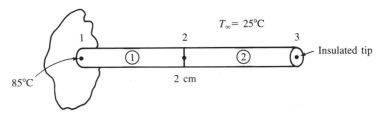

F I G U R E   16-21   Rod subjected to time-dependent temperature

Using Eq. (13.5.10), the stiffness matrix is

$$
\underline{k}^{(1)} = \underline{k}^{(2)} = \frac{AK_{xx}}{L} \overset{\begin{matrix}1 & 2\\2 & 3\end{matrix}}{\begin{bmatrix} 1 & -1 \\ -1 & 1 \end{bmatrix}} + \frac{hPL}{6} \overset{\begin{matrix}1 & 2\\2 & 3\end{matrix}}{\begin{bmatrix} 2 & 1 \\ 1 & 2 \end{bmatrix}}
$$

$$
\underline{k}^{(1)} = \underline{k}^{(2)} = \frac{\pi(0.004)^2(400)}{4(0.01)} \begin{bmatrix} 1 & -1 \\ -1 & 1 \end{bmatrix} + \frac{150(2\pi)(0.002)(0.01)}{6} \begin{bmatrix} 2 & 1 \\ 1 & 2 \end{bmatrix}
$$

$$(16.8.20)$$

Assembling the element stiffness matrices, Eq. (16.8.20), we obtain the global stiffness matrix as

$$
\underline{K} = \begin{bmatrix} 0.50894 & -0.49951 & 0 \\ -0.49951 & 1.01788 & -0.49951 \\ 0 & -0.49951 & 0.50894 \end{bmatrix} \qquad (16.8.21)
$$

Using Eq. (13.5.12), we obtain each element force matrix as

$$
\{f_h^{(1)}\} = \{f_h^{(2)}\} = \frac{hT_\infty PL}{2} \begin{Bmatrix} 1 \\ 1 \end{Bmatrix} = \frac{(150)(25°C)(2\pi)(0.002)(0.01)}{2} \begin{Bmatrix} 1 \\ 1 \end{Bmatrix}
$$

$$
\underline{f}_h^{(1)} = \underline{f}_h^{(2)} = \begin{Bmatrix} 0.23561 \\ 0.23561 \end{Bmatrix} \qquad (16.8.22)
$$

Using Eq. (16.8.22), the assembled global force matrix is

$$
\{F\} = \begin{Bmatrix} 0.23561 \\ 0.47122 \\ 0.23561 \end{Bmatrix} \text{W} \qquad (16.8.23)
$$

Next using Eq. (16.8.9), we obtain each element mass (capacitance) matrix as

$$
[m] = \frac{c\rho AL}{6} \begin{bmatrix} 2 & 1 \\ 1 & 2 \end{bmatrix}
$$

$$
\underline{m}^{(1)} = \underline{m}^{(2)} = \frac{(375)(8900)\dfrac{\pi(0.004)^2}{4}(0.01)}{6} \begin{bmatrix} 2 & 1 \\ 1 & 2 \end{bmatrix}
$$

$$
= 0.06990 \begin{bmatrix} 2 & 1 \\ 1 & 2 \end{bmatrix} \text{W} \cdot \text{s/°C} \qquad (16.8.24)
$$

Using Eq. (16.8.24), the assembled capacitance matrix is

$$
\underline{M} = \begin{bmatrix} 0.13980 & 0.06990 & 0 \\ 0.06990 & 0.27960 & 0.06990 \\ 0 & 0.06990 & 0.13980 \end{bmatrix} \qquad (16.8.25)
$$

Using Eq. (16.8.16) and Eqs. (16.8.21) and (16.8.25), we obtain

$$\left(\frac{1}{\Delta t}\underline{M} + \beta\underline{K}\right) = \begin{bmatrix} 1.7374 & 0.36603 & 0 \\ 0.36603 & 3.4747 & 0.36603 \\ 0 & 0.36603 & 1.7374 \end{bmatrix} \qquad (16.8.26)$$

and

$$\left[\frac{1}{\Delta t}\underline{M} - (1 - \beta)\underline{K}\right] = \begin{bmatrix} 1.2280 & 0.8655 & 0 \\ 0.8655 & 2.457 & 0.8655 \\ 0 & 0.8655 & 1.2280 \end{bmatrix} \qquad (16.8.27)$$

where $\beta = \frac{2}{3}$ and $\Delta t = 0.1$ s have been used to obtain Eqs. (16.8.26) and (16.8.27). For the first time step, $t = 0.1$ s, we then use Eqs. (16.8.23), (16.8.27), and (16.8.26) in Eq. (16.8.16) to obtain

$$\begin{bmatrix} 1.7374 & 0.36603 & 0 \\ 0.36603 & 3.4747 & 0.36603 \\ 0 & 0.36603 & 1.7374 \end{bmatrix} \begin{Bmatrix} 85°C \\ t_2 \\ t_3 \end{Bmatrix}$$

$$= \begin{bmatrix} 1.2280 & 0.8655 & 0 \\ 0.8655 & 2.457 & 0.8655 \\ 0 & 0.8655 & 1.2280 \end{bmatrix} \begin{Bmatrix} 25°C \\ 25°C \\ 25°C \end{Bmatrix} + \begin{Bmatrix} 0.23561 \\ 0.47122 \\ 0.23561 \end{Bmatrix} \qquad (16.8.28)$$

In Eq. (16.8.28), we should note that since $\underline{F}_i = \underline{F}_{i+1}$ for all time, the sum of the terms is $(1 - \beta)\underline{F}_i + \beta\underline{F}_{i+1} = \underline{F}_i$ for all time. This is the column matrix on the right side of Eq. (16.8.28). We now solve Eq. (16.8.28) in the usual manner by partitioning Eqs. 2 and 3 of Eq. (16.8.28) from Eq. 1 and solving Eqs. 2 and 3 simultaneously for t_2 and t_3. The results are

$$t_2 = 18.534°C \quad \text{and} \quad t_3 = 26.371°C$$

At time $t = 0.2$ s, Eq. (16.8.28) becomes

$$\begin{bmatrix} 1.7374 & 0.36603 & 0 \\ 0.36603 & 3.4747 & 0.36603 \\ 0 & 0.36603 & 1.7374 \end{bmatrix} \begin{Bmatrix} 85°C \\ t_2 \\ t_3 \end{Bmatrix}$$

$$= \begin{bmatrix} 1.2280 & 0.8655 & 0 \\ 0.8655 & 2.457 & 0.8655 \\ 0 & 0.8655 & 1.2280 \end{bmatrix} \begin{Bmatrix} 85°C \\ 18.534°C \\ 26.371°C \end{Bmatrix} + \begin{Bmatrix} 0.23561 \\ 0.47122 \\ 0.23561 \end{Bmatrix} \qquad (16.8.29)$$

Solving Eq. (16.8.29) for t_2 and t_3, we obtain

$$t_2 = 29.732°C \quad \text{and} \quad t_3 = 21.752°C$$

The results through a time of 3 s are tabulated in Table 16-6 and plotted in Figure 16-22.

T A B L E 16-6   Nodal temperatures at various times for Example 16.8

| Time, s | Temperature, °C, of Node Numbers | | |
	1	2	3
0.1	85	18.534	26.371
0.2	85	29.732	21.752
0.3	85	36.404	22.662
0.4	85	41.032	25.655
0.5	85	44.665	29.312
0.6	85	47.749	33.059
0.7	85	50.482	36.669
0.8	85	52.956	40.062
0.9	85	55.218	43.218
1.0	85	57.296	46.139
1.1	85	59.208	48.837
1.2	85	60.969	51.327
1.3	85	62.593	53.623
1.4	85	64.089	55.741
1.5	85	65.469	57.693
1.6	85	66.742	59.493
1.7	85	67.915	61.152
1.8	85	68.996	62.683
1.9	85	69.993	64.094
2.0	85	70.912	65.395
2.1	85	71.760	66.594
2.2	85	72.542	67.700
2.3	85	73.262	68.720
2.4	85	73.926	69.660
2.5	85	74.539	70.527
2.6	85	75.104	71.326
2.7	85	75.624	72.063
2.8	85	76.104	72.742
2.9	85	76.547	73.368
3.0	85	76.955	73.946

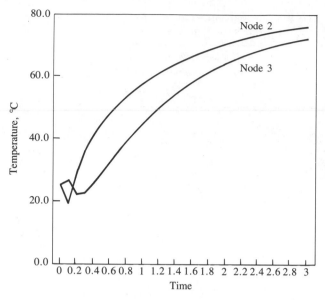

F I G U R E   16-22   Temperature as a function of time for nodes 2 and 3 of Example 16.8

REFERENCES

[1] Thompson, W. T., *Theory of Vibrations with Applications*, 2nd ed., Prentice Hall, Englewood Cliffs, New Jersey, 1981.

[2] Archer, J. S., "Consistent Matrix Formulations for Structural Analysis Using Finite Element Techniques," *Journal of the American Institute of Aeronautics and Astronautics*, Vol. 3, No. 10, pp. 1910–1918, 1965.

[3] James, M. L., Smith, G. M., and Wolford, J. C., *Applied Numerical Methods for Digital Computation*, 3rd ed., Harper & Row, New York, 1985.

[4] Biggs, J. M., *Introduction to Structural Dynamics*, McGraw-Hill, New York, 1964.

[5] Newmark, N. M., "A Method of Computation for Structural Dynamics," *Journal of the Engineering Mechanics Division*, American Society of Civil Engineers, Vol. 85, No. EM3, pp. 67–94, 1959.

[6] Clark, S. K., *Dynamics of Continuous Elements*, Prentice Hall, Englewood Cliffs, New Jersey, 1972.

[7] Bathe, K. J., *Finite Element Procedures in Engineering Analysis*, Prentice Hall, Englewood Cliffs, New Jersey, 1982.

[8] Bathe, K. J., and Wilson, E. L., *Numerical Methods in Finite Element Analysis*, Prentice Hall, Englewood Cliffs, New Jersey, 1976.

[9] Fujii, H., "Finite Element Schemes: Stability and Convergence," *Advances in Computational Methods in Structural Mechanics and Design*, J. T. Oden, R. W.

Clough, and Y. Yamamoto, Eds., University of Alabama Press, Alabama, pp. 201–218, 1972.

[10] Krieg, R. D., and Key, S. W., "Transient Shell Response by Numerical Time Integration," *International Journal of Numerical Methods in Engineering*, Vol. 17, pp. 273–286, 1973.

[11] Belytschko, T., "Transient Analysis," *Structural Mechanics Computer Programs, Surveys, Assessments, and Availability*, W. Pilkey, K. Saczalski, and H. Schaeffer, Eds., University of Virginia Press, Charlottesville, Virginia, pp. 255–276, 1974.

[12] Hughes, T. J. R., "Unconditionally Stable Algorithms for Nonlinear Heat Conduction," *Computational Methods in Applied Mechanical Engineering*, Vol. 10, No. 2, pp. 135–139, 1977.

[13] Paz, M., *Structural Dynamics Theory and Computation*, Van Nostrand Reinhold, New York, 1985.

PROBLEMS

16.1 Determine the consistent-mass matrix for the one-dimensional bar discretized into two elements as shown in Figure P16-1. Let the bar have modulus of elasticity E, mass density ρ, and cross-sectional area A.

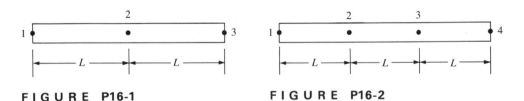

FIGURE P16-1   **FIGURE P16-2**

16.2 For the one-dimensional bar discretized into three elements as shown in Figure P16-2, determine the lumped- and consistent-mass matrices. Let the bar properties be E, ρ, and A throughout the bar.

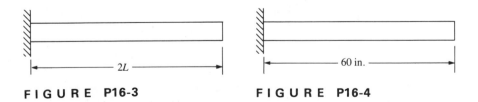

FIGURE P16-3   **FIGURE P16-4**

16.3 For the one-dimensional bar shown in Figure P16-3, determine the natural frequencies of vibration, ω's, using two elements of equal length. Use the consistent mass approach. Let the bar have modulus of elasticity E, mass density ρ, and cross-sectional area A. Compare your answers to those obtained using a lumped-mass matrix in Example 16.3.

16.4 For the one-dimensional bar shown in Figure P16-4, determine the natural frequencies of longitudinal vibration using first two and then three elements of equal length. Let the bar have $E = 30 \times 10^6$ psi, $\rho = 0.00073$ lb-s^2/in.^{4}, $A = 1$ in.^{2}, and $L = 60$ in.

16.5 For the spring-mass system shown in Figure P16-5, determine the mass displacement, velocity, and acceleration for five time steps using the central difference method. Let $k = 2000$ lb/ft and $m = 2$ slugs. Use a time step of $\Delta t = 0.03$ s. You might want to write a computer program to solve this problem.

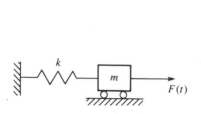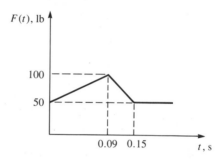

FIGURE P16-5

16.6 For the spring-mass system shown in Figure P16-6, determine the mass displacement, velocity, and acceleration for five time steps using (a) the central difference method and (b) Newmark's time integration method. Let $k = 1200$ lb/ft and $m = 2$ slugs.

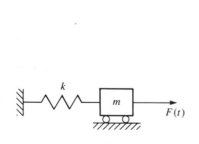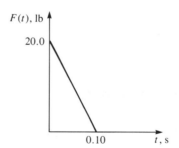

FIGURE P16-6

16.7 For the bar shown in Figure P16-7, determine the nodal displacements, velocities, and accelerations for five time steps using two finite elements. Let $E = 30 \times 10^6$ psi, $\rho = 0.00073$ lb-s^2/in.^{4}, $A = 1$ in.^{2}, and $L = 100$ in.

16.8 For the bar shown in Figure P16-8, determine the nodal displacements, velocities, and accelerations for five time steps using two finite elements. For simplicity of calculations, let $E = 1 \times 10^6$ psi, $\rho = 1$ lb-s^2/in.^{4}, $A = 1$ in.^{2}, and $L = 100$ in. Use Newmark's method.

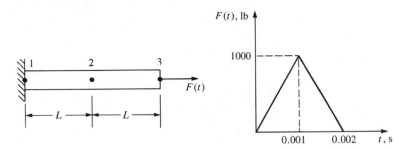

FIGURE P16-7

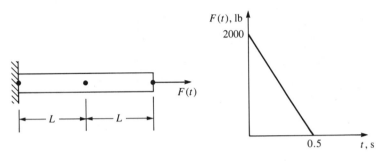

FIGURE P16-8

16.9 For the beams shown in Figure P16-9, determine the natural frequencies using first two and then three elements. Let E, ρ, and A be constant for the beams.

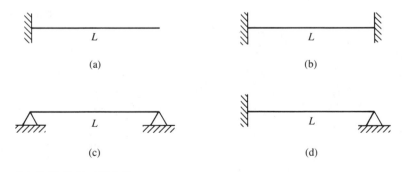

FIGURE P16-9

16.10, 16.11 For the beams in Figures P16-10 and P16-11 subjected to the forcing functions shown, determine the maximum deflections, velocities, and accelerations. Use program DFRAME or any other suitable program.

16.12, 16.13 For the rigid frames in Figures P16-12 and P16-13 subjected to the forcing functions shown, determine the maximum displacements, velocities, and accelerations. Use program DFRAME or any other suitable program.

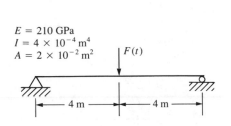

$E = 210$ GPa
$I = 4 \times 10^{-4}$ m^4
$A = 2 \times 10^{-2}$ m^2

4 m   4 m

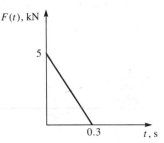

F I G U R E   P16-10

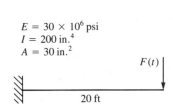

$E = 30 \times 10^6$ psi
$I = 200$ in.^{4}
$A = 30$ in.^{2}

20 ft

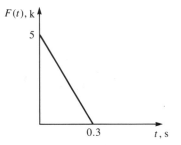

F I G U R E   P16-11

For elements 1 and 9,
  $A = 13$ in.^{2}, $I = 250$ in.^{4}
For elements 2, 3, 7, and 8,
  $A = 6$ in.^{2}, $I = 100$ in.^{4}
For elements 4, 5, and 6,
  $A = 14$ in.^{2}, $I = 800$ in.^{4}
For all elements,
  $E = 30 \times 10^6$ psi

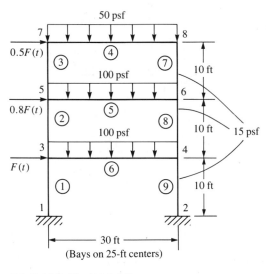

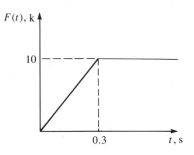

F I G U R E   P16-12

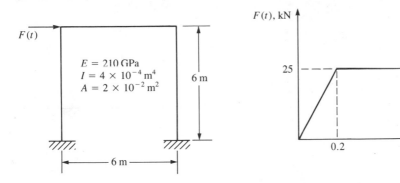

FIGURE P16-13

16.14 Develop the consistent-mass matrix for the plane frame element.

16.15 Describe how to develop the consistent-mass matrix for the constant-strain triangle of Chapter 7.

16.16 A marble slab with $k = 2$ W/(m·°C), $\rho = 2500$ kg/m³, and $c = 800$ W·s/(kg·°C) is 2 cm thick and at an initial uniform temperature of $T_i = 200°C$. The left surface is suddenly lowered to 0°C and is maintained at that temperature while the other surface is kept insulated. Determine the temperature distribution in the slab for 40 s. Use $\beta = \frac{2}{3}$ and a time step of 8 s.

16.17 A circular fin is made of pure copper with a thermal conductivity of $k = 400$ W/(m·°C), $h = 150$ W/(m²·°C), mass density $\rho = 8900$ kg/m³, and specific heat $c = 375$ J/(kg·°C). The initial temperature of the fin is 25°C. The fin length is 2 cm and the diameter is 0.4 cm. The right tip of the fin is insulated. See Figure P16-17. The base of the fin is then suddenly increased to a temperature of 85°C and maintained at this temperature. Use the lumped form of the capacitance matrix, a time step of 0.1 s, and $\beta = \frac{2}{3}$. Use two elements of equal length. Determine the temperature distribution up to 3 s. Compare your results with Example 16.8, which used the consistent form of the capacitance matrix.

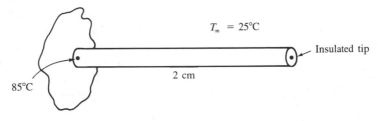

FIGURE P16-17

Matrix Algebra

INTRODUCTION

In this appendix, we provide an introduction to matrix algebra. We will consider the concepts relevant to the finite element method to provide an adequate background for the matrix algebra concepts used in this text.

$A.1$ Definition of a Matrix

A **matrix** *is an m × n array of numbers arranged in m rows and n columns.* The matrix is then described as being of order $m \times n$. Equation (A.1.1) illustrates a matrix with *m* rows and *n* columns.

$$[a] = \begin{bmatrix} a_{11} & a_{12} & a_{13} & a_{14} & \cdots & a_{1n} \\ a_{21} & a_{22} & a_{23} & a_{24} & \cdots & a_{2n} \\ a_{31} & a_{32} & a_{33} & a_{34} & \cdots & a_{3n} \\ \vdots & \vdots & \vdots & \vdots & & \vdots \\ a_{m1} & a_{m2} & a_{m3} & a_{m4} & \cdots & a_{mn} \end{bmatrix} \qquad (A.1.1)$$

If $m \neq n$ in matrix Eq. (A.1.1), the matrix is called **rectangular**. If $m = 1$ and $n > 1$, the elements of Eq. (A.1.1) form a single row called a **row matrix**. If $m > 1$ and $n = 1$, the elements form a single column called a **column matrix**. If $m = n$, the array is called a **square matrix**. Row matrices and rectangular matrices are denoted by using brackets [ ], and column matrices are denoted by using braces { }. For simplicity, matrices (row, column, or rectangular) are often denoted by using a line under a variable instead of surrounding it with brackets or braces. The order of the matrix should then be apparent from the context of its use. The force and displacement matrices used in structural analysis are column matrices, whereas the stiffness matrix is a square matrix.

To identify an element of matrix $\underline{a}$, we represent the element by a_{ij}, where the subscripts i and j indicate the row number and the column number, respectively, of $\underline{a}$. Hence, alternative notations for a matrix are given by

$$\underline{a} = [a] = [a_{ij}] \qquad (A.1.2)$$

Numerical examples of special types of matrices are given by Eqs. (A.1.3)–(A.1.6). A rectangle matrix $\underline{a}$ is given by

$$\underline{a} = \begin{bmatrix} 2 & 1 \\ 3 & 4 \\ 5 & 4 \end{bmatrix} \qquad (A.1.3)$$

where $\underline{a}$ has three rows and two columns. In matrix $\underline{a}$ of Eq. (A.1.1), if $m = 1$, a row matrix results, such as

$$\underline{a} = [2 \quad 3 \quad 4 \quad -1] \qquad (A.1.4)$$

If $n = 1$ in Eq. (A.1.1), a column matrix results, such as

$$\underline{a} = \begin{Bmatrix} 2 \\ 3 \end{Bmatrix} \qquad (A.1.5)$$

If $m = n$ in Eq. (A.1.1), a square matrix results, such as

$$\underline{a} = \begin{bmatrix} 2 & -1 \\ 3 & -2 \end{bmatrix} \qquad (A.1.6)$$

Matrices and matrix notation are often used to express algebraic equations in compact form, and are frequently used in the finite element formulation of equations. Matrix notation is also used to simplify the solution of a problem.

A.2 Matrix Operations

We will now present some common matrix operations that will be used in this text.

Multiplication of a Matrix by a Scalar

If we have a scalar k and a matrix $\underline{c}$, then the product $\underline{a} = k\underline{c}$ is given by

$$\underline{a}_{ij} = k\underline{c}_{ij} \qquad (A.2.1)$$

—that is, every element of the matrix $\underline{c}$ is multiplied by the scalar k. As a numerical example, consider

$$\underline{c} = \begin{bmatrix} 1 & 2 \\ 3 & 1 \end{bmatrix} \qquad k = 4$$

The product $\underline{a} = k\underline{c}$ is

$$\underline{a} = 4\begin{bmatrix} 1 & 2 \\ 3 & 1 \end{bmatrix} = \begin{bmatrix} 4 & 8 \\ 12 & 4 \end{bmatrix}$$

Note that if $\underline{c}$ is of order $m \times n$, then $\underline{a}$ is also of order $m \times n$.

Addition of Matrices

Matrices of the same order can be added together by summing corresponding elements of the matrices. Subtraction is performed in a similar manner. Matrices of unlike order cannot be added or subtracted. Matrices of the same order can be added (or subtracted) in any order (the commutative law for addition applies); that is

$$\underline{c} = \underline{a} + \underline{b} = \underline{b} + \underline{a} \qquad (A.2.2)$$

or, in subscript (index) notation, we have

$$[c_{ij}] = [a_{ij}] + [b_{ij}] = [b_{ij}] + [a_{ij}] \qquad (A.2.3)$$

As a numerical example, let

$$\underline{a} = \begin{bmatrix} -1 & 2 \\ -3 & 2 \end{bmatrix} \qquad \underline{b} = \begin{bmatrix} 1 & 2 \\ 3 & 1 \end{bmatrix}$$

The sum $\underline{a} + \underline{b} = \underline{c}$ is given by

$$\underline{c} = \begin{bmatrix} -1 & 2 \\ -3 & 2 \end{bmatrix} + \begin{bmatrix} 1 & 2 \\ 3 & 1 \end{bmatrix} = \begin{bmatrix} 0 & 4 \\ 0 & 3 \end{bmatrix}$$

Again, remember that the matrices $\underline{a}$, $\underline{b}$, and $\underline{c}$ must all be of the same order. For instance, a 2×2 matrix cannot be added to a 3×3 matrix.

Multiplication of Matrices

To multiply two matrices $\underline{a}$ and $\underline{b}$ in the order shown in Eq. (A.2.4), the number of columns in $\underline{a}$ must equal the number of rows in $\underline{b}$. For example, consider

$$\underline{c} = \underline{a}\underline{b} \qquad (A.2.4)$$

If $\underline{a}$ is an $m \times n$ matrix, then $\underline{b}$ must have n rows. Using subscript notation, we can write the product of matrices $\underline{a}$ and $\underline{b}$ as

$$[c_{ij}] = \sum_{e=1}^{n} a_{ie} b_{ej} \qquad (A.2.5)$$

where n is the total number of columns in $\underline{a}$ or of rows in $\underline{b}$. For matrix $\underline{a}$ of order 2×2 and matrix $\underline{b}$ of order 2×2, after multiplying the two matrices, we have

$$[c_{ij}] = \begin{bmatrix} a_{11}b_{11} + a_{12}b_{21} & a_{11}b_{12} + a_{12}b_{22} \\ a_{21}b_{11} + a_{22}b_{21} & a_{21}b_{12} + a_{22}b_{22} \end{bmatrix} \qquad (A.2.6)$$

For example, let

$$\underline{a} = \begin{bmatrix} 2 & 1 \\ 3 & 2 \end{bmatrix} \qquad \underline{b} = \begin{bmatrix} 1 & -1 \\ 2 & 0 \end{bmatrix}$$

The product of $\underline{a}\underline{b}$ is then

$$\underline{a}\underline{b} = \begin{bmatrix} 2(1) + 1(2) & 2(-1) + 1(0) \\ 3(1) + 2(2) & 3(-1) + 2(0) \end{bmatrix} = \begin{bmatrix} 4 & -2 \\ 7 & -3 \end{bmatrix}$$

In general, matrix multiplication is *not* commutative; that is,

$$\underline{a}\underline{b} \neq \underline{b}\underline{a} \qquad (A.2.7)$$

The validity of the product of two matrices $\underline{a}$ and $\underline{b}$ is commonly illustrated by

$$\begin{array}{ccc} \underline{a} & \underline{b} & = & \underline{c} \\ (i \times e)(e \times j) & & (i \times j) \end{array} \qquad (A.2.8)$$

where the product matrix $\underline{c}$ will be of order $i \times j$; that is, it will have the same number of rows as matrix $\underline{a}$ and the same number of columns as matrix $\underline{b}$.

Transpose of a Matrix

Any matrix, whether a row, column, or rectangular matrix, can be tranposed. This operation is frequently used in finite element equation formulations. The transpose of a matrix $\underline{a}$ is commonly denoted by $\underline{a}^T$. The superscript T will be used to denote the transpose of a matrix throughout this text. The transpose of a matrix is obtained by interchanging rows and columns; that is, the first row becomes the first column, the second row becomes the second column, and so on. For the transpose of matrix $\underline{a}$,

$$[a_{ij}] = [a_{ji}]^T \qquad (A.2.9)$$

For example, if we let

$$\underline{a} = \begin{bmatrix} 2 & 1 \\ 3 & 2 \\ 4 & 5 \end{bmatrix}$$

then

$$\underline{a}^T = \begin{bmatrix} 2 & 3 & 4 \\ 1 & 2 & 5 \end{bmatrix}$$

where we have interchanged the rows and columns of $\underline{a}$ to obtain its transpose.

Another important relationship using the transpose is

$$(\underline{a}\underline{b})^T = \underline{b}^T\underline{a}^T \qquad (A.2.10)$$

That is, the transpose of the product of matrices $\underline{a}$ and $\underline{b}$ is equal to the transpose of the latter matrix $\underline{b}$ multiplied by the transpose of matrix $\underline{a}$ in that order, provided the order of the initial matrices continues to satisfy the rule for matrix mutiplication, Eq. (A.2.8). In general, this property holds for any number of matrices; that is,

$$(\underline{a}\underline{b}\underline{c} \ldots \underline{k})^T = \underline{k}^T \ldots \underline{c}^T\underline{b}^T\underline{a}^T \qquad (A.2.11)$$

Note that the transpose of a column matrix is a row matrix.

As a numerical example of the use of Eq. (A.2.10), let

$$\underline{a} = \begin{bmatrix} 1 & 2 \\ 3 & 4 \end{bmatrix} \qquad \underline{b} = \begin{Bmatrix} 5 \\ 6 \end{Bmatrix}$$

First,
$$ab = \begin{bmatrix} 1 & 2 \\ 3 & 4 \end{bmatrix} \begin{Bmatrix} 5 \\ 6 \end{Bmatrix} = \begin{Bmatrix} 17 \\ 39 \end{Bmatrix}$$

Then,
$$(ab)^T = [17 \quad 39] \tag{A.2.12}$$

Since b^T and a^T can be multiplied according to the rule for matrix multiplication, we have

$$b^T a^T = [5 \quad 6] \begin{bmatrix} 1 & 3 \\ 2 & 4 \end{bmatrix} = [17 \quad 39] \tag{A.2.13}$$

Hence, on comparing Eqs. (A.2.12) and (A.2.13), we have shown (for this case) the validity of Eq. (A.2.10). A simple proof of the general validity of Eq. (A.2.10) is left to your discretion.

Symmetric Matrices

If a square matrix is equal to its transpose, it is called a **symmetric matrix**; that is, if

$$a = a^T$$

then a is a symmetric matrix. As an example,

$$a = \begin{bmatrix} 3 & 1 & 2 \\ 1 & 4 & 0 \\ 2 & 0 & 3 \end{bmatrix} \tag{A.2.14}$$

is a symmetric matrix because each element a_{ij} equals a_{ji} for $i \neq j$. In Eq. (A.2.14), note that the main diagonal running from the upper left corner to the lower right corner is the line of symmetry of the symmetric matrix a. Remember that only a square matrix can be symmetric.

Unit Matrix

The **unit** (or **identity**) **matrix** I is such that

$$aI = Ia = a \tag{A.2.15}$$

The unit matrix acts in the same way that one does in conventional multiplication. The unit matrix is always a square matrix of any possible order with each element of the main diagonal equal to one and all other elements equal to zero. For example, the 3×3 unit matrix is given by

$$I = \begin{bmatrix} 1 & 0 & 0 \\ 0 & 1 & 0 \\ 0 & 0 & 1 \end{bmatrix}$$

Inverse of a Matrix

The **inverse of a matrix** is a matrix such that

$$a^{-1}a = aa^{-1} = I \tag{A.2.16}$$

where the superscript, -1, denotes the inverse of $\underline{a}$ as $\underline{a}^{-1}$. Section A.3 provides more information regarding the properties of the inverse of a matrix and a method for its determination.

Differentiating a Matrix

A matrix is differentiated by differentiating every element in the matrix in the conventional manner. For example, if

$$\underline{a} = \begin{bmatrix} x^3 & 2x^2 & 3x \\ 2x^2 & x^4 & x \\ 3x & x & x^5 \end{bmatrix} \tag{A.2.17}$$

the derivative $d\underline{a}/dx$ is given by

$$\frac{d\underline{a}}{dx} = \begin{bmatrix} 3x^2 & 4x & 3 \\ 4x & 4x^3 & 1 \\ 3 & 1 & 5x^4 \end{bmatrix} \tag{A.2.18}$$

Similarly, the partial derivative of a matrix is illustrated as follows:

$$\frac{\partial \underline{a}}{\partial x} = \frac{\partial}{\partial x}\begin{bmatrix} x^2 & xy & xz \\ xy & y^2 & yz \\ xz & yz & z^2 \end{bmatrix} = \begin{bmatrix} 2x & y & z \\ y & 0 & 0 \\ z & 0 & 0 \end{bmatrix} \tag{A.2.19}$$

In structural analysis theory, we sometimes differentiate an expression of the form

$$U = \frac{1}{2}[x \;\; y]\begin{bmatrix} a_{11} & a_{12} \\ a_{12} & a_{22} \end{bmatrix}\begin{Bmatrix} x \\ y \end{Bmatrix} \tag{A.2.20}$$

where U might represent the strain energy in a bar. Expression (A.2.20) is known as a quadratic form. By matrix multiplication of Eq. (A.2.20), we obtain

$$U = \frac{1}{2}(a_{11}x^2 + 2a_{12}xy + a_{22}y^2) \tag{A.2.21}$$

Differentiating U now yields

$$\frac{\partial U}{\partial x} = a_{11}x + a_{12}y$$

$$\frac{\partial U}{\partial y} = a_{12}x + a_{22}y \tag{A.2.22}$$

Equation (A.2.22) in matrix form becomes

$$\begin{Bmatrix} \dfrac{\partial U}{\partial x} \\[2mm] \dfrac{\partial U}{\partial y} \end{Bmatrix} = \begin{bmatrix} a_{11} & a_{12} \\ a_{12} & a_{22} \end{bmatrix}\begin{Bmatrix} x \\ y \end{Bmatrix} \tag{A.2.23}$$

A general form of Eq. (A.2.20) is

$$U = \frac{1}{2}\{X\}^T[a]\{X\} \qquad (A.2.24)$$

Then by comparing Eq. (A.2.20) and (A.2.23), we obtain

$$\frac{\partial U}{\partial x_i} = [a]\{X\} \qquad (A.2.25)$$

where x_i denotes x and y. Here Eq. (A.2.25) depends on matrix $\underline{a}$ in Eq. (A.2.24) being symmetric.

Integrating a Matrix

Consistent with matrix differentation, to integrate a matrix, every element in the matrix must be integrated in the conventional manner. As an example, if

$$\underline{a} = \begin{bmatrix} 3x^2 & 4x & 3 \\ 4x & 4x^3 & 1 \\ 3 & 1 & 5x^4 \end{bmatrix}$$

we obtain the integration of $\underline{a}$ as

$$\int \underline{a}\, dx = \begin{bmatrix} x^3 & 2x^2 & 3x \\ 2x^2 & x^4 & x \\ 3x & x & x^5 \end{bmatrix}$$

In our finite element formulation of equations, we often integrate an expression of the form

$$\int\int [X]^T[A][X]\, dx\, dy \qquad (A.2.26)$$

The triple product in Eq. (A.2.26) will be symmetric if $\underline{A}$ is symmetric. The form $[X]^T[A][X]$ is also called a *quadratic form*. For example, letting

$$[A] = \begin{bmatrix} 9 & 2 & 3 \\ 2 & 8 & 0 \\ 3 & 0 & 5 \end{bmatrix} \qquad [X] = \begin{Bmatrix} x_1 \\ x_2 \\ x_3 \end{Bmatrix}$$

we obtain

$$\{X\}^T[A]\{X\} = [x_1 \quad x_2 \quad x_3]\begin{bmatrix} 9 & 2 & 3 \\ 2 & 8 & 0 \\ 3 & 0 & 5 \end{bmatrix}\begin{Bmatrix} x_1 \\ x_2 \\ x_3 \end{Bmatrix}$$

$$= 9x_1^2 + 4x_1 x_2 + 6x_1 x_3 + 8x_2^2 + 5x_3^2$$

which is in quadratic form.

A.3   Cofactor or Adjoint Method to Determine the Inverse of a Matrix

We will now introduce a method for finding the inverse of a matrix. This method is useful for longhand determination of the inverse of smaller-order square matrices (preferably of order 4×4 or less). A matrix $\underline{a}$ must be square to determine its inverse.

We must first define the determinant of a matrix. This concept is necessary to determine the inverse of a matrix by the cofactor method. *A **determinant** is a square array of elements expressed by*

$$|\underline{a}| = |a_{ij}| \qquad (A.3.1)$$

where the straight vertical bars, | |, on each side of the array denote the determinant. The resulting determinant of an array will be a single numerical value on evaluating the array.

To evaluate the determinant of $\underline{a}$, we must first determine the cofactors of $[a_{ij}]$. The cofactors of $[a_{ij}]$ are given by

$$C_{ij} = (-1)^{i+j}|\underline{d}| \qquad (A.3.2)$$

where the matrix $\underline{d}$, called the *first minor of* $[a_{ij}]$, is matrix $\underline{a}$ with row i and column j deleted. The inverse of matrix $\underline{a}$ is then given by

$$\underline{a}^{-1} = \frac{C^T}{|\underline{a}|} \qquad (A.3.3)$$

where $\underline{C}$ is the **cofactor matrix** and $|\underline{a}|$ is the determinant of $\underline{a}$. To illustrate the method of cofactors, we will determine the inverse of a matrix $\underline{a}$ given by

$$\underline{a} = \begin{bmatrix} -1 & 3 & -2 \\ 2 & -4 & 2 \\ 0 & 4 & 1 \end{bmatrix} \qquad (A.3.4)$$

The cofactors of matrix $\underline{a}$ are obtained on using Eq. (A.3.2) as

$$C_{11} = (-1)^{1+1} \begin{vmatrix} -4 & 2 \\ 4 & 1 \end{vmatrix} = -12$$

$$C_{12} = (-1)^{1+2} \begin{vmatrix} 2 & 2 \\ 0 & 1 \end{vmatrix} = -2$$

$$C_{13} = (-1)^{1+3} \begin{vmatrix} 2 & -4 \\ 0 & 4 \end{vmatrix} = 8$$

$$C_{21} = (-1)^{2+1} \begin{vmatrix} 3 & -2 \\ 4 & 1 \end{vmatrix} = -11 \qquad (A.3.5)$$

$$C_{22} = (-1)^{2+2} \begin{vmatrix} -1 & -2 \\ 0 & 1 \end{vmatrix} = -1$$

$$C_{23} = (-1)^{2+3} \begin{vmatrix} -1 & 3 \\ 0 & 4 \end{vmatrix} = 4$$

Similarly, $C_{31} = -2$ $C_{32} = -2$ $C_{33} = -2$ (*A.3.6*)

Therefore, from Eqs. (A.3.5) and (A.3.6), we have

$$\underline{C} = \begin{bmatrix} -12 & -2 & 8 \\ -11 & -1 & 4 \\ -2 & -2 & -2 \end{bmatrix}$$ (*A.3.7*)

The determinant of $\underline{a}$ is then

$$|\underline{a}| = \sum_{j=1}^{n} a_{ij} C_{ij} \qquad \text{with } i \text{ any row number } (1 \le i \le n) \qquad (A.3.8)$$

or $$|\underline{a}| = \sum_{j=1}^{n} a_{ji} C_{ji} \qquad \text{with } i \text{ any column number } (1 \le i \le n) \qquad (A.3.9)$$

For instance, if we choose the first rows of $\underline{a}$ and $\underline{C}$, then $i = 1$ in Eq. (A.3.8), and j is summed from 1 to 3 such that

$$\begin{aligned} |\underline{a}| &= a_{11} C_{11} + a_{12} C_{12} + a_{13} C_{13} \\ &= (-1)(-12) + (3)(-2) + (-2)(8) = -10 \end{aligned} \qquad (A.3.10)$$

Using the definition of the inverse given by Eq. (A.3.3), we have

$$\underline{a}^{-1} = \frac{\underline{C}^T}{|\underline{a}|} = \frac{1}{-10} \begin{bmatrix} -12 & -11 & -2 \\ -2 & -1 & -2 \\ 8 & 4 & -2 \end{bmatrix} \qquad (A.3.11)$$

We can then check that

$$\underline{a}\,\underline{a}^{-1} = \begin{bmatrix} 1 & 0 & 0 \\ 0 & 1 & 0 \\ 0 & 0 & 1 \end{bmatrix}$$

The transpose of the cofactor matrix is often defined as the **adjoint matrix**; that is,

$$\text{adj } \underline{a} = \underline{C}^T$$

Therefore, an alternative equation for the inverse of $\underline{a}$ is

$$\underline{a}^{-1} = \frac{\text{adj } \underline{a}}{|\underline{a}|} \qquad (A.3.12)$$

An important property associated with the determinant of a matrix is that if the determinant of a matrix is zero—that is, $|\underline{a}| = 0$—then the matrix is said to be **singular**. A singular matrix does not have an inverse. The stiffness matrices used in the finite element method are initially singular until sufficient boundary conditions (support conditions) are applied. This characteristic of the stiffness matrix is further discussed in the text.

A.4 Inverse of a Matrix by Row Reduction

The inverse of a nonsingular square matrix $\underline{a}$ can be found by the method of row reduction (sometimes called the *method of Gauss–Jordan*) by performing identical simultaneous operations on the matrix $\underline{a}$ and the identity matrix $\underline{I}$ (of the same order as $\underline{a}$) such that the matrix $\underline{a}$ becomes an identity matrix and the original identity matrix becomes the inverse of $\underline{a}$.

A numerical example will best illustrate the procedure. We begin by converting matrix $\underline{a}$ to an upper triangular form by setting all elements below the main diagonal equal to zero, starting with the first column and continuing with succeeding columns. We then proceed from the last column to the first, setting all elements above the main diagonal equal to zero.

We will invert the following matrix by row reduction.

$$\underline{a} = \begin{bmatrix} 2 & 2 & 1 \\ 2 & 1 & 0 \\ 1 & 1 & 1 \end{bmatrix} \tag{A.4.1}$$

To find $\underline{a}^{-1}$, we need to find $\underline{x}$ such that $\underline{a}\underline{x} = \underline{I}$, where

$$\underline{x} = \begin{bmatrix} x_{11} & x_{12} & x_{13} \\ x_{21} & x_{22} & x_{23} \\ x_{31} & x_{32} & x_{33} \end{bmatrix}$$

That is, solve

$$\begin{bmatrix} 2 & 2 & 1 \\ 2 & 1 & 0 \\ 1 & 1 & 1 \end{bmatrix} \underline{x} = \begin{bmatrix} 1 & 0 & 0 \\ 0 & 1 & 0 \\ 0 & 0 & 1 \end{bmatrix}$$

We begin by writing $\underline{a}$ and $\underline{I}$ side by side as

$$\left[\begin{array}{ccc|ccc} 2 & 2 & 1 & 1 & 0 & 0 \\ 2 & 1 & 0 & 0 & 1 & 0 \\ 1 & 1 & 1 & 0 & 0 & 1 \end{array} \right] \tag{A.4.2}$$

where the vertical dotted line separates $\underline{a}$ and $\underline{I}$.

1. Divide the first row of Eq. (A.4.2) by 2.

$$\left[\begin{array}{ccc|ccc} 1 & 1 & \frac{1}{2} & \frac{1}{2} & 0 & 0 \\ 2 & 1 & 0 & 0 & 1 & 0 \\ 1 & 1 & 1 & 0 & 0 & 1 \end{array} \right] \tag{A.4.3}$$

2. Multiply the first row of Eq. (A.4.3) by -2 and add the result to the second row.

$$\left[\begin{array}{ccc|ccc} 1 & 1 & \frac{1}{2} & \frac{1}{2} & 0 & 0 \\ 0 & -1 & -1 & -1 & 1 & 0 \\ 1 & 1 & 1 & 0 & 0 & 1 \end{array} \right] \tag{A.4.4}$$

3. Subtract the first row of Eq. (A.4.4) from the third row.

$$\left[\begin{array}{ccc|ccc} 1 & 1 & \frac{1}{2} & \frac{1}{2} & 0 & 0 \\ 0 & -1 & -1 & -1 & 1 & 0 \\ 0 & 0 & \frac{1}{2} & -\frac{1}{2} & 0 & 1 \end{array}\right] \qquad (A.4.5)$$

4. Multiply the second row of Eq. (A.4.5) by -1 and the third row by 2.

$$\left[\begin{array}{ccc|ccc} 1 & 1 & \frac{1}{2} & \frac{1}{2} & 0 & 0 \\ 0 & 1 & 1 & 1 & -1 & 0 \\ 0 & 0 & 1 & -1 & 0 & 2 \end{array}\right] \qquad (A.4.6)$$

5. Subtract the third row of Eq. (A.4.6) from the second row.

$$\left[\begin{array}{ccc|ccc} 1 & 1 & \frac{1}{2} & \frac{1}{2} & 0 & 0 \\ 0 & 1 & 0 & 2 & -1 & -2 \\ 0 & 0 & 1 & -1 & 0 & 2 \end{array}\right] \qquad (A.4.7)$$

6. Multiply the third row of Eq. (A.4.7) by $-1/2$ and add the result to the first row.

$$\left[\begin{array}{ccc|ccc} 1 & 1 & 0 & 1 & 0 & -1 \\ 0 & 1 & 0 & 2 & -1 & -2 \\ 0 & 0 & 1 & -1 & 0 & 2 \end{array}\right] \qquad (A.4.8)$$

7. Subtract the second row of Eq. (A.4.8) from the first row.

$$\left[\begin{array}{ccc|ccc} 1 & 0 & 0 & -1 & 1 & 1 \\ 0 & 1 & 0 & 2 & -1 & -2 \\ 0 & 0 & 1 & -1 & 0 & 2 \end{array}\right] \qquad (A.4.9)$$

The replacement of $\underline{a}$ by the inverse matrix is now complete. The inverse of $\underline{a}$ is then the right side of Eq. (A.4.9); that is,

$$\underline{a}^{-1} = \left[\begin{array}{ccc} -1 & 1 & 1 \\ 2 & -1 & -2 \\ -1 & 0 & 2 \end{array}\right] \qquad (A.4.10)$$

For additional information regarding matrix algebra, consult References [1] and [2].

REFERENCES

[1] Gere, J. M., and Weaver, W., Jr., *Matrix Algebra for Engineers*, Van Nostrand, Princeton, New Jersey, 1966.

[2] Jennings, A., *Matrix Computation for Engineers and Scientists*, Wiley, New York, 1977.

PROBLEMS

Solve Problems A.1–A.6 using matrices $\underline{A}$, $\underline{B}$, $\underline{C}$, $\underline{D}$, and $\underline{E}$ given by

$$\underline{A} = \begin{bmatrix} 1 & 0 \\ -1 & 4 \end{bmatrix} \quad \underline{B} = \begin{bmatrix} 2 & 0 \\ -2 & 8 \end{bmatrix} \quad \underline{C} = \begin{bmatrix} 3 & 1 & 0 \\ -1 & 0 & 3 \end{bmatrix}$$

$$\underline{D} = \begin{bmatrix} 3 & 1 & 2 \\ 1 & 4 & 0 \\ 2 & 0 & 3 \end{bmatrix} \quad \underline{E} = \begin{Bmatrix} 1 \\ 2 \\ 3 \end{Bmatrix}$$

(Write "nonsense" if the operation cannot be performed.)

A.1 (a) $\underline{A} + \underline{B}$ (b) $\underline{A} + \underline{C}$
(c) $\underline{A}\underline{C}^T$ (d) $\underline{D}\underline{E}$
(e) $\underline{D}\underline{C}$ (f) $\underline{C}\underline{D}$

A.2 Determine $\underline{A}^{-1}$ by the cofactor method.

A.3 Determine $\underline{D}^{-1}$ by the cofactor method.

A.4 Determine $\underline{C}^{-1}$.

A.5 Determine $\underline{B}^{-1}$ by row reduction.

A.6 Determine $\underline{D}^{-1}$ by row reduction.

A.7 Show that $(\underline{A}\underline{B})^T = \underline{B}^T\underline{A}^T$ by using

$$\underline{A} = \begin{bmatrix} a_{11} & a_{12} \\ a_{21} & a_{22} \end{bmatrix} \quad \underline{B} = \begin{bmatrix} b_{11} & b_{12} & b_{13} \\ b_{21} & b_{22} & b_{23} \end{bmatrix}$$

A.8 Find $\underline{T}^{-1}$ given that

$$\underline{T} = \begin{bmatrix} \cos\theta & \sin\theta \\ -\sin\theta & \cos\theta \end{bmatrix}$$

and show that $\underline{T}^{-1} = \underline{T}^T$, and hence, that $\underline{T}$ is an orthogonal matrix.

A.9 Given the matrices

$$\underline{X} = \begin{bmatrix} x & y \\ 1 & x \end{bmatrix}, \quad \underline{A} = \begin{bmatrix} a & b \\ b & c \end{bmatrix}$$

Show that the triple matrix product $\underline{X}^T\underline{A}\underline{X}$ is symmetric.

A.10 Evaluate the following integral in explicit form:

$$\underline{k} = \int_0^L \underline{B}^T\underline{E}\underline{B} \, dx$$

where

$$\underline{B} = \begin{bmatrix} -\dfrac{1}{L} & \dfrac{1}{L} \end{bmatrix}$$

(*Note:* This is the step needed to obtain Eq. (11.1.16) from (11.1.15).]

A.11   The following integral represents the strain energy in a bar:

$$U = \frac{A}{2} \int_0^L \underline{d}^T \underline{B} \underline{D} \underline{B} \underline{d} \; dx$$

where $\underline{d} = \begin{Bmatrix} d_1 \\ d_2 \end{Bmatrix}$ and $\underline{B} = \begin{bmatrix} -\dfrac{1}{L} & \dfrac{1}{L} \end{bmatrix}$ and $\underline{D} = E.$

Show that $dU/d\{d\}$ yields $\underline{kd}$, where $\underline{k}$ is the bar stiffness matrix given by

$$\underline{k} = \frac{AE}{L} \begin{bmatrix} 1 & -1 \\ -1 & 1 \end{bmatrix}$$

Methods for Solution

of Simultaneous

Linear Equations

INTRODUCTION

Many problems in engineering and mathematical physics require the solution of a system of simultaneous linear algebraic equations. Stress analysis, heat transfer, and vibration analysis are typical engineering problems for which the finite element formulation for solution involves the solving of simultaneous linear equations. This appendix introduces methods applicable to both longhand and computer solutions of simultaneous linear equations. Many methods are available for solution of equations; therefore, for brevity's sake, we will discuss only some of the more common methods.

B.1 General Form of the Equations

In general, the set of equations will have the form

$$a_{11}x_1 + a_{12}x_2 + \cdots + a_{1n}x_n = c_1$$
$$a_{21}x_1 + a_{22}x_2 + \cdots + a_{2n}x_n = c_2$$
$$\vdots \qquad \vdots \qquad \qquad \vdots \qquad \vdots \qquad (B.1.1)$$
$$a_{n1}x_1 + a_{n2}x_2 + \cdots + a_{nn}x_n = c_n$$

where the a_{ij}'s are the coefficients of the unknown x_j's, and the c_i's are the known right-side terms. In the structural analysis problem, the a_{ij}'s are the stiffness coefficients k_{ij}'s, the x_j's are the unknown nodal displacements d_i's, and the c_i's are the known nodal forces F_i's.

If the c's are not all zero, the set of equations is *nonhomogeneous* and all equations must be independent to obtain a unique solution. Stress analysis problems typically involve solving sets of nonhomogeneous equations.

If the c's are all zero, the set of equations is *homogeneous* and nontrivial solutions exist only if all equations are not independent. Buckling and vibration problems typically involve homogeneous sets of equations.

B.2   Uniqueness, Nonuniqueness, and Nonexistence of Solution

To solve a system of simultaneous linear equations means to determine a unique set of values (if they exist) for the unknowns that satisfy every equation of the set simultaneously. A unique solution exists if and only if the determinant of the square coefficient matrix is not equal to zero. (All of the engineering problems considered in this text will result in square coefficient matrices.) The problems in this text will usually result in a system of equations that has a unique solution. Here we will briefly illustrate the concepts of uniqueness, nonuniqueness, and nonexistence of solution for systems of equations.

Uniqueness of Solution

$$2x_1 + 1x_2 = 6$$
$$1x_1 + 4x_2 = 17$$

$$(B.2.1)$$

For Eqs. (B.2.1), the determinant of the coefficient matrix is not zero, and a unique solution exists as shown by the single common point of intersection of the two Eqs. (B.2.1) in Figure B-1.

Nonuniqueness of Solution

$$2x_1 + 1x_2 = 6$$
$$4x_1 + 2x_2 = 12$$

$$(B.2.2)$$

For Eqs. (B.2.2), the determinant of the coefficient matrix is zero; that is,

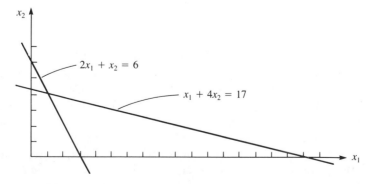

F I G U R E   B-1   Uniqueness of solution

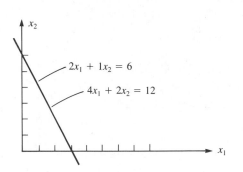

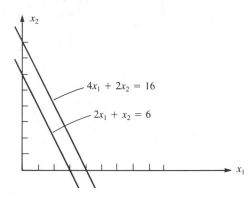

F I G U R E B-2   Nonuniqueness of solution

F I G U R E B-3   Nonexistence of solution

$$\begin{vmatrix} 2 & 1 \\ 4 & 2 \end{vmatrix} = 0$$

Hence, the equations are called *singular*, and either the solution is not unique or it does not exist. In this case, the solution is not unique, as shown in Figure B-2.

Nonexistence of Solution

$$2x_1 + x_2 = 6$$
$$4x_1 + 2x_2 = 16$$

(*B.2.3*)

Again, the determinant of the coefficient matrix is zero. In this case, no solution exists because we have parallel lines (no common point of intersection), as shown in Figure B-3.

B.3   Methods for Solving Linear Algebraic Equations

We will now present some common methods for solving systems of linear algebraic equations that have unique solutions. Some of these methods will work best for small sets of equations solved longhand, whereas others are well suited for computer application.

Cramer's Rule

We begin by introducing a method known as *Cramer's rule*, which is useful for the longhand solution of small numbers of simultaneous equations. Given the set of equations

$$\underline{a}\underline{x} = \underline{c}$$

(*B.3.1*)

or, in index notation,

$$\sum_{i=1}^{n} a_{ij}x_j = c_i \qquad (B.3.2)$$

We first let $\underline{d}^{(i)}$ be the matrix $\underline{a}$ with column i replaced by the column matrix $\underline{c}$. Then the unknown x_i's are determined by

$$x_i = \frac{|\underline{d}^{(i)}|}{|\underline{a}|} \qquad (B.3.3)$$

As an example of Cramer's rule, consider the following equations:

$$-x_1 + 3x_2 - 2x_3 = 2$$
$$2x_1 - 4x_2 + 2x_3 = 1 \qquad (B.3.4)$$
$$4x_2 + x_3 = 3$$

In matrix form, Eqs. (B.3.4) become

$$\begin{bmatrix} -1 & 3 & -2 \\ 2 & -4 & 2 \\ 0 & 4 & 1 \end{bmatrix} \begin{Bmatrix} x_1 \\ x_2 \\ x_3 \end{Bmatrix} = \begin{Bmatrix} 2 \\ 1 \\ 3 \end{Bmatrix} \qquad (B.3.5)$$

By Eq. (B.3.3), we can solve for the unknown x_i's as

$$x_1 = \frac{|\underline{d}^{(1)}|}{|\underline{a}|} = \frac{\begin{vmatrix} 2 & 3 & -2 \\ 1 & -4 & 2 \\ 3 & 4 & 1 \end{vmatrix}}{\begin{vmatrix} -1 & 3 & -2 \\ 2 & -4 & 2 \\ 0 & 4 & 1 \end{vmatrix}} = \frac{-41}{-10} = 4.1$$

$$x_2 = \frac{|\underline{d}^{(2)}|}{|\underline{a}|} = \frac{\begin{vmatrix} -1 & 2 & -2 \\ 2 & 1 & 2 \\ 0 & 3 & 1 \end{vmatrix}}{-10} = 1.1 \qquad (B.3.6)$$

$$x_3 = \frac{|\underline{d}^{(3)}|}{|\underline{a}|} = \frac{\begin{vmatrix} -1 & 3 & 2 \\ 2 & -4 & 1 \\ 0 & 4 & 3 \end{vmatrix}}{-10} = -1.4$$

In general, to find the determinant of an $n \times n$ matrix, we must evaluate the determinants of n matrices of order $(n-1) \times (n-1)$. It has been shown that the solution of n simultaneous equations by Cramer's rule, evaluating determinants by expansion by minors, requires $(n-1)(n+1)!$ multiplications. Hence, this method takes large amounts of computer time and therefore is not used in solving large systems of simultaneous equations either long-hand or by computer.

Inversion of the Coefficient Matrix

The set of equations $\underline{a}\underline{x} = \underline{c}$ can be solved for $\underline{x}$ by inverting the coefficient matrix $\underline{a}$ and premultiplying both sides of the original set of equations by $\underline{a}^{-1}$, such that

$$\underline{a}^{-1}\underline{a}\underline{x} = \underline{a}^{-1}\underline{c}$$

$$\underline{I}\underline{x} = \underline{a}^{-1}\underline{c} \qquad (B.3.7)$$

$$\underline{x} = \underline{a}^{-1}\underline{c}$$

Two methods for determining the inverse of a matrix (the cofactor method and row reduction) were discussed in Appendix A.

The inverse method is much more time consuming (because much time is required to determine the inverse of $\underline{a}$) than either the elimination or iteration methods, which are discussed subsequently. Therefore, inversion is practical only for small systems of equations.

However, the concept of inversion is often used during the formulation of the finite element equations, even though elimination or iteration is used in achieving the final solution for the unknowns (such as nodal displacements).

Besides the tedious calculations necessary to obtain the inverse, the method usually involves determining the inverse of sparse, banded matrices (stiffness matrices in structural analysis usually contain many zeros with the nonzero coefficients located in a band around the main diagonal). This sparsity and banded nature can be used to advantage in terms of storage requirements and solution algorithms on the computer. The inverse results in a dense, full matrix with loss of the advantages resulting from the sparse, banded nature of the original coefficient matrix.

To illustrate the solution of a system of equations by the inverse method, consider the same equations that we solved previously by Cramer's rule. For convenience' sake, we repeat the equations here.

$$\begin{bmatrix} -1 & 3 & -2 \\ 2 & -4 & 2 \\ 0 & 4 & 1 \end{bmatrix} \begin{Bmatrix} x_1 \\ x_2 \\ x_3 \end{Bmatrix} = \begin{Bmatrix} 2 \\ 1 \\ 3 \end{Bmatrix} \qquad (B.3.8)$$

The inverse of this coefficient matrix was found in Eq. (A.3.11) of Appendix A. The unknowns are then determined as

$$\begin{Bmatrix} x_1 \\ x_2 \\ x_3 \end{Bmatrix} = -\frac{1}{10} \begin{bmatrix} -12 & -11 & -2 \\ -2 & -1 & -2 \\ 8 & 4 & -2 \end{bmatrix} \begin{Bmatrix} 2 \\ 1 \\ 3 \end{Bmatrix} = \begin{Bmatrix} 4.1 \\ 1.1 \\ -1.4 \end{Bmatrix} \qquad (B.3.9)$$

Gaussian Elimination

We will now consider a commonly used method called *Gaussian elimination* that is easily adapted to the computer for solving systems of simultaneous equations. It is based on triangularization of the coefficient matrix and evaluation of the unknowns by back-substitution starting from the last equation.

The general system of n equations with n unknowns given by

$$
\begin{bmatrix}
a_{11} & a_{12} & \cdots & a_{1n} \\
a_{21} & a_{22} & \cdots & a_{2n} \\
\vdots & \vdots & & \vdots \\
a_{n1} & a_{n2} & \cdots & a_{nn}
\end{bmatrix}
\begin{Bmatrix}
x_1 \\
x_2 \\
\vdots \\
x_n
\end{Bmatrix}
=
\begin{Bmatrix}
c_1 \\
c_2 \\
\vdots \\
c_n
\end{Bmatrix}
\qquad (B.3.10)
$$

will be used to explain the Gaussian elimination method as follows:

1. Eliminate the coefficient of x_1 in every equation except the first one. To do this, select a_{11} as the pivot, and
    a. Add the multiple $-a_{21}/a_{11}$ of the first row to the second row.
    b. Add the multiple $-a_{31}/a_{11}$ of the first row to the third row.
    c. Continue this procedure through the nth row.
    The system of equations will then be reduced to the following form:

$$
\begin{bmatrix}
a_{11} & a_{12} & \cdots & a_{1n} \\
0 & a'_{22} & \cdots & a'_{2n} \\
\vdots & & & \vdots \\
0 & a'_{n2} & \cdots & a'_{nn}
\end{bmatrix}
\begin{Bmatrix}
x_1 \\
x_2 \\
\vdots \\
x_n
\end{Bmatrix}
=
\begin{Bmatrix}
c_1 \\
c'_2 \\
\vdots \\
c'_n
\end{Bmatrix}
\qquad (B.3.11)
$$

2. Eliminate the coefficient of x_2 in every equation below the second equation. To do this, select a'_{22} as the pivot, and
    a. Add the multiple $-a'_{32}/a'_{22}$ of the second row to the third row.
    b. Add the multiple $-a'_{42}/a'_{22}$ of the second row to the fourth row.
    c. Continue this procedure through the nth row.
    The system of equations will then be reduced to the following

form:

$$
\begin{bmatrix}
a_{11} & a_{12} & a_{13} & \cdots & a_{1n} \\
0 & a'_{22} & a'_{23} & \cdots & a'_{2n} \\
0 & 0 & a''_{33} & \cdots & a''_{3n} \\
\vdots & & & & \vdots \\
0 & 0 & a''_{n3} & \cdots & a''_{nn}
\end{bmatrix}
\begin{Bmatrix}
x_1 \\
x_2 \\
x_3 \\
\vdots \\
x_n
\end{Bmatrix}
=
\begin{Bmatrix}
c_1 \\
c'_2 \\
c''_3 \\
\vdots \\
c''_n
\end{Bmatrix}
\qquad (B.3.12)
$$

We repeat this process for the remaining rows until we have the system of equations (called *triangularized*) as

$$
\begin{bmatrix}
a_{11} & a_{12} & a_{13} & a_{14} & \cdots & a_{1n} \\
0 & a'_{22} & a'_{23} & a'_{24} & \cdots & a'_{2n} \\
0 & 0 & a''_{33} & a''_{34} & \cdots & a''_{3n} \\
0 & 0 & 0 & a'''_{44} & \cdots & a'''_{4n} \\
\vdots & \vdots & \vdots & \vdots & & \vdots \\
0 & 0 & 0 & 0 & \cdots & a^{n-1}_{nn}
\end{bmatrix}
\begin{Bmatrix}
x_1 \\
x_2 \\
x_3 \\
x_4 \\
\vdots \\
x_n
\end{Bmatrix}
=
\begin{Bmatrix}
c_1 \\
c'_2 \\
c''_3 \\
c'''_4 \\
\vdots \\
c^{n-1}_n
\end{Bmatrix}
\qquad (B.3.13)
$$

3. Determine x_n from the last equation as

$$
x_n = \frac{c^{n-1}_n}{a^{n-1}_{nn}}
\qquad (B.3.14)
$$

and determine the other unknowns by back-substitution. These steps are summarized in general form by

$$
a_{ij} = a_{ij} - a_{kj}\frac{a_{ik}}{a_{kk}} \qquad \begin{aligned} & k = 1, 2, \ldots, n-1 \\ & i = k+1, \ldots, n \\ & j = k, \ldots, n+1 \end{aligned}
$$

$$(B.3.15)$$

$$
x_i = \frac{1}{a_{ii}}\left(a_{i,n+1} - \sum_{r=i+1}^{n} a_{ir}x_r\right)
$$

where $a_{i,n+1}$ represent the latest right side c's given by Eq. (B.3.13).

We will solve the following example to illustrate the Gaussian elimination method.

E X A M P L E   B.1

$$
\begin{aligned}
2x_1 + 2x_2 + 1x_3 &= 9 \\
2x_1 + 1x_2 \qquad\;\; &= 4 \\
1x_1 + 1x_2 + 1x_3 &= 6
\end{aligned}
$$

$$(B.3.16)$$

Step 1

Eliminate the coefficient of x_1 in every equation except the first one. Select $a_{11} = 2$ as the pivot, and

    a. Add the multiple $-a_{21}/a_{11} = -2/2$ of the first row to the second row

    b. Add the multiple $-a_{31}/a_{11} = -1/2$ of the first row to the third row. We then obtain

$$
\begin{aligned}
2x_1 + 2x_2 + 1x_3 &= 9 \\
0x_1 - 1x_2 - 1x_3 &= 4 - 9 = -5 \\
0x_1 + 0x_2 + \tfrac{1}{2}x_3 &= 6 - \tfrac{9}{2} = \tfrac{3}{2}
\end{aligned}
$$

$$(B.3.17)$$

Step 2

Eliminate the coefficient of x_2 in every equation below the second equation. In this case, we accomplished this in Step 1.

Step 3

Solve for x_3 in the third of Eqs. (B.3.17) as

$$
x_3 = \frac{\left(\tfrac{3}{2}\right)}{\left(\tfrac{1}{2}\right)} = 3
$$

Solve for x_2 in the second of Eqs. (B.3.17) as

$$x_2 = \frac{-5 + 3}{-1} = 2$$

Solve for x_1 in the first of Eqs. (B.3.17) as

$$x_1 = \frac{9 - 2(2) - 3}{2} = 1$$

To illustrate the use of the index Eqs. (B.3.15), we resolve the same example as follows. The ranges of the indexes in Eqs. (B.3.15) are $k = 1, 2$; $i = 2, 3$; and $j = 1, 2, 3, 4$.

Step 1

For $k = 1$, $i = 2$, and j indexing from 1 to 4,

$$a_{21} = a_{21} - a_{11} \frac{a_{21}}{a_{11}} = 2 - 2\left(\frac{2}{2}\right) = 0$$

$$a_{22} = a_{22} - a_{12} \frac{a_{21}}{a_{11}} = 1 - 2\left(\frac{2}{2}\right) = -1$$

$$a_{23} = a_{23} - a_{13} \frac{a_{21}}{a_{11}} = 0 - 1\left(\frac{2}{2}\right) = -1 \qquad (B.3.18)$$

$$a_{24} = a_{24} - a_{14} \frac{a_{21}}{a_{11}} = 4 - 9\left(\frac{2}{2}\right) = -5$$

Note that these new coefficients correspond to those of the second of Eqs. (B.3.17), where the right-side a's of Eqs. (B.3.18) are those from the previous step [here from Eqs. (B.3.16)], the right-side a_{24} is really $c_2 = 4$, and the left-side a_{24} is the new $c_2 = -5$.

For $k = 1$, $i = 3$, and j indexing from 1 to 4,

$$a_{31} = a_{31} - a_{11} \frac{a_{31}}{a_{11}} = 1 - 2\left(\frac{1}{2}\right) = 0$$

$$a_{32} = a_{32} - a_{12} \frac{a_{31}}{a_{11}} = 1 - 2\left(\frac{1}{2}\right) = 0$$

$$a_{33} = a_{33} - a_{13} \frac{a_{31}}{a_{11}} = 1 - 1\left(\frac{1}{2}\right) = \frac{1}{2} \qquad (B.3.19)$$

$$a_{34} = a_{34} - a_{14} \frac{a_{31}}{a_{11}} = 6 - 9\left(\frac{1}{2}\right) = \frac{3}{2}$$

where these new coefficients correspond to those of the third of Eqs. (B.3.17) as previously explained.

Step 2

For $k = 2$, $i = 3$, and $j \, (= k)$ indexing from 2 to 4,

$$a_{32} = a_{32} - a_{22}\left(\frac{a_{32}}{a_{22}}\right) = 0 - (-1)\left(\frac{0}{-1}\right) = 0$$

$$a_{33} = a_{33} - a_{23}\left(\frac{a_{32}}{a_{22}}\right) = \frac{1}{2} - (-1)\left(\frac{0}{-1}\right) = \frac{1}{2} \qquad (B.3.20)$$

$$a_{34} = a_{34} - a_{24}\left(\frac{a_{32}}{a_{22}}\right) = \frac{3}{2} - (-5)\left(\frac{0}{-1}\right) = \frac{3}{2}$$

where the new coefficients again correspond to those of the third of Eqs. (B.3.17), because Step 1 already eliminated the coefficients of x_2 as observed in the third of Eqs. (B.3.17), and the a's on the right side of Eqs. (B.3.20) are taken from Eqs. (B.3.18) and (B.3.19).

Step 3

By Eqs. (B.3.15), for x_3, we have

$$x_3 = \frac{1}{a_{33}}(a_{34} - 0)$$

or, using a_{33} and a_{34} from Eqs. (B.3.20),

$$x_3 = \frac{1}{(\frac{1}{2})}\left(\frac{3}{2}\right) = 3$$

where the summation is interpreted as zero in the second of Eqs. (B.3.15) when $r > n$ (for x_3, $r = 4$, and $n = 3$). For x_2, we have

$$x_2 = \frac{1}{a_{22}}(a_{24} - a_{23}x_3)$$

or, using the appropriate a's from Eqs. (B.3.18),

$$x_2 = \frac{1}{-1}[-5 - (-1)(3)] = 2$$

and for x_1, we have

$$x_1 = \frac{1}{a_{11}}(a_{14} - a_{12}x_2 - a_{13}x_3)$$

or, using the a's from the first of Eqs. (B.3.16),

$$x_1 = \tfrac{1}{2}[9 - 2(2) - 1(3)] = 1$$

In summary, the latest a's from the previous steps have been used in Eqs. (B.3.15) to obtain the x's.   ∎

Note that the pivot element was the diagonal element in each step. However, the diagonal element must be nonzero because we divide by it in each step. An original matrix with all nonzero diagonal elements does not ensure that the pivots in each step will remain nonzero, because we are adding numbers to equations below the pivot in each following step. Therefore, a test is necessary to determine whether the pivot a_{kk} at each step is zero. If it is zero, the current row (equation) must be interchanged with one of the following rows—usually with the next row unless that row has a zero at the position that would next become the pivot. Remember that the right-side corresponding element in c must also be interchanged. After making this test and, if necessary, interchanging the equations, the procedure continues in the usual manner.

An example will now illustrate the method for treating the occurrence of a zero pivot element.

EXAMPLE B.2

Solve the following set of simultaneous equations.

$$2x_1 + 2x_2 + 1x_3 = 9$$
$$1x_1 + 1x_2 + 1x_3 = 6 \qquad (B.3.21)$$
$$2x_1 + 1x_2 \qquad\quad = 4$$

It will often be convenient to set up the solution procedure by considering the coefficient matrix a plus the right-side matrix c in one matrix without writing down the unknown matrix x. This new matrix is called the *augmented matrix*. For the set of Eqs. (B.3.21), we have the augmented matrix written as

$$\begin{bmatrix} 2 & 2 & 1 & 9 \\ 1 & 1 & 1 & 6 \\ 2 & 1 & 0 & 4 \end{bmatrix} \qquad (B.3.22)$$

We use the steps previously outlined as follows:

Step 1

We select $a_{11} = 2$ as the pivot, and

    a   Add the multiple $-a_{21}/a_{11} = -1/2$ of the first row to the second row of Eq. (B.3.22).

    b.   Add the multiple $-a_{31}/a_{11} = -2/2$ of the first row to the third row of Eq. (3.22) to obtain

$$\begin{bmatrix} 2 & 2 & 1 & 9 \\ 0 & 0 & \frac{1}{2} & \frac{3}{2} \\ 0 & -1 & -1 & -5 \end{bmatrix} \qquad (B.3.23)$$

At the end of Step 1, we would normally choose a_{22} as the next pivot. However, a_{22} is now equal to zero. If we interchange the second and third rows of Eq. (B.3.23), the new a_{22} will be nonzero and can be used as a pivot. The interchange of rows 2 and 3 results in

$$\begin{bmatrix} 2 & 2 & 1 & | & 9 \\ 0 & -1 & -1 & | & -5 \\ 0 & 0 & \frac{1}{2} & | & \frac{3}{2} \end{bmatrix} \qquad (B.3.24)$$

For this special set of only three equations, the interchange has resulted in an upper-triangular coefficient matrix and concludes the elimination procedure. The back-substitution process of Step 3 now yields

$$x_3 = 3 \qquad x_2 = 2 \qquad x_1 = 1 \qquad \blacksquare$$

A second problem when selecting the pivots in sequential manner without testing for the best possible pivot is that loss of accuracy due to rounding in the results can occur. In general, the pivots should be selected as the largest (in absolute value) of the elements in any column. For example, consider the set of equations given by

$$0.002x_1 + 2.00x_2 = 2.00$$
$$3.00x_1 + 1.50x_2 = 4.50 \qquad (B.3.25)$$

whose actual solution is given by

$$x_1 = 1.0005 \qquad x_2 = 0.999 \qquad (B.3.26)$$

The solution by Gaussian elimination without testing for the largest absolute value of the element in any column is

$$0.002x_1 + 2.00x_2 = 2.00$$
$$-2998.5x_2 = -995.5$$
$$x_2 = 0.3320$$
$$x_1 = 668 \qquad (B.3.27)$$

This solution does not satisfy the second of Eqs. (B.3.25). The solution by interchanging equations is

$$3.00x_1 + 1.50x_2 = 4.50$$
$$0.002x_1 + 2.00x_2 = 2.00$$

or

$$3.00x_1 + 1.50x_2 = 4.50$$
$$1.999x_2 = 1.997$$
$$x_2 = 0.999$$
$$x_1 = 1.0005 \qquad (B.3.28)$$

Equations (B.3.28) agree with the actual solution, Eqs. (B.3.26).

Hence, in general, the pivots should be selected as the largest (in absolute value) of the elements in any column. This process is called *partial pivoting*. Even better results can be obtained by choosing the pivot as the largest element in the whole matrix of the remaining equations and performing appropriate interchanging of rows. This is called *complete pivoting*. Complete pivoting requires a large amount of testing, and consequently, is not recommended in general.

The finite element equations generally involve coefficients with different orders of magnitude, so that Gaussian elimination with partial pivoting is a useful method for solving the equations.

Finally, it has been shown that for n simultaneous equations, the number of arithmetic operations required in Gaussian elimination is n divisions, $\frac{1}{3}n^3 + n^2$ multiplications, and $\frac{1}{3}n^3 + n$ additions. If partial pivoting is included, the number of comparisons needed to select pivots is $n(n + 1)/2$.

Other elimination methods, including the Gauss–Jordan and Cholesky methods, have some advantages over Gaussian elimination, and are sometimes used to solve large systems of equations. For descriptions of other methods, see References [1]–[3].

Gauss–Seidel Iteration

Another general class of methods (other than the elimination methods) used to solve systems of linear algebraic equations is the *iterative methods*. Iterative methods work well when the system of equations is large and sparse (many zero coefficients). The Gauss–Seidel method starts with the original set of equations $\underline{a}\underline{x} = \underline{c}$ written in the following form:

$$x_1 = \frac{1}{a_{11}}(c_1 - a_{12}x_2 - a_{13}x_3 - \cdots - a_{1n}x_n)$$

$$x_2 = \frac{1}{a_{22}}(c_2 - a_{21}x_1 - a_{23}x_3 - \cdots - a_{2n}x_n)$$

$$\vdots \qquad\qquad (B.3.29)$$

$$x_n = \frac{1}{a_{nn}}(c_n - a_{n1}x_1 - a_{n2}x_2 - \cdots - a_{n,n-1}x_{n-1})$$

The following steps are then applied.

1. Assume a set of initial values for the unknowns $x_1, x_2, \ldots, x_n$, and substitute them into the right side of the first of Eqs. (B.3.29) to solve for the new x_1.

2. Use the latest value for x_1 obtained from Step 1 and the initial values for $x_3, x_4, \ldots, x_n$ in the right side of the second of Eqs. (B.3.29) to solve for the new x_2.

3. Continue using the latest values of the x's obtained in the left side of Eqs. (B.3.29) as the next trial values in the right side for each succeeding step.

4. Iterate until convergence is satisfactory.

A good initial set of values (guesses) is often $x_i = c_i/a_{ii}$. An example will serve to illustrate the method.

E X A M P L E   B.3

Consider the set of linear simultaneous equations given by

$$
\begin{aligned}
4x_1 - x_2 \qquad\qquad &= 2 \\
-x_1 + 4x_2 - x_3 \qquad &= 5 \\
- x_2 + 4x_3 - x_4 &= 6 \\
- x_3 + 2x_4 &= -2
\end{aligned}
\qquad (B.3.30)
$$

Using the initial guesses given by $x_i = c_i/a_{ii}$, we have

$$
x_1 = \tfrac{2}{4} = \tfrac{1}{2} \qquad x_2 = \tfrac{5}{4} \approx 1 \qquad x_3 = \tfrac{6}{4} \approx 1 \qquad x_4 = -1
$$

Solving the first of Eqs. (B.3.30) for x_1 yields

$$
x_1 = \tfrac{1}{4}(2 + x_2) = \tfrac{1}{4}(2 + 1) = \tfrac{3}{4}
$$

Solving the second of Eqs. (B.3.30) for x_2, we have

$$
x_2 = \tfrac{1}{4}(5 + x_1 + x_3) = \tfrac{1}{4}(5 + \tfrac{3}{4} + 1) = 1.68
$$

Solving the third of Eqs. (B.3.30) for x_3, we have

$$
x_3 = \tfrac{1}{4}(6 + x_2 + x_4) = \tfrac{1}{4}[6 + 1.68 + (-1)] = 1.672
$$

Solving the fourth of Eqs. (B.3.30) for x_4, we obtain

$$
x_4 = \tfrac{1}{2}(-2 + x_3) = \tfrac{1}{2}(-2 + 1.67) = -0.16
$$

The first iteration has now been completed. The second iteration yields

$$
x_1 = \tfrac{1}{4}(2 + 1.68) = 0.922
$$

$$
x_2 = \tfrac{1}{4}(5 + 0.922 + 1.672) = 1.899
$$

$$
x_3 = \tfrac{1}{4}[6 + 1.899 + (-0.16)] = 1.944
$$

$$
x_4 = \tfrac{1}{2}(-2 + 1.944) = -0.028
$$

Table B-1 lists the results of four iterations of the Gauss–Seidel method and the exact solution. From Table B-1, we observe that convergence to the exact solution has proceeded rapidly by the fourth iteration, and the accuracy of the solution is dependent on the number of iterations. ■

In general, iteration methods are self-correcting, so that an error made in calculations at one iteration will be corrected by later iterations. However, there are certain systems of equations for which iterative methods are not

T A B L E   B-1   Results of four iterations of the Gauss–Seidel method for Eqs. (B.3.30)

Iteration	x_1	x_2	x_3	x_4
0	0.5	1.0	1.0	−1.0
1	0.75	1.68	1.672	−0.16
2	0.922	1.899	1.944	−0.028
3	0.975	1.979	1.988	−0.006
4	0.9985	1.9945	1.9983	−0.0008
Exact	1.0	2.0	2.00	0

convergent. When the equations can be arranged such that the diagonal terms are greater than the off-diagonal terms, the possibility of convergence is usually enhanced.

Finally, it has been shown that for n simultaneous equations, the number of arithmetic operations required by Gauss–Seidel iteration is n divisions, n^2 multiplications, and $n^2 - n$ additions for each iteration.

REFERENCES

[1] Southworth, R. W., and DeLeeuw, S. L., *Digital Computation and Numerical Methods*, McGraw-Hill, New York, 1965.

[2] James, M. L., Smith, G. M., and Wolford, J. C., *Applied Numerical Methods for Digital Computation*, 3rd ed., Harper & Row, New York, 1985.

[3] Bathe, K. J., and Wilson, E. L., *Numerical Methods in Finite Element Analysis*, Prentice Hall, Englewood Cliffs, New Jersey, 1976.

PROBLEMS

B.1 Determine the solution of the following simultaneous equations by Cramer's rule.

$$1x_1 + 3x_2 = 5$$
$$4x_1 - 1x_2 = 12$$

B.2 Determine the solution of the following simultaneous equations by the inverse method.

$$1x_1 + 3x_2 = 5$$
$$4x_1 - 1x_2 = 12$$

B.3 Solve the following system of simultaneous equations by Gaussian elimination.

$$x_1 - 4x_2 - 5x_3 = 4$$
$$3x_2 + 4x_3 = -1$$
$$-2x_1 - 1x_2 + 2x_3 = -3$$

B.4 Solve the following system of simultaneous equations by Gaussian elimination.

$$2x_1 + 1x_2 - 3x_3 = 11$$
$$4x_1 - 2x_2 + 3x_3 = 8$$
$$-2x_1 + 2x_2 - 1x_3 = -6$$

B.5 Given that

$$x_1 = 2y_1 - y_2 \qquad z_1 = -x_1 - x_2$$
$$x_2 = y_1 - y_2 \qquad z_2 = 2x_1 + x_2$$

a. Write these relationships in matrix form.
b. Express z in terms of y.
c. Express y in terms of z.

B.6 Starting with the initial guess $\underline{X}^T = [1 \ \ 1 \ \ 1 \ \ 1 \ \ 1]$, perform five iterations of the Gauss–Seidel method on the following system of equations. Based on the results of these five iterations, what is the exact solution?

$$2x_1 - 1x_2 \qquad\qquad\qquad = -1$$
$$-1x_1 + 6x_2 - 1x_3 \qquad\qquad = 4$$
$$-2x_2 + 4x_3 - 1x_4 \qquad = 4$$
$$-1x_3 + 4x_4 - 1x_5 = 6$$
$$-1x_4 + 2x_5 = -2$$

B.7 Solve Problem B.1 by Gauss–Seidel iteration.

B.8 Classify the solutions to the following systems of equations according to Section B.2 as unique, nonunique, or nonexistent.

a. $2x_1 - 4x_2 = 2$
$-9x_1 + 12x_2 = -6$

b. $10x_1 + 1x_2 = 0$
$5x_1 + \frac{1}{2}x_2 = 3$

c. $2x_1 + 1x_2 + 1x_3 = 6$
$3x_1 + 1x_2 - 1x_3 = 4$
$5x_1 + 2x_2 + 2x_3 = 8$

d. $1x_1 + 1x_2 + 1x_3 = 1$
$2x_1 + 2x_2 + 2x_3 = 2$
$3x_1 + 3x_2 + 3x_3 = 3$

Equations from
Elasticity Theory

INTRODUCTION

In this appendix, we will develop the basic equations of the theory of elasticity. These equations should be referred to frequently throughout the structural mechanics portions of this text.

There are three basic sets of equations included in theory of elasticity. These equations must be satisfied if an exact solution to a structural mechanics problem is to be obtained. These sets of equations are (1) the differential equations of equilibrium formulated here in terms of the stresses acting on a body, (2) the strain/displacement and compatibility differential equations, and (3) the stress/strain or material constitutive laws.

C.1 Differential Equations of Equilibrium

For simplicity, we initially consider the equilibrium of a plane element subjected to normal stresses σ_x and σ_y, in-plane shear stress τ_{xy} (in units of force per unit area), and body forces X_b and Y_b (in units of force per unit volume), as shown in Figure C-1. The stresses are assumed to be constant as they act on the width of each face. However, the stresses are assumed to vary from one face to the opposite. For example, we have σ_x acting on the left vertical face, whereas $\sigma_x + (\partial\sigma_x/\partial x)\, dx$ acts on the right vertical face. The element is assumed to have unit thickness.

Summing forces in the x direction, we have

$$\sum F_x = 0 = \left(\sigma_x + \frac{\partial\sigma_x}{\partial x}\, dx\right) dy(1) - \sigma_x\, dy(1) + X_b\, dx\, dy(1)$$
$$+ \left(\tau_{yx} + \frac{\partial\tau_{yx}}{\partial y}\, dy\right) dx(1) - \tau_{yx}\, dx(1) = 0 \qquad (C.1.1)$$

After simplifying and canceling terms in Eq. (C.1.1), we obtain

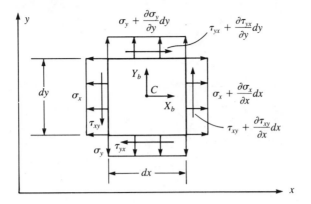

FIGURE   C-1   Plane differential element subjected to stresses

$$\frac{\partial \sigma_x}{\partial x} + \frac{\partial \tau_{yx}}{\partial y} + X_b = 0 \tag{C.1.2}$$

Similarly, summing forces in the y direction, we obtain

$$\frac{\partial \sigma_y}{\partial y} + \frac{\partial \tau_{xy}}{\partial x} + Y_b = 0 \tag{C.1.3}$$

Since we are considering only the planar element, three equilibrium equations must be satisfied. The third equation is equilibrium of moments about an axis normal to the x-y plane; that is, taking moments about point C in Figure C-1, we have

$$\sum M_z = 0 = \tau_{xy}\, dy(1)\frac{dx}{2} + \left(\tau_{xy} + \frac{\partial \tau_{xy}}{\partial x}\, dx\right)\frac{dx}{2}$$

$$- \tau_{yx}\, dx(1)\frac{dy}{2} - \left(\tau_{yx} + \frac{\partial \tau_{yx}}{\partial y}\, dy\right)\frac{dy}{2} = 0 \tag{C.1.4}$$

Simplifying Eq. (C.1.4) and neglecting higher-order terms yields

$$\tau_{xy} = \tau_{yx} \tag{C.1.5}$$

We now consider the three-dimensional state of stress shown in Figure C-2, which shows the additional stresses σ_z, τ_{xz}, and τ_{yz}. For clarity, we show only the stresses on three mutually perpendicular planes. With a straightforward procedure, we can extend the two-dimensional equations (C.1.2), (C.1.3), and (C.1.5) to three dimensions. The resulting total set of equilibrium equations is

$$\frac{\partial \sigma_x}{\partial x} + \frac{\partial \tau_{xy}}{\partial y} + \frac{\partial \tau_{xz}}{\partial z} + X_b = 0$$

$$\frac{\partial \tau_{xy}}{\partial x} + \frac{\partial \sigma_y}{\partial y} + \frac{\partial \tau_{yz}}{\partial z} + Y_b = 0 \tag{C.1.6}$$

$$\frac{\partial \tau_{xz}}{\partial x} + \frac{\partial \tau_{yz}}{\partial y} + \frac{\partial \sigma_z}{\partial z} + Z_b = 0$$

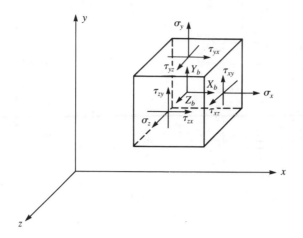

FIGURE C-2   Three-dimensional stress element

and $\qquad\tau_{xy} = \tau_{yx} \qquad \tau_{xz} = \tau_{zx} \qquad \tau_{yz} = \tau_{zy}$ $\qquad$ *(C.1.7)*

C.2   Strain/Displacement and Compatibility Equations

We first obtain the strain/displacement or kinematic differential relationships for the two-dimensional case. We begin by considering the differential element shown in Figure C-3, where the undeformed state is represented by the dotted lines and the deformed shape (after straining takes place) is represented by the solid lines.

Considering line element AB in the x direction, we can see that it becomes $A'B'$ after deformation, where u and v represent the displacements in the x and y directions. By the definition of engineering normal strain (that is, the change in length divided by the original length of a line), we have

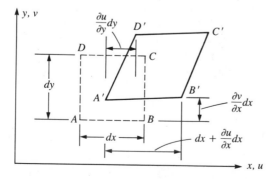

FIGURE C-3   Differential element before and after deformation

$$\varepsilon_x = \frac{A'B' - AB}{AB} \qquad (C.2.1)$$

Now $$AB = dx \qquad (C.2.2)$$

and $$(A'B')^2 = \left(dx + \frac{\partial u}{\partial x}\,dx\right)^2 + \left(\frac{\partial v}{\partial x}\,dx\right)^2 \qquad (C.2.3)$$

Therefore, evaluating $A'B'$ using the binomial theorem and neglecting the higher-order terms $(\partial u/\partial x)^2$ and $(\partial v/\partial x)^2$ (consistent with the assumption of small strains), we have

$$A'B' = dx + \frac{\partial u}{\partial x}\,dx \qquad (C.2.4)$$

Using Eqs. (C.2.2) and (C.2.4) in Eq. (C.2.1), we obtain

$$\varepsilon_x = \frac{\partial u}{\partial x} \qquad (C.2.5)$$

Similarly, considering line element AD in the y direction, we have

$$\varepsilon_y = \frac{\partial v}{\partial y} \qquad (C.2.6)$$

The shear strain γ_{xy} is defined to be the change in the angle between two lines, such as AB and AD, that originally formed a right angle. Hence, from Figure C-3, we can see that γ_{xy} is the sum of two angles and is given by

$$\gamma_{xy} = \frac{\partial u}{\partial y} + \frac{\partial v}{\partial x} \qquad (C.2.7)$$

Equations (C.2.5), (C.2.6), and (C.2.7) represent the strain/displacement relationships for in-plane behavior.

For three-dimensional situations, we have a displacement w in the z direction. It then becomes straightforward to extend the two-dimensional derivations to the three-dimensional case to obtain the additional strain/displacement equations as

$$\varepsilon_x = \frac{\partial w}{\partial z} \qquad (C.2.8)$$

$$\gamma_{xz} = \frac{\partial u}{\partial z} + \frac{\partial w}{\partial x} \qquad (C.2.9)$$

$$\gamma_{yz} = \frac{\partial v}{\partial z} + \frac{\partial w}{\partial y} \qquad (C.2.10)$$

Along with the strain/displacement equations, we need compatibility equations to ensure that the displacement components u, v, and w are single-valued continuous functions so that tearing or overlap of elements does not occur. For the planar-elastic case, we obtain the compatibility equation by differentiating γ_{xy} with respect to both x and y, and then using the definitions for ε_x and ε_y given by Eqs. (C.2.5) and (C.2.6). Hence,

$$\frac{\partial^2 \gamma_{xy}}{\partial x\, \partial y} = \frac{\partial^2}{\partial x\, \partial y} \frac{\partial u}{\partial y} + \frac{\partial^2}{\partial x\, \partial y} \frac{\partial v}{\partial x} = \frac{\partial^2 \varepsilon_x}{\partial y^2} + \frac{\partial^2 \varepsilon_y}{\partial x^2} \qquad (C.2.11)$$

where the second equation in terms of the strains on the right side is obtained by noting that single-valued continuity of displacements requires that the partial differentiations with respect to x and y be interchangeable in order. Therefore, we have $\partial^2/\partial x\, \partial y = \partial^2/\partial y\, \partial x$. Equation (C.2.11) is called the *condition of compatibility*, and it must be satisfied by the strain components to obtain unique expressions for u and v. Equations (C.2.5), (C.2.6), (C.2.7), and (C.2.11) together are then sufficient to obtain unique single-valued functions for u and v.

In three dimensions, we obtain five additional compatibility equations by differentiating γ_{xz} and γ_{yz} in a manner similar to that described above for γ_{xy}. We need not list these equations here; details of their derivation can be found in Reference [1].

In addition to the compatibility conditions that ensure single-valued continuous functions within the body, we must also satisfy displacement or kinematic boundary conditions. This simply means that the displacement functions must also satisfy prescribed or given displacements on the surface of the body. These conditions often occur as support conditions from rollers and/or pins. In general, we might have

$$u = u_0 \qquad v = v_0 \qquad w = w_0 \qquad (C.2.12)$$

at specified surface locations on the body. We may also have conditions other than displacements prescribed (for example, prescribed rotations).

C.3   Stress/Strain Relationships

We will now develop the three-dimensional stress/strain relationships for an isotropic body only. This is done by considering the response of a body to imposed stresses. We subject the body to the stresses σ_x, σ_y, and σ_z independently as shown in Figure C-4.

We first consider the change in length of the element in the x direction due to the independent stresses σ_x, σ_y, and σ_z. We assume the principle of superposition to hold; that is, we assume that the resultant strain in a system due to several forces is the algebraic sum of their individual effects.

Considering Figure C-4(b), the stress in the x direction produces a positive strain

$$\varepsilon_x' = \frac{\sigma_x}{E} \qquad (C.3.1)$$

where Hooke's law, $\sigma = E\varepsilon$, has been used in writing Eq. (C.3.1), and E is defined as the *modulus of elasticity*. Considering Figure C-4(c), the positive stress in the y direction produces a negative strain in the x direction as a result

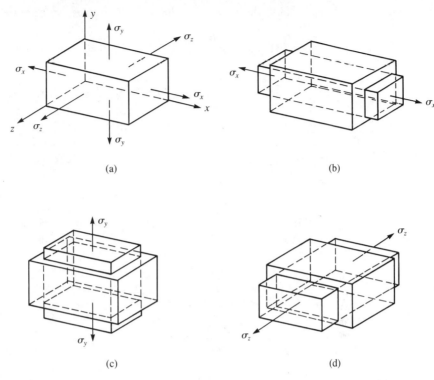

(a)

(b)

(c)

(d)

FIGURE C-4 Element subjected to normal stress acting in three mutually perpendicular directions

of Poisson's effect given by

$$\varepsilon_x'' = -\frac{v\sigma_y}{E} \qquad (C.3.2)$$

where v is Poisson's ratio. Similarly, considering Figure C-4(d), the stress in the z direction produces a negative strain in the x direction given by

$$\varepsilon_x''' = -\frac{v\sigma_z}{E} \qquad (C.3.3)$$

Using superposition of Eqs. (C.3.1), (C.3.2), and (C.3.3), we obtain

$$\varepsilon_x = \frac{\sigma_x}{E} - v\frac{\sigma_y}{E} - v\frac{\sigma_z}{E} \qquad (C.3.4)$$

The strains in the y and z directions can be determined in a manner similar to that used to obtain Eq. (C.3.4) for the x direction. They are

$$\varepsilon_y = -v\frac{\sigma_x}{E} + \frac{\sigma_y}{E} - v\frac{\sigma_z}{E}$$

$$\qquad (C.3.5)$$

$$\varepsilon_z = -v\frac{\sigma_x}{E} - v\frac{\sigma_y}{E} + \frac{\sigma_z}{E}$$

Solving Eqs. (C.3.4) and (C.3.5) for the normal stresses, we obtain

$$\sigma_x = \frac{E}{(1+v)(1-2v)}[\varepsilon_x(1-v) + v\varepsilon_y + v\varepsilon_z]$$

$$\sigma_y = \frac{E}{(1+v)(1-2v)}[v\varepsilon_x + (1-v)\varepsilon_y + v\varepsilon_z] \qquad (C.3.6)$$

$$\sigma_z = \frac{E}{(1+v)(1-2v)}[v\varepsilon_x + v\varepsilon_y + (1-v)\varepsilon_z]$$

The Hooke's law relationship, $\sigma = E\varepsilon$, used for normal stress also applies for shear stress and strain; that is,

$$\tau = G\gamma \qquad (C.3.7)$$

where G is the *shear modulus*. Hence, the expressions for the three different sets of shear strains are

$$\gamma_{xy} = \frac{\tau_{xy}}{G} \qquad \gamma_{yz} = \frac{\tau_{yz}}{G} \qquad \gamma_{zx} = \frac{\tau_{zx}}{G} \qquad (C.3.8)$$

Solving Eqs. (C.3.8) for the stresses, we have

$$\tau_{xy} = G\gamma_{xy} \qquad \tau_{yz} = G\gamma_{yz} \qquad \tau_{zx} = G\gamma_{zx} \qquad (C.3.9)$$

In matrix form, we can express the stresses in Eqs. (C.3.6) and (C.3.9) as

$$\begin{Bmatrix} \sigma_x \\ \sigma_y \\ \sigma_z \\ \tau_{xy} \\ \tau_{yz} \\ \tau_{zx} \end{Bmatrix} = \frac{E}{(1+v)(1-2v)}$$

$$\times \begin{bmatrix} 1-v & v & v & 0 & 0 & 0 \\ & 1-v & v & 0 & 0 & 0 \\ & & 1-v & 0 & 0 & 0 \\ & & & \dfrac{1-2v}{2} & 0 & 0 \\ & & & & \dfrac{1-2v}{2} & 0 \\ \text{Symmetry} & & & & & \dfrac{1-2v}{2} \end{bmatrix} \begin{Bmatrix} \varepsilon_x \\ \varepsilon_y \\ \varepsilon_z \\ \gamma_{xy} \\ \gamma_{yz} \\ \gamma_{zx} \end{Bmatrix}$$

$$(C.3.10)$$

where we note that the relationship

$$G = \frac{E}{2(1+v)}$$

has been used in Eq. (C.3.10). The square matrix on the right side of Eq. (C.3.10) is called the *stress/strain* or *constitutive matrix* and is defined by $\underline{D}$, where $\underline{D}$ is

$$
[D] = \frac{E}{(1+v)(1-2v)}
\begin{bmatrix}
1-v & v & v & 0 & 0 & 0 \\
 & 1-v & v & 0 & 0 & 0 \\
 & & 1-v & 0 & 0 & 0 \\
 & & & \dfrac{1-2v}{2} & 0 & 0 \\
 & & & & \dfrac{1-2v}{2} & 0 \\
 \text{Symmetry} & & & & & \dfrac{1-2v}{2}
\end{bmatrix}
$$

$$(C.3.11)$$

REFERENCE

[1] Timoshenko, S., and Goodier, J., *Theory of Elasticity*, 3rd ed., McGraw-Hill, New York, 1970.

Equivalent Nodal Forces

The equivalent nodal (or joint) forces for different types of loads on beam elements are shown in Table D-1 on p. 625.

PROBLEMS

D.1 Determine the equivalent joint or nodal forces for the beam elements shown in Figure PD-1.

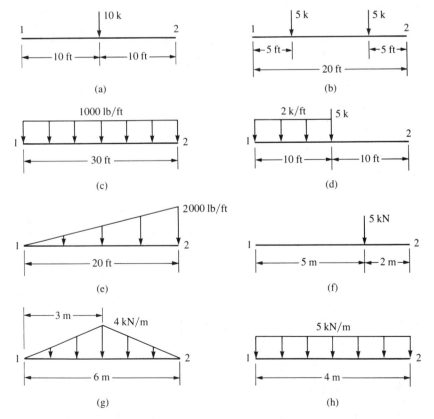

FIGURE PD-1

T A B L E   D-1   Equivalent joint forces f_0 for different types of loads

	f_{1y}	m_1	Loading case	f_{2y}	m_2
1.	$-\dfrac{P}{2}$	$-\dfrac{PL}{8}$		$-\dfrac{P}{2}$	$\dfrac{PL}{8}$
2.	$-\dfrac{Pb^2(L+2a)}{L^3}$	$-\dfrac{Pab^2}{L^2}$		$-\dfrac{Pa^2(L+2b)}{L^3}$	$\dfrac{Pa^2b}{L^2}$
3.	$-P$	$-\alpha(1-\alpha)PL$		$-P$	$\alpha(1-\alpha)PL$
4.	$-\dfrac{wL}{2}$	$-\dfrac{wL^2}{12}$		$-\dfrac{wL}{2}$	$\dfrac{wL^2}{12}$
5.	$-\dfrac{7wL}{20}$	$-\dfrac{wL^2}{20}$		$-\dfrac{3wL}{20}$	$\dfrac{wL^2}{30}$
6.	$-\dfrac{wL}{4}$	$-\dfrac{5wL^2}{96}$		$-\dfrac{wL}{4}$	$\dfrac{5wL^2}{96}$

T A B L E   D-1 (*Continued*)

	f_{1y}	m_1	Loading case	f_{2y}	m_2
7.	$-\dfrac{13wL}{32}$	$-\dfrac{11wL^2}{192}$		$-\dfrac{3wL}{32}$	$\dfrac{5wL^2}{192}$
8.	$-\dfrac{wL}{3}$	$-\dfrac{wL^2}{15}$		$-\dfrac{wL}{3}$	$\dfrac{wL^2}{15}$
9.	$-\dfrac{M(a^2 + b^2 - 4ab + L^2)}{L^3}$	$\dfrac{Mb(2a - b)}{L^2}$		$\dfrac{M(a^2 + b^2 - 4ab + L^2)}{L^3}$	$\dfrac{Ma(2b - a)}{L^2}$

Positive Nodal Force Conventions

Principle of Virtual Work

In this appendix, we will use the principle of virtual work to derive the general finite element equations for a dynamic system.

Even though the principle of virtual work strictly applies to a static system, through the introduction of D'Alembert's principle, we will be able to use the principle of virtual work to derive the finite element equations applicable for a dynamic system.

The principle of virtual work is stated as follows:

> If a deformable body in equilibrium is subjected to arbitrary virtual (imaginary) displacements associated with a compatible deformation of the body, the virtual work of external forces of the body is equal to the virtual strain energy of the internal stresses.

In the principle, *compatible displacements* are those that satisfy the boundary conditions and ensure that no discontinuities, such as voids or overlaps, occur within the body. Figure E-1 shows the hypothetical actual displacement, a compatible (admissible) displacement, and an incompatible (inadmissible) displacement for a simply supported beam. Here δv represents the variation in the transverse displacement function v. In the finite element formulation, δv would be replaced by nodal degrees of freedom δd_i. The inadmissible displacements shown in Figure E-1(b) are the result when the support condition at the right end of the beam and the continuity of displacement and slope within the beam are not satisfied. For more details of this principle, consult structural mechanics references such as Reference [1]. Also, for additional descriptions of strain energy and work done by external forces (as applied to a bar), consult Section 3.9.

Applying the principle to a finite element, we have

$$\delta U^{(e)} = \delta W^{(e)} \qquad (E.1)$$

where $\delta U^{(e)}$ is the virtual strain energy due to internal stresses and $\delta W^{(e)}$ is the virtual work of external forces on the element. We can express the internal virtual strain energy using matrix notation as

$$\delta U^{(e)} = \iiint\limits_{V} \delta \underline{\varepsilon}^T \underline{\sigma} \, dV \qquad (E.2)$$

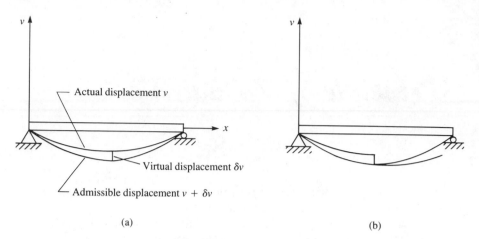

FIGURE E-1   (a) Admissible and (b) inadmissible virtual displacement functions

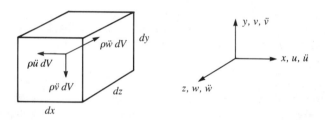

FIGURE E-2   Effective forces acting on an element

From Eq. (E.2), we can observe that internal strain energy is due to internal stresses moving through virtual strains $\delta\varepsilon$. The external virtual work is due to nodal, surface, and body forces. In addition, application of D'Alembert's principle yields effective or inertial forces $\rho\ddot{u}\, dV$, $\rho\ddot{v}\, dV$, and $\rho\ddot{w}\, dV$, where the double dots indicate second derivatives of the translations u, v, and w in the x, y, and z directions, respectively, with respect to time. These forces are shown in Figure E-2. According to D'Alembert's principle, these effective forces act in directions that are opposite to the assumed positive sense of the accelerations. We can now express the external virtual work as

$$\delta W^{(e)} = \delta\underline{d}^T\underline{P} + \iint_S \delta\underline{\psi}^T\underline{T}\, dS + \iiint_V \delta\underline{\psi}^T(\underline{X} - \rho\underline{\ddot{\psi}})\, dV \qquad (E.3)$$

where $\delta\underline{d}$ is the vector of virtual nodal displacements, $\delta\underline{\psi}$ is the vector of virtual displacement functions δu, δv, and δw, $\underline{P}$ is the nodal load matrix, $\underline{T}$ is the surface force per unit area matrix, and $\underline{X}$ is the body force per unit volume matrix.

Substituting Eqs. (E.2) and (E.3) into Eq. (E.1), we obtain

$$\iiint_V \delta\underline{\varepsilon}^T \underline{\sigma} \, dV = \delta\underline{d}^T\underline{P} + \iint_S \delta\underline{\psi}^T\underline{T} \, dS + \iiint_V \delta\underline{\psi}^T(\underline{X} - \rho\underline{\ddot{\psi}}) \, dV \qquad (E.4)$$

As shown throughout this text, shape functions are used to relate displacement functions to nodal displacements as

$$\underline{\psi} = \underline{N}\underline{d} \qquad (E.5)$$

Strains are related to nodal displacements as

$$\underline{\varepsilon} = \underline{B}\underline{d} \qquad (E.6)$$

and stresses are related to strains by

$$\underline{\sigma} = \underline{D}\underline{\varepsilon} \qquad (E.7)$$

Hence, substituting Eqs. (E.5), (E.6), and (E.7) for $\underline{\psi}, \underline{\varepsilon}$, and $\underline{\sigma}$ into Eq. (E.4), we obtain

$$\iiint_V \delta\underline{d}^T\underline{B}^T\underline{D}\underline{B}\underline{d} \, dV = \delta\underline{d}^T\underline{P} + \iint_S \delta\underline{d}^T\underline{N}^T\underline{T} \, dS$$

$$+ \iiint_V \delta\underline{d}^T\underline{N}^T(\underline{X} - \rho\underline{N}\underline{\ddot{d}}) \, dV \qquad (E.8)$$

Note that the shape functions are independent of time. Since $\underline{d}$ (or $\underline{d}^T$) is the matrix of nodal displacements, which is independent of spatial integration, we can simplify Eq. (E.8) by taking the $\underline{d}^T$ terms from the integrals to obtain

$$\delta\underline{d}^T \iiint_V \underline{B}^T\underline{D}\underline{B} \, dV \, \underline{d} = \delta\underline{d}^T\underline{P} + \delta\underline{d}^T \iint_S \underline{N}^T\underline{T} \, dS$$

$$+ \delta\underline{d}^T \iiint_V \underline{N}^T(\underline{X} - \rho\underline{N}\underline{\ddot{d}}) \, dV \qquad (E.9)$$

Since $\delta\underline{d}^T$ is an arbitrary virtual nodal displacement vector common to each term in Eq. (E.9), the following relationship must be true.

$$\iiint_V \underline{B}^T\underline{D}\underline{B} \, dV \, \underline{d} = \underline{P} + \iint_S \underline{N}^T\underline{T} \, dS + \iiint_V \underline{N}^T\underline{X} \, dV$$

$$- \iiint_V \rho\underline{N}^T\underline{N} \, dV \, \underline{\ddot{d}} \qquad (E.10)$$

We now define

$$\underline{m} = \iiint_V \rho\underline{N}^T\underline{N} \, dV \qquad (E.11)$$

$$\underline{k} = \iiint_V \underline{B}^T \underline{D} \underline{B} \, dV \qquad (E.12)$$

$$\underline{f}_s = \iint_S \underline{N}^T \underline{T} \, dS \qquad (E.13)$$

$$\underline{f}_b = \iiint_V \underline{N}^T \underline{X} \, dV \qquad (E.14)$$

Using Eqs. (E.11)–(E.14) in Eq. (E.10) and moving the last term of Eq. (E.10) to the left side, we obtain

$$\underline{m}\underline{\ddot{d}} + \underline{k}\underline{d} = \underline{P} + \underline{f}_s + \underline{f}_b \qquad (E.15)$$

The matrix $\underline{m}$ in Eq. (E.11) is the element consistent-mass matrix (also see Reference [2]), $\underline{k}$ in Eq. (E.12) is the element stiffness matrix, $\underline{f}_s$ in Eq. (E.13) is the matrix of element equivalent nodal loads due to surface forces, and $\underline{f}_b$ in Eq. (E.14) is the matrix of element equivalent nodal loads due to body forces.

Specific applications of Eq. (E.15) are given in Chapter 16 for bars and beams subjected to dynamic (time-dependent) forces. For static problems, we set $\underline{\ddot{d}}$ equal to zero in Eq. (E.15) to obtain

$$\underline{k}\underline{d} = \underline{P} + \underline{f}_s + \underline{f}_b \qquad (E.16)$$

Chapters 3–10 and 12 illustrate the use of Eq. (E.16) applied to bars, trusses, beams, frames, and plane stress, axisymmetric stress, and three-dimensional stress problems.

REFERENCES

[1] Oden, J. T., and Ripperger, E. A., *Mechanics of Elastic Structures*, 2nd ed., McGraw-Hill, New York, 1981.

[2] Archer, J. S., "Consistent Matrix Formulations for Structural Analysis Using Finite Element Techniques," *Journal of the American Institute of Aeronautics and Astronautics*, Vol. 3, No. 10, pp. 1910–1918, 1965.

Answers to Selected Problems

Chapter 2

2.1 **a.** $K = \begin{bmatrix} k_1 & 0 & -k_1 & 0 \\ 0 & k_3 & 0 & -k_3 \\ -k_1 & 0 & k_1 + k_2 & -k_2 \\ 0 & -k_3 & -k_2 & k_2 + k_3 \end{bmatrix}$

b. $d_{3x} = \dfrac{k_2 P}{k_1 k_2 + k_1 k_3 + k_2 k_3}$, $\quad d_{4x} = \dfrac{(k_1 + k_2)P}{k_1 k_2 + k_1 k_3 + k_2 k_3}$

c. $F_{1x} = \dfrac{-k_1 k_2 P}{k_1 k_2 + k_1 k_3 + k_2 k_3}$, $\quad F_{2x} = \dfrac{-k_3(k_1 + k_2)P}{k_1 k_2 + k_1 k_3 + k_2 k_3}$

2.2 $d_{2x} = 0.5$ in., $\quad F_{3x} = 500$ lb $\quad \hat{f}_{1x}^{(1)} = -\hat{f}_{2x}^{(1)} = -500$ lb,
$\hat{f}_{2x}^{(2)} = -\hat{f}_{3x}^{(2)} = -500$ lb

2.3 **a.** $K = \begin{bmatrix} k & -k & 0 & 0 & 0 \\ -k & 2k & -k & 0 & 0 \\ 0 & -k & 2k & -k & 0 \\ 0 & 0 & -k & 2k & -k \\ 0 & 0 & 0 & -k & k \end{bmatrix}$

b. $d_{2x} = \dfrac{P}{2k}$, $\quad d_{3x} = \dfrac{P}{k}$, $\quad d_{4x} = \dfrac{P}{2k}$ $\quad$ **c.** $F_{1x} = -\dfrac{P}{2}$, $\quad F_{5x} = -\dfrac{P}{2}$

2.4 **a.** K same as 2.3a. $\quad$ **b.** $d_{2x} = \dfrac{\delta}{4}$, $\quad d_{3x} = \dfrac{\delta}{2}$, $\quad d_{4x} = \dfrac{3\delta}{4}$

c. $F_{1x} = \dfrac{-k\delta}{4}$, $\quad F_{5x} = \dfrac{k\delta}{4}$

2.5 $d_{2x} = 2$ in., $\quad d_{3x} = 4$ in.
$\hat{f}_{1x}^{(1)} = -\hat{f}_{2x}^{(1)} = -1000$ lb, $\quad \hat{f}_{2x}^{(2)} = -\hat{f}_{3x}^{(2)} = -1000$ lb $\quad F_{1x} = -1000$ lb

2.6 $d_{1x} = 0$, $\quad d_{2x} = 3$ in., $\quad d_{3x} = 7$ in., $\quad d_{4x} = 11$ in.
$\hat{f}_{1x}^{(1)} = -\hat{f}_{2x}^{(1)} = -3000$ lb, $\quad \hat{f}_{2x}^{(2)} = -\hat{f}_{3x}^{(2)} = -4000$ lb
$\hat{f}_{3x}^{(3)} = -\hat{f}_{4x}^{(3)} = -4000$ lb, $\quad F_{1x} = -3000$ lb

2.7 $d_{2x} = -2$ in.
$\hat{f}_{1x}^{(1)} = -\hat{f}_{2x}^{(1)} = 2000$ lb, $\hat{f}_{2x}^{(2)} = -\hat{f}_{3x}^{(2)} = -1000$ lb
$\hat{f}_{2x}^{(3)} = -\hat{f}_{4x}^{(3)} = -1000$ lb, $F_{1x} = 2000$ lb, $F_{3x} = F_{4x} = 1000$ lb

2.8 $d_{2x} = 0.005$ m, $\hat{f}_{1x}^{(1)} = -\hat{f}_{2x}^{(1)} = -10$ N
$\hat{f}_{2x}^{(2)} = -\hat{f}_{3x}^{(2)} = -10$ N, $F_{1x} = -10$ N

2.9 $d_{2x} = 0.027$ m, $d_{3x} = 0.018$ m
$\hat{f}_{1x}^{(1)} = -\hat{f}_{2x}^{(1)} = -270$ N, $\hat{f}_{2x}^{(2)} = -\hat{f}_{3x}^{(2)} = 180$ N
$\hat{f}_{3x}^{(3)} = -\hat{f}_{4x}^{(3)} = 180$ N, $F_{1x} = -270$ N, $F_{4x} = -180$ N

2.10 $d_{2x} = 0.25$ m, $d_{3x} = 0.5$ m, $d_{4x} = 0.25$ m
$\hat{f}_{1x}^{(1)} = -\hat{f}_{2x}^{(1)} = -5$ kN, $\hat{f}_{2x}^{(2)} = -\hat{f}_{3x}^{(2)} = -5$ kN
$\hat{f}_{3x}^{(3)} = -\hat{f}_{4x}^{(3)} = 5$ kN, $\hat{f}_{4x}^{(4)} = -\hat{f}_{5x}^{(4)} = 5$ kN
$F_{1x} = -5$ kN, $F_{5x} = -5$ kN

2.11 $d_{2x} = -0.25$ m, $d_{3x} = -0.75$ m
$\hat{f}_{1x}^{(1)} = -\hat{f}_{2x}^{(1)} = 100$ N, $\hat{f}_{2x}^{(2)} = -\hat{f}_{3x}^{(2)} = 200$ N
$F_{1x} = 100$ N

2.12 $d_{3x} = 0.001$ m, $\hat{f}_{1x}^{(1)} = -\hat{f}_{3x}^{(1)} = -0.5$ kN
$\hat{f}_{2x}^{(2)} = -\hat{f}_{3x}^{(2)} = -0.5$ kN, $\hat{f}_{3x}^{(3)} = -\hat{f}_{4x}^{(3)} = 1$ kN
$F_{1x} = -0.5$ kN, $F_{2x} = -0.5$ kN, $F_{4x} = -1$ kN

2.13 **a.** $x = 0.5$ in. ↓, $\pi_{p_{min}} = -125$ lb-in.
  b. $x = 2.0$ in. ←, $\pi_{p_{min}} = -1000$ lb-in.
  c. $x = 1.962$ mm ↓, $\pi_{p_{min}} = -3849$ N · mm
  d. $x = 2.4525$ mm →, $\pi_{p_{min}} = -1203$ N · mm

2.14 $x = 2.0$ in. ↑

2.15 $x = 0.707$ in. ←, $\pi_{p_{min}} = -235.7$ in.-lb

2.16 Same as 2.7

2.17 Same as 2.12

Chapter 3

3.1 **a.** $\underline{K} = \begin{bmatrix} \dfrac{A_1 E_1}{L_1} & \dfrac{-A_1 E_1}{L_1} & 0 & 0 \\[2ex] \dfrac{-A_1 E_1}{L_1} & \dfrac{A_1 E_1}{L_1} + \dfrac{A_2 E_2}{L_2} & \dfrac{-A_2 E_2}{L_2} & 0 \\[2ex] 0 & \dfrac{-A_2 E_2}{L_2} & \dfrac{A_2 E_2}{L^2} + \dfrac{A_3 E_3}{L_3} & \dfrac{-A_3 E_3}{L_3} \\[2ex] 0 & 0 & \dfrac{-A_3 E_3}{L_3} & \dfrac{A_3 E_3}{L_3} \end{bmatrix}$

  b. $d_{2x} = \dfrac{PL}{3AE}$, $d_{3x} = \dfrac{2PL}{3AE}$

  c. **i.** $d_{2x} = 3.33 \times 10^{-4}$ in., $d_{3x} = 6.67 \times 10^{-4}$ in.
    ii. $F_{1x} = -333$ lb, $F_{4x} = -667$ lb
    iii. $\sigma^{(1)} = 333$ psi(T), $\sigma^{(2)} = 333$ psi(T), $\sigma^{(3)} = -667$ psi(C)

3.2 $d_{2x} = -1.19 \times 10^{-4}$ m, $d_{3x} = -2.38 \times 10^{-4}$ m, $F_{1x} = 10$ kN
$\hat{f}_{1x}^{(1)} = -\hat{f}_{2x}^{(1)} = 10$ kN, $\hat{f}_{2x}^{(2)} = -\hat{f}_{3x}^{(2)} = 10$ kN

3.3 $d_{2x} = 1.91 \times 10^{-3}$ in., $F_{1x} = -5715$ lb, $F_{3x} = -2286$ lb
$\hat{f}_{1x}^{(1)} = -\hat{f}_{2x}^{(1)} = -5715$ lb, $\hat{f}_{2x}^{(2)} = -\hat{f}_{3x}^{(2)} = 2286$ lb

3.4 $d_{2x} = -1.66 \times 10^{-4}$ in., $d_{3x} = -1.33 \times 10^{-3}$ in.
$F_{1x} = 667$ lb, $F_{4x} = 5333$ lb
$\hat{f}_{1x}^{(1)} = -\hat{f}_{2x}^{(1)} = 667$ lb, $\hat{f}_{2x}^{(2)} = \hat{f}_{3x}^{(2)} = 4667$ lb
$\hat{f}_{3x}^{(3)} = -\hat{f}_{4x}^{(3)} = -5333$ lb

3.5 $d_{2x} = 0.0015$ in., $d_{3x} = 0.0045$ in., $F_{1x} = -7500$ lb
$\hat{f}_{1x}^{(1)} = -\hat{f}_{2x}^{(1)} = \hat{f}_{2x}^{(2)} = -\hat{f}_{3x}^{(2)} = -7500$ lb

3.6 $d_{2x} = 3.16 \times 10^{-3}$ in., $F_{1x} = -3790$ lb, $F_{3x} = F_{4x} = -2105$ lb
$\hat{f}_{1x}^{(1)} = -\hat{f}_{2x}^{(1)} = -3790$ lb, $\hat{f}_{2x}^{(2)} = -\hat{f}_{3x}^{(2)} = \hat{f}_{2x}^{(3)} = -\hat{f}_{4x}^{(3)} = 2105$ lb

3.7 $d_{2x} = 2.21 \times 10^{-5}$ in., $d_{3x} = 6.65 \times 10^{-3}$ in.
$F_{1x} = -33.15$ lb, $F_{4x} = -9975$ lb
$\hat{f}_{1x}^{(1)} = -\hat{f}_{2x}^{(1)} = \hat{f}_{2x}^{(2)} = -\hat{f}_{3x}^{(2)} = -33.15$ lb $\quad \hat{f}_{3x}^{(3)} = -\hat{f}_{4x}^{(3)} = 9975$ lb

3.8 $d_{2x} = -0.125$ mm, $d_{3x} = -0.839$ mm

3.9 $d_{2x} = 0.01225$ m, $F_{1x} = -257.25$ kN, $F_{3x} = 267.75$ kN
$\hat{f}_{1x}^{(1)} = -\hat{f}_{2x}^{(1)} = -257.25$ kN, $\hat{f}_{2x}^{(2)} = -\hat{f}_{3x}^{(2)} = -267.75$ kN

3.10 $d_{2x} = 0.935 \times 10^{-3}$ m, $d_{3x} = 0.727 \times 10^{-3}$ m
$F_{1x} = -6.546$ kN, $F_{4x} = -1.455$ kN
$\hat{f}_{1x}^{(1)} = -\hat{f}_{2x}^{(1)} = -6.546$ kN, $\hat{f}_{2x}^{(2)} = -\hat{f}_{3x}^{(2)} = 1.455$ kN,
$\hat{f}_{3x}^{(3)} = -\hat{f}_{4x}^{(3)} = 1.455$ kN

3.11 $d_{2x} = 1.786 \times 10^{-4}$ m, $F_{1x} = -3.75$ kN, $F_{3x} = F_{4x} = F_{5x} = -3.75$ kN
$\hat{f}_{1x}^{(1)} = -\hat{f}_{2x}^{(1)} = -3.75$ kN,
$\hat{f}_{2x}^{(2)} = -\hat{f}_{3x}^{(2)} = \hat{f}_{2x}^{(3)} = -\hat{f}_{4x}^{(3)} = \hat{f}_{2x}^{(4)} = -\hat{f}_{5x}^{(4)} = 3.75$ kN

3.12 2 element solution, $d_{1x} = -0.686 \times 10^{-3}$ in.
1 element solution, $d_{1x} = -0.667 \times 10^{-3}$ in.

3.13 $B = \left[-\dfrac{1}{L} + \dfrac{4x}{L^2} \quad \dfrac{-8x}{L^2} \quad \dfrac{1}{L} + \dfrac{4x}{L^2} \right] \qquad k = A \displaystyle\int_{-L/2}^{L/2} B^T E B \, dx$

3.15 **a.** $k = 2.25 \times 10^6 \begin{bmatrix} 1 & 1 & -1 & -1 \\ 1 & 1 & -1 & -1 \\ -1 & -1 & 1 & 1 \\ -1 & -1 & 1 & 1 \end{bmatrix}$ lb/in.

b. $k = \dfrac{10^6}{4} \begin{bmatrix} 1 & -\sqrt{3} & -1 & \sqrt{3} \\ -\sqrt{3} & 3 & \sqrt{3} & -3 \\ -1 & \sqrt{3} & 1 & -\sqrt{3} \\ \sqrt{3} & -3 & -\sqrt{3} & 3 \end{bmatrix}$ lb/in.

c. $k = 7000 \begin{bmatrix} 3 & -\sqrt{3} & -3 & \sqrt{3} \\ -\sqrt{3} & 1 & \sqrt{3} & -1 \\ -3 & \sqrt{3} & 3 & -\sqrt{3} \\ \sqrt{3} & -1 & -\sqrt{3} & 1 \end{bmatrix}$ kN/m

$$\textbf{d. } \underline{k} = 1.4 \times 10^4 \begin{bmatrix} 0.883 & 0.321 & -0.883 & -0.321 \\ 0.321 & 0.117 & -0.321 & -0.117 \\ -0.883 & -0.321 & 0.883 & 0.321 \\ -0.321 & -0.117 & 0.321 & 0.117 \end{bmatrix} \text{kN/m}$$

3.16. a. $\hat{d}_{1x} = 0.433$ in., $\hat{d}_{2x} = 0.592$ in.
  b. $\hat{d}_{1x} = 0.433$ in., $\hat{d}_{2x} - -0.1585$ in.

3.17 a. $\hat{d}_{1x} = 2.165$ mm, $\hat{d}_{1y} = -1.25$ mm,
  $\hat{d}_{2x} = 0.098$ mm, $\hat{d}_{2y} = -5.83$ mm
  b. $\hat{d}_{1x} = -1.25$ mm, $\hat{d}_{1y} = 2.165$ mm,
  $\hat{d}_{2x} = 3.03$ mm, $\hat{d}_{2y} = 5.098$ mm

3.18 a. $\sigma = 10,600$ psi, **b.** 45.47 MPa

3.19 a. $\underline{K} = k \begin{bmatrix} 2 & 0 & -\frac{1}{2} & \frac{1}{2} & -1 & 0 & -\frac{1}{2} & -\frac{1}{2} \\ 0 & 1 & \frac{1}{2} & -\frac{1}{2} & 0 & 0 & -\frac{1}{2} & -\frac{1}{2} \\ -\frac{1}{2} & \frac{1}{2} & \frac{1}{2} & -\frac{1}{2} & 0 & 0 & 0 & 0 \\ \frac{1}{2} & -\frac{1}{2} & -\frac{1}{2} & \frac{1}{2} & 0 & 0 & 0 & 0 \\ -1 & 0 & 0 & 0 & 1 & 0 & 0 & 0 \\ 0 & 0 & 0 & 0 & 0 & 0 & 0 & 0 \\ -\frac{1}{2} & -\frac{1}{2} & 0 & 0 & 0 & 0 & \frac{1}{2} & \frac{1}{2} \\ -\frac{1}{2} & -\frac{1}{2} & 0 & 0 & 0 & 0 & \frac{1}{2} & \frac{1}{2} \end{bmatrix}$

  b. $d_{1x} = 0$, $d_{1y} = \dfrac{-10}{k}$

3.20 $d_{2x} = 0$, $d_{2y} = 0.285$ in., $\sigma^{(1)} = \sigma^{(2)} = 1414$ psi(T)

3.21 $d_{1x} = \dfrac{231L}{AE}$, $d_{1y} = \dfrac{43.5L}{AE}$

3.22 $d_{1x} = \dfrac{422L}{AE}$, $d_{1y} = \dfrac{1570L}{AE}$

  $\sigma^{(1)} = \dfrac{574}{A}$(C), $\sigma^{(2)} = \dfrac{422}{A}$(T), $\sigma^{(3)} = \dfrac{996}{A}$(T)

3.23 $d_{1x} = 0.12$ in., $d_{1y} = 0$, $\sigma^{(1)} = 6000$ psi

3.24 $d_{2x} = \dfrac{26,675}{AE}$, $d_{2y} = \dfrac{105,021}{AE}$, $d_{3x} = \dfrac{-26,675}{AE}$, $d_{3y} = \dfrac{105,021}{AE}$
  $\hat{f}_{1x}^{(1)} = -\hat{f}_{2x}^{(1)} = -1333$ lb, $\hat{f}_{1x}^{(2)} = -\hat{f}_{3x}^{(2)} = -1667$ lb
  $\hat{f}_{2x}^{(3)} = -\hat{f}_{4x}^{(3)} = 1667$ lb, $\hat{f}_{2x}^{(4)} = -\hat{f}_{3x}^{(4)} = 0$
  $\hat{f}_{3x}^{(5)} = -\hat{f}_{4x}^{(5)} = 1333$ lb, $\hat{f}_{1x}^{(6)} = -\hat{f}_{4x}^{(6)} = 0$

3.25 $d_{2x} = 0$, $d_{2y} = \dfrac{225,000}{AE}$, $d_{3x} = \dfrac{-53,340}{AE}$, $d_{3y} = \dfrac{210,000}{AE}$
  $\hat{f}_{1x}^{(1)} = -\hat{f}_{2x}^{(1)} = 0$, $\hat{f}_{1x}^{(2)} = -\hat{f}_{3x}^{(2)} = -3333$ lb
  $\hat{f}_{2x}^{(4)} = -\hat{f}_{3x}^{(4)} = 1000$ lb, $\hat{f}_{3x}^{(5)} = -\hat{f}_{4x}^{(5)} = 2667$ lb
  $\hat{f}_{1x}^{(6)} = -\hat{f}_{4x}^{(6)} = 0$

3.26 No, the truss is unstable, $|\underline{K}| = 0$.

3.27 $d_{3x} = 0.0463$ in., $d_{3y} = -0.0176$ in.
  $\hat{f}_{1x}^{(1)} = -\hat{f}_{3x}^{(1)} = -2.055$ k, $\hat{f}_{2x}^{(2)} = -\hat{f}_{3x}^{(2)} = 6.279$ k
  $\hat{f}_{3x}^{(3)} = -\hat{f}_{4x}^{(3)} = -6.6$ k

3.28 $\underline{T}^T = \begin{bmatrix} C & -S & 0 & 0 \\ S & C & 0 & 0 \\ 0 & 0 & C & -S \\ 0 & 0 & S & C \end{bmatrix}$ and $\underline{T}\underline{T}^T = \begin{bmatrix} 1 & 0 & 0 & 0 \\ 0 & 1 & 0 & 0 \\ 0 & 0 & 1 & 0 \\ 0 & 0 & 0 & 1 \end{bmatrix}$

$\therefore \underline{T}^T = \underline{T}^{-1}$

3.29 $d_{1x} = -0.893 \times 10^{-4}$ m, $d_{1y} = -4.46 \times 10^{-4}$ m
$\sigma^{(1)} = 31.2$ MPa (T), $\sigma^{(2)} = 26.5$ MPa (T), $\sigma^{(3)} = 6.25$ MPa (T)

3.30 $d_{1x} = 1.71 \times 10^{-4}$ m, $d_{1y} = -7.55 \times 10^{-4}$ m
$\sigma^{(1)} = 79.28$ MPa (T), $\sigma^{(2)} = 11.97$ MPa (T), $\sigma^{(3)} = -23.87$ MPa (C)

3.31 $d_{1x} = 8.25 \times 10^{-4}$ m, $d_{1y} = -3.65 \times 10^{-3}$ m
$\sigma^{(2)} = 57.74$ MPa (T), $\sigma^{(3)} = -115.5$ MPa (C)

3.32 $d_{2x} = 0.135 \times 10^{-2}$ m, $d_{2y} = -0.850 \times 10^{-2}$ m,
$d_{3y} = -0.137 \times 10^{-1}$ m $d_{4y} = -0.164 \times 10^{-1}$ m,
$\sigma^{(1)} = -198$ MPa (C), $\sigma^{(2)} = 0$, $\sigma^{(3)} = 44.6$ MPa (T)
$\sigma^{(4)} = -31.6$ MPa (C), $\sigma^{(5)} = -191$ MPa (C),
$\sigma^{(6)} = -63.1$ MPa (C)

3.33 $d_{1x} = -1.724 \times 10^{-3}$ m, $d_{1y} = -3.448 \times 10^{-3}$ m
$\sigma^{(1)} = 51.2$ MPa (T), $\sigma^{(2)} = -36.2$ MPa (C)

3.34 $d_{4x} = 9.93 \times 10^{-3}$ in., $d_{4y} = -2.46 \times 10^{-3}$ in.
$\sigma^{(1)} = 31.25$ ksi (T), $\sigma^{(2)} = 3.459$ ksi (T), $\sigma^{(3)} = -1.538$ ksi (C)
$\sigma^{(4)} = -3.103$ ksi (C), $\sigma^{(5)} = 0$

3.35 $d_{1y} = -0.5 \times 10^{-3}$ in., $\sigma^{(1)} = 250$ psi (T)

3.36 $\hat{d}_{1x} = 0.212$ in.

3.37 $\hat{d}_{1x} = 0.0397$ in.

3.38 $\hat{d}_{2x} = 16.98$ mm

3.39 $\hat{d}_{2x} = 1.71$ mm

3.40 $d_{1x} = -3.018 \times 10^{-5}$ m, $d_{1y} = -1.517 \times 10^{-5}$ m,
$d_{1x} = 2.684 \times 10^{-5}$ m, $\sigma^{(1)} = -338$ kN/m² (C),
$\sigma^{(2)} = -1690$ kN/m² (C), $\sigma^{(3)} = -7965$ kN/m² (C)
$\sigma^{(4)} = -2726$ kN/m² (C)

3.41 $d_{1x} = 1.383 \times 10^{-3}$ m, $d_{1y} = -5.119 \times 10^{-5}$ m
$d_{1x} = 6.015 \times 10^{-5}$ m, $\sigma^{(1)} = 20.51$ MPa (T),
$\sigma^{(2)} = 4.21$ MPa (T), $\sigma^{(3)} = -5.29$ MPa (C)

3.42 $d_{5x} = 0.0014$ in., $d_{5y} = 0$, $d_{5z} = -0.00042$ in.
$\sigma^{(1)} = \sigma^{(4)} = 180$ psi (T), $\sigma^{(2)} = \sigma^{(3)} = 140$ psi (C)

3.43 $d_{4x} = 0.00863$ in., $d_{4y} = 0$, $d_{4z} = -0.00683$ in.
$\sigma^{(1)} = -916$ psi (C)

3.46 $d_{1x}' = 0$, $d_{2y} = -0.00283$ in., $F_{2x} = 2000$ lb
$\sigma^{(1)} = 0$, $\sigma^{(2)} = 1414$ psi (T), $\sigma^{(3)} = 0$

3.47 $d_{2y} = -0.00283$ in.

3.48 $d_{2y}' = -0.002$ in., $f_{1x}' = -2800$ lb., $f_{2x}' = -2000$ lb
$f_{2y}' = -2000$ lb

3.49 **a.** $d_{1x} = 0.010$ in. $\downarrow$, $\pi_{p_{min}} = -100$ lb-in.
     b. $d_{1x} = 0.00833$ in. $\rightarrow$, $\pi_{p_{min}} = -41.67$ lb-in.

3.50 $\underline{k} = \dfrac{3A_0E}{2L}\begin{bmatrix} 1 & -1 \\ -1 & 1 \end{bmatrix}$

3.51 2-element solution: $d_{2x} = 0.00825$ in., $d_{3x} = 0.012$ in., $\sigma^{(1)} = 8250$ psi (T), $\sigma^{(2)} = 3750$ psi (T),

3.52 2-element solution: $d_{2x} = 6.75 \times 10^{-3}$ in., $d_{3x} = 0.009$ in. $\sigma^{(1)} = 6750$ psi (T), $\sigma^{(2)} = 2250$ psi (T)

3.55 **a.** $f_{1x} = 583.3$ lb, $f_{2x} = 666.7$ lb
     b. $f_{1x} = 26.7$ kN, $f_{2x} = 80$ kN

Chapter 4

4.1 $d_{2y} = -0.0192$ in., $d_{3y} = -0.0168$ in.
     $\sigma^{(1)} = -1688$ psi (C), $\sigma^{(2)} = 1332$ psi (T), $\sigma^{(3)} = 1000$ psi (T)

4.2 $d_{1x} = \dfrac{-110P}{AE}$ in., $d_{1y} = 0$, $d_{2x} = 0$, $d_{2y} = \dfrac{-405P}{AE}$ in.,

$d_{3x} = 0$, $d_{3y} = \dfrac{-433P}{AE}$ in., $d_{4x} = \dfrac{50P}{AE}$ in., $d_{4y} = \dfrac{-208P}{AE}$ in.

$\sigma^{(1)} = -0.156\dfrac{P}{A}$, $\sigma^{(2)} = -0.208\dfrac{P}{A}$, $\sigma^{(3)} = -1.16\dfrac{P}{A}$

$\sigma^{(4)} = 0.260\dfrac{P}{A}$, $\sigma^{(5)} = -0.573\dfrac{P}{A}$, $\sigma^{(6)} = 0.458\dfrac{P}{A}$

4.3 $d_{2y} = -0.955 \times 10^{-2}$ m, $d_{4y} = -1.03 \times 10^{-2}$ m,
     $\sigma^{(1)} = 67.1$ MPa (C), $\sigma^{(2)} = 60.0$ MPa (T), $\sigma^{(3)} = 22.4$ MPa (C)
     $\sigma^{(4)} = 44.7$ MPa (C), $\sigma^{(3)} = 20.0$ MPa (T)

4.4 **a.** $n_b = 8$ **b.** $n_b = 12$

4.5 $d_{2y} = -0.370 \times 10^{-3}$ m, $d_{3y} = -0.307 \times 10^{-3}$ m,
     $d_{4y} = -0.386 \times 10^{-3}$ m, $\sigma^{(1)} = 4.71$ MPa (C),
     $\sigma^{(2)} = \sigma^{(3)} = \sigma^{(6)} = 3.33$ MPa (T), $\sigma^{(4)} = 3.33$ MPa (C)

4.6 $d_{2x} = 0.0223$ in., $d_{2y} = 0.00528$ in., $\sigma^{(1)} = 880$ psi (T),
     $\sigma^{(2)} = 1033$ psi (T), $\sigma^{(3)} = 826$ psi (C), $\sigma^{(4)} = 1467$ psi (C)

4.7 $A = 0.444$ in.^{2}

4.8 $d_{1x} = 0.8342$ in., $\sigma^{(1)} = \sigma^{(3)} = -12.37$ ksi (C), $\sigma^{(2)} = \sigma^{(4)} = 9.74$ ksi (T)

4.10 $d_{1x} = -0.711 \times 10^{-1}$ in., $d_{1y} = 0$, $d_{1x} = -0.266$ in.
     $\sigma^{(3)} = 2868$ psi (C), $\sigma^{(1)} = \sigma^{(4)} = 948$ psi (C)
     $\sigma^{(2)} = \sigma^{(5)} = 1445$ psi (T)

4.11 $d_{1x} = 0.00391$ m, $d_{1y} = -0.0150$ m, $d_{1z} = -0.00458$ m
     $\sigma^{(1)} = 41.1$ MPa (C), $\sigma^{(2)} = 43.0$ MPa (C), $\sigma^{(3)} = 19.65$ MPa (C),
     $\sigma^{(4)} = 27.8$ MPa (T)

4.12 $d_{7x} = 0.102$ m, $d_{7y} = 0.0189$ m, $\sigma^{(1)} = 440$ MPa (T),
     $\sigma^{(2)} = 227$ MPa (T), $\sigma^{(3)} = -245$ MPa (C)

4.13 $d_{6x} = 1.14$ mm, $d_{6y} = -0.571$ mm, $d_{6z} = 0.286$ mm
     $\sigma^{(1)} = 0$, $\sigma^{(3)} = 20$ MPa (C), $\sigma^{(7)} = 40.0$ MPa (C)

Chapter 5

5.3 $d_{2y} = \dfrac{-7PL^3}{768EI}$, $\phi_1 = \dfrac{-PL^2}{32EI}$, $\phi_2 = \dfrac{PL^2}{128EI}$

$F_{1y} = \dfrac{5P}{16}$, $M_1 = 0$, $F_{3y} = \dfrac{11P}{16}$, $M_3 = \dfrac{-3PL}{16}$

5.4 $d_{1y} = \dfrac{-PL^3}{3EI}$, $\phi_1 = \dfrac{PL^3}{2EI}$, $F_{2y} = P$, $M_2 = -PL$

5.5 $d_{1y} = -2.688$ in., $\phi_1 = 0.0144$ rad, $\phi_2 = 0.0048$ rad
$F_{2y} = 2.5$ k, $F_{3y} = -1.5$ k, $M_3 = 10.0$ k-ft

5.8 $d_{2y} = -1.34 \times 10^{-4}$ m, $\phi_2 = 8.93 \times 10^{-5}$ rad
$F_{1y} = 10$ kN, $M_1 = 12.5$ kN·m, $F_{3y} = 1.87$ N, $M_3 = -2.5$ kN·m

5.9 $d_{3y} = -7.619 \times 10^{-4}$ m, $\phi_2 = -3.809 \times 10^{-4}$ rad, $\phi_1 = 1.904 \times 10^{-4}$ rad
$F_{1y} = -0.889$ kN, $F_{2y} = 4.889$ kN

5.10 $d_{2y} = -0.886$ in., $\phi_2 = -0.00554$ rad
$F_{1y} = 1115$ lb, $M_1 = -267$ k-in.

5.11 $d_{2y} = -7.934 \times 10^{-3}$ m, $\phi_1 = -2.975 \times 10^{-3}$ rad
$F_{1y} = 5.208$ kN, $F_{3y} = 5.208$ kN
$F_{\text{spring}} = 1.587$ kN

5.12 $d_{2y} = d_{4y} = \dfrac{-1wL^4}{607.5EI}$, $d_{3y} = \dfrac{-wL^4}{507EI}$

$\phi_2 = \dfrac{-1wL^3}{270EI}$, $\phi_4 = -\phi_2$

$F_{1y} = \dfrac{wL}{2}$, $M_1 = \dfrac{wL^2}{12}$

5.13 $d_{2y} = \dfrac{-wL^4}{384EI}$, $F_{1y} = \dfrac{wL}{2}$, $M_1 = \dfrac{wL^2}{12}$

5.14 $d_{2y} = \dfrac{-5wL^4}{384EI}$, $\phi_1 = -\phi_3 = \dfrac{-wL^3}{24EI}$, $F_{1y} = \dfrac{wL}{2}$

5.15 $d_{3y} = \dfrac{-wL^4}{4EI}$, $\phi_2 = \dfrac{-wL^3}{8EI}$, $\phi_3 = \dfrac{-7wL^3}{24EI}$

$F_{1y} = \dfrac{-3wL}{4}$, $M_1 = \dfrac{-wL^2}{4}$, $F_{2y} = \dfrac{7wL}{4}$

5.16 $\hat{f}_{1y} = \dfrac{-3wL}{20}$, $\hat{m}_1 = \dfrac{-wL^2}{30}$, $\hat{f}_{2y} = \dfrac{-7wL}{20}$, $\hat{m}_2 = \dfrac{wL^2}{20}$

5.17 $F_{1y} = \dfrac{3wL}{20}$, $M_1 = \dfrac{wL^2}{30}$, $F_{3y} = \dfrac{7wL}{20}$, $M_3 = \dfrac{-wL^2}{20}$

5.18 $\phi_2 = \dfrac{wL^3}{80EI}$, $F_{1y} = \dfrac{9wL}{40}$, $M_1 = \dfrac{7wL^2}{120}$, $F_{2y} = \dfrac{11wL}{40}$

5.19 $d_{3y} = -0.0244$ m, $\phi_3 = -0.0071$ rad, $\phi_2 = -0.00305$ rad
$F_{1y} = -24$ kN, $M_1 = -32$ kN·m, $F_{2y} = 56$ kN
$\hat{f}_{1y}^{(1)} = -\hat{f}_{2y}^{(1)} = -24$ kN, $\hat{m}_1^{(1)} = -32$ kN·m, $\hat{m}_2^{(1)} = -64$ kN·m
$\hat{f}_{2y}^{(2)} = 32$ kN, $\hat{m}_2^{(2)} = 64$ kN·m, $\hat{f}_{3y}^{(2)} = 0$, $\hat{m}_3^{(2)} = 0$

5.20 $\phi_1 = -0.0032$ rad, $d_{2y} = -0.0115$ m, $\phi_3 = 0.0032$ rad
$F_{1y} = 29.94$ kN, $F_{2y} = 0.1152$ kN, $F_{3y} = 29.94$ kN
$\hat{f}_{1y}^{(1)} = 29.94$ kN, $\hat{m}_1^{(1)} = 0$, $\hat{f}_{2y}^{(1)} = 0.058$ kN, $\hat{m}_2^{(1)} = 59.65$ kN·m

5.21 $d_{2y} = -2.514$ in., $\phi_2 = -0.00698$ rad, $\phi_3 = 0.0279$ rad
$F_{1y} = 37.5$ k, $M_1 = 225$ k-ft, $F_{3y} = 22.5$ k

5.22 $d_{3y} = -3.277$ in., $\phi_3 = -0.0323$ rad, $\phi_2 = -0.0130$ rad
$F_{1y} = -20.5$ k, $M_1 = -71.67$ k-ft, $F_{2y} = 60.5$ k

5.23 $d_{2y} = -2.34$ in., $F_{1y} = 5325$ lb $= F_{3y}$, $M_1 = 19,900$ lb-ft $= -M_3$

5.24 $\phi_1 = -3.596 \times 10^{-4}$ rad, $\phi_2 = 9.92 \times 10^{-5}$ rad, $\phi_3 = 1.091 \times 10^{-4}$ rad
$F_{1y} = 9875$ N, $F_{2y} = 28,406$ N, $F_{3y} = 6719$ N

5.25 $d_{2y} = \dfrac{-PL^3}{192EI} - \dfrac{wL^4}{384EI}$, $F_{1y} = \dfrac{P + wL}{2}$, $M_1 = \dfrac{PL}{8} + \dfrac{wL^2}{12}$

5.26 $d_{2y} = \dfrac{-5PL^3}{648EI}$

5.27 $d_{2y} = \dfrac{-(25P + 22wL)L^3}{240EI}$, $\phi_2 = \dfrac{-(PL^2 + wL^3)}{8EI}$

  $F_{1y} = P + \dfrac{wL}{2}$, $M_1 = \dfrac{PL}{2} + \dfrac{wL^2}{3}$

5.34 $\hat{\underline{k}} = EI \displaystyle\int_0^L [B]^T[B]\, d\hat{x} + k_f \int_0^L [N]^T[N]\, d\hat{x}$

Chapter 6

6.1 $d_{2x} = 0.0278$ in., $d_{2y} = 0$, $\phi_2 = -0.555 \times 10^{-4}$ rad
$\hat{f}_{1x}^{(1)} = -\hat{f}_{2x}^{(1)} = -8300$ lb, $\hat{f}_{1y}^{(1)} = -\hat{f}_{2y}^{(1)} = 4.6$ lb
$\hat{m}_1^{(1)} = 2775$ lb-in., $\hat{m}_2^{(1)} = 0$

6.2 $d_{2x} = d_{3x} = 0.688$ in., $d_{2y} = -d_{3y} = 0.00171$ in.
$\phi_2 = -\phi_3 = -0.00173$ rad
$\hat{f}_{1x}^{(1)} = -\hat{f}_{2x}^{(1)} = -2140$ lb, $\hat{f}_{1y}^{(1)} = -\hat{f}_{2y}^{(1)} = -2503$ lb
$\hat{m}_1^{(1)} = 343,600$ lb-in., $\hat{m}_2^{(1)} = 257,000$ lb-in.
$\hat{f}_{2x}^{(2)} = -\hat{f}_{3x}^{(2)} = 2497$ lb, $\hat{f}_{2y}^{(2)} = -\hat{f}_{3y}^{(2)} = -2140$ lb
$\hat{m}_2^{(2)} = -257,000$ lb-in., $\hat{m}_3^{(2)} = -256,600$ lb-in.
$\hat{f}_{3x}^{(3)} = -\hat{f}_{4x}^{(3)} = 2140$ lb, $\hat{f}_{3y}^{(3)} = -\hat{f}_{4y}^{(3)} = 2497$ lb
$\hat{m}_3^{(3)} = 256,600$ lb-in., $\hat{m}_4^{(3)} = 342,700$ lb-in.
$F_{1x} = F_{4x} = -2503$ lb, $F_{1y} = -F_{4y} = -2140$ lb
$M_1 = 343,600$ lb-in., $M_4 = 342,700$ lb-in.

6.3 Channel section 6 × 8.2 based on $M_{max} = 106,900$ lb-in.

6.4 $d_{4x} = 0.00445$ in., $d_{4y} = -0.0123$ in., $\phi_4 = -0.00290$ rad
$\hat{f}_{1x}^{(1)} = -\hat{f}_{4x}^{(1)} = 4.04$ k, $\hat{f}_{1y}^{(1)} = -\hat{f}_{4y}^{(1)} = -1.43$ k
$\hat{m}_1^{(1)} = -254$ k-in., $\hat{m}_4^{(1)} = -513$ k-in.
$\hat{f}_{2x}^{(2)} = -\hat{f}_{4x}^{(2)} = 5.82$ k, $\hat{f}_{2y}^{(2)} = -\hat{f}_{4y}^{(2)} = -1.45$ k
$\hat{m}_2^{(2)} = -260$ k-in., $\hat{m}_4^{(2)} = -519$ k-in.
$F_{1x} = 3.1$ k, $F_{1y} = 2.96$ k, $M_1 = -254$ k-in.
$F_{2x} = -1.31$ k, $F_{2y} = 5.86$ k, $M_2 = -260$ k-in.
$F_{3x} = -1.78$ k, $F_{3y} = 11.17$ k, $M_3 = -1736$ k-in.

6.5 $d_{2x} = 0.05618$ in., $d_{2y} = -0.1792$ in., $\phi_2 = -0.00965$ rad
$\hat{f}_{1x}^{(1)} = 90.07$ k, $\hat{f}_{1y}^{(1)} = 3.83$ k, $\hat{m}_1^{(1)} = 361$ k-in.
$\hat{f}_{2x}^{(1)} = -73.43$ k, $\hat{f}_{2y}^{(1)} = 7.27$ k, $\hat{m}_2^{(1)} = -1106$ k-in.
$\hat{f}_{2x}^{(2)} = -\hat{f}_{3x}^{(2)} = 46.8$ k, $\hat{f}_{2y}^{(2)} = 17.05$ k, $\hat{m}_2^{(2)} = 1107$ k-in.
$\hat{f}_{3y}^{(2)} = 22.95$ k, $\hat{m}_3^{(2)} = -2171$ k-in.
$F_{1x} = F_{3x} = 46.8$ k, $F_{1y} = 77.1$ k, $M_1 = 361$ k-in.
$F_{3y} = 22.95$ k, $M_3 = 2171$ k-in.

6.6 $d_{2x} = -0.000269$ in., $d_{2y} = -0.0363$ in., $\phi_2 = -0.00347$ rad
$\hat{f}_{1x}^{(1)} = 46.6$ k, $\hat{f}_{1y}^{(1)} = 6.07$ k, $\hat{m}_1^{(1)} = 491.3$ k-in.
$\hat{f}_{2x}^{(1)} = -32.4$ k, $\hat{f}_{2y}^{(1)} = 8.07$ k, $\hat{m}_2^{(1)} = -831.3$ k-in.
$\hat{f}_{2x}^{(2)} = -\hat{f}_{3x}^{(2)} = -0.28$ k, $\hat{f}_{2y}^{(2)} = 58.31$ k, $\hat{m}_2^{(2)} = 1123.9$ k-in.
$\hat{f}_{3y}^{(2)} = 21.69$ k, $\hat{m}_3^{(2)} = -1611.8$ k-in.
$\hat{f}_{4x}^{(3)} = -\hat{f}_{2x}^{(3)} = 50.2$ k, $\hat{f}_{4y}^{(3)} = -\hat{f}_{2y}^{(3)} = -1.49$ k, $\hat{m}_4^{(3)} = -154.2$ k-in.
$\hat{m}_2^{(3)} = -293.2$ k-in.
$F_{1x} = 28.65$ k, $F_{1y} = 37.24$ k, $M_1 = 491.3$ k-in.
$F_{3x} = 0.28$ k, $F_{3y} = 21.69$ k, $M_3 = -1611.8$ k-in.
$F_{4x} = -28.93$ k, $F_{4y} = 41.05$ k, $M_4 = -154.2$ k-in.

6.7 $d_{2x} = 0.4308 \times 10^{-4}$ m, $d_{2y} = -0.9067 \times 10^{-4}$ m,
$\phi_2 = -0.1403 \times 10^{-2}$ rad
$\hat{f}_{1x}^{(1)} = -\hat{f}_{2x}^{(1)} = 23.8$ kN, $\hat{f}_{1y}^{(1)} = 17.26$ kN, $\hat{m}_1^{(1)} = 32.77$ kN·m
$\hat{f}_{2y}^{(1)} = 22.74$ kN, $\hat{m}_2^{(1)} = -54.64$ kN·m
$\hat{f}_{2x}^{(2)} = -\hat{f}_{3x}^{(2)} = 11.31$ kN, $\hat{f}_{2y}^{(2)} = 37.19$ kN, $\hat{m}_2^{(2)} = 65.09$ kN·m
$\hat{f}_{3y}^{(2)} = 42.81$ kN, $\hat{m}_3^{(2)} = -87.54$ kN·m
$\hat{f}_{2x}^{(3)} = -\hat{f}_{4x}^{(3)} = 17.55$ kN, $\hat{f}_{2y}^{(3)} = -\hat{f}_{4y}^{(3)} = 1.40$ kN
$\hat{m}_2^{(3)} = -10.51$ kN·m, $\hat{m}_4^{(3)} = -5.30$ kN·m
$F_{1x} = -17.26$ kN, $F_{1y} = 23.80$ kN, $M_1 = 32.77$ kN·m
$F_{3x} = -11.31$ kN, $F_{3y} = 42.81$ kN, $M_3 = -87.54$ kN·m
$F_{4x} = -11.42$ kN, $F_{4y} = 13.40$ kN, $M_4 = -5.30$ kN·m

6.9 $d_{2x} = -4.95 \times 10^{-5}$ m, $d_{2y} = -2.56 \times 10^{-5}$ m, $\phi_2 = 2.66 \times 10^{-3}$ rad
$\hat{f}_{1x}^{(1)} = -\hat{f}_{2x}^{(1)} = 26.9$ kN, $\hat{f}_{1y}^{(1)} = -\hat{f}_{2y}^{(1)} = -42.0$ kN
$\hat{m}_1^{(1)} = 55.9$ kN·m, $\hat{m}_2^{(1)} = 111.7$ kN·m
$\hat{f}_{2x}^{(2)} = -\hat{f}_{3x}^{(2)} = -42.0$ kN, $\hat{f}_{2y}^{(2)} = -\hat{f}_{3y}^{(2)} = 26.9$ kN
$M_1 = 55.9$ kN·m, $M_3 = 44.7$ kN·m

6.10 $d_{2y} = -0.1423 \times 10^{-2}$ m, $\phi_2 = -0.5917 \times 10^{-3}$ rad
$\hat{f}_{1x}^{(1)} = 0$, $\hat{f}_{1y}^{(1)} = 10$ kN, $\hat{m}_1^{(1)} = 23.3$ kN·m, $\hat{f}_{2x}^{(1)} = 0$,
$\hat{f}_{2y}^{(1)} = -10$ kN, $\hat{m}_2^{(1)} = 6.7$ kN·m

6.11 $d_{2y} = -0.928 \times 10^{-5}$ m, $F_{1x} = 1860$ N, $F_{1y} = 2500$ N
$M_1 = 28$ N·m

6.13 $d_{2x} = 0.0559$ in., $d_{2y} = 0.00382$ in., $\phi_2 = -0.000150$ rad
$d_{3x} = 0.0558$ in., $d_{3y} = -0.000133$ in., $\phi_3 = 0.000149$ rad
$F_{1x} = -198$ lb, $F_{1y} = -4770$ lb, $M_1 = 27460$ lb·in.
$F_{4x} = -4802$ lb, $F_{4y} = 4770$ lb, $M_4 = 27430$ lb·in.

6.14 $d_{2x} = 0.0174$ in., $d_{2y} = -0.0481$ in., $\phi_2 = -0.00165$ rad
$\hat{f}_{1x}^{(1)} = 19160$ lb, $\hat{f}_{1y}^{(1)} = -1385$ lb, $\hat{m}_1^{(1)} = -59050$ lb·in.
$\hat{f}_{2x}^{(1)} = -19160$ lb, $\hat{f}_{2y}^{(1)} = 1385$ lb, $\hat{m}_2^{(1)} = -176{,}000$ lb·in.

6.15 $d_{2x} = -1.76 \times 10^{-2}$ m, $d_{2y} = -1.87 \times 10^{-5}$ m, $\phi_2 = 5.00 \times 10^{-3}$ rad
$d_{3x} = -1.76 \times 10^{-2}$ m, $\phi_3 = -2.49 \times 10^{-3}$ rad
$F_{1x} = 20.0$ kN, $F_{1y} = 13.1$ kN, $M_1 = -57.4$ kN·m, $F_{3y} = -13.1$ kN

6.16 $d_{3y} = -2.83 \times 10^{-5}$ m, $d_{4x} = 1.0 \times 10^{-5}$ m, $d_{4y} = -2.03 \times 10^{-5}$ m

6.17 $d_{3y} = -0.397$ in., $\phi_3 = 0$

6.18 $d_{2x} = d_{2y} = -5 \times 10^{-6}$ m, $\phi_2 = 8.83 \times 10^{-5}$ rad

6.19 $d_{1x} = 0.702$ in., $d_{1y} = 0.00797$ in., $\phi_1 = -0.00446$ rad
$\hat{f}_{3x}^{(1)} = -\hat{f}_{1x}^{(1)} = -19.93$ k, $\hat{f}_{3y}^{(1)} = -\hat{f}_{1y}^{(1)} = 18.1$ k, $\hat{m}_3^{(1)} = 1309$ k·in.
$\hat{m}_1^{(1)} = 863$ k·in.

6.20 $d_{3x} = 1.24$ in., $d_{3y} = 0.00203$ in., $\phi_3 = -0.000556$ rad
$\hat{f}_{1x}^{(1)} = -2.76$ k, $\hat{f}_{1y}^{(1)} = 1.79$ k, $\hat{m}_1^{(1)} = 0$, $\hat{f}_{2x}^{(1)} = 2.76$ k, $\hat{f}_{2y}^{(1)} = -1.79$ k,
$\hat{m}_2^{(1)} = 322$ k·in.

6.23 $d_{5x} = 0.0204$ in., $d_{5y} = 0.00122$ in., $\phi_5 = 0.000207$ rad

6.24 $d_{5x} = 2.82$ in., $d_{5y} = 0.00266$ in., $\phi_5 = -0.00139$ rad

6.26 $d_{2x} = 0.596 \times 10^{-5}$ in., $d_{2y} = -0.332 \times 10^{-2}$ in., $\phi_2 = -0.100 \times 10^{-3}$ rad
$F_{1x} = 130$ lb, $F_{1y} = 10360$ lb, $F_{4x} = -130$ lb, $F_{4y} = 10360$ lb

6.27 $d_{3y} = -0.0128$ in., $\hat{f}_{1x}^{(1)} = 25$ kN, $\hat{f}_{1y}^{(1)} = -5.55$ kN, $\hat{m}_1^{(1)} = 0$

6.28 $d_{2x} = 5.70$ mm, $d_{2y} = -0.0244$ mm, $\phi_2 = 0.00523$ rad

6.29 $d_{3y} = -1.83$ in., $d_{4y} = -1.22$ in.

6.30 $d_{3y} = 6.67$ in., $d_{4y} = -6.67$ in., $\phi_3 = -\phi_4 = -3.20$ rad
$F_{1x} = 11.69$ kN, $F_{1y} = 30$ kN, $M_1 = -1810$ kN·m
$F_{6x} = -11.69$ kN, $F_{6y} = 30$ kN, $M_6 = 1810$ kN·m

6.32 $d_{2x} = 4.30$ mm, $\phi_2 = -0.241 \times 10^{-3}$ rad
$F_{1x} = -8339$ N, $F_{1y} = -4995$ N, $M_1 = 26,700$ N·m,
$F_{4x} = -6661$ N, $F_{4y} = 4995$ N, $M_4 = 23,330$ N·m

6.33 $d_{7x} = 0.0264$ m, $d_{7y} = 0.463 \times 10^{-4}$ m, $\phi_7 = 0.171 \times 10^{-2}$ rad
$\hat{f}_{1x}^{(1)} = -21.1$ N, $\hat{f}_{1y}^{(1)} = 30.4$ N, $\hat{m}_1^{(1)} = 74.95$ N·m
$\hat{f}_{3x}^{(1)} = 21.1$ N, $\hat{f}_{3y}^{(1)} = -30.4$ N, $\hat{m}_3^{(1)} = 46.65$ N·m

6.35 $d_{9x} = 0.0174$ m, $\hat{f}_{1x}^{(1)} = -22.6$ kN, $\hat{f}_{1y}^{(1)} = 16.0$ kN, $\hat{m}_1^{(1)} = 53.6$ kN·m
$\hat{f}_{3x}^{(1)} = 22.6$ kN, $\hat{f}_{3y}^{(1)} = -16.0$ kN, $\hat{m}_3^{(1)} = 42.4$ kN·m

6.37 $d_{4x} = -0.495 \times 10^{-4}$ m, $d_{4y} = -0.290 \times 10^{-3}$ m, $\phi_4 = 0.1547 \times 10^{-2}$ rad

6.39 Truss: $d_{7x} = 0.0260$ m, $d_{7y} = 0.00566$ m,
Frame: $d_{7x} = 0.0180$ m, $d_{7y} = 0.00424$ m
Truss, element 1: $\hat{f}_{1x} = -49,730$ N, $\hat{f}_{1y} = 0$
Frame, element 1: $\hat{f}_{1x} = -43,060$ N, $\hat{f}_{1y} = 22670$ N

6.40 Tapered beam $n = 3$
1 element: $d_{1y} = -0.222 \times 10^{-1}$ in.
2 elements: $d_{1y} = -0.189 \times 10^{-1}$ in.
4 elements: $d_{1y} = -0.181 \times 10^{-1}$ in.
8 elements: $d_{1y} = -0.179 \times 10^{-1}$ in.

6.41 $\underline{K} = 15\dfrac{G J_0}{L}\begin{bmatrix} 1 & -1 \\ -1 & 1 \end{bmatrix}$

6.43 $d_{2y} = -0.214$ in.

6.44 $d_{2y} = -0.729$ in.

6.46 $d_{1y} = -0.690 \times 10^{-2}$ m

Chapter 7

7.1 Use Eq. (7.2.10) in Eq. (7.2.18) to show $N_i + N_j + N_m = 1$.

7.3 **a.** $\underline{k} = 4.0 \times 10^6$
$$\begin{bmatrix} 2.5 & 1.25 & -2.0 & -1.5 & -0.5 & 0.25 \\ & 4.375 & -1.0 & -0.75 & -0.25 & -3.625 \\ & & 4.0 & 0 & -2.0 & 1.0 \\ & & & 1.5 & 1.5 & -0.75 \\ & & & & 2.5 & -1.25 \\ \text{symmetry} & & & & & 4.375 \end{bmatrix} \text{lb/in.}$$

b. $\underline{k} = 13.33 \times 10^6$
$$\begin{bmatrix} 1.54 & 0.75 & -1.0 & -0.45 & -0.54 & -0.3 \\ & 1.815 & -0.3 & -0.375 & -0.45 & -1.44 \\ & & 1.0 & 0 & 0 & 0.3 \\ & & & 0.375 & 0.45 & 0 \\ & & & & 0.54 & 0 \\ \text{symmetry} & & & & & 1.44 \end{bmatrix} \text{lb/in.}$$

7.4 **a.** $\sigma_x = 19.2$ ksi, $\sigma_y = 4.8$ ksi, $\tau_{xy} = -15.0$ ksi
$\sigma_1 = 28.6$ ksi, $\sigma_2 = -4.64$ ksi, $\theta_p = -32.2°$
b. $\sigma_x = 32.0$ ksi, $\sigma_y = 8.0$ ksi, $\tau_{xy} = -25.0$ ksi
$\sigma_1 = 47.7$ ksi, $\sigma_2 = -7.73$ ksi, $\theta_p = -32.2°$

7.5 **a.** $\underline{k} = 2.074$
$\times 10^5$
$$\begin{bmatrix} 8437.5 & 1687.5 & -7762.5 & -337.5 & -675 & -1350 \\ 1687.5 & 3937.5 & 337.5 & -2137.5 & -2025 & -1800 \\ -7762.5 & 337.5 & 8437.5 & -1687.5 & -675 & 1350 \\ -337.5 & -2137.5 & -1687.5 & 3937.5 & 2025 & -1800 \\ -675 & -2025 & -675 & 2025 & 1350 & 0 \\ -1350 & -1800 & 1350 & -1800 & 0 & 3600 \end{bmatrix} \text{N/m}$$

b. $\underline{k} = 4.48$
$\times 10^7$
$$\begin{bmatrix} 25.0 & 0 & -12.5 & 6.25 & -12.5 & -6.25 \\ & 9.375 & 9.375 & -4.6875 & -9.375 & -4.6875 \\ & & 15.625 & -7.8125 & -3.125 & -1.5625 \\ & & & 27.343 & 1.5625 & -3.125 \\ & & & & 15.625 & 7.8125 \\ \text{symmetry} & & & & & 27.343 \end{bmatrix} \text{N/m.}$$

7.6 **a.** $\sigma_x = -5.289$ GPa, $\sigma_y = -0.156$ GPa, $\tau_{xy} = 0.233$ GPa
$\sigma_1 = -0.1459$ GPa, $\sigma_2 = -5.30$ GPa, $\theta_p = -2.59°$
b. $\sigma_x = 0$, $\sigma_y = 42.0$ MPa, $\tau_{xy} = 33.6$ MPa
$\sigma_1 = 60.6$ MPa, $\sigma_2 = -18.6$ MPa, $\theta_p = -29°$

7.7 **a.** $\sigma_x = -15.0$ ksi, $\sigma_y = -45.0$ ksi, $\tau_{xy} = -18.0$ ksi
$\sigma_1 = -6.57$ ksi, $\sigma_2 = -53.4$ ksi, $\theta_p = -25.1°$
b. $\sigma_x = -15.0$ ksi, $\sigma_y = -45$ ksi, $\tau_{xy} = -21.0$ ksi
$\sigma_1 = -4.19$ ksi, $\sigma_2 = -55.8$ ksi, $\theta_p = -27.2°$

7.8 **a.** $\sigma_x = -52.5$ MPa, $\sigma_y = -32.8$ MPa, $\tau_{xy} = -5.38$ MPa
$\sigma_1 = -31.4$ MPa, $\sigma_2 = -53.9$ MPa, $\theta_p = -14.3°$
b. $\sigma_x = -31.4$ MPa, $\sigma_y = -13.5$ MPa, $\tau_{xy} = 5.38$ MPa
$\sigma_1 = -12.0$ MPa, $\sigma_2 = -32.9$ MPa, $\theta_p = -15.5°$
c. $\sigma_x = -27.6$ MPa, $\sigma_y = -19.5$ MPa, $\tau_{xy} = 4.04$ MPa
$\sigma_1 = 17.9$ MPa, $\sigma_2 = -29.3$ MPa, $\theta_p = -22.5°$
d. $\sigma_x = -31.6$ MPa, $\sigma_y = -28.9$ MPa, $\tau_{xy} = -6.73$ MPa
$\sigma_1 = -23.0$ MPa, $\sigma_2 = -38.0$ MPa, $\theta_p = 39°$

7.9 **a.** $f_{s1x} = 0$, $f_{s1y} = 0$, $f_{s2x} = p_0 Lt/6$, $f_{s2y} = 0$
$f_{s3x} = p_0 Lt/3$, $f_{s3y} = 0$
b. $f_{s1x} = 0$, $f_{s2x} = p_0 Lt/12$, $f_{s3x} = p_0 Lt/4$

7.10 $d_{3x} = 0.5 \times 10^{-3}$ in., $d_{3y} = -0.275 \times 10^{-2}$ in.
$d_{4x} = -0.609 \times 10^{-3}$ in., $d_{4y} = -0.293 \times 10^{-2}$ in.
$\sigma_x^{(1)} = 824$ psi, $\sigma_y^{(1)} = 247$ psi, $\tau_{xy}^{(1)} = -1587$ psi
$\sigma_1^{(1)} = 2149$ psi, $\sigma_2^{(1)} = -1077$ psi, $\theta_p^{(1)} = -40°$
$\sigma_x^{(2)} = -826$ psi, $\sigma_y^{(2)} = 292$ psi, $\tau_{xy}^{(2)} = -411$ psi
$\sigma_1^{(2)} = 426$ psi, $\sigma_2^{(2)} = -960$ psi, $\theta_p^{(2)} = 18.15°$

7.11 **a.** $d_{2x} = 0.281 \times 10^{-4}$ m, $d_{2y} = -0.330 \times 10^{-4}$ m
$d_{5x} = 0.115 \times 10^{-4}$ m, $d_{5y} = -0.103 \times 10^{-4}$ m
$\sigma_x^{(2)} = 16.4$ MPa, $\sigma_y^{(2)} = 15.2$ MPa
$\tau_{xy}^{(2)} = -6.99$ MPa, $\sigma_1^{(2)} = 22.8$ MPa
$\sigma_2^{(2)} = 8.80$ MPa, $\theta_p^{(2)} = -42.7°$
$\sigma_x^{(1)} = 10.6$ MPa, $\sigma_y^{(1)} = 3.18$ MPa
$\tau_{xy}^{(1)} = -3.34$ MPa, $\sigma_1^{(1)} = 11.9$ MPa
$\sigma_2^{(1)} = 1.90$ MPa, $\theta_p^{(1)} = -21.0°$
b. $d_{1x} = -d_{2x} = -0.165 \times 10^{-5}$ m, $d_{1y} = d_{2y} = -0.125 \times 10^{-4}$ m
$d_{5x} = 0.274 \times 10^{-12}$ m, $d_{5y} = -0.163 \times 10^{-4}$ m
$\sigma_x^{(1)} = 5.99 \times 10^5$ N/m², $\sigma_y^{(1)} = -3.78 \times 10^6$ N/m²
$\tau_{xy}^{(1)} = 4.05 \times 10^{-1}$ N/m², $\sigma_1^{(1)} = 5.99 \times 10^5$ N/m²
$\sigma_2^{(1)} = -3.78 \times 10^6$ N/m², $\theta_p^{(1)} = 0°$, $\sigma_x^{(3)} = 5.64 \times 10^6$ N/m²
$\sigma_y^{(3)} = 1.88 \times 10^7$ N/m², $\tau_{xy}^{(3)} = -1.11 \times 10^{-1}$ N/m²
$\sigma_1^{(3)} = 1.88 \times 10^7$ N/m², $\sigma_2^{(3)} = 5.64 \times 10^6$ N/m², $\theta_p^{(3)} = -90°$

7.12 All f_{bx}'s are equal to 0.
$f_{b1y} = f_{b2y} = f_{b3y} = f_{b4y} = 10.28$ N, $f_{b5y} = 20.56$ N

7.13 **a.** $n_b = 8$, **b.** $n_b = 12$

Chapter 8

8.6 $d_{2x} = d_{3x} = 0.647 \times 10^{-3}$ in., $d_{2y} = 0.666 \times 10^{-4}$ in.
$d_{3y} = -0.666 \times 10^{-4}$ in., skew effect

8.7 Stress approaches 2.5 psi near edge of whole for model of 70 nodes, 54 elements.

8.8 At depth 4 in. equal to width stress approaches uniform $\sigma_y = -1000$ psi

8.11 For 106-element model at re-entrant corner, $\sigma_1 = 3500$ psi, $\sigma_2 = 670$ psi

8.14 For the model with 12 in. × $\frac{1}{2}$ in. size elements, finite element solution yields free end deflection of -0.499 in., exact solution is -1.15 in. (See Table 8-1 in text for other results.)

8.22 Largest principal stress 30 to 40 MPa in support region (72 element, 91 node model)

8.24 Largest principal stress $\sigma_1 = 1005$ MPa at narrowest width of member (70 element, 94 node model)

Chapter 9

9.2 $\varepsilon_x = \dfrac{1}{3b}(-u_1 + u_2 + 4u_4 - 4u_5), \quad \varepsilon_y = \dfrac{1}{3h}(-v_1 + v_3 + 4v_4 - 4v_6)$

$\gamma_{xy} = \dfrac{1}{3h}(-u_1 + u_3 + 4u_4 - 4u_6) + \dfrac{1}{3b}(-v_1 + v_3 + 4v_4 - 4v_6)$

$\sigma_x = \dfrac{E}{1 - v^2}(\varepsilon_x + v\varepsilon_y), \quad \sigma_y = \dfrac{E}{1 - v^2}(\varepsilon_y + v\varepsilon_x), \quad \tau_{xy} = G\gamma_{xy}$

9.3 $f_{s1x} = f_{s3x} = \dfrac{-pth}{6}, \qquad f_{s5x} = \dfrac{-2pth}{3}$

9.4 $f_{s1x} = 0, \quad f_{s3x} = \dfrac{-p_0 th}{6}, \qquad f_{s5x} = \dfrac{-p_0 th}{3}$

9.5 a. $\varepsilon_x = -5 \times 10^{-5}y + 2.5 \times 10^{-4}, \quad \varepsilon_y = -1.67 \times 10^{-4}x + 3.33 \times 10^{-5},$
$\gamma_{xy} = -5 \times 10^{-5}x - 1.11 \times 10^{-4}y + 4.17 \times 10^{-4}$
$\sigma_x = 3290$ psi, $\quad \sigma_y = -4850$ psi, $\quad \tau_{xy} = 1540$ psi

b. $\varepsilon_x = -5 \times 10^{-5}y + 1.67 \times 10^{-4}, \quad \varepsilon_y = -1.67 \times 10^{-4}x + 5 \times 10^{-5}$
$\gamma_{xy} = -5 \times 10^{-5}x - 4.17 \times 10^{-5}y + 2.08 \times 10^{-4}$
$\sigma_x = 928$ psi, $\quad \sigma_y = -8290$ psi, $\quad \tau_{xy} = 632$ psi

9.6 $\varepsilon_x = 2.54 \times 10^{-3}$
$\varepsilon_y = -7.62 \times 10^{-3}$
$\gamma_{xy} = -7.04 \times 10^{-3}$

9.7 $N_1 = 1 - \dfrac{x}{20} + \dfrac{x^2}{1800}, \quad N_2 = \dfrac{-x + y}{60} + \dfrac{x^2 + y^2}{1800} - \dfrac{xy}{900}$

$N_3 = \dfrac{-y}{60} + \dfrac{y^2}{1800}, \quad N_4 = \dfrac{xy}{900} - \dfrac{y^2}{900}, \quad N_5 = \dfrac{y}{15} - \dfrac{xy}{900},$ etc.

Chapter 10

10.1 a. $\underline{K} = 25.132 \times 10^6 \begin{bmatrix} 5 & 1 & 0 & -1 & 1 & 0 \\ 1 & 4 & -2 & -1 & -2 & -3 \\ 0 & -2 & 8 & 0 & 4 & 2 \\ -1 & -1 & 0 & 1 & 1 & 0 \\ 1 & -2 & 4 & 1 & 4 & 1 \\ 0 & -3 & 2 & 0 & 1 & 3 \end{bmatrix}$ lb/in.

b. $\underline{K} = 50.265 \times 10^6$ $\begin{bmatrix} 2.75 & 0 & -2.25 & 0.5 & 0.25 & -0.5 \\ 0 & 1 & 1 & -1 & -1 & 0 \\ -2.25 & 1 & 5.75 & -2.5 & 0.25 & 1.5 \\ 0.5 & -1 & -2.5 & 4 & 0.5 & -3 \\ 0.25 & -1 & 0.25 & 0.5 & 1.75 & 0.5 \\ -0.5 & 0 & 1.5 & -3 & 0.5 & 3 \end{bmatrix}$ lb/in.

10.2 $f_{s2x} = \dfrac{2\pi b p_0 h}{6}$, $f_{s3x} = \dfrac{2\pi b p_0 h}{3}$

10.3 $f_{b1r} = f_{b2r} = f_{b3r} = 0.382$ lb
$f_{b1z} = f_{b2rz} = f_{b3z} = -6.32$ lb

10.4 **a.** $\sigma_r = 8000$ psi, $\sigma_z = 0$, $\sigma_\theta = 8000$ psi, $\tau_{rz} = 1200$ psi
b. $\sigma_r = 5830$ psi, $\sigma_z = -3770$ psi, $\sigma_\theta = 3090$ psi, $\tau_{rz} = 400$ psi

10.6 **a.** $\underline{k} = 7.037$ $\begin{bmatrix} 3125 & 625 & 0 & -625 & 625 & 0 \\ & 2500 & -1250 & -625 & -1250 & -1875 \\ & & 5000 & 0 & 2500 & 1250 \\ & & & 625 & 625 & 0 \\ & & & & 2500 & 625 \\ \text{symmetry} & & & & & 1875 \end{bmatrix}$ kN/mm

b. $\underline{k} = 11.73$ $\begin{bmatrix} 2475 & 0 & -2025 & 450 & 225 & -450 \\ & 900 & 900 & -900 & -900 & 0 \\ & & 5175 & -2250 & 225 & 1350 \\ & & & 3600 & 450 & -2700 \\ & & & & 1575 & 450 \\ \text{symmetry} & & & & & 2700 \end{bmatrix}$ kN/mm

10.7 **a.** $\sigma_r = -84$ MPa, $\sigma_z = -84$ MPa, $\sigma_\theta = 252$ MPa, $\tau_{rz} = -101$ MPa
b. $\sigma_r = -103$ MPa, $\sigma_z = -103$ MPa, $\sigma_\theta = 112$ MPa, $\tau_{rz} = 73$ MPa.

Chapter 11

11.3 $N_1 = \dfrac{s(s-1)}{2}$, $N_2 = \dfrac{s(s+1)}{2}$, $N_3 = 1 - s^2$

$[B] = \begin{bmatrix} \dfrac{2s-1}{L} & \dfrac{2s+1}{L} & \dfrac{-4s}{L} \end{bmatrix}$

11.4 $N_1 = \dfrac{-2s^3}{3} + \dfrac{2s^2}{3} + \dfrac{s}{6} - \dfrac{1}{6}$, $N_2 = \dfrac{4s^3}{3} - \dfrac{2s^2}{3} - \dfrac{4s}{3} + \dfrac{2}{3}$

$N_3 = \dfrac{-4s^3}{3} - \dfrac{2s^2}{3} + \dfrac{4s}{3} + \dfrac{2}{3}$, $N_4 = \dfrac{2s^3}{3} + \dfrac{2s^2}{3} - \dfrac{s}{6} - \dfrac{1}{6}$

$[B] = \begin{bmatrix} \dfrac{-12s^2 + 8s + 1}{3L} & \dfrac{12s^2 - 4s - 4}{1.5L} & \dfrac{-12s^2 - 4s + 4}{1.5L} & \dfrac{12s^2 + 8s - 1}{3L} \end{bmatrix}$

11.6 $\varepsilon_x = 0.0009375$ in./in., $\varepsilon_y = -0.00125$ in./in., $\gamma_{xy} = -0.000625$ rad
$\sigma_x = 18.5$ ksi, $\sigma_y = -31.9$ ksi, $\tau_{xy} = -7.21$ ksi

11.11 **a.** $f_{s2s} = 500$ lb, $f_{s3s} = 500$ lb, **b.** $f_{s1t} = 83.33$ lb, $f_{s4t} = 41.67$ lb

11.12 **a.** 1.917, **b.** 0.667, **c.** 0.400, **d.** 2.87, **e.** 0.250, **f.** 0.382

Chapter 12

12.1 $\underline{B} = \dfrac{1}{8}\begin{bmatrix} 0 & 0 & 0 & 0 & 0 & 4 & 0 & 0 & -4 & 0 & 0 \\ 0 & 0 & 0 & 0 & 4 & 0 & 0 & 0 & 0 & -4 & 0 \\ 0 & 0 & 4 & 0 & 0 & 0 & 0 & 0 & 0 & 0 & -4 \\ 0 & 0 & 0 & 0 & 0 & 0 & 0 & 0 & -4 & -4 & 0 \\ 0 & 4 & 0 & 0 & 0 & 4 & 0 & 0 & 0 & -4 & -4 \\ 4 & 0 & 0 & 0 & 0 & 0 & 0 & 4 & -4 & 0 & -4 \end{bmatrix}$

12.2 $\sigma_x = 77.9$ ksi, $\sigma_y = 8.65$ ksi, $\sigma_z = -49.0$ ksi
$\tau_{xy} = 11.5$ ksi, $\tau_{yz} = -23.1$ ksi, $\tau_{zx} = 5.77$ ksi

12.4 $\underline{B} = \dfrac{1}{18750}$

$\times \begin{bmatrix} -625 & 0 & 0 & 0 & 0 & 0 & 0 & 0 & 0 & 625 & 0 & 0 \\ 0 & -375 & 0 & 0 & -375 & 0 & 0 & 0 & 0 & 0 & -375 & 0 \\ 0 & 0 & -375 & 0 & 0 & 0 & 0 & 0 & 750 & 0 & 0 & -375 \\ -375 & -625 & 0 & 750 & 0 & 0 & 0 & 0 & 0 & -375 & 625 & 0 \\ 0 & -375 & -375 & 0 & 0 & 750 & 0 & 750 & 0 & 0 & -375 & -375 \\ -375 & 0 & -625 & 0 & 0 & 0 & 750 & 0 & 0 & -375 & 0 & 625 \end{bmatrix}$

12.5 $\sigma_x = 72.7$ MPa, $\sigma_y = 169.6$ MPa, $\sigma_z = 72.7$ MPa
$\tau_{xy} = 59.2$ MPa, $\tau_{yz} = 32.3$ MPa, $\tau_{zx} = 91.5$ MPa

12.6 $N_2 = \dfrac{(1-s)(1-t)(1-z')}{8}$, $\quad N_3 = \dfrac{(1-s)(1+t)(1-z')}{8}$,

$N_4 = \dfrac{(1-s)(1+t)(1+z')}{8}$,

$N_5 = \dfrac{(1+s)(1-t)(1+z')}{8}$, $\quad N_6 = \dfrac{(1+s)(1-t)(1-z')}{8}$,

$N_7 = \dfrac{(1+s)(1+t)(1-z')}{8}$, $\quad N_8 = \dfrac{(1+s)(1+t)(1+z')}{8}$,

12.7 $N_1 = \dfrac{(1-s)(1-t)(1+z')(-s-t+z'-2)}{8}$,

$N_2 = \dfrac{(1-s)(1-t)(1-z')(-s-t-z'-2)}{8}$

Chapter 13

13.1 $t_2 = 166.7°C$, $t_3 = 233.3°C$

13.2 $t_2 = 150°F$, $t_3 = 100°F$, $t_4 = 50°F$

13.3 $t_2 = 875°F$, $t_3 = 1250°F$

13.4 $t_1 = 151°F$, $t_2 = 148°F$, $t_3 = 140°F$, $t_4 = 125°F$

13.5 $t_2 = 183°F$, $t_3 = 267°F$, $t_4 = 350°F$, $t_5 = 433°F$

13.6 $t_2 = 421°C$, $t_3 = 121°C$, $q^{(3)} = 3975$ W/m^2

13.7 $t_2 = 418.2°C$, $t_3 = 527.3°C$

13.9 $\underline{k} = \begin{bmatrix} 39.57 & 7.076 & -5.417 \\ & 35.82 & -1.667 \\ & & 7.083 \end{bmatrix}, \quad \underline{f} = \begin{Bmatrix} 2936 \\ 2936 \\ 50 \end{Bmatrix} \text{Btu/hr}$

13.10 $\underline{f} = \begin{Bmatrix} 1291 \\ 27.3 \\ 1254 \end{Bmatrix} \text{W}$

13.13 $t_3 = 75°\text{F}, \quad t_6 = 25°\text{F}$

Chapter 14

14.1 $p_2 = 4.545 \text{ m}, \quad p_3 = 1.818 \text{ m}, \quad v_x^{(1)} = 10.91 \text{ m/s}, \quad Q_f^{(1)} = 21.82 \text{ m}^3/\text{s}$

14.2 $p_2 = -15 \text{ m}, \quad p_3 = -40 \text{ m}, \quad p_4 = -65 \text{ m}, \quad v_x^{(1)} = 25 \text{ m/s}, \quad Q_1 = 50 \text{ m}^3/\text{s}$

14.3 $p_2 = 8.182 \text{ in.}, \quad p_3 = 5.455 \text{ in.}, \quad v_x^{(1)} = 0.182 \text{ in./s}, \quad v_x^{(2)} = 0.273 \text{ in./s},$
$v_x^{(3)} = 0.545 \text{ in./s}, \quad Q_f^{(1)} = 1.091 \text{ in.}^3/\text{s}$

14.4 $p_2 = -3 \text{ cm}, \quad p_3 = -8 \text{ cm}, \quad v_x^{(1)} = 1.2 \text{ cm/s}, \quad v_x^{(2)} = 2 \text{ cm/s},$
$Q_1 = Q_2 = 6 \text{ cm}^3/\text{s}$

14.6 $v^{(1)} = 2.0 \text{ in./s}, \quad v^{(2)} = 4.0 \text{ in./s}, \quad Q^{(1)} = Q^{(2)} = 4 \text{ in.}^3/\text{s}$

14.7 $\underline{f}_Q = \begin{Bmatrix} 54.76 \\ 28.57 \\ 16.67 \end{Bmatrix} \text{m}^3/\text{s}$

14.8 $f_1 = f_3 = 5 \text{ in.}^3/\text{s}, \quad f_2 = 0$

14.9 $p_2 = p_3 = 12 \text{ m}, \quad p_5 = 11 \text{ m}$

Chapter 15

15.1 $d_{2x} = 0.021 \text{ in.}, \quad d_{3x} = 0.042 \text{ in.}, \quad \sigma_x = 0$

15.2 $d_{2x} = 0, \quad \sigma_x = 50.4 \text{ MPa}$

15.3 $d_{1x} = d_{1y} = -0.0175 \text{ in.}, \quad \sigma^{(1)} = 4350 \text{ psi (T)}$
$\sigma^{(2)} = -6150 \text{ psi (C)}, \quad \sigma^{(3)} = 4350 \text{ psi (T)}$

15.4 $d_{1x} = -0.0291 \text{ in.}, \quad d_{1y} = -0.0095 \text{ in.}$
$\sigma^{(1)} = -1370 \text{ psi (C)}, \quad \sigma^{(2)} = 2375 \text{ psi (T)}, \quad \sigma^{(3)} = -1370 \text{ psi (C)}$

15.5 $d_{2x} = 1.44 \times 10^{-4} \text{ m}, \quad \sigma^{(1)} = -20.2 \text{ MPa (C)}, \quad \sigma^{(2)} = \sigma^{(3)} = -10.1 \text{ MPa (C)}$

15.6 $d_{1x} = 0, \quad d_{1y} = 6.0 \times 10^{-4} \text{ m}, \quad \sigma^{(1)} = \sigma^{(3)} = -10.5 \text{ MPa (C)}$
$\sigma^{(2)} = 18.2 \text{ MPa (T)}$

15.7 $d_{1x} = 0, \quad d_{1y} = -3.6 \times 10^{-4} \text{ m}, \quad \sigma^{(1)} = \sigma^{(2)} = 0$

15.8 $d_{2x} = 0.0173 \text{ in.}, \quad \sigma_{st} = 840 \text{ psi (T)}, \quad \sigma_{br} = 1680 \text{ psi (C)}$

15.9 $f_{T1x} = -4464 \text{ lb}, \quad f_{T1y} = -8929 \text{ lb}, \quad f_{T2x} = 4464 \text{ lb}$
$f_{T2y} = -8929 \text{ lb}, \quad f_{T3x} = 0, \quad f_{T3y} = 17,857 \text{ lb}$

15.10 $f_{T1x} = -43.125 \text{ kN}, \quad f_{T1y} = 0, \quad f_{T2x} = 43.125 \text{ kN}, \quad f_{T2y} = -86.250 \text{ kN}$
$f_{T3x} = 0, \quad f_{T3y} = 86.250 \text{ kN}$

15.11 $f_{T1x} = -60.0 \text{ k}, \quad f_{T1y} = -90 \text{ k}, \quad f_{T2x} = 60 \text{ k}, \quad f_{T2y} = 0$
$f_{T3x} = 0, \quad f_{T3y} = 90 \text{ k}$

15.12 $f_{T1x} = 134$ kN, $f_{T1y} = 134$ kN, $f_{T2x} = -134$ kN, $f_{T2y} = 0$
$f_{T3x} = 0$, $f_{T3y} = -134$ kN

15.13 $\sigma_x = \sigma_y = -8929$ psi (C), $\tau_{xy} = 0$

15.14 $\sigma_x = 67.2$ MPa, $\sigma_y = 67.2$ MPa, $\tau_{xy} = 0$

15.15 $\{f_T\} = \dfrac{AE\alpha_0}{6}\begin{Bmatrix} -4t_1 - 5t_2 \\ 4t_1 + 5t_2 \end{Bmatrix}$

15.16 $\dfrac{AE\alpha}{2}\begin{Bmatrix} -t_1 - t_2 \\ t_1 + t_2 \end{Bmatrix}$

15.17 $\{f_T\} = \dfrac{2\pi\bar{r}AE\alpha(\Delta T)[\bar{B}]^T}{1 - 2v}\begin{Bmatrix} 1 \\ 1 \\ 1 \\ 0 \end{Bmatrix}$

15.18 $d_{2x} = 0.8 \times 10^{-3}$ in., $d_{3x} = 0$, $d_{3y} = 0.8 \times 10^{-3}$ in.
$d_{4x} = d_{4y} = 0.8 \times 10^{-3}$ in., stresses are zero

15.19 $d_{2x} = 0.989 \times 10^{-3}$ in., $d_{3x} = -0.756 \times 10^{-3}$ in.,
$d_{3y} = 0.989 \times 10^{-3}$ in., $d_{4x} = 0.132 \times 10^{-2}$ in.,
$d_{4y} = 0.2045 \times 10^{-2}$ in., $\sigma_1^{(1)} = 17$ ksi, $\sigma_2^{(2)} = -17$ ksi

Chapter 16

16.1 $[M] = \dfrac{\rho AL}{6}\begin{bmatrix} 2 & 1 & 0 \\ 1 & 4 & 1 \\ 0 & 1 & 2 \end{bmatrix}$

16.2 a. $[M] = \dfrac{\rho AL}{2}\begin{bmatrix} 1 & 0 & 0 & 0 \\ 0 & 2 & 0 & 0 \\ 0 & 0 & 2 & 0 \\ 0 & 0 & 0 & 1 \end{bmatrix}$

b. $[M] = \dfrac{\rho AL}{6}\begin{bmatrix} 2 & 1 & 0 & 0 \\ 1 & 4 & 1 & 0 \\ 0 & 1 & 4 & 1 \\ 0 & 0 & 1 & 2 \end{bmatrix}$

16.3 $\omega_1 = 0.806\sqrt{u}$, $\omega_2 = 2.81\sqrt{\mu}$

16.4 $\omega_1 = 5.368 \times 10^3$ rad/sec, $\omega_2 = 17.556 \times 10^3$ rad/sec

16.5 a.

t, sec	d_i, ft	$\dot{d}_i$, ft/sec	$\ddot{d}_i$, ft/sec^2
0	0	0	25
0.03	0.01125	0.71	22.09
0.06	0.04238	1.03	-0.715
0.09	0.07287	0.67	-22.87
0.12	0.08278	-0.35	-45.28
0.15	0.05194	-1.43	-26.94

16.6 a.

t, sec	d_i, ft	$\dot{d}_i, ft/sec$	$\ddot{d}_i, ft/sec^2$
0	0	0	0
0.02	0.0020	0.168	6.80
0.04	0.00672	0.256	1.968
0.06	0.01223	0.242	−3.338
0.08	0.01640	0.130	−7.84
0.10	0.1743	−0.053	−10.46

b.

t, sec	d_i, ft	$\dot{d}_i, ft/sec$	$\ddot{d}_i, ft/sec^2$	F(t), lb
0.00	0.00000	0.000	10.000	20.0
0.02	0.00179	0.169	6.923	16.0
0.04	0.00625	0.263	2.248	12.0
0.06	0.0115	0.254	−2.945	8.0
0.08	0.0157	0.150	−7.458	4.0
0.10	0.0169	−0.0147	−10.251	0.0

16.7

Node	t, sec	$d_i, in.$	$\dot{d}_i, in./sec$	$\ddot{d}_i, in./sec^2$
2	0	0	0	0
	0.00025	2.6E-8	0	249.6
	0.00050	3.4E-5	0.284	1768.9
	0.00075	1.9E-4	1.085	4641.9
	0.0010	6.36E-4	2.605	7519.3
3	0	0	0	0
	0.00025	6.59E-5	0.791	6328.8
	0.00050	4.99E-4	2.817	9881.2
	0.00075	1.51E-3	5.265	9701.7
	0.0010	3.10E-3	7.369	7128.3

16.8 Using Newmark's method with $\gamma = \frac{1}{2}, \beta = \frac{1}{6}$

Node	t, sec	d_i, ft	$\dot{d}_i, ft/sec$	$\ddot{d}_i, ft/sec^2$	F(t), lb
2	0	0	0	0	0
	0.05	0.00172	0.103	4.131	0
	0.10	0.01544	0.513	12.27	0
3	0	0	0	40.0	2000
	0.05	0.0448	1.685	27.39	1800
	0.10	0.1536	2.479	4.37	1600

16.9 a. $\omega_1 = \dfrac{3.15}{L^2}\left(\dfrac{EI}{\rho A}\right)^{1/2}$, $\omega_2 = \dfrac{16.24}{L^2}\left(\dfrac{EI}{\rho A}\right)^{1/2}$, **c.** $\omega_1 = \dfrac{9.8}{L^2}\left(\dfrac{EI}{\rho A}\right)^{1/2}$

d. $\omega = \dfrac{14.8}{L^2}\left(\dfrac{EI}{\rho A}\right)^{1/2}$

16.14 $\quad [M] = \dfrac{\rho AL}{420}$
$$\begin{bmatrix} 140 & 0 & 0 & 70 & 0 & 0 \\ & 156 & 22L & 0 & 54 & -13L \\ & & 4L^2 & 0 & 13L & -3L^2 \\ & & & 140 & 0 & 0 \\ & & & & 156 & -22L \\ \text{symmetry} & & & & & 4L^2 \end{bmatrix}$$

16.16

Node:	1	2	3	4	5	6
i $\quad$ t, s			Temperatures, °C			
0 $\quad$ 0	200	200	200	200	200	200
1 $\quad$ 8	0	159.0095	191.4441	198.2110	199.6110	199.8444
2 $\quad$ 16	0	135.5852	178.1491	193.6620	198.2112	199.1445
3 $\quad$ 24	0	120.2309	165.7003	187.3485	195.5379	197.5152
4 $\quad$ 32	0	109.1993	154.9587	180.4038	191.7446	194.8115
5 $\quad$ 40	0	100.7600	145.7784	173.4129	187.1268	191.1242
6 $\quad$ 48	0	94.00311	137.8529	166.6182	181.9599	186.6590
7 $\quad$ 56	0	88.39929	130.9034	160.1012	176.4598	181.6395
8 $\quad$ 64	0	83.61745	124.7101	153.8759	170.7856	176.2620
9 $\quad$ 72	0	79.43935	119.1075	147.9316	165.0508	170.6822
10 $\quad$ 80	0	75.71603	113.9733	142.2502	159.3352	165.0171

16.17

Time (sec)		Node	
	1	2	3
		Temperature, °C	
0	25	25	25
0.1	85	18.53611	26.36189
0.2	85	29.61303	21.63526
0.3	85	36.18435	22.42717
0.4	85	40.72491	25.30428
0.5	85	44.27834	28.85201
0.6	85	47.29072	32.49614
0.7	85	49.95809	36.01157
0.8	85	52.37152	39.31761
0.9	85	54.57756	42.39278
1	85	56.60353	45.23933
1.1	85	58.46814	47.86852
1.2	85	60.1859	50.29457
1.3	85	61.76908	52.53218
1.4	85	63.22852	54.59557
1.5	85	64.574	56.49814
1.6	85	65.81448	58.25235
1.7	85	66.95818	59.86974

16.17

Time (sec)	1	2	3
		Temperature, °C	
0	25	25	25
1.8	85	68.01265	61.36096
1.9	85	68.98485	62.73586
2	85	69.88121	64.0035
2.1	85	70.70765	65.17226
2.2	85	71.46961	66.24984
2.3	85	72.17214	67.24336
2.4	85	72.81986	68.15938
2.5	85	73.41705	69.00393
2.6	85	73.96766	69.78261
2.7	85	74.47531	70.50053
2.8	85	74.94336	71.16246
2.9	85	75.3749	71.77274
3	85	75.77277	72.33542

Appendix A

A1. **a.** $\begin{bmatrix} 3 & 0 \\ -3 & 12 \end{bmatrix}$ **b.** Nonsense   **c.** Nonsense

d. $\left\{ \begin{array}{c} 11 \\ 9 \\ 11 \end{array} \right\}$ **e.** Nonsense   **f.** $\begin{bmatrix} 10 & 7 & 6 \\ 3 & -1 & 7 \end{bmatrix}$

A2. $\begin{bmatrix} 1 & 0 \\ \frac{1}{4} & \frac{1}{4} \end{bmatrix}$

A3. $\frac{1}{17} \begin{bmatrix} 12 & -3 & -8 \\ -3 & 5 & 2 \\ -8 & 2 & 11 \end{bmatrix}$

A4. Nonsense

A5. $\begin{bmatrix} \frac{1}{2} & 0 \\ \frac{1}{8} & \frac{1}{8} \end{bmatrix}$

A6. $\begin{bmatrix} \cos \theta & -\sin \theta \\ \sin \theta & \cos \theta \end{bmatrix}$

Appendix B

B1. $x_1 = 3.15, \quad x_2 = 0.62$

B2. $x_1 = 3.15, \quad x_2 = 0.62$

B3. $x_1 = 2.5, \quad x_2 = -1, \quad x_3 = 0.5$

B4. $x_1 = 3, \quad x_2 = -1, \quad x_3 = -2$

B5. **a.** $\begin{Bmatrix} x_1 \\ x_2 \end{Bmatrix} = \begin{bmatrix} 2 & -1 \\ 1 & -1 \end{bmatrix} \begin{Bmatrix} y_1 \\ y_2 \end{Bmatrix}$ **b.** $\begin{Bmatrix} z_1 \\ z_2 \end{Bmatrix} = \begin{bmatrix} -3 & 2 \\ 5 & -3 \end{bmatrix} \begin{Bmatrix} y_1 \\ y_2 \end{Bmatrix}$

B6. $x_1 = 0, \quad x_2 = 1, \quad x_3 = 2, \quad x_4 = 2, \quad x_5 = 0$

B7. $x_1 = 3.15, \quad x_2 = 0.62$

B8. **a.** Unique **b.** Nonexistent **c.** Unique **d.** Nonunique

Appendix D

D1. **a.** $f_{1y} = f_{2y} = -5 \text{ k}, \quad m_1 = -m_2 = -100 \text{ k-ft}$
b. $f_{1y} = f_{2y} = -5 \text{ k}, \quad m_1 = -m_2 = -18.75 \text{ k-ft}$
c. $f_{1y} = f_{2y} = -15 \text{ k}, \quad m_1 = -m_2 = -75 \text{ k-ft}$
d. $f_{1y} = -18.75 \text{ k}, \quad f_{2y} = -6.25 \text{ k}, \quad m_1 = -58.3 \text{ k-ft}, \quad m_2 = 33.3 \text{ k-ft}$
e. $f_{1y} = -6 \text{ k}, \quad f_{2y} = -14 \text{ k}, \quad m_1 = -26.67 \text{ k-ft}, \quad m_2 = 40 \text{ k-ft}$
f. $f_{1y} = -0.99 \text{ kN}, \quad f_{2y} = -4.0 \text{ kN}, \quad m_1 = -2.04 \text{ kN} \cdot \text{m}, \quad m_2 = 5.10 \text{ kN} \cdot \text{m}$
g. $f_{1y} = f_{2y} = -6 \text{ kN}, \quad m_1 = -m_2 = -7.5 \text{ kN} \cdot \text{m}$
h. $f_{1y} = f_{2y} = -10 \text{ kN}, \quad m_1 = -m_2 = -6.67 \text{ kN} \cdot \text{m}$

Index

WARNING: IF THE DISK PACKAGE SEAL IS BROKEN, THE PURCHASER FORFEITS ALL RETURNS, RIGHTS, AND PRIVILEGES TO THE SELLER.